Periodic Table of the Elements

Metals (main-group)
Metals (transition)
Metals (inner transition)
Metalloids
Nonmetals

MAIN–GROUP ELEMENTS | TRANSITION ELEMENTS | MAIN–GROUP ELEMENTS

Period	1A (1)	2A (2)	3B (3)	4B (4)	5B (5)	6B (6)	7B (7)	8B (8)	8B (9)	8B (10)	1B (11)	2B (12)	3A (13)	4A (14)	5A (15)	6A (16)	7A (17)	8A (18)
1	1 **H** 1.008																	2 **He** 4.003
2	3 **Li** 6.941	4 **Be** 9.012											5 **B** 10.81	6 **C** 12.01	7 **N** 14.01	8 **O** 16.00	9 **F** 19.00	10 **Ne** 20.18
3	11 **Na** 22.99	12 **Mg** 24.31											13 **Al** 26.98	14 **Si** 28.09	15 **P** 30.97	16 **S** 32.07	17 **Cl** 35.45	18 **Ar** 39.95
4	19 **K** 39.10	20 **Ca** 40.08	21 **Sc** 44.96	22 **Ti** 47.88	23 **V** 50.94	24 **Cr** 52.00	25 **Mn** 54.94	26 **Fe** 55.85	27 **Co** 58.93	28 **Ni** 58.69	29 **Cu** 63.55	30 **Zn** 65.41	31 **Ga** 69.72	32 **Ge** 72.61	33 **As** 74.92	34 **Se** 78.96	35 **Br** 79.90	36 **Kr** 83.80
5	37 **Rb** 85.47	38 **Sr** 87.62	39 **Y** 88.91	40 **Zr** 91.22	41 **Nb** 92.91	42 **Mo** 95.94	43 **Tc** (98)	44 **Ru** 101.1	45 **Rh** 102.9	46 **Pd** 106.4	47 **Ag** 107.9	48 **Cd** 112.4	49 **In** 114.8	50 **Sn** 118.7	51 **Sb** 121.8	52 **Te** 127.6	53 **I** 126.	
6	55 **Cs** 132.9	56 **Ba** 137.3	57 **La** 138.9	72 **Hf** 178.5	73 **Ta** 180.9	74 **W** 183.9	75 **Re** 186.2	76 **Os** 190.2	77 **Ir** 192.2	78 **Pt** 195.1	79 **Au** 197.0	80 **Hg** 200.6	81 **Tl** 204.4	82 **Pb** 207.2	83 **Bi** 209.0	84 **Po** (209)	85 **At** (21	
7	87 **Fr** (223)	88 **Ra** (226)	89 **Ac** (227)	104 **Rf** (263)	105 **Db** (262)	106 **Sg** (266)	107 **Bh** (267)	108 **Hs** (277)	109 **Mt** (268)	110 **Ds** (281)	111 **Rg** (272)	112 (285)	113 284)	114 (289)	115 288)	116 (292)		

As of
eleme
throu
not b

INNER TRANSITION ELEMENTS

Period															
6	Lanthanides	58 **Ce** 140.1	59 **Pr** 140.9	60 **Nd** 144.2	61 **Pm** (145)	62 **Sm** 150.4	63 **Eu** 152.0	64 **Gd** 157.3	65 **Tb** 158.9	66 **Dy** 162.5	67 **Ho** 164.9	68 **Er** 167.3	69 **Tm** 168.9	70 **Yb** 173.0	71 **Lu** 175.0
7	Actinides	90 **Th** 232.0	91 **Pa** (231)	92 **U** 238.0	93 **Np** (237)	94 **Pu** (242)	95 **Am** (243)	96 **Cm** (247)	97 **Bk** (247)	98 **Cf** (251)	99 **Es** (252)	100 **Fm** (257)	101 **Md** (258)	102 **No** (259)	103 **Lr** (260)

Useful Data and Information

McGraw-Hill's ARIS Online Registration Code

to accompany

McGraw-Hill's ARIS (Assessment, Review, and Instruction System) is a web-based supplement that allows you to access homework, quizzing, tutorials and self-study material. You will use ARIS to view and complete assignments. Once you submit your assignments, your results are saved online so you can review your progress. ARIS also provides tools for self-assessment and self-study to help you succeed in this course. Don't wait; use the enclosed online registration code to get started with ARIS.

STUDENTS: DO NOT THROW AWAY!

This is you personalized registration card for online access to textbook and assignment material.

The registration code on the back of this card allows you to access assignments and other material that your instructor requires for this course. The code is unique and for one specific McGraw-Hill product; it is not related to any other registration number or ID you may have. Once it is used, it cannot be re-used. Your student registration code protects valuable content. Your course materials contain much more than a textbook. You may be viewing animations, taking automatically graded assignments, or using the self-assessments to check your own understanding of the course content. The 20-character student registration code is the only way to safely and legally protect your access to this valuable content.

ARIS™ REGISTRATION PROCEDURES

To register and activate your ARIS account, simply follow these easy steps...

- Go to http://www.mharis.com and click on "Need to Register/I'm a student"
- If your instructor has provided you with a student section code, enter it under the "Join a Course" section and click "Next".
- When prompted, enter the registration code listed below.
- Complete the brief online registration form.
- Suggestion: Bookmark your instructor's ARIS course URL for fast, easy access.
- If your instructor has not provided you with a section code, click "Next" under "Register for Self-Study Resources"
- Suggestion: before "Registering for Self-Study" click on "Plug-In Checker
- Follow Steps 1 and 2 by selecting your book by author. Click on "Silberberg Principles of Chemistry, 2e."
- When prompted, enter the registration code listed below.
- Complete the brief online registration form.
- Suggestion: Bookmark the URL for fast, easy access.

REGISTRATION CODE	HU8C-JNDW-84XP-T3CY-BJVY

Record your information below:

Instructor URL:	
USERNAME:	
PASSWORD:	

If you have trouble with registration, please contact Customer Support at

www.mhhe.com/support

Telephone Support for instructors and students: **(800) 331-5094**
Monday through Thursday: 8 am–11 pm; Friday: 8 am–6 pm; Sunday: 6 pm–11 pm.
All hours are Central Standard Time (CST).

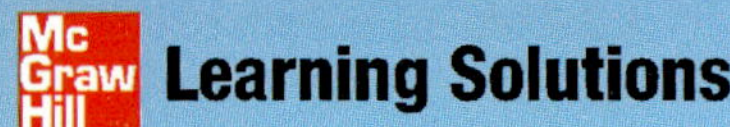

The McGraw-Hill Companies

ISBN-13: 978-0-07-747019-7 ISBN-10: 0-07-747019-2

Principles of GENERAL CHEMISTRY

Volume I

Second Edition

MARTIN S. SILBERBERG

UC SAN DIEGO

Boston Burr Ridge, IL Dubuque, IA New York San Francisco St. Louis
Bangkok Bogotá Caracas Lisbon London Madrid
Mexico City Milan New Delhi Seoul Singapore Sydney Taipei Toronto

The McGraw-Hill Companies

PRINCIPLES OF GENERAL CHEMISTRY, Second Edition
VOLUME I
UC SAN DIEGO

3 4 5 6 7 8 9 0 QDB QDB 12 11 10

ISBN-13: 978-0-07-747013-5
ISBN-10: 0-07-747013-3

Learning Solutions Manager: Danielle Meier
Production Editor: Jennifer Pickel
Cover Design: Fairfax Hutter
Printer/Binder: Quad/Graphics

To Ruth and Daniel,
with all my love and gratitude

Brief Contents

Contents

4 CHAPTER

Three Major Classes of Chemical Reactions 113

6 CHAPTER

Thermochemistry: Energy Flow and Chemical Change 185

7 CHAPTER

Quantum Theory and Atomic Structure 214

8 CHAPTER

Electron Configuration and Chemical Periodicity 245

9 CHAPTER

Models of Chemical Bonding 278

10 CHAPTER

The Shapes of Molecules 305

Keys to the Study of Chemistry

1

Key Principles
to focus on while studying this chapter

- *Matter* can undergo two kinds of change: *physical change* involves a change in *state*—gas, liquid, or solid—but not in ultimate makeup *(composition); chemical change (reaction)* is more fundamental because it does involve a change in composition. The changes we observe result ultimately from changes too small to observe. *(Section 1.1)*
- *Energy* occurs in different forms that are interconvertible, even as the total quantity of energy is conserved. When opposite charges are pulled apart, their *potential energy* increases; when they are released, potential energy is converted to the *kinetic energy* of the charges moving together. Matter consists of charged particles, so changes in energy accompany changes in matter. *(Section 1.1)*
- *Scientific thinking* involves making *observations* and gathering *data* to develop *hypotheses* that are tested by *controlled experiments* until enough results are obtained to create a *model (theory)* that explains how nature works. A sound theory can predict events but must be changed if new results conflict with it. *(Section 1.2)*
- Any *measured quantity* is expressed by a number together with a *unit*. *Conversion factors* are ratios of equivalent quantities having different units; they are used in calculations to change the units of quantities. *Decimal prefixes* and *exponential notation* are used to express very large or very small quantities. *(Section 1.3)*
- *The SI system* consists of seven *fundamental units,* each identifying a physical quantity such as length (meter), mass (kilogram), or temperature (kelvin). These are combined into many *derived units* used to identify quantities such as volume, density, and energy. *Extensive properties,* such as mass, depend on sample size; *intensive properties,* such as temperature, do not. *(Section 1.4)*
- *Uncertainty* characterizes every measurement and is indicated by the number of *significant figures*. We *round* the final answer of a calculation to the same number of digits as in the least certain measurement. *Accuracy* refers to how close a measurement is to the true value; *precision* refers to how close measurements are to one another. *(Section 1.5)*

A Molecular View of the World *Learning the principles of chemistry opens your mind to an amazing world a billion times smaller than the one you see every day, like this view of a lab burner. This chapter introduces some ideas and skills that prepare you to enter this new level of reality.*

Outline

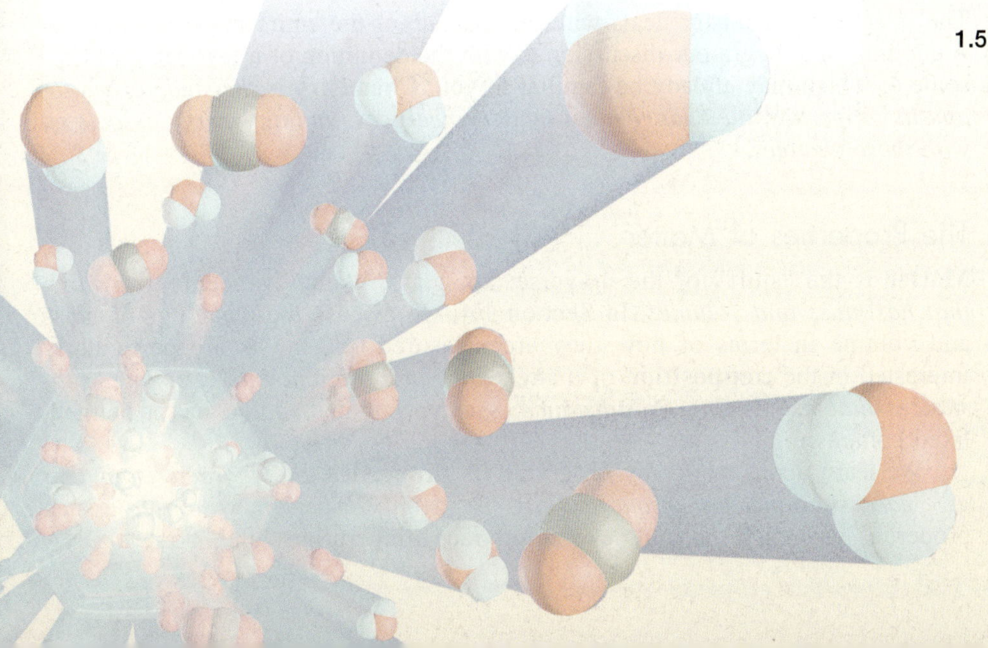

Concepts & Skills to Review before studying this chapter

- exponential (scientific) notation (Appendix A)

Today, as always, the science of chemistry, together with the other sciences that depend on it, stands at the forefront of discovery. Developing "greener" energy sources to power society and using our newfound knowledge of the human genome to cure diseases are but two of the tasks that will occupy researchers in the chemical, biological, and engineering sciences for much of the 21st century. Addressing these and countless other challenges and opportunities depends on an understanding of the concepts you will learn in this course.

The impact of chemistry on your personal, everyday life is mind-boggling. Consider what the beginning of a typical day might look like from a chemical point of view. Molecules align in the liquid crystal display of your alarm clock and electrons flow to create a noise. A cascade of neuronal activators triggers your brain's arousal center, and you throw off a thermal insulator of manufactured polymer. You jump in the shower to emulsify fatty substances on your skin and hair with purified water and formulated detergents. Then, you adorn yourself in an array of processed chemicals—pleasant-smelling pigmented materials suspended in cosmetic gels, dyed polymeric fibers, synthetic footwear, and metal-alloyed jewelry. Breakfast is a bowl of nutrient-enriched, spoilage-retarded cereal and milk, a piece of fertilizer-grown, pesticide-treated fruit, and a cup of a hot aqueous solution of stimulating alkaloid. After abrading your teeth with artificially flavored, dental-hardening agents in a colloidal dispersion, you're ready to leave. You grab your laptop—an electronic device containing ultrathin, microetched semiconductor layers powered by a series of voltaic cells; you collect some books—processed cellulose and plastic, electronically printed with light- and oxygen-resistant inks; you hop in your hydrocarbon-fueled, metal-vinyl-ceramic vehicle, electrically ignite a synchronized series of controlled gaseous explosions, and you're off to class!

This course comes with a bonus—the development of two mental skills you can apply to any science-related field. The first, common to all science courses, is the ability to solve quantitative problems systematically. The second is specific to chemistry, for as you comprehend its ideas, your mind's eye will learn to see a hidden level of the universe, one filled with incredibly minute particles hurtling at fantastic speeds, colliding billions of times a second, and interacting in ways that determine how everything inside and outside of you behaves. The first chapter holds the keys to help you enter this new world.

1.1 SOME FUNDAMENTAL DEFINITIONS

The science of chemistry deals with the makeup of the entire physical universe. A good place to begin our discussion is with the definition of a few central ideas, some of which may already be familiar to you. **Chemistry** is *the study of matter and its properties, the changes that matter undergoes, and the energy associated with those changes.*

The Properties of Matter

Matter is the "stuff" of the universe: air, glass, planets, students—*anything that has mass and volume.* (In Section 1.4, we discuss the meanings of mass and volume in terms of how they are measured.) Chemists are particularly interested in the **composition** of matter, *the types and amounts of simpler substances that make it up.* A *substance* is a type of matter that has a defined, fixed composition.

We learn about matter by observing its **properties,** *the characteristics that give each substance its unique identity.* To identify a person, we observe such properties as height, weight, eye color, race, fingerprints, and, now, a DNA

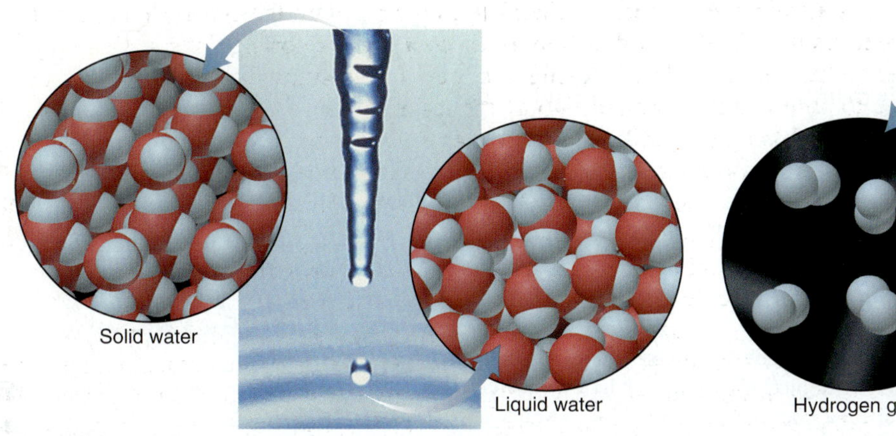

A Physical change:
Solid form of water becomes liquid form; composition does not change because particles are the same.

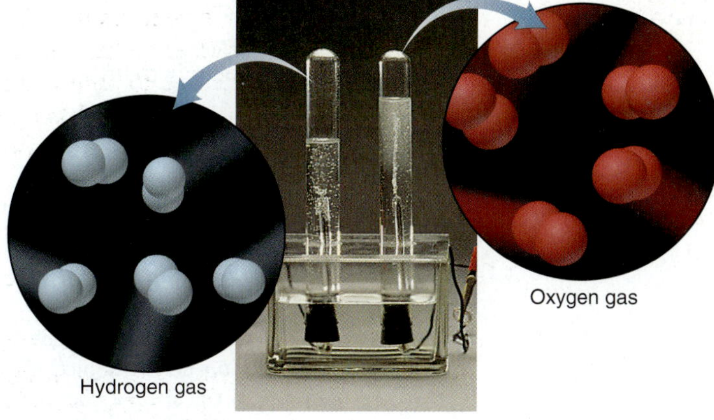

B Chemical change:
Electric current decomposes water into different substances (hydrogen and oxygen); composition does change because particles are different.

FIGURE 1.1 The distinction between physical and chemical change.

pattern, until we arrive at a unique identification. To identify a substance, chemists observe two types of properties, physical and chemical, which are closely related to two types of change that matter undergoes. **Physical properties** are those that a substance shows *by itself, without changing into or interacting with another substance.* Some physical properties are color, melting point, electrical conductivity, and density.

A **physical change** occurs when a substance *alters its physical form,* ***not*** *its composition.* Thus, a physical change results in different physical properties. For example, when ice melts, several physical properties change, such as hardness, density, and ability to flow. But the composition of the sample has *not* changed: the substance is still water. The photo in Figure 1.1A shows this change the way you would see it in everyday life. In your imagination, try to see the magnified view that appears in the "blow-up" circles. Here we see the particles that make up the sample; note that the same particles appear in solid and liquid water, even though they may be arranged differently.

Physical change (same substance before and after):

$$\text{Water (solid form)} \longrightarrow \text{water (liquid form)}$$

On the other hand, **chemical properties** are those that a substance shows *as it changes into or interacts with another substance (or substances).* Some examples of chemical properties are flammability, corrosiveness, and reactivity with acids. A **chemical change,** also called a **chemical reaction,** occurs when *a substance (or substances) is converted into a different substance (or substances).*

Figure 1.1B shows the chemical change (reaction) that occurs when you pass an electric current through water: the water decomposes (breaks down) into two other substances, hydrogen and oxygen, each with physical and chemical properties different from those of the other *and* from those of water. The sample *has* changed its composition: it is no longer water, as you can see from the different particles in the magnified view.

Chemical change (different substances before and after):

$$\text{Water} \xrightarrow{\text{electric current}} \text{hydrogen gas} + \text{oxygen gas}$$

Let's work through a sample problem so that you can visualize this important distinction between physical and chemical change.

SAMPLE PROBLEM 1.1 Visualizing Change on the Atomic Scale

Problem The scenes below represent an atomic-scale view of a sample of matter, A, undergoing two different changes, left to B and right to C:

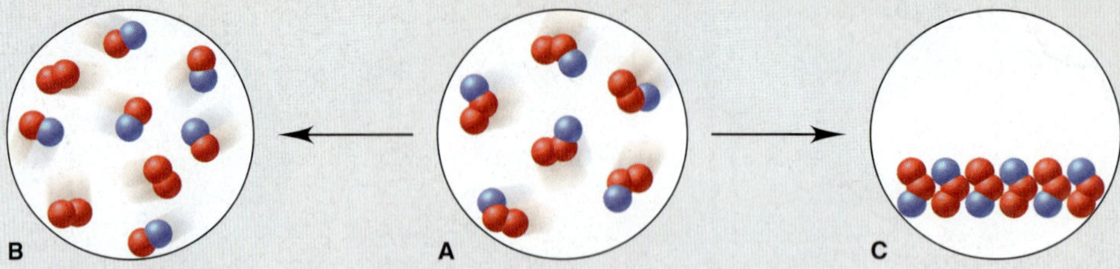

Decide whether each depiction shows a physical or chemical change.

Plan Given depictions of the changes, we have to determine whether each represents a physical or a chemical change. The number and color of the little spheres that make up each particle tell its "composition." Samples with particles of the *same* composition but in a different form depict a *physical* change, and those with particles of a *different* composition depict a *chemical* change.

Solution In A, each particle consists of one blue and two red spheres. The particles in A change into two types in B, one made of red and blue spheres and the other made of two red spheres; therefore, they have undergone a chemical change to form different particles in B. The particles in C are the same as those in A, though they are closer together and aligned; therefore, the conversion from A to C represents a physical change.

FOLLOW-UP PROBLEM 1.1 Is the following change chemical or physical?

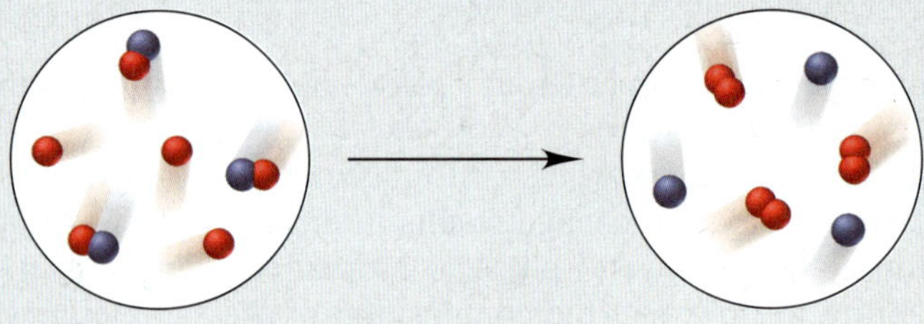

The Three States of Matter

Matter occurs commonly in three physical forms called **states:** solid, liquid, and gas. As shown in Figure 1.2 for a general substance, each state is defined by the way it fills a container. A **solid** *has a fixed shape that does not conform to the container shape*. A **liquid** *conforms to the container shape but fills the container only to the extent of the liquid's volume;* thus, a liquid forms a *surface*. A **gas** *conforms to the container shape also, but it fills the entire container,* and thus, does *not* form a surface. Now, look at the views within the blow-up circles of the figure. The particles in the solid lie next to each other in a regular, three-dimensional array with a definite pattern. Particles in the liquid also lie together but are jumbled and move randomly around one another. Particles in the gas usually have great distances between them, as they move randomly throughout the entire container.

Depending on the temperature and pressure of the surroundings, many substances can exist in each of the three physical states, and they can undergo changes in state as well. For example, as the temperature increases, solid water melts to liquid water and then boils to gaseous water (also called *water vapor*). Similarly, with decreasing temperature, water vapor condenses to liquid water, and the liquid freezes to ice. Benzene, iron, nitrogen, and many other substances behave similarly.

Thus, a physical change caused by heating can generally be reversed by cooling, and vice versa. This is *not* generally true for a chemical change. For example, heating iron in moist air causes a chemical reaction that slowly yields the brown, crumbly substance known as rust. Cooling does not reverse this change; rather, another chemical change (or series of them) is required.

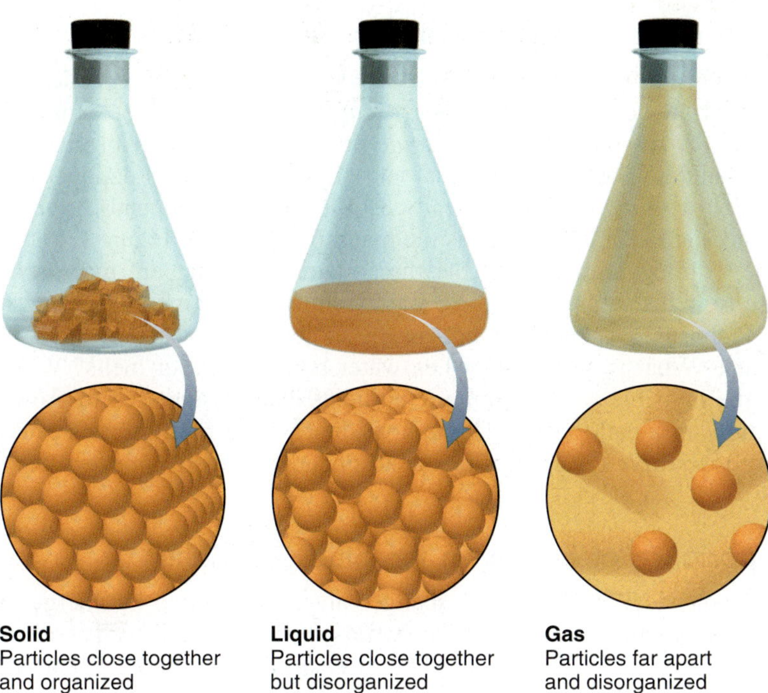

FIGURE 1.2 The physical states of matter. The magnified (blow-up) views show the atomic-scale arrangement of the particles in the three states of matter.

To summarize the key distinctions:

- A physical change leads to a different form of the same substance (same composition), whereas a chemical change leads to a different substance (different composition).
- A physical change caused by a temperature change can generally be reversed by the opposite temperature change, but this is not generally true of a chemical change.

SAMPLE PROBLEM 1.2 Distinguishing Between Physical and Chemical Change

Problem Decide whether each of the following processes is primarily a physical or a chemical change, and explain briefly:

(a) Frost forms as the temperature drops on a humid winter night.
(b) A cornstalk grows from a seed that is watered and fertilized.
(c) A match ignites to form ash and a mixture of gases.
(d) Perspiration evaporates when you relax after jogging.
(e) A silver fork tarnishes slowly in air.

Plan The basic question we ask to decide whether a change is chemical or physical is, "Does the substance change composition or just change form?"

Solution **(a)** Frost forming is a physical change: the drop in temperature changes water vapor (gaseous water) in humid air to ice crystals (solid water).
(b) A seed growing involves chemical change: the seed uses substances from air, fertilizer, soil, and water, and energy from sunlight to make complex changes in composition.
(c) The match burning is a chemical change: combustible substances in the match head are converted into other substances.
(d) Perspiration evaporating is a physical change: the water in sweat changes its form, from liquid to gas, but not its composition.
(e) Tarnishing is a chemical change: silver changes to silver sulfide by reacting with sulfur-containing substances in the air.

FOLLOW-UP PROBLEM 1.2 Decide whether each of the following processes is primarily a physical or a chemical change, and explain briefly:
(a) Purple iodine vapor appears when solid iodine is warmed.
(b) Gasoline fumes are ignited by a spark in an automobile engine cylinder.
(c) A scab forms over an open cut.

The Central Theme in Chemistry

Understanding the properties of a substance and the changes it undergoes leads to the central theme in chemistry: *macroscopic* properties and behavior, those we can see, are the results of *submicroscopic* properties and behavior that we cannot see. The distinction between chemical and physical change is defined by composition, which we study macroscopically. But it ultimately depends on the makeup of substances at the atomic scale, as the magnified views of Figure 1.1 show. Similarly, the defining properties of the three states of matter are macroscopic, but they arise from the submicroscopic behavior shown in the magnified views of Figure 1.2. Picturing a chemical event on the molecular scale helps clarify what is taking place. What is happening when water boils or copper melts? What events occur in the invisible world of minute particles that cause a seed to grow, a neon light to glow, or a nail to rust? Throughout the text, we return to this central idea: we study *observable* changes in matter to understand their *unobservable* causes.

The Importance of Energy in the Study of Matter

In general, physical and chemical changes are accompanied by energy changes. **Energy** is often defined as *the ability to do work.* Essentially, all work involves moving something. Work is done when your arm lifts a book, when an engine moves a car's wheels, or when a falling rock moves the ground as it lands. The object doing the work (arm, engine, rock) transfers some of the energy it possesses to the object on which the work is done (book, wheels, ground).

The total energy an object possesses is the sum of its potential energy and its kinetic energy. **Potential energy** is the *energy due to the* ***position*** *of the object.* **Kinetic energy** is the *energy due to the* ***motion*** *of the object.* Let's examine four systems that illustrate the relationship between these two forms of energy: (1) a weight raised above the ground, (2) two balls attached by a spring, (3) two electrically charged particles, and (4) a fuel and its waste products. A key concept illustrated by all four cases is that *energy is conserved: it may be converted from one form to the other, but it is not destroyed.*

Suppose you lift a weight off the ground, as in Figure 1.3A. The energy you use to move the weight against the gravitational attraction of Earth increases the weight's potential energy (energy due to its position). When the weight is dropped, this additional potential energy is converted to kinetic energy (energy due to motion). Some of this kinetic energy is transferred to the ground as the weight does work, such as driving a stake or simply moving dirt and pebbles. As you can see, the added potential energy does not disappear, but is converted to kinetic energy.

In nature, situations of lower energy are typically favored over those of higher energy: because the weight has less potential energy (and thus less total energy) at rest on the ground than held in the air, it will fall when released. Therefore, the situation with the weight elevated and higher in potential energy is *less stable,* and the situation after the weight has fallen and is lower in potential energy is *more stable.*

Next, consider the two balls attached by a relaxed spring in Figure 1.3B. When you pull the balls apart, the energy you exert to stretch the spring increases its potential energy. This change in potential energy is converted to kinetic energy when you release the balls and they move closer together. The system of balls and spring is less stable (has more potential energy) when the spring is stretched than when it is relaxed.

There are no springs in a chemical substance, of course, but the following situation is similar in terms of energy. Much of the matter in the universe is composed of positively and negatively charged particles. A well-known behavior of

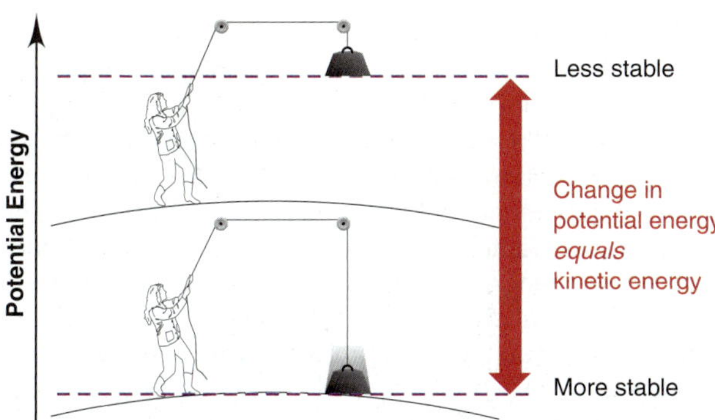

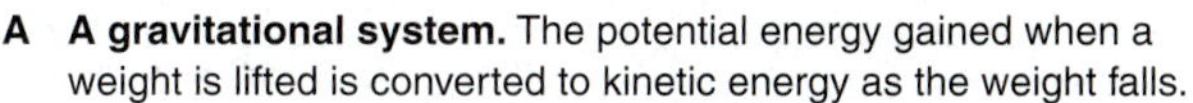

A A gravitational system. The potential energy gained when a weight is lifted is converted to kinetic energy as the weight falls.

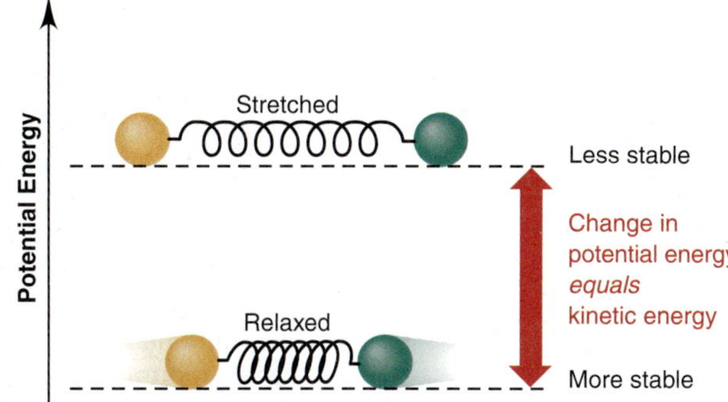

B A system of two balls attached by a spring. The potential energy gained when the spring is stretched is converted to the kinetic energy of the moving balls when it is released.

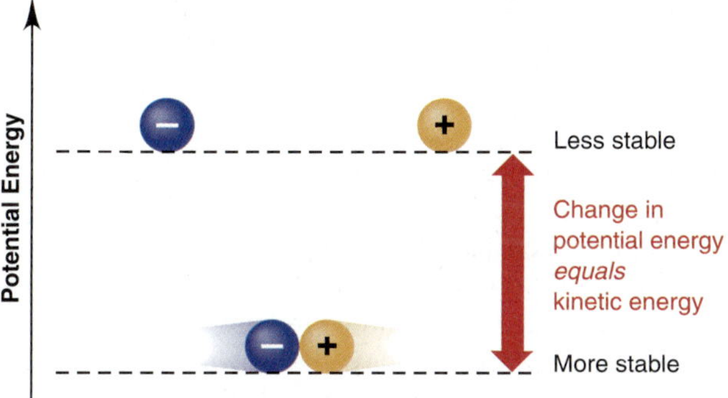

C A system of oppositely charged particles. The potential energy gained when the charges are separated is converted to kinetic energy as the attraction pulls them together.

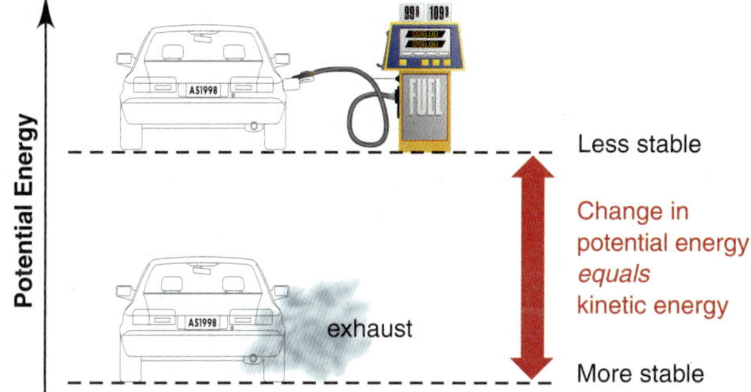

D A system of fuel and exhaust. A fuel is higher in chemical potential energy than the exhaust. As the fuel burns, some of its potential energy is converted to the kinetic energy of the moving car.

FIGURE 1.3 Potential energy is converted to kinetic energy. In all four parts of the figure, the dashed horizontal lines indicate the potential energy of the system in each situation.

charged particles (similar to the behavior of the poles of magnets) results from interactions known as *electrostatic forces: opposite charges attract each other, and like charges repel each other.* When work is done to separate a positive particle from a negative one, the potential energy of the particles increases. As Figure 1.3C shows, that increase in potential energy is converted to kinetic energy when the particles move together again. Also, when two positive (or two negative) particles are pushed toward each other, their potential energy increases, and when they are allowed to move apart, that increase in potential energy is changed into kinetic energy. Like the weight above the ground and the balls connected by a spring, charged particles move naturally toward a position of lower energy, which is more stable.

The chemical potential energy of a substance results from the relative positions and the attractions and repulsions among all its particles. Some substances are richer in this chemical potential energy than others. Fuels and foods, for example, contain more potential energy than the waste products they form. Figure 1.3D shows that when gasoline burns in a car engine, substances with higher chemical potential energy (gasoline and air) form substances with lower potential energy (exhaust gases). This difference in potential energy is eventually converted into the kinetic energy of the moving car; it also heats the passenger compartment, makes the lights shine, and so forth. Similarly, the difference in potential energy between the food and air we take in and the waste products we excrete is used to move, grow, keep warm, study chemistry, and so on. Note again the essential point: *energy is neither created nor destroyed—it is always conserved as it is converted from one form to the other.*

SECTION 1.1 SUMMARY

Chemists study the composition and properties of matter and how they change. • Each substance has a unique set of physical properties (attributes of the substance itself) and chemical properties (attributes of the substance as it interacts with or changes to other substances). • Changes in matter can be physical (different form of the same substance) or chemical (different substance). • Matter exists in three physical states—solid, liquid, and gas. The observable features that distinguish these states reflect the arrangement of their particles. • A change in physical state brought about by heating may be reversed by cooling. A chemical change can be reversed only by other chemical changes. Macroscopic changes result from submicroscopic changes. • Changes in matter are accompanied by changes in energy. • An object's potential energy is due to its position; an object's kinetic energy is due to its motion. • Energy used to lift a weight, stretch a spring, or separate opposite charges increases the system's potential energy, which is converted to kinetic energy as the system returns to its original condition. • Chemical potential energy arises from the positions and interactions of the particles in a substance. Higher energy substances are less stable than lower energy substances. When a less stable substance is converted into a more stable substance, some potential energy is converted into kinetic energy, which can do work.

1.2 THE SCIENTIFIC APPROACH: DEVELOPING A MODEL

The principles of chemistry have been modified over time and are *still* evolving. At the dawn of human experience, our ancestors survived through knowledge acquired by *trial and error:* which types of stone were hard enough to shape others, which plants were edible, and so forth. Today, the science of chemistry, with its powerful *quantitative theories,* helps us understand the essential nature of materials to make better use of them and create new ones: specialized drugs, advanced composites, synthetic polymers, and countless other new materials.

Is there something special about the way scientists think? If we could break down a "typical" modern scientist's thought processes, we could organize them into an approach called the **scientific method.** This approach is not a stepwise checklist, but rather a flexible process of creative thinking and testing aimed at objective, verifiable discoveries about how nature works. Note, however, that there is no typical scientist and no single method, and that luck or a "flash" of insight can and often has played a key role in scientific discovery. In general terms, the scientific approach includes the following parts (Figure 1.4):

1. ***Observations.*** These are the facts that our ideas must explain. Observation is basic to scientific thinking. The most useful observations are quantitative because they can be compared and allow trends to be seen. Pieces of quantitative information are **data.** When the same observation is made by many investigators in many situations with no clear exceptions, it is summarized, often in mathematical terms, and called a **natural law.**

2. ***Hypothesis.*** Whether derived from actual observation or from a "spark" of intuition, a hypothesis is a proposal made to explain an observation. A valid hypothesis need not be correct, but it must be *testable.* Thus, a hypothesis is often the reason for performing an experiment. If the hypothesis is inconsistent with the experimental results, it must be revised or discarded.

3. ***Experiment.*** An experiment is a clear set of procedural steps that tests a hypothesis. Often, hypothesis leads to experiment, which leads to revised hypothesis, and so forth. Hypotheses can be altered, but the results of an experiment cannot.

An experiment typically contains at least two **variables,** quantities that can have more than a single value. A well-designed experiment is **controlled** in that

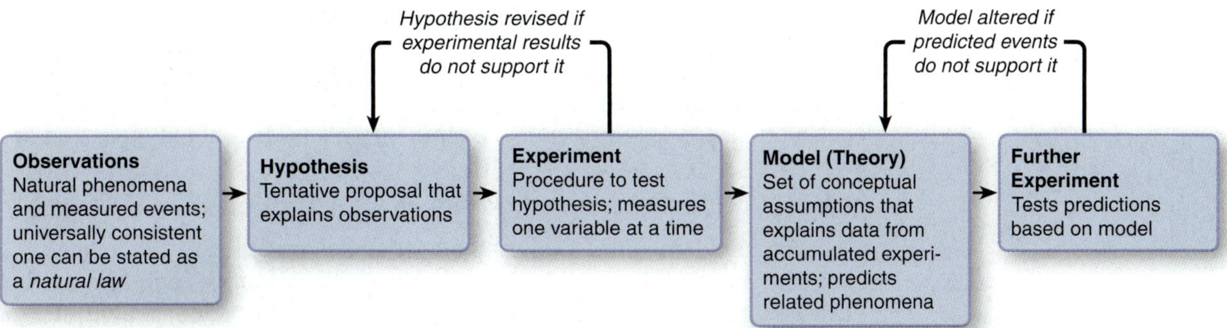

FIGURE 1.4 The scientific approach to understanding nature. Note that hypotheses and models are mental pictures that are changed to match observations and experimental results, *not* the other way around.

it measures the effect of one variable on another while keeping all others constant. For experimental results to be accepted, they must be *reproducible,* not only by the person who designed the experiment, but also by others. Both skill and creativity play a part in experimental design.

4. ***Model.*** Formulating conceptual models, or **theories,** *based on experiments* is what distinguishes scientific thinking from speculation. As hypotheses are revised according to experimental results, a model gradually emerges that describes how the observed phenomenon occurs. A model is not an exact representation of nature, but rather a simplified version of it that can be used to make *predictions* about related phenomena. Further investigation refines a model by testing its predictions and altering it to account for new facts.

The following short paragraph is the first of an occasional feature that will help you learn a concept through an analogy, a unifying idea, or a memorization aid.

THINK OF IT THIS WAY
Everyday Scientific Thinking

Consider this familiar scenario. While listening to an FM broadcast on your stereo system, you notice the sound is garbled (observation) and assume it is caused by poor reception (hypothesis). To isolate this variable, you play a CD (experiment): the sound is still garbled. If the problem is not poor reception, perhaps the speakers are at fault (new hypothesis). To isolate this variable, you play the CD and listen with headphones (experiment): the sound is clear. You conclude that the speakers need to be repaired (model). The repair shop says the speakers check out fine (new observation), but the power amplifier may be at fault (new hypothesis). Repairing the amplifier corrects the garbled sound (new experiment), so the power amplifier was the problem (revised model). Approaching a problem scientifically is a common practice, even if you're not aware of it.

SECTION 1.2 SUMMARY

The scientific method is not a rigid sequence of steps, but rather a dynamic process designed to explain and predict real phenomena. • Observations (sometimes expressed as natural laws) lead to hypotheses about how or why something occurs. • Hypotheses are tested in controlled experiments and adjusted if necessary. • If all the data collected support a hypothesis, a model (theory) can be developed to explain the observations. • A good model is useful in predicting related phenomena but must be refined if conflicting data appear.

1.3 CHEMICAL PROBLEM SOLVING

In many ways, learning chemistry is learning how to solve chemistry problems. In this section, we discuss the problem-solving approach. Most problems include calculations, so let's first go over some important ideas about measured quantities.

Units and Conversion Factors in Calculations

All measured quantities consist of a number *and* a unit; a person's height is "6 feet," not "6." Ratios of quantities have ratios of units, such as miles/hour. (We discuss the most important units in chemistry in the next section.) To minimize errors, try to make a habit of including units in all calculations. The arithmetic operations used with measured quantities are the same as those used with pure numbers; in other words, units can be multiplied, divided, and canceled:

- A carpet measuring 3 feet (ft) by 4 ft has an area of

$$\text{Area} = 3\text{ ft} \times 4\text{ ft} = (3 \times 4)\,(\text{ft} \times \text{ft}) = 12\text{ ft}^2$$

- A car traveling 350 miles (mi) in 7 hours (h) has a speed of

$$\text{Speed} = \frac{350\text{ mi}}{7\text{ h}} = \frac{50\text{ mi}}{1\text{ h}}\ \text{(often written 50 mi·h}^{-1}\text{)}$$

- In 3 hours, the car travels a distance of

$$\text{Distance} = 3\,\cancel{\text{h}} \times \frac{50\text{ mi}}{1\,\cancel{\text{h}}} = 150\text{ mi}$$

Conversion factors are ratios used to express a measured quantity in different units. Suppose we want to know the distance of that 150-mile car trip in feet. To convert the distance between miles and feet, we use equivalent quantities to construct the desired conversion factor. The equivalent quantities in this case are 1 mile and the number of feet in 1 mile:

$$1\text{ mi} = 5280\text{ ft}$$

We can construct two conversion factors from this equivalency. Dividing both sides by 5280 ft gives one conversion factor (shown in blue):

$$\frac{1\text{ mi}}{5280\text{ ft}} = \frac{5280\,\cancel{\text{ft}}}{5280\,\cancel{\text{ft}}} = 1$$

And, dividing both sides by 1 mi gives the other conversion factor (the inverse):

$$\frac{1\,\cancel{\text{mi}}}{1\,\cancel{\text{mi}}} = \frac{5280\text{ ft}}{1\text{ mi}} = 1$$

It's very important to see that, since the numerator and denominator of a conversion factor are equal, multiplying by a conversion factor is the same as multiplying by 1. Therefore, *even though the number and unit of the quantity change, the size of the quantity remains the same.*

In our example, we want to convert the distance in miles to the equivalent distance in feet. Therefore, we choose the conversion factor with units of feet in the numerator, because it cancels units of miles and gives units of feet:

$$\text{Distance (ft)} = 150\,\cancel{\text{mi}} \times \frac{5280\text{ ft}}{1\,\cancel{\text{mi}}} = 792{,}000\text{ ft}$$

$$\text{mi} \Longrightarrow \text{ft}$$

Choosing the correct conversion factor is made much easier if you think through the calculation to decide whether the answer expressed in the new units should have a larger or smaller number. In the previous case, we know that a foot is *smaller* than a mile, so the distance in feet should have a *larger* number (792,000) than the distance in miles (150). The conversion factor has the larger number (5280) in the numerator, so it gave a larger number in the answer. The main goal is that *the chosen conversion factor cancels all units except those required for the answer.* Set up the calculation so that the unit you are converting *from* (beginning unit) is in the *opposite position in the conversion factor* (numerator or denominator). It will then cancel and leave the unit you are converting *to* (final unit):

$$\cancel{\text{beginning unit}} \times \frac{\text{final unit}}{\cancel{\text{beginning unit}}} = \text{final unit} \qquad \text{as in} \qquad \cancel{\text{mi}} \times \frac{\text{ft}}{\cancel{\text{mi}}} = \text{ft}$$

Or, in cases that involve units raised to a power,

$$(\cancel{\text{beginning unit}} \times \cancel{\text{beginning unit}}) \times \frac{\text{final unit}^2}{\cancel{\text{beginning unit}^2}} = \text{final unit}^2$$

$$\text{as in} \qquad (\cancel{\text{ft}} \times \cancel{\text{ft}}) \times \frac{\text{mi}^2}{\cancel{\text{ft}^2}} = \text{mi}^2$$

Or, in cases that involve a ratio of units,

$$\frac{\cancel{\text{beginning unit}}}{\text{final unit}_1} \times \frac{\text{final unit}_2}{\cancel{\text{beginning unit}}} = \frac{\text{final unit}_2}{\text{final unit}_1} \qquad \text{as in} \qquad \frac{\cancel{\text{mi}}}{\text{h}} \times \frac{\text{ft}}{\cancel{\text{mi}}} = \frac{\text{ft}}{\text{h}}$$

We use the same procedure to convert between systems of units, for example, between the English (or American) unit system and the International System (a revised metric system discussed fully in the next section). Suppose we know the height of Angel Falls in Venezuela to be 3212 ft, and we find its height in miles as

$$\text{Height (mi)} = 3212\ \cancel{\text{ft}} \times \frac{1\ \text{mi}}{5280\ \cancel{\text{ft}}} = 0.6083\ \text{mi}$$

$$\text{ft} \Longrightarrow \text{mi}$$

Now, we want its height in kilometers (km). The equivalent quantities are

$$1.609\ \text{km} = 1\ \text{mi}$$

Because we are converting from miles to kilometers, we use the conversion factor with kilometers in the numerator in order to cancel miles:

$$\text{Height (km)} = 0.6083\ \cancel{\text{mi}} \times \frac{1.609\ \text{km}}{1\ \cancel{\text{mi}}} = 0.9788\ \text{km}$$

$$\text{mi} \Longrightarrow \text{km}$$

Notice that, because kilometers are *smaller* than miles, this conversion factor gave us a *larger* number (0.9788 is larger than 0.6083).

If we want the height of Angel Falls in meters (m), we use the equivalent quantities 1 km = 1000 m to construct the conversion factor:

$$\text{Height (m)} = 0.9788\ \cancel{\text{km}} \times \frac{1000\ \text{m}}{1\ \cancel{\text{km}}} = 978.8\ \text{m}$$

$$\text{km} \Longrightarrow \text{m}$$

In longer calculations, we often string together several conversion steps:

$$\text{Height (m)} = 3212\ \cancel{\text{ft}} \times \frac{1\ \cancel{\text{mi}}}{5280\ \cancel{\text{ft}}} \times \frac{1.609\ \cancel{\text{km}}}{1\ \cancel{\text{mi}}} \times \frac{1000\ \text{m}}{1\ \cancel{\text{km}}} = 978.8\ \text{m}$$

$$\text{ft} \Longrightarrow \text{mi} \Longrightarrow \text{km} \Longrightarrow \text{m}$$

The use of conversion factors in calculations is known by various names, such as the factor-label method or **dimensional analysis** (because units represent physical dimensions). We use this method in quantitative problems throughout the text.

A Systematic Approach to Solving Chemistry Problems

The approach we use in this text provides a systematic way to work through a problem. It emphasizes reasoning, not memorizing, and is based on a very simple idea: plan how to solve the problem *before* you go on to solve it, and then check your answer. Try to develop a similar approach on homework and exams.

In general, the sample problems consist of several parts:

1. **Problem.** This part states all the information you need to solve the problem (usually framed in some interesting context).
2. **Plan.** The overall solution is broken up into two parts, *plan* and *solution,* to make a point: *think* about how to solve the problem *before* juggling numbers. There is often more than one way to solve a problem, and the plan shown in a given problem is just one possibility; develop a plan that seems clearest to you. The plan will
   - Clarify the known and unknown. (What information do you have, and what are you trying to find?)
   - Suggest the steps from known to unknown. (What ideas, conversions, or equations are needed to solve the problem?)
   - Present a "roadmap" of the solution for many problems in early chapters (and in some later ones). The roadmap is a visual summary of the planned steps. Each step is shown by an arrow labeled with information about the conversion factor or operation needed.
3. **Solution.** In this part, the steps appear in the same order as in the plan.
4. **Check.** In most cases, a quick check is provided to see if the results make sense: Are the units correct? Does the answer seem to be the right size? Did the change occur in the expected direction? And, most important, is it reasonable chemically? We often do a rough calculation to see if the answer is "in the same ballpark" as the calculated result, just to make sure we didn't make a large error. Here's a typical "ballpark" calculation. You are at the music store and buy three CDs at $14.97 each. With a 5% sales tax, the bill comes to $47.16. In your mind, you round $14.97 to $15, and quickly compute that 3 times $15 is $45; given the sales tax, the cost should be a bit more. So, the amount of the bill *is* in the right ballpark. *Always* check your answers, especially in a multipart problem, where an error in an early step can affect all later steps.
5. **Comment.** This part is included occasionally to provide additional information, such as an application, an alternative approach, a common mistake to avoid, or an overview.
6. **Follow-up Problem.** This part consists of a problem statement only and applies the same ideas as the sample problem. Try to solve it *before* you look at the brief worked-out solution at the end of the chapter.

Of course, you can't learn to solve chemistry problems, any more than you can learn to swim, by reading about doing it. Practice is the key. Try to:

- Follow along in the sample problem with pencil, paper, and calculator.
- Do the follow-up problem as soon as you finish studying the sample problem. Check your calculation steps and answer against the *brief solution* at the end of the chapter.
- Read the sample problem and text explanations again if you have trouble.
- Work on as many of the problems at the end of the chapter as you can. They review and extend the concepts and skills in the text. Answers are given in the back of the book for problems with a colored number, but try to solve them yourself first. Now let's apply this approach in a unit-conversion problem.

SAMPLE PROBLEM 1.3 Converting Units of Length

Problem To wire your stereo equipment, you need 325 centimeters (cm) of speaker wire that sells for $0.15/ft. What is the price of the wire?

Plan We know the length of wire in centimeters (325 cm) and the cost in dollars per foot ($0.15/ft). We can find the unknown price of the wire by converting the length from centimeters to inches (in) and from inches to feet. Then, we use the cost as a conversion factor to convert feet of wire to price in dollars. The roadmap starts with the known and moves through the calculation steps to the unknown.

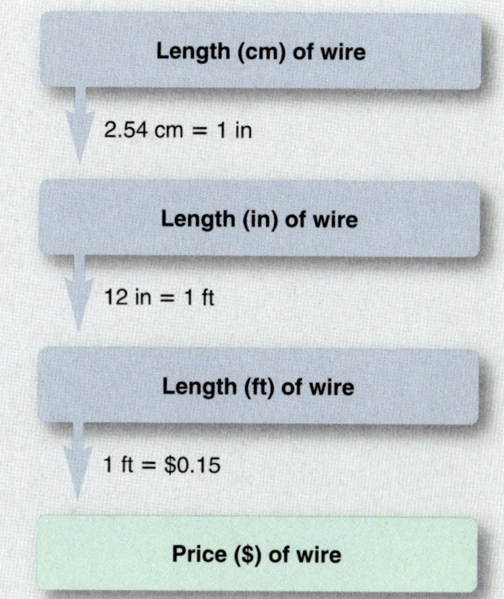

Solution Converting the known length from centimeters to inches: The equivalent quantities alongside the roadmap arrow are the ones needed to construct the conversion factor. We choose 1 in/2.54 cm, rather than the inverse, because it gives an answer in inches:

$$\text{Length (in)} = \text{length (cm)} \times \text{conversion factor} = 325\ \cancel{\text{cm}} \times \frac{1\ \text{in}}{2.54\ \cancel{\text{cm}}} = 128\ \text{in}$$

Converting the length from inches to feet:

$$\text{Length (ft)} = \text{length (in)} \times \text{conversion factor} = 128\ \cancel{\text{in}} \times \frac{1\ \text{ft}}{12\ \cancel{\text{in}}} = 10.7\ \text{ft}$$

Converting the length in feet to price in dollars:

$$\text{Price (\$)} = \text{length (ft)} \times \text{conversion factor} = 10.7\ \cancel{\text{ft}} \times \frac{\$0.15}{1\ \cancel{\text{ft}}} = \$1.60$$

Check The units are correct for each step. The conversion factors make sense in terms of the relative unit sizes: the number of inches is *smaller* than the number of centimeters (an inch is *larger* than a centimeter), and the number of feet is *smaller* than the number of inches. The total price seems reasonable: a little more than 10 ft of wire at $0.15/ft should cost a little more than $1.50.

Comment 1. We could also have strung the three steps together:

$$\text{Price (\$)} = 325\ \cancel{\text{cm}} \times \frac{1\ \cancel{\text{in}}}{2.54\ \cancel{\text{cm}}} \times \frac{1\ \cancel{\text{ft}}}{12\ \cancel{\text{in}}} \times \frac{\$0.15}{1\ \cancel{\text{ft}}} = \$1.60$$

2. There are usually alternative sequences in unit-conversion problems. Here, for example, we would get the same answer if we first converted the cost of wire from $/ft to $/cm and kept the wire length in cm. Try it yourself.

FOLLOW-UP PROBLEM 1.3 A furniture factory needs 31.5 ft^2 of fabric to upholster one chair. Its Dutch supplier sends the fabric in bolts of exactly 200 m^2. What is the maximum number of chairs that can be upholstered by 3 bolts of fabric (1 m = 3.281 ft)?

SECTION 1.3 SUMMARY

A measured quantity consists of a number and a unit. Conversion factors are used to express a quantity in different units and are constructed as a ratio of equivalent quantities. • The problem-solving approach used in this text usually has four parts: (1) devise a plan for the solution, (2) put the plan into effect in the calculations, (3) check to see if the answer makes sense, and (4) practice with similar problems.

1.4 MEASUREMENT IN SCIENTIFIC STUDY

Almost everything we own—clothes, house, food, vehicle—is manufactured with measured parts, sold in measured amounts, and paid for with measured currency. Measurement has a history characterized by the search for *exact, invariable standards.* Our current system of measurement began in 1790, when the newly formed National Assembly of France set up a committee to establish consistent unit standards. This effort led to the development of the *metric system.* In 1960, another international committee met in France to establish the International System of Units, a revised metric system now accepted by scientists throughout the world. The units of this system are called **SI units,** from the French **S**ystème **I**nternational d'Unités.

General Features of SI Units

As Table 1.1 shows, the SI system is based on a set of seven **fundamental units,** or **base units,** each of which is identified with a physical quantity. All other units, called **derived units,** are combinations of these seven base units. For example, the derived unit for speed, meters per second (m/s), is the base unit for length (m) divided by the base unit for time (s). (Derived units that occur as a ratio of two or more base units can be used as conversion factors.) For quantities that are much smaller or much larger than the base unit, we use decimal prefixes and exponential (scientific) notation. Table 1.2 shows the most important prefixes. (If you need a review of exponential notation, read Appendix A.) Because these prefixes are based on powers of 10, SI units are easier to use in calculations than are English units such as pounds and inches.

Table 1.1 SI Base Units

Physical Quantity (Dimension)	Unit Name	Unit Abbreviation
Mass	kilogram	kg
Length	meter	m
Time	second	s
Temperature	kelvin	K
Electric current	ampere	A
Amount of substance	mole	mol
Luminous intensity	candela	cd

Table 1.2 Common Decimal Prefixes Used with SI Units

Prefix*	Prefix Symbol	Word	Conventional Notation	Exponential Notation
tera	T	trillion	1,000,000,000,000	1×10^{12}
giga	G	billion	1,000,000,000	1×10^{9}
mega	M	million	1,000,000	1×10^{6}
kilo	k	thousand	1,000	1×10^{3}
hecto	h	hundred	100	1×10^{2}
deka	da	ten	10	1×10^{1}
—	—	one	1	1×10^{0}
deci	d	tenth	0.1	1×10^{-1}
centi	c	hundredth	0.01	1×10^{-2}
milli	m	thousandth	0.001	1×10^{-3}
micro	μ	millionth	0.000001	1×10^{-6}
nano	n	billionth	0.000000001	1×10^{-9}
pico	p	trillionth	0.000000000001	1×10^{-12}
femto	f	quadrillionth	0.000000000000001	1×10^{-15}

*The prefixes most frequently used by chemists appear in bold type.

Some Important SI Units in Chemistry

Let's discuss some of the SI units for quantities that we use early in the text: length, volume, mass, density, temperature, and time. (Units for other quantities are presented in later chapters, as they are used.) Table 1.3 shows some useful SI quantities for length, volume, and mass, along with their equivalents in the English system.

Table 1.3 Common SI-English Equivalent Quantities

Quantity	SI	SI Equivalents	English Equivalents	English to SI Equivalent
Length	1 kilometer (km)	1000 (10^3) meters	0.6214 mile (mi)	1 mile = 1.609 km
	1 meter (m)	100 (10^2) centimeters	1.094 yards (yd)	1 yard = 0.9144 m
		1000 millimeters (mm)	39.37 inches (in)	1 foot (ft) = 0.3048 m
	1 centimeter (cm)	0.01 (10^{-2}) meter	0.3937 inch	1 inch = 2.54 cm (exactly)
Volume	1 cubic meter (m^3)	1,000,000 (10^6) cubic centimeters	35.31 cubic feet (ft^3)	1 cubic foot = 0.02832 m^3
	1 cubic decimeter (dm^3)	1000 cubic centimeters	0.2642 gallon (gal)	1 gallon = 3.785 dm^3
			1.057 quarts (qt)	1 quart = 0.9464 dm^3
				1 quart = 946.4 cm^3
	1 cubic centimeter (cm^3)	0.001 dm^3	0.03381 fluid ounce	1 fluid ounce = 29.57 cm^3
Mass	1 kilogram (kg)	1000 grams	2.205 pounds (lb)	1 pound = 0.4536 kg
	1 gram (g)	1000 milligrams (mg)	0.03527 ounce (oz)	1 ounce = 28.35 g

Length The SI base unit of length is the **meter (m).** It is about 2.5 times the width of this textbook when open. The standard meter is defined as the distance light travels in a vacuum in 1/299,792,458 second. Biological cells are often measured in micrometers (1 μm = 10^{-6} m). On the atomic-size scale, nanometers and picometers are used (1 nm = 10^{-9} m; 1 pm = 10^{-12} m). Many proteins have diameters of around 2 nm; atomic diameters are around 200 pm (0.2 nm). An older unit still in use is the angstrom (1 Å = 10^{-10} m = 0.1 nm = 100 pm).

Volume Any sample of matter has a certain **volume (*V*),** the amount of space that the sample occupies. The SI unit of volume is the **cubic meter (m^3).** In chemistry, the most important volume units are non-SI units, the **liter (L)** and the **milliliter (mL)** (note the uppercase L). A liter is slightly larger than a quart (qt) (1 L = 1.057 qt; 1 qt = 946.4 mL). Physicians and other medical practitioners measure body fluids in cubic decimeters (dm^3), which is equivalent to liters:

$$1\text{ L} = 1\text{ dm}^3 = 10^{-3}\text{ m}^3$$

As the prefix *milli-* indicates, 1 mL is $\frac{1}{1000}$ of a liter, and it is equal to exactly 1 cubic centimeter (cm^3):

$$1\text{ mL} = 1\text{ cm}^3 = 10^{-3}\text{ dm}^3 = 10^{-3}\text{ L} = 10^{-6}\text{ m}^3$$

Figure 1.5 shows some of the types of laboratory glassware designed to contain liquids or measure their volumes.

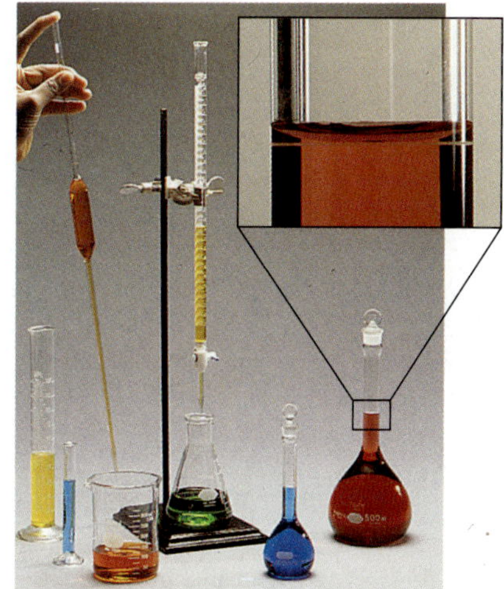

FIGURE 1.5 Common laboratory volumetric glassware. From left to right are two graduated cylinders, a pipet being emptied into a beaker, a buret delivering liquid to an Erlenmeyer flask, and two volumetric flasks. **Inset,** In contact with glass, this liquid forms a concave meniscus (curved surface).

SAMPLE PROBLEM 1.4 Converting Units of Volume

Problem The volume of an irregularly shaped solid can be determined from the volume of water it displaces. A graduated cylinder contains 19.9 mL of water. When a small piece of galena, an ore of lead, is added, it sinks and the volume increases to 24.5 mL. What is the volume of the piece of galena in cm^3 and in L?

Plan We have to find the volume of the galena from the change in volume of the cylinder contents. The volume of galena in mL is the difference in the known volumes before (19.9 mL) and after (24.5 mL) adding it. The units mL and cm^3 represent identical volumes, so the volume of the galena in mL equals the volume in cm^3. We construct a conversion factor to convert the volume from mL to L. The calculation steps are shown in the roadmap on the next page.

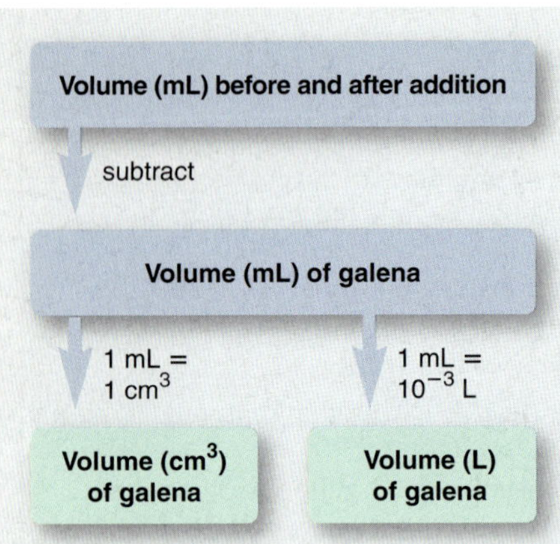

Solution Finding the volume of galena:

$$\text{Volume (mL)} = \text{volume after} - \text{volume before} = 24.5\ \text{mL} - 19.9\ \text{mL} = 4.6\ \text{mL}$$

Converting the volume from mL to cm^3:

$$\text{Volume (cm}^3) = 4.6\ \cancel{\text{mL}} \times \frac{1\ \text{cm}^3}{1\ \cancel{\text{mL}}} = 4.6\ \text{cm}^3$$

Converting the volume from mL to L:

$$\text{Volume (L)} = 4.6\ \cancel{\text{mL}} \times \frac{10^{-3}\ \text{L}}{1\ \cancel{\text{mL}}} = 4.6\times10^{-3}\ \text{L}$$

Check The units and magnitudes of the answers seem correct. It makes sense that the volume expressed in mL would have a number 1000 times larger than the volume expressed in L, because a milliliter is $\frac{1}{1000}$ of a liter.

FOLLOW-UP PROBLEM 1.4 Within a cell, proteins are synthesized on particles called ribosomes. Assuming ribosomes are generally spherical, what is the volume (in dm^3 and μL) of a ribosome whose average diameter is 21.4 nm (V of a sphere $= \frac{4}{3}\pi r^3$)?

Mass The **mass** of an object refers to the quantity of matter it contains. The SI unit of mass is the **kilogram (kg),** the only base unit whose standard is a physical object—a platinum-iridium cylinder kept in France. It is also the only base unit whose name has a prefix. (In contrast to the practice with other base units, however, we attach prefixes to the word "gram," rather than to the word "kilogram"; thus, 10^{-3} gram is 1 milligram, not 1 microkilogram.)

The terms *mass* and *weight* have distinct meanings. Because a given object's quantity of matter cannot change, its *mass is constant.* Its **weight,** on the other hand, depends on its mass *and* the strength of the local gravitational field pulling on it. Because the strength of this field varies with height above Earth's surface, the object's weight also varies. For instance, you actually weigh slightly less on a high mountaintop than at sea level.

SAMPLE PROBLEM 1.5 Converting Units of Mass

Problem International computer communications are often carried by optical fibers in cables laid along the ocean floor. If one strand of optical fiber weighs 1.19×10^{-3} lb/m, what is the mass (in kg) of a cable made of six strands of optical fiber, each long enough to link New York and Paris (8.84×10^3 km)?

Plan We have to find the mass of cable (in kg) from the given mass/length of fiber (1.19×10^{-3} lb/m), number of fibers/cable (6 fibers/cable), and the length (8.84×10^3 km, distance from New York to Paris). One way to do this (as shown in the roadmap) is to first find the mass of one fiber and then find the mass of cable. We convert the length of one fiber from km to m and then find its mass (in lb) by using the lb/m factor. The cable mass is six times the fiber mass, and finally we convert lb to kg.

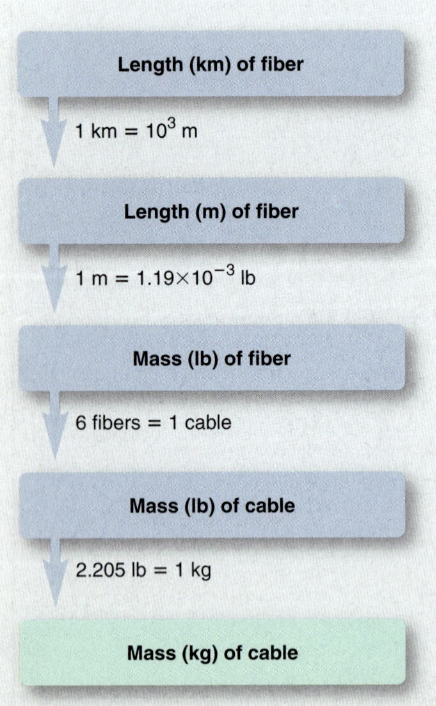

Solution Converting the fiber length from km to m:

$$\text{Length (m) of fiber} = 8.84\times10^3\ \cancel{\text{km}} \times \frac{10^3\ \text{m}}{1\ \cancel{\text{km}}} = 8.84\times10^6\ \text{m}$$

Converting the length of one fiber to mass (lb):

$$\text{Mass (lb) of fiber} = 8.84\times10^6\ \cancel{\text{m}} \times \frac{1.19\times10^{-3}\ \text{lb}}{1\ \cancel{\text{m}}} = 1.05\times10^4\ \text{lb}$$

Finding the mass of the cable (lb):

$$\text{Mass (lb) of cable} = \frac{1.05\times10^4\ \text{lb}}{1\ \cancel{\text{fiber}}} \times \frac{6\ \cancel{\text{fibers}}}{1\ \text{cable}} = 6.30\times10^4\ \text{lb/cable}$$

Converting the mass of cable from lb to kg:

$$\text{Mass (kg) of cable} = \frac{6.30\times10^4\ \cancel{\text{lb}}}{1\ \text{cable}} \times \frac{1\ \text{kg}}{2.205\ \cancel{\text{lb}}} = 2.86\times10^4\ \text{kg/cable}$$

Check The units are correct. Let's think through the relative sizes of the answers to see if they make sense: The number of m should be 10^3 larger than the number of km. If 1 m of fiber weighs about 10^{-3} lb, about 10^7 m should weigh about 10^4 lb. The cable mass should be six times as much, or about 6×10^4 lb. Since 1 lb is about $\frac{1}{2}$ kg, the number of kg should be about half the number of lb.

FOLLOW-UP PROBLEM 1.5 An intravenous bag delivers a nutrient solution to a hospital patient at a rate of 1.5 drops per second. If a drop weighs 65 mg on average, how many kilograms of solution are delivered in 8.0 h?

Density The **density (*d*)** of an object is its mass divided by its volume:

$$\text{Density} = \frac{\text{mass}}{\text{volume}} \tag{1.1}$$

Whenever needed, you can isolate mathematically each of the component variables by treating density as a conversion factor:

$$\text{Mass} = \text{volume} \times \text{density} = \cancel{\text{volume}} \times \frac{\text{mass}}{\cancel{\text{volume}}}$$

Or,

$$\text{Volume} = \text{mass} \times \frac{1}{\text{density}} = \cancel{\text{mass}} \times \frac{\text{volume}}{\cancel{\text{mass}}}$$

Because volume may change with temperature, density may change also. But, under given conditions of temperature and pressure, *density is a characteristic physical property of a substance* and has a specific value. Mass and volume are examples of **extensive properties,** those dependent on the amount of substance. Density, on the other hand, is an **intensive property,** one that is independent of the amount of substance. For example, the mass of a gallon of water is four times the mass of a quart of water, but its volume is also four times greater; therefore, the density of the water, the *ratio* of its mass to its volume, is constant at a particular temperature and pressure, regardless of the sample size.

The SI unit of density is the kilogram per cubic meter (kg/m^3), but in chemistry, density is typically given in units of g/L (g/dm^3) or g/mL (g/cm^3). As you might expect from the magnified views in Figure 1.2, at ordinary pressure and 20°C, the densities of gases (for example, 0.0000899 g/cm^3 for hydrogen) are much lower than the densities of either liquids (1.00 g/cm^3 for water) or solids (2.17 g/cm^3 for aluminum).

SAMPLE PROBLEM 1.6 Calculating Density from Mass and Length

Problem Lithium is a soft, gray solid that has the lowest density of any metal. It is an essential component of some advanced batteries, such as the one in your laptop. If a small rectangular slab of lithium weighs 1.49×10^3 mg and has sides that measure 20.9 mm by 11.1 mm by 11.9 mm, what is the density of lithium in g/cm^3?

Plan To find the density in g/cm^3, we need the mass of lithium in g and the volume in cm^3. The mass is given in mg (1.49×10^3 mg), so we convert mg to g. Volume data are not given, but we can convert the given side lengths (20.9 mm, 11.1 mm, 11.9 mm), from mm to cm, and then multiply them to find the volume in cm^3. Finally, we divide mass by volume to get density. The steps are shown in the roadmap.

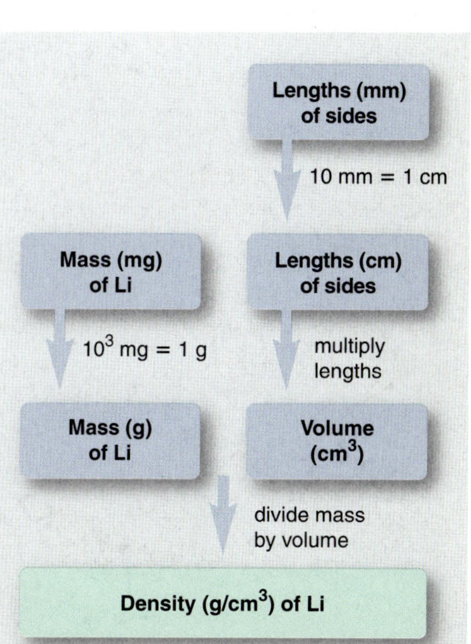

Solution Converting the mass from mg to g:

$$\text{Mass (g) of lithium} = 1.49\times10^3 \cancel{\text{mg}} \left(\frac{10^{-3}\text{ g}}{1 \cancel{\text{mg}}}\right) = 1.49 \text{ g}$$

Converting side lengths from mm to cm:

$$\text{Length (cm) of one side} = 20.9 \cancel{\text{mm}} \times \frac{1 \text{ cm}}{10 \cancel{\text{mm}}} = 2.09 \text{ cm}$$

Similarly, the other side lengths are 1.11 cm and 1.19 cm.

Finding the volume:

$$\text{Volume (cm}^3) = 2.09 \text{ cm} \times 1.11 \text{ cm} \times 1.19 \text{ cm} = 2.76 \text{ cm}^3$$

Calculating the density:

$$\text{Density of lithium} = \frac{\text{mass}}{\text{volume}} = \frac{1.49 \text{ g}}{2.76 \text{ cm}^3} = \boxed{0.540 \text{ g/cm}^3}$$

Check Since 1 cm = 10 mm, the number of cm in each length should be $\frac{1}{10}$ the number of mm. The units for density are correct, and the size of the answer (~0.5 g/cm^3) seems correct since the number of g (1.49) is about half the number of cm^3 (2.76). The problem states that lithium has a very low density, so this answer makes sense.

FOLLOW-UP PROBLEM 1.6 The piece of galena in Sample Problem 1.4 has a volume of 4.6 cm^3. If the density of galena is 7.5 g/cm^3, what is the mass (in kg) of that piece of galena?

Temperature There is a common misunderstanding about heat and temperature. **Temperature (*T*)** is a measure of how hot or cold a substance is *relative to another substance.* **Heat** is the energy that flows between objects that are at different temperatures. Temperature is related to the *direction* of that energy flow: when two objects at different temperatures touch, energy flows *from* the one with the higher temperature *to* the one with the lower temperature until their temperatures are equal. When you hold an ice cube, it feels cold because heat flows *from* your hand into the ice. (In Chapter 6, we will see how heat is measured and how it is related to chemical and physical change.) Energy is an *extensive* property (as is volume), but temperature is an *intensive* property (as is density): a vat of boiling water has more energy than a cup of boiling water, but the temperatures of the two water samples are the same.

In the laboratory, the most common means for measuring temperature is the **thermometer,** a device that contains a fluid that expands when it is heated. When the thermometer's fluid-filled bulb is immersed in a substance hotter than itself, heat flows from the substance through the glass and into the fluid, which expands and rises in the thermometer tube. If a substance is colder than the thermometer, heat flows outward from the fluid, which contracts and falls within the tube.

The three temperature scales most important for us to consider are the Celsius (°C, formerly called centigrade), the Kelvin (K), and the Fahrenheit (°F) scales. The SI base unit of temperature is the **kelvin (K);** note that the kelvin has no degree sign (°). The Kelvin scale, also known as the *absolute scale,* is preferred in all scientific work, although the Celsius scale is used frequently. In the United States, the Fahrenheit scale is still used for weather reporting, body temperature, and other everyday purposes. *The three scales differ in the size of the unit and/or the temperature of the zero point.* Figure 1.6 shows the freezing and boiling points of water in the three scales.

The **Celsius scale,** devised in the 18th century by the Swedish astronomer Anders Celsius, is based on changes in the physical state of water: 0°C is set at water's freezing point, and 100°C is set at its boiling point (at normal atmospheric pressure). The **Kelvin (absolute) scale** was devised by the English physicist William Thomson, known as Lord Kelvin, in 1854 during his experiments on the expansion and contraction of gases. *The Kelvin scale uses the same size degree unit as the Celsius scale*—$\frac{1}{100}$ of the difference between the freezing and boiling points of water—*but it differs in zero point.* The zero point in the Kelvin scale, 0 K, is called *absolute zero* and equals −273.15°C. In the Kelvin scale, *all temperatures have positive values.* Water freezes at +273.15 K (0°C) and boils at +373.15 K (100°C).

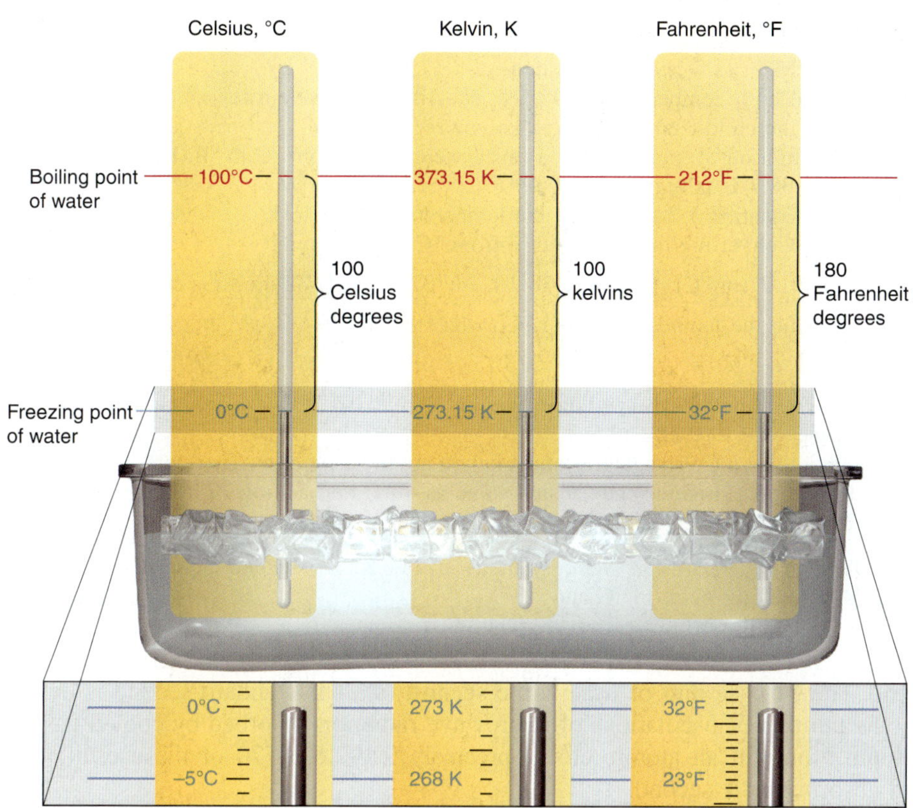

FIGURE 1.6 The freezing point and the boiling point of water in the Celsius, Kelvin (absolute), and Fahrenheit temperature scales. As you can see, this range consists of 100 degrees on the Celsius and Kelvin scales, but 180 degrees on the Fahrenheit scale. At the bottom of the figure, a portion of each of the three thermometer scales is expanded to show the sizes of the units. A Celsius degree (°C; *left*) and a kelvin (K; *center*) are the same size, and each is $\frac{9}{5}$ the size of a Fahrenheit degree (°F; *right*).

We can convert between the Celsius and Kelvin scales by remembering the difference in zero points: since 0°C = 273.15 K,

$$T \text{ (in K)} = T \text{ (in °C)} + 273.15 \qquad \textbf{(1.2)}$$

Solving Equation 1.2 for T (in °C) gives

$$T \text{ (in °C)} = T \text{ (in K)} - 273.15 \qquad \textbf{(1.3)}$$

The Fahrenheit scale differs from the other scales in its zero point *and* in the size of its unit. Water freezes at 32°F and boils at 212°F. Therefore, 180 Fahrenheit degrees (212°F − 32°F) represents the same temperature change as 100 Celsius degrees (or 100 kelvins). Because 100 Celsius degrees equal 180 Fahrenheit degrees,

$$1 \text{ Celsius degree} = \frac{180}{100} \text{ Fahrenheit degrees} = \frac{9}{5} \text{ Fahrenheit degrees}$$

To convert a temperature in °C to °F, first change the degree size and then adjust the zero point:

$$T \text{ (in °F)} = \tfrac{9}{5}T \text{ (in °C)} + 32 \qquad \textbf{(1.4)}$$

To convert a temperature in °F to °C, do the two steps in the opposite order; that is, first adjust the zero point and then change the degree size. In other words, solve Equation 1.4 for T (in °C):

$$T \text{ (in °C)} = [T \text{ (in °F)} - 32]\tfrac{5}{9} \qquad \textbf{(1.5)}$$

(The only temperature with the same numerical value in the Celsius and Fahrenheit scales is −40°; that is, −40°F = −40°C.)

SAMPLE PROBLEM 1.7 Converting Units of Temperature

Problem A child has a body temperature of 38.7°C.
(a) If normal body temperature is 98.6°F, does the child have a fever?
(b) What is the child's temperature in kelvins?

Plan **(a)** To find out if the child has a fever, we convert from °C to °F (Equation 1.4) and see whether 38.7°C is higher than 98.6°F.
(b) We use Equation 1.2 to convert the temperature in °C to K.

Solution **(a)** Converting the temperature from °C to °F:

$$T\text{ (in °F)} = \tfrac{9}{5}T\text{ (in °C)} + 32 = \tfrac{9}{5}(38.7\text{°C}) + 32 = 101.7\text{°F; yes, the child has a fever.}$$

(b) Converting the temperature from °C to K:

$$T\text{ (in K)} = T\text{ (in °C)} + 273.15 = 38.7\text{°C} + 273.15 = 311.8\text{ K}$$

Check **(a)** From everyday experience, you know that 101.7°F is a reasonable temperature for someone with a fever.
(b) We know that a Celsius degree and a kelvin are the same size. Therefore, we can check the math by approximating the Celsius value as 40°C and adding 273: 40 + 273 = 313, which is close to our calculation, so there is no large error.

FOLLOW-UP PROBLEM 1.7 Mercury melts at 234 K, lower than any other pure metal. What is its melting point in °C and °F?

Time The SI base unit of time is the **second (s).** The standard second is defined by the number of oscillations of microwave radiation absorbed by cooled gaseous cesium atoms in an atomic clock; precisely 9,192,631,770 of these oscillations are absorbed in 1 second.

In the laboratory, we study the speed (or *rate*) of a reaction by measuring the time it takes a fixed amount of substance to undergo a chemical change. The range of reaction rates is enormous: a fast reaction may be over in less than a nanosecond (10^{-9} s), whereas slow ones, such as rusting or aging, take years. Chemists now use lasers to study changes that occur in a few picoseconds (10^{-12} s) or femtoseconds (10^{-15} s).

SECTION 1.4 SUMMARY

SI units consist of seven base units and numerous derived units. • Exponential notation and prefixes based on powers of 10 are used to express very small and very large numbers. • The SI base unit of length is the meter (m). Length units on the atomic scale are the nanometer (nm) and picometer (pm). Volume units are derived from length units; the most important volume units are the cubic meter (m^3) and the liter (L). • The mass of an object, a measure of the quantity of matter in it, is constant. The SI unit of mass is the kilogram (kg). The weight of an object varies with the gravitational field influencing it. • Density (*d*) is the ratio of mass to volume of a substance and is one of its characteristic physical properties. • Temperature (*T*) is a measure of the relative hotness of an object. Heat is energy that flows from an object at higher temperature to one at lower temperature. Temperature scales differ in the size of the degree unit and/or the zero point. In chemistry, temperature is measured in kelvins (K) or degrees Celsius (°C). • Extensive properties, such as mass, volume, and energy, depend on the amount of a substance. Intensive properties, such as density and temperature, are independent of amount.

1.5 UNCERTAINTY IN MEASUREMENT: SIGNIFICANT FIGURES

We can never measure a quantity exactly, because measuring devices are made to limited specifications and we use our imperfect senses and skills to read them. Therefore, every measurement includes some **uncertainty.**

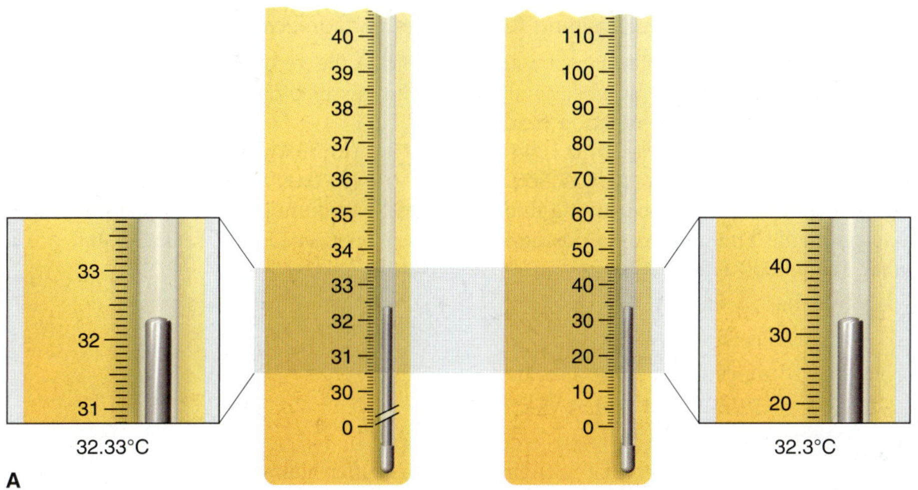

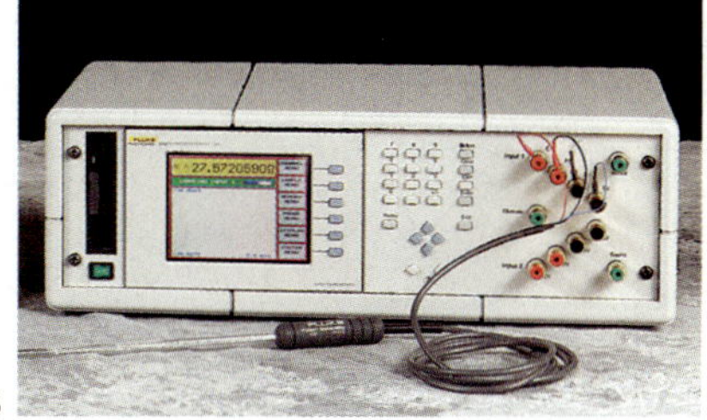

FIGURE 1.7 The number of significant figures in a measurement depends on the measuring device. A, Two thermometers measuring the same temperature are shown with expanded views. The thermometer on the left is graduated in 0.1°C and reads 32.33°C; the one on the right is graduated in 1°C and reads 32.3°C. Therefore, a reading with more significant figures (more certainty) can be made with the thermometer on the left. **B,** This modern electronic thermometer measures the resistance through a fine platinum wire in the probe to determine temperatures to the nearest microkelvin (10^{-6} K).

The measuring device we choose in a given situation depends on how much uncertainty we are willing to accept. When you buy potatoes, a supermarket scale that measures in 0.1-kg increments is perfectly acceptable; it tells you that the mass is, for example, 2.0 ± 0.1 kg. The term "± 0.1 kg" expresses the uncertainty in the measurement: the potatoes weigh between 1.9 and 2.1 kg. For a large-scale reaction, a chemist uses a lab balance that measures in 0.001-kg increments in order to obtain 2.036 ± 0.001 kg of a chemical, that is, between 2.035 and 2.037 kg. The greater number of digits in the mass of the chemical indicates that we know its mass with *more certainty* than we know the mass of the potatoes. The uncertainty of a measured quantity can be expressed with the ± sign, but generally we drop the sign and *assume an uncertainty of one unit in the rightmost digit.*

The digits we record in a measurement, both the certain and the uncertain ones, are called **significant figures.** There are four significant figures in 2.036 kg and two in 2.0 kg. *The greater the number of significant figures in a measurement, the greater is the certainty.* Figure 1.7 shows this point for two thermometers.

Determining Significant Figures

When you take measurements or use them in calculations, you must know the number of digits that are significant. In general, *all digits are significant, except zeros that are not measured but are used only to position the decimal point.* Here is a simple procedure that applies this general point:

1. Make sure that the measured quantity has a decimal point.
2. Start at the left, and move right to the first nonzero digit.
3. Count that digit and every digit to its right as significant.

A complication may arise with zeros that *end* a number. Zeros that end a number and lie either after or before the decimal point *are* significant; thus, 1.030 mL has four significant figures, and 5300. L has four significant figures also. If there is no decimal point, as in 5300 L, we assume that the zeros are *not* significant; exponential notation is needed to show which of the zeros, if any, were measured and therefore are significant. Thus, 5.300×10^3 L has four significant figures, 5.30×10^3 L has three, and 5.3×10^3 L has only two. A terminal decimal point is used to clarify the number of significant figures; thus, 500 mL has one significant figure, but 5.00×10^2 mL, 500. mL, and 0.500 L have three.

SAMPLE PROBLEM 1.8 Determining the Number of Significant Figures

Problem For each of the following quantities, underline the zeros that are significant figures (sf), and determine the number of significant figures in each quantity. For (d) to (f), express each in exponential notation first.

(a) 0.0030 L **(b)** 0.1044 g **(c)** 53,069 mL
(d) 0.00004715 m **(e)** 57,600. s **(f)** 0.0000007160 cm^3

Plan We determine the number of significant figures by counting digits, as just presented, paying particular attention to the position of zeros in relation to the decimal point.

Solution (a) $0.003\underline{0}$ L has 2 sf

(b) $0.1\underline{0}44$ g has 4 sf

(c) $53,\underline{0}69$ mL has 5 sf

(d) 0.00004715 m, or 4.715×10^{-5} m, has 4 sf

(e) $57,6\underline{00}.$ s, or $5.76\underline{00} \times 10^4$ s, has 5 sf

(f) $0.000000716\underline{0}$ cm^3, or $7.16\underline{0} \times 10^{-7}$ cm^3, has 4 sf

Check Be sure that every zero that is significant comes after nonzero digit(s) in the number.

FOLLOW-UP PROBLEM 1.8 For each of the following quantities, underline the zeros that are significant figures and determine the number of significant figures (sf) in each quantity. For (d) to (f), express each in exponential notation first.

(a) 31.070 mg **(b)** 0.06060 g **(c)** 850.°C
(d) 200.0 mL **(e)** 0.0000039 m **(f)** 0.000401 L

Significant Figures in Calculations

Measurements often contain differing numbers of significant figures. In a calculation, we keep track of the number of significant figures in each quantity so that we don't claim more significant figures (more certainty) in the answer than in the original data. If we have too many significant figures, we **round off** the answer to obtain the proper number of them.

The general rule for rounding is that *the least certain measurement sets the limit on certainty for the entire calculation and determines the number of significant figures in the final answer.* Suppose you want to find the density of a new ceramic. You measure the mass of a piece on a precise laboratory balance and obtain 3.8056 g; you measure its volume as 2.5 mL by displacement of water in a graduated cylinder. The mass has five significant figures, but the volume has only two. Should you report the density as 3.8056 g/2.5 mL = 1.5222 g/mL or as 1.5 g/mL? The answer with five significant figures implies more certainty than the answer with two. But you didn't measure the volume to five significant figures, so you can't possibly know the density with that much certainty. Therefore, you report the answer as 1.5 g/mL.

Significant Figures and Arithmetic Operations

The following two rules tell how many significant figures to show based on the arithmetic operation:

1. *For multiplication and division.* The answer contains the same number of *significant figures* as in the measurement with the fewest significant figures. Suppose you want to find the volume of a sheet of a new graphite composite. The length (9.2 cm) and width (6.8 cm) are obtained with a meterstick and the thickness (0.3744 cm) with a set of fine calipers. The volume calculation is

$$\text{Volume (cm}^3) = 9.2\text{ cm} \times 6.8\text{ cm} \times 0.3744\text{ cm} = 23\text{ cm}^3$$

The calculator shows 23.4225 cm^3, but you should report the answer as 23 cm^3, with two significant figures, because the length and width measurements determine the overall certainty, and they contain only two significant figures.

2. *For addition and subtraction.* The answer has the same number of *decimal places* as there are in the measurement with the fewest decimal places. Suppose you measure 83.5 mL of water in a graduated cylinder and add 23.28 mL of protein solution from a buret. The total volume is

$$\text{Volume (mL)} = 83.5\text{ mL} + 23.28\text{ mL} = 106.8\text{ mL}$$

Here the calculator shows 106.78 mL, but you report the volume as 106.8 mL, with one decimal place, because the measurement with fewer decimal places (83.5 mL) has one decimal place. (Appendix A covers significant figures in calculations involving logarithms, which arise later in the text.)

Rules for Rounding Off In most calculations, you need to round off the answer to obtain the proper number of significant figures or decimal places. Notice that in calculating the volume of the graphite composite above, we removed the extra digits, but in calculating the total protein solution volume, we removed the extra digit and increased the last digit by 1. Here are rules for rounding off:

1. If the digit removed is *more than 5,* the preceding number is increased by 1: 5.379 rounds to 5.38 if three significant figures are retained and to 5.4 if two significant figures are retained.

2. If the digit removed is *less than 5,* the preceding number is unchanged: 0.2413 rounds to 0.241 if three significant figures are retained and to 0.24 if two significant figures are retained.

3. If the digit removed *is 5,* the preceding number is increased by 1 if it is odd and remains unchanged if it is even: 17.75 rounds to 17.8, but 17.65 rounds to 17.6. If the 5 is followed only by zeros, rule 3 is followed; if the 5 is followed by nonzeros, rule 1 is followed: 17.6500 rounds to 17.6, but 17.6513 rounds to 17.7.

4. *Always carry one or two additional significant figures through a multistep calculation and round off the final answer* ***only.*** Don't be concerned if you string together a calculation to check a sample or follow-up problem and find that your answer differs in the last decimal place from the one in the book. To show you the correct number of significant figures in text calculations, *we round off intermediate steps,* and this process may sometimes change the last digit.

A calculator usually gives answers with too many significant figures. For example, if your calculator displays ten digits and you divide 15.6 by 9.1, it will show 1.714285714. Obviously, most of these digits are not significant; the answer should be rounded off to 1.7 so that it has two significant figures, the same as in 9.1.

Exact Numbers Some numbers are called **exact numbers** because *they have no uncertainty* associated with them. Some exact numbers are part of a unit definition: there are 60 minutes in 1 hour, 1000 micrograms in 1 milligram, and 2.54 centimeters in 1 inch. Other exact numbers result from actually counting individual items: there are exactly 3 quarters in my hand, 26 letters in the English alphabet, and so forth. Because they have no uncertainty, *exact numbers do not limit the number of significant figures in the answer.* Put another way, exact numbers have as many significant figures as a calculation requires.

SAMPLE PROBLEM 1.9 Significant Figures and Rounding

Problem Perform the following calculations and round the answer to the correct number of significant figures:

(a) $\dfrac{16.3521\ \text{cm}^2 - 1.448\ \text{cm}^2}{7.085\ \text{cm}}$ **(b)** $\dfrac{(4.80\times10^4\ \text{mg})\left(\dfrac{1\ \text{g}}{1000\ \text{mg}}\right)}{11.55\ \text{cm}^3}$

Plan We use the rules just presented in the text. In (a), we subtract before we divide. In (b), we note that the unit conversion involves an exact number.

Solution **(a)** $\dfrac{16.3521\ \text{cm}^2 - 1.448\ \text{cm}^2}{7.085\ \text{cm}} = \dfrac{14.904\ \text{cm}^{\cancel{2}}}{7.085\ \cancel{\text{cm}}} = 2.104\ \text{cm}$

(b) $\dfrac{(4.80\times10^4\ \cancel{\text{mg}})\left(\dfrac{1\ \text{g}}{1000\ \cancel{\text{mg}}}\right)}{11.55\ \text{cm}^3} = \dfrac{48.0\ \text{g}}{11.55\ \text{cm}^3} = 4.16\ \text{g/cm}^3$

Check Note that in (a) we lose a decimal place in the numerator, and in (b) we retain 3 sf in the answer because there are 3 sf in 4.80. Rounding to the nearest whole number is always a good way to check: **(a)** $(16 - 1)/7 \approx 2$; **(b)** $(5\times10^4/1\times10^3)/12 \approx 4$.

FOLLOW-UP PROBLEM 1.9 Perform the following calculation and round the answer to the correct number of significant figures: $\dfrac{25.65\text{ mL} + 37.4\text{ mL}}{73.55\text{ s}\left(\dfrac{1\text{ min}}{60\text{ s}}\right)}$

Precision, Accuracy, and Instrument Calibration

Precision and accuracy are two aspects of certainty. We often use these terms interchangeably in everyday speech, but in scientific measurements they have distinct meanings. **Precision,** or *reproducibility,* refers to how close the measurements in a series are to one another. **Accuracy** refers to how close a measurement is to the actual value.

Precision and accuracy are linked with two common types of error:

1. **Systematic error** produces values that are *either* all higher or all lower than the actual value. Such error is part of the experimental system, often caused by a faulty measuring device or by a consistent mistake in taking a reading.
2. **Random error,** in the absence of systematic error, produces values that are higher *and* lower than the actual value. Random error *always* occurs, but its size depends on the measurer's skill and the instrument's precision.

Precise measurements have low random error, that is, small deviations from the average. *Accurate measurements have low systematic error and, generally, low random error as well.* In some cases, when many measurements are taken that have a high random error, the *average* may still be accurate.

Suppose each of four students measures 25.0 mL of water in a pre-weighed graduated cylinder and then weighs the water *plus* cylinder on a balance. If the density of water is 1.00 g/mL at the temperature of the experiment, the actual mass of 25.0 mL of water is 25.0 g. Each student performs the operation four times, subtracts the mass of the empty cylinder, and obtains one of the four graphs shown in Figure 1.8. In graphs A and B, the random error is small; that is, the precision is high (the weighings are reproducible). In A, however, the accuracy is high as well (all the values are close to 25.0 g), whereas in B the accuracy is low (there is a systematic error). In graphs C and D, there is a large random error; that is, the precision is low. Large random error is often called large *scatter.* Note, however, that in D there is also a systematic error (all the values are high), whereas in C the average of the values is close to the actual value.

Systematic error can be avoided, or at least taken into account, through **calibration** of the measuring device, that is, by comparing it with a known standard. The systematic error in graph B, for example, might be caused by a poorly manufactured cylinder that reads "25.0" when it actually contains about 27 mL. If you detect such an error by means of a calibration procedure, you could adjust all volumes measured with that cylinder. Instrument calibration is an essential part of careful measurement.

SECTION 1.5 SUMMARY

The final digit of a measurement is always estimated. Thus, all measurements have a limit to their certainty, which is expressed by the number of significant figures. • The certainty of a calculated result depends on the certainty of the data, so the answer has as many significant figures as in the least certain measurement. Excess digits are rounded off in the final answer. Exact numbers have as many significant figures as the calculation requires. • Precision (how close values are to one another) and accuracy (how close values are to the actual value) are two aspects of certainty. • Systematic errors result in values that are either all higher or all lower than the actual value. Random errors result in some values that are higher and some values that are

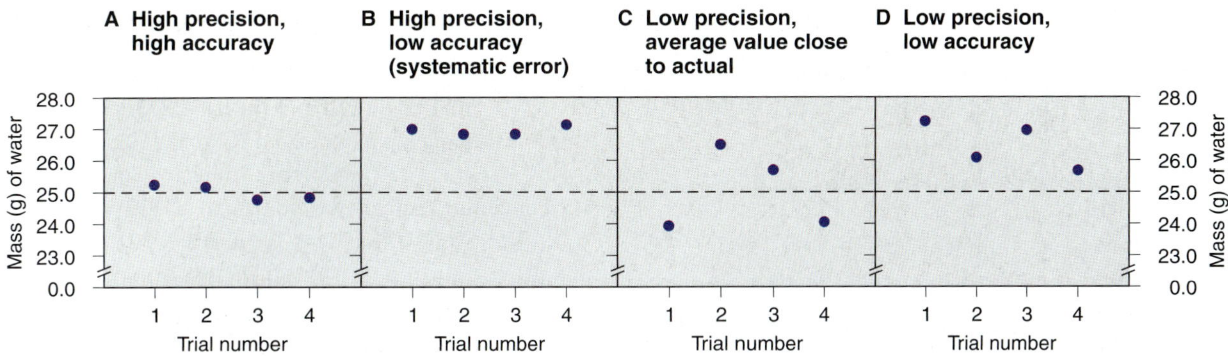

FIGURE 1.8 Precision and accuracy in a laboratory calibration. Each graph represents four measurements made with a graduated cylinder that is being calibrated (see text for details).

lower than the actual value. • Precise measurements have low random error; accurate measurements have low systematic error and often low random error. The size of random errors depends on the skill of the measurer and the precision of the instrument. A systematic error, however, is often caused by faulty equipment and can be compensated for by calibration.

CHAPTER REVIEW GUIDE

The following sections provide many aids to help you study this chapter. (Numbers in parentheses refer to pages, unless noted otherwise.)

• LEARNING OBJECTIVES *These are concepts and skills to review after studying this chapter.*

Related section (§), sample problem (SP), and upcoming end-of-chapter problem (EP) numbers are listed in parentheses.

1. Distinguish between physical and chemical properties and changes (§ 1.1) (SPs 1.1, 1.2) (EPs 1.1, 1.3–1.6)
2. Define the features of the states of matter (§ 1.1) (EP 1.2)
3. Understand the nature of potential and kinetic energy and their interconversion (§ 1.1) (EPs 1.7–1.8)
4. Understand the scientific approach to studying phenomena and distinguish between observation, hypothesis, experiment, and model (§ 1.2) (EPs 1.9–1.12)
5. Use conversion factors in calculations (§ 1.3) (SP 1.3) (EPs 1.13–1.15)
6. Distinguish between mass and weight, heat and temperature, and intensive and extensive properties (§ 1.4) (EPs 1.16, 1.17, 1.19)
7. Use numerical prefixes and common units of length, mass, volume, and temperature in unit-conversion calculations (§ 1.4) (SPs 1.4–1.7) (EPs 1.21–1.34, 1.36–1.40)
8. Understand scientific notation and the meaning of uncertainty; determine the number of significant figures and the number of digits after rounding (§ 1.5) (SPs 1.8, 1.9) (EPs 1.42–1.56)
9. Distinguish between accuracy and precision and between systematic and random error (§ 1.5) (EPs 1.57–1.59)

• KEY TERMS *These important terms appear in boldface in the chapter and are defined again in the Glossary.*

Section 1.1
chemistry (2)
matter (2)
composition (2)
property (2)
physical property (3)
physical change (3)
chemical property (3)
chemical change (chemical reaction) (3)
state of matter (4)
solid (4)
liquid (4)
gas (4)
energy (6)
potential energy (6)
kinetic energy (6)

Section 1.2
scientific method (8)
observation (8)
data (8)
natural law (8)
hypothesis (8)
experiment (8)
variable (8)
controlled experiment (8)
model (theory) (9)

Section 1.3
conversion factor (10)
dimensional analysis (11)

Section 1.4
SI units (13)
base (fundamental) unit (14)
derived unit (14)
meter (m) (15)
volume (V) (15)
cubic meter (m^3) (15)
liter (L) (15)
milliliter (mL) (15)
mass (16)
kilogram (kg) (16)
weight (16)
density (d) (17)
extensive property (17)
intensive property (17)
temperature (T) (18)
heat (18)
thermometer (18)
kelvin (K) (18)
Celsius scale (18)
Kelvin (absolute) scale (18)
second (s) (20)

Section 1.5
uncertainty (20)
significant figures (21)
round off (22)
exact number (23)
precision (24)
accuracy (24)
systematic error (24)
random error (24)
calibration (24)

• KEY EQUATIONS AND RELATIONSHIPS *Numbered and screened concepts are listed for you to refer to or memorize.*

1.1 Calculating density from mass and volume (17):

$$\text{Density} = \frac{\text{mass}}{\text{volume}}$$

1.2 Converting temperature from °C to K (19):

$$T\ (\text{in K}) = T\ (\text{in °C}) + 273.15$$

1.3 Converting temperature from K to °C (19):

$$T\ (\text{in °C}) = T\ (\text{in K}) - 273.15$$

1.4 Converting temperature from °C to °F (19):

$$T\ (\text{in °F}) = \tfrac{9}{5}T\ (\text{in °C}) + 32$$

1.5 Converting temperature from °F to °C (19):

$$T\ (\text{in °C}) = [T\ (\text{in °F}) - 32]\tfrac{5}{9}$$

• BRIEF SOLUTIONS TO ***FOLLOW-UP PROBLEMS*** *Compare your own solutions to these calculation steps and answers.*

1.1 Chemical. The red-and-blue and separate red particles on the left become paired red and separate blue particles on the right.

1.2 (a) Physical. Solid iodine changes to gaseous iodine.
(b) Chemical. Gasoline burns in air to form different substances.
(c) Chemical. In contact with air, torn skin and blood react to form different substances.

1.3 No. of chairs

$$= 3\ \cancel{\text{bolts}} \times \frac{200\ \cancel{\text{m}^2}}{1\ \cancel{\text{bolt}}} \times \frac{3.281\ \cancel{\text{ft}}}{1\ \cancel{\text{m}}} \times \frac{3.281\ \cancel{\text{ft}}}{1\ \cancel{\text{m}}} \times \frac{1\ \text{chair}}{31.5\ \cancel{\text{ft}^2}}$$

$$= 205\ \text{chairs}$$

1.4 Radius of ribosome (dm) $= \dfrac{21.4\ \cancel{\text{nm}}}{2} \times \dfrac{1\ \text{dm}}{10^8\ \cancel{\text{nm}}}$

$$= 1.07\times10^{-7}\ \text{dm}$$

Volume of ribosome (dm^3) $= \frac{4}{3}\pi r^3 = \frac{4}{3}(3.14)(1.07\times10^{-7}\ \text{dm})^3$

$$= 5.13\times10^{-21}\ \text{dm}^3$$

Volume of ribosome (μL) $= (5.13\times10^{-21}\ \cancel{\text{dm}^3})\left(\dfrac{1\ \cancel{\text{L}}}{1\ \cancel{\text{dm}^3}}\right)\left(\dfrac{10^6\ \mu\text{L}}{1\ \cancel{\text{L}}}\right)$

$$= 5.13\times10^{-15}\ \mu\text{L}$$

1.5 Mass (kg) of solution $= 8.0\ \cancel{\text{h}} \times \dfrac{60\ \cancel{\text{min}}}{1\ \cancel{\text{h}}} \times \dfrac{60\ \cancel{\text{s}}}{1\ \cancel{\text{min}}} \times \dfrac{1.5\ \cancel{\text{drops}}}{1\ \cancel{\text{s}}}$

$$\times \frac{65\ \cancel{\text{mg}}}{1\ \cancel{\text{drop}}} \times \frac{1\ \cancel{\text{g}}}{10^3\ \cancel{\text{mg}}} \times \frac{1\ \text{kg}}{10^3\ \cancel{\text{g}}}$$

$$= 2.8\ \text{kg}$$

1.6 Mass (kg) of sample $= 4.6\ \cancel{\text{cm}^3} \times \dfrac{7.5\ \cancel{\text{g}}}{1\ \cancel{\text{cm}^3}} \times \dfrac{1\ \text{kg}}{10^3\ \cancel{\text{g}}}$

$$= 0.034\ \text{kg}$$

1.7 T (in °C) = 234 K − 273.15 = −39°C
T (in °F) = $\frac{9}{5}$(−39°C) + 32 = −38°F
Answer contains two significant figures (see Section 1.5).

1.8 (a) $31.\underline{0}7\underline{0}$ mg, 5 sf (b) $0.06\underline{0}6\underline{0}$ g, 4 sf
(c) $85\underline{0}.$°C, 3 sf (d) $2.\underline{000}\times10^2$ mL, 4 sf
(e) 3.9×10^{-6} m, 2 sf (f) $4.\underline{0}1\times10^{-4}$ L, 3 sf

1.9 $\dfrac{25.65\ \text{mL} + 37.4\ \text{mL}}{73.55\ \cancel{\text{s}}\left(\dfrac{1\ \text{min}}{60\ \cancel{\text{s}}}\right)} = 51.4\ \text{mL/min}$

PROBLEMS

*Problems with **colored** numbers are answered in Appendix E. Sections match the text and provide the numbers of relevant sample problems. Bracketed problems are grouped in pairs (indicated by a short rule) that cover the same concept. Comprehensive Problems are based on material from any section.*

Some Fundamental Definitions

(Sample Problems 1.1 and 1.2)

1.1 Scenes A–D represent atomic-scale views of different samples of substances:

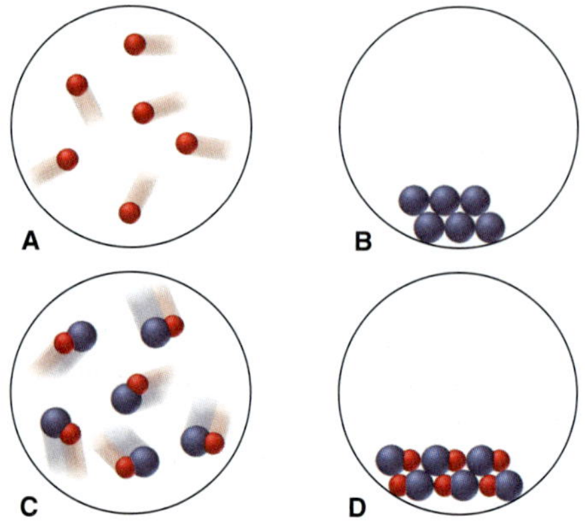

(a) Under one set of conditions, the substances in A and B mix and the result is depicted in C. Does this represent a chemical or a physical change?
(b) Under a second set of conditions, the same substances mix and the result is depicted in D. Does this represent a chemical or a physical change?
(c) Under a third set of conditions, the sample depicted in C changes to that in D. Does this represent a chemical or a physical change?
(d) When the change in part (c) occurs, does the sample have different chemical properties? Physical properties?

1.2 Describe solids, liquids, and gases in terms of how they fill a container. Use your descriptions to identify the physical state (at room temperature) of the following: (a) helium in a toy balloon; (b) mercury in a thermometer; (c) soup in a bowl.

1.3 Define *physical property* and *chemical property.* Identify each type of property in the following statements:
(a) Yellow-green chlorine gas attacks silvery sodium metal to form white crystals of sodium chloride (table salt).
(b) A magnet separates a mixture of black iron shavings and white sand.

1.4 Define *physical change* and *chemical change.* State which type of change occurs in each of the following statements:

(a) Passing an electric current through molten magnesium chloride yields molten magnesium and gaseous chlorine.
(b) The iron in discarded automobiles slowly forms reddish brown, crumbly rust.

1.5 Which of the following is a chemical change? Explain your reasoning: (a) boiling canned soup; (b) toasting a slice of bread; (c) chopping a log; (d) burning a log.

1.6 Which of the following changes can be reversed by changing the temperature (that is, which are physical changes): (a) dew condensing on a leaf; (b) an egg turning hard when it is boiled; (c) ice cream melting; (d) a spoonful of batter cooking on a hot griddle?

1.7 For each pair, which has higher potential energy?
(a) The fuel in your car or the products in its exhaust
(b) Wood in a fireplace or the ashes in the fireplace after the wood burns

1.8 For each pair, which has higher kinetic energy?
(a) A sled resting at the top of a hill or a sled sliding down the hill
(b) Water above a dam or water falling over the dam

The Scientific Approach: Developing a Model

1.9 How are the key elements of scientific thinking used in the following scenario? While making your breakfast toast, you notice it fails to pop out of the toaster. Thinking the spring mechanism is stuck, you notice that the bread is unchanged. Assuming you forgot to plug in the toaster, you check and find it is plugged in. When you take the toaster into the dining room and plug it into a different outlet, you find the toaster works. Returning to the kitchen, you turn on the switch for the overhead light and nothing happens.

1.10 Why is a quantitative observation more useful than a nonquantitative one? Which of the following are quantitative?
(a) The Sun rises in the east.
(b) An astronaut weighs one-sixth as much on the Moon as on Earth.
(c) Ice floats on water.
(d) An old-fashioned hand pump cannot draw water from a well more than 34 ft deep.

1.11 Describe the essential features of a well-designed experiment.

1.12 Describe the essential features of a scientific model.

Chemical Problem Solving

(Sample Problem 1.3)

1.13 When you convert feet to inches, how do you decide which portion of the conversion factor should be in the numerator and which in the denominator?

1.14 Write the conversion factor(s) for (a) in^2 to m^2; (b) km^2 to cm^2; (c) mi/h to m/s; (d) lb/ft^3 to g/cm^3.

1.15 Write the conversion factor(s) for (a) cm/min to in/s; (b) m^3 to in^3; (c) m/s^2 to km/h^2; (d) gallons/h to L/min.

Measurement in Scientific Study

(Sample Problems 1.4 to 1.7)

1.16 Describe the difference between intensive and extensive properties. Which of the following properties are intensive: (a) mass; (b) density; (c) volume; (d) melting point?

1.17 Explain the difference between mass and weight. Why is your weight on the Moon one-sixth that on Earth?

1.18 For each of the following cases, state whether the density of the object increases, decreases, or remains the same:
(a) A sample of chlorine gas is compressed.
(b) A lead weight is carried from sea level to the top of a high mountain.
(c) A sample of water is frozen.
(d) An iron bar is cooled.
(e) A diamond is submerged in water.

1.19 Explain the difference between heat and temperature. Does 1 L of water at 65°F have more, less, or the same quantity of energy as 1 L of water at 65°C?

1.20 A one-step conversion is sufficient to convert a temperature in the Celsius scale into the Kelvin scale, but not into the Fahrenheit scale. Explain.

1.21 The average radius of a molecule of lysozyme, an enzyme in tears, is 1430 pm. What is its radius in nanometers (nm)?

1.22 The radius of a barium atom is 2.22×10^{-10} m. What is its radius in angstroms (Å)?

1.23 A small hole in the wing of a space shuttle requires a 20.7-cm^2 patch. (a) What is the patch's area in square kilometers (km^2)? (b) If the patching material costs NASA \$3.25/in^2, what is the cost of the patch?

1.24 The area of a telescope lens is 7903 mm^2. (a) What is the area of the lens in square feet (ft^2)? (b) If it takes a technician 45 s to polish 135 mm^2, how long does it take her to polish the entire lens?

1.25 The average density of Earth is 5.52 g/cm^3. What is its density in (a) kg/m^3; (b) lb/ft^3?

1.26 The speed of light in a vacuum is 2.998×10^8 m/s. What is its speed in (a) km/h; (b) mi/min?

1.27 The volume of a certain bacterial cell is 2.56 μm^3. (a) What is its volume in cubic millimeters (mm^3)? (b) What is the volume of 10^5 cells in liters (L)?

1.28 (a) How many cubic meters of milk are in 1 qt (946.4 mL)? (b) How many liters of milk are in 835 gallons (1 gal = 4 qt)?

1.29 An empty vial weighs 55.32 g. (a) If the vial weighs 185.56 g when filled with liquid mercury (d = 13.53 g/cm^3), what is its volume? (b) How much would the vial weigh if it were filled with water (d = 0.997 g/cm^3 at 25°C)?

1.30 An empty Erlenmeyer flask weighs 241.3 g. When filled with water (d = 1.00 g/cm^3), the flask and its contents weigh 489.1 g. (a) What is the flask's volume? (b) How much does the flask weigh when filled with chloroform (d = 1.48 g/cm^3)?

1.31 A small cube of aluminum measures 15.6 mm on a side and weighs 10.25 g. What is the density of aluminum in g/cm^3?

1.32 A steel ball-bearing with a circumference of 32.5 mm weighs 4.20 g. What is the density of the steel in g/cm^3 (V of a sphere = $\frac{4}{3}\pi r^3$; circumference of a circle = $2\pi r$)?

1.33 Perform the following conversions:
(a) 72°F (a pleasant spring day) to °C and K
(b) −164°C (the boiling point of methane, the main component of natural gas) to K and °F
(c) 0 K (absolute zero, theoretically the coldest possible temperature) to °C and °F

1.34 Perform the following conversions:
(a) 106°F (the body temperature of many birds) to K and °C
(b) 3410°C (the melting point of tungsten, the highest for any element) to K and °F
(c) 6.1×10^3 K (the surface temperature of the Sun) to °F and °C

1.35 A 25.0-g sample of each of three unknown metals is added to 25.0 mL of water in graduated cylinders A, B, and C, and the final volumes are depicted in the circles below. Given their densities, identify the metal in each cylinder: zinc (7.14 g/mL), iron (7.87 g/mL), nickel (8.91 g/mL).

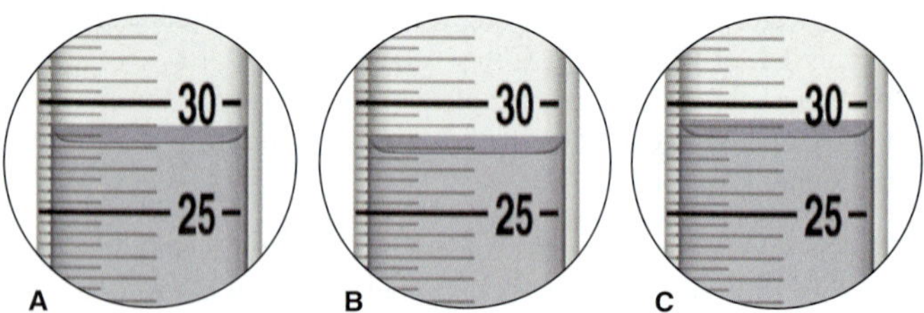

1.36 Anton van Leeuwenhoek, a 17th-century pioneer in the use of the microscope, described the microorganisms he saw as "animalcules" whose length was "25 thousandths of an inch." How long were the animalcules in meters?

1.37 The distance between two adjacent peaks on a wave is called the *wavelength.*
(a) The wavelength of a beam of ultraviolet light is 247 nanometers (nm). What is its wavelength in meters?
(b) The wavelength of a beam of red light is 6760 pm. What is its wavelength in angstroms (Å)?

1.38 In the early 20th century, thin metal foils were used to study atomic structure. (a) How many in^2 of gold foil with a thickness of 1.6×10^{-5} in could have been made from 2.0 troy oz? (b) If gold cost \$20.00/troy oz at that time, how many cm^2 of gold foil could have been made from \$75.00 worth of gold (1 troy oz = 31.1 g; d of gold = 19.3 g/cm^3)?

1.39 A cylindrical tube 9.5 cm high and 0.85 cm in diameter is used to collect blood samples. How many cubic decimeters (dm^3) of blood can it hold (V of a cylinder = $\pi r^2 h$)?

1.40 Copper can be drawn into thin wires. How many meters of 34-gauge wire (diameter = 6.304×10^{-3} in) can be produced from the copper in 5.01 lb of covellite, an ore of copper that is 66% copper by mass? (*Hint:* Treat the wire as a cylinder: V of cylinder = $\pi r^2 h$; d of copper = 8.95 g/cm^3.)

1.41 Each of the beakers depicted below contains two liquids that do not dissolve in each other. Three of the liquids are designated A, B, and C, and water is designated W.

(a) Which of the liquids is(are) more dense than water and which less dense?
(b) If the densities of W, C, and A are 1.0 g/mL, 0.88 g/mL, and 1.4 g/mL, respectively, which of the following densities is possible for liquid B: 0.79 g/mL, 0.86 g/mL, 0.94 g/mL, 1.2 g/mL?

Uncertainty in Measurement: Significant Figures

(Sample Problems 1.8 and 1.9)

1.42 What is an exact number? How are exact numbers treated differently from other numbers in a calculation?

1.43 All nonzero digits are significant. State a rule that tells which zeros are significant.

1.44 Underline the significant zeros in the following numbers:
(a) 0.41 (b) 0.041 (c) 0.0410 (d) 4.0100×10^4

1.45 Underline the significant zeros in the following numbers:
(a) 5.08 (b) 508 (c) 5.080×10^3 (d) 0.05080

1.46 Carry out the following calculations, making sure that your answer has the correct number of significant figures:
(a) $\dfrac{2.795 \text{ m} \times 3.10 \text{ m}}{6.48 \text{ m}}$
(b) $V = \frac{4}{3}\pi r^3$, where r = 17.282 cm
(c) 1.110 cm + 17.3 cm + 108.2 cm + 316 cm

1.47 Carry out the following calculations, making sure that your answer has the correct number of significant figures:
(a) $\dfrac{2.420 \text{ g} + 15.6 \text{ g}}{4.8 \text{ g}}$ (b) $\dfrac{7.87 \text{ mL}}{16.1 \text{ mL} - 8.44 \text{ mL}}$
(c) $V = \pi r^2 h$, where r = 6.23 cm and h = 4.630 cm

1.48 Write the following numbers in scientific notation:
(a) 131,000.0 (b) 0.00047 (c) 210,006 (d) 2160.5

1.49 Write the following numbers in scientific notation:
(a) 282.0 (b) 0.0380 (c) 4270.8 (d) 58,200.9

1.50 Write the following numbers in standard notation. Use a terminal decimal point when needed:
(a) 5.55×10^3 (b) 1.0070×10^4 (c) 8.85×10^{-7} (d) 3.004×10^{-3}

1.51 Write the following numbers in standard notation. Use a terminal decimal point when needed:
(a) 6.500×10^3 (b) 3.46×10^{-5} (c) 7.5×10^2 (d) 1.8856×10^2

1.52 Carry out each of the following calculations, paying special attention to significant figures, rounding, and units (J = joule, the SI unit of energy; mol = mole, the SI unit for amount of substance):
(a) $\dfrac{(6.626\times10^{-34} \text{ J·s})(2.9979\times10^8 \text{ m/s})}{489\times10^{-9} \text{ m}}$
(b) $\dfrac{(6.022\times10^{23} \text{ molecules/mol})(1.23\times10^2 \text{ g})}{46.07 \text{ g/mol}}$
(c) $(6.022\times10^{23} \text{ atoms/mol})(2.18\times10^{-18} \text{ J/atom})\left(\dfrac{1}{2^2} - \dfrac{1}{3^2}\right)$, where the numbers 2 and 3 in the last term are exact.

1.53 Carry out each of the following calculations, paying special attention to significant figures, rounding, and units:
(a) $\dfrac{4.32\times10^7 \text{ g}}{\frac{4}{3}(3.1416)(1.95\times10^2 \text{ cm})^3}$ (The term $\frac{4}{3}$ is exact.)
(b) $\dfrac{(1.84\times10^2 \text{ g})(44.7 \text{ m/s})^2}{2}$ (The term 2 is exact.)
(c) $\dfrac{(1.07\times10^{-4} \text{ mol/L})^2(3.8\times10^{-3} \text{ mol/L})}{(8.35\times10^{-5} \text{ mol/L})(1.48\times10^{-2} \text{ mol/L})^3}$

1.54 Which statements include exact numbers?
(a) Angel Falls in Venezuela is 3212 ft high.
(b) There are nine known planets in the Solar System.
(c) There are 453.59 g in 1 lb.
(d) There are 1000 mm in 1 m.

1.55 Which of the following include exact numbers?
(a) The speed of light in a vacuum is a physical constant; to six significant figures, it is 2.99792×10^8 m/s.
(b) The density of mercury at 25°C is 13.53 g/mL.
(c) There are 3600 s in 1 h.
(d) In 2007, the United States had 50 states.

1.56 How long is the metal strip shown below? Be sure to answer with the correct number of significant figures.

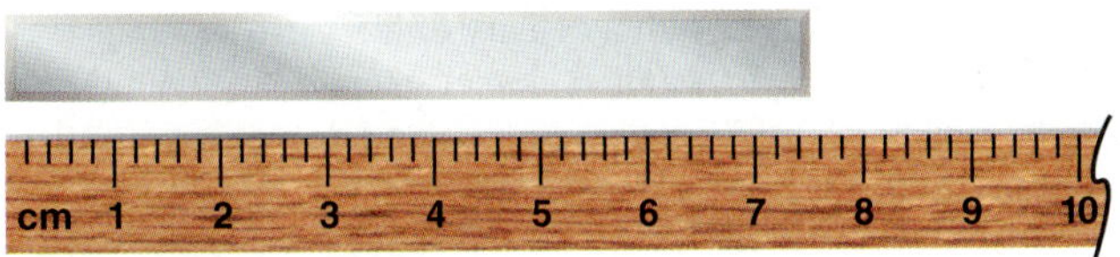

1.57 These organic solvents are used to clean compact discs:

Solvent	Density (g/mL) at 20°C
Chloroform	1.492
Diethyl ether	0.714
Ethanol	0.789
Isopropanol	0.785
Toluene	0.867

(a) If a 15.00-mL sample of CD cleaner weighs 11.775 g at 20°C, which solvent is most likely to be present?
(b) The chemist analyzing the cleaner calibrates her equipment and finds that the pipet is accurate to ±0.02 mL, and the balance is accurate to ±0.003 g. Is this equipment precise enough to distinguish between ethanol and isopropanol?

1.58 A laboratory instructor gives a sample of amino-acid powder to each of four students, I, II, III, and IV, and they weigh the samples. The true value is 8.72 g. Their results for three trials are
I: 8.72 g, 8.74 g, 8.70 g II: 8.56 g, 8.77 g, 8.83 g
III: 8.50 g, 8.48 g, 8.51 g IV: 8.41 g, 8.72 g, 8.55 g
(a) Calculate the average mass from each set of data, and tell which set is the most accurate.
(b) Precision is a measure of the average of the deviations of each piece of data from the average value. Which set of data is the most precise? Is this set also the most accurate?
(c) Which set of data is both the most accurate and most precise?
(d) Which set of data is both the least accurate and least precise?

1.59 The following dartboards illustrate the types of errors often seen in measurements. The bull's-eye represents the actual value, and the darts represent the data.

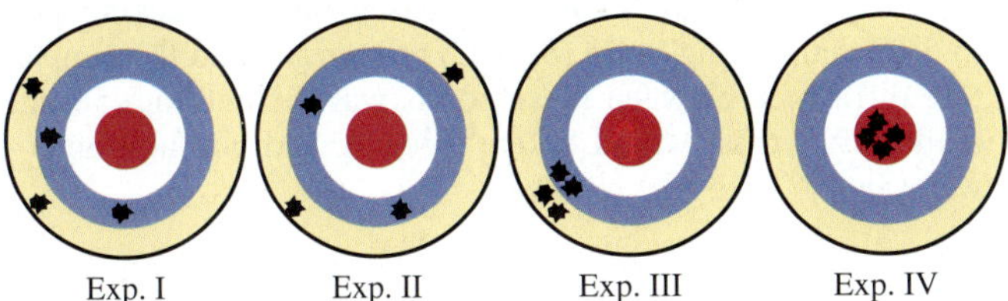

(a) Which experiments yield the same average result?
(b) Which experiment(s) display(s) high precision?
(c) Which experiment(s) display(s) high accuracy?
(d) Which experiment(s) show(s) a systematic error?

Comprehensive Problems

Problems with an asterisk (*) are more challenging.

1.60 To make 2.000 gal of a powdered sports drink, a group of students measure out 2.000 gal of water with 500.-mL, 50.-mL, and 5-mL graduated cylinders. Show how they could get closest to 2.000 gal of water, using these cylinders the fewest times.

1.61 Two blank potential energy diagrams (see Figure 1.3) appear below. Beneath each diagram are objects to place in the diagram. Draw the objects on the dashed lines to indicate higher or lower potential energy and label each case as more or less stable:

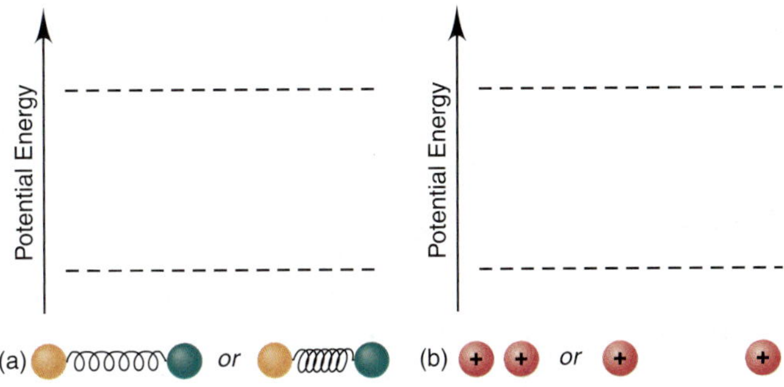

(a) Two balls attached to a relaxed *or* a compressed spring
(b) Two positive charges near *or* apart from each other

1.62 Soft drinks are about as dense as water (1.0 g/cm^3); many common metals, including iron, copper, and silver, have densities around 9.5 g/cm^3. (a) What is the mass of the liquid in a standard 12-oz bottle of diet cola? (b) What is the mass of a dime? (*Hint:* A stack of five dimes has a volume of about 1 cm^3.)

1.63 The molecular scenes below illustrate two different mixtures. When mixture A at 273 K is heated to 473 K, mixture B results.

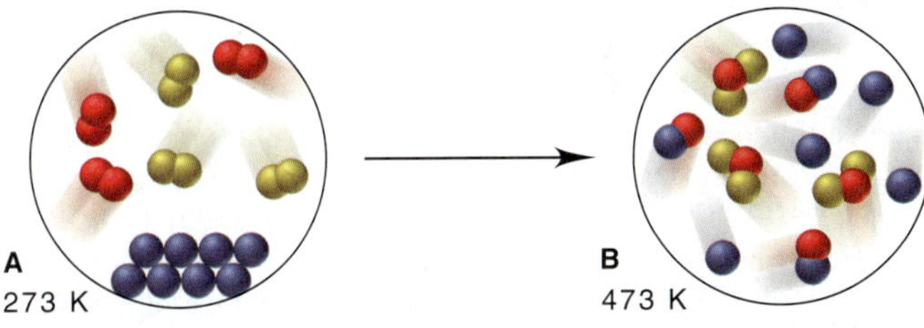

(a) How many different chemical changes occur?
(b) How many different physical changes occur?

1.64 Suppose your dorm room is 11 ft wide by 12 ft long by 8.5 ft high and has an air conditioner that exchanges air at a rate of 1200 L/min. How long would it take the air conditioner to exchange the air in your room once?

1.65 In 1933, the United States went off the international gold standard, and the price of gold increased from \$20.00 to \$35.00/troy oz. The twenty-dollar gold piece, known as the double eagle, weighed 33.436 g and was 90.0% gold by mass. (a) What was the value of the gold in the double eagle before and after the price change? (b) How many coins could be made from 50.0 troy oz of gold? (c) How many coins could be made from 2.00 in^3 of gold (1 troy oz = 31.1 g; *d* of gold = 19.3 g/cm^3)?

1.66 Bromine is used to prepare the pesticide methyl bromide and flame retardants for plastic electronic housings. It is recovered from seawater, underground brines, and the Dead Sea. The average concentrations of bromine in seawater (d = 1.024 g/mL) and the Dead Sea (d = 1.22 g/mL) are 0.065 g/L and 0.50 g/L, respectively. What is the mass ratio of bromine in the Dead Sea to that in seawater?

1.67 An Olympic-size pool is 50.0 m long and 25.0 m wide. (a) How many gallons of water (d = 1.0 g/mL) are needed to fill the pool to an average depth of 4.8 ft? (b) What is the mass (in kg) of water in the pool?

* **1.68** At room temperature (20°C) and pressure, the density of air is 1.189 g/L. An object will float in air if its density is less than that of air. In a buoyancy experiment with a new plastic, a chemist creates a rigid, thin-walled ball that weighs 0.12 g and has a volume of 560 cm^3.
(a) Will the ball float if it is evacuated?
(b) Will it float if filled with carbon dioxide (d = 1.830 g/L)?
(c) Will it float if filled with hydrogen (d = 0.0899 g/L)?
(d) Will it float if filled with oxygen (d = 1.330 g/L)?
(e) Will it float if filled with nitrogen (d = 1.165 g/L)?
(f) For any case that will float, how much weight must be added to make the ball sink?

1.69 Asbestos is a fibrous silicate mineral with remarkably high tensile strength. But it is no longer used because airborne asbestos particles can cause lung cancer. Grunerite, a type of asbestos, has a *tensile strength* of 3.5×10^2 kg/mm^2 (thus, a strand of grunerite with a 1-mm^2 cross-sectional area can hold up to 3.5×10^2 kg). The tensile strengths of aluminum and Steel No. 5137 are 2.5×10^4 lb/in^2 and 5.0×10^4 lb/in^2, respectively. Calculate the cross-sectional area (in mm^2) of wires of aluminum and of Steel No. 5137 that have the same tensile strength as a fiber of grunerite with a cross-sectional area of 1.0 μm^2.

1.70 According to the lore of ancient Greece, Archimedes discovered the displacement method of density determination while bathing and used it to find the composition of the king's crown. If a crown weighing 4 lb 13 oz displaces 186 mL of water, is the crown made of pure gold (d = 19.3 g/cm^3)?

1.71 Earth's oceans have an average depth of 3800 m, a total area of 3.63×10^8 km^2, and an average concentration of dissolved gold of 5.8×10^{-9} g/L. (a) How many grams of gold are in the oceans? (b) How many m^3 of gold are in the oceans? (c) Just a few years ago, the price of gold was $370.00/troy oz. What was the value of gold in the oceans (1 troy oz = 31.1 g; d of gold = 19.3 g/cm^3)?

1.72 For the year 2007, worldwide production of aluminum was 35.1 million metric tons (t). (a) How many pounds of aluminum were produced? (b) What was its volume in cubic feet (1 t = 1000 kg; d of aluminum = 2.70 g/cm^3)?

1.73 Liquid nitrogen is obtained from liquefied air and is used industrially to prepare frozen foods. It boils at 77.36 K. (a) What is this temperature in °C? (b) What is this temperature in °F? (c) At the boiling point, the density of the liquid is 809 g/L and that of the gas is 4.566 g/L. How many liters of liquid nitrogen are produced when 895.0 L of nitrogen gas is liquefied at 77.36 K?

1.74 The speed of sound varies according to the material through which it travels. Sound travels at 5.4×10^3 cm/s through rubber and at 1.97×10^4 ft/s through granite. Calculate each of these speeds in m/s.

1.75 If a raindrop weighs 0.52 mg on average and 5.1×10^5 raindrops fall on a lawn every minute, what mass (in kg) of rain falls on the lawn in 1.5 h?

* **1.76** The Environmental Protection Agency (EPA) proposed a safety standard for microparticulates in air: for particles up to 2.5 μm in diameter, the maximum allowable amount is 50. μg/m^3. If your 10.0 ft × 8.25 ft × 12.5 ft dorm room just meets the EPA standard, how many of these particles are in your room? How many are in each 0.500-L breath you take? (Assume the particles are spheres of diameter 2.5 μm and made primarily of soot, a form of carbon with a density of 2.5 g/cm^3.)

1.77 Molecular scenes A and B depict changes in matter at the atomic scale:

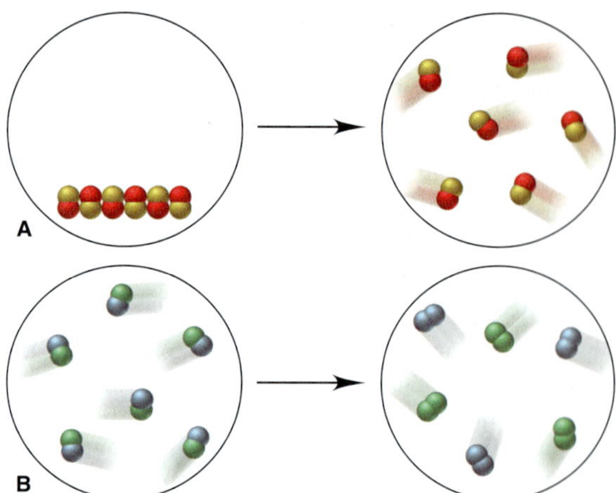

(a) Which show(s) a physical change?
(b) Which show(s) a chemical change?
(c) Which result(s) in different physical properties?
(d) Which result(s) in different chemical properties?
(e) Which result(s) in a change in state?

* **1.78** Earth's surface area is 5.10×10^8 km^2, and its crust has a mean thickness of 35 km and mean density of 2.8 g/cm^3. The two most abundant elements in the crust are oxygen (4.55×10^5 g/metric ton, t) and silicon (2.72×10^5 g/t), and the two rarest nonradioactive elements are ruthenium and rhodium, each with an abundance of 1×10^{-4} g/t. What is the total mass of each of these elements in Earth's crust (1 t = 1000 kg)?

* **1.79** The three states of matter differ greatly in their viscosity, a measure of their resistance to flow. Rank the three states from highest to lowest viscosity. Explain in submicroscopic terms.

* **1.80** If a temperature scale were based on the freezing point (5.5°C) and boiling point (80.1°C) of benzene and the temperature difference between these points was divided into 50 units (called °X), what would be the freezing and boiling points of water in °X? (See Figure 1.6.)

The Components of Matter

2

Key Principles
to focus on while studying this chapter

- A *substance* is matter with a *fixed composition.* The two types of substances are elements and compounds: An *element* consists of a single type of *atom,* and a *compound* consists of *molecules* (or formula units) made up of two or more atoms combined in a specific numerical ratio. A *mixture* consists of two or more substances intermingled physically and, thus, has a *variable composition*. A compound's properties differ from those of its components, but a mixture's properties do not. *(Section 2.1)*
- Three *mass laws* led to an atomic theory of matter: *mass is conserved* during a chemical change; any sample of a compound has its elements in the *same proportions by mass;* in different compounds consisting of the same two elements, the masses of one of the elements that combine with a given mass of the other can be expressed as a *ratio of small integers. (Section 2.2)*
- According to *Dalton's atomic theory,* atoms of a given element have a unique mass and other properties. Mass is conserved during a chemical reaction because the atoms of the reacting substances are just rearranged into different substances. *(Section 2.3)*
- Early 20th-century experiments showed that atoms are divisible, consisting of a positively charged *nucleus,* which contains nearly all the atom's mass but a tiny fraction of its volume, and negatively charged *electrons* that move continuously around the nucleus. *(Section 2.4)*
- Atoms have a structure made of three types of *subatomic particles:* positively charged *protons* and uncharged *neutrons* make up the *nucleus,* and negatively charged *electrons* exist outside the nucleus. *Atoms are neutral* because the number of protons equals the number of electrons. All the atoms of an element have the same number of protons *(atomic number, Z)* and thus the same chemical behavior, but *isotopes* of an element have atoms with different masses because they have different numbers of neutrons. The *atomic mass* of an element is the weighted *average* of the masses of its naturally occurring isotopes. *(Section 2.5)*
- In the *periodic table*, the elements are arranged by increasing atomic number into a grid of horizontal rows *(periods)* and vertical columns *(groups)*. *Metals* occupy most of the lower-left portion, and *nonmetals* are found in the upper-right corner, with *metalloids* in between. Elements in a group have similar properties. *(Section 2.6)*
- The electrons of atoms are involved in forming compounds. In *ionic bonding,* metal atoms *transfer* electrons to nonmetal atoms, and the resulting charged particles *(ions)* attract each other into solid arrays. In *covalent bonding,* nonmetal atoms *share* electrons and usually form individual molecules. Each compound has a unique name, formula, and mass based on its component elements. *(Sections 2.7 and 2.8)*
- Unlike compounds, mixtures can be *separated by physical means* into their components. A *heterogeneous mixture* has a nonuniform composition with visible boundaries between the components. A *homogeneous mixture (solution)* has a uniform composition because the components (elements and/or compounds) are mixed as individual atoms, ions, or molecules. *(Section 2.9)*

Taking It Apart *Like this fine watch, everyday matter consists of simpler components that are themselves made of even simpler parts. In this chapter, you'll learn the properties of matter and discover how chemists identify its components to see how they combine.*

Outline

Concepts & Skills to Review before studying this chapter

- physical and chemical change (Section 1.1)
- states of matter (Section 1.1)
- attraction and repulsion between charged particles (Section 1.1)
- meaning of a scientific model (Section 1.2)
- SI units and conversion factors (Section 1.4)
- significant figures in calculations (Section 1.5)

It may seem surprising, but questioning what something is made of is as common today as it was among the philosophers of ancient Greece, even though we approach the question differently. They believed that everything was made of one or, at most, a few elemental substances (elements). Some believed it to be water, others thought it was air, and still others believed there were four elements—fire, air, water, and earth.

Democritus (c. 460–370 BC), the father of atomism, focused on the ultimate components of *all* substances, and his reasoning went something like this: If you cut a piece of, say, copper smaller and smaller, you must eventually reach a particle of copper so small that it can no longer be cut. Therefore, matter is ultimately composed of indivisible particles, with nothing between them but empty space. He called the particles *atoms* (Greek *atomos,* "uncuttable"). However, Aristotle (384–322 BC) held that it was impossible for "nothing" to exist, and his influence suppressed the concept of atoms for 2000 years.

Finally, in the 17th century, the great English scientist Robert Boyle argued that an element is composed of "simple Bodies, . . . of which all mixed Bodies are compounded." Boyle's hypothesis is remarkably similar to today's idea of an element, in which the "simple Bodies" are atoms. Further studies in the 18th century gave rise to laws concerning the relative masses of substances that react with each other. Then, at the beginning of the 19th century, John Dalton proposed an atomic model that explained these mass laws. By that century's close, however, further observation exposed the need to revise Dalton's model. A burst of creativity in the early 20th century gave rise to a picture of the atom with a complex internal structure, which led to our current model.

2.1 ELEMENTS, COMPOUNDS, AND MIXTURES: AN ATOMIC OVERVIEW

Matter can be classified into three types based on composition—elements, compounds, and mixtures. An **element** is the simplest type of matter with unique physical and chemical properties. *An element consists of only one kind of atom.* Therefore, it cannot be broken down into a simpler type of matter by any physical or chemical methods. An element is one kind of **substance,** matter whose composition is fixed. Each element has a name, such as silicon, oxygen, or copper. A sample of silicon contains only silicon atoms. A key point to remember is that the *macroscopic* properties of a piece of silicon, such as color, density, and combustibility, are different from those of a piece of copper because silicon atoms are different from copper atoms; in other words, *each element is unique because the properties of its atoms are unique.*

Most elements exist in nature as populations of atoms. Figure 2.1A shows atoms of a gaseous element such as neon. Several elements occur naturally as molecules: a **molecule** is an independent structure consisting of two or more atoms chemically bound together (Figure 2.1B). Elemental oxygen, for example, occurs in air as *diatomic* (two-atom) molecules.

A **compound** is a type of matter composed of *two or more different elements that are chemically bound together* (Figure 2.1C). Ammonia, water, and carbon dioxide are some common compounds. One defining feature of a compound is that *the elements are present in fixed parts by mass* (fixed mass ratio). Because of this fixed composition, *a compound is also considered a substance.* Any molecule of the compound has the same fixed parts by mass because it consists of *fixed numbers* of atoms of the component elements. For example, any sample of ammonia is 14 parts nitrogen by mass plus 3 parts hydrogen by mass. Since

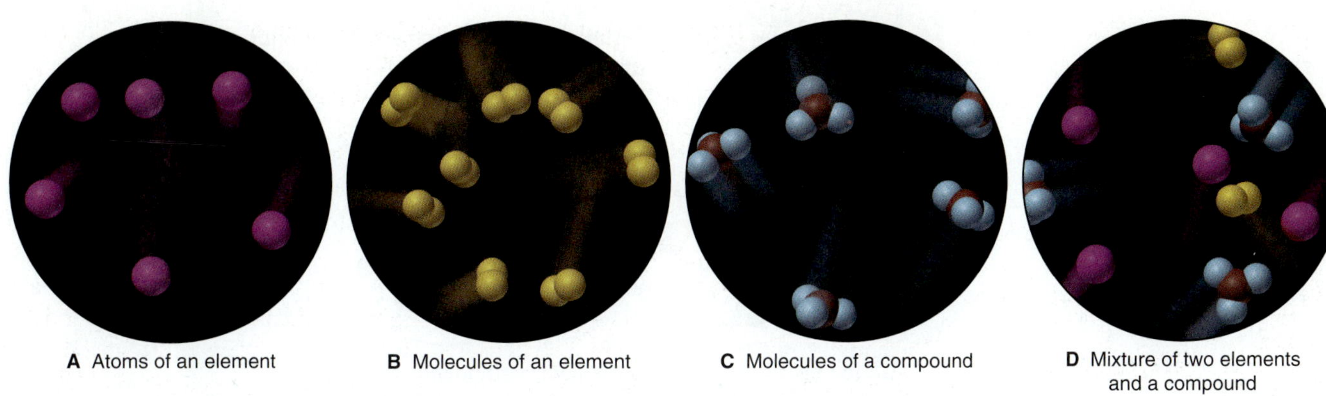

A Atoms of an element B Molecules of an element C Molecules of a compound D Mixture of two elements and a compound

FIGURE 2.1 Elements, compounds, and mixtures on the atomic scale. A, Most elements consist of a large collection of identical atoms. **B,** Some elements occur as molecules. **C,** A molecule of a compound consists of characteristic numbers of atoms of two or more elements chemically bound together. **D,** A mixture contains the individual units of two or more elements and/or compounds that are physically intermingled. The samples shown here are gases, but elements, compounds, and mixtures occur as liquids and solids also.

1 nitrogen atom has 14 times the mass of 1 hydrogen atom, ammonia must consist of 1 nitrogen atom for every 3 hydrogen atoms:

Ammonia is 14 parts N and 3 parts H by mass
1 N atom has 14 times the mass of 1 H atom
Therefore, ammonia has 1 N atom for every 3 H atoms.

Another defining feature of a compound is that *its properties are different from those of its component elements.* Table 2.1 shows a striking example. Soft, silvery sodium metal and yellow-green, poisonous chlorine gas have very different properties from the compound they form—white, crystalline sodium chloride, or common table salt! Unlike an element, a compound *can* be broken down into simpler substances—its component elements. For example, an electric current breaks down molten sodium chloride into metallic sodium and chlorine gas. Note that this breakdown is a *chemical change,* not a physical one.

Figure 2.1D depicts a **mixture,** a group of two or more substances (elements and/or compounds) that are physically intermingled. In contrast to a compound, *the components of a mixture* ***can*** *vary in their parts by mass.* Because its composition is not fixed, a mixture is *not* a substance. A mixture of the two compounds sodium chloride and water, for example, can have many different parts by mass of salt to water. At the atomic scale, a mixture is merely a group of the individual units that make up its component elements and/or compounds. Therefore, *a mixture retains many of the properties of its components.* Saltwater, for instance, is colorless like water and tastes salty like sodium chloride. Unlike compounds, mixtures can be separated into their components by *physical changes;* chemical changes are not needed. For example, the water in saltwater can be boiled off, a physical process that leaves behind solid sodium chloride. Sample Problem 2.1 will help you to differentiate the three types of matter.

Table 2.1 Some Properties of Sodium, Chlorine, and Sodium Chloride

Property	Sodium	+	Chlorine	⟶	Sodium Chloride
Melting point	97.8°C		−101°C		801°C
Boiling point	881.4°C		−34°C		1413°C
Color	Silvery		Yellow-green		Colorless (white)
Density	0.97 g/cm^3		0.0032 g/cm^3		2.16 g/cm^3
Behavior in water	Reacts		Dissolves slightly		Dissolves freely

SAMPLE PROBLEM 2.1 Distinguishing Elements, Compounds, and Mixtures at the Atomic Scale

Problem The scenes below represent an atomic-scale view of three samples of matter:

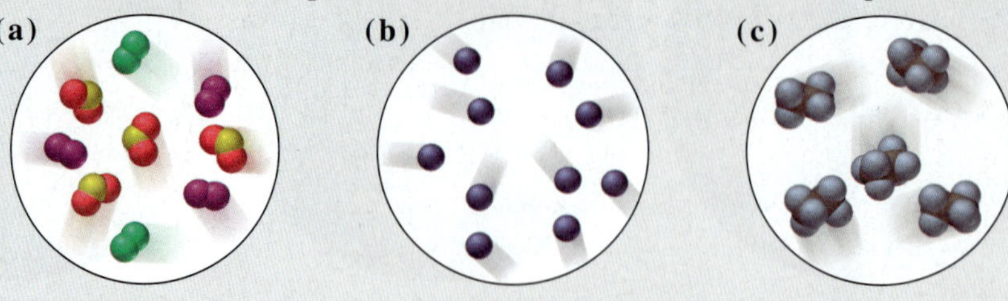

Describe each sample as an element, compound, or mixture.

Plan From depictions of the samples, we have to determine the type of matter by examining the component particles. If a sample contains only one type of particle, it is either an element or a compound; if it contains more than one type, it is a mixture. Particles of an element have only one kind of atom (one color), and particles of a compound have two or more kinds of atoms.

Solution **(a)** This sample is a mixture: there are three different types of particles, two types contain only one kind of atom, either green or purple, so they are elements, and the third type contains two red atoms for every one yellow, so it is a compound. **(b)** This sample is an element: it consists of only blue atoms, **(c)** This sample is a compound: it consists of molecules that each have two black and six blue atoms.

FOLLOW-UP PROBLEM 2.1 Describe this reaction in terms of elements, compounds, and mixtures.

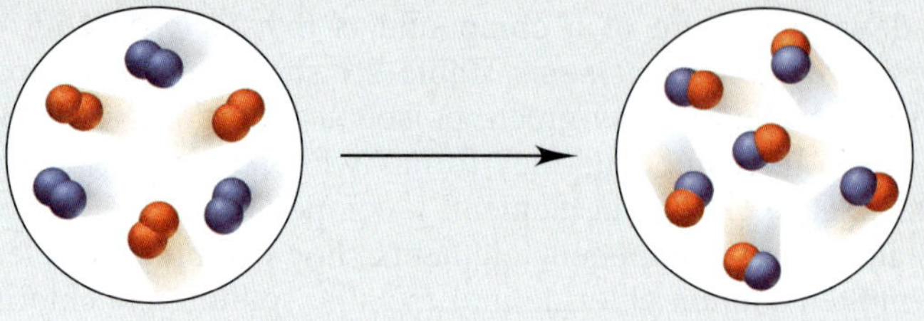

SECTION 2.1 SUMMARY

All matter exists as elements, compounds, or mixtures. • Elements and compounds are referred to as substances because their compositions are fixed. An element consists of only one type of atom. A compound contains two or more elements in chemical combination and exhibits different properties from its component elements. The elements of a compound occur in fixed parts by mass because each unit of the compound has fixed numbers of each type of atom. • A mixture consists of two or more substances mixed together, not chemically combined. The components retain their individual properties and can be present in any proportions.

2.2 THE OBSERVATIONS THAT LED TO AN ATOMIC VIEW OF MATTER

Any model of the composition of matter had to explain two extremely important chemical observations that were well established by the end of the 18th century: the *law of mass conservation* and the *law of definite (or constant) composition.* As you'll see, John Dalton's atomic theory explained these laws and another observation now known as the *law of multiple proportions.*

Mass Conservation

The most fundamental chemical observation of the 18th century was the **law of mass conservation:** *the total mass of substances does not change during a chemical reaction.* The *number* of substances may change and, by definition, their properties must, but the *total amount* of matter remains constant. Antoine Lavoisier (1743–1794), the great French chemist and statesman, had first stated this law on the basis of experiments in which he reacted mercury with oxygen. He found the mass of oxygen plus the mass of mercury always equaled the mass of mercuric oxide that formed.

Even in a complex biochemical change—such as the metabolism of the sugar glucose, which involves many reactions—mass is conserved:

$$\text{180 g glucose} + \text{192 g oxygen gas} \longrightarrow \text{264 g carbon dioxide} + \text{108 g water}$$
$$\text{372 g material before change} \longrightarrow \text{372 g material after change}$$

Mass conservation means that, based on all chemical experience, *matter cannot be created or destroyed.* (As you'll see later, however, mass *does* change in nuclear reactions, although *not* in chemical reactions.)

Definite Composition

Another fundamental chemical observation is summarized as the **law of definite (or constant) composition:** *no matter what its source, a particular compound is composed of the same elements in the same parts (fractions) by mass.* The **fraction by mass (mass fraction)** is that part of the compound's mass that each element contributes. It is obtained by dividing the mass of each element by the total mass of compound. The **percent by mass (mass percent, mass %)** is the fraction by mass expressed as a percentage.

Consider calcium carbonate, the major compound in marble. It is composed of three elements—calcium, carbon, and oxygen—and each is present in a fixed fraction (or percent) by mass. The results shown in the table below are obtained for the elemental mass composition of 20.0 g of calcium carbonate (for example, 8.0 g of calcium/20.0 g = 0.40 parts of calcium):

Analysis by Mass (grams/20.0 g)	Mass Fraction (parts/1.00 part)	Percent by Mass (parts/100 parts)
8.0 g calcium	0.40 calcium	40% calcium
2.4 g carbon	0.12 carbon	12% carbon
9.6 g oxygen	0.48 oxygen	48% oxygen
20.0 g	1.00 part by mass	100% by mass

FIGURE 2.2 The law of definite composition. Calcium carbonate is found naturally in many forms, including marble *(top)*, coral *(bottom)*, chalk, and seashells. The mass percents of its component elements do not change regardless of the compound's source.

As you can see, the sum of the mass fractions (or mass percents) equals 1.00 part (or 100%) by mass. The law of definite composition tells us that pure samples of calcium carbonate, no matter where they come from, always contain these elements in the same percents by mass (Figure 2.2).

Because a given element always constitutes the same mass fraction of a given compound, we can use that mass fraction to find the actual mass of the element in any sample of the compound:

$$\text{Mass of element} = \text{mass of compound} \times \frac{\text{part by mass of element}}{\text{one part by mass of compound}}$$

Or, more simply, mass analysis tells us the parts by mass, so we can use that directly with *any* mass unit and skip the need to find the mass fraction first:

$$\text{Mass of element in sample} = \text{mass of compound in sample} \times \frac{\text{mass of element in compound}}{\text{mass of compound}} \quad (2.1)$$

SAMPLE PROBLEM 2.2 Calculating the Mass of an Element in a Compound

Problem Pitchblende is the most commercially important compound of uranium. Analysis shows that 84.2 g of pitchblende contains 71.4 g of uranium, with oxygen as the only other element. How many grams of uranium can be obtained from 102 kg of pitchblende?

Plan We have to find the mass of uranium in a known mass of pitchblende (102 kg), given the mass of uranium (71.4 g) in a different mass of pitchblende (84.2 g). The mass ratio of uranium/pitchblende is the same for any sample of pitchblende. Therefore, as shown

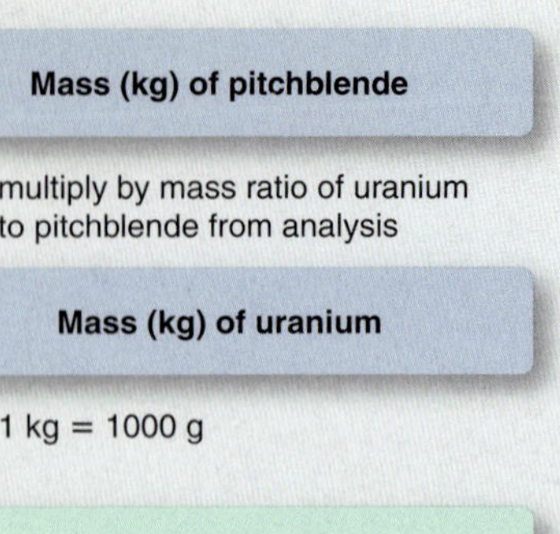

by Equation 2.1, we multiply the mass (in kg) of pitchblende by the ratio of uranium to pitchblende that we construct from the mass analysis. This gives the mass (in kg) of uranium, and we just convert kilograms to grams.

Solution Finding the mass (kg) of uranium in 102 kg of pitchblende:

$$\text{Mass (kg) of uranium} = \text{mass (kg) of pitchblende} \times \frac{\text{mass (kg) of uranium in pitchblende}}{\text{mass (kg) of pitchblende}}$$

$$\text{Mass (kg) of uranium} = 102\ \cancel{\text{kg pitchblende}} \times \frac{71.4\ \text{kg uranium}}{84.2\ \cancel{\text{kg pitchblende}}} = 86.5\ \text{kg uranium}$$

Converting the mass of uranium from kg to g:

$$\text{Mass (g) of uranium} = 86.5\ \cancel{\text{kg}}\ \text{uranium} \times \frac{1000\ \text{g}}{1\ \cancel{\text{kg}}} = 8.65\times10^4\ \text{g uranium}$$

Check The analysis showed that most of the mass of pitchblende is due to uranium, so the large mass of uranium makes sense. Rounding off to check the math gives:

$$\sim 100\ \text{kg pitchblende} \times \frac{70}{85} = 82\ \text{kg uranium}$$

FOLLOW-UP PROBLEM 2.2 How many metric tons (t) of oxygen are combined in a sample of pitchblende that contains 2.3 t of uranium? (*Hint:* Remember that oxygen is the only other element present.)

Multiple Proportions

Dalton described a phenomenon that occurs when two elements form more than one compound. His observation is now called the **law of multiple proportions:** *if elements A and B react to form two compounds, the different masses of B that combine with a fixed mass of A can be expressed as a ratio of small whole numbers.* Consider two compounds that form from carbon and oxygen; for now, let's call them carbon oxides I and II. They have very different properties. For example, measured at the same temperature and pressure, the density of carbon oxide I is 1.25 g/L, whereas that of II is 1.98 g/L. Moreover, I is poisonous and flammable, but II is not. Analysis shows that their compositions by mass are

Carbon oxide I: 57.1 mass % oxygen and 42.9 mass % carbon
Carbon oxide II: 72.7 mass % oxygen and 27.3 mass % carbon

To see the phenomenon of multiple proportions, we use the mass percents of oxygen and of carbon in each compound to find the masses of these elements in a given mass, for example, 100 g, of each compound. Then we divide the mass of oxygen by the mass of carbon in each compound to obtain the mass of oxygen that combines with a fixed mass of carbon:

	Carbon Oxide I	Carbon Oxide II
g oxygen/100 g compound	57.1	72.7
g carbon/100 g compound	42.9	27.3
g oxygen/g carbon	$\frac{57.1}{42.9} = 1.33$	$\frac{72.7}{27.3} = 2.66$

If we then divide the grams of oxygen per gram of carbon in II by that in I, we obtain a ratio of small whole numbers:

$$\frac{2.66\ \text{g oxygen/g carbon in II}}{1.33\ \text{g oxygen/g carbon in I}} = \frac{2}{1}$$

The law of multiple proportions tells us that in two compounds of the same elements, the mass fraction of one element relative to the other element changes in *increments based on ratios of small whole numbers.* In this case, the ratio is 2/1 for a given mass of carbon, II contains *2 times* as much oxygen as I, not 1.583 times, 1.716 times, or any other intermediate amount. As you'll see next, Dalton's theory allows us to explain the composition of carbon oxides I and II on the atomic scale.

SECTION 2.2 SUMMARY

Three fundamental observations are known as the mass laws. The law of mass conservation states that the total mass remains constant during a chemical reaction. • The law of definite composition states that any sample of a given compound has the same elements present in the same parts by mass. • The law of multiple proportions states that in different compounds of the same elements, the masses of one element that combine with a fixed mass of the other can be expressed as a ratio of small whole numbers.

2.3 DALTON'S ATOMIC THEORY

With more than 200 years of hindsight, it may be easy to see how the mass laws could be explained by an atomic model—matter existing in indestructible units, each with a particular mass—but it was a major breakthrough in 1808 when John Dalton (1766–1844) presented his atomic theory of matter in *A New System of Chemical Philosophy.*

Postulates of the Atomic Theory

Dalton expressed his theory in a series of postulates. Like most great thinkers, Dalton incorporated the ideas of others into his own to create the new theory. As we go through the postulates, which are presented here in modern terms, let's see which were original and which came from others.

1. All matter consists of **atoms,** tiny indivisible particles of an element that cannot be created or destroyed. (Derives from the "eternal, indestructible atoms" proposed by Democritus more than 2000 years earlier and conforms to mass conservation as stated by Lavoisier.)
2. Atoms of one element *cannot* be converted into atoms of another element. In chemical reactions, the atoms of the original substances recombine to form different substances. (Rejects the earlier belief by alchemists that one element could be magically transformed into another, such as lead into gold.)
3. Atoms of an element are identical in mass and other properties and are different from atoms of any other element. (Contains Dalton's major new ideas: unique mass and properties for all the atoms of a given element.)
4. Compounds result from the chemical combination of a specific ratio of atoms of different elements. (Follows directly from the fact of definite composition.)

How the Theory Explains the Mass Laws

Let's see how Dalton's postulates explain the mass laws:

- *Mass conservation.* Atoms cannot be created or destroyed (postulate 1) or converted into other types of atoms (postulate 2). Since each type of atom has a fixed mass (postulate 3), a chemical reaction, in which atoms are just combined differently with each other, cannot possibly result in a mass change.
- *Definite composition.* A compound is a combination of a *specific* ratio of different atoms (postulate 4), each of which has a particular mass (postulate 3). Thus, each element in a compound constitutes a fixed fraction of the total mass.
- *Multiple proportions.* Atoms of an element have the same mass (postulate 3) and are indivisible (postulate 1). The masses of element B that combine with a fixed mass of element A give a small, whole-number ratio because different numbers of B atoms combine with each A atom in different compounds.

The *simplest* arrangement consistent with the mass data for carbon oxides I and II in our earlier example is that one atom of oxygen combines with one atom of carbon in compound I (carbon monoxide) and that two atoms of oxygen combine with one atom of carbon in compound II (carbon dioxide) (Figure 2.3).

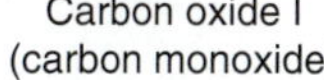

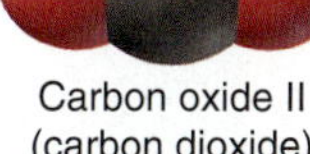

FIGURE 2.3 The atomic basis of the law of multiple proportions. Carbon and oxygen combine to form carbon oxide I (carbon monoxide) and carbon oxide II (carbon dioxide). The masses of oxygen in the two compounds relative to a fixed mass of carbon are in a ratio of small whole numbers.

SAMPLE PROBLEM 2.3 Visualizing the Mass Laws

Problem The scenes below represent an atomic-scale view of a chemical reaction:

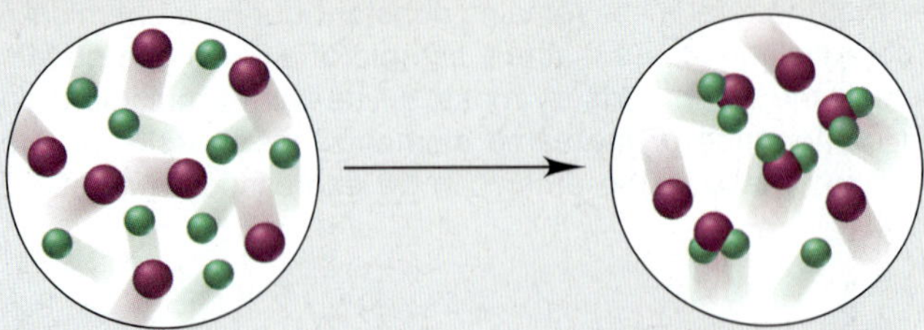

Which of the mass laws—mass conservation, definite composition, or multiple proportions—is (are) illustrated?

Plan From the depictions, we note the number, color, and combinations of atoms (spheres) to see which mass laws pertain. If the numbers of each atom are the same before and after the reaction, the total mass did not change (mass conservation). If a compound forms that always has the same atom ratio, the elements are present in fixed parts by mass (definite composition). When the same elements form different compounds and the ratio of the atoms of one element that combine with one atom of the other element is a small whole number, the ratio of their masses is a small whole number as well (multiple proportions).

Solution There are seven purple and nine green atoms in each circle, so mass is conserved. The compound formed has one purple and two green atoms, so it has definite composition. Only one compound forms, so the law of multiple proportions does not pertain.

FOLLOW-UP PROBLEM 2.3 Which sample(s) best display(s) the fact that compounds of bromine (orange) and fluorine (yellow) exhibit the law of multiple proportions? Explain.

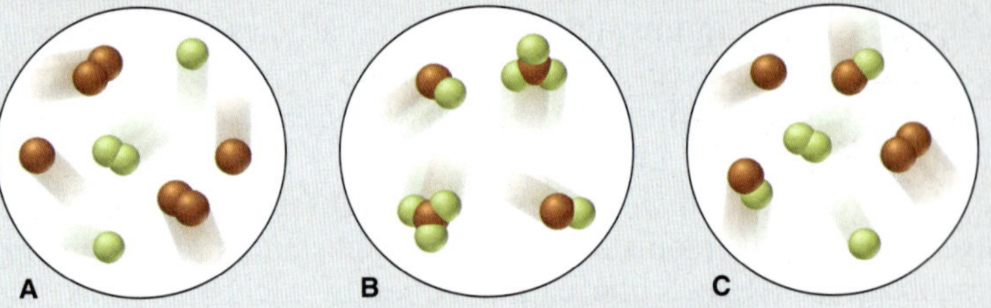

Dalton's atomic model was crucial to the idea that masses of reacting elements could be explained in terms of atoms, and it led to experiments to learn the relative masses of atoms in compounds. However, the model did not explain why atoms bond as they do: for example, why do two, and not three, hydrogen atoms bond with one oxygen atom in water? Also, Dalton's model did not account for the charged particles that were being observed in experiments. Clearly, a more complex atomic model needed to be developed, as we discuss in the next section.

SECTION 2.3 SUMMARY

Dalton's atomic theory explained the mass laws by proposing that all matter consists of indivisible, unchangeable atoms of fixed, unique mass. • Mass is constant during a reaction because atoms form new combinations. • Each compound has a fixed mass fraction of each of its elements because it is composed of a fixed number of each type of atom. • Different compounds of the same elements exhibit multiple proportions because they each consist of whole atoms.

2.4 THE OBSERVATIONS THAT LED TO THE NUCLEAR ATOM MODEL

The path of discovery is often winding and unpredictable. Basic research into the nature of electricity eventually led to the discovery of *electrons,* negatively charged particles that are part of all atoms. Soon thereafter, other experiments revealed that the atom has a *nucleus*—a tiny, central core of mass and positive charge. In this section, we examine some key experiments that led to our current model of the atom.

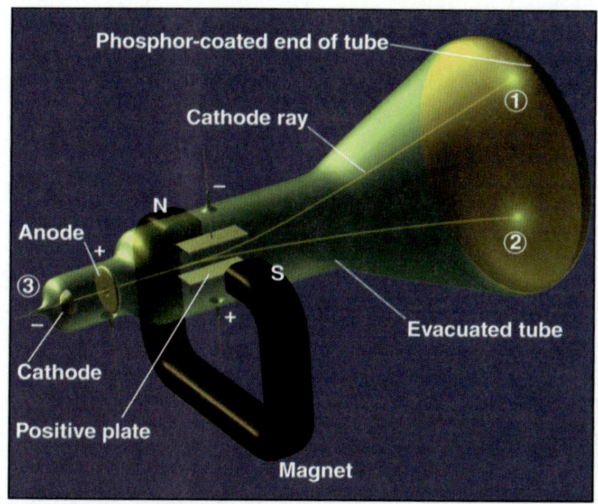

OBSERVATION	CONCLUSION
1. Ray bends in magnetic field	Consists of charged particles
2. Ray bends toward positive plate in electric field	Consists of negative particles
3. Ray is identical for any cathode	Particles found in all matter

FIGURE 2.4 Experiments to determine the properties of cathode rays. A cathode ray forms when high voltage is applied to a partially evacuated tube. The ray passes through a hole in the anode and hits the coated end of the tube to produce a glow.

Animation: Cathode Ray Tube

Discovery of the Electron and Its Properties

Nineteenth-century investigators of electricity knew that matter and electric charge were somehow related, but they did not know what an electric current consists of. Some investigators tried passing current from a high-voltage source through nearly evacuated glass tubes fitted with metal electrodes that were sealed in place and connected to an external source of electricity. When the power was turned on, a "ray" could be seen striking the phosphor-coated end of the tube and emitting a glow. The rays were called **cathode rays** because they originated at the negative electrode (cathode) and moved to the positive electrode (anode). Cathode rays travel in a straight line, but in a magnetic field the path is bent, indicating that the particles are charged, and in an electric field the path bends toward the positive plate. Moreover, the ray is identical no matter what metal is used as the cathode. It was concluded that cathode rays consist of negatively charged particles found in all matter (Figure 2.4). The rays appear when these particles collide with the few remaining gas molecules in the evacuated tube. Cathode ray particles were later named *electrons.*

In 1897, the British physicist J. J. Thomson (1856–1940) used magnetic and electric fields to measure the ratio of the cathode ray particle's mass to its charge. By comparing this value with the mass/charge ratio for other particles, Thomson estimated that the cathode ray particle weighed less than $\frac{1}{1000}$ as much as hydrogen, the lightest atom! He was shocked because this implied that, contrary to Dalton's atomic theory, *atoms are divisible into even smaller particles.* Fellow scientists reacted with disbelief, and some even thought Thomson was joking.

In 1909, the American physicist Robert Millikan (1868–1953) measured the *charge* of the electron. He did so by observing the movement of tiny droplets of the "highest grade clock oil" in an apparatus that contained electrically charged plates and an x-ray source (Figure 2.5, next page). X-rays knocked electrons from gas molecules in the air, and as an oil droplet fell through a hole in the positive (upper) plate, the electrons stuck to the drop, giving it a negative charge. With the electric field off, Millikan measured the mass of the droplet from its rate of fall. By turning on the field and varying its strength, he could make the drop fall more slowly, rise, or pause suspended. From these data, Millikan calculated the total charge of the droplet.

After studying many droplets, Millikan calculated that the various charges of the droplets were always some *whole-number multiple of a minimum charge.* He reasoned that different oil droplets picked up different numbers of electrons, so this minimum charge must be that of the electron itself. The value, which he calculated around 100 years ago, is within 1% of the modern value of the electron's

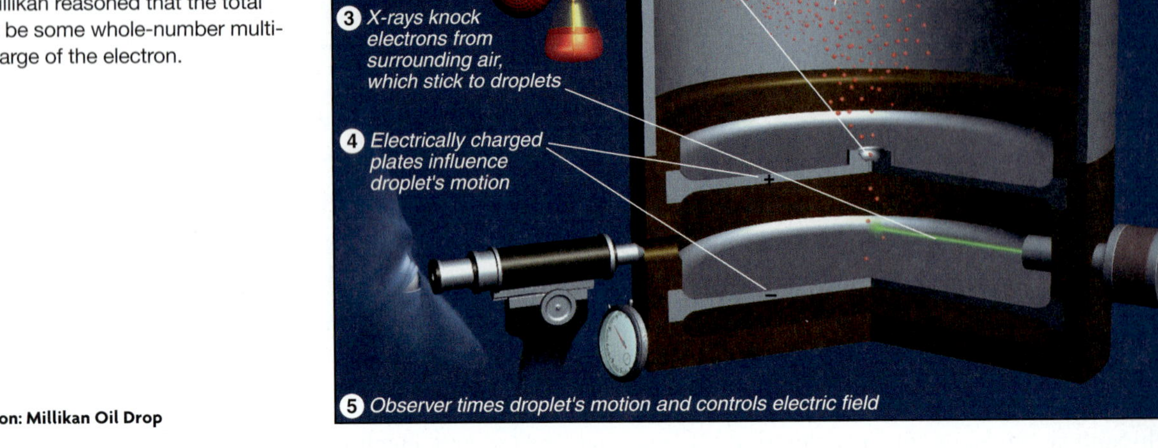

FIGURE 2.5 Millikan's oil-drop experiment for measuring an electron's charge. The motion of a given oil droplet depends on the variation in electric field and the total charge on the droplet, which depends in turn on the number of attached electrons. Millikan reasoned that the total charge must be some whole-number multiple of the charge of the electron.

Animation: Millikan Oil Drop

charge, -1.602×10^{-19} C (C stands for *coulomb,* the SI unit of charge). Using the electron's mass/charge ratio from work by Thomson and others and this value for the electron's charge, let's calculate the electron's *extremely* small mass the way Millikan did:

$$\text{Mass of electron} = \frac{\text{mass}}{\text{charge}} \times \text{charge} = \left(-5.686\times10^{-12}\ \frac{\text{kg}}{\cancel{\text{C}}}\right)(-1.602\times10^{-19}\cancel{\text{C}})$$
$$= 9.109\times10^{-31}\ \text{kg} = 9.109\times10^{-28}\ \text{g}$$

Discovery of the Atomic Nucleus

The properties of the electron posed major questions about the inner structure of atoms. If everyday matter is electrically neutral, the atoms that make it up must be neutral also. But if atoms contain negatively charged electrons, what positive charges balance them? And if an electron has such an incredibly tiny mass, what accounts for an atom's much larger mass? To address these questions, Thomson proposed a model of a spherical atom composed of diffuse, positively charged matter, in which electrons were embedded like "raisins in a plum pudding."

Near the turn of the 20th century, French scientists discovered radioactivity, the emission of particles and/or radiation from atoms of certain elements. Just a few years later, in 1910, the New Zealand–born physicist Ernest Rutherford (1871–1937) used one type of radioactive particle in a series of experiments that solved this dilemma of atomic structure.

Figure 2.6 is a three-part representation of Rutherford's experiment. Tiny, dense, positively charged alpha (α) particles emitted from radium were aimed, like minute projectiles, at thin gold foil. The figure illustrates (A) the "plum pudding" hypothesis, (B) the apparatus used to measure the deflection (scattering) of the α particles from the light flashes created when the particles struck a circular, coated screen, and (C) the actual result.

With Thomson's model in mind, Rutherford expected only minor, if any, deflections of the α particles because they should act as tiny, dense, positively charged "bullets" and go right through the gold atoms. According to the model, the embedded electrons could not deflect the α particles any more than a Ping-Pong ball could deflect a speeding baseball. Initial results confirmed this, but soon the unexpected happened. As Rutherford recalled: "Then I remember two or three days later Geiger [one of his coworkers] coming to me in great excitement and

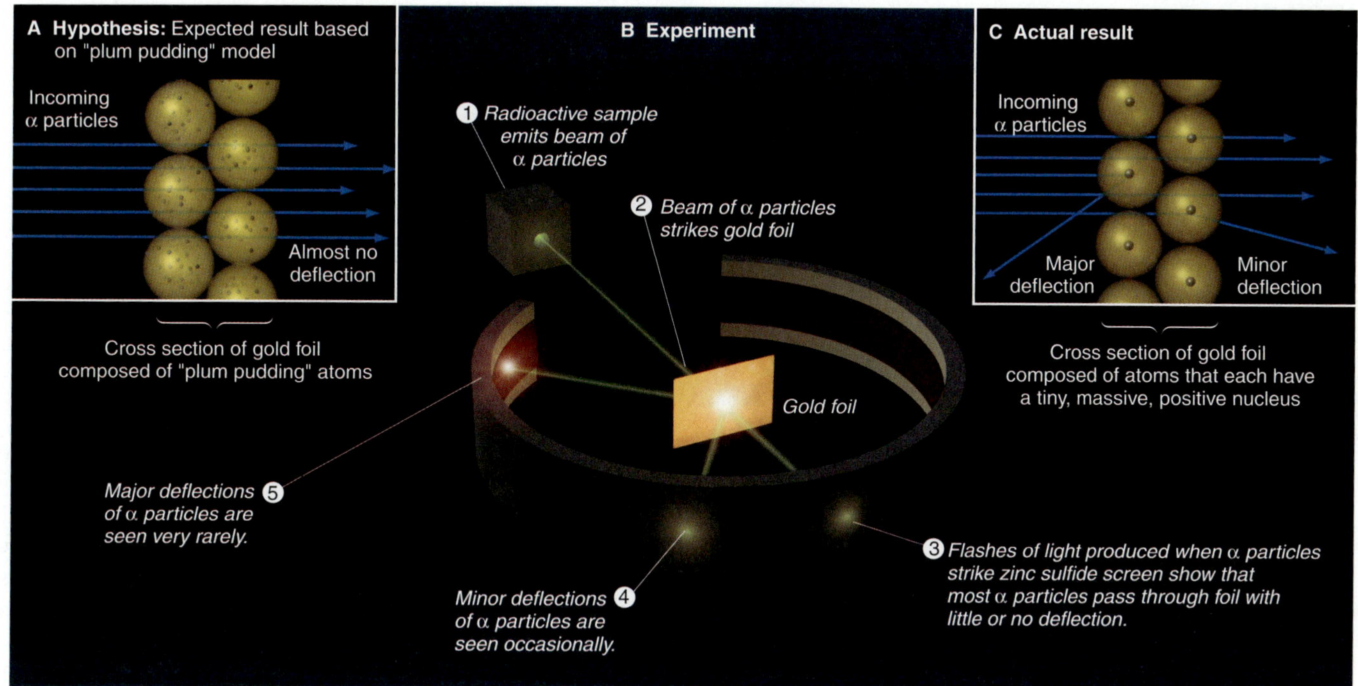

FIGURE 2.6 Rutherford's α-scattering experiment and discovery of the atomic nucleus. A, HYPOTHESIS: Atoms consist of electrons embedded in diffuse, positively charged matter, so the speeding α particles should pass through the gold foil with, at most, minor deflections. **B,** EXPERIMENT: α particles emit a flash of light when they pass through the gold atoms and hit a phosphor-coated screen. **C,** RESULTS: Occasional minor deflections and very infrequent major deflections are seen. This means very high mass and positive charge are concentrated in a tiny region within the atom, the nucleus.

Animation: Rutherford's Experiment

saying, 'We have been able to get some of the α particles coming backwards . . . .' It was quite the most incredible event that has ever happened to me in my life. It was almost as incredible as if you fired a 15-inch shell at a piece of tissue paper and it came back and hit you."

The data showed that very few α particles were deflected at all, and that only 1 in 20,000 was deflected by more than 90° ("coming backwards"). It seemed that these few α particles were being repelled by something small, dense, and positive within the gold atoms. From the data, Rutherford calculated that *an atom is mostly space occupied by electrons,* but in the center of that space is a tiny region, which he called the **nucleus,** that contains *all the positive charge and nearly all the mass of the atom.* He proposed that positive particles lay within the nucleus and called them *protons*. Rutherford's model explained the charged nature of matter, but it could not account for all the atom's mass. After more than 20 years, this issue was resolved when, in 1932, James Chadwick discovered the *neutron,* an uncharged dense particle that also resides in the nucleus.

SECTION 2.4 SUMMARY

Several major discoveries at the turn of the 20th century led to our current model of atomic structure. • Cathode rays were shown to consist of negative particles (electrons) that exist in all matter. J. J. Thomson measured their mass/charge ratio and concluded that they are much smaller and lighter than atoms. • Robert Millikan determined the charge of the electron, which he combined with other data to calculate its mass. • Ernest Rutherford proposed that atoms consist of a tiny, massive, positive nucleus surrounded by electrons.

2.5 THE ATOMIC THEORY TODAY

For over 200 years, scientists have known that all matter consists of atoms, and they have learned astonishing things about them. Dalton's tiny indivisible particles have given way to atoms with "fuzzy," indistinct boundaries and an elaborate internal architecture of subatomic particles. In this section, we examine our current model and begin to see how the properties of subatomic particles affect the properties of atoms.

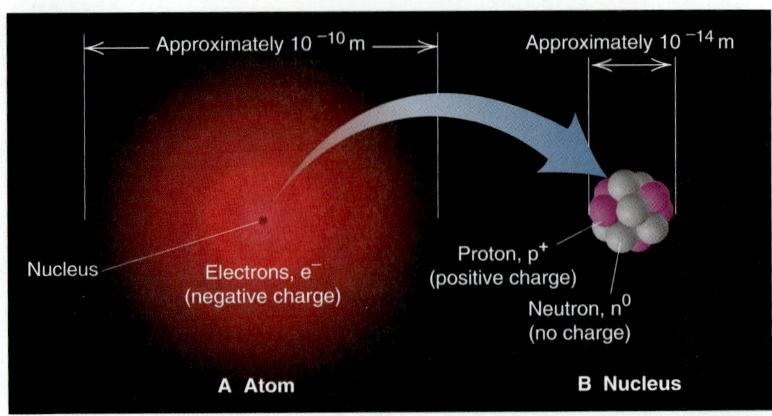

FIGURE 2.7 General features of the atom. A, A "cloud" of rapidly moving, negatively charged electrons occupies virtually all the atomic volume and surrounds the tiny, central nucleus. **B,** The nucleus contains virtually all the mass of the atom and consists of positively charged protons and uncharged neutrons.

Structure of the Atom

An *atom* is an electrically neutral, spherical entity composed of a positively charged central nucleus surrounded by one or more negatively charged electrons (Figure 2.7). The electrons move rapidly within the available atomic volume, held there by the attraction of the nucleus. The nucleus is incredibly dense: it contributes 99.97% of the atom's mass but occupies only about 1 ten-trillionth of its volume. An atom's diameter ($\sim 10^{-10}$ m) is about 10,000 times the diameter of its nucleus ($\sim 10^{-14}$ m).

THINK OF IT THIS WAY
The Tiny, Massive Nucleus

A few analogies will help you appreciate the incredible properties of the atomic nucleus. A nucleus the size of the period at the end of this sentence would weigh about 100 tons, as much as 50 cars. An atom the size of the Houston Astrodome would have a nucleus the size of a marble that would contain virtually all the atom's mass. If a nucleus were actually the size shown in Figure 2.7 (about 1 cm across), the atom would be about 100 m across, or slightly more than the length of a football field!

An atomic nucleus consists of protons and neutrons (the only exception is the simplest hydrogen nucleus, which is a single proton). The **proton (p^+)** has a positive charge, and the **neutron (n^0)** has no charge; thus, the positive charge of the nucleus results from its protons. The *magnitude* of the charge possessed by a proton is equal to that of an **electron (e^-),** but the *signs* of the charges are opposite. *An atom is neutral because the number of protons in the nucleus equals the number of electrons surrounding the nucleus.* Some properties of these three subatomic particles are listed in Table 2.2.

Table 2.2 Properties of the Three Key Subatomic Particles

	Charge		Mass		
Name (Symbol)	**Relative**	**Absolute (C)***	**Relative (amu)†**	**Absolute (g)**	**Location in Atom**
Proton (p^+)	1+	$+1.60218\times10^{-19}$	1.00727	1.67262×10^{-24}	Nucleus
Neutron (n^0)	0	0	1.00866	1.67493×10^{-24}	Nucleus
Electron (e^-)	1−	-1.60218×10^{-19}	0.00054858	9.10939×10^{-28}	Outside nucleus

*The coulomb (C) is the SI unit of charge.
†The atomic mass unit (amu; discussed later in this section) equals 1.66054×10^{-24} g.

Atomic Number, Mass Number, and Atomic Symbol

The **atomic number (Z)** of an element equals the number of protons in the nucleus of each of its atoms. *All atoms of a particular element have the same atomic number, and each element has a different atomic number from that of any other element.* All carbon atoms ($Z = 6$) have 6 protons, all oxygen atoms ($Z = 8$) have 8 protons, and all uranium atoms ($Z = 92$) have 92 protons. There are currently 116 known elements, of which 90 occur in nature; the remaining 26 have been synthesized by nuclear scientists.

The total number of protons and neutrons in the nucleus of an atom is its **mass number (A).** Each proton and each neutron contributes one unit to the mass number. Thus, a carbon atom with 6 protons and 6 neutrons in its nucleus has a mass number of 12, and a uranium atom with 92 protons and 146 neutrons in its nucleus has a mass number of 238.

The nuclear mass number and charge are often written with the **atomic symbol** (or *element symbol*). Every element has a symbol based on its English, Latin, or Greek name, such as C for carbon, O for oxygen, S for sulfur, and Na for sodium (Latin *natrium*). The atomic number (Z) is written as a left *sub*script and the mass number (A) as a left *super*script to the symbol, so element X would be ${}^{A}_{Z}\text{X}$. The mass number is the sum of protons and neutrons, so the number of neutrons (N) equals the mass number minus the atomic number:

$$\text{Number of neutrons} = \text{mass number} - \text{atomic number}, \quad \text{or} \quad N = A - Z \qquad (2.2)$$

Thus, a chlorine atom, which is symbolized as ${}^{35}_{17}\text{Cl}$, has $A = 35$, $Z = 17$, and $N = 35 - 17 = 18$. Each element has its own atomic number, so we know the atomic number from the symbol. For example, every carbon atom has 6 protons. Therefore, instead of writing ${}^{12}_{6}\text{C}$ for carbon with mass number 12, we can write ${}^{12}\text{C}$ (spoken "carbon twelve"), with $Z = 6$ understood. Another way to write this atom is carbon-12.

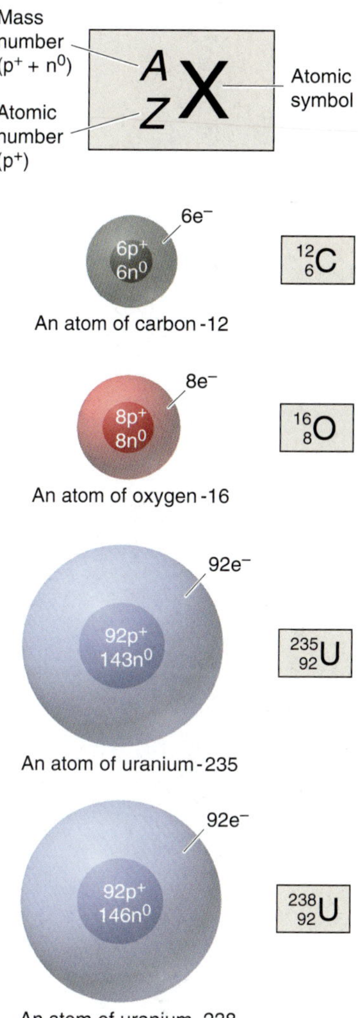

FIGURE 2.8 Depicting the atom. Atoms of carbon-12, oxygen-16, uranium-235, and uranium-238 are shown (nuclei not drawn to scale) with their symbolic representations. The sum of the number of protons (Z) and the number of neutrons (N) equals the mass number (A). An atom is neutral, so the number of protons in the nucleus equals the number of electrons around the nucleus. The two uranium atoms are isotopes of the element.

Isotopes and Atomic Masses of the Elements

All atoms of an element are identical in atomic number but not in mass number. **Isotopes** of an element are atoms that have *different numbers of neutrons* and therefore different mass numbers. For example, all carbon atoms ($Z = 6$) have 6 protons and 6 electrons, but only 98.89% of naturally occurring carbon atoms have 6 neutrons in the nucleus ($A = 12$). A small percentage (1.11%) have 7 neutrons in the nucleus ($A = 13$), and even fewer (less than 0.01%) have 8 ($A = 14$). These are carbon's three naturally occurring isotopes—${}^{12}\text{C}$, ${}^{13}\text{C}$, and ${}^{14}\text{C}$. Five other carbon isotopes—${}^{9}\text{C}$, ${}^{10}\text{C}$, ${}^{11}\text{C}$, ${}^{15}\text{C}$, and ${}^{16}\text{C}$—have been created in the laboratory. Figure 2.8 depicts the atomic number, mass number, and symbol for four atoms, two of which are isotopes of the same element.

A key point is that the chemical properties of an element are primarily determined by the number of electrons, so *all isotopes of an element have nearly identical chemical behavior,* even though they have different masses.

SAMPLE PROBLEM 2.4 Determining the Number of Subatomic Particles in the Isotopes of an Element

Problem Silicon (Si) is essential to the computer industry as a major component of semiconductor chips. It has three naturally occurring isotopes: ${}^{28}\text{Si}$, ${}^{29}\text{Si}$, and ${}^{30}\text{Si}$. Determine the numbers of protons, neutrons, and electrons in each silicon isotope.

Plan The mass number (A) of each of the three isotopes is given, so we know the sum of protons and neutrons. From the elements list on the text's inside front cover, we find the atomic number (Z, number of protons), which equals the number of electrons. We obtain the number of neutrons from Equation 2.2.

Solution From the elements list, the atomic number of silicon is 14. Therefore,

^{28}Si has $14p^+$, $14e^-$, and $14n^0$ (28 − 14)
^{29}Si has $14p^+$, $14e^-$, and $15n^0$ (29 − 14)
^{30}Si has $14p^+$, $14e^-$, and $16n^0$ (30 − 14)

FOLLOW-UP PROBLEM 2.4 How many protons, neutrons, and electrons are in **(a)** $^{11}_{5}Q$? **(b)** $^{41}_{20}X$? **(c)** $^{131}_{53}Y$? What element symbols do Q, X, and Y represent?

The mass of an atom is measured *relative* to the mass of an atomic standard. The modern atomic mass standard is the carbon-12 atom. Its mass is defined as *exactly* 12 atomic mass units. Thus, the **atomic mass unit (amu)** is $\frac{1}{12}$ the mass of a carbon-12 atom. Based on this standard, the ^1H atom has a mass of 1.008 amu; in other words, a ^{12}C atom has almost 12 times the mass of an ^1H atom. We will continue to use the term *atomic mass unit* in the text, although the name of the unit has been changed to the **dalton (Da);** thus, one ^{12}C atom has a mass of 12 daltons (12 Da, or 12 amu). The atomic mass unit, which is a unit of relative mass, has an absolute mass of 1.66054×10^{-24} g.

The isotopic makeup of an element is determined by **mass spectrometry,** a method for measuring the relative masses and abundances of atomic-scale particles very precisely. In one type of mass spectrometer, atoms of a sample of, say, elemental neon are bombarded by a high-energy electron beam (Figure 2.9A). As a result, one electron is knocked off each Ne atom, and each resulting particle has one positive charge. Thus, its mass/charge ratio (*m/e*) equals the mass of an Ne atom divided by 1+. The *m/e* values can be measured in the mass spectrometer to identify the masses of different isotopes of the element. The positively charged Ne particles are attracted toward a series of negatively charged plates with slits in them, and some of the particles pass through into an evacuated tube exposed to a magnetic field. As the particles zoom through this region, they are deflected (their paths are bent) according to their *m/e* values: the lightest particles are deflected most and the heaviest particles least. At the end of the magnetic region, the particles strike a detector, which records their relative positions and abundances (Figure 2.9B).

Mass spectrometry is now used to measure the mass of virtually any atom or molecule. In 2002, the Nobel Prize in chemistry was awarded for the study of proteins by mass spectrometry.

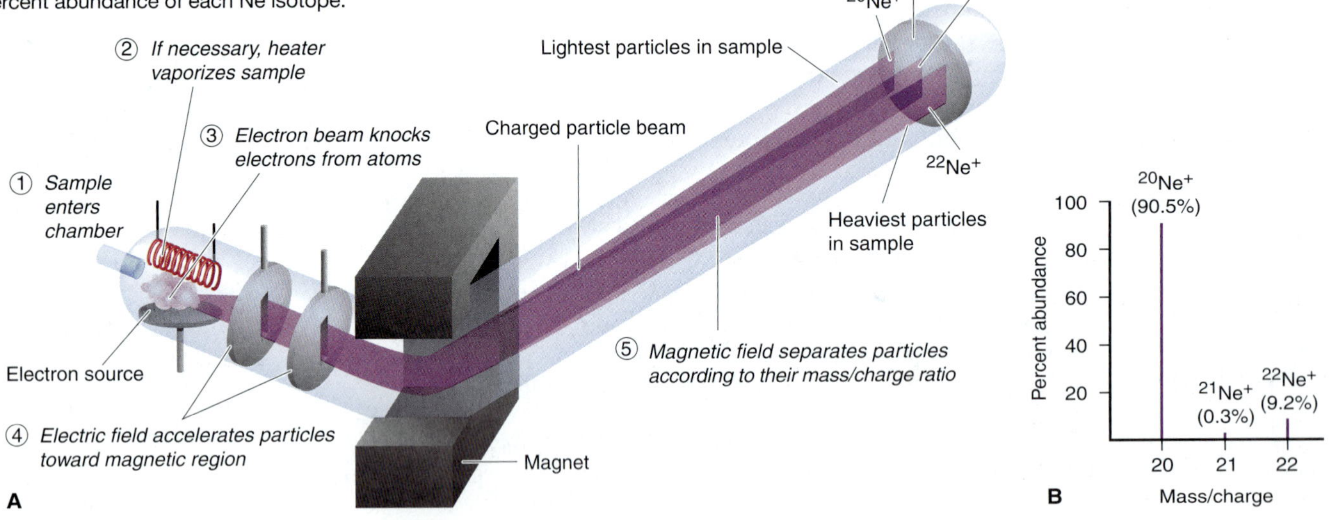

FIGURE 2.9 The mass spectrometer and its data. A, Charged particles are separated on the basis of their *m/e* values. Ne is the sample here. **B,** Data show the percent abundance of each Ne isotope.

Let's see how data obtained with this instrument give us key information. Using a mass spectrometer, we measure the mass ratio of, say, ^{28}Si to ^{12}C as

$$\frac{\text{Mass of } ^{28}\text{Si atom}}{\text{Mass of } ^{12}\text{C standard}} = 2.331411$$

From this mass ratio, we find the **isotopic mass** of the ^{28}Si atom, the mass of the isotope relative to the mass of the standard carbon-12 isotope:

$$\begin{aligned}\text{Isotopic mass of } ^{28}\text{Si} &= \text{measured mass ratio} \times \text{mass of } ^{12}\text{C}\\ &= 2.331411 \times 12 \text{ amu} = 27.97693 \text{ amu}\end{aligned}$$

Along with the isotopic mass, the mass spectrometer gives the relative abundance (fraction) of each isotope in a sample of the element. For example, the percent abundance of ^{28}Si is 92.23%. Such data allow us to calculate the **atomic mass** (also called *atomic weight*) of an element, the *average* of the masses of its naturally occurring isotopes weighted according to their abundances.

Each naturally occurring isotope of an element contributes a certain portion to the atomic mass. For instance, as just noted, 92.23% of Si atoms are ^{28}Si. Using this percent abundance as a fraction and multiplying by the isotopic mass of ^{28}Si gives the portion of the atomic mass of Si contributed by ^{28}Si:

$$\text{Portion of Si atomic mass from } ^{28}\text{Si} = 27.97693 \text{ amu} \times 0.9223 = 25.8031 \text{ amu}$$

(retaining two additional significant figures)

Similar calculations give the portions contributed by ^{29}Si (28.976495 amu × 0.0467 = 1.3532 amu) and by ^{30}Si (29.973770 amu × 0.0310 = 0.9292 amu), and adding the three portions together (rounding to two decimal places at the end) gives the atomic mass of silicon:

$$\begin{aligned}\text{Atomic mass of Si} &= 25.8031 \text{ amu} + 1.3532 \text{ amu} + 0.9292 \text{ amu}\\ &= 28.0855 \text{ amu} = 28.09 \text{ amu}\end{aligned}$$

Note that this atomic mass is an average value, and averages must be interpreted carefully. Although the average number of children in an American family in 1985 was 2.4, no family actually had 2.4 children; similarly, no individual silicon atom has a mass of 28.09 amu. But for most laboratory purposes, we consider a sample of silicon to consist of atoms with this average mass.

SAMPLE PROBLEM 2.5 Calculating the Atomic Mass of an Element

Problem Silver (Ag; $Z = 47$) has 46 known isotopes, but only two occur naturally, ^{107}Ag and ^{109}Ag. Given the following mass spectrometric data, calculate the atomic mass of Ag:

Isotope	Mass (amu)	Abundance (%)
^{107}Ag	106.90509	51.84
^{109}Ag	108.90476	48.16

Plan From the mass and abundance of the two Ag isotopes, we have to find the atomic mass of Ag (weighted average of the isotopic masses). We multiply each isotopic mass by its fractional abundance to find the portion of the atomic mass contributed by each isotope. The sum of the isotopic portions is the atomic mass.

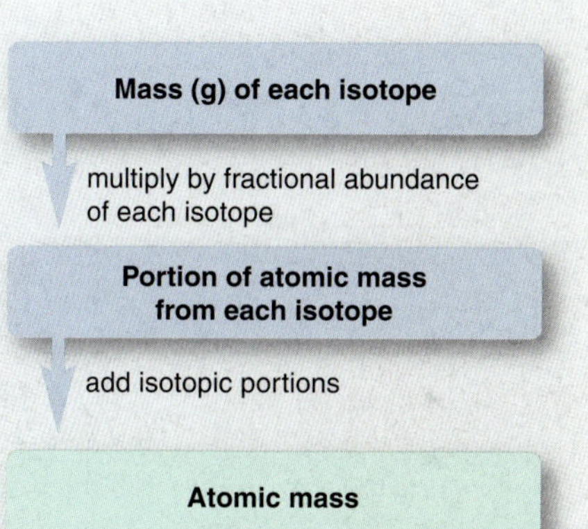

Solution Finding the portion of the atomic mass from each isotope:

$$\begin{aligned}\text{Portion of atomic mass from } ^{107}\text{Ag:} &= \text{isotopic mass} \times \text{fractional abundance}\\ &= 106.90509 \text{ amu} \times 0.5184 = 55.42 \text{ amu}\\ \text{Portion of atomic mass from } ^{109}\text{Ag:} &= 108.90476 \text{ amu} \times 0.4816 = 52.45 \text{ amu}\end{aligned}$$

Finding the atomic mass of silver:

$$\text{Atomic mass of Ag} = 55.42 \text{ amu} + 52.45 \text{ amu} = 107.87 \text{ amu}$$

Check The individual portions seem right: ~100 amu × 0.50 = 50 amu. The portions should be almost the same because the two isotopic abundances are almost the same. We rounded each portion to four significant figures because that is the number of significant figures in the abundance values. This is the correct atomic mass (to two decimal places), as shown in the list of elements *(inside front cover).*

FOLLOW-UP PROBLEM 2.5 Boron (B; Z = 5) has two naturally occurring isotopes. Find the percent abundances of ^{10}B and ^{11}B given the atomic mass of B = 10.81 amu, the isotopic mass of ^{10}B = 10.0129 amu, and the isotopic mass of ^{11}B = 11.0093 amu. (*Hint:* The sum of the fractional abundances is 1. If x = abundance of ^{10}B, then $1 - x$ = abundance of ^{11}B.)

SECTION 2.5 SUMMARY

An atom has a central nucleus, which contains positively charged protons and uncharged neutrons and is surrounded by negatively charged electrons. An atom is neutral because the number of electrons equals the number of protons. • An atom is represented by the notation $^{A}_{Z}X$, in which Z is the atomic number (number of protons), A the mass number (sum of protons and neutrons), and X the atomic symbol. • An element occurs naturally as a mixture of isotopes, atoms with the same number of protons but different numbers of neutrons. Each isotope has a mass relative to the ^{12}C mass standard. • The atomic mass of an element is the average of the masses of its isotopes weighted according to their natural abundances and is determined by mass spectrometry.

2.6 ELEMENTS: A FIRST LOOK AT THE PERIODIC TABLE

At the end of the 18th century, Lavoisier compiled a list of the 23 elements known at that time; by 1870, 65 were known; by 1925, 88; today, there are 116 and still counting! These elements combine to form millions of compounds, so we clearly need some way to organize what we know about their behavior. By the mid-19th century, enormous amounts of information concerning reactions, properties, and atomic masses of the elements led researchers to note recurring, or *periodic,* patterns of element behavior. In 1871, the Russian chemist Dmitri Mendeleev organized this information into a table that listed the elements by increasing atomic mass, arranged so that elements with similar chemical properties fell in the same column. The modern **periodic table of the elements,** based on Mendeleev's earlier version—but arranged by atomic number, not mass—is one of the great classifying schemes in science and has become an indispensable tool to chemists.

Organization of the Periodic Table A modern version of the periodic table appears in Figure 2.10 and inside the front cover. It is formatted as follows:

1. Each element has a box that contains its atomic number, atomic symbol, and atomic mass. The boxes lie in order of *increasing atomic number* (number of protons) as you move from left to right.
2. The boxes are arranged into a grid of **periods** (horizontal rows) and **groups** (vertical columns). Each period has a number from 1 to 7. Each group has a number from 1 to 8 *and* either the letter A or B. A new system, with group numbers from 1 to 18 but no letters, appears in parentheses under the number-letter designations. (Most chemists still use the number-letter system, so the text retains it and shows the new numbering system in parentheses.)
3. The eight A groups (two on the left and six on the right) contain the *main-group,* or *representative, elements.* The ten B groups, located between Groups 2A(2) and 3A(13), contain the *transition elements.* Two horizontal series of *inner transition elements,* the lanthanides and the actinides, fit *between* the elements in Group 3B(3) and Group 4B(4) and are usually placed below the main body of the table.

Metals (main-group)
Metals (transition)
Metals (inner transition)
Metalloids
Nonmetals

MAIN–GROUP ELEMENTS | TRANSITION ELEMENTS | MAIN–GROUP ELEMENTS

Period	1A (1)	2A (2)	3B (3)	4B (4)	5B (5)	6B (6)	7B (7)	8B (8)	8B (9)	8B (10)	1B (11)	2B (12)	3A (13)	4A (14)	5A (15)	6A (16)	7A (17)	8A (18)
1	1 **H** 1.008																	2 **He** 4.003
2	3 **Li** 6.941	4 **Be** 9.012											5 **B** 10.81	6 **C** 12.01	7 **N** 14.01	8 **O** 16.00	9 **F** 19.00	10 **Ne** 20.18
3	11 **Na** 22.99	12 **Mg** 24.31											13 **Al** 26.98	14 **Si** 28.09	15 **P** 30.97	16 **S** 32.07	17 **Cl** 35.45	18 **Ar** 39.95
4	19 **K** 39.10	20 **Ca** 40.08	21 **Sc** 44.96	22 **Ti** 47.88	23 **V** 50.94	24 **Cr** 52.00	25 **Mn** 54.94	26 **Fe** 55.85	27 **Co** 58.93	28 **Ni** 58.69	29 **Cu** 63.55	30 **Zn** 65.41	31 **Ga** 69.72	32 **Ge** 72.61	33 **As** 74.92	34 **Se** 78.96	35 **Br** 79.90	36 **Kr** 83.80
5	37 **Rb** 85.47	38 **Sr** 87.62	39 **Y** 88.91	40 **Zr** 91.22	41 **Nb** 92.91	42 **Mo** 95.94	43 **Tc** (98)	44 **Ru** 101.1	45 **Rh** 102.9	46 **Pd** 106.4	47 **Ag** 107.9	48 **Cd** 112.4	49 **In** 114.8	50 **Sn** 118.7	51 **Sb** 121.8	52 **Te** 127.6	53 **I** 126.9	54 **Xe** 131.3
6	55 **Cs** 132.9	56 **Ba** 137.3	57 **La** 138.9	72 **Hf** 178.5	73 **Ta** 180.9	74 **W** 183.9	75 **Re** 186.2	76 **Os** 190.2	77 **Ir** 192.2	78 **Pt** 195.1	79 **Au** 197.0	80 **Hg** 200.6	81 **Tl** 204.4	82 **Pb** 207.2	83 **Bi** 209.0	84 **Po** (209)	85 **At** (210)	86 **Rn** (222)
7	87 **Fr** (223)	88 **Ra** (226)	89 **Ac** (227)	104 **Rf** (263)	105 **Db** (262)	106 **Sg** (266)	107 **Bh** (267)	108 **Hs** (277)	109 **Mt** (268)	110 **Ds** (281)	111 **Rg** (272)	112 (285)	113 (284)	114 (289)	115 (288)	116 (292)		

INNER TRANSITION ELEMENTS

Period	Series														
6	Lanthanides	58 **Ce** 140.1	59 **Pr** 140.9	60 **Nd** 144.2	61 **Pm** (145)	62 **Sm** 150.4	63 **Eu** 152.0	64 **Gd** 157.3	65 **Tb** 158.9	66 **Dy** 162.5	67 **Ho** 164.9	68 **Er** 167.3	69 **Tm** 168.9	70 **Yb** 173.0	71 **Lu** 175.0
7	Actinides	90 **Th** 232.0	91 **Pa** (231)	92 **U** 238.0	93 **Np** (237)	94 **Pu** (242)	95 **Am** (243)	96 **Cm** (247)	97 **Bk** (247)	98 **Cf** (251)	99 **Es** (252)	100 **Fm** (257)	101 **Md** (258)	102 **No** (259)	103 **Lr** (260)

FIGURE 2.10 The modern periodic table. Element boxes are arranged by *increasing* atomic number into vertical *groups* and horizontal *periods*. Each box contains the atomic number, atomic symbol, and atomic mass. (A mass in parentheses is the mass number of the most stable isotope of that element.) The periods are numbered 1 to 7. The groups (sometimes called *families*) have a number-letter designation and a new group number in parentheses. The A groups are the main-group elements; the B groups are the transition elements. Two series of inner transition elements are placed below the main body of the table. Metals lie below and to the left of the thick "staircase" line [top of 3A(13) to bottom of 6A(16) in Period 6] and include main-group metals *(purple-blue)*, transition elements *(blue)*, and inner transition elements *(gray-blue)*. Nonmetals *(yellow)* lie to the right of the line. Metalloids *(green)* lie along the line. As of mid-2008, elements 112 through 116 had not yet been named.

At this point in the text, the clearest distinction among the elements is their classification as metals, nonmetals, or metalloids. The "staircase" line that runs from the top of Group 3A(13) to the bottom of Group 6A(16) in Period 6 is a dividing line for this classification. The **metals** (three shades of blue) appear in the large lower-left portion of the table. About three-quarters of the elements are metals, including many main-group elements and all the transition and inner transition elements. They are generally shiny solids at room temperature (mercury is the only liquid) that conduct heat and electricity well and can be tooled into sheets (malleable) and wires (ductile). The **nonmetals** (yellow) appear in the small upper-right portion of the table. They are generally gases or dull, brittle solids at room temperature (bromine is the only liquid) and conduct heat and electricity poorly. Along the staircase line lie the **metalloids** (green; also called **semimetals**), which have properties between those of metals and nonmetals. Several metalloids, such as silicon (Si) and germanium (Ge), play major roles in modern electronics.

It is important to learn some of the group (family) names. Group 1A(1), except for hydrogen, consists of the *alkali metals,* and Group 2A(2) consists of the *alkaline earth metals.* Both groups consist of highly reactive elements. The *halogens,* Group 7A(17), are highly reactive nonmetals, whereas the *noble gases,* Group 8A(18), are relatively unreactive nonmetals. Other main groups [3A(13) to 6A(16)] are often named for the first element in the group; for example, Group 6A is the *oxygen family.*

A key point that we return to many times is that, in general, *elements in a group have **similar** chemical properties and elements in a period have **different** chemical properties.* We begin applying the organizing power of the periodic table in the next section, where we discuss how elements combine to form compounds.

SECTION 2.6 SUMMARY

In the periodic table, the elements are arranged by atomic number into horizontal periods and vertical groups. • Because of the periodic recurrence of certain key properties, elements within a group have similar behavior, whereas elements in a period have dissimilar behavior. • Nonmetals appear in the upper-right portion of the table, metalloids lie along a staircase line, and metals fill the rest of the table.

Animation: Formation of an Ionic Compound

2.7 COMPOUNDS: INTRODUCTION TO BONDING

The overwhelming majority of elements occur in chemical combination with other elements. In fact, only a few elements occur free in nature. The noble gases—helium (He), neon (Ne), argon (Ar), krypton (Kr), xenon (Xe), and radon (Rn)—occur in air as separate atoms. In addition to occurring in compounds, oxygen (O), nitrogen (N), and sulfur (S) occur in the most common elemental form as the molecules O_2, N_2, and S_8, and carbon (C) occurs in vast, nearly pure deposits of coal. Some of the metals, such as copper (Cu), silver (Ag), gold (Au), and platinum (Pt), may also occur uncombined with other elements. But these few exceptions reinforce the general rule that elements occur combined in compounds.

It is the electrons of the atoms of interacting elements that are involved in compound formation. Elements combine in two general ways:

1. *Transferring electrons* from the atoms of one element to those of another to form **ionic compounds** (Figure 2.11)
2. *Sharing electrons* between atoms of different elements to form **covalent compounds**

These processes generate **chemical bonds,** the forces that hold the atoms of elements together in a compound. We'll introduce compound formation next and have much more to say about it in later chapters.

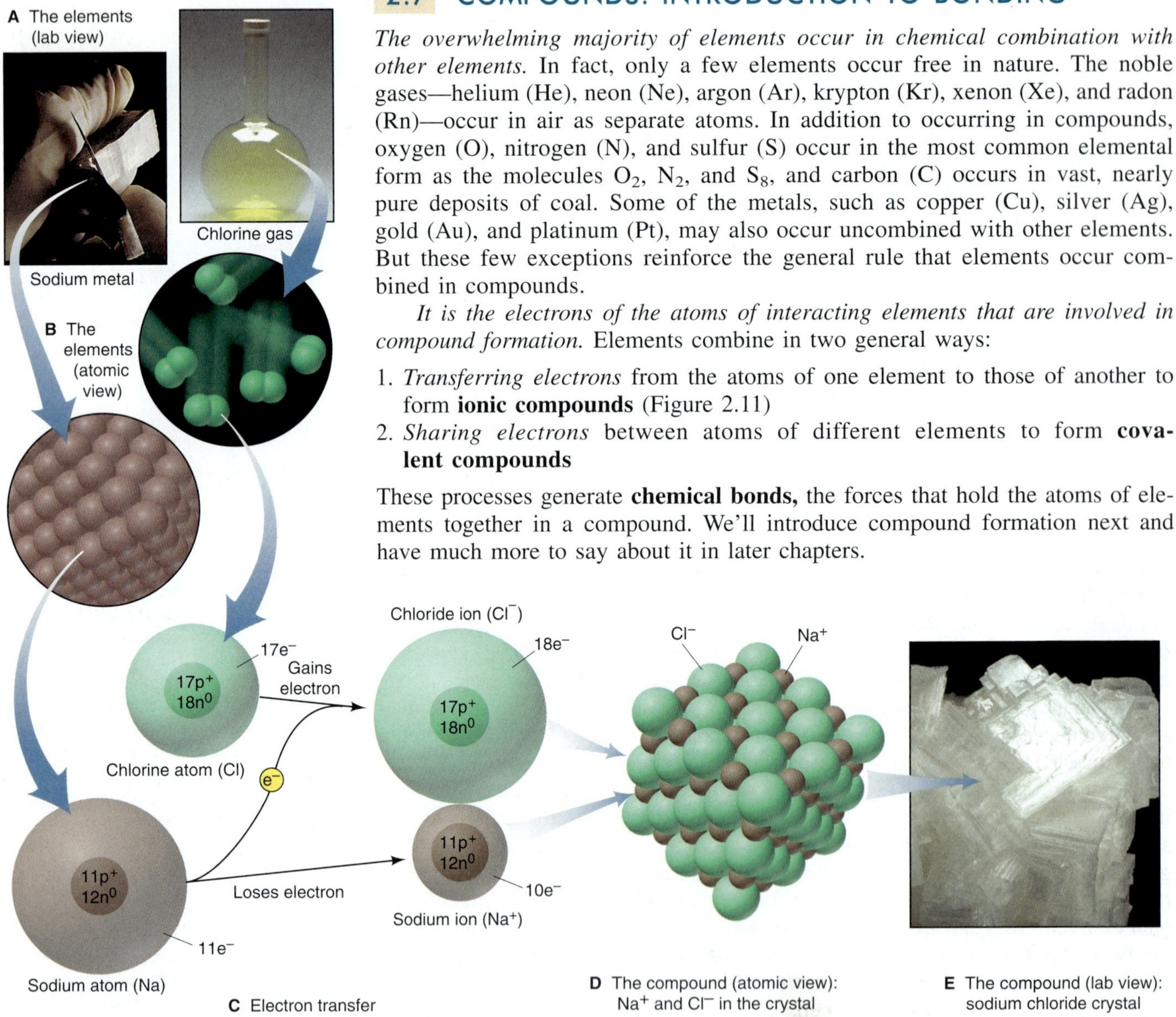

FIGURE 2.11 The formation of an ionic compound. A, The two elements as seen in the laboratory. **B,** The elements as they might appear on the atomic scale. **C,** The neutral sodium atom loses one electron to become a sodium cation (Na^+), and the chlorine atom gains one electron to become a chloride anion (Cl^-). (Note that when atoms lose electrons, they become ions that are smaller, and when they gain electrons, they become ions that are larger.) **D,** Na^+ and Cl^- ions attract each other and form a regular three-dimensional array. **E,** This arrangement of the ions is reflected in the structure of crystalline NaCl, which occurs naturally as the mineral halite, hence the name *halogens* for the Group 7A(17) elements.

The Formation of Ionic Compounds

Ionic compounds are composed of **ions,** charged particles that form when an atom (or small group of atoms) gains or loses one or more electrons. The simplest type of ionic compound is a **binary ionic compound,** one composed of just two elements. It typically forms *when a metal reacts with a nonmetal.* Each metal atom loses a certain number of its electrons and becomes a **cation,** a positively charged ion. The nonmetal atoms gain the electrons lost by the metal atoms and become **anions,** negatively charged ions. In effect, the metal atoms *transfer electrons* to the nonmetal atoms. The resulting cations and anions attract each other through electrostatic forces and form the ionic compound. *All binary ionic compounds are solids.* A cation or anion derived from a single atom is called a **monatomic ion;** we'll discuss polyatomic ions, those derived from a small group of atoms, later in this section.

The formation of the binary ionic compound sodium chloride, common table salt, is depicted in Figure 2.11, from the elements through the atomic-scale electron transfer to the compound. In the electron transfer, a sodium atom, which is neutral because it has the same number of protons as electrons, *loses* 1 electron and forms a sodium cation, Na^+. Note that the charge on an ion is written as a *right superscript.* A chlorine atom *gains* the electron and becomes a chloride anion, Cl^-. (The name change from the nonmetal atom to the ion is discussed in the next section.) Even the tiniest visible grain of table salt contains an enormous number of sodium and chloride ions. The oppositely charged ions (Na^+ and Cl^-) attract each other, and the similarly charged ions (Na^+ and Na^+, or Cl^- and Cl^-) repel each other. The resulting solid aggregation is a regular array of alternating Na^+ and Cl^- ions that extends in all three dimensions.

The strength of the ionic bonding depends to a great extent on the net strength of these attractions and repulsions and is described by *Coulomb's law,* which can be expressed as follows: *the energy of attraction (or repulsion) between two particles is directly proportional to the product of the charges and inversely proportional to the distance between them.*

$$\text{Energy} \propto \frac{\text{charge 1} \times \text{charge 2}}{\text{distance}}$$

In other words, ions with higher charges attract (or repel) each other more strongly than ions with lower charges. Likewise, smaller ions attract (or repel) each other more strongly than larger ions because the distance between their charges is shorter. These effects are summarized in Figure 2.12.

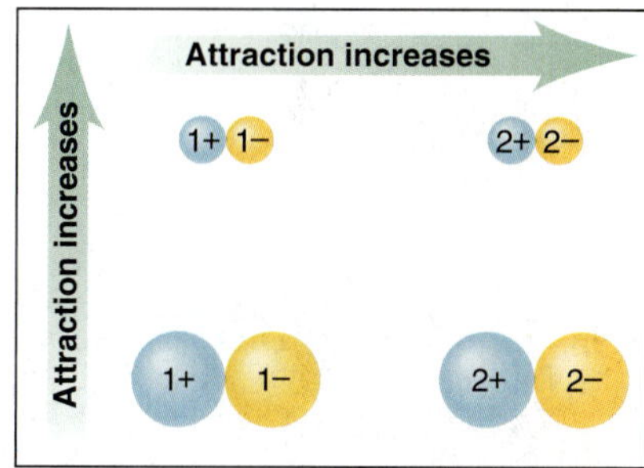

FIGURE 2.12 Factors that influence the strength of ionic bonding. For ions of a given size, strength of attraction *(arrows)* increases with higher ionic charge *(left to right).* For ions of a given charge, strength of attraction increases with smaller ionic size *(bottom to top).*

Ionic compounds are neutral; that is, they possess no net charge. For this to occur, they must contain equal numbers of positive and negative *charges*—not necessarily equal numbers of positive and negative *ions.* Because Na^+ and Cl^- each bear a unit charge (1+ or 1−), equal numbers of these ions are present in sodium chloride; but in sodium oxide, for example, there are two Na^+ ions needed to balance the 2− charge of each oxide ion, O^{2-}.

Can we predict the number of electrons a given atom will lose or gain when it forms an ion? In the formation of sodium chloride, for example, why does each sodium atom give up only 1 of its 11 electrons? Why doesn't each chlorine atom gain two electrons, instead of just one? For A-group elements, the periodic table provides an answer. We generally find that metals lose electrons and nonmetals gain electrons to *form ions with the same number of electrons as in an atom of the nearest noble gas* [Group 8A(18)]. Noble gases have a stability (low reactivity) that is related to their number (and arrangement) of electrons. A sodium atom ($11e^-$) can attain the stability of neon ($10e^-$), the nearest noble gas, by losing one electron. Similarly, by gaining one electron, a chlorine atom ($17e^-$) attains the stability of argon ($18e^-$), its nearest noble gas. Thus, when an element located near a noble gas forms a monatomic ion, *it gains or loses enough electrons to attain the same number as that noble gas.* Specifically, the elements in Group 1A(1) lose one electron, those in Group 2A(2) lose two, and aluminum in Group

3A(13) loses three; the elements in Group 7A(17) gain one electron, oxygen and sulfur in Group 6A(16) gain two, and nitrogen in Group 5A(15) gains three.

SAMPLE PROBLEM 2.6 Predicting the Ion an Element Forms

Problem What monatomic ions do the following elements form?
(a) Iodine (Z = 53) **(b)** Calcium (Z = 20) **(c)** Aluminum (Z = 13)

Plan We use the given Z value to find the element in the periodic table and see whether an element loses or gains electrons to attain the same number as the nearest noble gas. Elements in Groups 1A, 2A, and 3A *lose* electrons and become positive ions; those in Groups 5A, 6A, and 7A *gain* electrons and become negative ions.

Solution (a) I^- Iodine ($_{53}I$) is a nonmetal in Group 7A(17), one of the halogens. Like any member of this group, it gains 1 electron to have the same number as the nearest Group 8A(18) member, in this case $_{54}Xe$.

(b) Ca^{2+} Calcium ($_{20}Ca$) is a member of Group 2A(2), the alkaline earth metals. Like any Group 2A member, it loses 2 electrons to attain the same number as the nearest noble gas, in this case, $_{18}Ar$.

(c) Al^{3+} Aluminum ($_{13}Al$) is a metal in the boron family [Group 3A(13)] and thus loses 3 electrons to attain the same number as its nearest noble gas, $_{10}Ne$.

FOLLOW-UP PROBLEM 2.6 What monatomic ion does each of the following elements form: **(a)** $_{16}S$; **(b)** $_{37}Rb$; **(c)** $_{56}Ba$?

A No interaction

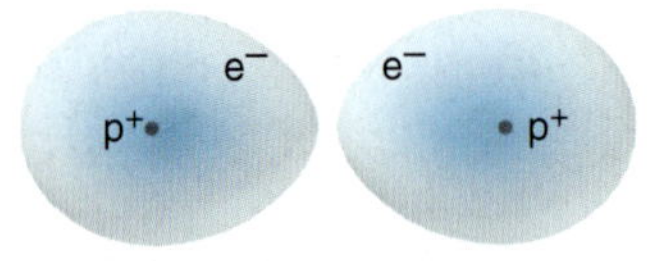

B Attraction begins

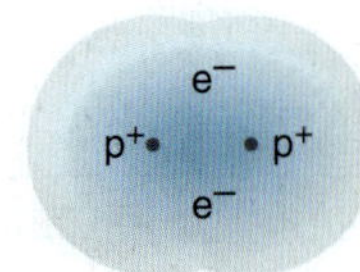

C Covalent bond

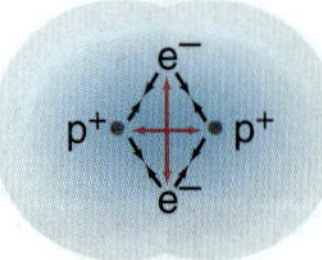

D Interaction of forces

FIGURE 2.13 Formation of a covalent bond between two H atoms. A, The distance is too great for the atoms to affect each other. **B,** As the distance decreases, the nucleus of each atom begins to attract the electron of the other. **C,** The covalent bond forms when the two nuclei mutually attract the pair of electrons at some optimum distance. **D,** The H_2 molecule is more stable than the separate atoms because the attractive forces *(black arrows)* between each nucleus and the two electrons are greater than the repulsive forces *(red arrows)* between the electrons and between the nuclei.

The Formation of Covalent Compounds

Covalent compounds form when elements share electrons, which usually occurs between nonmetals. Even though relatively few nonmetals exist, they interact in many combinations to form a very large number of covalent compounds.

The simplest case of electron sharing occurs not in a compound but between two hydrogen atoms (H; Z = 1). Imagine two separated H atoms approaching each other, as in Figure 2.13. As they get closer, the nucleus of each atom attracts the electron of the other atom more and more strongly, and the separated atoms begin to interpenetrate each other. At some optimum distance between the nuclei, the two atoms form a **covalent bond,** a pair of electrons mutually attracted by the two nuclei. The result is a hydrogen molecule, in which each electron no longer "belongs" to a particular H atom: the two electrons are *shared* by the two nuclei. Repulsions between the nuclei and between the electrons also occur, but the net attraction is greater than the net repulsion. (We discuss the properties of covalent bonds in great detail in Chapter 9.)

A sample of hydrogen gas consists of these diatomic molecules (H_2)—pairs of atoms that are chemically bound and behave as an independent unit—*not* separate H atoms. Other nonmetals that exist as diatomic molecules at room temperature are nitrogen (N_2), oxygen (O_2), and the halogens [fluorine (F_2), chlorine (Cl_2), bromine (Br_2), and iodine (I_2)]. Phosphorus exists as tetratomic molecules (P_4), and sulfur and selenium as octatomic molecules (S_8 and Se_8) (Figure 2.14). At room temperature, covalent substances may be gases, liquids, or solids.

Atoms of different elements share electrons to form the molecules of a covalent compound. A sample of hydrogen fluoride, for example, consists of molecules in which one H atom forms a covalent bond with one F atom; water consists of molecules in which one O atom forms covalent bonds with two H atoms:

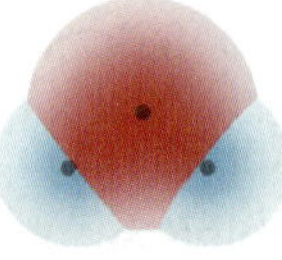

(As you'll see in Chapter 9, covalent bonding provides another way for atoms to attain the same number of electrons as the nearest noble gas.)

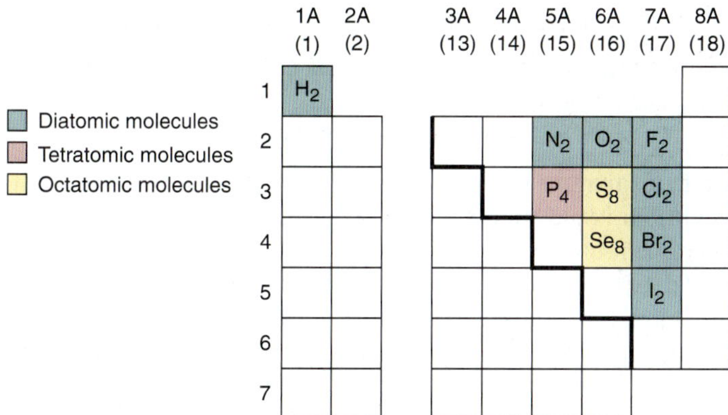

FIGURE 2.14 Elements that occur as molecules.

Distinguishing the Entities in Covalent and Ionic Substances There is a key distinction between the chemical entities in covalent and ionic substances. *Most covalent substances consist of molecules.* A cup of water, for example, consists of individual water molecules lying near each other. In contrast, under ordinary conditions, *no molecules exist in a sample of an ionic compound.* A piece of sodium chloride, for example, is a continuous array of oppositely charged sodium and chloride ions, *not* a collection of individual "sodium chloride molecules."

Another key distinction concerns the nature of the particles attracting each other. Covalent bonding involves the mutual attraction between two (positively charged) nuclei and the two (negatively charged) electrons between them. Ionic bonding involves the mutual attraction between positive and negative ions.

Polyatomic Ions: Covalent Bonds Within Ions Many ionic compounds contain **polyatomic ions,** which consist of two or more atoms bonded *covalently* and have a net positive or negative charge. For example, the ionic compound calcium carbonate is an array of polyatomic carbonate anions and monatomic calcium cations attracted to each other. The carbonate ion consists of a carbon atom covalently bonded to three oxygen atoms, and two additional electrons give the ion its 2− charge (Figure 2.15). In many reactions, a polyatomic ion stays together as a unit.

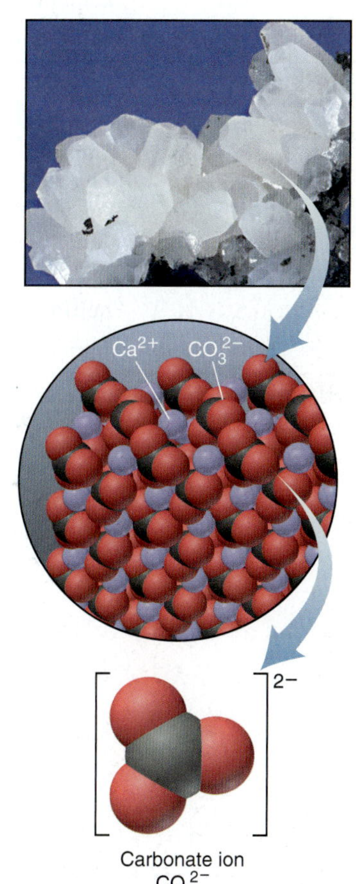

FIGURE 2.15 A polyatomic ion. Calcium carbonate is a three-dimensional array of monatomic calcium cations *(purple spheres)* and polyatomic carbonate anions. As the bottom structure shows, each carbonate ion consists of four covalently bonded atoms.

SECTION 2.7 SUMMARY

Although a few elements occur uncombined in nature, the great majority exist in compounds. • Ionic compounds form when a metal *transfers electrons* to a nonmetal, and the resulting positive and negative ions attract each other to form a three-dimensional array. In many cases, metal atoms lose and nonmetal atoms gain enough electrons to attain the same number of electrons as in atoms of the nearest noble gas. • Covalent compounds form when elements, usually nonmetals, *share electrons.* Each covalent bond is an electron pair mutually attracted by two atomic nuclei. • Monatomic ions are derived from single atoms. Polyatomic ions consist of two or more covalently bonded atoms that have a net positive or negative charge due to a deficit or excess of electrons.

2.8 COMPOUNDS: FORMULAS, NAMES, AND MASSES

Names and formulas of compounds form the vocabulary of the chemical language. In this section, you'll learn the names and formulas of ionic and simple covalent compounds and how to calculate the mass of a unit of a compound from its formula.

Table 2.3 Common Monatomic Ions*

Charge	Formula	Name
Cations		
1+	H^+	hydrogen
	Li^+	**lithium**
	Na^+	**sodium**
	K^+	**potassium**
	Cs^+	cesium
	Ag^+	**silver**
2+	Mg^{2+}	**magnesium**
	Ca^{2+}	**calcium**
	Sr^{2+}	strontium
	Ba^{2+}	**barium**
	Zn^{2+}	**zinc**
	Cd^{2+}	cadmium
3+	Al^{3+}	**aluminum**
Anions		
1−	H^-	hydride
	F^-	**fluoride**
	Cl^-	**chloride**
	Br^-	**bromide**
	I^-	**iodide**
2−	O^{2-}	**oxide**
	S^{2-}	**sulfide**
3−	N^{3-}	nitride

*Listed by charge; those in **boldface** are most common.

Types of Chemical Formulas

In a **chemical formula,** element symbols and numerical subscripts show the type and number of each atom present in the smallest unit of the substance. There are several types of chemical formulas for a compound:

1. The **empirical formula** shows the *relative* number of atoms of each element in the compound. It is the simplest type of formula and is derived from the masses of the component elements. For example, in hydrogen peroxide, there is 1 part by mass of hydrogen for every 16 parts by mass of oxygen. Therefore, the empirical formula of hydrogen peroxide is HO: one H atom for every O atom.
2. The **molecular formula** shows the *actual* number of atoms of each element in a molecule of the compound. The molecular formula of hydrogen peroxide is H_2O_2; there are two H atoms and two O atoms in each molecule.
3. A **structural formula** shows the number of atoms and *the bonds between them;* that is, the relative placement and connections of atoms in the molecule. The structural formula of hydrogen peroxide is H—O—O—H; each H is bonded to an O, and the O's are bonded to each other.

Names and Formulas of Ionic Compounds

All ionic compound names give the positive ion (cation) first and the negative ion (anion) second. Here are some points to note about ion charges:

- Members of a periodic table group have the same ionic charge; for example, Li, Na, and K are all in Group 1A and all have a 1+ charge.
- For A-group cations, ion charge = group number: for example, Na^+ is in Group 1A, Ba^{2+} in Group 2A. (Exceptions in Figure 2.16 are Sn^{2+} and Pb^{2+}.)
- For anions, ion charge = group number minus 8: for example, S is in Group 6A ($6 - 8 = -2$), so the ion is S^{2-}. Table 2.3 lists some of the more common monatomic ions.

Compounds Formed from Monatomic Ions Let's first consider how to name binary ionic compounds, those composed of ions of two elements.

- *The name of the cation is the same as the name of the metal.* Many metal names end in *-ium*.
- *The name of the anion takes the root of the nonmetal name and adds the suffix -ide.*

For example, the anion formed from brom*ine* is named brom*ide* (brom+ide). Therefore, the compound formed from the metal calcium and the nonmetal bromine is named *calcium bromide*.

SAMPLE PROBLEM 2.7 Naming Binary Ionic Compounds

Problem Name the ionic compound formed from the following pairs of elements:
(a) Magnesium and nitrogen **(b)** Iodine and cadmium
(c) Strontium and fluorine **(d)** Sulfur and cesium

Plan The key to naming a binary ionic compound is to recognize which element is the metal and which is the nonmetal. When in doubt, check the periodic table. We place the cation name first, add the suffix *-ide* to the nonmetal root, and place the anion name last.

Solution **(a)** Magnesium is the metal; *nitr-* is the nonmetal root: magnesium nitride

(b) Cadmium is the metal; *iod-* is the nonmetal root: cadmium iodide

(c) Strontium is the metal; *fluor-* is the nonmetal root: strontium fluoride (Note the spelling is fl*uo*ride, not fl*ou*ride.)

(d) Cesium is the metal; *sulf-* is the nonmetal root: cesium sulfide

FOLLOW-UP PROBLEM 2.7 For the following ionic compounds, give the name and periodic table group number of each of the elements present: **(a)** zinc oxide; **(b)** silver bromide; **(c)** lithium chloride; **(d)** aluminum sulfide.

FIGURE 2.16 Some common monatomic ions of the elements. Main-group elements usually form a single monatomic ion. Note that members of a group have ions with the same charge. [Hydrogen is shown as both the cation H^+ in Group 1A(1) and the anion H^- in Group 7A(17).] Many transition elements form two different monatomic ions. (Although Hg_2^{2+} is a diatomic ion, it is included for comparison with Hg^{2+}.)

Period	1A (1)	2A (2)	3B (3)	4B (4)	5B (5)	6B (6)	7B (7)	8B (8)	8B (9)	8B (10)	1B (11)	2B (12)	3A (13)	4A (14)	5A (15)	6A (16)	7A (17)	8A (18)
1	H^+																H^-	
2	Li^+														N^{3-}	O^{2-}	F^-	
3	Na^+	Mg^{2+}											Al^{3+}			S^{2-}	Cl^-	
4	K^+	Ca^{2+}				Cr^{2+} Cr^{3+}	Mn^{2+}	Fe^{2+} Fe^{3+}	Co^{2+} Co^{3+}		Cu^+ Cu^{2+}	Zn^{2+}					Br^-	
5	Rb^+	Sr^{2+}									Ag^+	Cd^{2+}		Sn^{2+} Sn^{4+}			I^-	
6	Cs^+	Ba^{2+}										Hg_2^{2+} Hg^{2+}		Pb^{2+} Pb^{4+}				
7																		

Ionic compounds are arrays of oppositely charged ions rather than separate molecular units. Therefore, we write a formula for the **formula unit,** which gives the *relative* numbers of cations and anions in the compound. Thus, ionic compounds generally have only empirical formulas.* The compound has zero net charge, so the positive charges of the cations must balance the negative charges of the anions. For example, calcium bromide is composed of Ca^{2+} ions and Br^- ions; therefore, two Br^- balance each Ca^{2+}. The formula is $CaBr_2$, not Ca_2Br. In this and all other formulas,

- The subscript refers to the element *preceding* it.
- The *subscript 1 is understood* from the presence of the element symbol alone (that is, we do not write Ca_1Br_2).
- The charge (without the sign) of one ion becomes the subscript of the other:

$$Ca^{2+} \quad Br^{1-} \qquad \text{gives} \qquad Ca_1Br_2 \qquad \text{or} \qquad CaBr_2$$

Reduce the subscripts to the smallest whole numbers that retain the ratio of ions. Thus, for example, from the ions Ca^{2+} and O^{2-} we have Ca_2O_2, which we reduce to the formula CaO (but see the footnote).

SAMPLE PROBLEM 2.8 Determining Formulas of Binary Ionic Compounds

Problem Write empirical formulas for the compounds named in Sample Problem 2.7.

Plan We write the empirical formula by finding the smallest number of each ion that gives the neutral compound. These numbers appear as *right subscripts* to the element symbol.

Solution

(a) Mg^{2+} and N^{3-}; three Mg^{2+} ions (6+) balance two N^{3-} ions (6−): Mg_3N_2

(b) Cd^{2+} and I^-; one Cd^{2+} ion (2+) balances two I^- ions (2−): CdI_2

(c) Sr^{2+} and F^-; one Sr^{2+} ion (2+) balances two F^- ions (2−): SrF_2

(d) Cs^+ and S^{2-}; two Cs^+ ions (2+) balance one S^{2-} ion (2−): Cs_2S

*Compounds of the mercury(I) ion, such as Hg_2Cl_2, and peroxides of the alkali metals, such as Na_2O_2, are the only two common exceptions. Their empirical formulas are HgCl and NaO, respectively.

Comment Note that ion charges do *not* appear in the compound formula. That is, for cadmium iodide, we do *not* write $Cd^{2+}I_2^{-}$.

FOLLOW-UP PROBLEM 2.8 Write the formulas of the compounds named in Follow-up Problem 2.7.

Compounds with Metals That Can Form More Than One Ion Many metals, particularly the transition elements (B groups), can form more than one ion, each with its own particular charge. Table 2.4 lists some examples, and Figure 2.16 shows their placement in the periodic table. Names of compounds containing these elements include a *Roman numeral within parentheses* immediately after the metal ion's name to indicate its ionic charge. For example, iron can form Fe^{2+} and Fe^{3+} ions. The two compounds that iron forms with chlorine are $FeCl_2$, named iron(II) chloride (spoken "iron two chloride"), and $FeCl_3$, named iron(III) chloride.

In common names, the Latin root of the metal is followed by either of two suffixes:

- The suffix *-ous* for the ion with the lower charge
- The suffix *-ic* for the ion with the higher charge

Thus, iron(II) chloride is also called ferr*ous* chloride and iron(III) chloride is fer*ric* chloride. (You can easily remember this naming relationship because there is an ***o*** in *-ous* and *lower,* and an ***i*** in *-ic* and *higher.*)

Table 2.4 Some Metals That Form More Than One Monatomic Ion*

Element	Ion Formula	Systematic Name	Common (Trivial) Name
Chromium	Cr^{2+}	chromium(II)	chromous
	$\mathbf{Cr^{3+}}$	**chromium(III)**	chromic
Cobalt	Co^{2+}	cobalt(II)	
	Co^{3+}	cobalt(III)	
Copper	$\mathbf{Cu^{+}}$	**copper(I)**	cuprous
	$\mathbf{Cu^{2+}}$	**copper(II)**	cupric
Iron	$\mathbf{Fe^{2+}}$	**iron(II)**	ferrous
	$\mathbf{Fe^{3+}}$	**iron(III)**	ferric
Lead	$\mathbf{Pb^{2+}}$	**lead(II)**	
	Pb^{4+}	lead(IV)	
Mercury	Hg_2^{2+}	mercury(I)	mercurous
	$\mathbf{Hg^{2+}}$	**mercury(II)**	mercuric
Tin	$\mathbf{Sn^{2+}}$	**tin(II)**	stannous
	Sn^{4+}	tin(IV)	stannic

*Listed alphabetically by metal name; those in **boldface** are most common.

SAMPLE PROBLEM 2.9 Determining Names and Formulas of Ionic Compounds of Elements That Form More Than One Ion

Problem Give the systematic names for the formulas or the formulas for the names of the following compounds: **(a)** tin(II) fluoride; **(b)** CrI_3; **(c)** ferric oxide; **(d)** CoS.

Solution **(a)** Tin(II) is Sn^{2+}; fluoride is F^{-}. Two F^{-} ions balance one Sn^{2+} ion: tin(II) fluoride is SnF_2. (The common name is stannous fluoride.)

(b) The anion is I^{-}, iodide, and the formula shows three I^{-}. Therefore, the cation must be Cr^{3+}, chromium(III): CrI_3 is chromium(III) iodide. (The common name is chromic iodide.)

(c) Ferric is the common name for iron(III), Fe^{3+}; oxide ion is O^{2-}. To balance the ionic charges, the formula of ferric oxide is Fe_2O_3. [The systematic name is iron(III) oxide.] **(d)** The anion is sulfide, S^{2-}, which requires that the cation be Co^{2+}. The name is cobalt(II) sulfide.

FOLLOW-UP PROBLEM 2.9 Give the systematic names for the formulas or the formulas for the names of the following compounds: **(a)** lead(IV) oxide; **(b)** Cu_2S; **(c)** $FeBr_2$; **(d)** mercuric chloride.

Compounds Formed from Polyatomic Ions Ionic compounds in which one or both of the ions are polyatomic are very common. Table 2.5 gives the formulas and the names of some common polyatomic ions. Remember that *the polyatomic ion stays together as a charged unit.* The formula for potassium nitrate is KNO_3: each K^+ balances one NO_3^-. The formula for sodium carbonate is Na_2CO_3: two Na^+ balance one CO_3^{2-}. *When two or more of the same polyatomic ion are present in the formula unit, that ion appears in parentheses with the subscript written outside.* For example, calcium nitrate, which contains one Ca^{2+} and two NO_3^- ions, has the formula $Ca(NO_3)_2$. Parentheses and a subscript are *not* used unless *more than one* of the polyatomic ions is present; thus, sodium nitrate is $NaNO_3$, *not* $Na(NO_3)$.

Families of Oxoanions As Table 2.5 shows, most polyatomic ions are **oxoanions,** those in which an element, usually a nonmetal, is bonded to one or more oxygen atoms. There are several families of two or four oxoanions that differ only in the number of oxygen atoms. The following simple naming convention is used with these ions.

With two oxoanions in the family:

- The ion with *more* O atoms takes the nonmetal root and the suffix *-ate*.
- The ion with *fewer* O atoms takes the nonmetal root and the suffix *-ite*.

For example, SO_4^{2-} is the sulf*ate* ion; SO_3^{2-} is the sulf*ite* ion; similarly, NO_3^- is nitr*ate*, and NO_2^- is nitr*ite*.

With four oxoanions in the family (usually a halogen bonded to varying numbers of O atoms), as Figure 2.17 shows:

- The ion with *most* O atoms has the prefix *per-*, the nonmetal root, and the suffix *-ate*.
- The ion with *one fewer* O atom has just the root and the suffix *-ate*.
- The ion with *two fewer* O atoms has just the root and the suffix *-ite*.
- The ion with *least (three fewer)* O atoms has the prefix *hypo-*, the root, and the suffix *-ite*.

For example, for the four chlorine oxoanions,

ClO_4^- is *per*chlor*ate*, ClO_3^- is chlor*ate*, ClO_2^- is chlor*ite*, ClO^- is *hypo*chlor*ite*

Hydrated Ionic Compounds Ionic compounds called **hydrates** have a specific number of water molecules associated with each formula unit. In their formulas, this number is shown after a centered dot. It is indicated in the systematic name by a Greek numerical prefix before the word *hydrate.* Table 2.6 on the next page shows these prefixes. For example, Epsom salt has the formula $MgSO_4{\cdot}7H_2O$ and the name magnesium sulfate *hepta*hydrate. Similarly, the mineral gypsum has the formula $CaSO_4{\cdot}2H_2O$ and the name calcium sulfate *di*hydrate. The water molecules, referred to as "waters of hydration," are part of the hydrate's structure. Heating can remove some or all of them, leading to a different substance. For example, when heated strongly, blue copper(II) sulfate pentahydrate ($CuSO_4{\cdot}5H_2O$) is converted to white copper(II) sulfate ($CuSO_4$).

Table 2.5 Common Polyatomic Ions*

Formula	Name
Cations	
NH_4^+	**ammonium**
H_3O^+	**hydronium**
Anions	
CH_3COO^- (or $C_2H_3O_2^-$)	**acetate**
CN^-	cyanide
OH^-	**hydroxide**
ClO^-	hypochlorite
ClO_2^-	chlorite
ClO_3^-	**chlorate**
ClO_4^-	**perchlorate**
NO_2^-	nitrite
NO_3^-	**nitrate**
MnO_4^-	**permanganate**
CO_3^{2-}	**carbonate**
HCO_3^-	**hydrogen carbonate** (or **bicarbonate**)
CrO_4^{2-}	chromate
$Cr_2O_7^{2-}$	**dichromate**
O_2^{2-}	peroxide
PO_4^{3-}	**phosphate**
HPO_4^{2-}	hydrogen phosphate
$H_2PO_4^-$	dihydrogen phosphate
SO_3^{2-}	sulfite
SO_4^{2-}	**sulfate**
HSO_4^-	hydrogen sulfate (or bisulfate)

***Boldface** ions are most common.

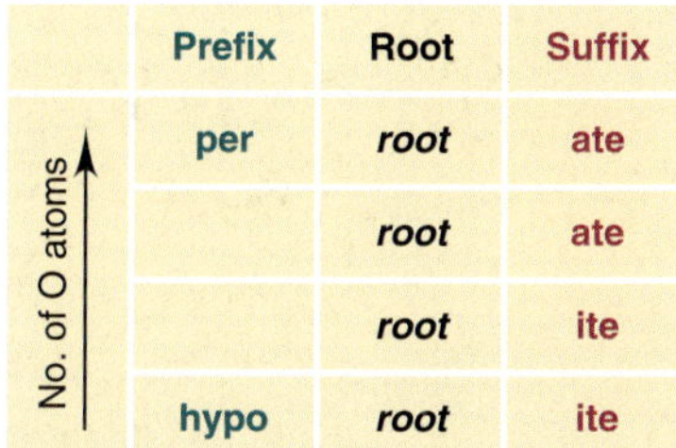

FIGURE 2.17 Naming oxoanions. Prefixes and suffixes indicate the number of O atoms in the anion.

Table 2.6 Numerical Prefixes for Hydrates and Binary Covalent Compounds

Number	Prefix
1	mono-
2	di-
3	tri-
4	tetra-
5	penta-
6	hexa-
7	hepta-
8	octa-
9	nona-
10	deca-

SAMPLE PROBLEM 2.10 Determining Names and Formulas of Ionic Compounds Containing Polyatomic Ions

Problem Give the systematic names for the formulas or the formulas for the names of the following compounds:
(a) $Fe(ClO_4)_2$ **(b)** Sodium sulfite **(c)** $Ba(OH)_2 \cdot 8H_2O$

Solution (a) ClO_4^- is perchlorate, which has a 1− charge, so the cation must be Fe^{2+}. The name is iron(II) perchlorate. (The common name is ferrous perchlorate.)
(b) Sodium is Na^+; sulfite is SO_3^{2-}. Therefore, two Na^+ ions balance one SO_3^{2-} ion. The formula is Na_2SO_3. **(c)** Ba^{2+} is barium; OH^- is hydroxide. There are eight (*octa-*) water molecules in each formula unit. The name is barium hydroxide octahydrate.

FOLLOW-UP PROBLEM 2.10 Give the systematic names for the formulas or the formulas for the names of the following compounds:
(a) Cupric nitrate trihydrate **(b)** Zinc hydroxide **(c)** LiCN

SAMPLE PROBLEM 2.11 Recognizing Incorrect Names and Formulas of Ionic Compounds

Problem Something is wrong with the second part of each statement. Provide the correct name or formula.
(a) $Ba(C_2H_3O_2)_2$ is called barium diacetate.
(b) Sodium sulfide has the formula $(Na)_2SO_3$.
(c) Iron(II) sulfate has the formula $Fe_2(SO_4)_3$.
(d) Cesium carbonate has the formula $Cs_2(CO_3)$.

Solution (a) The charge of the Ba^{2+} ion *must* be balanced by *two* $C_2H_3O_2^-$ ions, so the prefix *di-* is unnecessary. For ionic compounds, we do not indicate the number of ions with numerical prefixes. The correct name is barium acetate.
(b) Two mistakes occur here. The sodium ion is monatomic, so it does *not* require parentheses. The sulfide ion is S^{2-}, *not* SO_3^{2-} (called "sulfite"). The correct formula is Na_2S.
(c) The Roman numeral refers to the charge of the ion, *not* the number of ions in the formula. Fe^{2+} is the cation, so it requires one SO_4^{2-} to balance its charge. The correct formula is $FeSO_4$.
(d) Parentheses are *not* required when only one polyatomic ion of a kind is present. The correct formula is Cs_2CO_3.

FOLLOW-UP PROBLEM 2.11 State why the second part of each statement is incorrect, and correct it:
(a) Ammonium phosphate is $(NH_3)_4PO_4$.
(b) Aluminum hydroxide is $AlOH_3$.
(c) $Mg(HCO_3)_2$ is manganese(II) carbonate.
(d) $Cr(NO_3)_3$ is chromic(III) nitride.
(e) $Ca(NO_2)_2$ is cadmium nitrate.

Acid Names from Anion Names Acids are an important group of hydrogen-containing compounds that have been used in chemical reactions for centuries. In the laboratory, acids are typically used in water solution. When naming them and writing their formulas, we consider them as anions connected to the number of hydrogen ions (H^+) needed for charge neutrality. The two common types of acids are binary acids and oxoacids:

1. *Binary acid* solutions form when certain gaseous compounds dissolve in water. For example, when gaseous hydrogen chloride (HCl) dissolves in water, it forms a solution whose name consists of the following parts:

   Prefix *hydro-* + nonmetal *root* + suffix *-ic* + separate word *acid*
   hydro + chlor + ic + acid

   or *hydrochloric acid.* This naming pattern holds for many compounds in which hydrogen combines with an anion that has an *-ide* suffix.

2. *Oxoacid* names are similar to those of the oxoanions, except for two suffix changes:
   - *-ate* in the anion becomes *-ic* in the acid
   - *-ite* in the anion becomes *-ous* in the acid

   The oxoanion prefixes *hypo-* and *per-* are kept. Thus,

   BrO_4^- is *per*brom*ate*, and $HBrO_4$ is *per*brom*ic* acid

   IO_2^- is iod*ite*, and HIO_2 is iod*ous* acid

SAMPLE PROBLEM 2.12 Determining Names and Formulas of Anions and Acids

Problem Name the following anions and give the names and formulas of the acids derived from them: **(a)** Br^-; **(b)** IO_3^-; **(c)** CN^-; **(d)** SO_4^{2-}; **(e)** NO_2^-.

Solution **(a)** The anion is bromide; the acid is hydrobromic acid, HBr.

(b) The anion is iodate; the acid is iodic acid, HIO_3.

(c) The anion is cyanide; the acid is hydrocyanic acid, HCN.

(d) The anion is sulfate; the acid is sulfuric acid, H_2SO_4. (In this case, the suffix is added to the element name *sulfur,* not to the root, *sulf-*.)

(e) The anion is nitrite; the acid is nitrous acid, HNO_2.

Comment We added *two* H^+ ions to the sulfate ion to obtain sulfuric acid because it has a 2− charge.

FOLLOW-UP PROBLEM 2.12 Write the formula for the name or name for the formula of each acid: **(a)** chloric acid; **(b)** HF; **(c)** acetic acid; **(d)** sulfurous acid; **(e)** HBrO.

Names and Formulas of Binary Covalent Compounds

Binary covalent compounds are formed by the combination of two elements, usually nonmetals. Several are so familiar, such as ammonia (NH_3), methane (CH_4), and water (H_2O), that we use their common names, but most are named in a systematic way:

1. The element with the lower group number in the periodic table is the first word in the name; the element with the higher group number is the second word. (*Exception:* When the compound contains oxygen and any of the halogens chlorine, bromine, and iodine, the halogen is named first.)
2. If both elements are in the same group, the one with the higher period number is named first.
3. The second element is named with its root and the suffix *-ide*.
4. Covalent compounds have Greek numerical prefixes (see Table 2.6) to indicate the number of atoms of each element in the compound. The first word has a prefix *only* when more than one atom of the element is present; the second word *usually* has a numerical prefix.

SAMPLE PROBLEM 2.13 Determining Names and Formulas of Binary Covalent Compounds

Problem **(a)** What is the formula of carbon disulfide? **(b)** What is the name of PCl_5? **(c)** Give the name and formula of the compound whose molecules each consist of two N atoms and four O atoms.

Solution **(a)** The prefix *di-* means "two." The formula is CS_2.

(b) P is the symbol for phosphorus; there are five chlorine atoms, which is indicated by the prefix *penta-*. The name is phosphorus pentachloride.

(c) Nitrogen (N) comes first in the name (lower group number). The compound is dinitrogen tetraoxide, N_2O_4.

FOLLOW-UP PROBLEM 2.13 Give the name or the formula for **(a)** SO_3; **(b)** SiO_2; **(c)** dinitrogen monoxide; **(d)** selenium hexafluoride.

SAMPLE PROBLEM 2.14 Recognizing Incorrect Names and Formulas of Binary Covalent Compounds

Problem Explain what is wrong with the name or formula in the second part of each statement and correct it: **(a)** SF_4 is monosulfur pentafluoride. **(b)** Dichlorine heptaoxide is Cl_2O_6. **(c)** N_2O_3 is dinitrotrioxide.

Solution **(a)** There are two mistakes. *Mono-* is not needed if there is only one atom of the first element, and the prefix for four is *tetra-*, not *penta-*. The correct name is sulfur tetrafluoride.

(b) The prefix *hepta-* indicates seven, not six. The correct formula is Cl_2O_7.

(c) The full name of the first element is needed, and a space separates the two element names. The correct name is dinitrogen trioxide.

FOLLOW-UP PROBLEM 2.14 Explain what is wrong with the second part of each statement and correct it: **(a)** S_2Cl_2 is disulfurous dichloride. **(b)** Nitrogen monoxide is N_2O. **(c)** $BrCl_3$ is trichlorine bromide.

Naming Alkanes

Organic compounds typically have complex structural formulas that consist of chains, branches, and/or rings of carbon atoms bonded to hydrogen atoms and, often, to atoms of oxygen, nitrogen, and a few other elements. At this point, we'll see how the simplest organic compounds are named. Much more on the rules of organic nomenclature appears in Chapter 15.

Hydrocarbons, the simplest type of organic compound, contain *only* carbon and hydrogen. *Alkanes* are the simplest type of hydrocarbon; many function as important fuels, such as methane, propane, butane, and the mixture of alkanes in gasoline. The simplest alkanes to name are the *straight-chain alkanes* because the carbon chains have no branches. Alkanes are named with a *root,* based on the number of C atoms in the chain, followed by the suffix *-ane*. Table 2.7 gives the names, molecular formulas, and space-filling models (discussed shortly) of the first 10 straight-chain alkanes. Note that the roots of the four smallest ones are new, but those for the larger ones are the same as the Greek prefixes in Table 2.6.

Table 2.7 The First 10 Straight-Chain Alkanes

Name (Formula)	Model
Methane (CH_4)	
Ethane (C_2H_6)	
Propane (C_3H_8)	
Butane (C_4H_{10})	
Pentane (C_5H_{12})	
Hexane (C_6H_{14})	
Heptane (C_7H_{16})	
Octane (C_8H_{18})	
Nonane (C_9H_{20})	
Decane ($C_{10}H_{22}$)	

Molecular Masses from Chemical Formulas

In Section 2.5, we calculated the atomic mass of an element. Using the periodic table and the formula of a compound to see the number of atoms of each element present, we calculate the **molecular mass** (also called *molecular weight*) of a formula unit of the compound as the sum of the atomic masses:

$$\text{Molecular mass} = \text{sum of atomic masses} \quad (2.3)$$

The molecular mass of a water molecule (using atomic masses to four significant figures from the periodic table) is

$$\begin{aligned}\text{Molecular mass of } H_2O &= (2 \times \text{atomic mass of H}) + (1 \times \text{atomic mass of O}) \\ &= (2 \times 1.008 \text{ amu}) + 16.00 \text{ amu} = 18.02 \text{ amu}\end{aligned}$$

Ionic compounds are treated the same, but because they do not consist of molecules, we use the term **formula mass** for an ionic compound. To calculate its formula mass, *the number of atoms of each element inside the parentheses is multiplied by the subscript outside the parentheses.* For barium nitrate, $Ba(NO_3)_2$,

$$\begin{aligned}&\text{Formula mass of } Ba(NO_3)_2 \\ &= (1 \times \text{atomic mass of Ba}) + (2 \times \text{atomic mass of N}) + (6 \times \text{atomic mass of O}) \\ &= 137.3 \text{ amu} + (2 \times 14.01 \text{ amu}) + (6 \times 16.00 \text{ amu}) = 261.3 \text{ amu}\end{aligned}$$

Atomic, not ionic, masses are used to find a formula mass because electron loss equals electron gain in the compound; thus, electron mass is balanced. In the next two sample problems, a name or molecular-scale depiction is used to find a compound's formula and molecular mass.

SAMPLE PROBLEM 2.15 Calculating the Molecular Mass of a Compound

Problem Using data in the periodic table, calculate each molecular (or formula) mass:
(a) Tetraphosphorus trisulfide **(b)** Ammonium nitrate

Plan We first write the formula, then multiply the number of atoms (or ions) of each element by its atomic mass, and find the sum.

Solution (a) The formula is P_4S_3.

$$\text{Molecular mass} = (4 \times \text{atomic mass of P}) + (3 \times \text{atomic mass of S})$$
$$= (4 \times 30.97\text{ amu}) + (3 \times 32.07\text{ amu}) = \boxed{220.09\text{ amu}}$$

(b) The formula is NH_4NO_3. We count the total number of N atoms even though they belong to different ions:

Formula mass

$$= (2 \times \text{atomic mass of N}) + (4 \times \text{atomic mass of H}) + (3 \times \text{atomic mass of O})$$
$$= (2 \times 14.01\text{ amu}) + (4 \times 1.008\text{ amu}) + (3 \times 16.00\text{ amu}) = \boxed{80.05\text{ amu}}$$

Check You can often find large errors by rounding atomic masses to the nearest 5 and adding: **(a)** $(4 \times 30) + (3 \times 30) = 210 \approx 220.09$. The sum has two decimal places because the atomic masses have two. **(b)** $(2 \times 15) + 4 + (3 \times 15) = 79 \approx 80.05$.

FOLLOW-UP PROBLEM 2.15 What is the formula and molecular (or formula) mass of each of the following compounds: **(a)** hydrogen peroxide; **(b)** cesium chloride; **(c)** sulfuric acid; **(d)** potassium sulfate?

SAMPLE PROBLEM 2.16 Using Molecular Depictions to Determine Formula, Name, and Mass

Problem Each circle contains a representation of a binary compound. Determine its formula, name, and molecular (formula) mass.

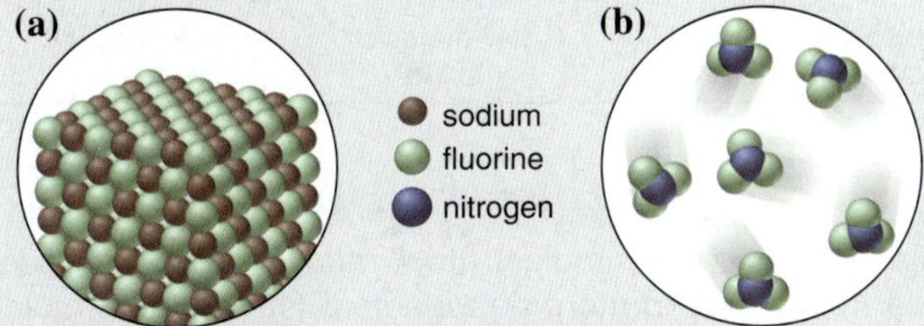

Plan Each of the compounds contains only two elements, so to find the formula, we find the simplest whole-number ratio of one atom to the other. Then we determine the name (see Sample Problems 2.7, 2.8, and 2.13) and the molecular or formula mass (see Sample Problem 2.14).

Solution (a) There is one brown (sodium) for each green (fluorine), so the formula is NaF. A metal and nonmetal form an ionic compound, in which the metal is named first: sodium fluoride.

$$\text{Formula mass} = (1 \times \text{atomic mass of Na}) + (1 \times \text{atomic mass of F})$$
$$= 22.99\text{ amu} + 19.00\text{ amu} = \boxed{41.99\text{ amu}}$$

(b) There are three green (fluorine) for each blue (nitrogen), so the formula is NF_3. Two nonmetals form a covalent compound. Nitrogen has a lower group number, so it is named first: nitrogen trifluoride.

$$\text{Molecular mass} = (1 \times \text{atomic mass of N}) + (3 \times \text{atomic mass of F})$$
$$= 14.01\text{ amu} + (3 \times 19.00\text{ amu}) = \boxed{71.01\text{ amu}}$$

Check (a) For binary ionic compounds, we predict ionic charges from the periodic table (see Figure 2.10). Na forms a 1+ ion, and F forms a 1− ion, so the charges balance with one Na^+ per F^-. Also, ionic compounds are solids, consistent with the picture. **(b)** Covalent compounds often occur as individual molecules, as in the picture. Rounding in (a) gives 25 + 20 = 45; in (b), we get 15 + (3 × 20) = 75, so there are no large errors.

FOLLOW-UP PROBLEM 2.16 Each circle contains a representation of a binary compound. Determine its name, formula, and molecular (formula) mass.

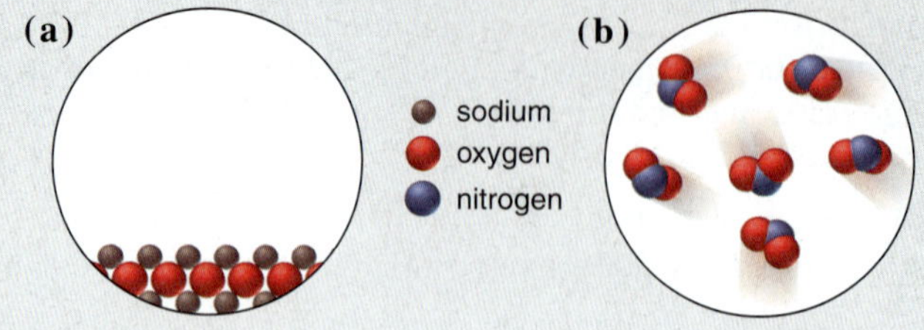

Picturing Molecules

Molecules are depicted in a variety of useful ways, as shown in Figure 2.18 for the water molecule. A *chemical formula* shows only the relative numbers of atoms. *Electron-dot* and *bond-line formulas* show a bond between atoms as either a pair of dots or a line. A *ball-and-stick model* shows atoms as spheres and bonds as sticks, with accurate angles and relative sizes, but distances are exaggerated. A *space-filling model* is an accurately scaled-up version of a molecule, but it does not show bonds. An *electron-density model* shows the ball-and-stick model within the space-filling shape and colors the regions of high *(red)* and low *(blue)* electron charge.

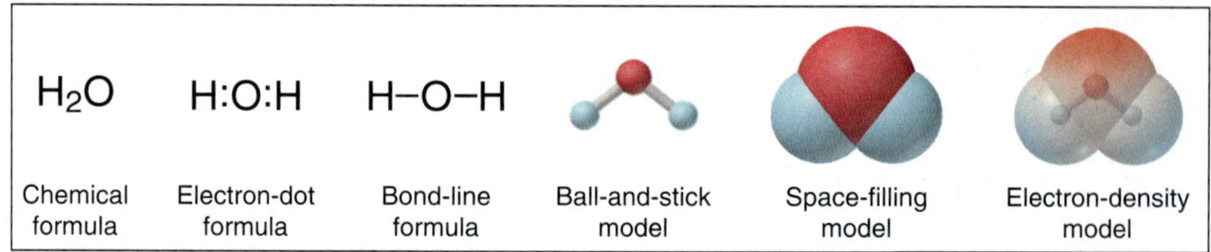

FIGURE 2.18 Representations of a water molecule.

SECTION 2.8 SUMMARY

Chemical formulas describe the simplest atom ratio (empirical formula), actual atom number (molecular formula), and atom arrangement (structural formula) of one unit of a compound. • An ionic compound is named with cation first and anion second. For metals that can form more than one ion, the charge is shown with a Roman numeral. • Oxoanions have suffixes, and sometimes prefixes, attached to the element root name to indicate the number of oxygen atoms. • Names of hydrates give the number of associated water molecules with a numerical prefix. • Acid names are based on anion names. • Covalent compounds have the first word of the name for the element that is leftmost or lower in the periodic table, and prefixes show the number of each atom. • The molecular (or formula) mass of a compound is the sum of the atomic masses in the formula. • Molecules are depicted by various types of formulas and models.

2.9 CLASSIFICATION OF MIXTURES

Although chemists pay a great deal of attention to pure substances, this form of matter almost never occurs around us. In the natural world, *matter usually occurs as mixtures*, such as air, seawater, soil, and organisms.

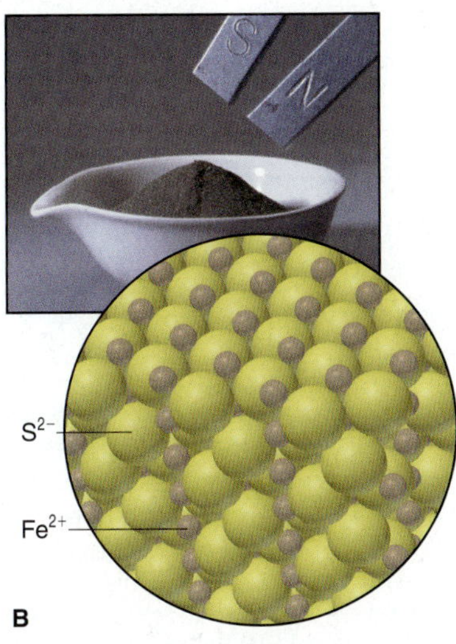

FIGURE 2.19 The distinction between mixtures and compounds. A, A *mixture* of iron and sulfur can be separated with a magnet because only the iron is magnetic. The blow-up shows separate regions of the two elements. **B,** After strong heating, the *compound* iron(II) sulfide forms, which is no longer magnetic. The blow-up shows the structure of the compound.

There are two broad classes of mixtures. A **heterogeneous mixture** has one or more visible boundaries between the components. Thus, its composition is *not* uniform; it varies from one region to another. Many rocks are heterogeneous, showing individual grains and flecks of different minerals. In some cases, as in milk and blood, the boundaries can be seen only with a microscope. A **homogeneous mixture** has no visible boundaries because the components are mixed as individual atoms, ions, and molecules. Thus, its composition *is* uniform. A mixture of sugar dissolved in water is homogeneous, for example, because the sugar molecules and water molecules are uniformly intermingled on the molecular level. We have no way to tell visually whether an object is a substance (element or compound) or a homogeneous mixture.

A homogeneous mixture is also called a **solution.** Although we usually think of solutions as liquid, they can exist in all three physical states. For example, air is a gaseous solution of mostly oxygen and nitrogen molecules, and wax is a solid solution of several fatty substances. Solutions in water, called **aqueous solutions,** are especially important in chemistry and comprise a major portion of the environment and of all organisms.

Recall that mixtures differ fundamentally from compounds in three ways: (1) the proportions of the components can vary; (2) the individual properties of the components are observable; and (3) the components can be separated by physical means. In some cases, as in a mixture of iron and sulfur (Figure 2.19), if we apply enough energy to the components, they react with each other and form a compound, after which their individual properties are no longer observable. The characteristics of mixtures and pure substances that we covered in this chapter are summarized in Figure 2.20 on the next page.

SECTION 2.9 SUMMARY

Heterogeneous mixtures have visible boundaries between the components. • Homogeneous mixtures have no visible boundaries because mixing occurs at the molecular level. A solution is a homogeneous mixture and can occur in any physical state. • Mixtures (not compounds) can have variable proportions, can be separated physically, and retain their components' properties.

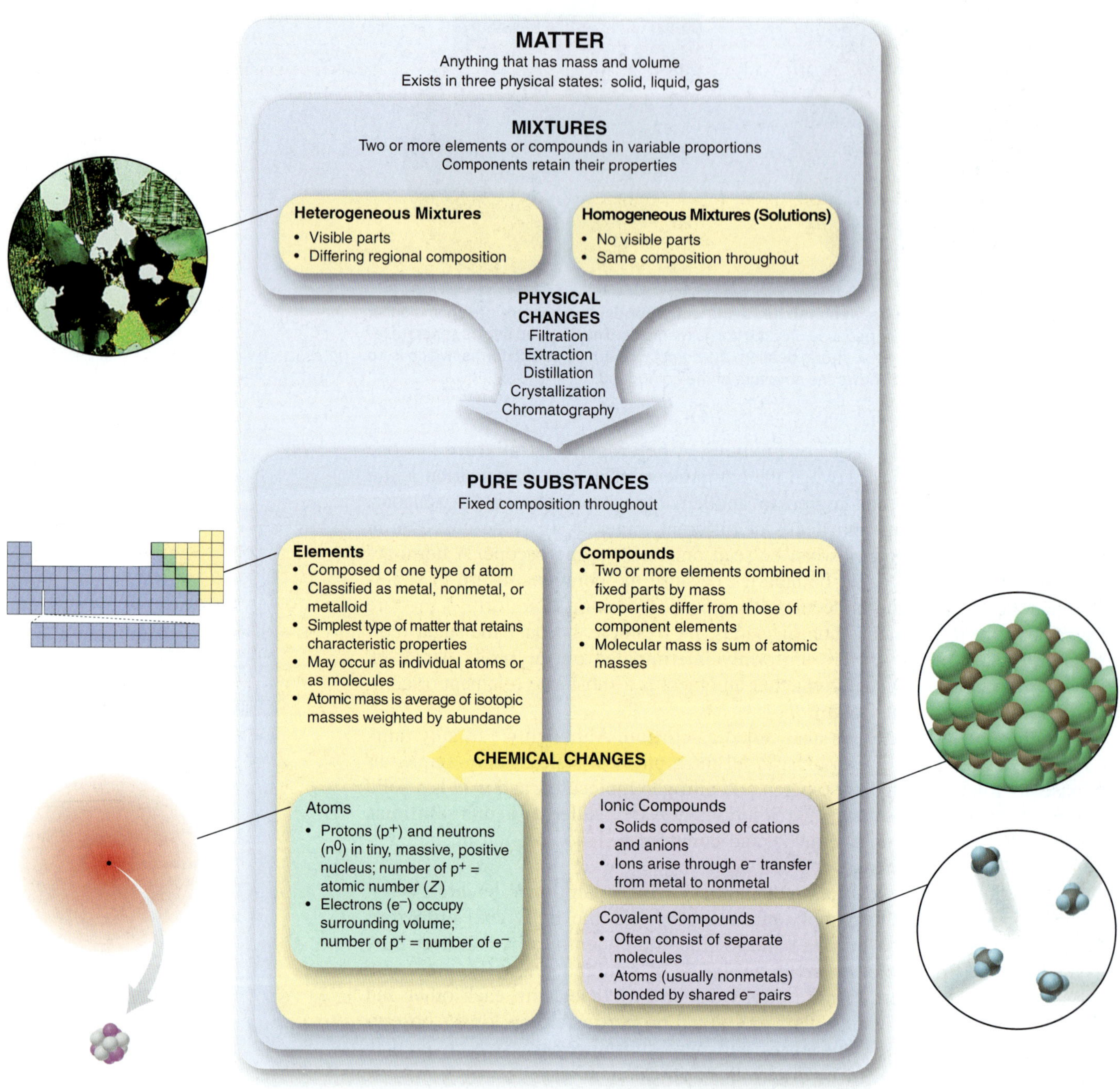

FIGURE 2.20 The classification of matter from a chemical point of view. Mixtures are separated by physical changes into elements and compounds. Chemical changes are required to convert elements into compounds, and vice versa.

CHAPTER REVIEW GUIDE

The following sections provide many aids to help you study this chapter. (Numbers in parentheses refer to pages, unless noted otherwise.)

• LEARNING OBJECTIVES *These are concepts and skills to review after studying this chapter.*

Related section (§), sample problem (SP), and upcoming end-of-chapter problem (EP) numbers are listed in parentheses.

1. Define the characteristics of the three types of matter—element, compound, and mixture—on the macroscopic and atomic levels (§ 2.1) (SP 2.1) (EPs 2.1–2.5)
2. Understand the laws of mass conservation, definite composition, and multiple proportions; use the mass ratio of element-to-compound to find the mass of an element in a compound (§ 2.2) (SP 2.2) (EPs 2.6–2.19)
3. Understand Dalton's atomic theory and how it explains the mass laws (§ 2.3) (SP 2.3) (EP 2.20)
4. Describe the results of the key experiments by Thomson, Millikan, and Rutherford concerning atomic structure (§ 2.4) (EPs 2.21, 2.22)
5. Explain the structure of the atom, the main features of the subatomic particles, and the significance of isotopes; use atomic notation to express the subatomic makeup of an isotope; calculate the atomic mass of an element from its isotopic composition (§ 2.5) (SPs 2.4, 2.5) (EPs 2.23–2.35)
6. Describe the format of the periodic table and the general location and characteristics of metals, metalloids, and nonmetals (§ 2.6) (EPs 2.36–2.42)
7. Explain the essential features of ionic and covalent bonding and distinguish between them; predict the monatomic ion formed from a main-group element (§ 2.7) (SP 2.6) (EPs 2.43–2.55)
8. Name, write the formula, and calculate the moleculer (or formula) mass of ionic and binary covalent compounds (§ 2.8) (SPs 2.7–2.16) (EPs 2.56–2.79)
9. Describe the types of mixtures and their properties (§ 2.9) (EPs 2.80–2.84)

• KEY TERMS *These important terms appear in boldface in the chapter and are defined again in the Glossary.*

Section 2.1
element (32)
substance (32)
molecule (32)
compound (32)
mixture (33)

Section 2.2
law of mass conservation (34)
law of definite (or constant) composition (35)
fraction by mass (mass fraction) (35)
percent by mass (mass percent, mass %) (35)
law of multiple proportions (36)

Section 2.3
atom (37)

Section 2.4
cathode ray (39)
nucleus (41)

Section 2.5
proton (p^+) (42)
neutron (n^0) (42)
electron (e^-) (42)
atomic number (Z) (43)
mass number (A) (43)
atomic symbol (43)
isotope (43)
atomic mass unit (amu) (44)
dalton (Da) (44)
mass spectrometry (44)
isotopic mass (45)
atomic mass (45)

Section 2.6
periodic table of the elements (46)
period (46)
group (46)
metal (47)
nonmetal (47)
metalloid (semimetal) (47)

Section 2.7
ionic compound (48)
covalent compound (48)
chemical bond (48)
ion (49)
binary ionic compound (49)
cation (49)
anion (49)
monatomic ion (49)
covalent bond (50)
polyatomic ion (51)

Section 2.8
chemical formula (52)
empirical formula (52)
molecular formula (52)
structural formula (52)
formula unit (53)
oxoanion (55)
hydrate (55)
binary covalent compound (57)
molecular mass (58)
formula mass (58)

Section 2.9
heterogeneous mixture (61)
homogeneous mixture (61)
solution (61)
aqueous solution (61)

• KEY EQUATIONS AND RELATIONSHIPS *Numbered and screened concepts are listed for you to refer to or memorize.*

2.1 Finding the mass of an element in a given mass of compound (35):

Mass of element in sample

$$= \text{mass of compound in sample} \times \frac{\text{mass of element}}{\text{mass of compound}}$$

2.2 Calculating the number of neutrons in an atom (43):

$$\text{Number of neutrons} = \text{mass number} - \text{atomic number}$$

or

$$N = A - Z$$

2.3 Determining the molecular mass of a formula unit of a compound (58):

$$\text{Molecular mass} = \text{sum of atomic masses}$$

BRIEF SOLUTIONS TO *FOLLOW-UP PROBLEMS* *Compare your own solutions to these calculation steps and answers.*

2.1 There are two types of particles reacting (left circle), one with two blue atoms and the other with two orange, so the depiction shows a mixture of two elements. In the product (right circle), all the particles have one blue atom and one orange; this is a compound.

2.2 Mass (t) of pitchblende

$$= 2.3 \cancel{\text{t uranium}} \times \frac{84.2 \text{ t pitchblende}}{71.4 \cancel{\text{t uranium}}} = 2.7 \text{ t pitchblende}$$

Mass (t) of oxygen

$$= 2.7 \cancel{\text{t pitchblende}} \times \frac{(84.2 - 71.4 \text{ t oxygen})}{84.2 \cancel{\text{t pitchblende}}} = 0.41 \text{ t oxygen}$$

2.3 Sample B. Two bromine-fluorine compounds appear. In one, there are three fluorine atoms for each bromine; in the other, there is one fluorine for each bromine. Therefore, in the two compounds, the ratio of fluorines combining with one bromine is 3/1.

2.4 (a) Q = B; $5p^+$, $6n^0$, $5e^-$
(b) X = Ca; $20p^+$, $21n^0$, $20e^-$
(c) Y = I; $53p^+$, $78n^0$, $53e^-$

2.5 $10.0129x + [11.0093(1 - x)] = 10.81$; $0.9964x = 0.1993$; $x = 0.2000$ and $1 - x = 0.8000$; % abundance of ^{10}B = 20.00%; % abundance of ^{11}B = 80.00%

2.6 (a) S^{2-}; (b) Rb^+; (c) Ba^{2+}

2.7 (a) Zinc [Group 2B(12)] and oxygen [Group 6A(16)]
(b) Silver [Group 1B(11)] and bromine [Group 7A(17)]
(c) Lithium [Group 1A(1)] and chlorine [Group 7A(17)]
(d) Aluminum [Group 3A(13)] and sulfur [Group 6A(16)]

2.8 (a) ZnO; (b) AgBr; (c) LiCl; (d) Al_2S_3

2.9 (a) PbO_2; (b) copper(I) sulfide (cuprous sulfide); (c) iron(II) bromide (ferrous bromide); (d) $HgCl_2$

2.10 (a) $Cu(NO_3)_2 \cdot 3H_2O$; (b) $Zn(OH)_2$; (c) lithium cyanide

2.11 (a) $(NH_4)_3PO_4$; ammonium is NH_4^+ and phosphate is PO_4^{3-}.
(b) $Al(OH)_3$; parentheses are needed around the polyatomic ion OH^-.
(c) Magnesium hydrogen carbonate; Mg^{2+} is magnesium and can have only a 2+ charge, so it does not need (II); HCO_3^- is hydrogen carbonate (or bicarbonate).
(d) Chromium(III) nitrate; the *-ic* ending is not used with Roman numerals; NO_3^- is nitr*ate*.
(e) Calcium nitrite; Ca^{2+} is calcium and NO_2^- is nitr*ite*.

2.12 (a) $HClO_3$; (b) hydrofluoric acid; (c) CH_3COOH (or $HC_2H_3O_2$); (d) H_2SO_3; (e) hypobromous acid

2.13 (a) Sulfur trioxide; (b) silicon dioxide; (c) N_2O; (d) SeF_6

2.14 (a) Disulfur dichloride; the *-ous* suffix is not used.
(b) NO; the name indicates one nitrogen.
(c) Bromine trichloride; Br is in a higher period in Group 7A(17), so it is named first.

2.15 (a) H_2O_2, 34.02 amu; (b) CsCl, 168.4 amu; (c) H_2SO_4, 98.09 amu; (d) K_2SO_4, 174.27 amu

2.16 (a) Na_2O. This is an ionic compound, so the name is sodium oxide.

Formula mass
$$= (2 \times \text{atomic mass of Na}) + (1 \times \text{atomic mass of O})$$
$$= (2 \times 22.99 \text{ amu}) + 16.00 \text{ amu} = 61.98 \text{ amu}$$

(b) NO_2. This is a covalent compound, and N has the lower group number, so the name is nitrogen dioxide.

Molecular mass
$$= (1 \times \text{atomic mass of N}) + (2 \times \text{atomic mass of O})$$
$$= 14.01 \text{ amu} + (2 \times 16.00 \text{ amu}) = 46.01 \text{ amu}$$

PROBLEMS

Problems with ***colored*** *numbers are answered in Appendix E. Sections match the text and provide the numbers of relevant sample problems. Bracketed problems are grouped in pairs (indicated by a short rule) that cover the same concept. Comprehensive Problems are based on material from any section or previous chapter.*

Elements, Compounds, and Mixtures: An Atomic Overview
(Sample Problem 2.1)

2.1 What is the key difference between an element and a compound?

2.2 List two differences between a compound and a mixture.

2.3 Which of the following are pure substances? Explain.
(a) Calcium chloride, used to melt ice on roads, consists of two elements, calcium and chlorine, in a fixed mass ratio.
(b) Sulfur consists of sulfur atoms combined into octatomic molecules.
(c) Baking powder, a leavening agent, contains 26% to 30% sodium hydrogen carbonate and 30% to 35% calcium dihydrogen phosphate by mass.
(d) Cytosine, a component of DNA, consists of H, C, N, and O atoms bonded in a specific arrangement.

2.4 Classify each substance in Problem 2.3 as an element, compound, or mixture, and explain your answers.

2.5 Each molecular-scale scene below represents a mixture. Describe each one in terms of the number of elements and/or compounds present.

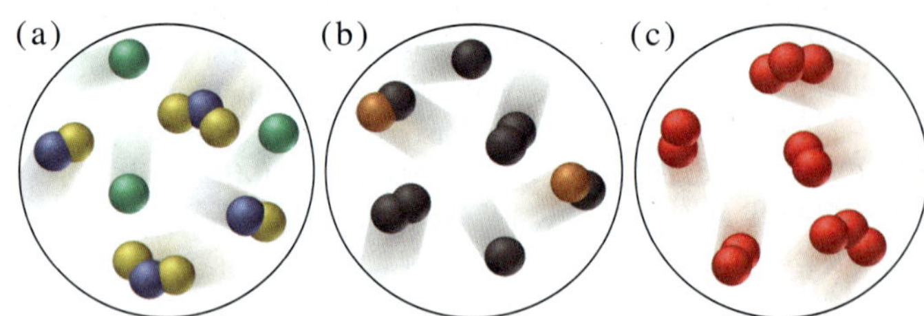

The Observations That Led to an Atomic View of Matter
(Sample Problem 2.2)

2.6 To which classes of matter—element, compound, and/or mixture—do the following apply: (a) law of mass conservation; (b) law of definite composition; (c) law of multiple proportions?

2.7 Identify the mass law that each of the following observations demonstrates, and explain your reasoning:
(a) A sample of potassium chloride from Chile contains the same percent by mass of potassium as one from Poland.

(b) A flashbulb contains magnesium and oxygen before use and magnesium oxide afterward, but its mass does not change.
(c) Arsenic and oxygen form one compound that is 65.2 mass % arsenic and another that is 75.8 mass % arsenic.

2.8 Which of the following scenes illustrate(s) the fact that compounds of chlorine (green) and oxygen (red) exhibit the law of multiple proportions? Name the compounds.

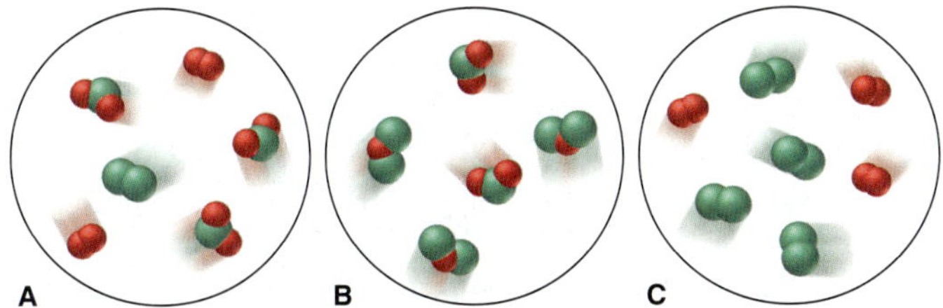

2.9 (a) Does the percent by mass of each element in a compound depend on the amount of compound? Explain.
(b) Does the mass of each element in a compound depend on the amount of compound? Explain.
(c) Does the percent by mass of each element in a compound depend on the amount of that element used to make the compound? Explain.

2.10 State the mass law(s) demonstrated by the following experimental results, and explain your reasoning:
Experiment 1: A student heats 1.00 g of a blue compound and obtains 0.64 g of a white compound and 0.36 g of a colorless gas.
Experiment 2: A second student heats 3.25 g of the same blue compound and obtains 2.08 g of a white compound and 1.17 g of a colorless gas.

2.11 State the mass law(s) demonstrated by the following experimental results, and explain your reasoning:
Experiment 1: A student heats 1.27 g of copper and 3.50 g of iodine to produce 3.81 g of a white compound, and 0.96 g of iodine remains.
Experiment 2: A second student heats 2.55 g of copper and 3.50 g of iodine to form 5.25 g of a white compound, and 0.80 g of copper remains.

2.12 Fluorite, a mineral of calcium, is a compound of the metal with fluorine. Analysis shows that a 2.76-g sample of fluorite contains 1.42 g of calcium. Calculate the (a) mass of fluorine in the sample; (b) mass fractions of calcium and fluorine in fluorite; (c) mass percents of calcium and fluorine in fluorite.

2.13 Galena, a mineral of lead, is a compound of the metal with sulfur. Analysis shows that a 2.34-g sample of galena contains 2.03 g of lead. Calculate the (a) mass of sulfur in the sample; (b) mass fractions of lead and sulfur in galena; (c) mass percents of lead and sulfur in galena.

2.14 A compound of copper and sulfur contains 88.39 g of metal and 44.61 g of nonmetal. How many grams of copper are in 5264 kg of compound? How many grams of sulfur?

2.15 A compound of iodine and cesium contains 63.94 g of metal and 61.06 g of nonmetal. How many grams of cesium are in 38.77 g of compound? How many grams of iodine?

2.16 Show, with calculations, how the following data illustrate the law of multiple proportions:
Compound 1: 47.5 mass % sulfur and 52.5 mass % chlorine
Compound 2: 31.1 mass % sulfur and 68.9 mass % chlorine

2.17 Show, with calculations, how the following data illustrate the law of multiple proportions:
Compound 1: 77.6 mass % xenon and 22.4 mass % fluorine
Compound 2: 63.3 mass % xenon and 36.7 mass % fluorine

2.18 Dolomite is a carbonate of magnesium and calcium. Analysis shows that 7.81 g of dolomite contains 1.70 g of Ca. Calculate the mass percent of Ca in dolomite. On the basis of the mass percent of Ca, and neglecting all other factors, which is the richer source of Ca, dolomite or fluorite (see Problem 2.12)?

2.19 The mass percent of sulfur in a sample of coal is a key factor in the environmental impact of the coal because the sulfur combines with oxygen when the coal is burned and the oxide can then be incorporated into acid rain. Which of the following coals would have the smallest environmental impact?

	Mass (g) of Sample	Mass (g) of Sulfur in Sample
Coal A	378	11.3
Coal B	495	19.0
Coal C	675	20.6

Dalton's Atomic Theory
(Sample Problem 2.3)

2.20 Use Dalton's theory to explain why potassium nitrate from India or Italy has the same mass percents of K, N, and O.

The Observations That Led to the Nuclear Atom Model

2.21 Thomson was able to determine the mass/charge ratio of the electron but not its mass. (a) How did Millikan's experiment allow determination of the electron's mass? (b) The following charges on individual oil droplets were obtained during an experiment similar to Millikan's. Determine a charge for the electron (in C, coulombs), and explain your answer: -3.204×10^{-19} C; -4.806×10^{-19} C; -8.010×10^{-19} C; -1.442×10^{-18} C.

2.22 When Rutherford's coworkers bombarded gold foil with α particles, they obtained results that overturned the existing (Thomson) model of the atom. Explain.

The Atomic Theory Today
(Sample Problems 2.4 and 2.5)

2.23 Choose the correct answer. The difference between the mass number of an isotope and its atomic number is (a) directly related to the identity of the element; (b) the number of electrons; (c) the number of neutrons; (d) the number of isotopes.

2.24 Argon has three naturally occurring isotopes, ^{36}Ar, ^{38}Ar, and ^{40}Ar. What is the mass number of each? How many protons, neutrons, and electrons are present in each?

2.25 Chlorine has two naturally occurring isotopes, ^{35}Cl and ^{37}Cl. What is the mass number of each isotope? How many protons, neutrons, and electrons are present in each?

2.26 Do both members of the following pairs have the same number of protons? Neutrons? Electrons?
(a) $^{16}_{8}O$ and $^{17}_{8}O$ (b) $^{40}_{18}Ar$ and $^{41}_{19}K$ (c) $^{60}_{27}Co$ and $^{60}_{28}Ni$
Which pair(s) consist(s) of atoms with the same Z value? N value? A value?

2.27 Do both members of the following pairs have the same number of protons? Neutrons? Electrons?
(a) ${}^{3}_{1}H$ and ${}^{3}_{2}He$ (b) ${}^{14}_{6}C$ and ${}^{15}_{7}N$ (c) ${}^{19}_{9}F$ and ${}^{18}_{9}F$
Which pair(s) consist(s) of atoms with the same *Z* value? *N* value? *A* value?

2.28 Write the ${}^{A}_{Z}X$ notation for each atomic depiction:

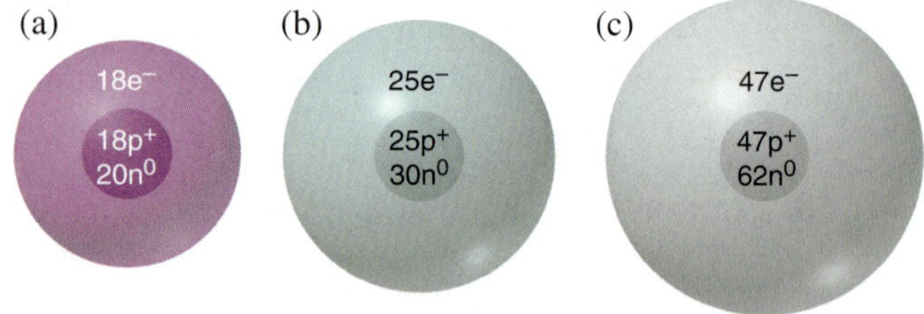

2.29 Write the ${}^{A}_{Z}X$ notation for each atomic depiction:

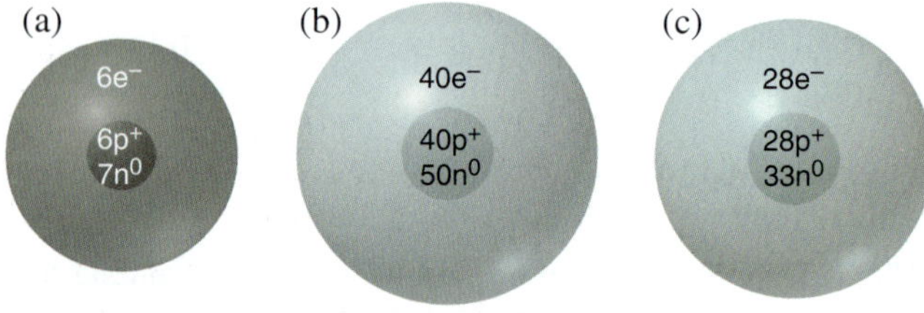

2.30 Draw atomic depictions similar to those in Problem 2.28 for (a) ${}^{48}_{22}Ti$; (b) ${}^{79}_{34}Se$; (c) ${}^{11}_{5}B$.

2.31 Draw atomic depictions similar to those in Problem 2.28 for (a) ${}^{207}_{82}Pb$; (b) ${}^{9}_{4}Be$; (c) ${}^{75}_{33}As$.

2.32 Gallium has two naturally occurring isotopes, ${}^{69}Ga$ (isotopic mass 68.9256 amu, abundance 60.11%) and ${}^{71}Ga$ (isotopic mass 70.9247 amu, abundance 39.89%). Calculate the atomic mass of gallium.

2.33 Magnesium has three naturally occurring isotopes, ${}^{24}Mg$ (isotopic mass 23.9850 amu, abundance 78.99%), ${}^{25}Mg$ (isotopic mass 24.9858 amu, abundance 10.00%), and ${}^{26}Mg$ (isotopic mass 25.9826 amu, abundance 11.01%). Calculate the atomic mass of magnesium.

2.34 Chlorine has two naturally occurring isotopes, ${}^{35}Cl$ (isotopic mass 34.9689 amu) and ${}^{37}Cl$ (isotopic mass 36.9659 amu). If chlorine has an atomic mass of 35.4527 amu, what is the percent abundance of each isotope?

2.35 Copper has two naturally occurring isotopes, ${}^{63}Cu$ (isotopic mass 62.9396 amu) and ${}^{65}Cu$ (isotopic mass 64.9278 amu). If copper has an atomic mass of 63.546 amu, what is the percent abundance of each isotope?

Elements: A First Look at the Periodic Table

2.36 Correct each of the following statements:
(a) In the modern periodic table, the elements are arranged in order of increasing atomic mass.
(b) Elements in a period have similar chemical properties.
(c) Elements can be classified as either metalloids or nonmetals.

2.37 What class of elements lies along the "staircase" line in the periodic table? How do their properties compare with those of metals and nonmetals?

2.38 What are some characteristic properties of elements to the left of the elements along the "staircase"? To the right?

2.39 Give the name, atomic symbol, and group number of the element with the following *Z* value, and classify it as a metal, metalloid, or nonmetal:
(a) $Z = 32$ (b) $Z = 15$ (c) $Z = 2$ (d) $Z = 3$ (e) $Z = 42$

2.40 Give the name, atomic symbol, and group number of the element with the following *Z* value, and classify it as a metal, metalloid, or nonmetal:
(a) $Z = 33$ (b) $Z = 20$ (c) $Z = 35$ (d) $Z = 19$ (e) $Z = 13$

2.41 Fill in the blanks:
(a) The symbol and atomic number of the heaviest alkaline earth metal are ________ and ________.
(b) The symbol and atomic number of the lightest metalloid in Group 4A(14) are ________ and ________.
(c) Group 1B(11) consists of the *coinage metals.* The symbol and atomic mass of the coinage metal whose atoms have the fewest electrons are ________ and ________.
(d) The symbol and atomic mass of the halogen in Period 4 are ________ and ________.

2.42 Fill in the blanks:
(a) The symbol and atomic number of the heaviest nonradioactive noble gas are ________ and ________.
(b) The symbol and group number of the Period 5 transition element whose atoms have the fewest protons are ________ and ________.
(c) The elements in Group 6A(16) are sometimes called the *chalcogens.* The symbol and atomic number of the only metallic chalcogen are ________ and ________.
(d) The symbol and number of protons of the Period 4 alkali metal atom are ________ and ________.

Compounds: Introduction to Bonding

(Sample Problem 2.6)

2.43 Describe the type and nature of the bonding that occurs between reactive metals and nonmetals.

2.44 Describe the type and nature of the bonding that often occurs between two nonmetals.

2.45 Given that the ions in LiF and in MgO are of similar size, which compound has stronger ionic bonding? Use Coulomb's law in your explanation.

2.46 Describe the formation of solid magnesium chloride ($MgCl_2$) from large numbers of magnesium and chlorine atoms.

2.47 Does potassium nitrate (KNO_3) incorporate ionic bonding, covalent bonding, or both? Explain.

2.48 What monatomic ions do potassium ($Z = 19$) and iodine ($Z = 53$) form?

2.49 What monatomic ions do barium ($Z = 56$) and selenium ($Z = 34$) form?

2.50 For each ionic depiction, give the name of the parent atom, its mass number, and its group and period numbers:

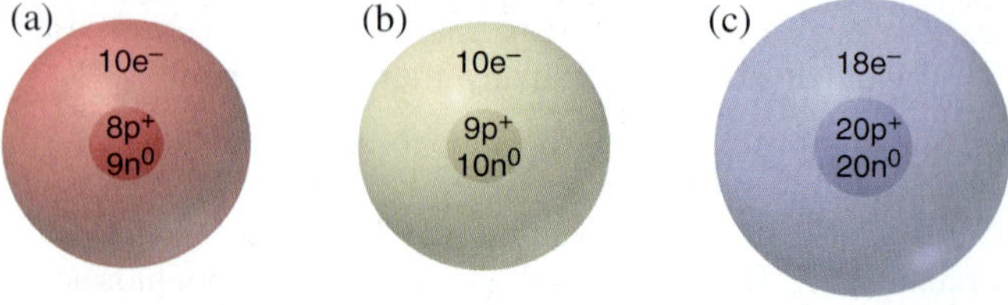

2.51 For each ionic depiction, give the name of the parent atom, its mass number, and its group and period numbers:

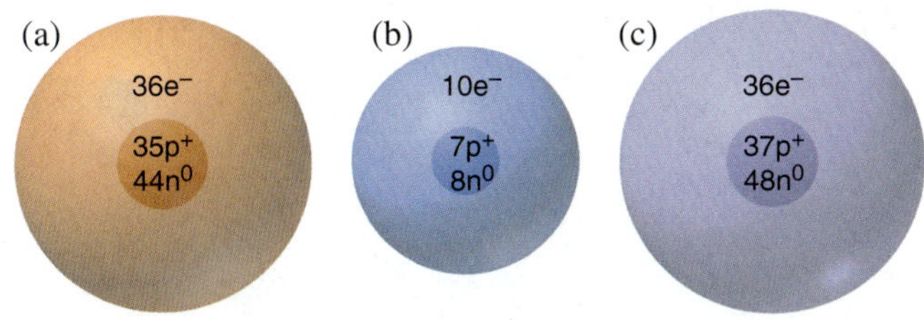

2.52 An ionic compound forms when lithium ($Z = 3$) reacts with oxygen ($Z = 8$). If a sample of the compound contains 8.4×10^{21} lithium ions, how many oxide ions does it contain?

2.53 An ionic compound forms when calcium ($Z = 20$) reacts with iodine ($Z = 53$). If a sample of the compound contains 7.4×10^{21} calcium ions, how many iodide ions does it contain?

2.54 The radii of the sodium and potassium ions are 102 pm and 138 pm, respectively. Which compound has stronger ionic attractions, sodium chloride or potassium chloride?

2.55 The radii of the lithium and magnesium ions are 76 pm and 72 pm, respectively. Which compound has stronger ionic attractions, lithium oxide or magnesium oxide?

Compounds: Formulas, Names, and Masses

(Sample Problems 2.7 to 2.16)

2.56 What is the difference between an empirical formula and a molecular formula? Can they ever be the same?

2.57 Consider a mixture of 10 billion O_2 molecules and 10 billion H_2 molecules. In what way is this mixture similar to a sample containing 10 billion hydrogen peroxide (H_2O_2) molecules? In what way is it different?

2.58 Write an empirical formula for each of the following:
(a) Hydrazine, a rocket fuel, molecular formula N_2H_4
(b) Glucose, a sugar, molecular formula $C_6H_{12}O_6$

2.59 Write an empirical formula for each of the following:
(a) Ethylene glycol, car antifreeze, molecular formula $C_2H_6O_2$
(b) Peroxodisulfuric acid, a compound used to make bleaching agents, molecular formula $H_2S_2O_8$

2.60 Give the name and formula of the compound formed from the following elements: (a) sodium and nitrogen; (b) oxygen and strontium; (c) aluminum and chlorine.

2.61 Give the name and formula of the compound formed from the following elements: (a) cesium and bromine; (b) sulfur and barium; (c) calcium and fluorine.

2.62 Give the name and formula of the compound formed from the following elements:
(a) $_{12}L$ and $_{9}M$ (b) $_{30}L$ and $_{16}M$ (c) $_{17}L$ and $_{38}M$

2.63 Give the name and formula of the compound formed from the following elements:
(a) $_{37}Q$ and $_{35}R$ (b) $_{8}Q$ and $_{13}R$ (c) $_{20}Q$ and $_{53}R$

2.64 Give the systematic names for the formulas or the formulas for the names:
(a) tin(IV) chloride (b) $FeBr_3$
(c) cuprous bromide (d) Mn_2O_3

2.65 Give the systematic names for the formulas or the formulas for the names:
(a) Na_2HPO_4
(b) potassium carbonate dihydrate
(c) $NaNO_2$
(d) ammonium perchlorate

2.66 Correct each of the following formulas:
(a) Barium oxide is BaO_2.
(b) Iron(II) nitrate is $Fe(NO_3)_3$.
(c) Magnesium sulfide is $MnSO_3$.

2.67 Correct each of the following names:
(a) CuI is cobalt(II) iodide.
(b) $Fe(HSO_4)_3$ is iron(II) sulfate.
(c) $MgCr_2O_7$ is magnesium dichromium heptaoxide.

2.68 Give the name and formula for the acid derived from each of the following anions:
(a) hydrogen sulfate (b) IO_3^- (c) cyanide (d) HS^-

2.69 Give the name and formula for the acid derived from each of the following anions:
(a) perchlorate (b) NO_3^- (c) bromite (d) F^-

2.70 Give the name and formula of the compound whose molecules consist of two sulfur atoms and four fluorine atoms.

2.71 Give the name and formula of the compound whose molecules consist of two chlorine atoms and one oxygen atom.

2.72 Give the number of atoms of the specified element in a formula unit of each of the following compounds, and calculate the molecular (formula) mass:
(a) Oxygen in aluminum sulfate, $Al_2(SO_4)_3$
(b) Hydrogen in ammonium hydrogen phosphate, $(NH_4)_2HPO_4$
(c) Oxygen in the mineral azurite, $Cu_3(OH)_2(CO_3)_2$

2.73 Give the number of atoms of the specified element in a formula unit of each of the following compounds, and calculate the molecular (formula) mass:
(a) Hydrogen in ammonium benzoate, $C_6H_5COONH_4$
(b) Nitrogen in hydrazinium sulfate, $N_2H_6SO_4$
(c) Oxygen in the mineral leadhillite, $Pb_4SO_4(CO_3)_2(OH)_2$

2.74 Write the formula of each compound, and determine its molecular (formula) mass: (a) ammonium sulfate; (b) sodium dihydrogen phosphate; (c) potassium bicarbonate.

2.75 Write the formula of each compound, and determine its molecular (formula) mass: (a) sodium dichromate; (b) ammonium perchlorate; (c) magnesium nitrite trihydrate.

2.76 Give the name, empirical formula, and molecular mass of the molecule depicted in Figure P2.76.

2.77 Give the name, empirical formula, and molecular mass of the molecule depicted in Figure P2.77.

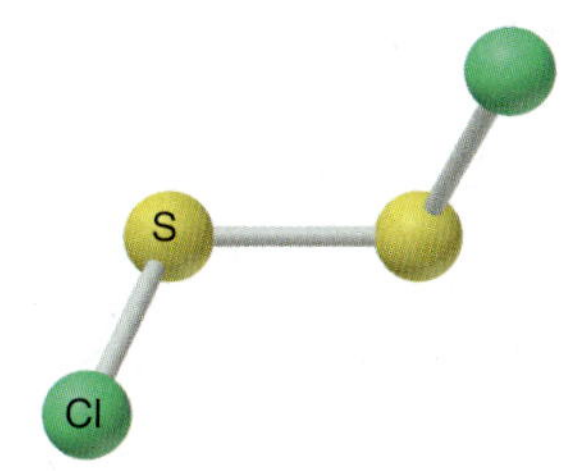

Figure P2.76

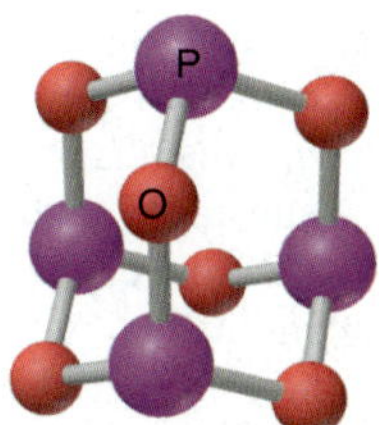

Figure P2.77

2.78 Give the formula, name, and molecular mass of the following molecules:

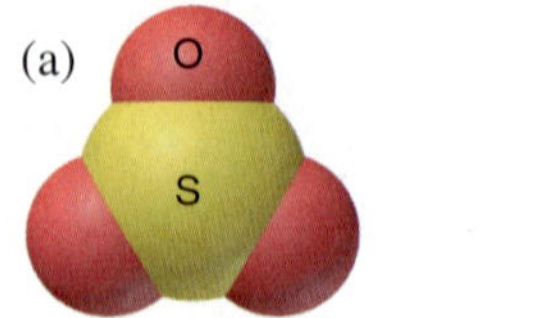

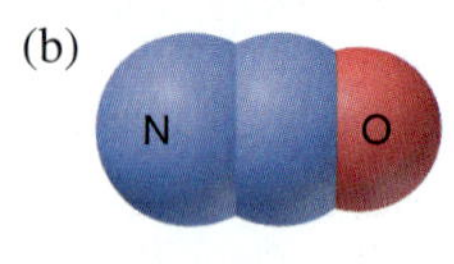

2.79 Before the use of systematic names, many compounds had common names. Give the systematic name for each of the following: (a) blue vitriol, $CuSO_4{\cdot}5H_2O$; (b) slaked lime, $Ca(OH)_2$; (c) oil of vitriol, H_2SO_4; (d) washing soda, Na_2CO_3; (e) muriatic acid, HCl; (f) Epsom salts, $MgSO_4{\cdot}7H_2O$; (g) chalk, $CaCO_3$; (h) dry ice, CO_2; (i) baking soda, $NaHCO_3$; (j) lye, NaOH.

Mixtures: Classification and Separation

2.80 In what main way is separating the components of a mixture different from separating the components of a compound?

2.81 What is the difference between a homogeneous and a heterogeneous mixture?

2.82 Is a solution a homogeneous or a heterogeneous mixture? Give an example of an aqueous solution.

2.83 Classify each of the following as a compound, a homogeneous mixture, or a heterogeneous mixture: (a) distilled water; (b) gasoline; (c) beach sand; (d) wine; (e) air.

2.84 Classify each of the following as a compound, a homogeneous mixture, or a heterogeneous mixture: (a) orange juice; (b) vegetable soup; (c) cement; (d) calcium sulfate; (e) tea.

Comprehensive Problems

Problems with an asterisk (*) are more challenging.

2.85 Helium is the lightest noble gas and the second most abundant element (after hydrogen) in the universe.

(a) The radius of a helium atom is 3.1×10^{-11} m; the radius of its nucleus is 2.5×10^{-15} m. What fraction of the spherical atomic volume is occupied by the nucleus (V of a sphere $= \frac{4}{3}\pi r^3$)?

(b) The mass of a helium-4 atom is 6.64648×10^{-24} g, and each of its two electrons has a mass of 9.10939×10^{-28} g. What fraction of this atom's mass is contributed by its nucleus?

2.86 Scenes A–I depict various types of matter on the atomic scale. Choose the correct scene(s) for each of the following:

(a) A mixture that fills its container
(b) A substance that cannot be broken down into simpler ones
(c) An element with a very high resistance to flow
(d) A homogeneous mixture
(e) An element that conforms to the walls of its container and displays a surface
(f) A gas consisting of diatomic particles
(g) A gas that can be broken down into simpler substances
(h) A substance with a 2/1 ratio of its component atoms
(i) Matter that can be separated into its component substances by physical means
(j) A heterogeneous mixture
(k) Matter that obeys the law of definite composition

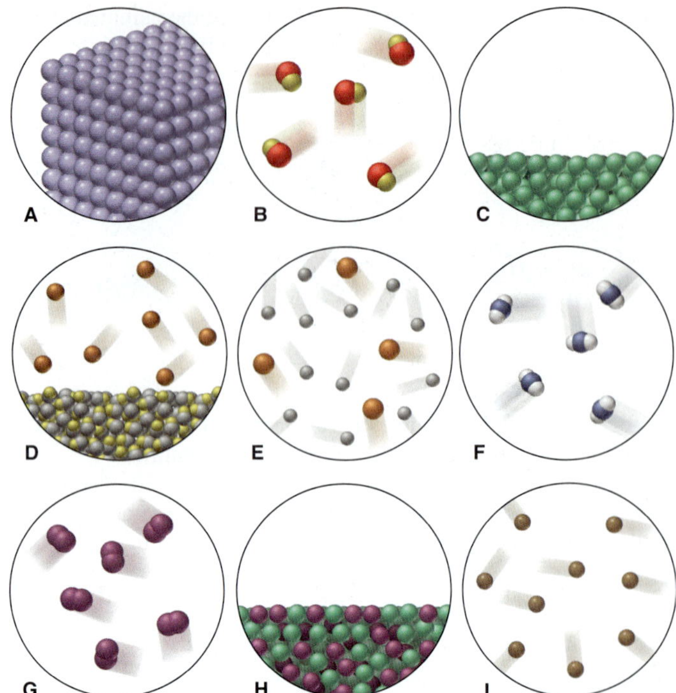

2.87 Nitrogen forms more oxides than any other element. The percents by mass of N in three different nitrogen oxides are (I) 46.69%, (II) 36.85%, (III) 25.94%. (a) Determine the empirical formula of each compound. (b) How many grams of oxygen per 1.00 g of nitrogen are in each compound?

2.88 Give the molecular mass of each compound depicted below, and provide a correct name for any that are named incorrectly.

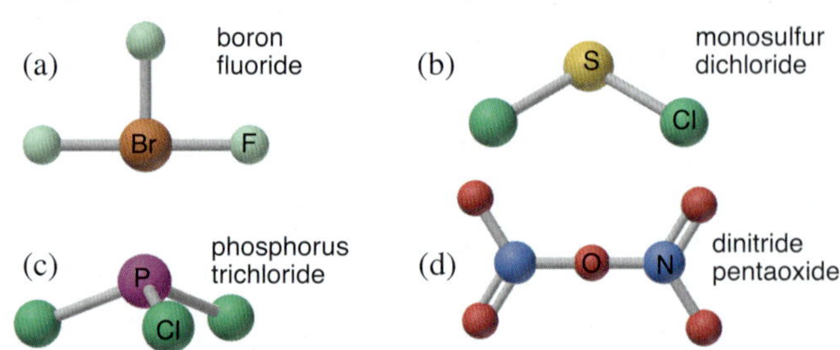

2.89 Dinitrogen monoxide (N_2O; nitrous oxide) is a greenhouse gas that enters the atmosphere principally from natural fertilizer breakdown. Some studies have shown that the isotope ratios of ^{15}N to ^{14}N and of ^{18}O to ^{16}O in N_2O depend on the source, which can thus be determined by measuring the relative abundance of molecular masses in a sample of N_2O.

(a) What different molecular masses are possible for N_2O?

(b) The percent abundance of ^{14}N is 99.6%, and that of ^{16}O is 99.8%. Which molecular mass of N_2O is least common, and which is most common?

2.90 The seven most abundant ions in seawater make up more than 99% by mass of the dissolved compounds. They are listed in units of mg ion/kg seawater: chloride, 18,980; sodium, 10,560; sulfate, 2650; magnesium, 1270; calcium, 400; potassium, 380; hydrogen carbonate, 140.

(a) What is the mass % of each ion in seawater?

(b) What percent of the total mass of ions is sodium ion?

(c) How does the total mass % of alkaline earth metal ions compare with the total mass % of alkali metal ions?
(d) Which makes up the larger mass fraction of dissolved components, anions or cations?

2.91 The scenes below represent a mixture of two monatomic gases undergoing a reaction when heated. Which mass law(s) is (are) illustrated by this change?

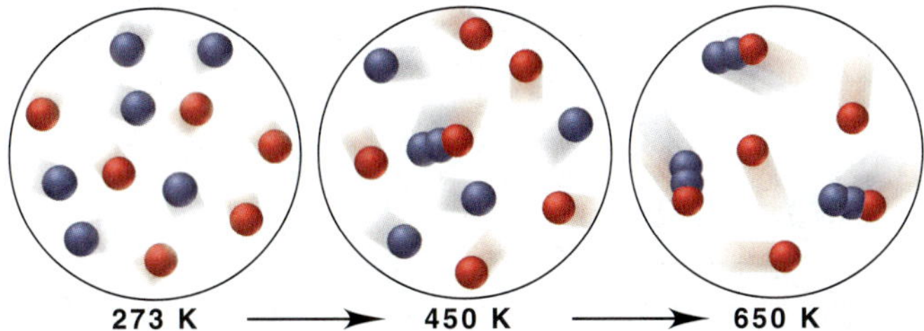

* **2.92** When barium (Ba) reacts with sulfur (S) to form barium sulfide (BaS), each Ba atom reacts with an S atom. If 2.50 cm^3 of Ba reacts with 1.75 cm^3 of S, are there enough Ba atoms to react with the S atoms (d of Ba = 3.51 g/cm^3; d of S = 2.07 g/cm^3)?

2.93 Succinic acid *(below)* is an important metabolite in biological energy production. Give the molecular formula, empirical formula, and molecular mass of succinic acid, and calculate the mass percent of each element.

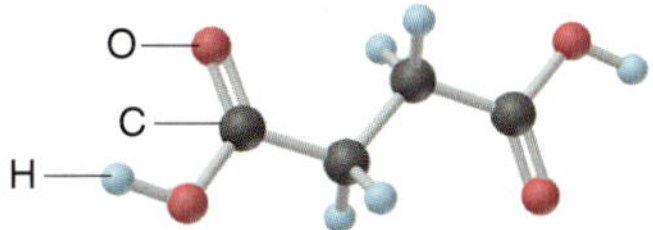

2.94 Which of the following models represent compounds having the same empirical formula? What is the molecular mass of this common empirical formula?

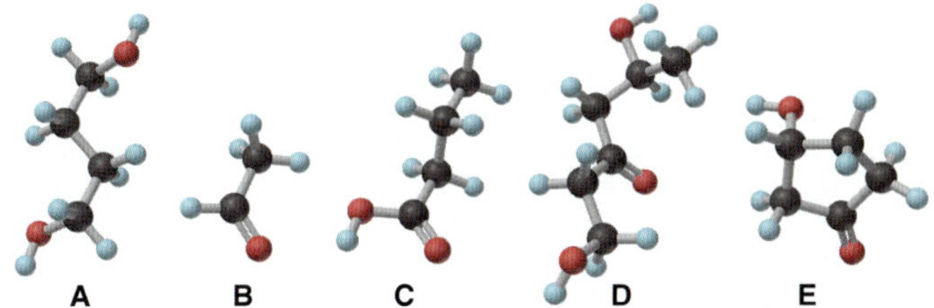

2.95 Antimony has many uses—for example, in semiconductor infrared devices and as part of an alloy in lead storage batteries. The element has two naturally occurring isotopes, one with mass 120.904 amu, the other with mass 122.904 amu. (a) Write the $^{A}_{Z}X$ notation for each isotope. (b) Use the atomic mass of antimony from the periodic table to calculate the natural abundance of each isotope.

2.96 The two isotopes of potassium with significant abundance in nature are ^{39}K (isotopic mass 38.9637 amu, 93.258%) and ^{41}K (isotopic mass 40.9618 amu, 6.730%). Fluorine has only one naturally occurring isotope, ^{19}F (isotopic mass = 18.9984 amu). Calculate the formula mass of potassium fluoride.

2.97 Dimercaprol ($HSCH_2CHSHCH_2OH$) is a complexing agent developed during World War I as an antidote to arsenic-based poison gas and used today to treat heavy-metal poisoning. Such an agent binds and removes the toxic element from the body.
(a) If each molecule binds one arsenic (As) atom, how many atoms of As could be removed by 250. mg of dimercaprol?
(b) If one molecule binds one metal atom, calculate the mass % of each of the following metals in a metal-dimercaprol complex: mercury, thallium, chromium.

* **2.98** TNT (trinitrotoluene; *below*) is used as an explosive in construction. Calculate the mass of each element in 1.00 lb of TNT.

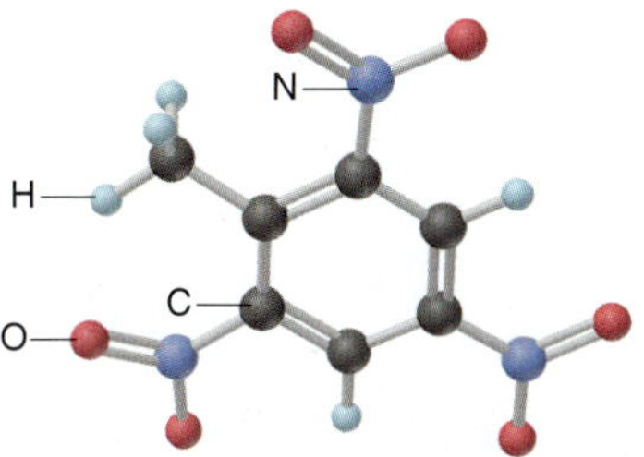

2.99 The anticancer drug Platinol (cisplatin), $Pt(NH_3)_2Cl_2$, reacts with the cancer cell's DNA and interferes with its growth. (a) What is the mass % of platinum (Pt) in Platinol? (b) If Pt costs \$19/g, how many grams of Platinol can be made for \$1.00 million (assume that the cost of Pt determines the cost of the drug)?

2.100 Choose the box color(s) in the periodic table below that match(es) the following:

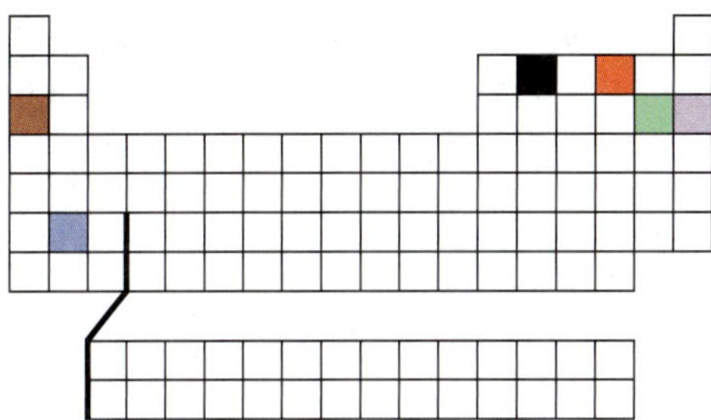

(a) Four elements that are nonmetals
(b) Two elements that are metals
(c) Three elements that are gases at room temperature
(d) Three elements that are solid at room temperature
(e) One pair of elements likely to form a covalent compound
(f) Another pair of elements likely to form a covalent compound
(g) One pair of elements likely to form an ionic compound with formula MX
(h) Another pair of elements likely to form an ionic compound with formula MX
(i) Two elements likely to form an ionic compound with formula M_2X
(j) Two elements likely to form an ionic compound with formula MX_2
(k) An element that forms no compounds
(l) A pair of elements whose compounds exhibit the law of multiple proportions

2.101 From the following ions and their radii (in pm), choose a pair that gives the strongest ionic bonding and a pair that gives the weakest: Mg^{2+}, 72; K^+, 138; Rb^+, 152; Ba^{2+}, 135; Cl^-, 181; O^{2-}, 140; I^-, 220.

2.102 A rock is 5.0% by mass fayalite (Fe_2SiO_4), 7.0% by mass forsterite (Mg_2SiO_4), and the remainder silicon dioxide. What is the mass percent of each element in the rock?

2.103 Fluoride ion is poisonous in relatively low amounts: 0.2 g of F^- per 70 kg of body weight can cause death. Nevertheless, in order to prevent tooth decay, F^- ions are added to drinking water at a concentration of 1 mg of F^- ion per L of water. How many liters of fluoridated drinking water would a 70-kg person have to consume in one day to reach this toxic level? How many kilograms of sodium fluoride would be needed to fluoridate the water in a 7.00×10^7-gal reservoir?

2.104 Nitrogen monoxide (NO) is a bioactive molecule in blood. Low NO concentrations cause respiratory distress and the formation of blood clots. Doctors prescribe nitroglycerin, $C_3H_5N_3O_9$, and isoamyl nitrate, $(CH_3)_2CHCH_2CH_2ONO_2$, to increase NO. If each compound releases one molecule of NO per atom of N, calculate the mass percent of NO in each medicine.

2.105 Which of the following steps in an overall process involve(s) a physical change and which involve(s) a chemical change?

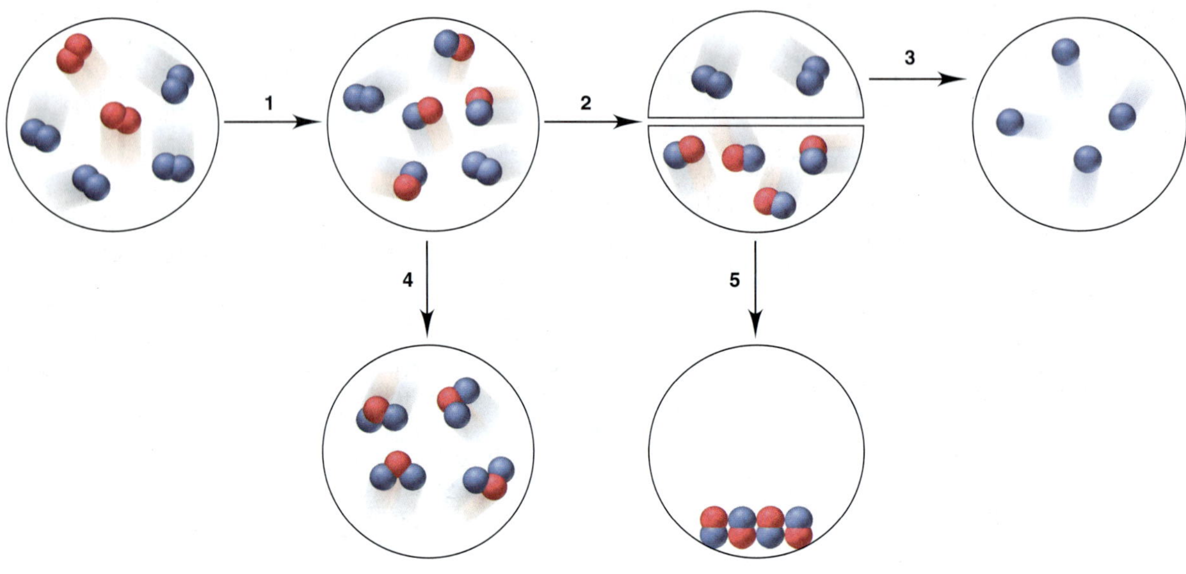

Stoichiometry of Formulas and Equations

3

Key Principles
to focus on while studying this chapter

- The *mole* (mol) is the standard unit for *amount of substance* and contains *Avogadro's number* (6.022×10^{23}) of chemical entities (atoms, molecules, or ions). It has the same numerical value in grams as a single entity of the substance has in atomic mass units; for example, 1 molecule of H_2O weighs 18.02 amu and 1 mol of H_2O molecules weighs 18.02 g. Therefore, if the amount of a substance is expressed in moles, we know the number of entities in a given mass of it, which means that amount, mass, and number are interconvertible *(Section 3.1)*.
- The subscripts in the *chemical formula* for a compound provide quantitative information about the amounts of each element in one mole of the compound. In an *empirical* formula, the subscripts show the *relative* numbers of moles of each element in the compound; in a *molecular* formula, they show the *actual* numbers. Isomers are different compounds with the same molecular formula *(Section 3.2)*.
- In a *balanced equation,* chemical formulas preceded by integer (whole-number) coefficients give the same number of each kind of atom on the left *(reactants)* as on the right *(products)* but with atoms in different combinations *(Section 3.3)*.
- Using molar ratios from the balanced equation, we calculate the amount of one substance from the amount of any other involved in the reaction. During a typical reaction, one substance (the *limiting reactant*) is used up, so it limits the amount of product that can form; the other reactant(s) are in excess. The *theoretical yield,* the amount indicated by the balanced equation, is never obtained in the lab because of competing *side reactions* and losses incurred while isolating the product *(Section 3.4)*.
- For reactions in solution, we determine amounts of substances from their *concentration (molarity)* and the solution volume. To dilute a solution, we add *solvent,* which lowers the amount of *solute* dissolved in each unit volume *(Section 3.5)*.

Weighing the Matter *By knowing the mass of table salt, you also know the number of sodium and chloride ions in the sample. In this chapter, you'll learn the arithmetic of chemical formulas and reactions.*

Outline

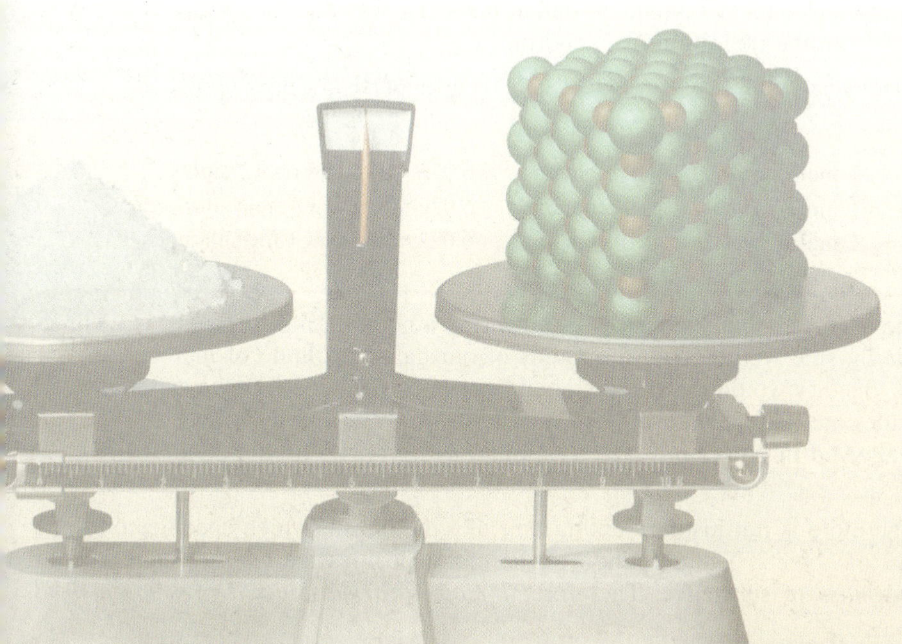

Concepts & Skills to Review before studying this chapter

- isotopes and atomic mass (Section 2.5)
- names and formulas of compounds (Section 2.8)
- molecular mass of a compound (Section 2.8)
- empirical and molecular formulas (Section 2.8)
- mass laws in chemical reactions (Section 2.2)

Chemistry is a practical science. Just imagine how useful it could be to determine the formula of a compound from the masses of its elements or to predict the amounts of substances consumed and produced in a reaction. Suppose you are a polymer chemist preparing a new plastic: how much of this new material will a given polymerization reaction yield? Or suppose you're a chemical engineer studying rocket engine thrust: what amount of exhaust gases will a test of this fuel mixture produce? Perhaps you are on a team of environmental chemists examining coal samples: what quantity of greenhouse gases will this sample release when burned? Or, maybe you're a biomedical researcher who has extracted a new cancer-preventing substance from a tropical plant: what is its formula, and what quantity of metabolic products will establish a safe dosage level? You can answer countless questions like these with a knowledge of **stoichiometry** (pronounced "stoy-key-AHM-uh-tree"; from the Greek *stoicheion,* "element or part," and *metron,* "measure"), the study of the quantitative aspects of chemical formulas and reactions. All the ideas and skills discussed in this chapter depend on an understanding of the *mole* concept, so the first section introduces this essential unit.

3.1 THE MOLE

In daily life, we often measure things by counting or by weighing, with the choice based on convenience. It is more convenient to weigh beans or rice than to count individual pieces, and it is more convenient to count eggs or pencils than to weigh them. To measure such things, we use mass units (a kilogram of rice) or counting units (a dozen pencils). Similarly, daily life in the laboratory involves measuring substances to prepare a solution or run a reaction. However, an obvious problem arises when we try to do this. The atoms, ions, molecules, or formula units are the entities that react with one another, so we would like to know the numbers of them that we mix together. But, how can we possibly count entities that are so small? To do this, chemists have devised a unit called the mole to *count chemical entities by weighing them.*

Defining the Mole

The **mole** (abbreviated **mol**) is the SI unit for amount of substance. It is defined as *the amount of a substance that contains the same number of entities as there are atoms in exactly 12 g of carbon-12.* This number is called **Avogadro's number,** in honor of the 19th-century Italian physicist Amedeo Avogadro, and as you can tell from the definition, it is enormous:

$$\text{One mole (1 mol) contains } 6.022\times10^{23} \text{ entities (to four significant figures)} \quad (3.1)$$

Thus,

1 mol of carbon-12	contains	6.022×10^{23} carbon-12 atoms
1 mol of H_2O	contains	6.022×10^{23} H_2O molecules
1 mol of NaCl	contains	6.022×10^{23} NaCl formula units

THINK OF IT THIS WAY
Imagine a Mole of . . .

A mole of any ordinary object is a staggering amount: a mole of periods (.) lined up side by side would equal the radius of our galaxy; a mole of marbles stacked tightly together would cover the United States 70 miles deep. However, atoms and molecules are not ordinary objects: a mole of water molecules (about 18 mL) can be swallowed in one gulp!

The mole is not just a counting unit like the dozen, which specifies only the *number* of objects. The definition of the mole specifies the *number* of objects in a fixed *mass* of substance. Therefore, *1 mole of a substance represents a fixed*

FIGURE 3.1 Counting objects of fixed relative mass. A, If marbles had a fixed mass, we could count them by weighing them. Each red marble weighs 7 g, and each yellow marble weighs 4 g, so 84 g of red marbles and 48 g of yellow marbles each contains 12 marbles. Equal numbers of the two types of marbles always have a 7/4 mass ratio of red/yellow marbles. **B,** Because atoms of a substance have a fixed mass, we can weigh the substance to count the atoms; 55.85 g of Fe (left pan) and 32.07 g of S (right pan) each contains 6.022×10^{23} atoms (1 mol of atoms). Any two samples of Fe and S that contain equal numbers of atoms have a 55.85/32.07 mass ratio of Fe/S.

number of chemical entities ***and*** *has a fixed mass.* To see why this is important, consider the marbles in Figure 3.1A, which we'll use as an analogy for atoms. Suppose you have large groups of red marbles and yellow marbles; each red marble weighs 7 g and each yellow marble weighs 4 g. Right away you know that there are 12 marbles in 84 g of red marbles or in 48 g of yellow marbles. Moreover, because one red marble weighs $\frac{7}{4}$ as much as one yellow marble, any given *number* of red and of yellow marbles always has this 7/4 *mass* ratio. By the same token, any given *mass* of red and of yellow marbles always has a 4/7 *number* ratio. For example, 280 g of red marbles contains 40 marbles, and 280 g of yellow marbles contains 70 marbles. As you can see, the fixed masses of the marbles allow you to count marbles by weighing them.

Atoms have fixed masses also, and the mole unit allows us to determine the number of atoms, molecules, or formula units in a sample by weighing it. Let's focus on elements first and recall a key point from Chapter 2: the atomic mass of an element (which appears on the periodic table) is the weighted average of the masses of its naturally occurring isotopes. For purposes of weighing, *all atoms of an element are considered to have this mass.* That is, all iron (Fe) atoms have an atomic mass of 55.85 amu, all sulfur (S) atoms have an atomic mass of 32.07 amu, and so forth.

The central relationship between the mass of one atom and the mass of 1 mole of those atoms is that *the atomic mass of an element expressed in* ***amu*** *is numerically the same as the mass of 1 mole of atoms of the element expressed in* ***grams.*** You can see this from the definition of the mole, which referred to the number of atoms in "**12** g of carbon-**12**." Thus,

1 Fe atom	has a mass of	55.85 amu	and	1 mol of Fe atoms	has a mass of	55.85 g
1 S atom	has a mass of	32.07 amu	and	1 mol of S atoms	has a mass of	32.07 g
1 O atom	has a mass of	16.00 amu	and	1 mol of O atoms	has a mass of	16.00 g
1 O_2 molecule	has a mass of	32.00 amu	and	1 mol of O_2 molecules	has a mass of	32.00 g

Moreover, because of their fixed atomic masses, we know that 55.85 g of Fe atoms and 32.07 g of S atoms each contains 6.022×10^{23} atoms. As with marbles of fixed mass, one Fe atom weighs $\frac{55.85}{32.07}$ as much as one S atom, and 1 mol of Fe atoms weighs $\frac{55.85}{32.07}$ as much as 1 mol of S atoms (Figure 3.1B).

A similar relationship holds for compounds: *the molecular mass (or formula mass) of a compound expressed in* ***amu*** *is numerically the same as the mass of 1 mole of the compound expressed in* ***grams.*** Thus, for example,

1 molecule of H_2O	has a mass of	18.02 amu	and	1 mol of H_2O (6.022×10^{23} molecules)	has a mass of	18.02 g
1 formula unit of NaCl	has a mass of	58.44 amu	and	1 mol of NaCl (6.022×10^{23} formula units)	has a mass of	58.44 g

To summarize the two key points about the usefulness of the mole concept:

- The mole maintains the *same mass relationship* between macroscopic samples as exists between individual chemical entities.
- The mole relates the *number* of chemical entities to the *mass* of a sample of those entities.

FIGURE 3.2 One mole of some familiar substances. One mole of a substance is the amount that contains 6.022×10^{23} atoms, molecules, or formula units. From left to right: 1 mol (172.19 g) of writing chalk (calcium sulfate dihydrate), 1 mol (32.00 g) of gaseous O_2, 1 mol (63.55 g) of copper, and 1 mol (18.02 g) of liquid H_2O.

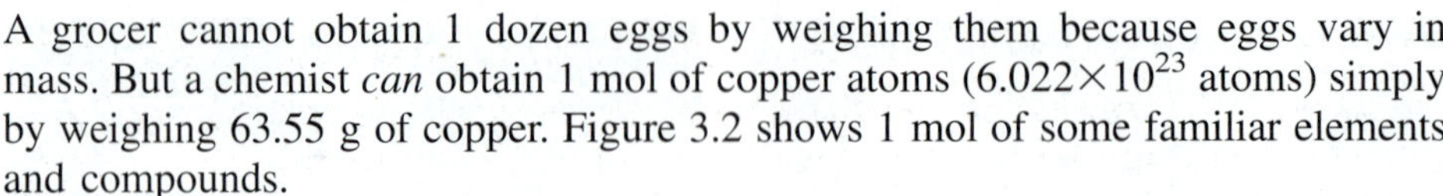

A grocer cannot obtain 1 dozen eggs by weighing them because eggs vary in mass. But a chemist *can* obtain 1 mol of copper atoms (6.022×10^{23} atoms) simply by weighing 63.55 g of copper. Figure 3.2 shows 1 mol of some familiar elements and compounds.

Molar Mass

The **molar mass ($\mathcal{M}$)** of a substance is the mass per mole of its entities (atoms, molecules, or formula units). Thus, molar mass has units of grams per mole (g/mol). The periodic table is indispensable for calculating the molar mass of a substance. Here's how the calculations are done:

1. **Elements.** You find the molar mass of an element simply by looking up its atomic mass in the periodic table and then noting whether the element occurs naturally as individual atoms or as molecules.

- *Monatomic elements.* For elements that occur as individual atoms, the molar mass is the numerical value from the periodic table expressed in units of grams per mole.* Thus, the molar mass of neon is 20.18 g/mol, the molar mass of iron is 55.85 g/mol, and the molar mass of gold is 197.0 g/mol.
- *Molecular elements.* For elements that occur as molecules, you must know the molecular formula to determine the molar mass. For example, oxygen exists normally in air as diatomic molecules, so the molar mass of O_2 molecules is twice that of O atoms:

$$\text{Molar mass } (\mathcal{M}) \text{ of } O_2 = 2 \times \mathcal{M} \text{ of O} = 2 \times 16.00 \text{ g/mol} = 32.00 \text{ g/mol}$$

The most common form of sulfur exists as octatomic molecules, S_8:

$$\mathcal{M} \text{ of } S_8 = 8 \times \mathcal{M} \text{ of S} = 8 \times 32.07 \text{ g/mol} = 256.6 \text{ g/mol}$$

2. **Compounds.** *The molar mass of a compound is the sum of the molar masses of the atoms of the elements in the formula.* For example, the formula of sulfur dioxide (SO_2) tells us that 1 mol of SO_2 molecules contains 1 mol of S atoms and 2 mol of O atoms:

$$\begin{aligned}\mathcal{M} \text{ of } SO_2 &= \mathcal{M} \text{ of S} + (2 \times \mathcal{M} \text{ of O}) = 32.07 \text{ g/mol} + (2 \times 16.00 \text{ g/mol}) \\ &= 64.07 \text{ g/mol}\end{aligned}$$

Similarly, for ionic compounds, such as potassium sulfide (K_2S), we have

$$\begin{aligned}\mathcal{M} \text{ of } K_2S &= (2 \times \mathcal{M} \text{ of K}) + \mathcal{M} \text{ of S} = (2 \times 39.10 \text{ g/mol}) + 32.07 \text{ g/mol} \\ &= 110.27 \text{ g/mol}\end{aligned}$$

A key point to note is that *the subscripts in a formula refer to individual atoms (or ions), as well as to moles of atoms (or ions).* Table 3.1 presents this idea for glucose ($C_6H_{12}O_6$; *margin*), the essential sugar in energy metabolism.

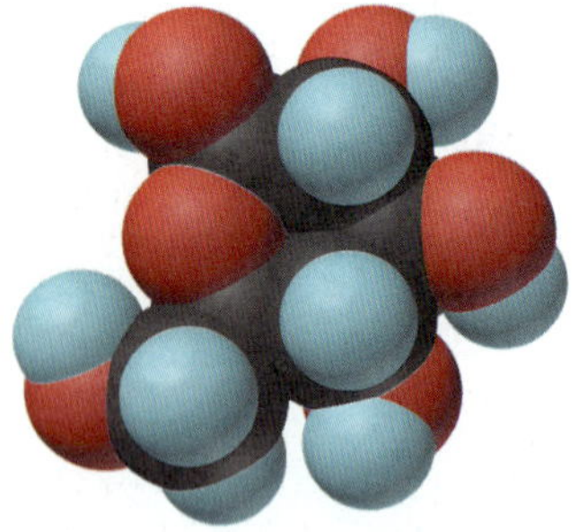

Glucose

*The mass value in the periodic table has no units because it is a *relative* atomic mass, given by the atomic mass (in amu) divided by 1 amu ($\frac{1}{12}$ mass of one ^{12}C atom in amu):

$$\text{Relative atomic mass} = \frac{\text{atomic mass } (\cancel{\text{amu}})}{\frac{1}{12} \text{ mass of } ^{12}\text{C } (\cancel{\text{amu}})}$$

Therefore, you use the same number for the atomic mass (weighted average mass of one atom in amu) and the molar mass (mass of 1 mole of atoms in grams).

Table 3.1 Information Contained in the Chemical Formula of Glucose, $C_6H_{12}O_6$ ($\mathcal{M}$ = 180.16 g/mol)

	Carbon (C)	Hydrogen (H)	Oxygen (O)
Atoms/molecule of compound	6 atoms	12 atoms	6 atoms
Moles of atoms/mole of compound	6 mol of atoms	12 mol of atoms	6 mol of atoms
Atoms/mole of compound	$6(6.022 \times 10^{23})$ atoms	$12(6.022 \times 10^{23})$ atoms	$6(6.022 \times 10^{23})$ atoms
Mass/molecule of compound	6(12.01 amu) = 72.06 amu	12(1.008 amu) = 12.10 amu	6(16.00 amu) = 96.00 amu
Mass/mole of compound	72.06 g	12.10 g	96.00 g

Interconverting Moles, Mass, and Number of Chemical Entities

One of the reasons the mole is such a convenient unit for laboratory work is that it allows you to calculate the mass or number of entities of a substance in a sample if you know the amount (number of moles) of the substance. Conversely, if you know the mass or number of entities of a substance, you can calculate the number of moles.

The molar mass, which expresses the equivalent relationship between 1 mole of a substance and its mass in grams, can be used as a conversion factor. We multiply by the molar mass of an element or compound ($\mathcal{M}$, in g/mol) to convert a given amount (in moles) to mass (in grams):

$$\text{Mass (g)} = \text{no. of moles} \times \frac{\text{no. of grams}}{1 \text{ mol}} \quad (3.2)$$

Or, we divide by the molar mass (multiply by $1/\mathcal{M}$) to convert a given mass (in grams) to amount (in moles):

$$\text{No. of moles} = \text{mass (g)} \times \frac{1 \text{ mol}}{\text{no. of grams}} \quad (3.3)$$

In a similar way, we use Avogadro's number, which expresses the equivalent relationship between 1 mole of a substance and the number of entities it contains, as a conversion factor. We multiply by Avogadro's number to convert amount of substance (in moles) to the number of entities (atoms, molecules, or formula units):

$$\text{No. of entities} = \text{no. of moles} \times \frac{6.022\times10^{23} \text{ entities}}{1 \text{ mol}} \quad (3.4)$$

Or, we divide by Avogadro's number to do the reverse:

$$\text{No. of moles} = \text{no. of entities} \times \frac{1 \text{ mol}}{6.022\times10^{23} \text{ entities}} \quad (3.5)$$

Converting Moles of Elements For problems involving mass-mole-number relationships of elements, keep these points in mind:

- To convert between amount (mol) and mass (g), use the molar mass ($\mathcal{M}$ in g/mol).
- To convert between amount (mol) and number of entities, use Avogadro's number (6.022×10^{23} entities/mol). For elements that occur as molecules, use the molecular formula to find atoms/mol.
- Mass and number of entities relate directly to number of moles, *not* to each other. Therefore, to convert between number of entities and mass, *first convert to number of moles.* For example, to find the number of atoms in a given mass,

$$\text{No. of atoms} = \text{mass } \cancel{\text{(g)}} \times \frac{1 \cancel{\text{ mol}}}{\text{no. } \cancel{\text{of grams}}} \times \frac{6.022\times10^{23} \text{ atoms}}{1 \cancel{\text{ mol}}}$$

These relationships are summarized in Figure 3.3 and demonstrated in Sample Problem 3.1.

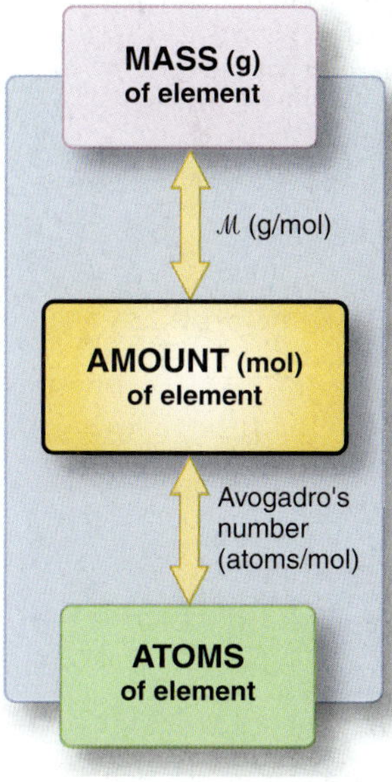

MASS (g) of element

$\mathcal{M}$ (g/mol)

AMOUNT (mol) of element

Avogadro's number (atoms/mol)

ATOMS of element

FIGURE 3.3 Summary of the mass-mole-number relationships for elements. The amount (mol) of an element is related to its mass (g) through the molar mass ($\mathcal{M}$ in g/mol) and to its number of atoms through Avogadro's number (6.022×10^{23} atoms/mol). For elements that occur as molecules, Avogadro's number gives *molecules* per mole.

SAMPLE PROBLEM 3.1 Calculating the Mass and Number of Atoms in a Given Number of Moles of an Element

Problem **(a)** Silver (Ag) is used in jewelry and tableware but no longer in U.S. coins. How many grams of Ag are in 0.0342 mol of Ag?

(b) Iron (Fe), the main component of steel, is the most important metal in industrial society. How many Fe atoms are in 95.8 g of Fe?

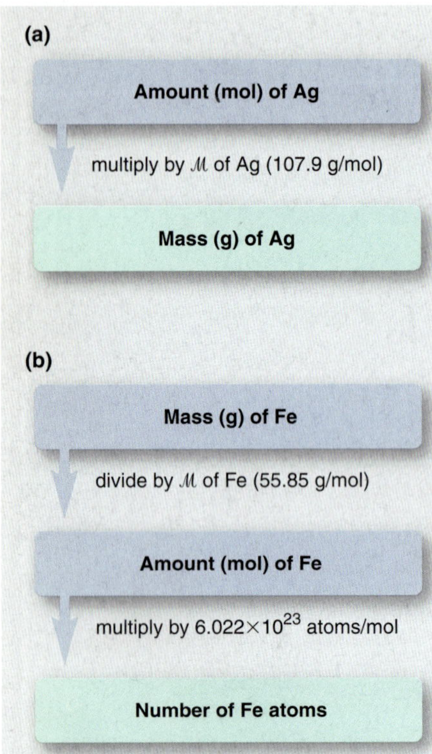

(a) Determining the mass (g) of Ag

Plan We know the number of moles of Ag (0.0342 mol) and have to find the mass (in g). To convert *moles* of Ag to *grams* of Ag, we multiply by the *molar mass* of Ag, which we find in the periodic table (see roadmap a).

Solution Converting from moles of Ag to grams:

$$\text{Mass (g) of Ag} = 0.0342 \cancel{\text{mol Ag}} \times \frac{107.9 \text{ g Ag}}{1 \cancel{\text{mol Ag}}} = 3.69 \text{ g Ag}$$

Check We rounded the mass to three significant figures because the number of moles has three. The units are correct. About 0.03 mol × 100 g/mol gives 3 g; the small mass makes sense because 0.0342 is a small fraction of a mole.

(b) Determining the number of Fe atoms

Plan We know the grams of Fe (95.8 g) and need the number of Fe atoms. We cannot convert directly from grams to atoms, so we first convert to moles by dividing grams of Fe by its molar mass. [This is the reverse of the step in part (a).] Then, we multiply number of moles by Avogadro's number to find number of atoms (see roadmap b).

Solution Converting from grams of Fe to moles:

$$\text{Moles of Fe} = 95.8 \cancel{\text{g Fe}} \times \frac{1 \text{ mol Fe}}{55.85 \cancel{\text{g Fe}}} = 1.72 \text{ mol Fe}$$

Converting from moles of Fe to number of atoms:

$$\text{No. of Fe atoms} = 1.72 \cancel{\text{mol Fe}} \times \frac{6.022\times10^{23} \text{ atoms Fe}}{1 \cancel{\text{mol Fe}}}$$

$$= 10.4\times10^{23} \text{ atoms Fe} = 1.04\times10^{24} \text{ atoms Fe}$$

Check When we approximate the mass of Fe and the molar mass of Fe, we have ~100 g/(~50 g/mol) = 2 mol. Therefore, the number of atoms should be about twice Avogadro's number: $2(6\times10^{23}) = 1.2\times10^{24}$.

FOLLOW-UP PROBLEM 3.1 **(a)** Graphite is the crystalline form of carbon used in "lead" pencils. How many moles of carbon are in 315 mg of graphite?
(b) Manganese (Mn) is a transition element essential for the growth of bones. What is the mass in grams of 3.22×10^{20} Mn atoms, the number found in 1 kg of bone?

Converting Moles of Compounds Solving mass-mole-number problems involving compounds requires a very similar approach to the one for elements. We need the chemical formula to find the molar mass and to determine the moles of a given element in the compound. These relationships are shown in Figure 3.4, and an example is worked through in Sample Problem 3.2.

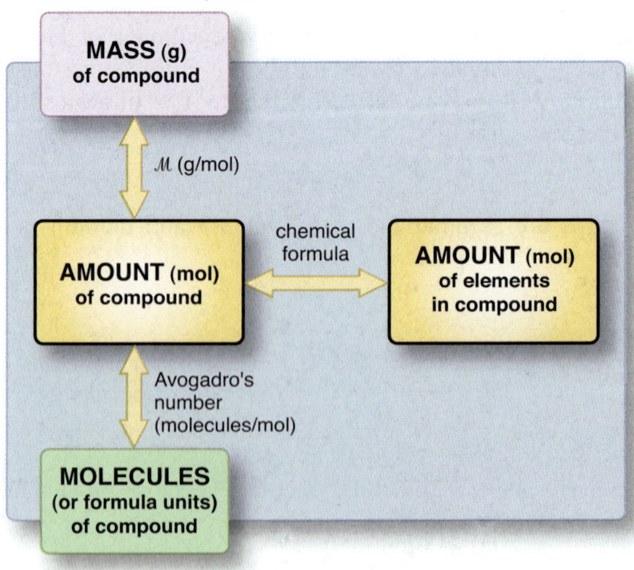

FIGURE 3.4 Summary of the mass-mole-number relationships for compounds. Moles of a compound are related to grams of the compound through the molar mass (ℳ in g/mol) and to the number of molecules (or formula units) through Avogadro's number (6.022×10^{23} molecules/mol). To find the number of molecules (or formula units) in a given mass, or vice versa, convert the information to moles first. With the chemical formula, you can calculate mass-mole-number information about each component element.

SAMPLE PROBLEM 3.2 Calculating the Moles and Number of Formula Units in a Given Mass of a Compound

Problem Ammonium carbonate is a white solid that decomposes with warming. Among its many uses, it is a component of baking powder, fire extinguishers, and smelling salts. How many formula units are in 41.6 g of ammonium carbonate?

Plan We know the mass of compound (41.6 g) and need to find the number of formula units. As we saw in Sample Problem 3.1(b), to convert grams to number of entities, we have to find number of moles first, so we must divide the grams by the molar mass ($\mathcal{M}$). For this, we need $\mathcal{M}$, so we determine the formula (see Table 2.5) and take the sum of the elements' molar masses. Once we have the number of moles, we multiply by Avogadro's number to find the number of formula units.

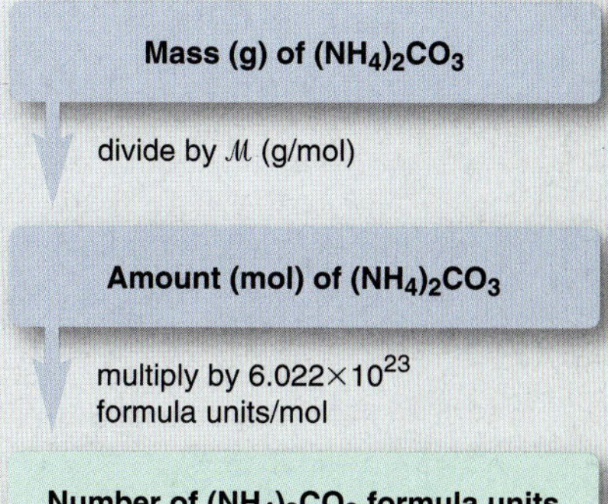

Solution The formula is $(NH_4)_2CO_3$. Calculating molar mass:

$$\begin{aligned}\mathcal{M} &= (2 \times \mathcal{M} \text{ of N}) + (8 \times \mathcal{M} \text{ of H}) + (1 \times \mathcal{M} \text{ of C}) + (3 \times \mathcal{M} \text{ of O})\\ &= (2 \times 14.01 \text{ g/mol}) + (8 \times 1.008 \text{ g/mol}) + 12.01 \text{ g/mol} + (3 \times 16.00 \text{ g/mol})\\ &= 96.09 \text{ g/mol}\end{aligned}$$

Converting from grams to moles:

$$\text{Moles of } (NH_4)_2CO_3 = 41.6 \cancel{\text{g } (NH_4)_2CO_3} \times \frac{1 \text{ mol } (NH_4)_2CO_3}{96.09 \cancel{\text{g } (NH_4)_2CO_3}} = 0.433 \text{ mol } (NH_4)_2CO_3$$

Converting from moles to formula units:

$$\begin{aligned}\text{Formula units of } (NH_4)_2CO_3 &= 0.433 \cancel{\text{mol } (NH_4)_2CO_3}\\ &\quad \times \frac{6.022\times10^{23} \text{ formula units } (NH_4)_2CO_3}{1 \cancel{\text{mol } (NH_4)_2CO_3}}\\ &= 2.61\times10^{23} \text{ formula units } (NH_4)_2CO_3\end{aligned}$$

Check The units are correct. The mass is less than half the molar mass ($\sim42/96 < 0.5$), so the number of formula units should be less than half Avogadro's number ($\sim2.6\times10^{23}/6.0\times10^{23} < 0.5$).

Comment A *common mistake* is to forget the subscript 2 outside the parentheses in $(NH_4)_2CO_3$, which would give a much lower molar mass.

FOLLOW-UP PROBLEM 3.2 Tetraphosphorus decaoxide reacts with water to form phosphoric acid, a major industrial acid. In the laboratory, the oxide is used as a drying agent. **(a)** What is the mass (in g) of 4.65×10^{22} molecules of tetraphosphorus decaoxide? **(b)** How many P atoms are present in this sample?

Mass Percent from the Chemical Formula

Each element in a compound constitutes its own particular portion of the compound's mass. For an individual molecule (or formula unit), we use the molecular (or formula) mass and chemical formula to find the mass percent of any element X in the compound:

$$\text{Mass \% of element X} = \frac{\text{atoms of X in formula} \times \text{atomic mass of X (amu)}}{\text{molecular (or formula) mass of compound (amu)}} \times 100$$

The formula also tells the number of *moles* of each element in the compound, so we can use the molar mass to find the mass percent of each element on a mole basis:

$$\text{Mass \% of element X} = \frac{\text{moles of X in formula} \times \text{molar mass of X (g/mol)}}{\text{mass (g) of 1 mol of compound}} \times 100 \qquad \textbf{(3.6)}$$

As always, the individual mass percents of the elements in the compound must add up to 100% (within rounding). As Sample Problem 3.3 demonstrates, an important practical use of mass percent is to determine the amount of an element in any size sample of a compound.

SAMPLE PROBLEM 3.3 Calculating Mass Percents and Masses of Elements in a Sample of a Compound

Problem In mammals, lactose (milk sugar) is metabolized to glucose ($C_6H_{12}O_6$), the key nutrient for generating chemical potential energy.
(a) What is the mass percent of each element in glucose?
(b) How many grams of carbon are in 16.55 g of glucose?

Amount (mol) of element X in 1 mol of compound
↓ multiply by $\mathcal{M}$ (g/mol) of X
Mass (g) of X in 1 mol of compound
↓ divide by mass (g) of 1 mol of compound
Mass fraction of X
↓ multiply by 100
Mass % of X

(a) Determining the mass percent of each element

Plan We know the relative numbers of moles of the elements in glucose from the formula (6 C, 12 H, 6 O). We multiply the number of moles of each element by its molar mass to find grams. Dividing each element's mass by the mass of 1 mol of glucose gives the mass fraction of each element, and multiplying each fraction by 100 gives the mass percent. The calculation steps for any element X are shown in the roadmap.

Solution Calculating the mass of 1 mol of $C_6H_{12}O_6$:

$$\begin{aligned}\mathcal{M} &= (6 \times \mathcal{M} \text{ of C}) + (12 \times \mathcal{M} \text{ of H}) + (6 \times \mathcal{M} \text{ of O})\\ &= (6 \times 12.01 \text{ g/mol}) + (12 \times 1.008 \text{ g/mol}) + (6 \times 16.00 \text{ g/mol})\\ &= 180.16 \text{ g/mol}\end{aligned}$$

Converting moles of C to grams: There are 6 mol of C in 1 mol of glucose, so

$$\text{Mass (g) of C} = 6 \cancel{\text{mol C}} \times \frac{12.01 \text{ g C}}{1 \cancel{\text{mol C}}} = 72.06 \text{ g C}$$

Finding the mass fraction of C in glucose:

$$\text{Mass fraction of C} = \frac{\text{total mass C}}{\text{mass of 1 mol glucose}} = \frac{72.06 \text{ g}}{180.16 \text{ g}} = 0.4000$$

Finding the mass percent of C:

$$\text{Mass \% of C} = \text{mass fraction of C} \times 100 = 0.4000 \times 100 = \boxed{40.00 \text{ mass \% C}}$$

Combining the steps for each of the other two elements in glucose:

$$\text{Mass \% of H} = \frac{\text{mol H} \times \mathcal{M} \text{ of H}}{\text{mass of 1 mol glucose}} \times 100 = \frac{12 \cancel{\text{mol H}} \times \dfrac{1.008 \text{ g H}}{1 \cancel{\text{mol H}}}}{180.16 \cancel{\text{g}}} \times 100 = \boxed{6.714 \text{ mass \% H}}$$

$$\text{Mass \% of O} = \frac{\text{mol O} \times \mathcal{M} \text{ of O}}{\text{mass of 1 mol glucose}} \times 100 = \frac{6 \cancel{\text{mol O}} \times \dfrac{16.00 \text{ g O}}{1 \cancel{\text{mol O}}}}{180.16 \cancel{\text{g}}} \times 100 = \boxed{53.29 \text{ mass \% O}}$$

Check The answers make sense: even though there are equal numbers of moles of O and C in the compound, the mass % of O is greater than the mass % of C because the molar mass of O is greater than the molar mass of C. The mass % of H is small because the molar mass of H is small. The total of the mass percents is 100.00%.

(b) Determining the mass (g) of carbon

Plan To find the mass of C in the glucose sample, we multiply the mass of the sample by the mass fraction of C from part (a).

Solution Finding the mass of C in a given mass of glucose (with units for mass fraction):

$$\text{Mass (g) of C} = \text{mass of glucose} \times \text{mass fraction of C} = 16.55 \cancel{\text{g glucose}} \times \frac{0.4000 \text{ g C}}{1 \cancel{\text{g glucose}}} = \boxed{6.620 \text{ g C}}$$

Check Rounding shows that the answer is "in the right ballpark": 16 g times less than 0.5 parts by mass should be less than 8 g.

Comment 1. A *more direct approach* to finding the mass of element in any mass of compound is similar to the approach we used in Sample Problem 2.2 and eliminates the need to calculate the mass fraction. Just multiply the given mass of compound by the ratio of the total mass of element to the mass of 1 mol of compound:

$$\text{Mass (g) of C} = 16.55 \text{ g glucose} \times \frac{72.06 \text{ g C}}{180.16 \text{ g glucose}} = 6.620 \text{ g C}$$

2. From here on, you should be able to determine the molar mass of a compound, so that calculation will no longer be shown.

FOLLOW-UP PROBLEM 3.3 Ammonium nitrate is a common fertilizer. Agronomists base the effectiveness of fertilizers on their nitrogen content.
(a) Calculate the mass percent of N in ammonium nitrate.
(b) How many grams of N are in 35.8 kg of ammonium nitrate?

SECTION 3.1 SUMMARY

A mole of substance is the amount that contains Avogadro's number (6.022×10^{23}) of chemical entities (atoms, molecules, or formula units). • The mass (in grams) of a mole has the same numerical value as the mass (in amu) of the entity. Thus, the mole allows us to count entities by weighing them. • Using the molar mass ($\mathcal{M}$, g/mol) of an element (or compound) and Avogadro's number as conversion factors, we can convert among amount (mol), mass (g), and number of entities. • The mass fraction of element X in a compound is used to find the mass of X in any amount of the compound.

3.2 DETERMINING THE FORMULA OF AN UNKNOWN COMPOUND

In Sample Problem 3.3, we knew the formula and used it to find the mass percent (or mass fraction) of an element in a compound *and* the mass of the element in a given mass of the compound. In this section, we do the reverse: use the masses of elements in a compound to find its formula. We'll present the mass data in several ways and then look briefly at molecular structures.

Empirical Formulas

An analytical chemist investigating a compound decomposes it into simpler substances, finds the mass of each component element, converts these masses to numbers of moles, and then arithmetically converts the moles to whole-number (integer) subscripts. This procedure yields the empirical formula, the *simplest whole-number ratio* of moles of each element in the compound (see Section 2.8). Let's see how to obtain the subscripts from the moles of each element.

Analysis of an unknown compound shows that the sample contains 0.21 mol of zinc, 0.14 mol of phosphorus, and 0.56 mol of oxygen. Because the subscripts in a formula represent individual atoms or moles of atoms, we write a preliminary formula that contains fractional subscripts: $Zn_{0.21}P_{0.14}O_{0.56}$. Next, we convert these fractional subscripts to whole numbers using one or two simple arithmetic steps (rounding when needed):

1. Divide each subscript by the smallest subscript:

$$Zn_{\frac{0.21}{0.14}}P_{\frac{0.14}{0.14}}O_{\frac{0.56}{0.14}} \longrightarrow Zn_{1.5}P_{1.0}O_{4.0}$$

   This step alone often gives integer subscripts.
2. If any of the subscripts is still not an integer, multiply through by the *smallest integer* that will turn all subscripts into integers. Here, we multiply by 2, the smallest integer that will make 1.5 (the subscript for Zn) into an integer:

$$Zn_{(1.5\times2)}P_{(1.0\times2)}O_{(4.0\times2)} \longrightarrow Zn_{3.0}P_{2.0}O_{8.0}, \text{ or } Zn_3P_2O_8$$

   Notice that the *relative* number of moles has not changed because we multiplied *all* the subscripts by 2.

Always check that the subscripts are the smallest set of integers with the same ratio as the original numbers of moles; that is, 3/2/8 is *in the same ratio* as 0.21/0.14/0.56. A more conventional way to write this formula is $Zn_3(PO_4)_2$; the compound is zinc phosphate, a dental cement.

Sample Problems 3.4, 3.5, and 3.6 demonstrate how other types of compositional data are used to determine chemical formulas. In the first problem, the empirical formula is found from data given as grams of each element rather than as moles.

SAMPLE PROBLEM 3.4 Determining an Empirical Formula from Masses of Elements

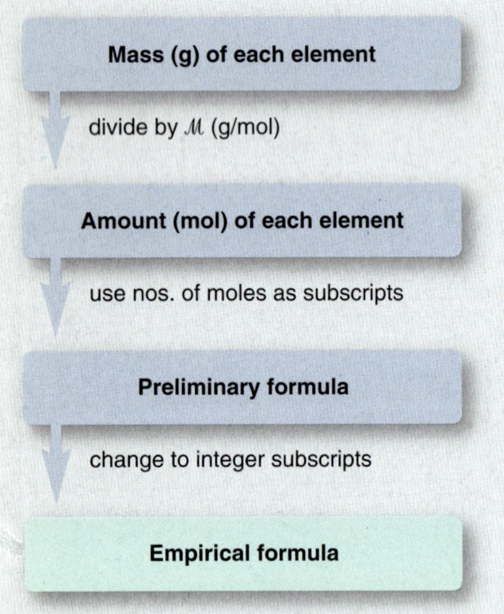

Mass (g) of each element

divide by $\mathcal{M}$ (g/mol)

Amount (mol) of each element

use nos. of moles as subscripts

Preliminary formula

change to integer subscripts

Empirical formula

Problem Elemental analysis of a sample of an ionic compound showed 2.82 g of Na, 4.35 g of Cl, and 7.83 g of O. What is the empirical formula and name of the compound?

Plan This problem is similar to the one we just discussed, except that we are given element *masses,* so we must convert the masses into integer subscripts. We first divide each mass by the element's molar mass to find *number of moles.* Then we construct a preliminary formula and convert the numbers of moles to integers.

Solution Finding moles of elements:

$$\text{Moles of Na} = 2.82\ \cancel{\text{g Na}} \times \frac{1 \text{ mol Na}}{22.99\ \cancel{\text{g Na}}} = 0.123 \text{ mol Na}$$

$$\text{Moles of Cl} = 4.35\ \cancel{\text{g Cl}} \times \frac{1 \text{ mol Cl}}{35.45\ \cancel{\text{g Cl}}} = 0.123 \text{ mol Cl}$$

$$\text{Moles of O} = 7.83\ \cancel{\text{g O}} \times \frac{1 \text{ mol O}}{16.00\ \cancel{\text{g O}}} = 0.489 \text{ mol O}$$

Constructing a preliminary formula: $Na_{0.123}Cl_{0.123}O_{0.489}$

Converting to integer subscripts (dividing all by the smallest subscript):

$$Na_{\frac{0.123}{0.123}}Cl_{\frac{0.123}{0.123}}O_{\frac{0.489}{0.123}} \longrightarrow Na_{1.00}Cl_{1.00}O_{3.98} \approx Na_1Cl_1O_4, \text{ or } NaClO_4$$

We rounded the subscript of O from 3.98 to 4. The empirical formula is $NaClO_4$; the name is sodium perchlorate.

Check The moles seem correct because the masses of Na and Cl are slightly more than 0.1 of their molar masses. The mass of O is greatest and its molar mass is smallest, so it should have the greatest number of moles. The ratio of subscripts, 1/1/4, is the same as the ratio of moles, 0.123/0.123/0.489 (within rounding).

FOLLOW-UP PROBLEM 3.4 An unknown metal M reacts with sulfur to form a compound with the formula M_2S_3. If 3.12 g of M reacts with 2.88 g of S, what are the names of M and M_2S_3? (*Hint:* Determine the number of moles of S and use the formula to find the number of moles of M.)

Molecular Formulas

If we know the molar mass of a compound, we can use the empirical formula to obtain the molecular formula, the *actual* number of moles of each element in 1 mol of compound. In some cases, such as water (H_2O), ammonia (NH_3), and methane (CH_4), the empirical and molecular formulas are identical, but in many others the molecular formula is a *whole-number multiple* of the empirical formula. Hydrogen peroxide, for example, has the empirical formula HO and the molecular formula H_2O_2. Dividing the molar mass of H_2O_2 (34.02 g/mol) by the empirical formula mass (17.01 g/mol) gives the whole-number multiple:

$$\text{Whole-number multiple} = \frac{\text{molar mass (g/mol)}}{\text{empirical formula mass (g/mol)}} = \frac{34.02\ \cancel{\text{g/mol}}}{17.01\ \cancel{\text{g/mol}}} = 2.000 = 2$$

Instead of giving compositional data in terms of masses of each element, analytical laboratories provide it as mass percents. From this, we determine the empirical formula by (1) assuming 100.0 g of compound, which allows us to express mass percent directly as mass, (2) converting the mass to number of moles, and (3) constructing the empirical formula. With the molar mass, we can also find the whole-number multiple and then the molecular formula.

SAMPLE PROBLEM 3.5 Determining a Molecular Formula from Elemental Analysis and Molar Mass

Problem During excessive physical activity, lactic acid ($\mathcal{M}$ = 90.08 g/mol) forms in muscle tissue and is responsible for muscle soreness. Elemental analysis shows that this compound contains 40.0 mass % C, 6.71 mass % H, and 53.3 mass % O.
(a) Determine the empirical formula of lactic acid.
(b) Determine the molecular formula.

(a) Determining the empirical formula
Plan We know the mass % of each element and must convert each to an integer subscript. Although the mass of lactic acid is not given, mass % is the same for any mass of compound, so we can assume 100.0 g of lactic acid and express each mass % directly as grams. Then, we convert grams to moles and construct the empirical formula as we did in Sample Problem 3.4.
Solution Expressing mass % as grams, assuming 100.0 g of lactic acid:

$$\text{Mass (g) of C} = \frac{40.0 \text{ parts C by mass}}{100 \text{ parts by mass}} \times 100.0 \text{ g} = 40.0 \text{ g C}$$

Similarly, we have 6.71 g of H and 53.3 g of O.
Converting from grams of each element to moles:

$$\text{Moles of C} = \text{mass of C} \times \frac{1}{\mathcal{M} \text{ of C}} = 40.0 \cancel{\text{g C}} \times \frac{1 \text{ mol C}}{12.01 \cancel{\text{g C}}} = 3.33 \text{ mol C}$$

Similarly, we have 6.66 mol of H and 3.33 mol of O.
Constructing the preliminary formula: $C_{3.33}H_{6.66}O_{3.33}$
Converting to integer subscripts:

$$C_{\frac{3.33}{3.33}}H_{\frac{6.66}{3.33}}O_{\frac{3.33}{3.33}} \longrightarrow C_{1.00}H_{2.00}O_{1.00} \approx C_1H_2O_1; \text{ the empirical formula is } \boxed{CH_2O}$$

Check The numbers of moles seem correct: the masses of C and O are each slightly more than 3 times their molar masses (e.g., for C, 40 g/(12 g/mol) > 3 mol), and the mass of H is over 6 times its molar mass.

(b) Determining the molecular formula
Plan The molecular formula subscripts are whole-number multiples of the empirical formula subscripts. To find this whole number, we divide the given molar mass (90.08 g/mol) by the empirical formula mass, which we find from the sum of the elements' molar masses. Then we multiply the whole number by each subscript in the empirical formula.
Solution The empirical-formula molar mass is 30.03 g/mol. Finding the whole-number multiple:

$$\text{Whole-number multiple} = \frac{\mathcal{M} \text{ of lactic acid}}{\mathcal{M} \text{ of empirical formula}} = \frac{90.08 \cancel{\text{g/mol}}}{30.03 \cancel{\text{g/mol}}} = 3.000 = 3$$

Determining the molecular formula:

$$C_{(1\times3)}H_{(2\times3)}O_{(1\times3)} = \boxed{C_3H_6O_3}$$

Check The calculated molecular formula has the same ratio of moles of elements (3/6/3) as the empirical formula (1/2/1) and corresponds to the given molar mass:

$$\begin{aligned}\mathcal{M} \text{ of lactic acid} &= (3 \times \mathcal{M} \text{ of C}) + (6 \times \mathcal{M} \text{ of H}) + (3 \times \mathcal{M} \text{ of O}) \\ &= (3 \times 12.01) + (6 \times 1.008) + (3 \times 16.00) = 90.08 \text{ g/mol}\end{aligned}$$

FOLLOW-UP PROBLEM 3.5 One of the most widespread environmental carcinogens (cancer-causing agents) is benzo[*a*]pyrene ($\mathcal{M}$ = 252.30 g/mol). It is found in coal dust, in cigarette smoke, and even in charcoal-grilled meat. Analysis of this hydrocarbon shows 95.21 mass % C and 4.79 mass % H. What is the molecular formula of benzo[*a*]pyrene?

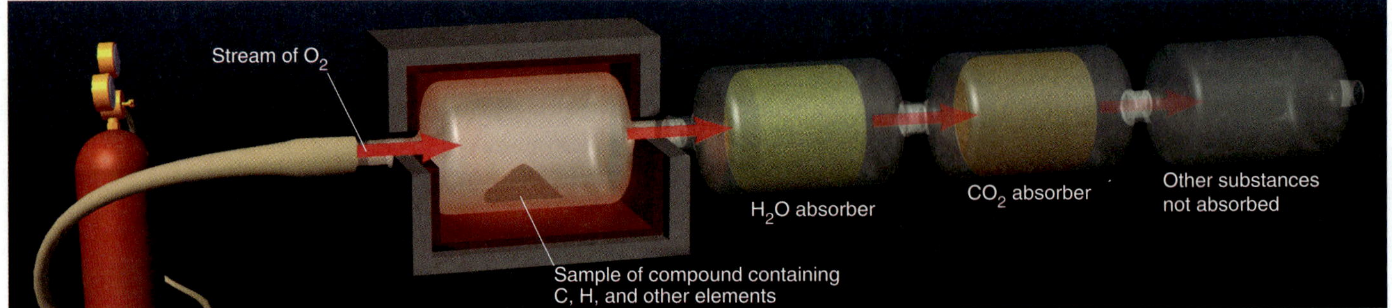

FIGURE 3.5 Combustion apparatus for determining formulas of organic compounds. A sample of compound that contains C and H (and perhaps other elements) is burned in a stream of O_2 gas. The CO_2 and H_2O formed are absorbed separately, while any other element oxides are carried through by the O_2 gas stream. H_2O is absorbed by $Mg(ClO_4)_2$; CO_2 is absorbed by NaOH. The increases in mass of the absorbers are used to calculate the amounts (mol) of C and H in the sample.

Combustion Analysis of Organic Compounds Still another type of compositional data is obtained through **combustion analysis,** a method used to measure the amounts of carbon and hydrogen in a combustible organic compound. The unknown compound is burned in pure O_2 in an apparatus that consists of a combustion furnace and chambers containing compounds that absorb either H_2O or CO_2 (Figure 3.5). All the H in the unknown is converted to H_2O, which is absorbed in the first chamber, and all the C is converted to CO_2, which is absorbed in the second. By weighing the absorbers before and after combustion, we find the masses of CO_2 and H_2O and use them to calculate the masses of C and H in the compound, from which we find the empirical formula.

Many organic compounds also contain at least one other element, such as oxygen, nitrogen, or a halogen. As long as the third element doesn't interfere with the absorption of CO_2 and H_2O, we calculate its mass by subtracting the masses of C and H from the original mass of the compound.

SAMPLE PROBLEM 3.6 Determining a Molecular Formula from Combustion Analysis

Problem Vitamin C ($\mathcal{M}$ = 176.12 g/mol) is a compound of C, H, and O found in many natural sources, especially citrus fruits. When a 1.000-g sample of vitamin C is placed in a combustion chamber and burned, the following data are obtained:

$$\text{Mass of } CO_2 \text{ absorber after combustion} = 85.35 \text{ g}$$
$$\text{Mass of } CO_2 \text{ absorber before combustion} = 83.85 \text{ g}$$
$$\text{Mass of } H_2O \text{ absorber after combustion} = 37.96 \text{ g}$$
$$\text{Mass of } H_2O \text{ absorber before combustion} = 37.55 \text{ g}$$

What is the molecular formula of vitamin C?

Plan We find the masses of CO_2 and H_2O by subtracting the masses of the absorbers before the reaction from the masses after. From the mass of CO_2, we use the mass fraction of C in CO_2 to find the mass of C (see Comment in Sample Problem 3.3). Similarly, we find the mass of H from the mass of H_2O. The mass of vitamin C (1.000 g) minus the sum of the C and H masses gives the mass of O, the third element present. Then, we proceed as in Sample Problem 3.5: calculate numbers of moles using the elements' molar masses, construct the empirical formula, determine the whole-number multiple from the given molar mass, and construct the molecular formula.

Solution Finding the masses of combustion products:

$$\begin{aligned}\text{Mass (g) of } CO_2 &= \text{mass of } CO_2 \text{ absorber after} - \text{mass before}\\ &= 85.35 \text{ g} - 83.85 \text{ g} = 1.50 \text{ g } CO_2\\ \text{Mass (g) of } H_2O &= \text{mass of } H_2O \text{ absorber after} - \text{mass before}\\ &= 37.96 \text{ g} - 37.55 \text{ g} = 0.41 \text{ g } H_2O\end{aligned}$$

Calculating masses of C and H using their mass fractions:

$$\text{Mass of element} = \text{mass of compound} \times \frac{\text{mass of element in compound}}{\text{mass of 1 mol of compound}}$$

$$\text{Mass (g) of C} = \text{mass of } CO_2 \times \frac{1 \text{ mol C} \times \mathcal{M} \text{ of C}}{\text{mass of 1 mol } CO_2} = 1.50 \cancel{\text{g } CO_2} \times \frac{12.01 \text{ g C}}{44.01 \cancel{\text{g } CO_2}}$$
$$= 0.409 \text{ g C}$$

$$\text{Mass (g) of H} = \text{mass of } H_2O \times \frac{2 \text{ mol H} \times \mathcal{M} \text{ of H}}{\text{mass of 1 mol } H_2O} = 0.41 \cancel{\text{g } H_2O} \times \frac{2.016 \text{ g H}}{18.02 \cancel{\text{g } H_2O}}$$
$$= 0.046 \text{ g H}$$

Calculating the mass of O:

$$\text{Mass (g) of O} = \text{mass of vitamin C sample} - (\text{mass of C} + \text{mass of H})$$
$$= 1.000 \text{ g} - (0.409 \text{ g} + 0.046 \text{ g}) = 0.545 \text{ g O}$$

Finding the amounts (mol) of elements: Dividing the mass in grams of each element by its molar mass gives 0.0341 mol of C, 0.046 mol of H, and 0.0341 mol of O.
Constructing the preliminary formula: $C_{0.0341}H_{0.046}O_{0.0341}$
Determining the empirical formula: Dividing through by the smallest subscript gives

$$C_{\frac{0.0341}{0.0341}}H_{\frac{0.046}{0.0341}}O_{\frac{0.0341}{0.0341}} = C_{1.00}H_{1.3}O_{1.00}$$

By trial and error, we find that 3 is the smallest integer that will make all subscripts approximately into integers:

$$C_{(1.00\times3)}H_{(1.3\times3)}O_{(1.00\times3)} = C_{3.00}H_{3.9}O_{3.00} \approx C_3H_4O_3$$

Determining the molecular formula:

$$\text{Whole-number multiple} = \frac{\mathcal{M} \text{ of vitamin C}}{\mathcal{M} \text{ of empirical formula}} = \frac{176.12 \cancel{\text{g/mol}}}{88.06 \cancel{\text{g/mol}}} = 2.000 = 2$$
$$C_{(3\times2)}H_{(4\times2)}O_{(3\times2)} = C_6H_8O_6$$

Check The element masses seem correct: carbon makes up slightly more than 0.25 of the mass of CO_2 (12 g/44 g > 0.25), as do the masses in the problem (0.409 g/1.50 g > 0.25). Hydrogen makes up slightly more than 0.10 of the mass of H_2O (2 g/18 g > 0.10), as do the masses in the problem (0.046 g/0.41 g > 0.10). The molecular formula has the same ratio of subscripts (6/8/6) as the empirical formula (3/4/3) and adds up to the given molar mass:

$$(6 \times \mathcal{M} \text{ of C}) + (8 \times \mathcal{M} \text{ of H}) + (6 \times \mathcal{M} \text{ of O}) = \mathcal{M} \text{ of vitamin C}$$
$$(6 \times 12.01) + (8 \times 1.008) + (6 \times 16.00) = 176.12 \text{ g/mol}$$

Comment In determining the subscript for H, if we string the calculation steps together, we obtain the subscript 4.0, rather than 3.9, and don't need to round:

$$\text{Subscript of H} = 0.41 \text{ g } H_2O \times \frac{2.016 \text{ g H}}{18.02 \text{ g } H_2O} \times \frac{1 \text{ mol H}}{1.008 \text{ g H}} \times \frac{1}{0.0341 \text{ mol}} \times 3 = 4.0$$

FOLLOW-UP PROBLEM 3.6 A dry-cleaning solvent ($\mathcal{M}$ = 146.99 g/mol) that contains C, H, and Cl is suspected to be a cancer-causing agent. When a 0.250-g sample was studied by combustion analysis, 0.451 g of CO_2 and 0.0617 g of H_2O formed. Find the molecular formula.

Isomers A molecular formula tells the *actual* number of each type of atom, providing as much information as possible from mass analysis. Yet *different compounds can have the **same** molecular formula* because the atoms can bond to each other in different arrangements to give more than one *structural formula.* **Isomers** are two or more compounds with the same molecular formula but different properties. The simplest type of isomerism, called *constitutional,* or *structural, isomerism,* occurs when the atoms link together in different arrangements. The pair of constitutional isomers shown in Table 3.2 (next page) share the molecular formula

Table 3.2 Constitutional Isomers of C_2H_6O

Property	Ethanol	Dimethyl Ether
$\mathcal{M}$ (g/mol)	46.07	46.07
Boiling point	78.5°C	−25°C
Density (at 20°C)	0.789 g/mL (liquid)	0.00195 g/mL (gas)
Structural formula	H H H—C—C—O—H H H	H H H—C—O—C—H H H
Space-filling model		

C_2H_6O but have very different properties because they are different compounds. In this case, they are even different *types* of compounds—one is an alcohol, and the other an ether.

As the number and kinds of atoms increase, the number of constitutional isomers—that is, the number of structural formulas that can be written for a given molecular formula—also increases: C_2H_6O has the two that you've seen, C_3H_8O has three, and $C_4H_{10}O$, seven. (We'll discuss this and other types of isomerism later in the text.)

SECTION 3.2 SUMMARY

From the masses of elements in an unknown compound, the relative amounts (in moles) are found and the empirical formula determined. • If the molar mass is known, the molecular formula can also be determined. • Methods such as combustion analysis provide data on the masses of elements in a compound, which are used to obtain the formula. • Because atoms can bond in different arrangements, more than one compound may have the same molecular formula (constitutional isomers).

3.3 WRITING AND BALANCING CHEMICAL EQUATIONS

Perhaps the most important reason for thinking in terms of moles is because it greatly clarifies the amounts of substances taking part in a reaction. Comparing masses doesn't tell the ratio of substances reacting but comparing numbers of moles does. It allows us to view substances as large populations of interacting particles rather than as grams of material. To clarify this idea, consider the formation of hydrogen fluoride gas from H_2 and F_2, a reaction that occurs explosively at room temperature. If we weigh the gases, we find that

2.016 g of H_2 and 38.00 g of F_2 react to form 40.02 g of HF

This information tells us little except that mass is conserved. However, if we convert these masses (in grams) to amounts (in moles), we find that

1 mol of H_2 and 1 mol of F_2 react to form 2 mol of HF

This information reveals that equal-size populations of H_2 and F_2 molecules combine to form twice as large a population of HF molecules. Dividing through by Avogadro's number shows us the chemical event that occurs between individual molecules:

1 H_2 molecule and 1 F_2 molecule react to form 2 HF molecules

Figure 3.6 shows that when we express the reaction in terms of moles, *the macroscopic (molar) change corresponds to the submicroscopic (molecular) change.* As you'll see, a balanced chemical equation shows both changes.

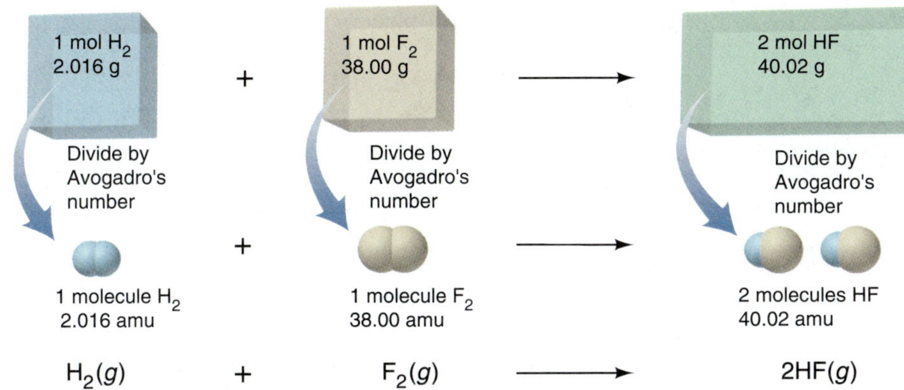

FIGURE 3.6 The formation of HF gas on the macroscopic and molecular levels. When 1 mol of H_2 (2.016 g) and 1 mol of F_2 (38.00 g) react, 2 mol of HF (40.02 g) forms. Dividing by Avogadro's number shows the change at the molecular level.

A **chemical equation** is a statement in formulas that expresses the identities and quantities of the substances involved in a chemical or physical change. Equations are the "sentences" of chemistry, just as chemical formulas are the "words" and atomic symbols the "letters." The left side of an equation shows the amount of each substance present before the change, and the right side shows the amounts present afterward. *For an equation to depict these amounts accurately, it must be balanced; that is, the same number of each type of atom must appear on both sides of the equation.* This requirement follows directly from the mass laws and the atomic theory:

- In a chemical process, atoms cannot be created, destroyed, or changed, only rearranged into different combinations.
- A formula represents a fixed ratio of the elements in a compound, so a different ratio represents a different compound.

Consider the chemical change that occurs in an old-fashioned photographic flashbulb, in many fireworks, and in a common lecture demo: a magnesium strip burns in oxygen gas to yield powdery magnesium oxide. (Light and heat are produced as well, but we're only concerned with the substances involved.) Let's convert this chemical statement into a balanced equation through the following steps:

1. *Translating the statement.* We first translate the chemical statement into a "skeleton" equation: chemical formulas arranged in an equation format. All the substances that react during the change, called **reactants,** are placed to the left of a "yield" arrow, which points to all the substances produced, called **products:**

$$\underbrace{\_\_Mg \quad + \quad \_\_O_2}_{\text{reactants}} \xrightarrow{\text{yield}} \underset{\text{product}}{\_\_MgO}$$

magnesium and oxygen yield magnesium oxide

At the beginning of the balancing process, we put a blank in front of each substance to remind us that we have to account for its atoms.

2. *Balancing the atoms.* The next step involves shifting our attention back and forth from right to left in order to *match the number of each type of atom on each side.* At the end of this step, each blank will contain a **balancing (stoichiometric) coefficient,** a numerical multiplier of *all the atoms* in the formula that follows it. In general, balancing is easiest when we

- Start with the most complex substance, the one with the largest number of atoms or different types of atoms.
- End with the least complex substance, such as an element by itself.

In this case, MgO is the most complex, so we place a coefficient 1 *in front of* the compound:

$$\_\_Mg + \_\_O_2 \longrightarrow \underline{1}\,MgO$$

To balance the Mg in MgO on the right, we place a 1 in front of Mg on the left:

$$\underline{1}\,Mg + \underline{}\,O_2 \longrightarrow \underline{1}\,MgO$$

The O atom on the right must be balanced by one O atom on the left. One-half an O_2 molecule provides one O atom:

$$\underline{1}\,Mg + \underline{\tfrac{1}{2}}\,O_2 \longrightarrow \underline{1}\,MgO$$

In terms of number and type of atom, the equation is balanced.

3. *Adjusting the coefficients.* There are several conventions about the final form of the coefficients:

- In most cases, *the smallest whole-number coefficients are preferred.* Whole numbers allow entities such as O_2 molecules to be treated as intact particles. One-half of an O_2 molecule cannot exist, so we multiply the equation by 2:

$$2Mg + 1O_2 \longrightarrow 2MgO$$

- We used the coefficient 1 to remind us to balance each substance. In the final form, a coefficient of 1 is implied just by the presence of the formula of the substance, so we don't need to write it:

$$2Mg + O_2 \longrightarrow 2MgO$$

(This convention is similar to not writing a subscript 1 in a formula.)

4. *Checking.* After balancing and adjusting the coefficients, always check that the equation is balanced:

$$\text{Reactants (2 Mg, 2 O)} \longrightarrow \text{products (2 Mg, 2 O)}$$

5. *Specifying the states of matter.* The final equation also indicates the physical state of each substance or whether it is dissolved in water. The abbreviations that are used for these states are solid (*s*), liquid (*l*), gas (*g*), and aqueous solution (*aq*). From the original statement, we know that the Mg strip is solid, O_2 is a gas, and powdery MgO is also solid. The balanced equation, therefore, is

$$2Mg(s) + O_2(g) \longrightarrow 2MgO(s)$$

Of course, the key point to realize is, as was pointed out in Figure 3.6, *the balancing coefficients refer to both individual chemical entities and moles of chemical entities.* Thus, 2 mol of Mg and 1 mol of O_2 yield 2 mol of MgO. Figure 3.7 shows this reaction from three points of view—as you see it on the macroscopic level, as chemists (and you!) can imagine it on the atomic level (darker colored atoms represent the stoichiometry), and on the symbolic level of the chemical equation.

Keep in mind these other key points about the balancing process:

- A coefficient operates on *all the atoms in the formula* that follows it: 2MgO means 2 × (MgO), or 2 Mg atoms and 2 O atoms; $2Ca(NO_3)_2$ means 2 × $[Ca(NO_3)_2]$, or 2 Ca atoms, 4 N atoms, and 12 O atoms.
- In balancing an equation, *chemical formulas cannot be altered.* In step 2 of the example, we *cannot* balance the O atoms by changing MgO to MgO_2 because MgO_2 has a different elemental composition and thus is a different compound.
- We *cannot add other reactants or products* to balance the equation because this would represent a different reaction. For example, we *cannot* balance the O atoms by changing O_2 to O or by adding one O atom to the products, because the chemical statement does not say that the reaction involves O atoms.

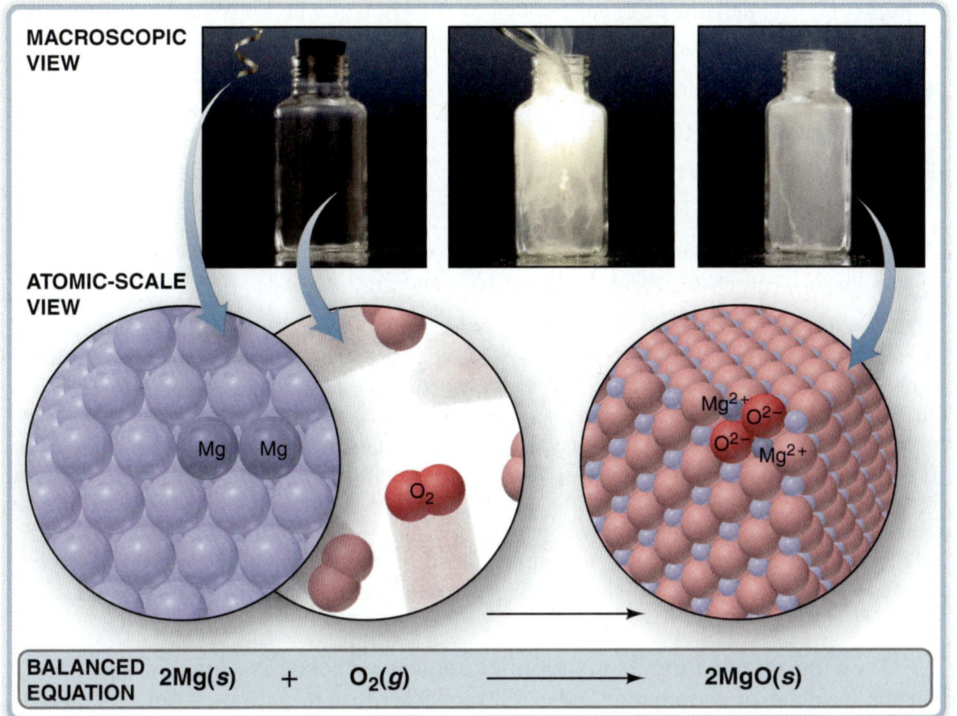

FIGURE 3.7 A three-level view of the reaction between magnesium and oxygen. The photos present the macroscopic view that you see in the lab. Before the reaction occurs, a piece of magnesium ribbon is held above a container of oxygen gas *(left)*. After the reaction, white, powdery magnesium oxide coats the container's inner surface *(right)*. The blow-up arrows lead to an atomic-scale view, a representation of the chemist's mental picture of the reaction. The darker colored spheres show the stoichiometry. By knowing the substances before and after a reaction, we can write a balanced equation *(bottom),* the chemist's symbolic shorthand for the change.

- A balanced equation remains balanced even if you multiply all the coefficients by the same number. For example,

$$4\text{Mg}(s) + 2\text{O}_2(g) \longrightarrow 4\text{MgO}(s)$$

is also balanced: it is just the balanced equation we obtained above, multiplied by 2. However, we balance an equation with the *smallest* whole-number coefficients.

SAMPLE PROBLEM 3.7 Balancing Chemical Equations

Problem Within the cylinders of a car's engine, the hydrocarbon octane (C_8H_{18}), one of many components of gasoline, mixes with oxygen from the air and burns to form carbon dioxide and water vapor. Write a balanced equation for this reaction.

Solution

1. *Translate* the statement into a skeleton equation (with coefficient blanks). Octane and oxygen are reactants; "oxygen from the air" implies molecular oxygen, O_2. Carbon dioxide and water vapor are products:

$$\_\text{C}_8\text{H}_{18} + \_\text{O}_2 \longrightarrow \_\text{CO}_2 + \_\text{H}_2\text{O}$$

2. *Balance the atoms.* We start with the most complex substance, C_8H_{18}, and balance O_2 last:

$$\underline{1}\,\text{C}_8\text{H}_{18} + \_\text{O}_2 \longrightarrow \_\text{CO}_2 + \_\text{H}_2\text{O}$$

The C atoms in C_8H_{18} end up in CO_2. Each CO_2 contains one C atom, so 8 molecules of CO_2 are needed to balance the 8 C atoms in each C_8H_{18}:

$$\underline{1}\,\text{C}_8\text{H}_{18} + \_\text{O}_2 \longrightarrow \underline{8}\,\text{CO}_2 + \_\text{H}_2\text{O}$$

The H atoms in C_8H_{18} end up in H_2O. The 18 H atoms in C_8H_{18} require the coefficient 9 in front of H_2O:

$$\underline{1}\,\text{C}_8\text{H}_{18} + \_\text{O}_2 \longrightarrow \underline{8}\,\text{CO}_2 + \underline{9}\,\text{H}_2\text{O}$$

There are 25 atoms of O on the right (16 in $8CO_2$ plus 9 in $9H_2O$), so we place the coefficient $\frac{25}{2}$ in front of O_2:

$$\underline{1}\,\text{C}_8\text{H}_{18} + \underline{\tfrac{25}{2}}\,\text{O}_2 \longrightarrow \underline{8}\,\text{CO}_2 + \underline{9}\,\text{H}_2\text{O}$$

3. *Adjust the coefficients.* Multiply through by 2 to obtain whole numbers:

$$2C_8H_{18} + 25O_2 \longrightarrow 16CO_2 + 18H_2O$$

4. *Check* that the equation is balanced:

$$\text{Reactants (16 C, 36 H, 50 O)} \longrightarrow \text{products (16 C, 36 H, 50 O)}$$

5. *Specify* states of matter. C_8H_{18} is liquid; O_2, CO_2, and H_2O vapor are gases:

$$2C_8H_{18}(l) + 25O_2(g) \longrightarrow 16CO_2(g) + 18H_2O(g)$$

Comment This is an example of a combustion reaction. *Any* compound containing C and H that burns in an excess of air produces CO_2 and H_2O.

FOLLOW-UP PROBLEM 3.7 Write a balanced equation for each chemical statement:
(a) A characteristic reaction of Group 1A(1) elements: chunks of sodium react violently with water to form hydrogen gas and sodium hydroxide solution.
(b) The destruction of marble statuary by acid rain: aqueous nitric acid reacts with calcium carbonate to form carbon dioxide, water, and aqueous calcium nitrate.
(c) Halogen compounds exchanging bonding partners: phosphorus trifluoride is prepared by the reaction of phosphorus trichloride and hydrogen fluoride; hydrogen chloride is the other product. The reaction involves gases only.
(d) Explosive decomposition of dynamite: liquid nitroglycerine ($C_3H_5N_3O_9$) explodes to produce a mixture of gases—carbon dioxide, water vapor, nitrogen, and oxygen.

Viewing an equation as a simplified molecular scene is a great way to focus on the essence of the change—the rearrangement of the atoms from reactants to products. Here's a simple representation of the combustion of octane:

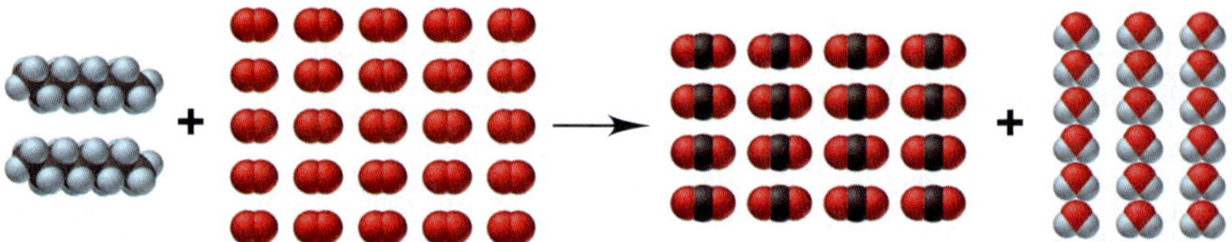

Now, let's work through a sample problem to do the reverse—derive a balanced equation from a molecular scene.

SAMPLE PROBLEM 3.8 Balancing an Equation from a Molecular Depiction

Problem The following molecular scene depicts an important reaction in nitrogen chemistry (nitrogen is blue; oxygen is red):

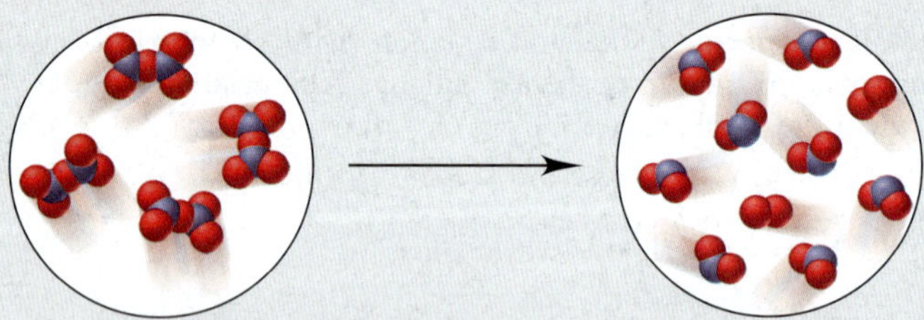

Write a balanced equation for this reaction.

Plan To write a balanced equation from the depiction, we first have to determine the formulas of the molecules and obtain coefficients by counting the number of each molecule. Then, we arrange this information into the correct equation format, using the smallest whole-number coefficients and including states of matter.

Solution The reactant circle shows only one type of molecule. It has two N and five O atoms, so the formula is N_2O_5; there are four of these molecules. The product circle shows two different molecules, one with one N and two O atoms, and the other with two O atoms; there are eight NO_2 and two O_2. Thus, we have:

$$4N_2O_5 \longrightarrow 8NO_2 + 2O_2$$

Writing the balanced equation with the smallest whole-number coefficients and all substances as gases:

$$2N_2O_5(g) \longrightarrow 4NO_2(g) + O_2(g)$$

Check Reactant (4 N, 10 O) $\longrightarrow$ products (4 N, 8 + 2 = 10 O)

FOLLOW-UP PROBLEM 3.8 Write a balanced equation for the important atmospheric reaction depicted below (carbon is black; oxygen is red):

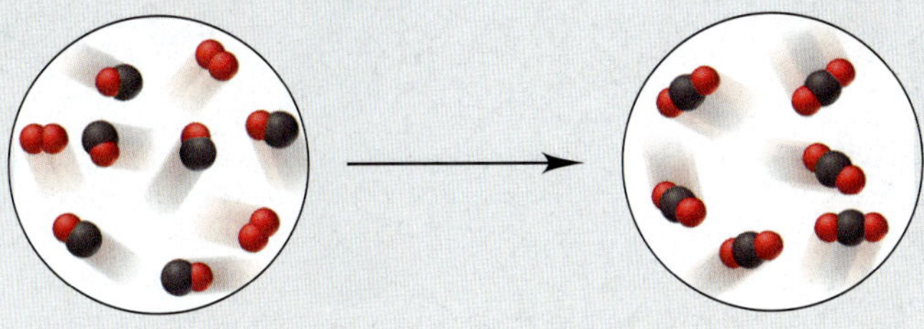

SECTION 3.3 SUMMARY

To conserve mass and maintain the fixed composition of compounds, a chemical equation must be balanced in terms of number and type of each atom. • A balanced equation has reactant formulas on the left of a yield arrow and product formulas on the right. • Balancing coefficients are integer multipliers for *all* the atoms in a formula and apply to the individual entity or to moles of entities.

3.4 CALCULATING AMOUNTS OF REACTANT AND PRODUCT

A balanced equation contains a wealth of quantitative information relating individual chemical entities, amounts of chemical entities, and masses of substances. It is essential for all calculations involving amounts of reactants and products: *if you know the number of moles of one substance, the balanced equation for the reaction tells you the number of moles of all the others.*

Stoichiometrically Equivalent Molar Ratios from the Balanced Equation

In a balanced equation, *the number of moles of one substance is stoichiometrically equivalent to the number of moles of any of the other substances.* The term *stoichiometrically equivalent* means that a definite amount of one substance is formed from, produces, or reacts with a definite amount of the other, and these quantitative relationships are expressed as *stoichiometrically equivalent molar ratios.* As you'll see shortly, we use these ratios as conversion factors to calculate the amounts. For example, let's consider the equation for the combustion of propane, a hydrocarbon fuel used in cooking and water heating:

$$C_3H_8(g) + 5O_2(g) \longrightarrow 3CO_2(g) + 4H_2O(g)$$

If we view the reaction quantitatively in terms of C_3H_8, we see that

1 mol of C_3H_8 reacts with 5 mol of O_2
1 mol of C_3H_8 produces 3 mol of CO_2
1 mol of C_3H_8 produces 4 mol of H_2O

Therefore, in this reaction,

1 mol of C_3H_8 is stoichiometrically equivalent to 5 mol of O_2
1 mol of C_3H_8 is stoichiometrically equivalent to 3 mol of CO_2
1 mol of C_3H_8 is stoichiometrically equivalent to 4 mol of H_2O

Table 3.3 Information Contained in a Balanced Equation

Viewed in Terms of	Reactants $C_3H_8(g) + 5O_2(g)$		Products $3CO_2(g) + 4H_2O(g)$
Molecules	1 molecule C_3H_8 + 5 molecules O_2	$\longrightarrow$	3 molecules CO_2 + 4 molecules H_2O
	+	$\longrightarrow$	+
Amount (mol)	1 mol C_3H_8 + 5 mol O_2	$\longrightarrow$	3 mol CO_2 + 4 mol H_2O
Mass (amu)	44.09 amu C_3H_8 + 160.00 amu O_2	$\longrightarrow$	132.03 amu CO_2 + 72.06 amu H_2O
Mass (g)	44.09 g C_3H_8 + 160.00 g O_2	$\longrightarrow$	132.03 g CO_2 + 72.06 g H_2O
Total mass (g)	204.09 g	$\longrightarrow$	204.09 g

We chose to look at C_3H_8, but any two of the substances are stoichiometrically equivalent to each other. Thus,

3 mol of CO_2 is stoichiometrically equivalent to 4 mol of H_2O
5 mol of O_2 is stoichiometrically equivalent to 3 mol of CO_2

and so on. Table 3.3 presents various ways to view the quantitative information contained in this equation.

Here's a typical problem that shows how stoichiometric equivalence is used to create conversion factors: in the combustion of propane, how many moles of O_2 are consumed when 10.0 mol of H_2O are produced? To solve this problem, we have to find the molar ratio between O_2 and H_2O. From the balanced equation, we see that for every 5 mol of O_2 consumed, 4 mol of H_2O is formed:

5 mol of O_2 is stoichiometrically equivalent to 4 mol of H_2O

We can construct two conversion factors from this equivalence, depending on the quantity we want to find:

$$\frac{5\ \text{mol}\ O_2}{4\ \text{mol}\ H_2O} \quad \text{or} \quad \frac{4\ \text{mol}\ H_2O}{5\ \text{mol}\ O_2}$$

Since we want to find moles of O_2 and we are given moles of H_2O, we choose "5 mol O_2/4 mol H_2O" to cancel "mol H_2O":

$$\text{Moles of } O_2 \text{ consumed} = 10.0\ \cancel{\text{mol}\ H_2O} \times \frac{5\ \text{mol}\ O_2}{4\ \cancel{\text{mol}\ H_2O}} = 12.5\ \text{mol}\ O_2$$

mol H_2O $\Longrightarrow$ (molar ratio as conversion factor) mol O_2

Obviously, we could not have solved this problem without the balanced equation. Here is a general approach for solving *any* stoichiometry problem that involves a chemical reaction:

1. Write a balanced equation for the reaction.
2. Convert the given mass (or number of entities) of the first substance to amount (mol).

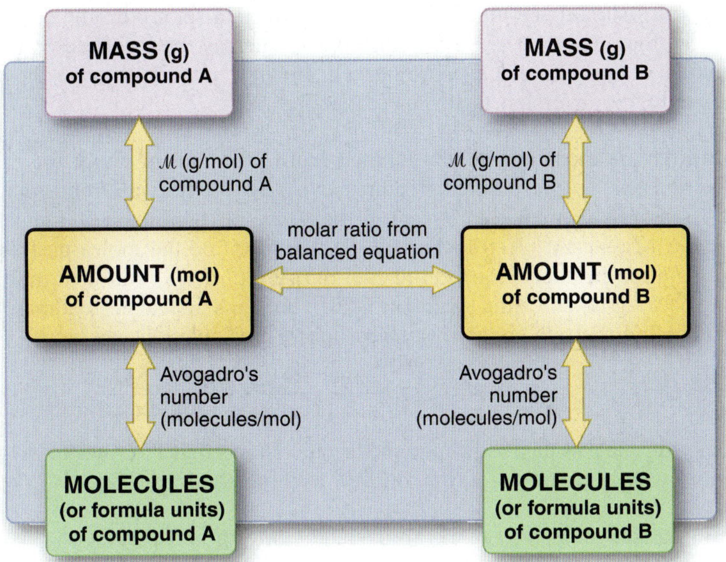

FIGURE 3.8 Summary of the mass-mole-number relationships in a chemical reaction. The amount of one substance in a reaction is related to that of any other. Quantities are expressed in terms of grams, moles, or number of entities (atoms, molecules, or formula units). Start at any box in the diagram (known) and move to any other box (unknown) by using the information on the arrows as conversion factors. As an example, if you know the mass (in g) of A and want to know the number of molecules of B, the path involves three calculation steps:
1. Grams of A to moles of A, using the molar mass ($\mathcal{M}$) of A
2. Moles of A to moles of B, using the molar ratio from the balanced equation
3. Moles of B to molecules of B, using Avogadro's number

Steps 1 and 3 refer to calculations discussed in Section 3.1 (see Figure 3.4).

3. Use the appropriate molar ratio from the balanced equation to calculate the amount (mol) of the second substance.
4. Convert the amount of the second substance to the desired mass (or number of entities).

Figure 3.8 summarizes the various relationships, and Sample Problem 3.9 applies three of them in an industrial setting.

SAMPLE PROBLEM 3.9 Calculating Amounts of Reactants and Products

Problem In a lifetime, the average American uses about 1750 lb (794 kg) of copper in coins, plumbing, and wiring. Copper is obtained from sulfide ores, such as chalcocite, or copper(I) sulfide, by a multistep process. After an initial grinding, the first step is to "roast" the ore (heat it strongly with oxygen gas) to form powdered copper(I) oxide and gaseous sulfur dioxide.

(a) How many moles of oxygen are required to roast 10.0 mol of copper(I) sulfide?

(b) How many grams of sulfur dioxide are formed when 10.0 mol of copper(I) sulfide is roasted?

(c) How many kilograms of oxygen are required to form 2.86 kg of copper(I) oxide?

(a) Determining the moles of O_2 needed to roast 10.0 mol of Cu_2S

Plan We *always* write the balanced equation first. The formulas of the reactants are Cu_2S and O_2, and the formulas of the products are Cu_2O and SO_2, so we have

$$2Cu_2S(s) + 3O_2(g) \longrightarrow 2Cu_2O(s) + 2SO_2(g)$$

We are given the *moles* of Cu_2S (10.0 mol) and need to find the *moles* of O_2. The balanced equation shows that 3 mol of O_2 is needed for every 2 mol of Cu_2S consumed. To cancel "mol Cu_2S," the conversion factor we construct is "3 mol O_2/2 mol Cu_2S" (see roadmap a).

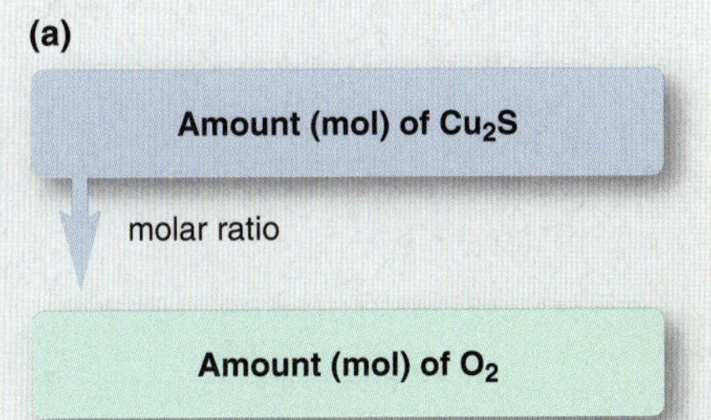

Solution Calculating number of moles of O_2:

$$\text{Moles of } O_2 = 10.0\ \cancel{\text{mol } Cu_2S} \times \frac{3 \text{ mol } O_2}{2\ \cancel{\text{mol } Cu_2S}} = 15.0 \text{ mol } O_2$$

Check The units are correct, and the answer is reasonable because this O_2/Cu_2S molar ratio (15/10) is equivalent to the ratio in the balanced equation (3/2).

Comment A *common mistake* is to use the incorrect conversion factor; the calculation would then be

$$\text{Moles of } O_2 = 10.0 \text{ mol } Cu_2S \times \frac{2 \text{ mol } Cu_2S}{3 \text{ mol } O_2} = \frac{6.67 \text{ mol}^2\ Cu_2S}{1 \text{ mol } O_2}$$

Such strange units should signal that you made an error in setting up the conversion factor. In addition, the answer, 6.67, is *less* than 10.0, whereas the balanced equation shows that *more* moles of O_2 than of Cu_2S are needed. Be sure to think through the calculation when setting up the conversion factor and canceling units.

(b)

Amount (mol) of Cu_2S

↓ molar ratio

Amount (mol) of SO_2

↓ multiply by $\mathcal{M}$ (g/mol)

Mass (g) of SO_2

(b) Determining the mass (g) of SO_2 formed from 10.0 mol of Cu_2S

Plan The second part of the problem requires two steps to convert from amount of one substance to mass of another. Here we need the *grams* of product (SO_2) that form from the given *moles* of reactant (10.0 mol of Cu_2S). We first find the moles of SO_2 using the molar ratio from the balanced equation (2 mol SO_2/2 mol Cu_2S) and then multiply by its molar mass (64.07 g/mol) to find grams of SO_2. The steps appear in roadmap b.

Solution Combining the two conversion steps into one calculation, we have

$$\text{Mass (g) of } SO_2 = 10.0\ \cancel{\text{mol } Cu_2S} \times \frac{2\ \cancel{\text{mol } SO_2}}{2\ \cancel{\text{mol } Cu_2S}} \times \frac{64.07\ \text{g } SO_2}{1\ \cancel{\text{mol } SO_2}} = \boxed{641\ \text{g } SO_2}$$

Check The answer makes sense, since the molar ratio shows that 10.0 mol of SO_2 is formed and each mole weighs about 64 g. We rounded to three significant figures.

(c)

Mass (kg) of Cu_2O

↓ $1\ \text{kg} = 10^3\ \text{g}$

Mass (g) of Cu_2O

↓ divide by $\mathcal{M}$ (g/mol)

Amount (mol) of Cu_2O

↓ molar ratio

Amount (mol) of O_2

↓ multiply by $\mathcal{M}$ (g/mol)

Mass (g) of O_2

↓ $10^3\ \text{g} = 1\ \text{kg}$

Mass (kg) of O_2

(c) Determining the mass (kg) of O_2 that yields 2.86 kg of Cu_2O

Plan In the final part, aside from unit conversions, we need three steps to convert from mass of one substance to mass of another. Here the mass of product (2.86 kg of Cu_2O) is known, and we need the mass of reactant (O_2) that reacts to form it. We first convert the quantity of Cu_2O from *kilograms* to *moles* (in two steps, as shown in roadmap c). Then, we use the molar ratio (3 mol O_2/2 mol Cu_2O) to find the *moles* of O_2 required. Finally, we convert *moles* of O_2 to *kilograms* (in two steps).

Solution Converting from kilograms of Cu_2O to moles of Cu_2O: Combining the mass unit conversion with the mass-to-mole conversion gives

$$\text{Moles of } Cu_2O = 2.86\ \cancel{\text{kg } Cu_2O} \times \frac{10^3\ \cancel{\text{g}}}{1\ \cancel{\text{kg}}} \times \frac{1\ \text{mol } Cu_2O}{143.10\ \cancel{\text{g } Cu_2O}} = 20.0\ \text{mol } Cu_2O$$

Converting from moles of Cu_2O to moles of O_2:

$$\text{Moles of } O_2 = 20.0\ \cancel{\text{mol } Cu_2O} \times \frac{3\ \text{mol } O_2}{2\ \cancel{\text{mol } Cu_2O}} = 30.0\ \text{mol } O_2$$

Converting from moles of O_2 to kilograms of O_2: Combining the mole-to-mass conversion with the mass unit conversion gives

$$\text{Mass (kg) of } O_2 = 30.0\ \cancel{\text{mol } O_2} \times \frac{32.00\ \cancel{\text{g } O_2}}{1\ \cancel{\text{mol } O_2}} \times \frac{1\ \text{kg}}{10^3\ \cancel{\text{g}}} = \boxed{0.960\ \text{kg } O_2}$$

Check The units are correct. Round off to check the math: for example, in the final step, $\sim 30\ \text{mol} \times 30\ \text{g/mol} \times 1\ \text{kg}/10^3\ \text{g} = 0.90\ \text{kg}$. The answer seems reasonable: even though the amount (mol) of O_2 is greater than the amount (mol) of Cu_2O, the mass of O_2 is less than the mass of Cu_2O because $\mathcal{M}$ of O_2 is less than $\mathcal{M}$ of Cu_2O.

Comment This problem highlights a key point for solving stoichiometry problems: *convert the information given into moles.* Then, use the appropriate molar ratio and any other conversion factors to complete the solution.

FOLLOW-UP PROBLEM 3.9 Thermite is a mixture of iron(III) oxide and aluminum powders that was once used to weld railroad tracks. It undergoes a spectacular reaction to yield solid aluminum oxide and molten iron.

(a) How many grams of iron form when 135 g of aluminum reacts?

(b) How many atoms of aluminum react for every 1.00 g of aluminum oxide formed?

Chemical Reactions That Involve a Limiting Reactant

In the problems we've considered up to now, the amount of *one* reactant was given, and we assumed there was enough of any other reactant for the first reactant to be completely used up. For example, to find the amount of SO_2 that forms when 100 g of Cu_2S reacts, we convert the grams of Cu_2S to moles and assume that the Cu_2S reacts with as much O_2 as needed. Because all the Cu_2S is used up, its initial amount determines, or limits, how much SO_2 can form. We call Cu_2S

the **limiting reactant** (or *limiting reagent*) because the product stops forming once the Cu_2S is gone, no matter how much O_2 is present.

Animation: Limiting Reagent

Suppose, however, that the amounts of Cu_2S *and* O_2 are given in the problem, and we need to find how much SO_2 forms. We first have to determine whether Cu_2S *or* O_2 is the limiting reactant (that is, which one is completely used up) because the amount of that reactant limits how much SO_2 can form. The other reactant is *in excess,* and whatever amount of it is not used is left over.

To clarify the idea of a limiting reactant, let's consider a situation from real life. A car assembly plant has 1500 car bodies and 4000 tires. How many cars can be made with the supplies on hand? Does the plant manager need to order more car bodies or more tires? Obviously, 4 tires are required for each car body, so the "balanced equation" is

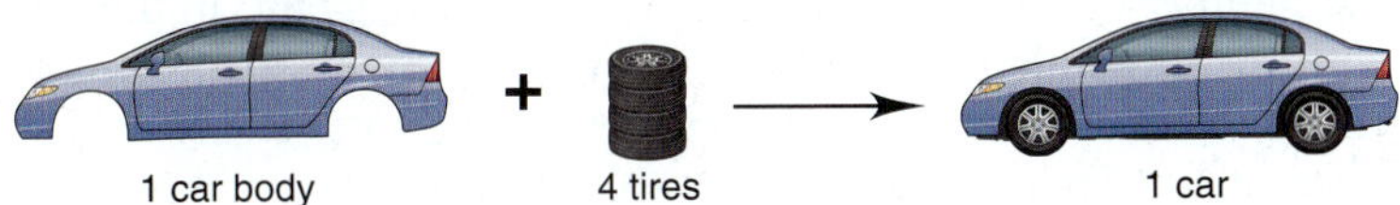

How much "product" (cars) can we make from the amount of each "reactant"?

$$1500 \text{ car bodies} \times \frac{1 \text{ car}}{1 \text{ car body}} = 1500 \text{ cars}$$

$$4000 \text{ tires} \times \frac{1 \text{ car}}{4 \text{ tires}} = 1000 \text{ cars}$$

The number of tires limits the number of cars because less "product" (fewer cars) can be produced from the available tires. There will be $1500 - 1000 = 500$ car bodies in excess, and they cannot be turned into cars until more tires are delivered.

A good way to keep track of the quantities in a limiting-reactant problem is with a *reaction table.* It shows the initial amounts of reactants and products, the changes in their amounts due to the reaction, and their final amounts. For example, for the car assembly "reaction," the reaction table is

Quantity	1 car body	+ 4 tires	⟶ 1 car
Initial	1500	4000	0
Change	−1000	−4000	+1000
Final	500	0	1000

At the top is the balanced equation, which provides column heads for the table. The first line of the table shows the initial amounts of reactants and products before the "reaction" starts. No "product" has yet formed, which is indicated in the table by a zero in the car column. The next line shows the changes in reactants and products as a result of the "reaction." Notice that since car bodies and tires were used to make the cars, the amounts of reactants decreased and their changes have a negative sign, while the amount of product increased so its change has a positive sign. We add the changes to the initial amounts to obtain the bottom line, the final amounts after the reaction is over. Now we can see that some car bodies are in excess, and the tires, the limiting reactant, have been used up.

THINK OF IT THIS WAY
Limiting Reactants in Everyday Life

In addition to an industrial setting, such as a car assembly plant, limiting-"reactant" situations arise in daily life all the time. A muffin recipe calls for 2 cups of flour and 1 cup of sugar, but you have 3 cups of flour and only $\frac{3}{4}$ cup of sugar. Clearly, the flour is in excess, and the sugar limits the number of muffins you can make. Or, you're making cheeseburgers for a picnic, and you have 10 buns, 12 meat patties, and 8 slices of cheese. Here, the cheese limits the number of cheeseburgers you can make. Or, there are 16 students in a cell biology lab but only 13 microscopes. The number of times limiting-reactant situations arise is almost limitless.

Now let's apply these ideas to solving chemical problems. In limiting-reactant problems, the amounts of two (or more) reactants are given, and we must first determine which is limiting. To do this, just as we did with the cars, we first note how much of each reactant *should* be present to completely use up the other, and then we compare it with the amount that is *actually* present. Simply put, the limiting reactant is the one there is not enough of; that is, *it is the reactant that limits the amount of the other reactant that can react, and thus the amount of product that can form.* In mathematical terms, *the limiting reactant is the one that yields the* ***lower*** *amount of product.*

We'll examine limiting reactants in the following two sample problems. Sample Problem 3.10 has two parts, and in both we have to identify the limiting reactant. In the first part, we look at a simple molecular view of a reaction and compare the number of molecules to find the limiting reactant; in the second part, we start with the amounts (mol) of two reactants and perform two calculations, each of which assumes an excess of one of the reactants, to see which reactant forms less product. Then, in Sample Problem 3.11, we go through a similar process but start with the masses of the two reactants.

SAMPLE PROBLEM 3.10 Using Molecular Depictions to Solve a Limiting-Reactant Problem

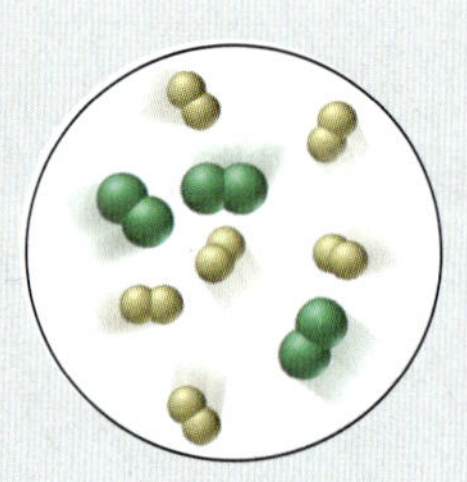

Problem Nuclear engineers use chlorine trifluoride in the processing of uranium fuel for power plants. This extremely reactive substance is formed as a gas in special metal containers by the reaction of elemental chlorine and fluorine.

(a) Suppose the circle shown at left represents a container of the reactant mixture before the reaction occurs (with chlorine colored green). Name the limiting reactant and draw the container contents after the reaction is complete.

(b) When the reaction is run again with 0.750 mol of Cl_2 and 3.00 mol of F_2, what mass of chlorine trifluoride will be prepared?

(a) Determining the limiting reactant and drawing the container contents

Plan We first write the balanced equation. From its name, we know that chlorine trifluoride consists of one Cl atom bonded to three F atoms, ClF_3. Elemental chlorine and fluorine refer to the diatomic molecules Cl_2 and F_2. All the substances are gases. To find the limiting reactant, we compare the number of molecules we have of each reactant, with the number we need for the other to react completely. The limiting reactant limits the amount of the other reactant that can react and the amount of product that will form.

Solution The balanced equation is

$$Cl_2(g) + 3F_2(g) \longrightarrow 2ClF_3(g)$$

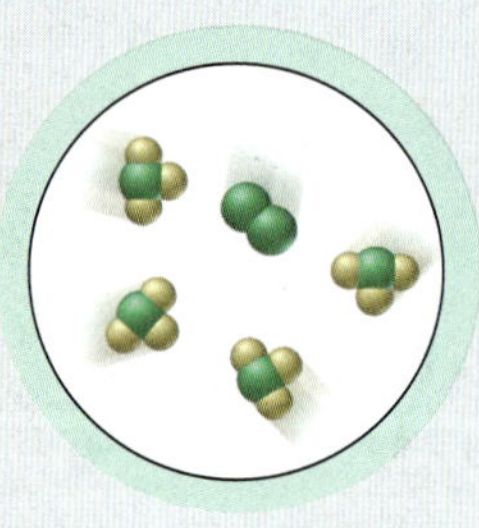

The equation shows that two ClF_3 molecules are formed for every one Cl_2 molecule and three F_2 molecules that react. Before the reaction, there are three Cl_2 molecules (six Cl atoms). For all the Cl_2 to react, we need three times three, or nine, F_2 molecules (18 F atoms). But there are only six F_2 molecules (12 F atoms). Therefore, F_2 is the limiting reactant because it limits the amount of Cl_2 that can react, and thus the amount of ClF_3 that can form. After the reaction, as the circle at left depicts, all 12 F atoms and four of the six Cl atoms make four ClF_3 molecules, and one Cl_2 molecule remains in excess.

Check The equation is balanced: reactants (2 Cl, 6 F) $\longrightarrow$ products (2 Cl, 6 F), and, in the circles, the number of each type of atom before the reaction equals the number after the reaction. You can check the choice of limiting reactant by examining the reaction from the perspective of Cl_2: Two Cl_2 molecules are enough to react with the six F_2 molecules in the container. But there are three Cl_2 molecules, so there is not enough F_2.

(b) Calculating the mass of ClF_3 formed

Plan We first determine the limiting reactant by using the molar ratios from the balanced equation to convert the moles of each reactant to moles of ClF_3 formed, assuming an excess of the other reactant. Whichever reactant forms fewer moles of ClF_3 is the limiting reactant. Then we use the molar mass of ClF_3 to convert this lower number of moles to grams.

Solution Determining the limiting reactant:
Finding moles of ClF_3 from moles of Cl_2 (assuming F_2 is in excess):

$$\text{Moles of } ClF_3 = 0.750 \cancel{\text{mol } Cl_2} \times \frac{2 \text{ mol } ClF_3}{1 \cancel{\text{mol } Cl_2}} = 1.50 \text{ mol } ClF_3$$

Finding moles of ClF_3 from moles of F_2 (assuming Cl_2 is in excess):

$$\text{Moles of } ClF_3 = 3.00 \cancel{\text{mol } F_2} \times \frac{2 \text{ mol } ClF_3}{3 \cancel{\text{mol } F_2}} = 2.00 \text{ mol } ClF_3$$

In this experiment, Cl_2 is limiting because it forms fewer moles of ClF_3.
Calculating grams of ClF_3 formed:

$$\text{Mass (g) of } ClF_3 = 1.50 \cancel{\text{mol } ClF_3} \times \frac{92.45 \text{ g } ClF_3}{1 \cancel{\text{mol } ClF_3}} = 139 \text{ g } ClF_3$$

Check Let's check our reasoning that Cl_2 is the limiting reactant by assuming, for the moment, that F_2 is limiting. In that case, all 3.00 mol of F_2 would react to form 2.00 mol of ClF_3. Based on the balanced equation, however, that amount of product would require that 1.00 mol of Cl_2 reacted. But that is impossible because only 0.750 mol of Cl_2 is present.
Comment Note that a reactant can be limiting even though it is present in the greater amount. It is the *reactant molar ratio in the balanced equation* that is the determining factor. In both parts (a) and (b), F_2 is present in greater amount than Cl_2. However, in (a), the F_2/Cl_2 ratio is 6/3, or 2/1, which is less than the required molar ratio of 3/1, so F_2 is limiting; in (b), the F_2/Cl_2 ratio is 3.00/0.750, greater than the required 3/1, so F_2 is in excess. When we write a reaction table for part (b), this fact is revealed clearly:

Amount (mol)	$Cl_2(g)$	+	$3F_2(g)$	$\longrightarrow$	$2ClF_3(g)$
Initial	0.750		3.00		0
Change	−0.750		−2.25		+1.50
Final	0		0.75		1.50

FOLLOW-UP PROBLEM 3.10 B_2 (red spheres) reacts with AB as shown below:

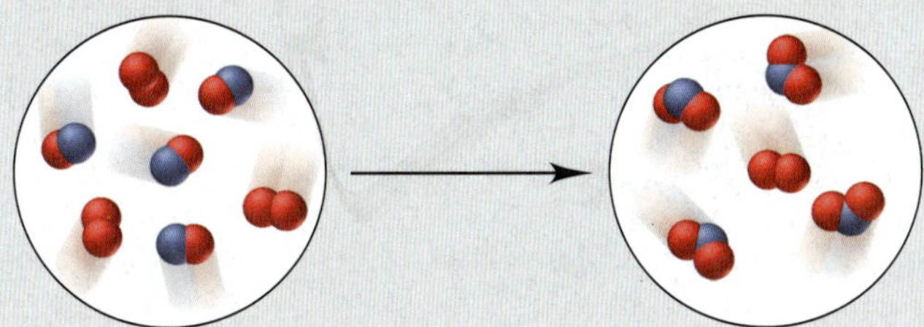

(a) Write a balanced equation for the reaction, and determine the limiting reactant.
(b) How many moles of product can form from the reaction of 1.5 mol of each reactant?

SAMPLE PROBLEM 3.11 Calculating Amounts of Reactant and Product in a Limiting-Reactant Problem

Problem A fuel mixture used in the early days of rocketry is composed of two liquids, hydrazine (N_2H_4) and dinitrogen tetraoxide (N_2O_4), which ignite on contact to form nitrogen gas and water vapor. How many grams of nitrogen gas form when 1.00×10^2 g of N_2H_4 and 2.00×10^2 g of N_2O_4 are mixed?
Plan We first write the balanced equation. *Because the amounts of two reactants are given, we know this is a limiting-reactant problem.* To determine which reactant is limiting, we calculate the mass of N_2 formed from each reactant *assuming an excess of the other.* We convert the grams of each reactant to moles and use the appropriate molar ratio to find the moles of N_2 each forms. Whichever yields *less* N_2 is the limiting reactant. Then, we convert this lower number of moles of N_2 to mass. The roadmap shows the steps.

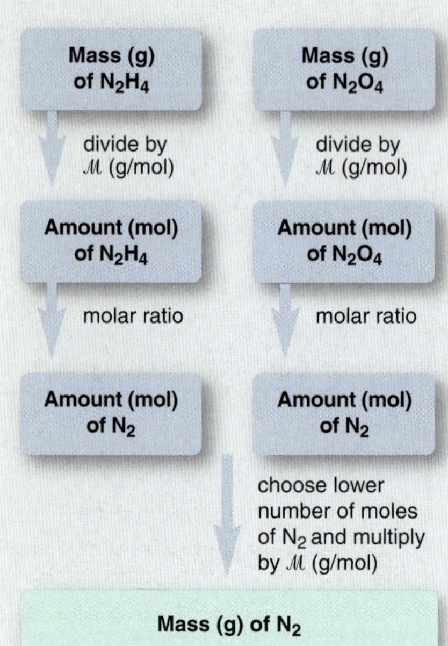

Solution Writing the balanced equation:

$$2N_2H_4(l) + N_2O_4(l) \longrightarrow 3N_2(g) + 4H_2O(g)$$

Finding the moles of N_2 from the moles of N_2H_4 (if N_2H_4 is limiting):

$$\text{Moles of } N_2H_4 = 1.00\times10^2 \text{ g } N_2H_4 \times \frac{1 \text{ mol } N_2H_4}{32.05 \text{ g } N_2H_4} = 3.12 \text{ mol } N_2H_4$$

$$\text{Moles of } N_2 = 3.12 \text{ mol } N_2H_4 \times \frac{3 \text{ mol } N_2}{2 \text{ mol } N_2H_4} = \mathbf{4.68 \text{ mol } N_2}$$

Finding the moles of N_2 from the moles of N_2O_4 (if N_2O_4 is limiting):

$$\text{Moles of } N_2O_4 = 2.00\times10^2 \text{ g } N_2O_4 \times \frac{1 \text{ mol } N_2O_4}{92.02 \text{ g } N_2O_4} = 2.17 \text{ mol } N_2O_4$$

$$\text{Moles of } N_2 = 2.17 \text{ mol } N_2O_4 \times \frac{3 \text{ mol } N_2}{1 \text{ mol } N_2O_4} = \mathbf{6.51 \text{ mol } N_2}$$

Thus, N_2H_4 is the limiting reactant because it yields fewer moles of N_2. Converting from moles of N_2 to grams:

$$\text{Mass (g) of } N_2 = 4.68 \text{ mol } N_2 \times \frac{28.02 \text{ g } N_2}{1 \text{ mol } N_2} = 131 \text{ g } N_2$$

Check The mass of N_2O_4 is greater than that of N_2H_4, but there are fewer moles of N_2O_4 because its $\mathcal{M}$ is much higher. Round off to check the math: for N_2H_4, 100 g N_2H_4 × 1 mol/32 g ≈ 3 mol; ~3 mol × $\frac{3}{2}$ ≈ 4.5 mol N_2; ~4.5 mol × 30 g/mol ≈ 135 g N_2.

Comment 1. Here are two *common mistakes* in solving limiting-reactant problems:

- The limiting reactant is not the *reactant* present in fewer moles (2.17 mol of N_2O_4 vs. 3.12 mol of N_2H_4). Rather, it is the reactant that forms fewer moles of *product.*
- Similarly, the limiting reactant is not the *reactant* present in lower mass. Rather, it is the reactant that forms the lower mass of *product.*

2. Here is an *alternative approach* to finding the limiting reactant. Find the moles of each reactant that would be needed to react with the other reactant. Then see which amount actually given in the problem is sufficient. That substance is in excess, and the other substance is limiting. For example, the balanced equation shows that 2 mol of N_2H_4 reacts with 1 mol of N_2O_4. The moles of N_2O_4 needed to react with the given moles of N_2H_4 are

$$\text{Moles of } N_2O_4 \text{ needed} = 3.12 \text{ mol } N_2H_4 \times \frac{1 \text{ mol } N_2O_4}{2 \text{ mol } N_2H_4} = 1.56 \text{ mol } N_2O_4$$

The moles of N_2H_4 needed to react with the given moles of N_2O_4 are

$$\text{Moles of } N_2H_4 \text{ needed} = 2.17 \text{ mol } N_2O_4 \times \frac{2 \text{ mol } N_2H_4}{1 \text{ mol } N_2O_4} = 4.34 \text{ mol } N_2H_4$$

We are given 2.17 mol of N_2O_4, which is *more* than the amount of N_2O_4 needed (1.56 mol) to react with the given amount of N_2H_4, and we are given 3.12 mol of N_2H_4, which is *less* than the amount of N_2H_4 needed (4.34 mol) to react with the given amount of N_2O_4. Therefore, N_2H_4 is limiting, and N_2O_4 is in excess. Once we determine this, we continue with the final calculation to find the amount of N_2.

3. Once again, a reaction table reveals the amounts of all the reactants and products before and after the reaction:

Amount (mol)	$2N_2H_4(l)$	+	$N_2O_4(l)$	⟶	$3N_2(g)$	+	$4H_2O(g)$
Initial	3.12		2.17		0		0
Change	−3.12		−1.56		+4.68		+6.24
Final	0		0.61		4.68		6.24

FOLLOW-UP PROBLEM 3.11 How many grams of solid aluminum sulfide can be prepared by the reaction of 10.0 g of aluminum and 15.0 g of sulfur? How much of the nonlimiting reactant is in excess?

Chemical Reactions in Practice: Theoretical, Actual, and Percent Yields

Up until now, we've been optimistic about the amount of product obtained from a reaction. We have assumed that 100% of the limiting reactant becomes product, that ideal separation and purification methods exist for isolating the product, and that we use perfect lab technique to collect all the product formed. In other words, we have assumed that we obtain the **theoretical yield,** the amount indicated by the stoichiometrically equivalent molar ratio in the balanced equation.

It's time to face reality. The theoretical yield is *never* obtained, for reasons that are largely uncontrollable. For one thing, although the major reaction predominates, many reactant mixtures also proceed through one or more **side reactions** that form smaller amounts of different products (Figure 3.9). In the rocket fuel reaction in Sample Problem 3.11, for example, the reactants might form some NO in the following side reaction:

$$N_2H_4(l) + 2N_2O_4(l) \longrightarrow 6NO(g) + 2H_2O(g)$$

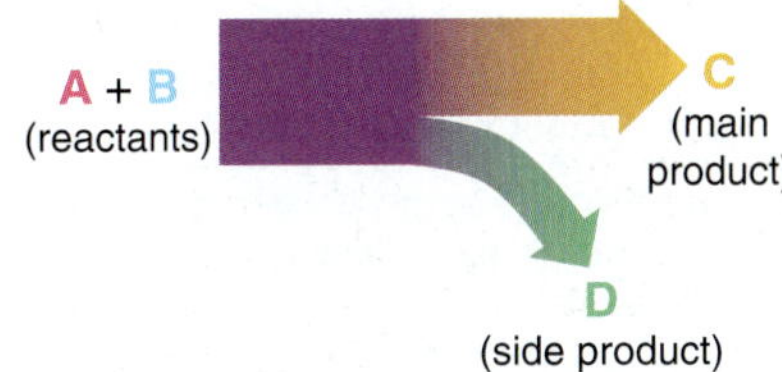

FIGURE 3.9 The effect of side reactions on yield. One reason the theoretical yield is never obtained is that other reactions lead some of the reactants along side paths to form undesired products.

This reaction decreases the amounts of reactants available for N_2 production (see Problem 3.84 at the end of the chapter). Even more important, as we'll discuss in later chapters, many reactions seem to stop before they are complete, which leaves some limiting reactant unused. But, even when a reaction does go completely to product, losses occur in virtually every step of the separation procedure used to isolate the product from the reaction mixture. With careful technique, you can minimize these losses but never eliminate them.

The amount of product that you actually obtain is the **actual yield.** Theoretical and actual yields are expressed in units of amount (moles) or mass (grams). The **percent yield (% yield)** is the actual yield expressed as a percentage of the theoretical yield:

$$\%\ \text{yield} = \frac{\text{actual yield}}{\text{theoretical yield}} \times 100 \qquad (3.7)$$

Because the actual yield *must* be less than the theoretical yield, the percent yield is *always* less than 100%. In multistep reaction sequences, the percent yield of each step is expressed as a fraction and multiplied by the others to find the overall yield. The result may sometimes be surprising. For example, suppose a six-step reaction sequence has a 90.0% yield for each step, which is quite high. Even so, the overall percent yield would be

$$\text{Overall}\ \%\ \text{yield} = (0.900 \times 0.900 \times 0.900 \times 0.900 \times 0.900 \times 0.900) \times 100 = 53.1\%$$

SAMPLE PROBLEM 3.12 Calculating Percent Yield

Problem Silicon carbide (SiC) is an important ceramic material that is made by allowing sand (silicon dioxide, SiO_2) to react with powdered carbon at high temperature. Carbon monoxide is also formed. When 100.0 kg of sand is processed, 51.4 kg of SiC is recovered. What is the percent yield of SiC from this process?

Plan We are given the actual yield of SiC (51.4 kg), so we need the theoretical yield to calculate the percent yield. After writing the balanced equation, we convert the given mass of SiO_2 (100.0 kg) to amount (mol). We use the molar ratio to find the amount of SiC formed and convert that amount to mass (kg) to obtain the theoretical yield [see Sample Problem 3.9(c)]. Then, we use Equation 3.7 to find the percent yield (see the roadmap).

Solution Writing the balanced equation:

$$SiO_2(s) + 3C(s) \longrightarrow SiC(s) + 2CO(g)$$

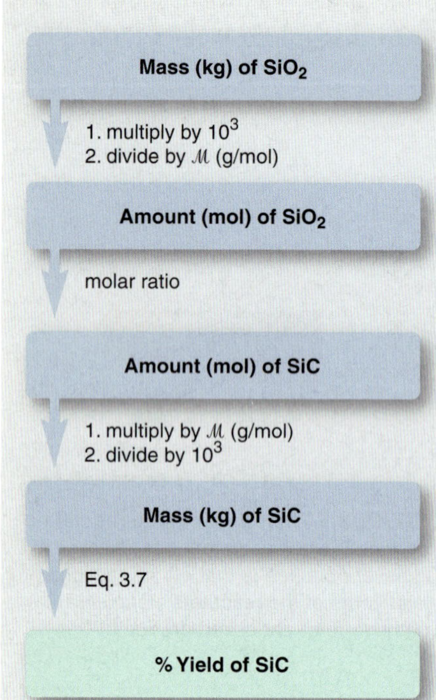

Converting from kilograms of SiO_2 to moles:

$$\text{Moles of SiO}_2 = 100.0\ \cancel{\text{kg SiO}_2} \times \frac{1000\ \cancel{\text{g}}}{1\ \cancel{\text{kg}}} \times \frac{1\ \text{mol SiO}_2}{60.09\ \cancel{\text{g SiO}_2}} = 1664\ \text{mol SiO}_2$$

Converting from moles of SiO_2 to moles of SiC: The molar ratio is 1 mol SiC/1 mol SiO_2, so

$$\text{Moles of SiO}_2 = \text{moles of SiC} = 1664\ \text{mol SiC}$$

Converting from moles of SiC to kilograms:

$$\text{Mass (kg) of SiC} = 1664\ \cancel{\text{mol SiC}} \times \frac{40.10\ \text{g SiC}}{1\ \cancel{\text{mol SiC}}} \times \frac{1\ \text{kg}}{1000\ \text{g}} = 66.73\ \text{kg SiC}$$

Calculating the percent yield:

$$\%\ \text{yield of SiC} = \frac{\text{actual yield}}{\text{theoretical yield}} \times 100 = \frac{51.4\ \cancel{\text{kg}}}{66.73\ \cancel{\text{kg}}} \times 100 = 77.0\%$$

Check Rounding shows that the mass of SiC seems correct: $\sim$1500 mol $\times$ 40 g/mol $\times$ 1 kg/1000 g = 60 kg. The molar ratio of SiC/SiO_2 is 1/1, and the $\mathcal{M}$ of SiC is about two-thirds ($\sim\frac{40}{60}$) the $\mathcal{M}$ of SiO_2, so 100 kg of SiO_2 should form about 66 kg of SiC.

FOLLOW-UP PROBLEM 3.12 Marble (calcium carbonate) reacts with hydrochloric acid solution to form calcium chloride solution, water, and carbon dioxide. What is the percent yield of carbon dioxide if 3.65 g of the gas is collected when 10.0 g of marble reacts?

SECTION 3.4 SUMMARY

The substances in a balanced equation are related to each other by stoichiometrically equivalent molar ratios, which are used as conversion factors to find the moles of one substance given the moles of another. • In limiting-reactant problems, the amounts of two (or more) reactants are given, and one of them limits the amount of product that forms. The limiting reactant is the one that forms the lower amount of product. • In practice, side reactions, incomplete reactions, and physical losses result in an actual yield of product that is less than the theoretical yield, the amount based on the molar ratio. The percent yield is the actual yield expressed as a percentage of the theoretical yield. In multistep reaction sequences, the overall yield is found by multiplying the percent yields for each step.

3.5 FUNDAMENTALS OF SOLUTION STOICHIOMETRY

Many environmental reactions and almost all biochemical reactions occur in solution, so an understanding of reactions in solution is extremely important in chemistry and related sciences. We'll discuss solution chemistry at many places in the text, but here we focus on solution stoichiometry. Only one aspect of the stoichiometry of dissolved substances is different from what we've seen so far. We know the amounts of pure substances by converting their masses directly into moles. For dissolved substances, we must know the *concentration*—the number of moles present in a certain volume of solution—to find the volume that contains a given number of moles. Of the various ways to express concentration, the most important is *molarity,* so we discuss it here (and wait until Chapter 13 to discuss the other ways). Then, we see how to prepare a solution of a specific molarity and how to use solutions in stoichiometric calculations.

Expressing Concentration in Terms of Molarity

A typical solution consists of a smaller amount of one substance, the **solute,** dissolved in a larger amount of another substance, the **solvent.** When a solution forms, the solute's individual chemical entities become evenly dispersed throughout the available volume and surrounded by solvent molecules. The **concentration** of a solution is usually expressed as *the amount of solute dissolved in a given amount of solution.* Concentration is an *intensive* quantity (like density or temperature) and thus independent of the volume of solution: a 50-L tank of a given solution has the *same concentration* (solute amount/solution amount) as a 50-mL beaker of the solution. **Molarity (*M*)** expresses the concentration in units of *moles of solute per liter of solution:*

$$\text{Molarity} = \frac{\text{moles of solute}}{\text{liters of solution}} \quad \text{or} \quad M = \frac{\text{mol solute}}{\text{L soln}} \tag{3.8}$$

SAMPLE PROBLEM 3.13 Calculating the Molarity of a Solution

Problem Glycine (H_2NCH_2COOH) is the simplest amino acid. What is the molarity of an aqueous solution that contains 0.715 mol of glycine in 495 mL?

Plan The molarity is the number of moles of solute in each liter of solution. We are given the number of moles (0.715 mol) and the volume (495 mL), so we divide moles by volume and convert the volume to liters to find the molarity (see the roadmap).

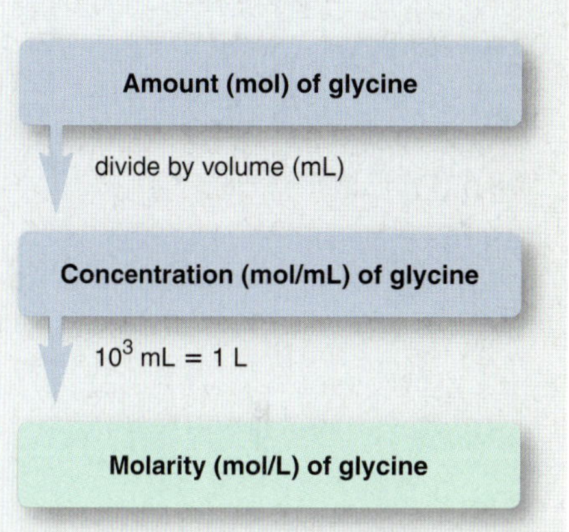

Solution

$$\text{Molarity} = \frac{0.715 \text{ mol glycine}}{495 \cancel{\text{mL soln}}} \times \frac{1000 \cancel{\text{mL}}}{1 \text{ L}} = 1.44 \ M \text{ glycine}$$

Check A quick look at the math shows about 0.7 mol of glycine in about 0.5 L of solution, so the concentration should be about 1.4 mol/L, or 1.4 *M*.

FOLLOW-UP PROBLEM 3.13 How many moles of KI are in 84 mL of 0.50 *M* KI?

Mole-Mass-Number Conversions Involving Solutions

Molarity can be thought of as a conversion factor used to convert between volume of solution and amount (mol) of solute, from which we then find the mass or the number of entities of solute. Figure 3.10 shows this new stoichiometric relationship, and Sample Problem 3.14 applies it.

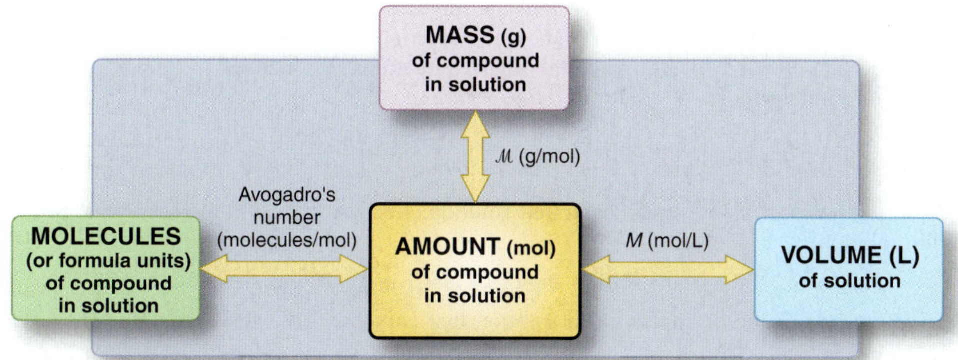

FIGURE 3.10 Summary of mass-mole-number-volume relationships in solution. The amount (in moles) of a compound in solution is related to the volume of solution in liters through the molarity (*M*) in moles per liter. The other relationships shown are identical to those in Figure 3.4, except that here they refer to the quantities *in solution*. As in previous cases, to find the quantity of substance expressed in one form or another, convert the given information to moles first.

SAMPLE PROBLEM 3.14 Calculating Mass of Solute in a Given Volume of Solution

Problem A *buffered* solution maintains acidity as a reaction occurs. In living cells, phosphate ions play a key buffering role, so biochemists often study reactions in such solutions. How many grams of solute are in 1.75 L of 0.460 *M* sodium monohydrogen phosphate?

Plan We know the solution volume (1.75 L) and molarity (0.460 *M*), and we need the mass of solute. We use the known quantities to find the amount (mol) of solute and then convert moles to grams with the solute molar mass, as shown in the roadmap.

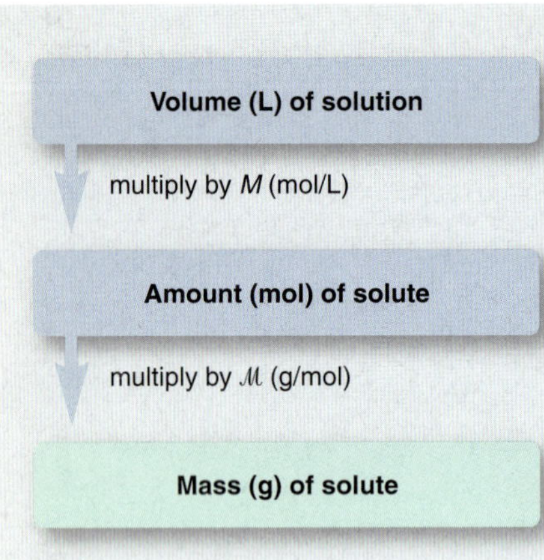

Solution Calculating moles of solute in solution:

$$\text{Moles of Na}_2\text{HPO}_4 = 1.75\ \text{L soln} \times \frac{0.460\ \text{mol Na}_2\text{HPO}_4}{1\ \text{L soln}} = 0.805\ \text{mol Na}_2\text{HPO}_4$$

Converting from moles of solute to grams:

$$\text{Mass (g) Na}_2\text{HPO}_4 = 0.805\ \text{mol Na}_2\text{HPO}_4 \times \frac{141.96\ \text{g Na}_2\text{HPO}_4}{1\ \text{mol Na}_2\text{HPO}_4} = 114\ \text{g Na}_2\text{HPO}_4$$

Check The answer seems to be correct: ~1.8 L of 0.5 mol/L contains 0.9 mol, and 150 g/mol × 0.9 mol = 135 g, which is close to 114 g of solute.

FOLLOW-UP PROBLEM 3.14 In biochemistry laboratories, solutions of sucrose (table sugar, $C_{12}H_{22}O_{11}$) are used in high-speed centrifuges to separate the parts of a biological cell. How many liters of 3.30 *M* sucrose contain 135 g of solute?

Dilution of Solutions

A concentrated solution (higher molarity) is converted to a dilute solution (lower molarity) by *adding solvent* to it. The solution volume increases while the number of moles of solute remains the same. Thus, a given volume of the final (dilute) solution contains fewer solute particles and has a lower concentration than the initial (concentrated) solution (Figure 3.11). If various lower concentrations of a solution are needed, it is common practice to prepare a more concentrated solution (called a *stock solution*), which is stored and diluted as needed.

SAMPLE PROBLEM 3.15 Preparing a Dilute Solution from a Concentrated Solution

Problem Isotonic saline is a 0.15 *M* aqueous solution of NaCl that simulates the total concentration of ions found in many cellular fluids. Its uses range from a cleansing rinse for contact lenses to a washing medium for red blood cells. How would you prepare 0.80 L of isotonic saline from a 6.0 *M* stock solution?

Volume (L) of dilute solution
multiply by *M* (mol/L) of dilute solution
Amount (mol) of NaCl in dilute solution = Amount (mol) of NaCl in concentrated solution
divide by *M* (mol/L) of concentrated solution
Volume (L) of concentrated solution

Plan To dilute a concentrated solution, we add only solvent, so the *moles of solute are the same in both solutions.* We know the volume (0.80 L) and molarity (0.15 *M*) of the dilute (dil) NaCl solution we need, so we find the moles of NaCl it contains and then find the volume of concentrated (conc; 6.0 *M*) NaCl solution that contains the same number of moles. Then, we add solvent *up to* the final volume (see roadmap).

Solution Finding moles of solute in dilute solution:

$$\text{Moles of NaCl in dil soln} = 0.80\ \text{L soln} \times \frac{0.15\ \text{mol NaCl}}{1\ \text{L soln}} = 0.12\ \text{mol NaCl}$$

Finding moles of solute in concentrated solution: Because we add only solvent to dilute the solution,

$$\text{Moles of NaCl in dil soln} = \text{moles of NaCl in conc soln} = 0.12\ \text{mol NaCl}$$

Finding the volume of concentrated solution that contains 0.12 mol of NaCl:

$$\text{Volume (L) of conc NaCl soln} = 0.12\ \text{mol NaCl} \times \frac{1\ \text{L soln}}{6.0\ \text{mol NaCl}} = 0.020\ \text{L soln}$$

To prepare 0.80 L of dilute solution, place 0.020 L of 6.0 *M* NaCl in a 1.0-L cylinder, add distilled water (~780 mL) to the 0.80-L mark, and stir thoroughly.

Check The answer seems reasonable because a small volume of concentrated solution is used to prepare a large volume of dilute solution. Also, the ratio of volumes (0.020 L/0.80 L) is the same as the ratio of concentrations (0.15 *M*/6.0 *M*).

FOLLOW-UP PROBLEM 3.15 To prepare a fertilizer, an engineer dilutes a stock solution of sulfuric acid by adding 25.0 m^3 of 7.50 *M* acid to enough water to make 500. m^3. What is the mass (in g) of sulfuric acid per milliliter of the diluted solution?

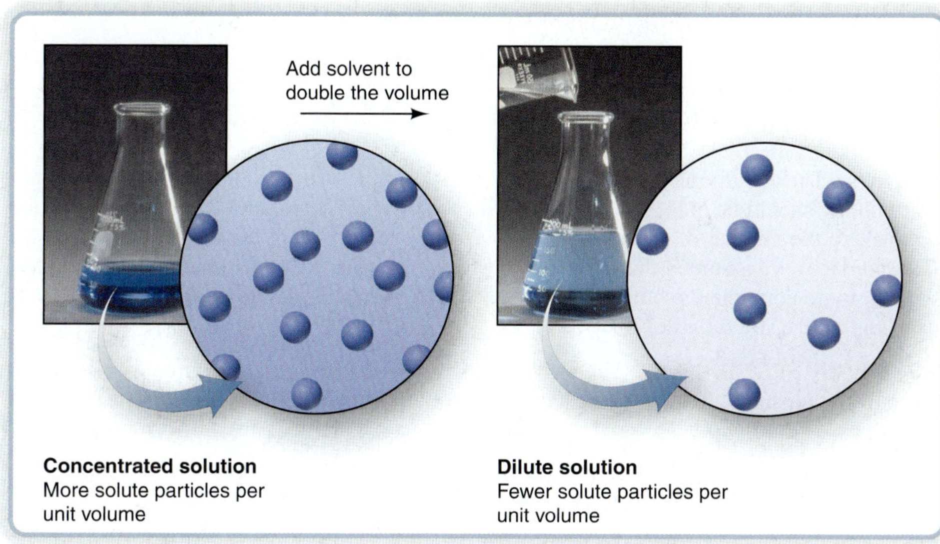

FIGURE 3.11 Converting a concentrated solution to a dilute solution. When a solution is diluted, only solvent is added. The solution volume increases while the total number of moles of solute remains the same. Therefore, as shown in the blow-up views, a unit volume of concentrated solution contains more solute particles than the same unit volume of dilute solution.

A very useful way to solve dilution problems, and others involving a change in concentration, applies the following relationship:

$$M_{\text{dil}} \times V_{\text{dil}} = \text{number of moles} = M_{\text{conc}} \times V_{\text{conc}} \qquad (3.9)$$

where the M and V terms are the molarity and volume of the *dil*ute and *conc*entrated solutions. In Sample Problem 3.15, for example, we could find the volume of concentrated solution, V_{conc}, using Equation 3.9:

$$V_{\text{conc}} = \frac{M_{\text{dil}} \times V_{\text{dil}}}{M_{\text{conc}}} = \frac{0.15\ \cancel{M} \times 0.80\ \text{L}}{6.0\ \cancel{M}} = 0.020\ \text{L}$$

The method worked out in the solution to Sample Problem 3.15 is actually the same calculation broken into two parts to emphasize the thinking process:

$$V_{\text{conc}} = 0.80\ \cancel{\text{L}} \times \frac{0.15\ \cancel{\text{mol NaCl}}}{1\ \cancel{\text{L}}} \times \frac{1\ \text{L}}{6.0\ \cancel{\text{mol NaCl}}} = 0.020\ \text{L}$$

In Sample Problem 3.16, we'll use a variation of this relationship to visualize changes in concentration.

SAMPLE PROBLEM 3.16 Visualizing Changes in Concentration

Problem The top circle at right represents a unit volume of a solution. Draw a circle representing a unit volume of the solution after each of these changes:

(a) For every 1 mL of solution, 1 mL of solvent is added.

(b) One third of the solution's total volume is boiled off.

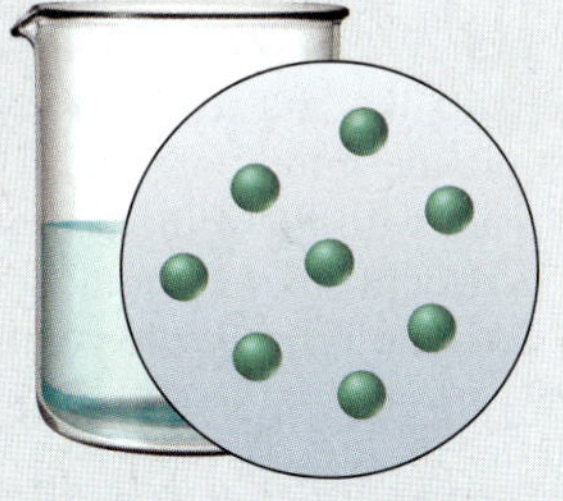

Plan Given the starting solution, we have to find the number of solute particles in a unit volume after each change. The number of particles per unit volume, N, is directly related to moles per unit volume, M, so we can use a relationship similar to Equation 3.9 to find the number of particles to show in each circle. In (a), the volume increases, so the final solution is more dilute—fewer particles per unit volume. In (b), some solvent is lost, so the final solution is more concentrated—more particles per unit volume.

Solution (a) Finding the number of particles in the dilute solution, N_{dil}:

$$N_{\text{dil}} \times V_{\text{dil}} = N_{\text{conc}} \times V_{\text{conc}}$$

thus, $$N_{\text{dil}} = N_{\text{conc}} \times \frac{V_{\text{conc}}}{V_{\text{dil}}} = 8\ \text{particles} \times \frac{1\ \text{mL}}{2\ \text{mL}} = 4\ \text{particles}$$

(a)

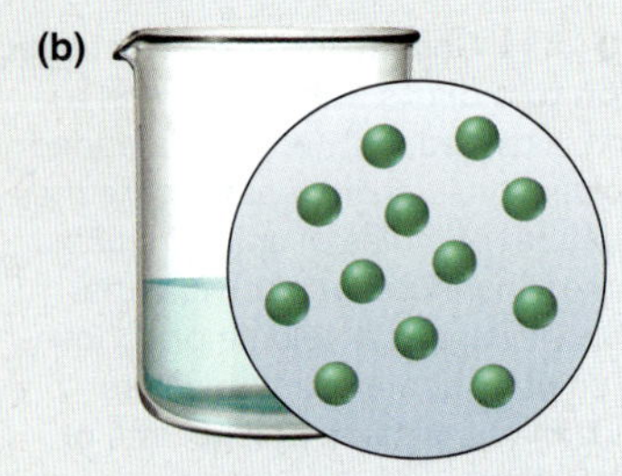

(b) Finding the number of particles in the concentrated solution, N_{conc}:

$$N_{dil} \times V_{dil} = N_{conc} \times V_{conc}$$

thus,
$$N_{conc} = N_{dil} \times \frac{V_{dil}}{V_{conc}} = 8 \text{ particles} \times \frac{1 \text{ mL}}{\frac{2}{3} \text{ mL}} = 12 \text{ particles}$$

Check (a) The volume is doubled (from 1 mL to 2 mL), so the number of particles per unit volume should be half of the original; $\frac{1}{2}$ of 8 is 4. **(b)** The volume is reduced to $\frac{2}{3}$ of the original, so the number of particles per unit volume should be $\frac{3}{2}$ of the original; $\frac{3}{2}$ of 8 is 12.

Comment (b) We assumed that only solvent boils off. This is true with nonvolatile solutes, such as ionic compounds, but in Chapter 13, we'll encounter solutions in which both solvent *and* solute are volatile.

FOLLOW-UP PROBLEM 3.16 The circle labeled A represents a unit volume of a solution. Explain the changes that must be made to A to obtain the solutions depicted in B and C.

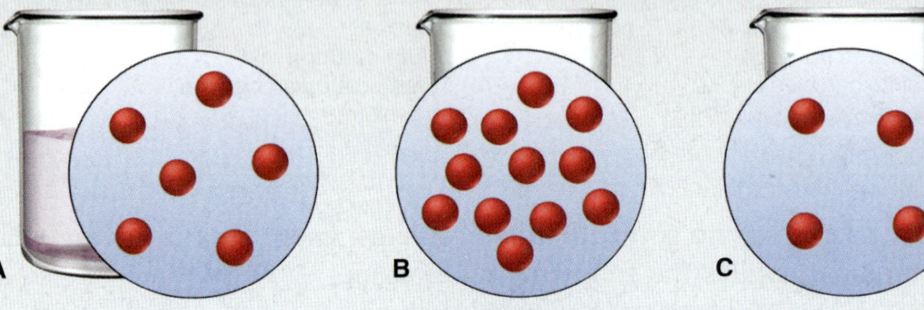

Stoichiometry of Chemical Reactions in Solution

Solving stoichiometry problems for reactions in solution requires the same approach as before, with the additional step of converting the volume of reactant or product to moles: (1) balance the equation, (2) find the number of moles of one substance, (3) relate it to the stoichiometrically equivalent number of moles of another substance, and (4) convert to the desired units.

SAMPLE PROBLEM 3.17 Calculating Amounts of Reactants and Products for a Reaction in Solution

Problem Specialized cells in the stomach release HCl to aid digestion. If they release too much, the excess can be neutralized with an antacid to avoid discomfort. A common antacid contains magnesium hydroxide, $Mg(OH)_2$, which reacts with the acid to form water and magnesium chloride solution. As a government chemist testing commercial antacids, you use 0.10 *M* HCl to simulate the acid concentration in the stomach. How many liters of "stomach acid" react with a tablet containing 0.10 g of $Mg(OH)_2$?

Plan We know the mass of $Mg(OH)_2$ (0.10 g) that reacts and the acid concentration (0.10 *M*), and we must find the acid volume. After writing the balanced equation, we convert the grams of $Mg(OH)_2$ to moles, use the molar ratio to find the moles of HCl that react with these moles of $Mg(OH)_2$, and then use the molarity of HCl to find the volume that contains this number of moles. The steps appear in the roadmap.

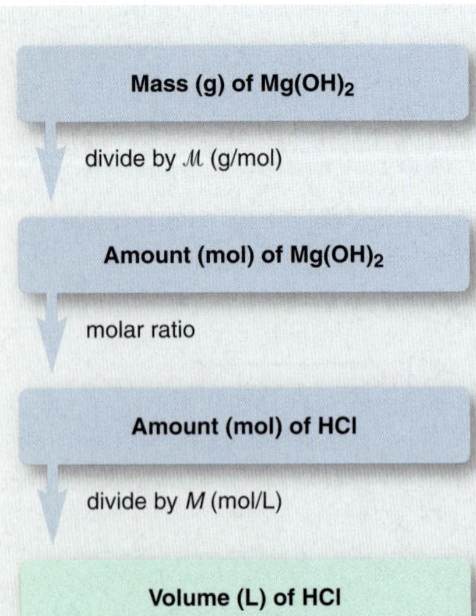

Solution Writing the balanced equation:

$$Mg(OH)_2(s) + 2HCl(aq) \longrightarrow MgCl_2(aq) + 2H_2O(l)$$

Converting from grams of $Mg(OH)_2$ to moles:

$$\text{Moles of } Mg(OH)_2 = 0.10 \cancel{\text{ g } Mg(OH)_2} \times \frac{1 \text{ mol } Mg(OH)_2}{58.33 \cancel{\text{ g } Mg(OH)_2}} = 1.7\times10^{-3} \text{ mol } Mg(OH)_2$$

Converting from moles of $Mg(OH)_2$ to moles of HCl:

$$\text{Moles of HCl} = 1.7\times10^{-3} \cancel{\text{ mol } Mg(OH)_2} \times \frac{2 \text{ mol HCl}}{1 \cancel{\text{ mol } Mg(OH)_2}} = 3.4\times10^{-3} \text{ mol HCl}$$

Converting from moles of HCl to liters:

$$\text{Volume (L) of HCl} = 3.4\times10^{-3} \cancel{\text{ mol HCl}} \times \frac{1 \text{ L}}{0.10 \cancel{\text{ mol HCl}}} = 3.4\times10^{-2} \text{ L}$$

Check The size of the answer seems reasonable: a small volume of dilute acid (0.034 L of 0.10 *M*) reacts with a small amount of antacid (0.0017 mol).

Comment The reaction as written is an oversimplification; in reality, HCl and $MgCl_2$ exist as separated ions in solution. This point will be covered in great detail in Chapters 4 and 18.

FOLLOW-UP PROBLEM 3.17 Another active ingredient in some antacids is aluminum hydroxide. Which is more effective at neutralizing stomach acid, magnesium hydroxide or aluminum hydroxide? [*Hint:* Effectiveness refers to the amount of acid that reacts with a given mass of antacid. You already know the effectiveness of 0.10 g of $Mg(OH)_2$.]

In limiting-reactant problems for reactions in solution, we determine which reactant is limiting and then determine the yield, as in Sample Problem 3.18.

SAMPLE PROBLEM 3.18 Solving Limiting-Reactant Problems for Reactions in Solution

Problem Mercury and its compounds have many uses, from fillings for teeth (as a mixture with silver, copper, and tin) to the industrial production of chlorine. Because of their toxicity, however, soluble mercury compounds, such as mercury(II) nitrate, must be removed from industrial wastewater. One removal method reacts the wastewater with sodium sulfide solution to produce solid mercury(II) sulfide and sodium nitrate solution. In a laboratory simulation, 0.050 L of 0.010 *M* mercury(II) nitrate reacts with 0.020 L of 0.10 *M* sodium sulfide. How many grams of mercury(II) sulfide form?

Plan This is a limiting-reactant problem because *the amounts of two reactants are given.* After balancing the equation, we must determine the limiting reactant. The molarity (0.010 *M*) and volume (0.050 L) of the mercury(II) nitrate solution and the molarity (0.10 *M*) and volume (0.020 L) of the sodium sulfide solution tell us the moles of the two reactants. Then, as in Sample Problem 3.11, we use the molar ratio to find the moles of HgS that form from each reactant, *assuming the other reactant is present in excess.* The limiting reactant is the one that forms fewer moles of HgS, which we convert to mass using the HgS molar mass. The roadmap shows the process.

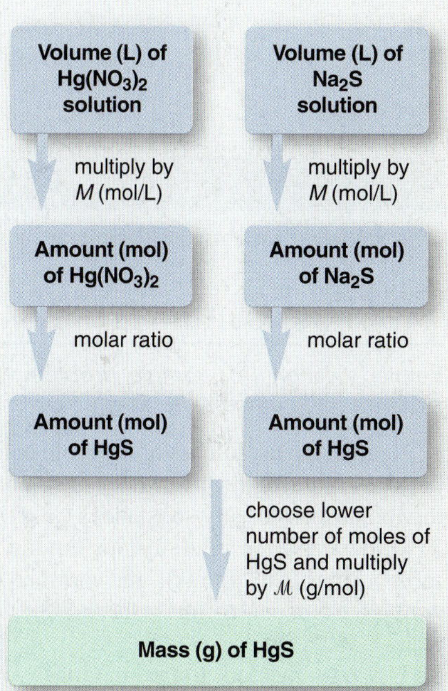

Solution Writing the balanced equation:

$$Hg(NO_3)_2(aq) + Na_2S(aq) \longrightarrow HgS(s) + 2NaNO_3(aq)$$

Finding moles of HgS assuming $Hg(NO_3)_2$ is limiting: Combining the steps gives

$$\text{Moles of HgS} = 0.050\ \cancel{\text{L soln}} \times \frac{0.010\ \cancel{\text{mol Hg(NO}_3)_2}}{1\ \cancel{\text{L soln}}} \times \frac{1\ \text{mol HgS}}{1\ \cancel{\text{mol Hg(NO}_3)_2}}$$
$$= \mathbf{5.0\times10^{-4}\ mol\ HgS}$$

Finding moles of HgS assuming Na_2S is limiting: Combining the steps gives

$$\text{Moles of HgS} = 0.020\ \cancel{\text{L soln}} \times \frac{0.10\ \cancel{\text{mol Na}_2\text{S}}}{1\ \cancel{\text{L soln}}} \times \frac{1\ \text{mol HgS}}{1\ \cancel{\text{mol Na}_2\text{S}}} = \mathbf{2.0\times10^{-3}\ mol\ HgS}$$

$Hg(NO_3)_2$ is the limiting reactant because it forms fewer moles of HgS.
Converting the moles of HgS formed from $Hg(NO_3)_2$ to grams:

$$\text{Mass (g) of HgS} = 5.0\times10^{-4}\ \cancel{\text{mol HgS}} \times \frac{232.7\ \text{g HgS}}{1\ \cancel{\text{mol HgS}}} = 0.12\ \text{g HgS}$$

Check As a check, let's use the alternative method for finding the limiting reactant (see Comment in Sample Problem 3.11). Finding moles of reactants available:

$$\text{Moles of Hg(NO}_3)_2 = 0.050\ \cancel{\text{L soln}} \times \frac{0.010\ \text{mol Hg(NO}_3)_2}{1\ \cancel{\text{L soln}}} = 5.0\times10^{-4}\ \text{mol Hg(NO}_3)_2$$

$$\text{Moles of Na}_2\text{S} = 0.020\ \cancel{\text{L soln}} \times \frac{0.10\ \text{mol Na}_2\text{S}}{1\ \cancel{\text{L soln}}} = 2.0\times10^{-3}\ \text{mol Na}_2\text{S}$$

The molar ratio of the reactants is 1 $Hg(NO_3)_2$/1 Na_2S. Therefore, $Hg(NO_3)_2$ is limiting because there are fewer moles of it than are needed to react with the moles of Na_2S.
Finding grams of product from moles of limiting reactant and the molar ratio:

$$\text{Mass (g) of HgS} = 5.0\times10^{-4}\ \cancel{\text{mol Hg(NO}_3)_2} \times \frac{1\ \cancel{\text{mol HgS}}}{1\ \cancel{\text{mol Hg(NO}_3)_2}} \times \frac{232.7\ \text{g HgS}}{1\ \cancel{\text{mol HgS}}}$$
$$= 0.12\ \text{g HgS}$$

Let's use these amounts to prepare a reaction table:

Amount (mol)	$Hg(NO_3)_2(aq)$	+	$Na_2S(aq)$	$\longrightarrow$	$HgS(s)$	+	$2NaNO_3(aq)$
Initial	5.0×10^{-4}		2.0×10^{-3}		0		0
Change	-5.0×10^{-4}		-5.0×10^{-4}		$+5.0\times10^{-4}$		$+1.0\times10^{-3}$
Final	0		1.5×10^{-3}		5.0×10^{-4}		1.0×10^{-3}

Note the large excess of Na_2S that remains after the reaction.

FOLLOW-UP PROBLEM 3.18 Even though gasoline sold in the United States no longer contains lead, this metal persists in the environment as a poison. Despite their toxicity, many compounds of lead are still used to make pigments.
(a) What volume of 1.50 *M* lead(II) acetate contains 0.400 mol of Pb^{2+} ions?
(b) When this volume reacts with 125 mL of 3.40 *M* sodium chloride, how many grams of solid lead(II) chloride can form? (Sodium acetate solution also forms.)

SECTION 3.5 SUMMARY

When reactions occur in solution, reactant and product amounts are given in terms of concentration and volume. • Molarity is the number of moles of solute dissolved in one liter of solution. • A concentrated (higher molarity) solution is converted to a dilute (lower molarity) solution by adding solvent. • Using molarity as a conversion factor, we apply the principles of stoichiometry to all aspects of reactions in solution.

CHAPTER REVIEW GUIDE

The following sections provide many aids to help you study this chapter. (Numbers in parentheses refer to pages, unless noted otherwise.)

LEARNING OBJECTIVES *These are concepts and skills to review after studying this chapter.*

Related section (§), sample problem (SP), and end-of-chapter problem (EP) numbers are listed in parentheses.

1. Realize the usefulness of the mole concept, and use the relation between molecular (or formula) mass and molar mass to calculate the molar mass of any substance (§ 3.1) (EPs 3.1–3.5, 3.7–3.10)
2. Understand the relationships among amount of substance (in moles), mass (in grams), and number of chemical entities and convert from one to any other (§ 3.1) (SPs 3.1, 3.2) (EPs 3.6, 3.11–3.16, 3.19)
3. Use mass percent to find the mass of element in a given mass of compound (§ 3.1) (SP 3.3) (EPs 3.17, 3.18, 3.20–3.23)
4. Determine the empirical and molecular formulas of a compound from mass analysis of its elements (§ 3.2) (SPs 3.4–3.6) (EPs 3.24–3.34)
5. Balance an equation given formulas or names, and use molar ratios to calculate amounts of reactants and products for reactions of pure or dissolved substances (§ 3.3, 3.5) (SPs 3.7, 3.8, 3.9, 3.17) (EPs 3.35–3.47, 3.64, 3.74, 3.75)
6. Understand why one reactant limits the yield of product, and solve limiting-reactant problems for reactions of pure or dissolved substances (§ 3.4, 3.5) (SPs 3.10, 3.11, 3.18) (EPs 3.48–3.55, 3.62, 3.76, 3.77)
7. Explain the reasons for lower-than-expected yields and the distinction between theoretical and actual yields, and calculate percent yield (§ 3.4) (SP 3.12) (EPs 3.56–3.61, 3.65)
8. Understand the meaning of concentration and the effect of dilution, and calculate molarity or mass of dissolved solute (§ 3.5) (SPs 3.13–3.16) (EPs 3.66–3.73, 3.78)

KEY TERMS *These important terms appear in boldface in the chapter and are defined again in the Glossary.*

stoichiometry (72)

Section 3.1
mole (mol) (72)
Avogadro's number (72)
molar mass ($\mathcal{M}$) (74)

Section 3.2
combustion analysis (82)
isomer (83)

Section 3.3
chemical equation (85)
reactant (85)
product (85)
balancing (stoichiometric) coefficient (85)

Section 3.4
limiting reactant (93)
theoretical yield (97)
side reaction (97)
actual yield (97)
percent yield (% yield) (97)

Section 3.5
solute (98)
solvent (98)
concentration (99)
molarity (M) (99)

KEY EQUATIONS AND RELATIONSHIPS *Numbered and screened concepts are listed for you to refer to or memorize.*

3.1 Number of entities in one mole (72):

1 mole contains 6.022×10^{23} entities (to 4 sf)

3.2 Converting amount (mol) to mass using $\mathcal{M}$ (75):

$$\text{Mass (g)} = \text{no. of moles} \times \frac{\text{no. of grams}}{1 \text{ mol}}$$

3.3 Converting mass to amount (mol) using $1/\mathcal{M}$ (75):

$$\text{No. of moles} = \text{mass (g)} \times \frac{1 \text{ mol}}{\text{no. of grams}}$$

3.4 Converting amount (mol) to number of entities (75):

$$\text{No. of entities} = \text{no. of moles} \times \frac{6.022\times10^{23} \text{ entities}}{1 \text{ mol}}$$

3.5 Converting number of entities to amount (mol) (75):

$$\text{No. of moles} = \text{no. of entities} \times \frac{1 \text{ mol}}{6.022\times10^{23} \text{ entities}}$$

3.6 Calculating mass % (77):

Mass % of element X

$$= \frac{\text{moles of X in formula} \times \text{molar mass of X (g/mol)}}{\text{mass (g) of 1 mol of compound}} \times 100$$

3.7 Calculating percent yield (97):

$$\%\text{ yield} = \frac{\text{actual yield}}{\text{theoretical yield}} \times 100$$

3.8 Defining molarity (99):

$$\text{Molarity} = \frac{\text{moles of solute}}{\text{liters of solution}} \quad \text{or} \quad M = \frac{\text{mol solute}}{\text{L soln}}$$

3.9 Diluting a concentrated solution (101):

$$M_{\text{dil}} \times V_{\text{dil}} = \text{number of moles} = M_{\text{conc}} \times V_{\text{conc}}$$

BRIEF SOLUTIONS TO *FOLLOW-UP PROBLEMS* *Compare your own solutions to these calculation steps and answers.*

3.1 (a) $\text{Moles of C} = 315 \cancel{\text{mg C}} \times \frac{1 \cancel{\text{g}}}{10^3 \cancel{\text{mg}}} \times \frac{1 \text{ mol C}}{12.01 \cancel{\text{g C}}}$

$= 2.62\times10^{-2}$ mol C

(b) $\text{Mass (g) of Mn} = 3.22\times10^{20} \cancel{\text{Mn atoms}} \times \frac{1 \cancel{\text{mol Mn}}}{6.022\times10^{23} \cancel{\text{Mn atoms}}} \times \frac{54.94 \text{ g Mn}}{1 \cancel{\text{mol Mn}}}$

$= 2.94\times10^{-2}$ g Mn

3.2 (a) Mass (g) of P_4O_{10}

$= 4.65\times10^{22} \cancel{\text{molecules } P_4O_{10}} \times \frac{1 \cancel{\text{mol } P_4O_{10}}}{6.022\times10^{23} \cancel{\text{molecules } P_4O_{10}}} \times \frac{283.88 \text{ g } P_4O_{10}}{1 \cancel{\text{mol } P_4O_{10}}}$

$= 21.9$ g P_4O_{10}

(b) $\text{No. of P atoms} = 4.65\times10^{22} \cancel{\text{molecules } P_4O_{10}} \times \frac{4 \text{ atoms P}}{1 \cancel{\text{molecule } P_4O_{10}}}$

$= 1.86\times10^{23}$ P atoms

3.3 (a) $$\text{Mass \% of N} = \frac{2 \cancel{\text{mol N}} \times \frac{14.01 \cancel{\text{g N}}}{1 \cancel{\text{mol N}}}}{80.05 \cancel{\text{g } NH_4NO_3}} \times 100$$

$= 35.00$ mass % N

(b) $\text{Mass (g) of N} = 35.8 \cancel{\text{kg } NH_4NO_3} \times \frac{10^3 \cancel{\text{g}}}{1 \cancel{\text{kg}}} \times \frac{0.3500 \text{ g N}}{1 \cancel{\text{g } NH_4NO_3}}$

$= 1.25\times10^4$ g N

3.4 $\text{Moles of S} = 2.88 \cancel{\text{g S}} \times \frac{1 \text{ mol S}}{32.07 \cancel{\text{g S}}} = 0.0898 \text{ mol S}$

$\text{Moles of M} = 0.0898 \cancel{\text{mol S}} \times \frac{2 \text{ mol M}}{3 \cancel{\text{mol S}}} = 0.0599 \text{ mol M}$

$\text{Molar mass of M} = \frac{3.12 \text{ g M}}{0.0599 \text{ mol M}} = 52.1 \text{ g/mol}$

M is chromium, and M_2S_3 is chromium(III) sulfide.

3.5 Assuming 100.00 g of compound, we have 95.21 g of C and 4.79 g of H:

$$\text{Moles of C} = 95.21 \cancel{\text{g C}} \times \frac{1 \text{ mol C}}{12.01 \cancel{\text{g C}}} = 7.928 \text{ mol C}$$

Also, 4.75 mol H

Preliminary formula = $C_{7.928}H_{4.75} \approx C_{1.67}H_{1.00}$

Empirical formula = C_5H_3

$\text{Whole-number multiple} = \frac{252.30 \cancel{\text{g/mol}}}{63.07 \cancel{\text{g/mol}}} = 4$

Molecular formula = $C_{20}H_{12}$

3.6 $\text{Mass (g) of C} = 0.451 \cancel{\text{g } CO_2} \times \frac{12.01 \text{ g C}}{44.01 \cancel{\text{g } CO_2}}$

$= 0.123$ g C

Also, 0.00690 g H

Mass (g) of Cl = 0.250 g − (0.123 g + 0.00690 g) = 0.120 g Cl

Moles of elements

= 0.0102 mol C; 0.00685 mol H; 0.00339 mol Cl

Empirical formula = C_3H_2Cl; multiple = 2

Molecular formula = $C_6H_4Cl_2$

3.7 (a) $2Na(s) + 2H_2O(l) \longrightarrow H_2(g) + 2NaOH(aq)$

(b) $2HNO_3(aq) + CaCO_3(s) \longrightarrow H_2O(l) + CO_2(g) + Ca(NO_3)_2(aq)$

(c) $PCl_3(g) + 3HF(g) \longrightarrow PF_3(g) + 3HCl(g)$

(d) $4C_3H_5N_3O_9(l) \longrightarrow 12CO_2(g) + 10H_2O(g) + 6N_2(g) + O_2(g)$

3.8 From the depiction, we have $6CO + 3O_2 \longrightarrow 6CO_2$

Or, $2CO(g) + O_2(g) \longrightarrow 2CO_2(g)$

3.9 $Fe_2O_3(s) + 2Al(s) \longrightarrow Al_2O_3(s) + 2Fe(l)$

(a) Mass (g) of Fe

$= 135 \cancel{\text{g Al}} \times \frac{1 \cancel{\text{mol Al}}}{26.98 \cancel{\text{g Al}}} \times \frac{2 \cancel{\text{mol Fe}}}{2 \cancel{\text{mol Al}}} \times \frac{55.85 \text{ g Fe}}{1 \cancel{\text{mol Fe}}}$

$= 279$ g Fe

(b) $\text{No. of Al atoms} = 1.00 \cancel{\text{g } Al_2O_3} \times \frac{1 \cancel{\text{mol } Al_2O_3}}{101.96 \cancel{\text{g } Al_2O_3}} \times \frac{2 \cancel{\text{mol Al}}}{1 \cancel{\text{mol } Al_2O_3}} \times \frac{6.022\times10^{23} \text{ Al atoms}}{1 \cancel{\text{mol Al}}}$

$= 1.18\times10^{22}$ Al atoms

3.10 (a) $2AB + B_2 \longrightarrow 2AB_2$

In the boxes, the AB/B_2 ratio is 4/3, which is less than the 2/1 ratio in the equation. Thus, there is not enough AB, so it is the limiting reactant; note that one B_2 is in excess.

BRIEF SOLUTIONS TO *FOLLOW-UP PROBLEMS* (continued)

(b) $\text{Moles of AB}_2 = 1.5\ \cancel{\text{mol AB}} \times \frac{2\ \text{mol AB}_2}{2\ \cancel{\text{mol AB}}} = 1.5\ \text{mol AB}_2$

$\text{Moles of AB}_2 = 1.5\ \cancel{\text{mol B}_2} \times \frac{2\ \text{mol AB}_2}{1\ \cancel{\text{mol B}_2}} = 3.0\ \text{mol AB}_2$

3.11 $2\text{Al}(s) + 3\text{S}(s) \longrightarrow \text{Al}_2\text{S}_3(s)$

Mass (g) of Al_2S_3 formed from Al

$= 10.0\ \cancel{\text{g Al}} \times \frac{1\ \cancel{\text{mol Al}}}{26.98\ \cancel{\text{g Al}}} \times \frac{1\ \cancel{\text{mol Al}_2\text{S}_3}}{2\ \cancel{\text{mol Al}}} \times \frac{150.17\ \text{g Al}_2\text{S}_3}{1\ \cancel{\text{mol Al}_2\text{S}_3}}$

$= 27.8\ \text{g Al}_2\text{S}_3$

Similarly, mass (g) of Al_2S_3 formed from S = 23.4 g Al_2S_3.

Therefore, S is limiting reactant, and 23.4 g of Al_2S_3 can form.

Mass (g) of Al in excess

= total mass of Al − mass of Al used

$= 10.0\ \text{g Al} - \left(15.0\ \cancel{\text{g S}} \times \frac{1\ \cancel{\text{mol S}}}{32.07\ \cancel{\text{g S}}} \times \frac{2\ \cancel{\text{mol Al}}}{3\ \cancel{\text{mol S}}} \times \frac{26.98\ \text{g Al}}{1\ \cancel{\text{mol Al}}}\right)$

$= 1.6\ \text{g Al}$

(We would obtain the same answer if sulfur were shown more correctly as S_8.)

3.12 $\text{CaCO}_3(s) + 2\text{HCl}(aq) \longrightarrow \text{CaCl}_2(aq) + \text{H}_2\text{O}(l) + \text{CO}_2(g)$

Theoretical yield (g) of CO_2

$= 10.0\ \cancel{\text{g CaCO}_3} \times \frac{1\ \cancel{\text{mol CaCO}_3}}{100.09\ \cancel{\text{g CaCO}_3}} \times \frac{1\ \cancel{\text{mol CO}_2}}{1\ \cancel{\text{mol CaCO}_3}}$

$\times \frac{44.01\ \text{g CO}_2}{1\ \cancel{\text{mol CO}_2}} = 4.40\ \text{g CO}_2$

$\%\ \text{yield} = \frac{3.65\ \cancel{\text{g CO}_2}}{4.40\ \cancel{\text{g CO}_2}} \times 100 = 83.0\%$

3.13 $\text{Moles of KI} = 84\ \cancel{\text{mL soln}} \times \frac{1\ \cancel{\text{L}}}{10^3\ \cancel{\text{mL}}} \times \frac{0.50\ \text{mol KI}}{1\ \cancel{\text{L soln}}}$

$= 0.042\ \text{mol KI}$

3.14 Volume (L) of soln

$= 135\ \cancel{\text{g sucrose}} \times \frac{1\ \cancel{\text{mol sucrose}}}{342.30\ \cancel{\text{g sucrose}}} \times \frac{1\ \text{L soln}}{3.30\ \cancel{\text{mol sucrose}}}$

$= 0.120\ \text{L soln}$

3.15 $M_{\text{dil}}\ \text{of H}_2\text{SO}_4 = \frac{7.50\ M \times 25.0\ \cancel{\text{m}^3}}{500.\ \cancel{\text{m}^3}} = 0.375\ M\ \text{H}_2\text{SO}_4$

Mass (g) of H_2SO_4/mL soln

$= \frac{0.375\ \cancel{\text{mol H}_2\text{SO}_4}}{1\ \cancel{\text{L soln}}} \times \frac{1\ \cancel{\text{L}}}{10^3\ \text{mL}} \times \frac{98.09\ \text{g H}_2\text{SO}_4}{1\ \cancel{\text{mol H}_2\text{SO}_4}}$

$= 3.68\times10^{-2}\ \text{g/mL soln}$

3.16 To obtain B, the total volume of solution A was reduced by half:

$$V_{\text{conc}} = V_{\text{dil}} \times \frac{N_{\text{dil}}}{N_{\text{conc}}} = 1.0\ \text{mL} \times \frac{6\ \text{particles}}{12\ \text{particles}} = 0.50\ \text{mL}$$

To obtain C, $\frac{1}{2}$ of a volume of solvent was added for every volume of A:

$$V_{\text{dil}} = V_{\text{conc}} \times \frac{N_{\text{conc}}}{N_{\text{dil}}} = 1.0\ \text{mL} \times \frac{6\ \text{particles}}{4\ \text{particles}} = 1.5\ \text{mL}$$

3.17 $\text{Al(OH)}_3(s) + 3\text{HCl}(aq) \longrightarrow \text{AlCl}_3(aq) + 3\text{H}_2\text{O}(l)$

Volume (L) of HCl consumed

$= 0.10\ \cancel{\text{g Al(OH)}_3} \times \frac{1\ \cancel{\text{mol Al(OH)}_3}}{78.00\ \cancel{\text{g Al(OH)}_3}}$

$\times \frac{3\ \cancel{\text{mol HCl}}}{1\ \cancel{\text{mol Al(OH)}_3}} \times \frac{1\ \text{L soln}}{0.10\ \cancel{\text{mol HCl}}}$

$= 3.8\times10^{-2}\ \text{L soln}$

Therefore, $Al(OH)_3$ is more effective than $Mg(OH)_2$.

3.18 (a) Volume (L) of soln

$= 0.400\ \cancel{\text{mol Pb}^{2+}}$

$\times \frac{1\ \cancel{\text{mol Pb(C}_2\text{H}_3\text{O}_2)_2}}{1\ \cancel{\text{mol Pb}^{2+}}} \times \frac{1\ \text{L soln}}{1.50\ \cancel{\text{mol Pb(C}_2\text{H}_3\text{O}_2)_2}}$

$= 0.267\ \text{L soln}$

(b) $\text{Pb(C}_2\text{H}_3\text{O}_2)_2(aq) + 2\text{NaCl}(aq) \longrightarrow \text{PbCl}_2(s) + 2\text{NaC}_2\text{H}_3\text{O}_2(aq)$

Mass (g) of $PbCl_2$ from $Pb(C_2H_3O_2)_2$ soln = 111 g $PbCl_2$

Mass (g) of $PbCl_2$ from NaCl soln = 59.1 g $PbCl_2$

Thus, NaCl is the limiting reactant, and 59.1 g of $PbCl_2$ can form.

PROBLEMS

*Problems with **colored** numbers are answered in Appendix E. Sections match the text and provide the numbers of relevant sample problems. Bracketed problems are grouped in pairs (indicated by a short rule) that cover the same concept. Comprehensive Problems are based on material from any section or previous chapter.*

The Mole

(Sample Problems 3.1 to 3.3)

3.1 The atomic mass of Cl is 35.45 amu, and the atomic mass of Al is 26.98 amu. What are the masses in grams of 3 mol of Al atoms and of 2 mol of Cl atoms?

3.2 (a) How many moles of C atoms are in 1 mol of sucrose ($C_{12}H_{22}O_{11}$)?
(b) How many C atoms are in 2 mol of sucrose?

3.3 Why might the expression "1 mol of chlorine" be confusing? What change would remove any uncertainty? For what other elements might a similar confusion exist? Why?

3.4 How is the molecular mass of a compound the same as the molar mass, and how is it different?

3.5 What advantage is there to using a counting unit (the mole) in chemistry rather than a mass unit?

3.6 Each of the following balances weighs the indicated numbers of atoms of two elements:

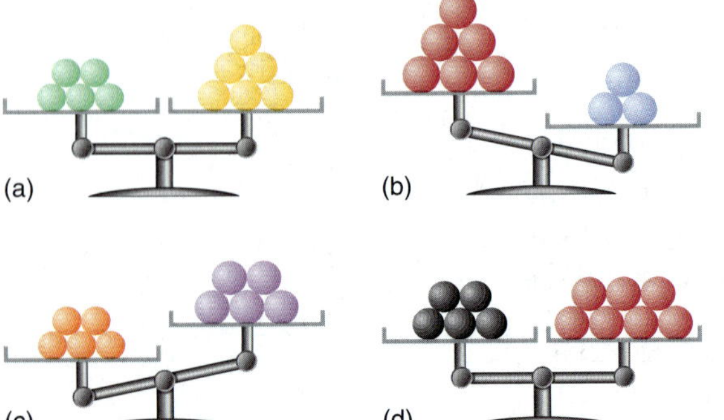

Which element—left, right, or neither: (a) Has the higher molar mass? (b) Has more atoms per gram? (c) Has fewer atoms per gram? (d) Has more atoms per mole?

3.7 Calculate the molar mass of each of the following:
(a) $Sr(OH)_2$ (b) N_2O_3 (c) $NaClO_3$ (d) Cr_2O_3

3.8 Calculate the molar mass of each of the following:
(a) $(NH_4)_3PO_4$ (b) CH_2Cl_2 (c) $CuSO_4 \cdot 5H_2O$ (d) BrF_3

3.9 Calculate the molar mass of each of the following:
(a) SnO (b) BaF_2 (c) $Al_2(SO_4)_3$ (d) $MnCl_2$

3.10 Calculate the molar mass of each of the following:
(a) N_2O_4 (b) C_8H_{10} (c) $MgSO_4 \cdot 7H_2O$ (d) $Ca(C_2H_3O_2)_2$

3.11 Calculate each of the following quantities:
(a) Mass in grams of 0.68 mol of $KMnO_4$
(b) Moles of O atoms in 8.18 g of $Ba(NO_3)_2$
(c) Number of O atoms in 7.3×10^{-3} g of $CaSO_4 \cdot 2H_2O$

3.12 Calculate each of the following quantities:
(a) Mass in kilograms of 4.6×10^{21} molecules of NO_2
(b) Moles of Cl atoms in 0.0615 g of $C_2H_4Cl_2$
(c) Number of H^- ions in 5.82 g of SrH_2

3.13 Calculate each of the following quantities:
(a) Mass in grams of 6.44×10^{-2} mol of $MnSO_4$
(b) Moles of compound in 15.8 kg of $Fe(ClO_4)_3$
(c) Number of N atoms in 92.6 mg of NH_4NO_2

3.14 Calculate each of the following quantities:
(a) Total number of ions in 38.1 g of SrF_2
(b) Mass in kilograms of 3.58 mol of $CuCl_2 \cdot 2H_2O$
(c) Mass in milligrams of 2.88×10^{22} formula units of $Bi(NO_3)_3 \cdot 5H_2O$

3.15 Calculate each of the following quantities:
(a) Mass in grams of 8.35 mol of copper(I) carbonate
(b) Mass in grams of 4.04×10^{20} molecules of dinitrogen pentaoxide
(c) Number of moles and formula units in 78.9 g of sodium perchlorate
(d) Number of sodium ions, perchlorate ions, Cl atoms, and O atoms in the mass of compound in part (c)

3.16 Calculate each of the following quantities:
(a) Mass in grams of 8.42 mol of chromium(III) sulfate decahydrate
(b) Mass in grams of 1.83×10^{24} molecules of dichlorine heptaoxide
(c) Number of moles and formula units in 6.2 g of lithium sulfate
(d) Number of lithium ions, sulfate ions, S atoms, and O atoms in the mass of compound in part (c)

3.17 Calculate each of the following:
(a) Mass % of H in ammonium bicarbonate
(b) Mass % of O in sodium dihydrogen phosphate heptahydrate

3.18 Calculate each of the following:
(a) Mass % of I in strontium periodate
(b) Mass % of Mn in potassium permanganate

3.19 Cisplatin *(right),* or Platinol, is a powerful drug used in the treatment of certain cancers. Calculate (a) the moles of compound in 285.3 g of cisplatin; (b) the number of hydrogen atoms in 0.98 mol of cisplatin.

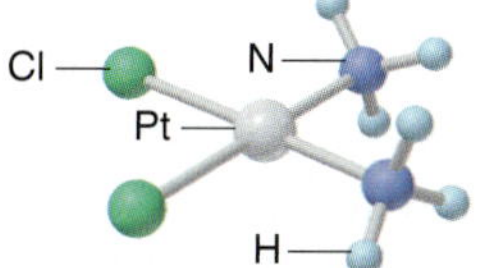

3.20 Propane is widely used in liquid form as a fuel for barbecue grills and camp stoves. For 85.5 g of propane, calculate (a) the moles of compound; (b) the grams of carbon.

3.21 The effectiveness of a nitrogen fertilizer is determined mainly by its mass % N. Rank the following fertilizers in terms of their effectiveness: potassium nitrate; ammonium nitrate; ammonium sulfate; urea, $CO(NH_2)_2$.

3.22 The mineral galena is composed of lead(II) sulfide and has an average density of 7.46 g/cm^3. (a) How many moles of lead(II) sulfide are in 1.00 ft^3 of galena? (b) How many lead atoms are in 1.00 dm^3 of galena?

3.23 Hemoglobin, a protein in red blood cells, carries O_2 from the lungs to the body's cells. Iron (as ferrous ion, Fe^{2+}) makes up 0.33 mass % of hemoglobin. If the molar mass of hemoglobin is 6.8×10^4 g/mol, how many Fe^{2+} ions are in one molecule?

Determining the Formula of an Unknown Compound

(Sample Problems 3.4 to 3.6)

3.24 Which of the following sets of information allows you to obtain the molecular formula of a covalent compound? In each case that allows it, explain how you would proceed (write a solution plan).
(a) Number of moles of each type of atom in a given sample of the compound
(b) Mass % of each element and the total number of atoms in a molecule of the compound
(c) Mass % of each element and the number of atoms of one element in a molecule of the compound
(d) Empirical formula and the mass % of each element in the compound
(e) Structural formula of the compound

3.25 What is the empirical formula and empirical formula mass for each of the following compounds?
(a) C_2H_4 (b) $C_2H_6O_2$ (c) N_2O_5 (d) $Ba_3(PO_4)_2$ (e) Te_4I_{16}

3.26 What is the empirical formula and empirical formula mass for each of the following compounds?
(a) C_4H_8 (b) $C_3H_6O_3$ (c) P_4O_{10} (d) $Ga_2(SO_4)_3$ (e) Al_2Br_6

3.27 What is the molecular formula of each compound?
(a) Empirical formula CH_2 ($\mathcal{M}$ = 42.08 g/mol)
(b) Empirical formula NH_2 ($\mathcal{M}$ = 32.05 g/mol)
(c) Empirical formula NO_2 ($\mathcal{M}$ = 92.02 g/mol)
(d) Empirical formula CHN ($\mathcal{M}$ = 135.14 g/mol)

3.28 What is the molecular formula of each compound?
(a) Empirical formula CH ($\mathcal{M}$ = 78.11 g/mol)
(b) Empirical formula $C_3H_6O_2$ ($\mathcal{M}$ = 74.08 g/mol)
(c) Empirical formula HgCl ($\mathcal{M}$ = 472.1 g/mol)
(d) Empirical formula $C_7H_4O_2$ ($\mathcal{M}$ = 240.20 g/mol)

3.29 Determine the empirical formulas of the following compounds:
(a) 0.063 mol of chlorine atoms combined with 0.22 mol of oxygen atoms
(b) 2.45 g of silicon combined with 12.4 g of chlorine
(c) 27.3 mass % carbon and 72.7 mass % oxygen

3.30 Determine the empirical formulas of the following compounds:
(a) 0.039 mol of iron atoms combined with 0.052 mol of oxygen atoms
(b) 0.903 g of phosphorus combined with 6.99 g of bromine
(c) A hydrocarbon with 79.9 mass % carbon

3.31 A sample of 0.600 mol of a metal M reacts completely with excess fluorine to form 46.8 g of MF_2.
(a) How many moles of F are in the sample of MF_2 that forms?
(b) How many grams of M are in this sample of MF_2?
(c) What element is represented by the symbol M?

3.32 A 0.370-mol sample of a metal oxide (M_2O_3) weighs 55.4 g.
(a) How many moles of O are in the sample?
(b) How many grams of M are in the sample?
(c) What element is represented by the symbol M?

3.33 Cortisol ($\mathcal{M}$ = 362.47 g/mol), one of the major steroid hormones, is used in the treatment of rheumatoid arthritis. Cortisol is 69.6% C, 8.34% H, and 22.1% O by mass. What is its molecular formula?

3.34 Menthol ($\mathcal{M}$ = 156.3 g/mol), a strong-smelling substance used in cough drops, is a compound of carbon, hydrogen, and oxygen. When 0.1595 g of menthol was subjected to combustion analysis, it produced 0.449 g of CO_2 and 0.184 g of H_2O. What is menthol's molecular formula?

Writing and Balancing Chemical Equations

(Sample Problems 3.7 and 3.8)

3.35 In the process of balancing the equation

$$Al + Cl_2 \longrightarrow AlCl_3$$

Student I writes: $Al + Cl_2 \longrightarrow AlCl_2$
Student II writes: $Al + Cl_2 + Cl \longrightarrow AlCl_3$
Student III writes: $2Al + 3Cl_2 \longrightarrow 2AlCl_3$
Is the approach of Student I valid? Student II? Student III? Explain.

3.36 The molecular scenes below represent a chemical reaction between elements A (red) and B (green):

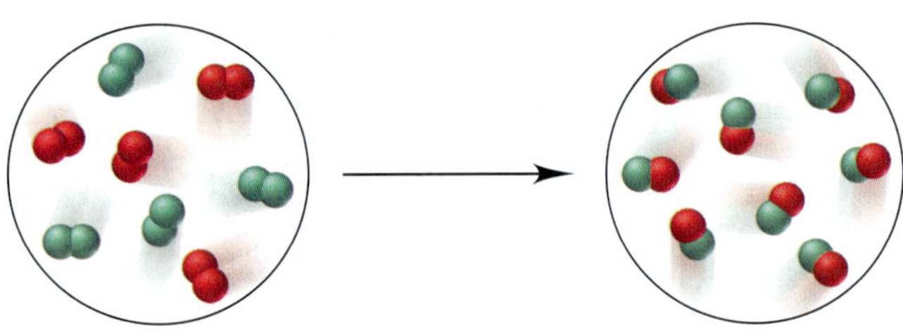

Which of the following best represents the balanced equation for the reaction?
(a) $2A + 2B \longrightarrow A_2 + B_2$ (b) $A_2 + B_2 \longrightarrow 2AB$
(c) $B_2 + 2AB \longrightarrow 2B_2 + A_2$ (d) $4A_2 + 4B_2 \longrightarrow 8AB$

3.37 Write balanced equations for each of the following by inserting the correct coefficients in the blanks:
(a) __$Cu(s)$ + __$S_8(s) \longrightarrow$ __$Cu_2S(s)$
(b) __$P_4O_{10}(s)$ + __$H_2O(l) \longrightarrow$ __$H_3PO_4(l)$
(c) __$B_2O_3(s)$ + __$NaOH(aq) \longrightarrow$ __$Na_3BO_3(aq)$ + __$H_2O(l)$
(d) __$CH_3NH_2(g)$ + __$O_2(g) \longrightarrow$ __$CO_2(g)$ + __$H_2O(g)$ + __$N_2(g)$

3.38 Write balanced equations for each of the following by inserting the correct coefficients in the blanks:
(a) __$Cu(NO_3)_2(aq)$ + __$KOH(aq) \longrightarrow$ __$Cu(OH)_2(s)$ + __$KNO_3(aq)$
(b) __$BCl_3(g)$ + __$H_2O(l) \longrightarrow$ __$H_3BO_3(s)$ + __$HCl(g)$
(c) __$CaSiO_3(s)$ + __$HF(g) \longrightarrow$ __$SiF_4(g)$ + __$CaF_2(s)$ + __$H_2O(l)$
(d) __$(CN)_2(g)$ + __$H_2O(l) \longrightarrow$ __$H_2C_2O_4(aq)$ + __$NH_3(g)$

3.39 Convert the following into balanced equations:
(a) When gallium metal is heated in oxygen gas, it melts and forms solid gallium(III) oxide.
(b) Liquid hexane burns in oxygen gas to form carbon dioxide gas and water vapor.
(c) When solutions of calcium chloride and sodium phosphate are mixed, solid calcium phosphate forms and sodium chloride remains in solution.

3.40 Convert the following into balanced equations:
(a) When lead(II) nitrate solution is added to potassium iodide solution, solid lead(II) iodide forms and potassium nitrate solution remains.
(b) Liquid disilicon hexachloride reacts with water to form solid silicon dioxide, hydrogen chloride gas, and hydrogen gas.
(c) When nitrogen dioxide is bubbled into water, a solution of nitric acid forms and gaseous nitrogen monoxide is released.

Calculating Amounts of Reactant and Product

(Sample Problems 3.9 to 3.12)

3.41 The molecular scene at right represents a mixture of A_2 (blue) and B_2 (green) before they react to form AB_3.
(a) What is the limiting reactant?
(b) How many molecules of product can form?

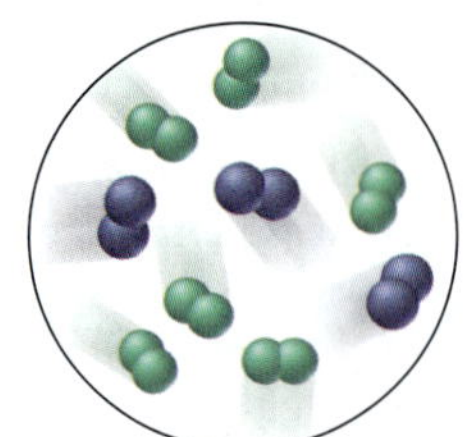

3.42 Potassium nitrate decomposes on heating, producing potassium oxide and gaseous nitrogen and oxygen:

$$4KNO_3(s) \longrightarrow 2K_2O(s) + 2N_2(g) + 5O_2(g)$$

To produce 56.6 kg of oxygen, how many (a) moles of KNO_3 must be heated? (b) Grams of KNO_3 must be heated?

3.43 Chromium(III) oxide reacts with hydrogen sulfide (H_2S) gas to form chromium(III) sulfide and water:

$$Cr_2O_3(s) + 3H_2S(g) \longrightarrow Cr_2S_3(s) + 3H_2O(l)$$

To produce 421 g of Cr_2S_3, (a) how many moles of Cr_2O_3 are required? (b) How many grams of Cr_2O_3 are required?

3.44 Calculate the mass of each product formed when 43.82 g of diborane (B_2H_6) reacts with excess water:

$$B_2H_6(g) + H_2O(l) \longrightarrow H_3BO_3(s) + H_2(g) \quad \text{[unbalanced]}$$

3.45 Calculate the mass of each product formed when 174 g of silver sulfide reacts with excess hydrochloric acid:

$$Ag_2S(s) + HCl(aq) \longrightarrow AgCl(s) + H_2S(g) \quad \text{[unbalanced]}$$

3.46 Elemental phosphorus occurs as tetratomic molecules, P_4. What mass of chlorine gas is needed to react completely with 455 g of phosphorus to form phosphorus pentachloride?

3.47 Elemental sulfur occurs as octatomic molecules, S_8. What mass of fluorine gas is needed to react completely with 17.8 g of sulfur to form sulfur hexafluoride?

3.48 Many metals react with oxygen gas to form the metal oxide. For example, calcium reacts as follows:

$$2Ca(s) + O_2(g) \longrightarrow 2CaO(s)$$

You wish to calculate the mass of calcium oxide that can be prepared from 4.20 g of Ca and 2.80 g of O_2. (a) How many moles of CaO can form from the given mass of Ca? (b) How many moles of CaO can form from the given mass of O_2? (c) Which is the limiting reactant? (d) How many grams of CaO can form?

3.49 Metal hydrides react with water to form hydrogen gas and the metal hydroxide. For example,

$$SrH_2(s) + 2H_2O(l) \longrightarrow Sr(OH)_2(s) + 2H_2(g)$$

You wish to calculate the mass of hydrogen gas that can be prepared from 5.70 g of SrH_2 and 4.75 g of H_2O.
(a) How many moles of H_2 can form from the given mass of SrH_2?
(b) How many moles of H_2 can form from the given mass of H_2O?
(c) Which is the limiting reactant?
(d) How many grams of H_2 can form?

3.50 Calculate the maximum numbers of moles and grams of iodic acid (HIO_3) that can form when 635 g of iodine trichloride reacts with 118.5 g of water:

$$ICl_3 + H_2O \longrightarrow ICl + HIO_3 + HCl \text{ [unbalanced]}$$

What mass of the excess reactant remains?

3.51 Calculate the maximum numbers of moles and grams of H_2S that can form when 158 g of aluminum sulfide reacts with 131 g of water:

$$Al_2S_3 + H_2O \longrightarrow Al(OH)_3 + H_2S \text{ [unbalanced]}$$

What mass of the excess reactant remains?

3.52 When 0.100 mol of carbon is burned in a closed vessel with 8.00 g of oxygen, how many grams of carbon dioxide can form? Which reactant is in excess, and how many grams of it remain after the reaction?

3.53 A mixture of 0.0375 g of hydrogen and 0.0185 mol of oxygen in a closed container is sparked to initiate a reaction. How many grams of water can form? Which reactant is in excess, and how many grams of it remain after the reaction?

3.54 Aluminum nitrite and ammonium chloride react to form aluminum chloride, nitrogen, and water. What mass of each substance is present after 72.5 g of aluminum nitrite and 58.6 g of ammonium chloride react completely?

3.55 Calcium nitrate and ammonium fluoride react to form calcium fluoride, dinitrogen monoxide, and water vapor. What mass of each substance is present after 16.8 g of calcium nitrate and 17.50 g of ammonium fluoride react completely?

3.56 Two successive reactions, A $\longrightarrow$ B and B $\longrightarrow$ C, have yields of 73% and 68%, respectively. What is the overall percent yield for conversion of A to C?

3.57 Two successive reactions, D $\longrightarrow$ E and E $\longrightarrow$ F, have yields of 48% and 73%, respectively. What is the overall percent yield for conversion of D to F?

3.58 What is the percent yield of a reaction in which 45.5 g of tungsten(VI) oxide (WO_3) reacts with excess hydrogen gas to produce metallic tungsten and 9.60 mL of water (d = 1.00 g/mL)?

3.59 What is the percent yield of a reaction in which 200. g of phosphorus trichloride reacts with excess water to form 128 g of HCl and aqueous phosphorous acid (H_3PO_3)?

3.60 When 20.5 g of methane and 45.0 g of chlorine gas undergo a reaction that has a 75.0% yield, what mass of chloromethane (CH_3Cl) forms? Hydrogen chloride also forms.

3.61 When 56.6 g of calcium and 30.5 g of nitrogen gas undergo a reaction that has a 93.0% yield, what mass of calcium nitride forms?

3.62 Cyanogen, $(CN)_2$, has been observed in the atmosphere of Titan, Saturn's largest moon, and in the gases of interstellar nebulas. On Earth, it is used as a welding gas and a fumigant. In its reaction with fluorine gas, carbon tetrafluoride and nitrogen trifluoride gases are produced. What mass of carbon tetrafluoride forms when 60.0 g of each reactant is used?

3.63 Gaseous dichlorine monoxide decomposes readily to chlorine and oxygen gases. (a) Which of the following scenes best depicts the product mixture after the decomposition?

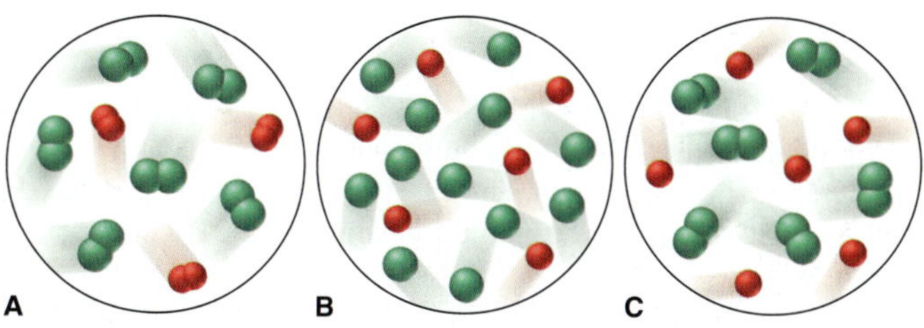

(b) Write the balanced equation for the decomposition.
(c) If each oxygen atom represents 0.050 mol, how many molecules of dichlorine monoxide were present before the decomposition?

3.64 Compressed butane gas is used as a liquid fuel in disposable cigarette lighters and lightweight camping stoves. Suppose a lighter contains 5.50 mL of butane (d = 0.579 g/mL). (a) What mass of oxygen is needed to burn the butane completely? (b) How many moles of H_2O form when all the butane burns? (c) How many total molecules of gas form when the butane burns completely?

3.65 Sodium borohydride ($NaBH_4$) is used industrially in many organic syntheses. One way to prepare it is by reacting sodium hydride with gaseous diborane (B_2H_6). Assuming an 88.5% yield, how many grams of $NaBH_4$ can be prepared by reacting 7.98 g of sodium hydride and 8.16 g of diborane?

Fundamentals of Solution Stoichiometry

(Sample Problems 3.13 to 3.18)

3.66 Six different aqueous solutions (with solvent molecules omitted for clarity) are represented in the beakers below, and their total volumes are noted.

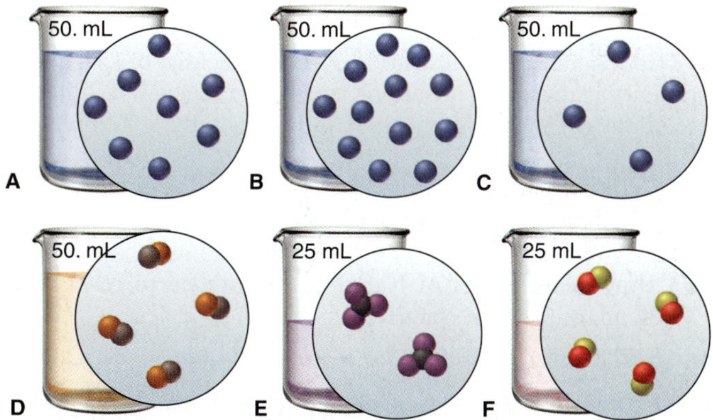

(a) Which solution has the highest molarity? (b) Which solutions have the same molarity? (c) If you mix solutions A and C, does the resulting solution have a higher, a lower, or the same molarity as solution B? (d) After 50. mL of water is added to solution D, is its molarity higher, lower, or the same as that of solution F after 75 mL of water is added to it? (e) How much solvent must be evaporated from solution E for it to have the same molarity as solution A?

3.67 Box A represents a unit volume of a solution. Choose from boxes B and C the one representing the same unit volume of solution that has (a) more solute added; (b) more solvent added; (c) higher molarity; (d) lower concentration.

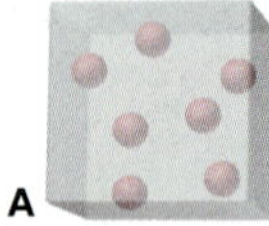

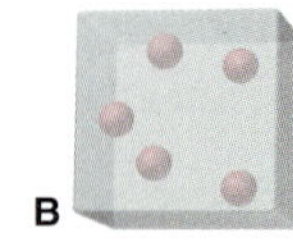

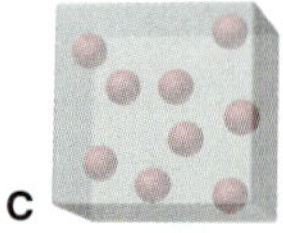

3.68 Calculate each of the following quantities:
(a) Grams of solute in 185.8 mL of 0.267 *M* calcium acetate
(b) Molarity of 500. mL of solution containing 21.1 g of potassium iodide
(c) Moles of solute in 145.6 L of 0.850 *M* sodium cyanide

3.69 Calculate each of the following quantities:
(a) Volume in milliliters of 2.26 *M* potassium hydroxide that contains 8.42 g of solute
(b) Number of Cu^{2+} ions in 52 L of 2.3 *M* copper(II) chloride
(c) Molarity of 275 mL of solution containing 135 mmol of glucose

3.70 Calculate each of the following quantities:
(a) Molarity of a solution prepared by diluting 37.00 mL of 0.250 *M* potassium chloride to 150.00 mL
(b) Molarity of a solution prepared by diluting 25.71 mL of 0.0706 *M* ammonium sulfate to 500.00 mL
(c) Molarity of sodium ion in a solution made by mixing 3.58 mL of 0.348 *M* sodium chloride with 500. mL of 6.81×10^{-2} *M* sodium sulfate (assume volumes are additive)

3.71 Calculate each of the following quantities:
(a) Volume of 2.050 *M* copper(II) nitrate that must be diluted with water to prepare 750.0 mL of a 0.8543 *M* solution
(b) Volume of 1.63 *M* calcium chloride that must be diluted with water to prepare 350. mL of a 2.86×10^{-2} *M* chloride ion solution
(c) Final volume of a 0.0700 *M* solution prepared by diluting 18.0 mL of 0.155 *M* lithium carbonate with water

3.72 A sample of concentrated nitric acid has a density of 1.41 g/mL and contains 70.0% HNO_3 by mass. (a) What mass of HNO_3 is present per liter of solution? (b) What is the molarity of the solution?

3.73 Concentrated sulfuric acid (18.3 *M*) has a density of 1.84 g/mL. (a) How many moles of sulfuric acid are present per milliliter of solution? (b) What is the mass % of H_2SO_4 in the solution?

3.74 How many milliliters of 0.383 *M* HCl are needed to react with 16.2 g of $CaCO_3$?

$$2HCl(aq) + CaCO_3(s) \longrightarrow CaCl_2(aq) + CO_2(g) + H_2O(l)$$

3.75 How many grams of NaH_2PO_4 are needed to react with 43.74 mL of 0.285 *M* NaOH?

$$NaH_2PO_4(s) + 2NaOH(aq) \longrightarrow Na_3PO_4(aq) + 2H_2O(l)$$

3.76 How many grams of solid barium sulfate form when 35.0 mL of 0.160 *M* barium chloride reacts with 58.0 mL of 0.065 *M* sodium sulfate? Aqueous sodium chloride also forms.

3.77 Which reactant is in excess and by how many moles when 350.0 mL of 0.210 *M* sulfuric acid reacts with 0.500 L of 0.196 *M* sodium hydroxide to form water and aqueous sodium sulfate?

3.78 Muriatic acid, an industrial grade of concentrated (11.7 *M*) HCl, is used to clean masonry and cement. (a) Write instructions for diluting the concentrated acid to make 3.0 gallons of 3.5 *M* acid for routine use (1 gal = 4 qt; 1 qt = 0.946 L). (b) How many milliliters of the muriatic acid solution contain 9.66 g of HCl?

Comprehensive Problems

Problems with an asterisk (*) are more challenging.

3.79 Narceine is a narcotic in opium. It crystallizes from water solution as a hydrate that contains 10.8 mass % water. If the molar mass of narceine hydrate is 499.52 g/mol, determine *x* in narceine·xH_2O.

3.80 Hydrogen-containing fuels have a "fuel value" based on their mass % H. Rank the following compounds from highest mass % H to lowest: ethane, propane, benzene, ethanol, cetyl palmitate (whale oil, $C_{32}H_{64}O_2$).

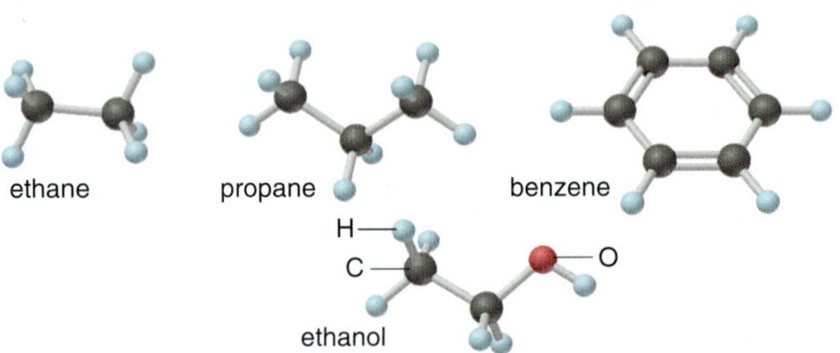

3.81 Convert the descriptions of reactions into balanced equations:
(a) In a gaseous reaction, hydrogen sulfide burns in oxygen to form sulfur dioxide and water vapor.
(b) When crystalline potassium chlorate is heated to just above its melting point, it reacts to form two different crystalline compounds, potassium chloride and potassium perchlorate.
(c) When hydrogen gas is passed over powdered iron(III) oxide, iron metal and water vapor form.
(d) The combustion of gaseous ethane in air forms carbon dioxide and water vapor.
(e) Iron(II) chloride is converted to iron(III) fluoride by treatment with chlorine trifluoride gas. Chlorine gas is also formed.

3.82 Isobutylene is a hydrocarbon used in the manufacture of synthetic rubber. When 0.847 g of isobutylene was analyzed by combustion (using an apparatus similar to that in Figure 3.5), the gain in mass of the CO_2 absorber was 2.657 g and that of the H_2O absorber was 1.089 g. What is the empirical formula of isobutylene?

3.83 One of the compounds used to increase the octane rating of gasoline is toluene *(right)*. Suppose 20.0 mL of toluene (*d* = 0.867 g/mL) is consumed when a sample of gasoline burns in air. (a) How many grams of oxygen are needed for complete combustion of the toluene? (b) How many total moles of gaseous products form? (c) How many molecules of water vapor form?

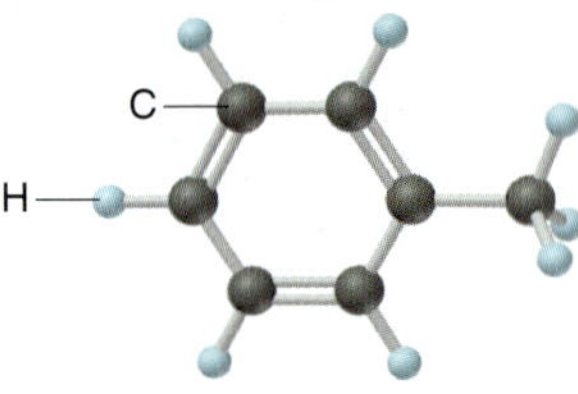

3.84 During studies of the reaction in Sample Problem 3.11,

$$2N_2H_4(l) + N_2O_4(l) \longrightarrow 3N_2(g) + 4H_2O(g)$$

a chemical engineer measured a less-than-expected yield of N_2 and discovered that the following side reaction occurs:

$$N_2H_4(l) + 2N_2O_4(l) \longrightarrow 6NO(g) + 2H_2O(g)$$

In one experiment, 10.0 g of NO formed when 100.0 g of each reactant was used. What is the highest percent yield of N_2 that can be expected?

3.85 These scenes depict a chemical reaction between AB_2 and B_2:

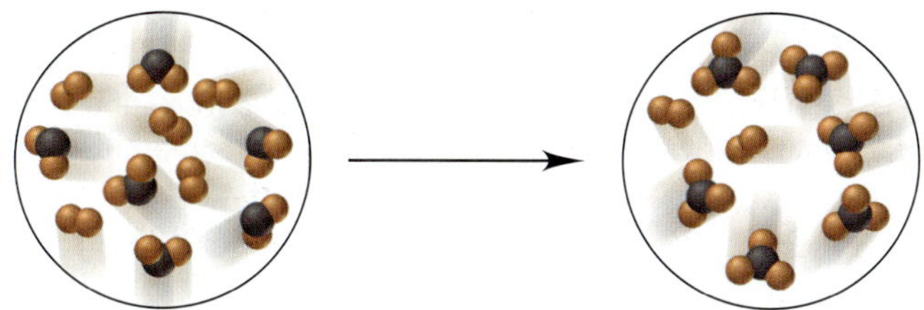

(a) Write a balanced equation for the reaction. (b) What is the limiting reactant in this reaction? (c) How many moles of product can be made from 3.0 mol of B_2 and 5.0 mol of AB_2? (d) How many moles of excess reactant remain after the reaction in part (c)?

3.86 Seawater is approximately 4.0% by mass dissolved ions. About 85% of the mass of the dissolved ions is from NaCl. (a) Calculate the mass percent of NaCl in seawater. (b) Calculate the mass percent of Na^+ ions and of Cl^- ions in seawater. (c) Calculate the molarity of NaCl in seawater at 15°C (*d* of seawater at 15°C = 1.025 g/mL).

3.87 Is each statement true or false? Correct any that are false:
(a) A mole of one substance has the same number of atoms as a mole of any other substance.
(b) The theoretical yield for a reaction is based on the balanced chemical equation.
(c) A limiting-reactant problem is presented when the quantity of available material is given in moles for one of the reactants.
(d) The concentration of a solution is an intensive property, but the amount of solute in a solution is an extensive property.

3.88 Box A represents one unit volume of solution A. Which box—B, C, or D—represents one unit volume after adding enough solvent to solution A to (a) triple its volume; (b) double its volume; (c) quadruple its volume?

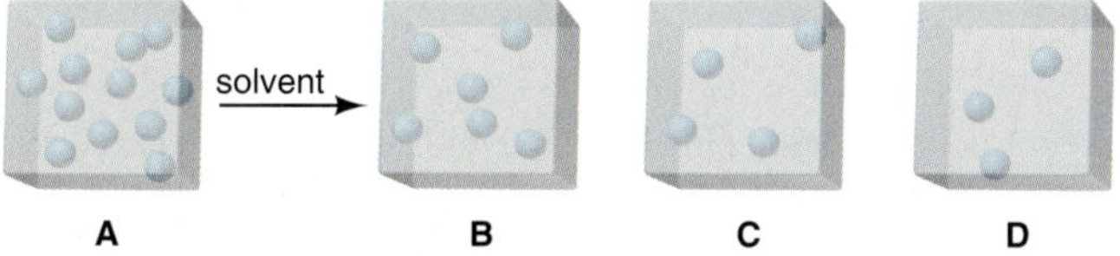

3.89 In each pair, which quantity is larger, or are they equal?
(a) Entities: 0.4 mol of O_3 molecules or 0.4 mol of O atoms
(b) Grams: 0.4 mol of O_3 molecules or 0.4 mol of O atoms
(c) Moles: 4.0 g of N_2O_4 or 3.3 g of SO_2
(d) Grams: 0.6 mol of C_2H_4 or 0.6 mol of F_2
(e) Total ions: 2.3 mol of sodium chlorate or 2.2 mol of magnesium chloride
(f) Molecules: 1.0 g of H_2O or 1.0 g of H_2O_2
(g) Na^+ ions: 0.500 L of 0.500 *M* NaBr or 0.0146 kg of NaCl
(h) Mass: 6.02×10^{23} atoms of ^{235}U or 6.02×10^{23} atoms of ^{238}U

3.90 Balance the equation for the reaction between solid tetraphosphorus trisulfide and oxygen gas to form solid tetraphosphorus decaoxide and gaseous sulfur dioxide. Tabulate the equation (see Table 3.3) in terms of (a) molecules, (b) moles, and (c) grams.

3.91 Hydrogen gas has been suggested as a clean fuel because it produces only water vapor when it burns. If the reaction has a 98.8% yield, what mass of hydrogen forms 105 kg of water?

3.92 Assuming that the volumes are additive, what is the concentration of KBr in a solution prepared by mixing 0.200 L of 0.053 *M* KBr with 0.550 L of 0.078 *M* KBr?

3.93 Calculate each of the following quantities:
(a) Moles of compound in 0.588 g of ammonium bromide
(b) Number of potassium ions in 88.5 g of potassium nitrate
(c) Mass in grams of 5.85 mol of glycerol ($C_3H_8O_3$)
(d) Volume of 2.85 mol of chloroform ($CHCl_3$; d = 1.48 g/mL)
(e) Number of sodium ions in 2.11 mol of sodium carbonate
(f) Number of atoms in 25.0 μg of cadmium
(g) Number of atoms in 0.0015 mol of fluorine gas

3.94 Elements X (green) and Y (purple) react according to the following equation: $X_2 + 3Y_2 \longrightarrow 2XY_3$. Which molecular scene represents the product of the reaction?

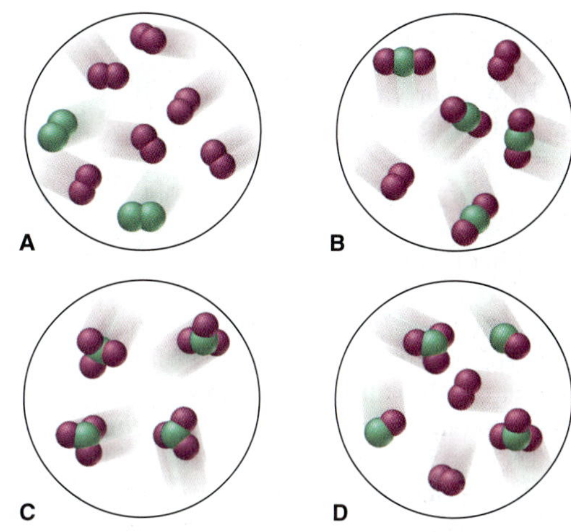

3.95 Hydrocarbon mixtures are used as fuels. (a) How many grams of $CO_2(g)$ are produced by the combustion of 200. g of a mixture that is 25.0% CH_4 and 75.0% C_3H_8 by mass? (b) A 252-g gaseous mixture of CH_4 and C_3H_8 burns in excess O_2, and 748 g of CO_2 gas is collected. What is the mass % of CH_4 in the initial mixture?

* **3.96** Nitrogen (N), phosphorus (P), and potassium (K) are the main nutrients in plant fertilizers. By convention, the numbers on the label refer to the mass percents of N, P_2O_5, and K_2O, in that order. Calculate the N/P/K ratio of a 30/10/10 fertilizer in terms of moles of each element, and express it as *x*/*y*/1.0.

* **3.97** What mass % of ammonium sulfate, ammonium hydrogen phosphate, and potassium chloride would you use to prepare 10/10/10 plant fertilizer (see Problem 3.96)?

3.98 A 0.652-g sample of a pure strontium halide reacts with excess sulfuric acid, and the solid strontium sulfate formed is separated, dried, and found to weigh 0.755 g. What is the formula of the original halide?

* **3.99** When carbon-containing compounds are burned in a limited amount of air, some CO(*g*) as well as $CO_2(g)$ is produced. A gaseous product mixture is 35.0 mass % CO and 65.0 mass % CO_2. What is the mass % C in the mixture?

3.100 Write a balanced equation for the reaction depicted below:

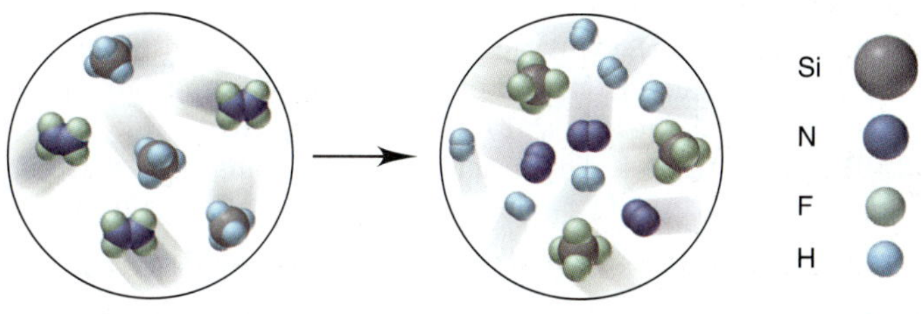

If each reactant molecule represents 1.25×10^{-2} mol and the reaction yield is 87%, how many grams of Si-containing product form?

3.101 Ferrocene, first synthesized in 1951, was the first organic iron compound with Fe—C bonds. An understanding of the structure of ferrocene gave rise to new ideas about chemical bonding and led to the preparation of many useful compounds. In the combustion analysis of ferrocene, which contains only Fe, C, and H, a 0.9437-g sample produced 2.233 g of CO_2 and 0.457 g of H_2O. What is the empirical formula of ferrocene?

* **3.102** Citric acid *(right)* is concentrated in citrus fruits and plays a central metabolic role in animal and plant cells. (a) What are the molar mass and formula of citric acid? (b) How many moles of citric acid are in 1.50 qt of lemon juice (d = 1.09 g/mL) that is 6.82% citric acid by mass?

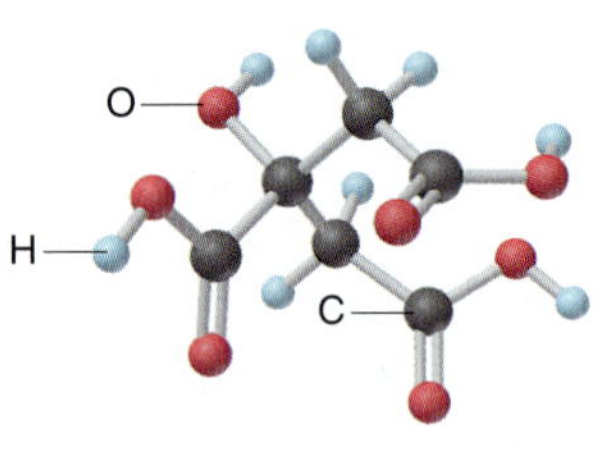

* **3.103** Fluorine is so reactive that it forms compounds with materials inert to other treatments.
(a) When 0.327 g of platinum is heated in fluorine, 0.519 g of a dark red, volatile solid forms. What is its empirical formula?
(b) When 0.265 g of this red solid reacts with excess xenon gas, 0.378 g of an orange-yellow solid forms. What is the empirical formula of this compound, the first noble gas compound formed?
(c) Fluorides of xenon can be formed by direct reaction of the elements at high pressure and temperature. Depending on conditions, the product mixture may include the difluoride, the tetrafluoride, and the hexafluoride. Under conditions that produce only the tetra- and hexafluorides, 1.85×10^{-4} mol of xenon reacted with 5.00×10^{-4} mol of fluorine, and 9.00×10^{-6} mol of xenon was found in excess. What are the mass percents of each xenon fluoride in the product mixture?

3.104 Hemoglobin is 6.0% heme ($C_{34}H_{32}FeN_4O_4$) by mass. To remove the heme, hemoglobin is treated with acetic acid and NaCl to form hemin ($C_{34}H_{32}N_4O_4FeCl$). At a crime scene, a blood sample contains 0.65 g of hemoglobin. (a) How many grams of heme are in the sample? (b) How many moles of heme? (c) How many grams of Fe? (d) How many grams of hemin could be formed for a forensic chemist to measure?

* **3.105** Manganese is a key component of extremely hard steel. The element occurs naturally in many oxides. A 542.3-g sample of a manganese oxide has an Mn/O ratio of 1.00/1.42 and consists of braunite (Mn_2O_3) and manganosite (MnO).
(a) What masses of braunite and manganosite are in the ore?
(b) What is the ratio Mn^{3+}/Mn^{2+} in the ore?

3.106 Hydroxyapatite, $Ca_5(PO_4)_3(OH)$, is the main mineral component of dental enamel, dentin, and bone, and thus has many medical uses. Coating it on metallic implants (such as titanium alloys and stainless steels) helps the body accept the implant. In the form of powder and beads, it is used to fill bone voids, which encourages natural bone to grow into the void. Hydroxyapatite is prepared by adding aqueous phosphoric acid to a dilute slurry of calcium hydroxide. (a) Write a balanced equation for this preparation. (b) What mass of hydroxyapatite could form from 100. g of 85% phosphoric acid and 100. g of calcium hydroxide?

3.107 Aspirin (acetylsalicylic acid, $C_9H_8O_4$) is made by reacting salicylic acid ($C_7H_6O_3$) with acetic anhydride [$(CH_3CO)_2O$]:

$$C_7H_6O_3(s) + (CH_3CO)_2O(l) \longrightarrow C_9H_8O_4(s) + CH_3COOH(l)$$

In one reaction, 3.077 g of salicylic acid and 5.50 mL of acetic anhydride react to form 3.281 g of aspirin.(a) Which is the limiting reactant (d of acetic anhydride = 1.080 g/mL)? (b) What is the percent yield of this reaction?

3.108 The human body excretes nitrogen in the form of urea, NH_2CONH_2. The key biochemical step in urea formation is the reaction of water with arginine to produce urea and ornithine:

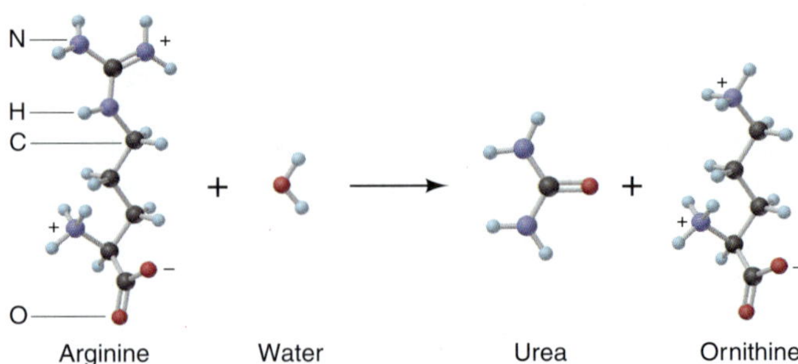

(a) What is the mass percent of nitrogen in urea, arginine, and ornithine? (b) How many grams of nitrogen can be excreted as urea when 135.2 g of ornithine is produced?

3.109 Nitrogen monoxide reacts with elemental oxygen to form nitrogen dioxide. The scene at right represents an initial mixture of reactants. If the reaction has a 66% yield, which of the scenes below (A, B, or C) best represents the final product mixture?

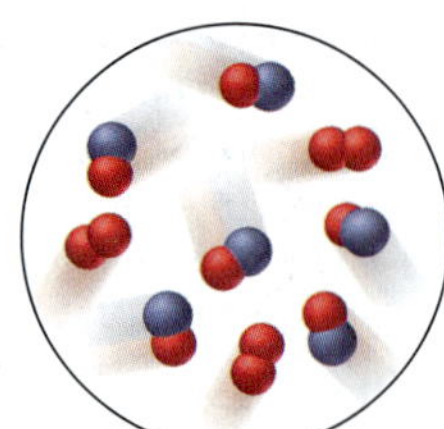

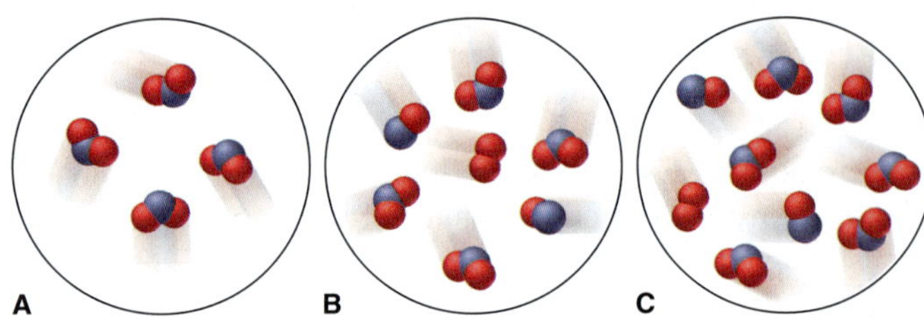

* **3.110** When powdered zinc is heated with sulfur, a violent reaction occurs, and zinc sulfide forms:

$$Zn(s) + S_8(s) \longrightarrow ZnS(s) \text{ [unbalanced]}$$

Some of the reactants also combine with oxygen in air to form zinc oxide and sulfur dioxide. When 83.2 g of Zn reacts with 52.4 g of S_8, 104.4 g of ZnS forms. What is the percent yield of ZnS? (b) If all the remaining reactants combine with oxygen, how many grams of each of the two oxides form?

* **3.111** High-temperature superconducting oxides hold great promise in the utility, transportation, and computer industries. (a) One superconductor is $La_{2-x}Sr_xCuO_4$. Calculate the molar mass of this oxide when $x = 0$, $x = 1$, and $x = 0.163$ (the last characterizes the compound with optimum superconducting properties).
(b) Another common superconducting oxide is made by heating a mixture of barium carbonate, copper(II) oxide, and yttrium(III) oxide, followed by further heating in O_2:

$$4BaCO_3(s) + 6CuO(s) + Y_2O_3(s) \longrightarrow 2YBa_2Cu_3O_{6.5}(s) + 4CO_2(g)$$

$$2YBa_2Cu_3O_{6.5}(s) + \tfrac{1}{2}O_2(g) \longrightarrow 2YBa_2Cu_3O_7(s)$$

When equal masses of the three reactants are heated, which reactant is limiting?
(c) After the product in part (b) is removed, what is the mass percent of each reactant in the solid mixture remaining?

Three Major Classes of Chemical Reactions

4

Key Principles
to focus on while studying this chapter

- *Aqueous* chemical reactions are those that occur in water. Because of its *shape* and *distribution of electrons,* water dissolves a variety of ionic and covalent substances. In water, ionic compounds (and a few simple covalent compounds, such as HCl) *dissociate* into ions *(Section 4.1).*
- Two types of ionic equations describe an aqueous ionic reaction. A *total ionic equation* shows ions for all soluble substances. A *net ionic equation* is more useful because it omits *spectator ions* (those not involved in the reaction) and shows the *actual chemical change* taking place *(Section 4.2).*
- *Precipitation reactions* occur when soluble ionic compounds exchange ions *(metathesis)* and form an insoluble product *(precipitate),* in which the ions attract each other so strongly that their attraction to water molecules cannot pull them apart *(Section 4.3).*
- An *acid* produces H^+ ions in solution, and a *base* produces OH^- ions. In an *acid-base (neutralization) reaction,* the H^+ and the OH^- ions form water. Another way to view this process is that an acid *transfers a proton* to a base. An acid-base titration is used to measure the amount (mol) of acid (or base) *(Section 4.4).*
- *Oxidation* is defined as *electron loss,* and *reduction* as *electron gain*. In an *oxidation-reduction (redox) reaction,* electrons move from one reactant to the other: the *reducing agent* is oxidized (loses the electrons), and the *oxidizing agent* is reduced (gains the electrons). Chemists use *oxidation number,* the number of electrons "owned" by each atom in a substance, to follow the change. *(Section 4.5).*
- Many common redox reactions (which are sometimes classified as *combination, decomposition, displacement,* or *combustion*) involve *elements* as reactants or products. In an *activity series,* metals are ranked by their ability to reduce H^+ or displace the ion of a different metal from an aqueous solution *(Section 4.6).*

Classifying the Countless *These silver halide particles (seen in colorized microscopy) were formed through one of the three major classes discussed in this chapter.*

Outline

Concepts & Skills to Review before studying this chapter

- names and formulas of compounds (Section 2.8)
- nature of ionic and covalent bonding (Section 2.7)
- mole-mass-number conversions (Section 3.1)
- molarity and mole-volume conversions (Section 3.5)
- balancing chemical equations (Section 3.3)
- calculating amounts of reactants and products (Section 3.4)

Rapid chemical changes occur among gas molecules as sunlight bathes the atmosphere or lightning rips through a stormy sky. Aqueous reactions go on unceasingly in the gigantic containers we know as oceans. And, in every cell of your body, thousands of reactions taking place right now allow you to function. Indeed, the amazing variety that we see in nature is largely a consequence of the amazing variety of chemical reactions.

Of the millions of reactions occurring in and around you, we have examined only a tiny fraction so far, and it would be impossible to examine them all. Fortunately, it isn't necessary to catalog every one, because when we survey even a small percentage of reactions, especially those occurring in aqueous solution, a few major patterns emerge. In this chapter, we examine the underlying nature of the three most common reaction processes. Because so many reactions occur in aqueous solution, first we'll highlight the importance of water.

4.1 THE ROLE OF WATER AS A SOLVENT

Our first step toward comprehending aqueous reactions is to understand how water acts as a solvent. Some solvents play a passive role, dispersing the dissolved substances into individual molecules but doing nothing further. Water plays a much more active role, interacting strongly with the substances and, in some cases, even reacting with them. To understand this active role, we'll examine the structure of water and how it interacts with ionic and covalent solutes.

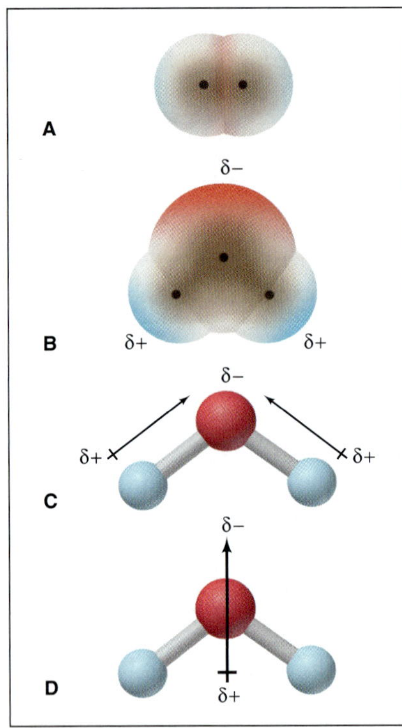

FIGURE 4.1 Electron distribution in molecules of H_2 and H_2O. A, In H_2, the identical nuclei attract the electrons equally. The central region of higher electron density (red) is balanced by the two outer regions of lower electron density (blue). **B,** In H_2O, the O nucleus attracts the shared electrons more strongly than the H nucleus. **C,** In this ball-and-stick model, a polar arrow points to the negative end of each O—H bond. **D,** The two polar O—H bonds and the bent shape give rise to the polar H_2O molecule.

The Polar Nature of Water

Of the many thousands of reactions that occur in the environment and in organisms, nearly all take place in water. Water's remarkable power as a solvent results from two features of its molecules: *the distribution of the bonding electrons* and *the overall shape.*

Recall from Section 2.7 that the electrons in a covalent bond are shared between the bonded atoms. In a covalent bond between identical atoms (as in H_2, Cl_2, O_2, etc.), the sharing is equal, so the electron charge is balanced between the atoms (Figure 4.1A). On the other hand, in covalent bonds between nonidentical atoms, the sharing is unequal: one atom attracts the electron pair more strongly than the other, so there is a charge imbalance. For reasons discussed in Chapter 9, an O atom attracts electrons more strongly than an H atom. Therefore, in each O—H bond in water, the shared electrons spend more time closer to the O atom.

This unequal distribution of negative charge creates partially charged "poles" at the ends of each O—H bond (Figure 4.1B). The O end acts as a slightly negative pole (represented by the red shading and the $\delta-$), and the H end acts as a slightly positive pole (represented by the blue shading and the $\delta+$). Figure 4.1C indicates the bond's polarity with a *polar arrow* (the arrowhead points to the negative pole and the tail is crossed to make a "plus").

The H—O—H arrangement forms an angle, so the water molecule is bent. The combined effects of its bent shape and its polar bonds make water a **polar molecule:** the O portion of the molecule is the partially negative pole, and the region midway between the H atoms on the other side of the molecule is the partially positive pole (Figure 4.1D).

Ionic Compounds in Water

In an ionic solid, the oppositely charged ions are held next to each other by electrostatic attraction (Figure 1.3C, p. 7). *Water separates the ions by replacing that attraction with one between the water molecules and the ions.* Imagine a granule of an ionic compound surrounded by bent, polar water molecules. The negative

FIGURE 4.2 The dissolution of an ionic compound. When an ionic compound dissolves in water, H_2O molecules separate, surround, and disperse the ions into the liquid. The negative ends of the H_2O molecules face the positive ions, and the positive ends face the negative ions.

ends of some water molecules are attracted to the cations, and the positive ends of others are attracted to the anions (Figure 4.2). Gradually, the attraction between each ion and the nearby water molecules outweighs the attraction of the ions for each other. By this process, the ions separate (dissociate) and become **solvated,** surrounded tightly by solvent molecules, as they move randomly in the solution. A similar scene occurs whenever an ionic compound dissolves in water.

Although many ionic compounds dissolve in water, many others do not. In the latter cases, the electrostatic attraction among ions in the compound remains greater than the attraction between ions and water molecules, so the solid stays largely intact. Actually, these so-called insoluble substances *do* dissolve to a very small extent, usually several orders of magnitude less than so-called soluble substances. Compare, for example, the solubilities of NaCl (a "soluble" compound) and AgCl (an "insoluble" compound):

$$\text{Solubility of NaCl in H}_2\text{O at }20^\circ\text{C} = 365\text{ g/L}$$
$$\text{Solubility of AgCl in H}_2\text{O at }20^\circ\text{C} = 0.009\text{ g/L}$$

The process of dissolving is in reality more complex than a contest between the relative attractions of the particles for each other and for the solvent. In Chapter 13, we'll see that it also involves the greater freedom of motion of the particles as they disperse throughout the solution.

Animation: Dissolution of an Ionic and a Covalent Compound

When an ionic compound dissolves, an important change occurs in the nature of the solution. Figure 4.3 (next page) shows this change with a simple apparatus that demonstrates *electrical conductivity,* the flow of electric current. When the electrodes are immersed in pure water or pushed into an ionic solid, such as potassium bromide (KBr), no current flows. In an aqueous KBr solution, however, a significant current flows, as shown by the brightly lit bulb. This current flow arises from the *movement of charged particles:* when KBr dissolves in water, the K^+ and Br^- ions dissociate, become solvated, and move toward the electrode of opposite charge. A substance that conducts a current when dissolved in water is an **electrolyte.** Soluble ionic compounds are called *strong* electrolytes because they dissociate completely into ions and create a large current. We express the dissociation of KBr into solvated ions in water as follows:

$$\text{KBr}(s) \xrightarrow{\text{H}_2\text{O}} \text{K}^+(aq) + \text{Br}^-(aq)$$

The "H_2O" above the arrow indicates that water is required as the solvent but is not a reactant in the usual sense.

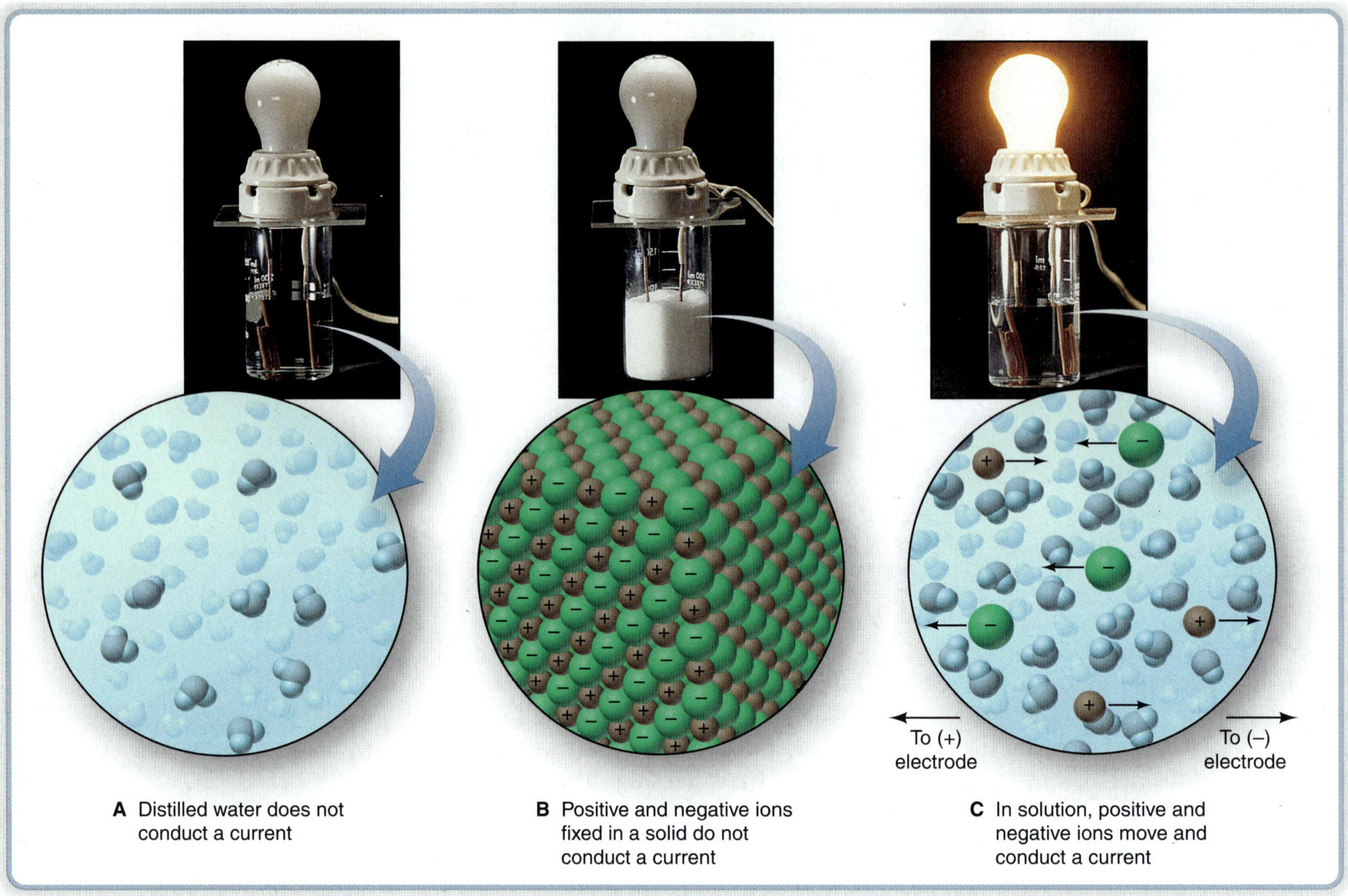

To (+) electrode

To (−) electrode

A Distilled water does not conduct a current

B Positive and negative ions fixed in a solid do not conduct a current

C In solution, positive and negative ions move and conduct a current

FIGURE 4.3 The electrical conductivity of ionic solutions. A, When electrodes connected to a power source are placed in distilled water, no current flows and the bulb is unlit. **B,** A solid ionic compound, such as KBr, conducts no current because the ions are bound tightly together. **C,** When KBr dissolves in H_2O, the ions separate and move through the solution toward the oppositely charged electrodes, thereby conducting a current.

Animation: Strong Electrolytes, Weak Electrolytes, and Nonelectrolytes

The formula of the compound tells the number of moles of different ions that result when the compound dissolves. Thus, 1 mol of KBr dissociates into 2 mol of ions—1 mol of K^+ and 1 mol of Br^-. Sample Problem 4.1 goes over this idea.

SAMPLE PROBLEM 4.1 Determining Moles of Ions in Aqueous Ionic Solutions

Problem For these soluble compounds, how many moles of each ion are in each solution?
(a) 5.0 mol of ammonium sulfate dissolved in water
(b) 78.5 g of cesium bromide dissolved in water
(c) 7.42×10^{22} formula units of copper(II) nitrate dissolved in water
(d) 35 mL of 0.84 *M* zinc chloride

Plan We write an equation that shows 1 mol of compound dissociating into ions. **(a)** We multiply the moles of ions by 5.0. **(b)** We first convert grams to moles. **(c)** We first convert formula units to moles. In (d), we first convert molarity and volume to moles.

Solution (a) $(NH_4)_2SO_4(s) \xrightarrow{H_2O} 2NH_4^+(aq) + SO_4^{2-}(aq)$

Remember that, in general, *polyatomic ions remain as intact units in solution.* Calculating moles of NH_4^+ ions:

$$\text{Moles of } NH_4^+ = 5.0\ \cancel{\text{mol }(NH_4)_2SO_4} \times \frac{2\text{ mol }NH_4^+}{1\ \cancel{\text{mol }(NH_4)_2SO_4}} = 10.\text{ mol }NH_4^+$$

5.0 mol of SO_4^{2-} is also present.

(b) $CsBr(s) \xrightarrow{H_2O} Cs^+(aq) + Br^-(aq)$

Converting from grams to moles:

$$\text{Moles of CsBr} = 78.5\ \cancel{\text{g CsBr}} \times \frac{1\text{ mol CsBr}}{212.8\ \cancel{\text{g CsBr}}} = 0.369\text{ mol CsBr}$$

Thus, 0.369 mol of Cs^+ and 0.369 mol of Br^- are present.

(c) $Cu(NO_3)_2(s) \xrightarrow{H_2O} Cu^{2+}(aq) + 2NO_3^-(aq)$
Converting from formula units to moles:

$$\text{Moles of } Cu(NO_3)_2 = 7.42\times10^{22} \text{ ~~formula units } Cu(NO_3)_2~~ \times \frac{1 \text{ mol } Cu(NO_3)_2}{6.022\times10^{23} \text{ ~~formula units } Cu(NO_3)_2~~} = 0.123 \text{ mol } Cu(NO_3)_2$$

$$\text{Moles of } NO_3^- = 0.123 \text{ ~~mol } Cu(NO_3)_2~~ \times \frac{2 \text{ mol } NO_3^-}{1 \text{ ~~mol } Cu(NO_3)_2~~} = 0.246 \text{ mol } NO_3^-$$

0.123 mol of Cu^{2+} is also present.

(d) $ZnCl_2(aq) \longrightarrow Zn^{2+}(aq) + 2Cl^-(aq)$
Converting from liters to moles:

$$\text{Moles of } ZnCl_2 = 35 \text{ ~~mL~~} \times \frac{1 \text{ ~~L~~}}{10^3 \text{ ~~mL~~}} \times \frac{0.84 \text{ mol } ZnCl_2}{1 \text{ ~~L~~}} = 2.9\times10^{-2} \text{ mol } ZnCl_2$$

$$\text{Moles of } Cl^- = 2.9\times10^{-2} \text{ ~~mol } ZnCl_2~~ \times \frac{2 \text{ mol } Cl^-}{1 \text{ ~~mol } ZnCl_2~~} = 5.8\times10^{-2} \text{ mol } Cl^-$$

2.9×10^{-2} mol of Zn^{2+} is also present.

Check After you round off to check the math, see if the relative moles of ions are consistent with the formula. For instance, in (a), 10 mol NH_4^+/5.0 mol SO_4^{2-} = 2 NH_4^+/1 SO_4^{2-}, or $(NH_4)_2SO_4$. In (d), 0.029 mol Zn^{2+}/0.058 mol Cl^- = 1 Zn^{2+}/2 Cl^-, or $ZnCl_2$.

FOLLOW-UP PROBLEM 4.1 How many moles of each ion are in each solution?
(a) 2 mol of potassium perchlorate dissolved in water
(b) 354 g of magnesium acetate dissolved in water
(c) 1.88×10^{24} formula units of ammonium chromate dissolved in water
(d) 1.32 L of 0.55 *M* sodium bisulfate

Covalent Compounds in Water

Water dissolves many covalent compounds also. Table sugar (sucrose, $C_{12}H_{22}O_{11}$), beverage (grain) alcohol (ethanol, CH_3CH_2OH), and automobile antifreeze (ethylene glycol, $HOCH_2CH_2OH$) are some familiar examples. All contain their own polar bonds, which interact with those of water. However, even though these substances dissolve, they do not dissociate into ions but remain as intact molecules. As a result, their aqueous solutions do not conduct an electric current, and these substances are called **nonelectrolytes.** (A small, but extremely important, group of H-containing covalent compounds interacts so strongly with water that their molecules *do* dissociate into ions. In aqueous solution, these substances are *acids,* as you'll see later in this chapter.) Many other covalent substances, such as benzene (C_6H_6) and octane (C_8H_{18}), do not contain polar bonds, and these substances do not dissolve appreciably in water.

SECTION 4.1 SUMMARY

Water plays an active role in dissolving ionic compounds because it consists of polar molecules that are attracted to the ions. • When an ionic compound dissolves in water, the ions dissociate from each other and become solvated by water molecules. Because the ions are free to move, their solutions conduct electricity. • Water also dissolves many covalent substances with polar bonds; however, the molecules remain intact, so their solutions do not conduct electricity. • Covalent compounds without polar bonds are mostly insoluble in water.

4.2 WRITING EQUATIONS FOR AQUEOUS IONIC REACTIONS

Chemists use different types of equations to represent aqueous ionic reactions: molecular, total ionic, and net ionic equations. As you'll see in the two types of ionic equations, by balancing the atoms, we also balance the charges.

Let's examine a reaction to see what each of these equations shows. When solutions of silver nitrate and sodium chromate are mixed, the brick-red solid silver chromate (Ag_2CrO_4) forms. Figure 4.4 depicts three views of this reaction: the change you would see if you mixed these solutions in the lab, how you might imagine the change at the atomic level among the ions, and how you can symbolize the change with the three types of equations. (The ions that are reacting are shown in red type.)

The **molecular equation** *(top)* reveals the least about the species in solution and is actually somewhat misleading because *it shows all the reactants and products as if they were intact, undissociated compounds:*

$$2AgNO_3(aq) + Na_2CrO_4(aq) \longrightarrow Ag_2CrO_4(s) + 2NaNO_3(aq)$$

Only by examining the state-of-matter designations (*s*) and (*aq*), can you tell the change that has occurred.

The **total ionic equation** *(middle)* is a much more accurate representation of the reaction because *it shows all the soluble ionic substances dissociated into ions.* Now the $Ag_2CrO_4(s)$ stands out as the only undissociated substance:

$$2Ag^+(aq) + 2NO_3^-(aq) + 2Na^+(aq) + CrO_4^{2-}(aq) \longrightarrow Ag_2CrO_4(s) + 2Na^+(aq) + 2NO_3^-(aq)$$

Notice that charges balance: there are four positive and four negative charges on the left for a net zero charge, and there are two positive and two negative charges on the right for a net zero charge.

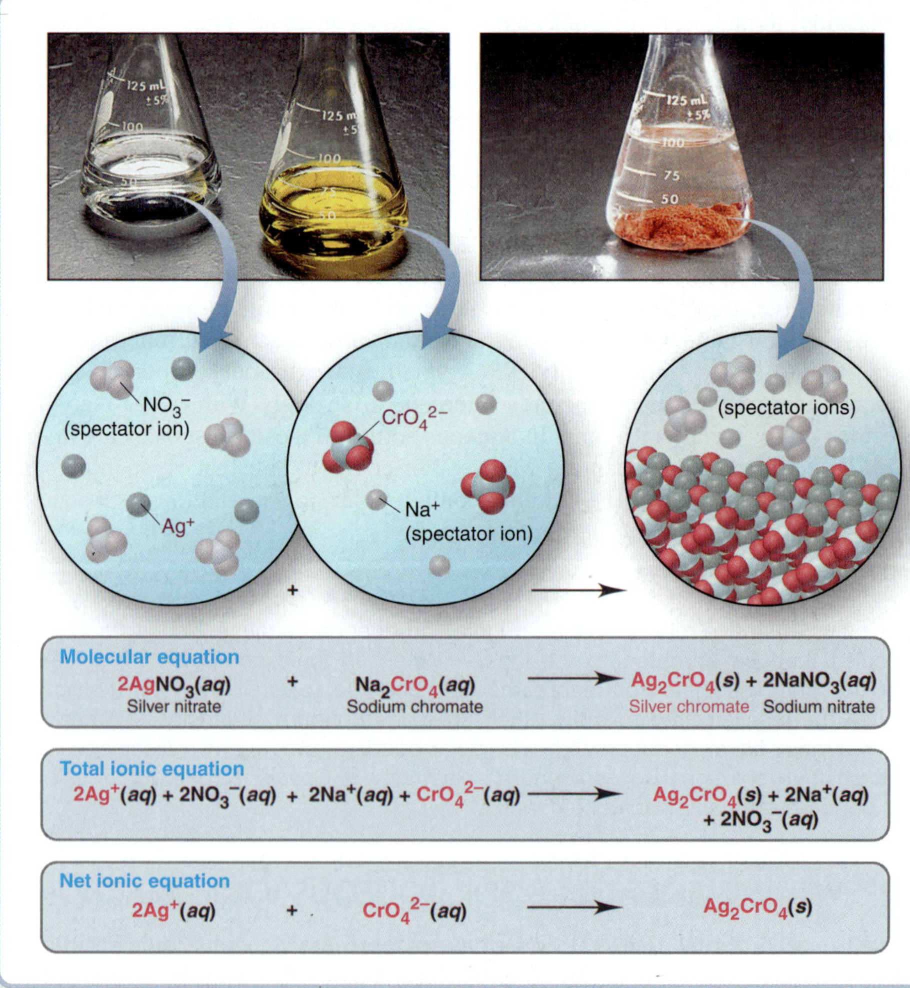

FIGURE 4.4 An aqueous ionic reaction and its equations. When silver nitrate and sodium chromate solutions are mixed, a reaction occurs that forms solid silver chromate and a solution of sodium nitrate. The photos present the macroscopic view of the reaction, the view the chemist sees in the lab. The blow-up arrows lead to an atomic-scale view, a representation of the chemist's mental picture of the reactants and products. (The pale ions are spectator ions, present for electrical neutrality, but not involved in the reaction.) Three equations represent the reaction in symbols. (The ions that are reacting are shown in red type.) The *molecular equation* shows all substances intact. The *total ionic equation* shows all *soluble* substances as separate, solvated ions. The *net ionic equation* omits the spectator ions to show only the reacting species.

Note that $Na^+(aq)$ and $NO_3^-(aq)$ appear in the same form on both sides of the equation. They are called **spectator ions** because they are not involved in the actual chemical change. These ions are present as part of the reactants to balance the charge. That is, we can't add an Ag^+ ion without also adding an anion, in this case, NO_3^- ion.

The **net ionic equation** *(bottom)* is the most useful because *it omits the spectator ions and shows the actual chemical change taking place:*

$$2Ag^+(aq) + CrO_4^{2-}(aq) \longrightarrow Ag_2CrO_4(s)$$

The formation of solid silver chromate from silver ions and chromate ions *is* the only change. In fact, if we had originally mixed solutions of potassium chromate, $K_2CrO_4(aq)$, and silver acetate, $AgC_2H_3O_2(aq)$, instead of sodium chromate and silver nitrate, the same change would have occurred. Only the spectator ions would differ—$K^+(aq)$ and $C_2H_3O_2^-(aq)$ instead of $Na^+(aq)$ and $NO_3^-(aq)$.

Now, let's apply these types of equations to three important classes of chemical reactions—precipitation, acid-base, and oxidation-reduction.

SECTION 4.2 SUMMARY

A molecular equation for an aqueous ionic reaction shows undissociated substances. • A total ionic equation shows all soluble ionic compounds as separate, solvated ions. Spectator ions appear unchanged on both sides of the equation. • The net ionic equation shows the actual chemical change, omitting all spectator ions.

4.3 PRECIPITATION REACTIONS

Precipitation reactions are common in both nature and commerce. Many geological formations, including coral reefs, some gems and minerals, and deep-sea structures form, in part, through this type of chemical process. And the chemical industry employs precipitation methods to produce several important inorganic compounds.

The Key Event: Formation of a Solid from Dissolved Ions

In **precipitation reactions,** two soluble ionic compounds react to form an insoluble product, a **precipitate.** The reaction you just saw between silver nitrate and sodium chromate is an example. Precipitates form for the same reason that some ionic compounds do not dissolve: the electrostatic attraction between the ions outweighs the tendency of the ions to remain solvated and move randomly throughout the solution. When solutions of such ions are mixed, the ions collide and stay together, and the resulting substance "comes out of solution" as a solid, as shown in Figure 4.5 (next page) for calcium fluoride. Thus, the key event in a precipitation reaction is *the formation of an insoluble product through the net removal of solvated ions from solution.*

Predicting Whether a Precipitate Will Form

If you mix aqueous solutions of two ionic compounds, can you predict if a precipitate will form? Consider this example. When solid sodium iodide and potassium nitrate are each dissolved in water, each solution consists of separated ions dispersed throughout the solution:

$$NaI(s) \xrightarrow{H_2O} Na^+(aq) + I^-(aq)$$
$$KNO_3(s) \xrightarrow{H_2O} K^+(aq) + NO_3^-(aq)$$

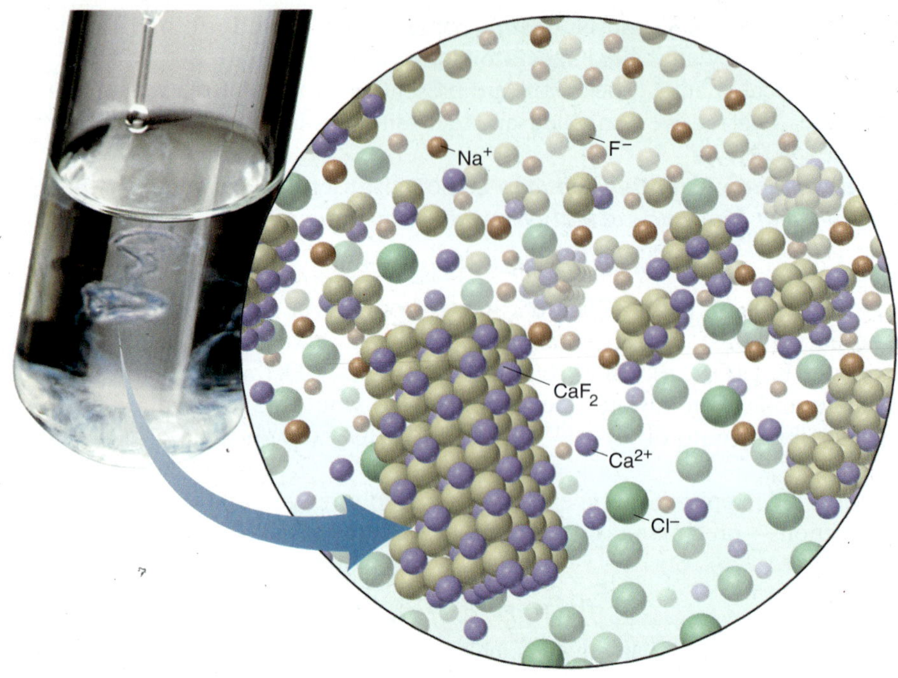

FIGURE 4.5 The precipitation of calcium fluoride. When an aqueous solution of NaF is added to a solution of $CaCl_2$, Ca^{2+} and F^- ions form particles of solid CaF_2.

Let's follow three steps to predict whether a precipitate will form:

1. *Note the ions present in the reactants.* The reactant ions are

$$Na^+(aq) + I^-(aq) + K^+(aq) + NO_3^-(aq) \longrightarrow ?$$

2. *Consider the possible cation-anion combinations.* In addition to the two original ones, NaI and KNO_3, which you know are soluble, the other possible cation-anion combinations are $NaNO_3$ and KI.
3. *Decide whether any of the combinations is insoluble.* A reaction does *not* occur when you mix these starting solutions because all the combinations—NaI, KNO_3, $NaNO_3$, and KI—are soluble. All the ions remain in solution. (You'll see shortly a set of rules for deciding if a product is soluble or not.)

Now, what happens if you substitute a solution of lead(II) nitrate, $Pb(NO_3)_2$, for the KNO_3? The reactant ions are Na^+, I^-, Pb^{2+}, and NO_3^-. In addition to the two soluble reactants, NaI and $Pb(NO_3)_2$, the other two possible cation-anion combinations are $NaNO_3$ and PbI_2. Lead(II) iodide is insoluble, so a reaction *does* occur because ions are removed from solution (Figure 4.6):

$$2Na^+(aq) + 2I^-(aq) + Pb^{2+}(aq) + 2NO_3^-(aq) \longrightarrow 2Na^+(aq) + 2NO_3^-(aq) + PbI_2(s)$$

A close look (with color) at the molecular equation shows that *the ions are exchanging partners:*

$$2NaI(aq) + Pb(NO_3)_2(aq) \longrightarrow PbI_2(s) + 2NaNO_3(aq)$$

Such reactions are called double-displacement reactions, or **metathesis** (pronounced *meh-TA-thuh-sis*) **reactions.** Several are important in industry, such as the preparation of silver bromide for the manufacture of black-and-white film:

$$AgNO_3(aq) + KBr(aq) \longrightarrow AgBr(s) + KNO_3(aq)$$

As we said, there is no simple way to decide whether any given ion combination is soluble or not, so Table 4.1 provides a short list of solubility rules to memorize. They allow you to predict the outcome of many precipitation reactions.

FIGURE 4.6 The reaction of $Pb(NO_3)_2$ and NaI. When aqueous solutions of these ionic compounds are mixed, the yellow solid PbI_2 forms.

Table 4.1 Solubility Rules for Ionic Compounds in Water

Soluble Ionic Compounds	Insoluble Ionic Compounds
1. All common compounds of Group 1A(1) ions (Li^+, Na^+, K^+, etc.) and ammonium ion (NH_4^+) are soluble.	1. All common metal hydroxides are insoluble, *except* those of Group 1A(1) and the larger members of Group 2A(2) (beginning with Ca^{2+}).
2. All common nitrates (NO_3^-), acetates (CH_3COO^- or $C_2H_3O_2^-$), and most perchlorates (ClO_4^-) are soluble.	2. All common carbonates (CO_3^{2-}) and phosphates (PO_4^{3-}) are insoluble, *except* those of Group 1A(1) and NH_4^+.
3. All common chlorides (Cl^-), bromides (Br^-), and iodides (I^-) are soluble, *except* those of Ag^+, Pb^{2+}, Cu^+, and Hg_2^{2+}. All common fluorides (F^-) are soluble, *except* those of Pb^{2+} and Group 2A(2).	3. All common sulfides are insoluble *except* those of Group 1A(1), Group 2A(2), and NH_4^+.
4. All common sulfates (SO_4^{2-}) are soluble, *except* those of Ca^{2+}, Sr^{2+}, Ba^{2+}, Ag^+, and Pb^{2+}.	

SAMPLE PROBLEM 4.2 Predicting Whether a Precipitation Reaction Occurs; Writing Ionic Equations

Problem Predict whether a reaction occurs when each of the following pairs of solutions are mixed. If a reaction does occur, write balanced molecular, total ionic, and net ionic equations, and identify the spectator ions.

(a) Potassium fluoride(*aq*) + strontium nitrate(*aq*) $\longrightarrow$

(b) Ammonium perchlorate(*aq*) + sodium bromide(*aq*) $\longrightarrow$

Plan For each pair of solutions, we note the ions present in the reactants, write the cation-anion combinations, and refer to Table 4.1 to see if any are insoluble. For the molecular equation, we predict the products. For the total ionic equation, we write the soluble compounds as separate ions. For the net ionic equation, we eliminate the spectator ions.

Solution **(a)** In addition to the reactants, the two other ion combinations are strontium fluoride and potassium nitrate. Table 4.1 shows that strontium fluoride is insoluble, so a reaction *does* occur. Writing the molecular equation:

$$2KF(aq) + Sr(NO_3)_2(aq) \longrightarrow SrF_2(s) + 2KNO_3(aq)$$

Writing the total ionic equation:

$$2K^+(aq) + F^-(aq) + Sr^{2+}(aq) + 2NO_3^-(aq) \longrightarrow SrF_2(s) + 2K^+(aq) + 2NO_3^-(aq)$$

Writing the net ionic equation:

$$Sr^{2+}(aq) + F^-(aq) \longrightarrow SrF_2(s)$$

The spectator ions are K^+ and NO_3^-.

(b) The other ion combinations are ammonium bromide and sodium perchlorate. Table 4.1 shows that all ammonium, sodium, and most perchlorate compounds are soluble, and all bromides are soluble except those of Ag^+, Pb^{2+}, Cu^+, and Hg_2^{2+}. Therefore, ***no*** reaction occurs. The compounds remain dissociated in solution as solvated ions.

FOLLOW-UP PROBLEM 4.2 Predict whether a reaction occurs, and write balanced total and net ionic equations:

(a) Iron(III) chloride(*aq*) + cesium phosphate(*aq*) $\longrightarrow$

(b) Sodium hydroxide(*aq*) + cadmium nitrate(*aq*) $\longrightarrow$

(c) Magnesium bromide(*aq*) + potassium acetate(*aq*) $\longrightarrow$

(d) Silver sulfate(*aq*) + barium chloride(*aq*) $\longrightarrow$

In Sample Problem 4.3, we'll use molecular depictions to examine a precipitation reaction quantitatively.

SAMPLE PROBLEM 4.3 Using Molecular Depictions to Understand a Precipitation Reaction

Problem Consider these molecular views of the reactant solutions in a precipitation reaction (with ions represented as simple spheres and solvent molecules omitted for clarity):

(a) Which compound is dissolved in solution A: KCl, Na_2SO_4, $MgBr_2$, or Ag_2SO_4?
(b) Which compound is dissolved in solution B: NH_4NO_3, $MgSO_4$, $Ba(NO_3)_2$, or CaF_2?
(c) Name the precipitate and the spectator ions that result when solutions A and B are mixed, and write balanced molecular, total ionic, and net ionic equations for the reaction.
(d) If each particle represents 0.010 mol of ions, what is the maximum mass of precipitate that can form (assuming complete reaction)?

Plan **(a)** and **(b)** From the depictions of the solutions in the beakers, we note the charge and number of each kind of ion and use Table 4.1 to determine the likely compounds. **(c)** Once we know the possible ion combinations, Table 4.1 helps us determine which two make up the solid. The other two are spectator ions. **(d)** We use the formula of the solid in part (c) and count the number of each kind of ion, to see which ion is in excess, which means the amount of the other ion limits the amount of precipitate that forms. We multiply the number of limiting ion particles by 0.010 mol and then use the molar mass of the precipitate to find the mass in grams.

Solution **(a)** In solution A, there are two 1+ particles for each 2− particle. Therefore, the dissolved compound cannot be KCl or $MgBr_2$. Of the remaining two choices, Ag_2SO_4 is insoluble, so it must be Na_2SO_4.
(b) In solution B, there are two 1− particles for each 2+ particle. Therefore, the dissolved compound cannot be NH_4NO_3 or $MgSO_4$. Of the remaining two choices, CaF_2 is insoluble, so it must be $Ba(NO_3)_2$.
(c) Of the remaining two ion combinations, the precipitate must be barium sulfate, and Na^+ and NO_3^- are the spectator ions.

Molecular: $Ba(NO_3)_2(aq) + Na_2SO_4(aq) \longrightarrow BaSO_4(s) + 2NaNO_3(aq)$

Total ionic: $Ba^{2+}(aq) + 2NO_3^-(aq) + 2Na^+(aq) + SO_4^{2-}(aq) \longrightarrow BaSO_4(s) + 2NO_3^-(aq) + 2Na^+(aq)$

Net ionic: $Ba^{2+}(aq) + SO_4^{2-}(aq) \longrightarrow BaSO_4(s)$

(d) The molar mass of $BaSO_4$ is 233.4 g/mol. There are four Ba^{2+} particles, so the maximum mass of $BaSO_4$ that can form is

$$\text{Mass (g) of BaSO}_4 = 4\ \text{Ba}^{2+}\text{particles} \times \frac{0.010\ \text{mol Ba}^{2+}\text{ions}}{1\ \text{particle}} \times \frac{1\ \text{mol BaSO}_4}{1\ \text{mol Ba}^{2+}\text{ions}} \times \frac{233.4\ \text{g BaSO}_4}{1\ \text{mol}} = 9.3\ \text{g BaSO}_4$$

Check Let's use the method for finding the limiting reactant that was introduced in Sample Problems 3.10 and 3.11; that is, see which reactant gives fewer moles of product:

$$\text{Amount (mol) of BaSO}_4 = 4\ \text{Ba}^{2+}\ \text{particles} \times 0.010\ \text{mol Ba}^{2+}\ \text{ions/particle} \times 1\ \text{mol of BaSO}_4/1\ \text{mol Ba}^{2+}\ \text{ions} = 0.040\ \text{mol BaSO}_4$$

$$\text{Amount (mol) of BaSO}_4 = 5\ \text{SO}_4^{2-}\ \text{particles} \times 0.010\ \text{mol SO}_4^{2-}\ \text{ions/particle} \times 1\ \text{mol of BaSO}_4/1\ \text{mol SO}_4^{2-}\ \text{ions} = 0.050\ \text{mol BaSO}_4$$

Therefore, Ba^{2+} is the limiting reactant ion, and the mass, after rounding, is 0.040 mol × 230 g/mol = 9.2 g, close to our calculated answer.

FOLLOW-UP PROBLEM 4.3 Molecular views of the reactant solutions in a precipitation reaction appear below (with ions shown as spheres and solvent molecules omitted):

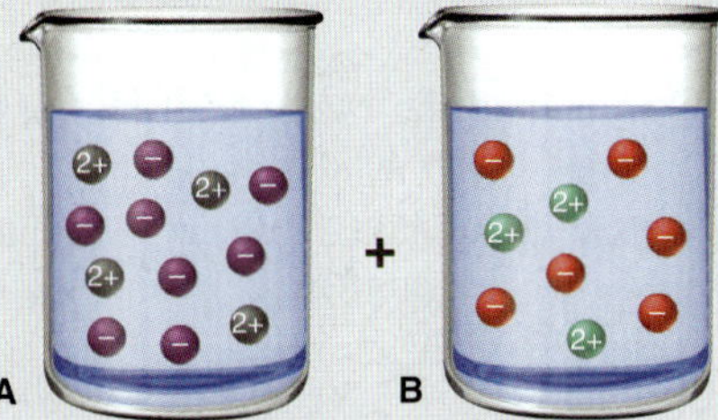

(a) Which compound is dissolved in beaker A: $Zn(NO_3)_2$, KCl, Na_2SO_4, or $PbCl_2$?
(b) Which compound is dissolved in beaker B: $(NH_4)_2SO_4$, $Cd(OH)_2$, $Ba(OH)_2$, or KNO_3?
(c) Name the precipitate and the spectator ions that result when the solutions in beakers A and B are mixed, and write balanced molecular, total ionic, and net ionic equations for the reaction.
(d) If each particle represents 0.050 mol of ions, what is the maximum mass of precipitate that can form (assuming complete reaction)?

SECTION 4.3 SUMMARY

Precipitation reactions involve the formation of an insoluble ionic compound from two soluble ones. These reactions occur because electrostatic attractions among certain pairs of solvated ions are strong enough to cause their removal from solution. • Formation of a precipitate is predicted by noting whether any possible ion combinations are insoluble, based on a set of solubility rules.

4.4 ACID-BASE REACTIONS

Aqueous acid-base reactions involve water not only as solvent but also in the more active roles of reactant and product. These reactions occur in processes as diverse as the biochemical synthesis of proteins, the industrial production of fertilizer, and some of the methods for revitalizing lakes damaged by acid rain.

Obviously, an **acid-base reaction** (also called a **neutralization reaction**) occurs when an acid reacts with a base, but the definitions of these terms and the scope of this reaction class have changed considerably over the years. For our purposes at this point, we'll use definitions that apply to chemicals you commonly encounter in the lab:

- *An **acid** is a substance that produces H^+ ions when dissolved in water.*

$$HX \xrightarrow{H_2O} H^+(aq) + X^-(aq)$$

- *A **base** is a substance that produces OH^- ions when dissolved in water.*

$$MOH \xrightarrow{H_2O} M^+(aq) + OH^-(aq)$$

(Other definitions of *acid* and *base* are presented later in this section and again in Chapter 18, along with a fuller meaning of *neutralization*.)

The Solvated Proton; Acids and Bases as Electrolytes Acidic solutions arise from a group of particular covalent molecules that *do* dissociate into ions in water. In every case, these molecules contain a polar bond to hydrogen in which the atom bonded to H pulls more strongly on the shared electron pair. A good example is hydrogen chloride gas. The Cl end of the HCl molecule is partially negative, and the H end is partially positive. When HCl dissolves in water, the partially charged poles of H_2O molecules are attracted to the oppositely charged poles of HCl.

The H—Cl bond breaks, with the H becoming the solvated cation $H^+(aq)$ and the Cl becoming the solvated anion $Cl^-(aq)$. Hydrogen bromide behaves similarly when it dissolves in water:

$$HBr(g) \xrightarrow{H_2O} H^+(aq) + Br^-(aq)$$

SAMPLE PROBLEM 4.4 Determining the Molarity of H^+ Ions in an Aqueous Solution of an Acid

Problem Nitric acid is a major chemical in the fertilizer and explosives industries. In aqueous solution, each molecule dissociates and the H becomes a solvated H^+ ion. What is the molarity of $H^+(aq)$ in 1.4 *M* nitric acid?

Plan We know the molarity of acid (1.4 *M*), so we just need the formula to find the number of moles of $H^+(aq)$ present in 1 L of solution.

Solution Nitrate ion is NO_3^-, so nitric acid is HNO_3. Thus, 1 mol of $H^+(aq)$ is released per mole of acid:

$$HNO_3(l) \xrightarrow{H_2O} H^+(aq) + NO_3^-(aq)$$

Therefore, 1.4 *M* HNO_3 contains 1.4 mol of $H^+(aq)$ per liter and is 1.4 *M* $H^+(aq)$.

FOLLOW-UP PROBLEM 4.4 How many moles of $H^+(aq)$ are present in 451 mL of 3.20 *M* hydrobromic acid?

Table 4.2 Strong and Weak Acids and Bases

Acids
Strong
Hydrochloric acid, HCl
Hydrobromic acid, HBr
Hydriodic acid, HI
Nitric acid, HNO_3
Sulfuric acid, H_2SO_4
Perchloric acid, $HClO_4$
Weak
Hydrofluoric acid, HF
Phosphoric acid, H_3PO_4
Acetic acid, CH_3COOH (or $HC_2H_3O_2$)
Bases
Strong
Sodium hydroxide, NaOH
Potassium hydroxide, KOH
Calcium hydroxide, $Ca(OH)_2$
Strontium hydroxide, $Sr(OH)_2$
Barium hydroxide, $Ba(OH)_2$
Weak
Ammonia, NH_3

Strong acids and bases are strong electrolytes.

Weak acids and bases are weak electrolytes.

Water interacts strongly with many ions, but most strongly with the hydrogen cation, H^+, a very unusual species. The H atom is a proton surrounded by an electron, so the H^+ ion is just a proton. Because its full positive charge is concentrated in such a tiny volume, H^+ attracts the negative pole of surrounding water molecules so strongly that it actually forms a covalent bond to one of them. We usually show this interaction by writing the aqueous H^+ ion as H_3O^+ (hydronium ion). Thus, to show more accurately what takes place when HBr(*g*) dissolves, we should write

$$HBr(g) + H_2O(l) \longrightarrow H_3O^+(aq) + Br^-(aq)$$

Acids and bases are electrolytes. Table 4.2 lists some acids and bases categorized in terms of their "strength"—the degree to which they dissociate into ions in aqueous solution. In water, *strong acids and strong bases dissociate completely* into ions. Therefore, like soluble ionic compounds, they are *strong* electrolytes and conduct a current well (see left photo in margin). (Note that nitric acid, featured in Sample Problem 4.4, is one of the strong acids.) In contrast, *weak acids and weak bases dissociate into ions very little, and most of their molecules remain intact.* As a result, they conduct only a small current and are *weak* electrolytes (see right photo in margin).

Strong and weak acids have one or more H atoms as part of their structure. Strong bases have either the OH^- or the O^{2-} ion as part of their structure. Soluble ionic oxides, such as K_2O, are strong bases because the oxide ion is not stable in water and reacts immediately to form hydroxide ion:

$$K_2O(s) + H_2O(l) \longrightarrow 2K^+(aq) + 2OH^-(aq)$$

Weak bases, such as ammonia, do not contain OH^- ions, but they produce them in a reaction with water that occurs to a small extent:

$$NH_3(g) + H_2O(l) \rightleftharpoons NH_4^+(aq) + OH^-(aq)$$

Weak Acids and Bases and the Equilibrium State The double reaction arrow in the equation above for ammonia's reaction with water indicates that *the reaction proceeds in both directions.* Indeed, as we'll discuss later in the text, most reactions behave this way: they seem to stop before they are complete (that is, before the limiting reactant is used up) because another reaction, the reverse of the first one, is taking place just as fast. As a result, *no further change in the amounts of reactants and products occurs,* and we say the reaction has reached a state of *equilibrium.*

The reversibility of reactions explains why some acids and bases are weak, that is, why they dissociate into ions to only a small extent: the dissociation becomes balanced by a reassociation. For example, when acetic acid dissolves in water, some of the CH_3COOH molecules react with water and form H_3O^+ and CH_3COO^- ions. As more of these ions form, they react with each other more often to re-form acetic acid and water, and we show this fact with the special (equilibrium) arrow:

$$CH_3COOH(aq) + H_2O(l) \rightleftharpoons H_3O^+(aq) + CH_3COO^-(aq)$$

In fact, in 0.1 *M* CH_3COOH at 25°C, only about 1.3% of the molecules dissociate into ions. A similarly small percentage of ammonia molecules form ions when 0.1 *M* NH_3 reacts with water at 25°C. We discuss the central idea of equilibrium and its applications for chemical and physical systems in Chapters 12, 13, and 17 through 21.

The Key Event: Formation of H_2O from H^+ and OH^-

Let's use ionic equations (and colored type) to see what occurs in acid-base reactions. We begin with the molecular equation for the reaction between the strong acid HCl and the strong base $Ba(OH)_2$:

$$2HCl(aq) + Ba(OH)_2(aq) \longrightarrow BaCl_2(aq) + 2H_2O(l)$$

Because HCl and $Ba(OH)_2$ dissociate completely and H_2O remains undissociated, the total ionic equation is

$$2H^+(aq) + 2Cl^-(aq) + Ba^{2+}(aq) + 2OH^-(aq) \longrightarrow Ba^{2+}(aq) + 2Cl^-(aq) + 2H_2O(l)$$

In the net ionic equation, we eliminate the spectator ions $Ba^{2+}(aq)$ and $Cl^-(aq)$ and see the actual reaction:

$$2H^+(aq) + 2OH^-(aq) \longrightarrow 2H_2O(l)$$

or

$$H^+(aq) + OH^-(aq) \longrightarrow H_2O(l)$$

Thus, *the essential change in all aqueous reactions between a strong acid and a strong base is that an H^+ ion from the acid and an OH^- ion from the base form a water molecule.* In fact, only the spectator ions differ from one strong acid–strong base reaction to another.

Like precipitation reactions, acid-base reactions occur through *the electrostatic attraction of ions and their removal from solution* as the product. In this case, the ions are H^+ and OH^- and the product is H_2O, which consists almost entirely of undissociated molecules. (Actually, water molecules *do* dissociate, but *very* slightly. As you'll see in Chapter 18, this slight dissociation is very important, but the formation of water in a neutralization reaction nevertheless represents an enormous net removal of H^+ and OH^- ions.)

Evaporate the water from the above reaction mixture, and the ionic solid barium chloride remains. An ionic compound that results from the reaction of an acid and a base is called a **salt.** Thus, in a typical aqueous neutralization reaction, *the reactants are an acid and a base, and the products are a salt solution and water:*

$$\underset{\text{acid}}{HX(aq)} + \underset{\text{base}}{MOH(aq)} \longrightarrow \underset{\text{salt}}{MX(aq)} + \underset{\text{water}}{H_2O(l)}$$

Note that *the cation of the salt comes from the base and the anion comes from the acid.*

As you can see, acid-base reactions, like precipitation reactions, are metathesis (double-displacement) reactions. The molecular equation for the reaction of aluminum hydroxide, the active ingredient in some antacid tablets, with HCl, the major component of stomach acid, shows this clearly:

$$3HCl(aq) + Al(OH)_3(s) \longrightarrow AlCl_3(aq) + 3H_2O(l)$$

Acid-base reactions occur frequently in the synthesis and breakdown of large biological molecules.

SAMPLE PROBLEM 4.5 Writing Ionic Equations for Acid-Base Reactions

Problem Write balanced molecular, total ionic, and net ionic equations for each of the following acid-base reactions and identify the spectator ions:
(a) Strontium hydroxide(*aq*) + perchloric acid(*aq*) ⟶
(b) Barium hydroxide(*aq*) + sulfuric acid(*aq*) ⟶

Plan All are strong acids and bases (see Table 4.2), so the essential reaction in both cases is between H^+ and OH^-. In (a), the products are H_2O and a salt solution consisting of the spectator ions. In (b), however, the salt ($BaSO_4$) is insoluble (see Table 4.1), so virtually all ions are removed from solution.

Solution (a) Writing the molecular equation:

$$Sr(OH)_2(aq) + 2HClO_4(aq) \longrightarrow Sr(ClO_4)_2(aq) + 2H_2O(l)$$

Writing the total ionic equation:

$$Sr^{2+}(aq) + 2OH^-(aq) + 2H^+(aq) + 2ClO_4^-(aq) \longrightarrow Sr^{2+}(aq) + 2ClO_4^-(aq) + 2H_2O(l)$$

Writing the net ionic equation:

$$2OH^-(aq) + 2H^+(aq) \longrightarrow 2H_2O(l) \quad \text{or} \quad OH^-(aq) + H^+(aq) \longrightarrow H_2O(l)$$

$Sr^{2+}(aq)$ and $ClO_4^-(aq)$ are the spectator ions.

(b) Writing the molecular equation:

$$Ba(OH)_2(aq) + H_2SO_4(aq) \longrightarrow BaSO_4(s) + 2H_2O(l)$$

Writing the total ionic equation:

$$Ba^{2+}(aq) + 2OH^-(aq) + 2H^+(aq) + SO_4^{2-}(aq) \longrightarrow BaSO_4(s) + 2H_2O(l)$$

The net ionic equation is the same as the total ionic equation. This is a precipitation *and* a neutralization reaction. There are no spectator ions because all the ions are used to form the two products.

FOLLOW-UP PROBLEM 4.5 Write balanced molecular, total ionic, and net ionic equations for the reaction between aqueous solutions of calcium hydroxide and nitric acid.

Acid-Base Titrations

Chemists study acid-base reactions quantitatively through titrations. In any **titration,** *one solution of known concentration is used to determine the concentration of another solution through a monitored reaction.*

In the acid-base titration shown in Figure 4.7, which is typical, a *standardized* solution of base, one whose concentration is *known,* is added slowly to an acid solution of *unknown* concentration. A known volume of the acid solution is placed in a flask, and a few drops of indicator solution are added. An *acid-base indicator* is a substance whose color is different in acid than in base. (We examine indicators in Chapters 18 and 19.) The standardized solution of base is added slowly to the flask from a buret. As the titration is close to its end, indicator molecules near a drop of added base change color due to the temporary excess of OH^- ions there. As soon as the solution is swirled, however, the indicator's acidic color returns. The **equivalence point** in the titration occurs when *all the moles of H^+ ions present in the original volume of acid solution have reacted with an equivalent number of moles of OH^- ions added from the buret:*

$$\text{Moles of } H^+ \text{ (originally in flask)} = \text{moles of } OH^- \text{ (added from buret)}$$

FIGURE 4.7 An acid-base titration. **A,** In this procedure, a measured volume of the unknown acid solution is placed in a flask beneath a buret containing the known (standardized) base solution. A few drops of indicator are added to the flask; the indicator used here is phenolphthalein, which is colorless in acid and pink in base. After an initial buret reading, base (OH^- ions) is added slowly to the acid (H^+ ions). **B,** Near the end of the titration, the indicator momentarily changes to its base color but reverts to its acid color with swirling. **C,** When the end point is reached, a tiny excess of OH^- is present, shown by the permanent change in color of the indicator. The difference between the final buret reading and the initial buret reading gives the volume of base used.

The **end point** of the titration occurs when a tiny excess of OH^- ions changes the indicator permanently to its color in base. In calculations, we assume this tiny excess is insignificant, and therefore *the amount of base needed to reach the end point is the same as the amount needed to reach the equivalence point.*

SAMPLE PROBLEM 4.6 Finding the Concentration of Acid from an Acid-Base Titration

Problem You perform an acid-base titration to standardize an HCl solution by placing 50.00 mL of HCl in a flask with a few drops of indicator solution. You put 0.1524 *M* NaOH into the buret, and the initial reading is 0.55 mL. At the end point, the buret reading is 33.87 mL. What is the concentration of the HCl solution?

Plan We must find the molarity of acid from the volume of acid (50.00 mL), the initial (0.55 mL) and final (33.87 mL) volumes of base, and the molarity of base (0.1524 *M*). First, we balance the equation. We find the volume of base added from the difference in buret readings and use the base's molarity to calculate the amount (mol) of base added. Then, we use the molar ratio from the balanced equation to find the amount (mol) of acid originally present and divide by the acid's original volume to find the molarity.

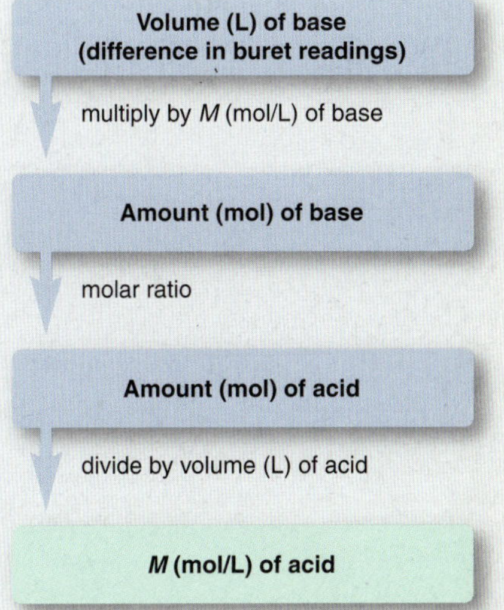

Solution Writing the balanced equation:

$$\mathrm{NaOH}(aq) + \mathrm{HCl}(aq) \longrightarrow \mathrm{NaCl}(aq) + \mathrm{H_2O}(l)$$

Finding volume (L) of NaOH solution added:

$$\text{Volume (L) of solution} = (33.87\ \cancel{\text{mL}}\ \text{soln} - 0.55\ \cancel{\text{mL}}\ \text{soln}) \times \frac{1\ \text{L}}{1000\ \cancel{\text{mL}}}$$
$$= 0.03332\ \text{L soln}$$

Finding amount (mol) of NaOH added:

$$\text{Moles of NaOH} = 0.03332\ \cancel{\text{L soln}} \times \frac{0.1524\ \text{mol NaOH}}{1\ \cancel{\text{L soln}}}$$
$$= 5.078\times10^{-3}\ \text{mol NaOH}$$

Finding amount (mol) of HCl originally present: Because the molar ratio is 1/1,

$$\text{Moles of HCl} = 5.078\times10^{-3}\ \cancel{\text{mol NaOH}} \times \frac{1\ \text{mol HCl}}{1\ \cancel{\text{mol NaOH}}} = 5.078\times10^{-3}\ \text{mol HCl}$$

Calculating molarity of HCl:

$$\text{Molarity of HCl} = \frac{5.078\times10^{-3}\ \text{mol HCl}}{50.00\ \cancel{\text{mL}}} \times \frac{1000\ \cancel{\text{mL}}}{1\ \text{L}}$$
$$= 0.1016\ M\ \text{HCl}$$

Check The answer makes sense: a larger volume of less concentrated acid neutralized a smaller volume of more concentrated base. Rounding shows that the moles of H^+ and OH^- are about equal: 50 mL $\times$ 0.1 M H^+ = 0.005 mol = 33 mL $\times$ 0.15 M OH^-.

FOLLOW-UP PROBLEM 4.6 What volume of 0.1292 M $Ba(OH)_2$ would neutralize 50.00 mL of the HCl solution standardized in the preceding sample problem?

Proton Transfer: A Closer Look at Acid-Base Reactions

We gain deeper insight into acid-base reactions if we look closely at the species in solution. Let's see what takes place when HCl gas dissolves in water. Polar water molecules pull apart each HCl molecule, and the H^+ ion ends up bonded to a water molecule. In essence, HCl *transfers its proton* to H_2O:

H^+ transfer

$$HCl(g) + H_2O(l) \longrightarrow H_3O^+(aq) + Cl^-(aq)$$

Thus, hydrochloric acid (an aqueous solution of HCl gas) actually consists of solvated H_3O^+ and Cl^- ions.

When sodium hydroxide solution is added, the H_3O^+ ion transfers a proton to the OH^- ion of the base (with the product water shown here as HOH):

H^+ transfer

$$[H_3O^+(aq) + Cl^-(aq)] + [Na^+(aq) + OH^-(aq)] \longrightarrow$$
$$H_2O(l) + Cl^-(aq) + Na^+(aq) + HOH(l)$$

Without the spectator ions, the transfer of a proton from H_3O^+ to OH^- is obvious:

H^+ transfer

$$H_3O^+(aq) + OH^-(aq) \longrightarrow H_2O(l) + HOH(l) \quad [\text{or } 2H_2O(l)]$$

This net ionic equation is identical with the one we saw earlier (see p. 125),

$$H^+(aq) + OH^-(aq) \longrightarrow H_2O(l)$$

with the additional H_2O molecule coming from the H_3O^+. Thus, *an acid-base reaction is a proton-transfer process.* In this case, the Cl^- and Na^+ ions remain in solution, and if the water is evaporated, they crystallize as the salt NaCl. Figure 4.8 shows this process on the atomic level. We'll discuss the proton-transfer concept thoroughly in Chapter 18.

Reactions of Weak Acids Ionic equations are written differently for the reactions of weak acids. When solutions of sodium hydroxide and acetic acid (CH_3COOH) are mixed, the molecular, total ionic, and net ionic equations are

Molecular equation:

$$CH_3COOH(aq) + NaOH(aq) \longrightarrow CH_3COONa(aq) + H_2O(l)$$

Total ionic equation:

$$CH_3COOH(aq) + Na^+(aq) + OH^-(aq) \longrightarrow CH_3COO^-(aq) + Na^+(aq) + H_2O(l)$$

Net ionic equation:

H^+ transfer

$$CH_3COOH(aq) + OH^-(aq) \longrightarrow CH_3COO^-(aq) + H_2O(l)$$

Acetic acid is a weak acid because it dissociates very little. To show this, *it appears undissociated in both ionic equations.* Note that H_3O^+ does not appear; rather, the proton is transferred from CH_3COOH. Therefore, only $Na^+(aq)$ is a spectator ion; $CH_3COO^-(aq)$ is not.

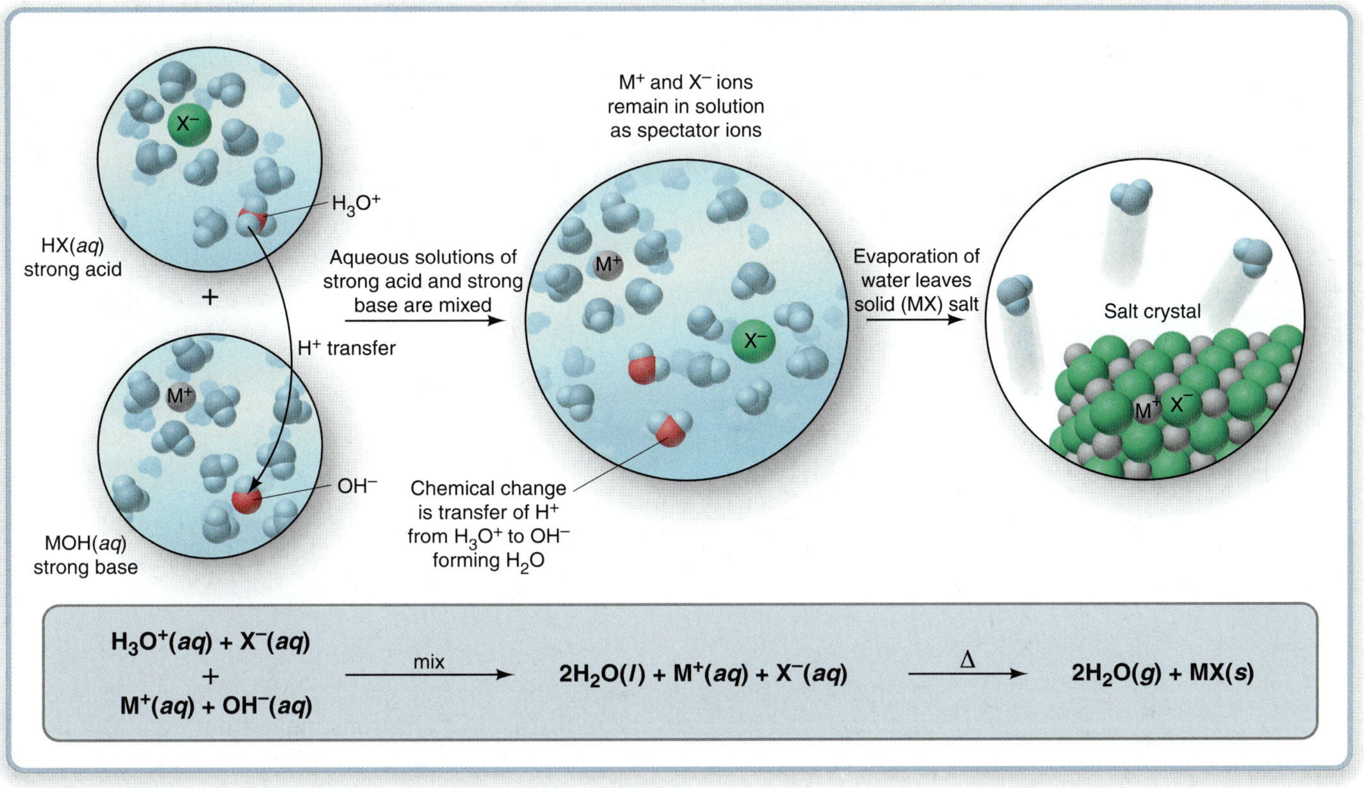

HX(*aq*) strong acid

H_3O^+

+

H^+ transfer

MOH(*aq*) strong base

OH^-

Aqueous solutions of strong acid and strong base are mixed

M^+ and X^- ions remain in solution as spectator ions

Chemical change is transfer of H^+ from H_3O^+ to OH^- forming H_2O

Evaporation of water leaves solid (MX) salt

Salt crystal

$$H_3O^+(aq) + X^-(aq) + M^+(aq) + OH^-(aq) \xrightarrow{\text{mix}} 2H_2O(l) + M^+(aq) + X^-(aq) \xrightarrow{\Delta} 2H_2O(g) + MX(s)$$

FIGURE 4.8 An aqueous strong acid–strong base reaction on the atomic scale. When solutions of a strong acid (HX) and a strong base (MOH) are mixed, the H_3O^+ from the acid transfers a proton to the OH^- from the base to form an H_2O molecule. Evaporation of the water leaves the spectator ions, X^- and M^+, as a solid ionic compound called a *salt.*

SECTION 4.4 SUMMARY

Acid-base (neutralization) reactions occur when an acid (an H^+-yielding substance) and a base (an OH^--yielding substance) react and the H^+ and OH^- ions form a water molecule. • Strong acids and bases dissociate completely in water; weak acids and bases dissociate slightly. • In a titration, a known concentration of one reactant is used to determine the concentration of the other. • An acid-base reaction can also be viewed as the transfer of a proton from an acid to a base. • Weak acids dissociate very little, so equations involving them show the acid as an intact molecule.

4.5 OXIDATION-REDUCTION (REDOX) REACTIONS

Redox reactions are a third and, perhaps, the most important class of chemical process. They include the formation of a compound from its elements (and vice versa), all combustion reactions, the reactions that generate electricity in batteries, the reactions that produce cellular energy, and many others. As you'll see, this class of reactions is so broad that many do not occur in aqueous solution. In this section, we examine the redox process and introduce some essential terminology.

The Key Event: Net Movement of Electrons Between Reactants

In **oxidation-reduction** (or **redox**) **reactions,** the key chemical event is the *net movement of electrons from one reactant to the other.* This movement of electrons occurs *from the reactant (or atom in the reactant) with less attraction for electrons to the reactant (or atom) with more attraction for electrons.*

Such movement of electron charge occurs in the formation of both ionic and covalent compounds. As an example, let's reconsider the reaction (Figure 3.7, p. 87) in which an ionic compound, MgO, forms from its elements:

$$2Mg(s) + O_2(g) \longrightarrow 2MgO(s)$$

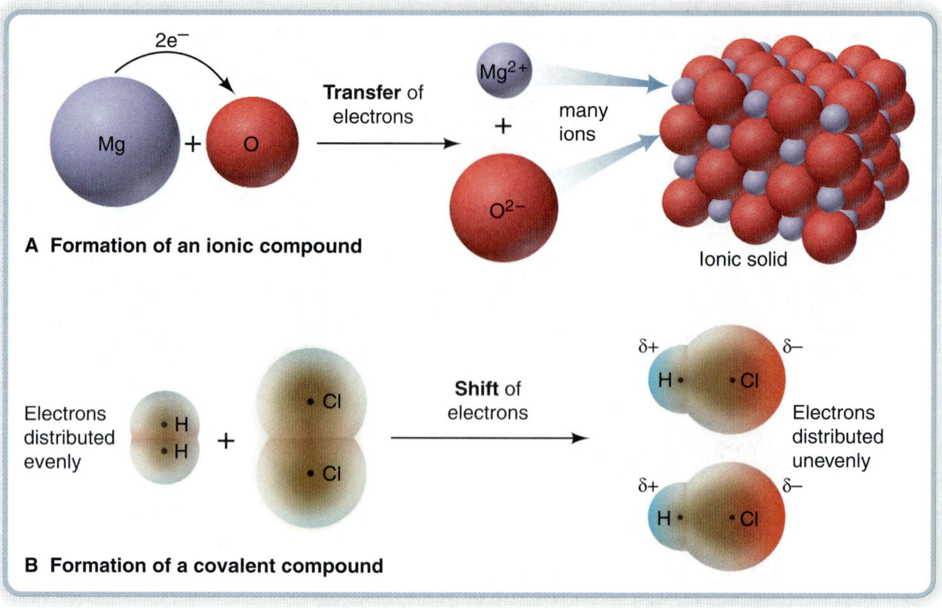

FIGURE 4.9 The redox process in compound formation. A, In forming the ionic compound MgO, each Mg atom transfers two electrons to each O atom. (Note that atoms become smaller when they lose electrons and larger when they gain electrons.) The resulting Mg^{2+} and O^{2-} ions aggregate with many others to form an ionic solid. **B,** In the reactants H_2 and Cl_2, the electron pairs are shared equally (indicated by even electron density shading). In the covalent product HCl, Cl attracts the shared electrons more strongly than H does. In effect, the H electron shifts toward Cl, as shown by higher electron density *(red)* near the Cl end of the molecule and lower electron density *(blue)* near the H end.

Figure 4.9A shows that during the reaction, each Mg atom loses two electrons and each O atom gains them; that is, two electrons move from each Mg atom to each O atom. This change represents a ***transfer** of electron charge* away from each Mg atom toward each O atom, resulting in the formation of Mg^{2+} and O^{2-} ions. The ions aggregate and form an ionic solid.

During the formation of a covalent compound from its elements, there is again a net movement of electrons, but it is more of a *shift* in electron charge than a full transfer. Thus, *ions do not form.* Consider the formation of HCl gas:

$$H_2(g) + Cl_2(g) \longrightarrow 2HCl(g)$$

To see the electron movement here, compare the electron charge distributions in the reactant bonds and in the product bonds. As Figure 4.9B shows, H_2 and Cl_2 molecules are each held together by covalent bonds in which the electrons are shared equally between the atoms (the tan shading is symmetrical). In the HCl molecule, the electrons are shared unequally because the Cl atom attracts them more strongly than the H atom does. Thus, in HCl, the H has less electron charge *(blue shading)* than it had in H_2, and the Cl has more charge *(red shading)* than it had in Cl_2. In other words, in the formation of HCl, there has been a relative ***shift** of electron charge* away from the H atom toward the Cl atom. This electron *shift* is not nearly as extreme as the electron *transfer* during MgO formation. In fact, in some reactions, the net movement of electrons may be very slight, but the reaction is still a redox process.

Some Essential Redox Terminology

Chemists use some important terminology to describe the movement of electrons in oxidation-reduction reactions. **Oxidation** is the *loss* of electrons, and **reduction** is the *gain* of electrons. (The original meaning of *reduction* comes from the process of reducing large amounts of metal ore to smaller amounts of metal, but you'll see shortly why we use the term "reduction" for the act of gaining.)

For example, during the formation of magnesium oxide, Mg undergoes oxidation (electron loss) and O_2 undergoes reduction (electron gain). The loss and gain are simultaneous, but we can imagine them occurring in separate steps:

Oxidation (electron loss by Mg): $Mg \longrightarrow Mg^{2+} + 2e^-$

Reduction (electron gain by O_2): $\frac{1}{2}O_2 + 2e^- \longrightarrow O^{2-}$

One reactant acts on the other. Thus, we say that *O_2 oxidizes Mg,* and that O_2 is the **oxidizing agent,** the species doing the oxidizing. Similarly, *Mg reduces O_2,* so Mg is the **reducing agent,** the species doing the reducing.

Note especially that O_2 takes the electrons that Mg loses or, put the other way around, Mg gives up the electrons that O_2 gains. This give-and-take of electrons means that *the oxidizing agent is reduced* because it takes the electrons (and thus gains them), and *the reducing agent is oxidized* because it gives up the electrons (and thus loses them). In the formation of HCl, Cl_2 oxidizes H_2 (H loses some electron charge and Cl gains it), which is the same as saying that H_2 reduces Cl_2. The reducing agent, H_2, is oxidized and the oxidizing agent, Cl_2, is reduced.

Using Oxidation Numbers to Monitor the Movement of Electron Charge

Chemists have devised a useful "bookkeeping" system to monitor which atom loses electron charge and which atom gains it. Each atom in a molecule (or ionic compound) is assigned an **oxidation number (O.N.),** or *oxidation state,* the charge the atom would have *if* electrons were not shared but were transferred completely. Thus, the oxidation number for each element in a binary *ionic* compound equals the ionic charge. However, the oxidation number for each element in a *covalent* compound (or in a polyatomic ion) is not as obvious because the atoms don't have whole charges. In general, *oxidation numbers are determined by the set of rules* in Table 4.3. (Oxidation numbers are assigned according to the relative attraction of an atom for electrons, so they are ultimately based on atomic properties, as you'll see in Chapters 8 and 9.) An O.N. has the sign *before* the number (e.g., +2), whereas an ionic charge has the sign *after* the number (e.g., 2+). Also, unlike a 1+ ionic charge in a chemical formula, an O.N. of +1 or −1 retains the numeral.

Table 4.3 Rules for Assigning an Oxidation Number (O.N.)

General Rules

1. For an atom in its elemental form (Na, O_2, Cl_2, etc.): O.N. = 0
2. For a monatomic ion: O.N. = ion charge
3. The sum of O.N. values for the atoms in a molecule or formula unit of a compound equals zero. The sum of O.N. values for the atoms in a polyatomic ion equals the ion's charge.

Rules for Specific Atoms or Periodic Table Groups

1. For Group 1A(1):	O.N. = +1 in all compounds
2. For Group 2A(2):	O.N. = +2 in all compounds
3. For hydrogen:	O.N. = +1 in combination with nonmetals
	O.N. = −1 in combination with metals and boron
4. For fluorine:	O.N. = −1 in all compounds
5. For oxygen:	O.N. = −1 in peroxides
	O.N. = −2 in all other compounds (except with F)
6. For Group 7A(17):	O.N. = −1 in combination with metals, nonmetals (except O), and other halogens lower in the group

SAMPLE PROBLEM 4.7 Determining the Oxidation Number of an Element

Problem Determine the oxidation number (O.N.) of each element in these compounds:
(a) Zinc chloride **(b)** Sulfur trioxide **(c)** Nitric acid

Plan We apply Table 4.3, noting the general rules that the O.N. values in a compound add up to zero, and the O.N. values in a polyatomic ion add up to the ion's charge.

Solution (a) $ZnCl_2$. The sum of O.N.s for the monatomic ions in the compound must equal zero. The O.N. of the Zn^{2+} ion is +2. The O.N. of each Cl^- ion is −1, for a total of −2. The sum of O.N.s is +2 + (−2), or 0.

(b) SO_3. The O.N. of each oxygen is -2, for a total of -6. The O.N.s must add up to zero, so the O.N. of S is $+6$.

(c) HNO_3. The O.N. of H is $+1$, so the O.N.s of the NO_3 group must add up to -1 to give zero for the compound. The O.N. of each O is -2 for a total of -6. Therefore, the O.N. of N is $+5$.

FOLLOW-UP PROBLEM 4.7 Determine the O.N. of each element in the following:
(a) Scandium oxide (Sc_2O_3) **(b)** Gallium chloride ($GaCl_3$)
(c) Hydrogen phosphate ion **(d)** Iodine trifluoride

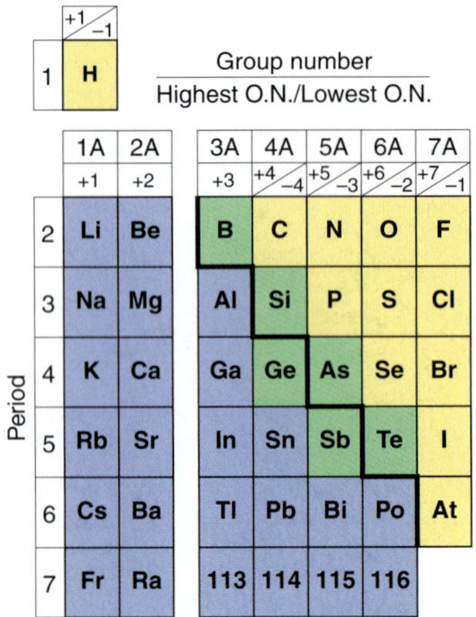

FIGURE 4.10 Highest and lowest oxidation numbers of reactive main-group elements. The A-group number shows the highest possible oxidation number (O.N.) for a main-group element. (Two important exceptions are O, which never has an O.N. of +6, and F, which never has an O.N. of +7.) For nonmetals *(yellow)* and metalloids *(green)*, the A-group number minus 8 gives the lowest possible oxidation number.

The periodic table is a great help in learning the highest and lowest oxidation numbers of most main-group elements, as Figure 4.10 shows:

- For most main-group elements, the A-group number (1A, 2A, and so on) is the *highest* oxidation number (always positive) of any element in the group. The exceptions are O and F (see Table 4.3).
- For main-group nonmetals and some metalloids, the A-group number minus 8 is the *lowest* oxidation number (always negative) of any element in the group.

For example, the highest oxidation number of S (Group 6A) is $+6$, as in SF_6, and the lowest is $(6 - 8)$, or -2, as in FeS and other metal sulfides.

Thus, another way to define a redox reaction is one in which *the oxidation numbers of the species change,* and the most important use of oxidation numbers is to monitor these changes:

- If a given atom has a higher (more positive or less negative) oxidation number in the product than it had in the reactant, the reactant species that contains the atom was oxidized (lost electrons) and is the reducing agent. Thus, *oxidation is represented by an increase in oxidation number.*
- If an atom has a lower (more negative or less positive) oxidation number in the product than it had in the reactant, the reactant species that contains the atom was reduced (gained electrons) and is the oxidizing agent. Thus, *the gain of electrons is represented by a decrease (a "reduction") in oxidation number.*

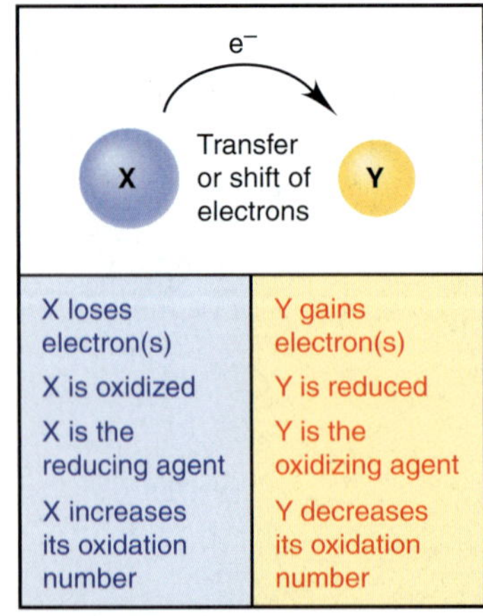

FIGURE 4.11 A summary of terminology for oxidation-reduction (redox) reactions.

It is essential to realize that *the transferred electrons are never free: the reducing agent loses electrons and the oxidizing agent gains them simultaneously.* In other words, a complete reaction *must* be an "oxidation-reduction," *not* an "oxidation" *or* a "reduction." Figure 4.11 summarizes redox terminology. (For the remainder of this chapter, blue oxidation numbers represent oxidation, and red oxidation numbers indicate reduction.)

SAMPLE PROBLEM 4.8 Recognizing Oxidizing and Reducing Agents

Problem Identify the oxidizing agent and reducing agent in each of the following:
(a) $2Al(s) + 3H_2SO_4(aq) \longrightarrow Al_2(SO_4)_3(aq) + 3H_2(g)$
(b) $PbO(s) + CO(g) \longrightarrow Pb(s) + CO_2(g)$
(c) $2H_2(g) + O_2(g) \longrightarrow 2H_2O(g)$

Plan We first assign an oxidation number (O.N.) to each atom (or ion) based on the rules in Table 4.3. The reactant is the reducing agent if it contains an atom that is oxidized (O.N. increased from left side to right side of the equation). The reactant is the oxidizing agent if it contains an atom that is reduced (O.N. decreased).

Solution (a) Assigning oxidation numbers:

$$\overset{0}{2Al}(s) + 3\overset{+1}{H_2}\overset{+6}{S}\overset{-2}{O_4}(aq) \longrightarrow \overset{+3}{Al_2}(\overset{+6}{S}\overset{-2}{O_4})_3(aq) + 3\overset{0}{H_2}(g)$$

The O.N. of Al increased from 0 to $+3$ (Al lost electrons), so Al was oxidized; Al is the reducing agent.

The O.N. of H decreased from $+1$ to 0 (H gained electrons), so H^+ was reduced; H_2SO_4 is the oxidizing agent.

(b) Assigning oxidation numbers:

$$\overset{+2\ -2}{PbO(s)} + \overset{+2\ -2}{CO(g)} \longrightarrow \overset{0}{Pb(s)} + \overset{+4\ -2}{CO_2(g)}$$

Pb decreased its O.N. from +2 to 0, so PbO was reduced; PbO is the oxidizing agent.
C increased its O.N. from +2 to +4, so CO was oxidized; CO is the reducing agent.
In general, when a substance (such as CO) becomes one with more O atoms (as in CO_2), it is oxidized; and when a substance (such as PbO) becomes one with fewer O atoms (as in Pb), it is reduced.

(c) Assigning oxidation numbers:

$$\overset{0}{2H_2(g)} + \overset{0}{O_2(g)} \longrightarrow \overset{+1\ -2}{2H_2O(g)}$$

O_2 was reduced (O.N. of O decreased from 0 to −2); O_2 is the oxidizing agent.
H_2 was oxidized (O.N. of H increased from 0 to +1); H_2 is the reducing agent.
Oxygen is always the oxidizing agent in a combustion reaction.

Comment **1.** Compare the O.N. values in (c) with those in another common reaction that forms water—the net ionic equation for an acid-base reaction:

$$\overset{+1}{H^+(aq)} + \overset{-2\ +1}{OH^-(aq)} \longrightarrow \overset{+1\ -2}{H_2O(l)}$$

Note that the O.N. values remain the same on both sides of the acid-base equation. Therefore, *an acid-base reaction is* **not** *a redox reaction.*

2. If a substance occurs in its elemental form on one side of an equation, it can't possibly be in its elemental form on the other side, so the reaction must be a redox process. Notice that elements appear in all three parts of this problem.

FOLLOW-UP PROBLEM 4.8 Identify each oxidizing agent and each reducing agent:
(a) $2Fe(s) + 3Cl_2(g) \longrightarrow 2FeCl_3(s)$ **(b)** $2C_2H_6(g) + 7O_2(g) \longrightarrow 4CO_2(g) + 6H_2O(g)$
(c) $5CO(g) + I_2O_5(s) \longrightarrow I_2(s) + 5CO_2(g)$

SECTION 4.5 SUMMARY

When one reactant has a greater attraction for electrons than another, there is a net movement of electron charge, and a redox reaction takes place. • Electron gain (reduction) and electron loss (oxidation) occur simultaneously. • The redox process is tracked by assigning oxidation numbers to each atom in a reaction. The species that is oxidized (contains an atom that increases in oxidation number) is the reducing agent; the species that is reduced (contains an atom that decreases in oxidation number) is the oxidizing agent.

4.6 ELEMENTS IN REDOX REACTIONS

As you saw in Comment 2 of Sample Problem 4.8, *whenever atoms appear in the form of a free element on one side of an equation and as part of a compound on the other, the reaction is a redox process.* And, while there are many redox reactions that do *not* involve free elements, we'll focus here on the many others that do. One way to classify these is by comparing the numbers of reactants and products. By that approach, we have three types:

- *Combination reactions:* two or more reactants form one product:

$$X + Y \longrightarrow Z$$

- *Decomposition reactions:* one reactant forms two or more products:

$$Z \longrightarrow X + Y$$

- *Displacement reactions:* the number of substances is the same but atoms (or ions) exchange places:

$$X + YZ \longrightarrow XZ + Y$$

Combining Two Elements Two elements may react to form binary ionic or covalent compounds. Here are some important examples:

1. *Metal and nonmetal form an ionic compound.* A metal, such as aluminum, reacts with a nonmetal, such as oxygen. The change in O.N.s shows that the metal is oxidized, so it is the reducing agent; the nonmetal is reduced, so it is the oxidizing agent.

$$4\overset{0}{\text{Al}}(s) + 3\overset{0}{\text{O}}_2(g) \longrightarrow 2\overset{+3}{\text{Al}}_2\overset{-2}{\text{O}}_3(s)$$

Figure 3.7 (p. 87) shows the redox reaction between magnesium metal and oxygen on the macroscopic and atomic scales.

2. *Two nonmetals form a covalent compound.* In one of thousands of examples, ammonia forms from nitrogen and hydrogen in a reaction that is carried out on an enormous scale in industry:

$$\overset{0}{\text{N}}_2(g) + 3\overset{0}{\text{H}}_2(g) \longrightarrow 2\overset{-3}{\text{N}}\overset{+1}{\text{H}}_3(g)$$

Combining Compound and Element Many binary covalent compounds react with nonmetals to form larger compounds. Many nonmetal oxides react with additional O_2 to form *higher oxides* (those with more O atoms in each molecule). For example,

$$2\overset{+2}{\text{N}}\overset{-2}{\text{O}}(g) + \overset{0}{\text{O}}_2(g) \longrightarrow 2\overset{+4}{\text{N}}\overset{-2}{\text{O}}_2(g)$$

Similarly, many nonmetal halides combine with additional halogen:

$$\overset{+3}{\text{P}}\overset{-1}{\text{Cl}}_3(l) + \overset{0}{\text{Cl}}_2(g) \longrightarrow \overset{+5}{\text{P}}\overset{-1}{\text{Cl}}_5(s)$$

Decomposing Compounds into Elements A decomposition reaction occurs when a reactant absorbs enough energy for one or more of its bonds to break. The energy can take many forms; we'll focus in this discussion on heat and electricity. The products are either elements or elements and smaller compounds. Following are several common examples:

1. *Thermal decomposition.* When the energy absorbed is heat, the reaction is a thermal decomposition. (A Greek *delta,* Δ, above a reaction arrow indicates that significant heat is required for the reaction.) Many metal oxides, chlorates, and perchlorates release oxygen when strongly heated. Heating potassium chlorate is a method for forming small amounts of oxygen in the laboratory; the same reaction occurs in some explosives and fireworks:

$$2\overset{+1}{\text{K}}\overset{+5}{\text{Cl}}\overset{-2}{\text{O}}_3(s) \xrightarrow{\Delta} 2\overset{+1}{\text{K}}\overset{-1}{\text{Cl}}(s) + 3\overset{0}{\text{O}}_2(g)$$

Notice that the lone reactant is the oxidizing *and* the reducing agent.

2. *Electrolytic decomposition.* In the process of *electrolysis,* a compound absorbs electrical energy and decomposes into its elements. Observing the electrolysis of water was crucial in the establishment of atomic masses:

$$2\overset{+1}{\text{H}}_2\overset{-2}{\text{O}}(l) \xrightarrow{\text{electricity}} 2\overset{0}{\text{H}}_2(g) + \overset{0}{\text{O}}_2(g)$$

Many active metals, such as sodium, magnesium, and calcium, are produced industrially by electrolysis of their molten halides:

$$\overset{+2}{\text{Mg}}\overset{-1}{\text{Cl}}_2(l) \xrightarrow{\text{electricity}} \overset{0}{\text{Mg}}(l) + \overset{0}{\text{Cl}}_2(g)$$

(We'll examine the details of electrolysis in Chapter 21.)

Displacing One Element by Another; Activity Series As we said, displacement reactions have the same number of reactants as products. We mentioned double-displacement (metathesis) reactions in discussing precipitation and acid-base reactions. The other type, *single-displacement reactions*, are all oxidation-reduction processes. They occur when one atom displaces the ion of a different atom from solution. When the reaction involves metals, the atom reduces the ion; when it involves nonmetals (specifically halogens), the atom oxidizes the ion. Chemists rank various elements into activity series—one for metals and one for halogens—in order of their ability to displace one another.

1. *The activity series of the metals*. Metals can be ranked by their ability to displace H_2 (actually reduce H^+) from various sources or by their ability to displace one another from solution.

- *A metal displaces H_2 from water or acid*. The most reactive metals, such as those from Group 1A(1) and Ca, Sr, and Ba from Group 2A(2), displace H_2 from water, and they do so vigorously. Figure 4.12 shows this reaction for lithium. Heat is needed to speed the reaction of slightly less reactive metals, such as Al and Zn, so these metals displace H_2 from steam:

$$\overset{0}{2Al}(s) + 6\overset{+1}{H_2}\overset{-2}{O}(g) \longrightarrow 2\overset{+3}{Al}(\overset{-2}{O}\overset{+1}{H})_3(s) + 3\overset{0}{H_2}(g)$$

Still less reactive metals, such as nickel, do not react with water but *do* react with acids. Because the concentration of H^+ is higher in acid solutions than in water, H_2 is displaced more easily. Here is the net ionic equation:

$$\overset{0}{Ni}(s) + 2\overset{+1}{H^+}(aq) \longrightarrow \overset{+2}{Ni^{2+}}(aq) + \overset{0}{H_2}(g)$$

In all such reactions, the metal is the reducing agent (O.N. of metal increases), and water or acid is the oxidizing agent (O.N. of H decreases). The least reactive metals, such as silver and gold, cannot displace H_2 from any source.

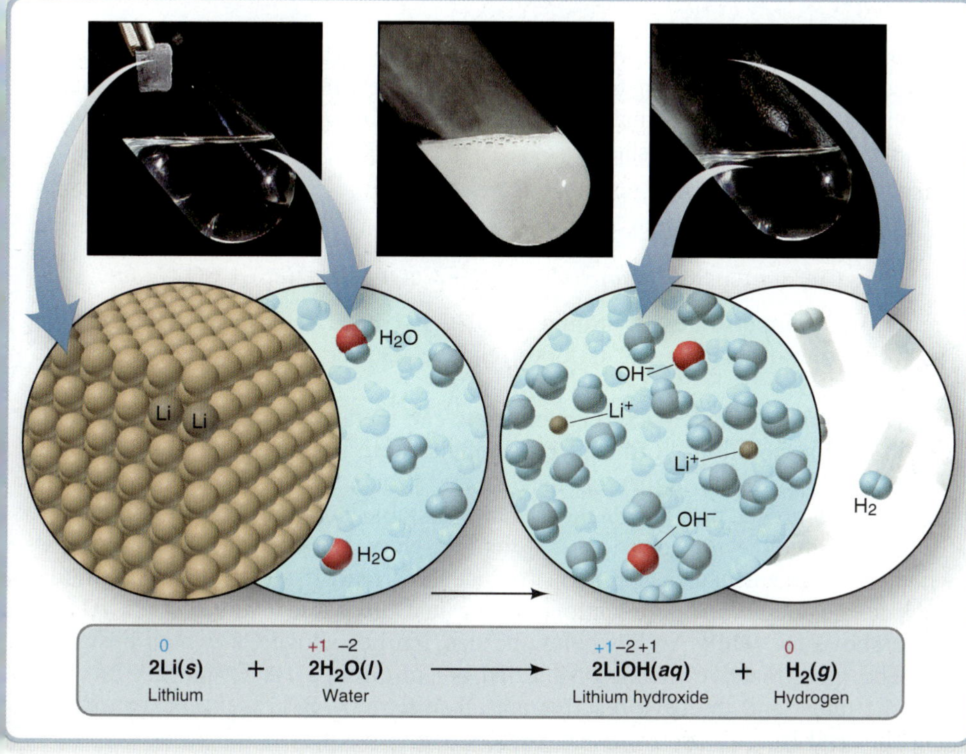

FIGURE 4.12 An active metal displacing hydrogen from water. Lithium displaces hydrogen from water in a vigorous reaction that yields an aqueous solution of lithium hydroxide and hydrogen gas, as shown on the macroscopic scale *(top)*, at the atomic scale *(middle)*, and as a balanced equation *(bottom)*. (For clarity, the atomic-scale view of water has been greatly simplified, and only water molecules involved in the reaction are colored red and blue.)

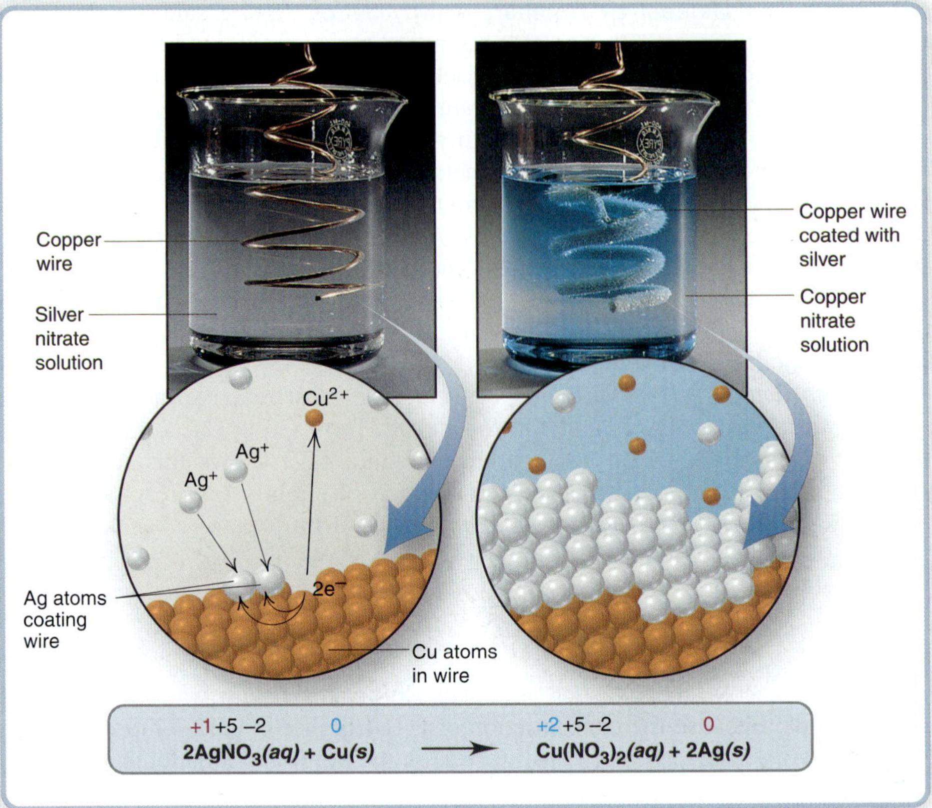

FIGURE 4.13 Displacing one metal by another. More reactive metals displace less reactive metals from solution. In this reaction, Cu atoms each give up two electrons as they become Cu^{2+} ions and leave the wire. The electrons are transferred to two Ag^{+} ions that become Ag atoms and deposit on the wire. With time, a coating of crystalline silver coats the wire. Thus, copper has displaced silver (reduced silver ion) from solution. The reaction is depicted as the laboratory view *(top),* the atomic-scale view *(middle),* and the balanced redox equation *(bottom).*

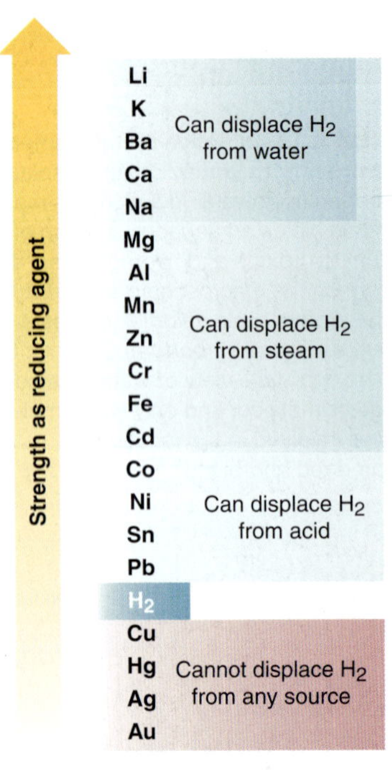

FIGURE 4.14 The activity series of the metals. This list of metals (and H_2) is arranged with the most active metal (strongest reducing agent) at the top and the least active metal (weakest reducing agent) at the bottom. The four metals below H_2 cannot displace it from any source.

- *A metal displaces another metal ion from solution.* Direct comparisons of metal reactivity are clearest in these reactions. For example, zinc metal displaces copper(II) ion from (actually reduces Cu^{2+} in) copper(II) sulfate solution, as the total ionic equation shows:

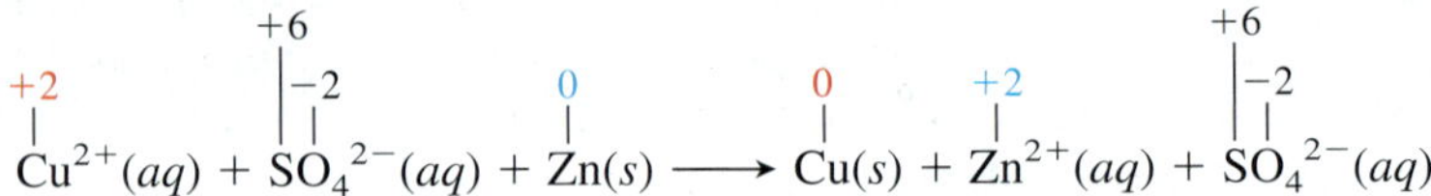

$$\overset{+2}{Cu^{2+}}(aq) + \overset{+6\,-2}{SO_4^{\;2-}}(aq) + \overset{0}{Zn}(s) \longrightarrow \overset{0}{Cu}(s) + \overset{+2}{Zn^{2+}}(aq) + \overset{+6\,-2}{SO_4^{\;2-}}(aq)$$

Figure 4.13 demonstrates in atomic detail that copper metal can displace silver ion from solution. Thus, zinc is more reactive than copper, which is more reactive than silver.

The results of many such reactions between metals and water, aqueous acids, and metal-ion solutions form the basis of the **activity series of the metals.** In Figure 4.14, elements higher on the list are stronger reducing agents than elements lower down; that is, for those that are stable in water, elements higher on the list can reduce aqueous ions of elements lower down. The list also shows whether the metal can displace H_2 (reduce H^+) and, if so, from which source. Look at the metals in the equations we've just discussed. Note that Li, Al, and Ni lie above H_2, while Ag lies below it; also, Zn lies above Cu, which lies above Ag. The most reactive metals on the list are in Groups 1A(1) and 2A(2) of the periodic table, and the least reactive are all in Group 1B(11), except for mercury (Hg) in 2B(12).

2. *The activity series of the halogens.* Reactivity decreases down Group 7A(17), so we can arrange the halogens into their own activity series:

$$F_2 > Cl_2 > Br_2 > I_2$$

A halogen higher in the periodic table is a stronger oxidizing agent than one lower down. Thus, chlorine can oxidize bromide ions or iodide ions from solution, and bromine can oxidize iodide ions. Here, chlorine displaces bromine:

$$\overset{-1}{2Br^-}(aq) + \overset{0}{Cl_2}(aq) \longrightarrow \overset{0}{Br_2}(aq) + \overset{-1}{2Cl^-}(aq)$$

Combustion Reactions *Combustion* is the process of combining with oxygen, often with the release of heat and sometimes light, as in a flame. Combustion reactions do not fall neatly into classes based on the number of reactants and products, but *all are redox processes* because elemental oxygen is a reactant:

$$2CO(g) + O_2(g) \longrightarrow 2CO_2(g)$$

The combustion reactions that we commonly use to produce energy involve organic mixtures such as coal, gasoline, and natural gas as reactants. These mixtures consist of substances with many carbon-carbon and carbon-hydrogen bonds. During the reaction, these bonds break, and each C and H atom combines with oxygen. Therefore, the major products are CO_2 and H_2O. The combustion of the hydrocarbon butane, which is used in camp stoves, is typical:

$$2C_4H_{10}(g) + 13O_2(g) \longrightarrow 8CO_2(g) + 10H_2O(g)$$

In modern *fuel cells,* in which H_2 reacts with O_2, and in many biochemical combustions, light is not produced. Biological *respiration* is a multistep combustion process that occurs within our cells when we "burn" organic foodstuffs, such as glucose, for energy:

$$C_6H_{12}O_6(s) + 6O_2(g) \longrightarrow 6CO_2(g) + 6H_2O(g)$$

SAMPLE PROBLEM 4.9 Identifying the Type of Redox Reaction

Problem Classify each of the following redox reactions as a combination, decomposition, or displacement reaction, write a balanced molecular equation for each, as well as total and net ionic equations for part (c), and identify the oxidizing and reducing agents:
(a) Magnesium(*s*) + nitrogen(*g*) ⟶ magnesium nitride(*s*)
(b) Hydrogen peroxide(*l*) ⟶ water + oxygen gas
(c) Aluminum(*s*) + lead(II) nitrate(*aq*) ⟶ aluminum nitrate(*aq*) + lead(*s*)

Plan To decide on reaction type, recall that combination reactions produce fewer products than reactants, decomposition reactions produce more products, and displacement reactions have the same number of reactants and products. The oxidation number (O.N.) becomes more positive for the reducing agent and less positive for the oxidizing agent.

Solution (a) Combination: two substances form one. This reaction occurs, along with formation of magnesium oxide, when magnesium burns in air:

$$3\overset{0}{Mg}(s) + \overset{0}{N_2}(g) \longrightarrow \overset{+2}{Mg_3}\overset{-3}{N_2}(s)$$

Mg is the reducing agent; N_2 is the oxidizing agent.

(b) Decomposition: one substance forms two. This reaction occurs within every bottle of this common household antiseptic. Hydrogen peroxide is very unstable and breaks down from heat, light, or just shaking:

$$2\overset{+1}{H_2}\overset{-1}{O_2}(l) \longrightarrow 2\overset{+1}{H_2}\overset{-2}{O}(l) + \overset{0}{O_2}(g)$$

(The upper -1 above O_2 in H_2O_2 also appears printed in blue above the red -1.)

H_2O_2 is both the oxidizing *and* the reducing agent. The O.N. of O in peroxides is -1. It increases to 0 in O_2 and decreases to -2 in H_2O.

(c) Displacement: two substances form two others. As Figure 4.14 shows, Al is more active than Pb and, thus, displaces it from aqueous solution:

$$\overset{0}{2Al(s)} + 3\overset{+2}{Pb}(\overset{+5}{N}\overset{-2}{O_3})_2(aq) \longrightarrow 2\overset{+3}{Al}(\overset{+5}{N}\overset{-2}{O_3})_3(aq) + 3\overset{0}{Pb}(s)$$

Al is the reducing agent; $Pb(NO_3)_2$ is the oxidizing agent.

The total ionic equation is

$$2Al(s) + 3Pb^{2+}(aq) + 6NO_3^-(aq) \longrightarrow 2Al^{3+}(aq) + 6NO_3^-(aq) + 3Pb(s)$$

The net ionic equation is

$$2Al(s) + 3Pb^{2+}(aq) \longrightarrow 2Al^{3+}(aq) + 3Pb(s)$$

FOLLOW-UP PROBLEM 4.9 Classify each of the following redox reactions as a combination, decomposition, or displacement reaction, write a balanced molecular equation for each, as well as total and net ionic equations for parts (b) and (c), and identify the oxidizing and reducing agents:

(a) $S_8(s) + F_2(g) \longrightarrow SF_4(g)$ **(b)** $CsI(aq) + Cl_2(aq) \longrightarrow CsCl(aq) + I_2(aq)$

(c) $Ni(NO_3)_2(aq) + Cr(s) \longrightarrow Ni(s) + Cr(NO_3)_3(aq)$

SECTION 4.6 SUMMARY

Any reaction that has the same atoms in elemental form and in a compound is a redox reaction. • In combination reactions, elements combine to form a compound, or a compound and an element combine. • Decomposition of compounds by absorption of heat or electricity can form elements or a compound and an element. • In displacement reactions, one element displaces another from solution. • Activity series rank elements in order of reactivity. The activity series of the metals ranks metals by their ability to displace H_2 from water, steam, or acid, or to displace one another from solution. Combustion releases heat (and sometimes light) through reaction of a substance with O_2.

CHAPTER REVIEW GUIDE

The following sections provide many aids to help you study this chapter. (Numbers in parentheses refer to pages, unless noted otherwise.)

• LEARNING OBJECTIVES *These are concepts and skills to review after studying this chapter.*

Related section (§), sample problem (SP), and end-of-chapter problem (EP) numbers are listed in parentheses.

1. Understand how water dissolves an ionic compound compared to a covalent compound and which solution contains an electrolyte; use a compound's formula to find moles of ions in solution (§ 4.1) (SP 4.1) (EPs 4.1–4.19)
2. Understand the key events in precipitation and acid-base reactions and use ionic equations to describe them; distinguish between strong and weak acids and bases and calculate an unknown concentration from a titration (§ 4.2–4.4) (SPs 4.2–4.6) (EPs 4.20–4.47)
3. Understand the key event in the redox process; determine the oxidation number of any element in a compound; identify the oxidizing and reducing agents in a reaction (§ 4.5) (SPs 4.7, 4.8) (EPs 4.48–4.62)
4. Identify three important types of redox reactions that include elements: combination, decomposition, displacement (§ 4.6) (SP 4.9) (EPs 4.63–4.77)

• KEY TERMS *These important terms appear in boldface in the chapter and are defined again in the Glossary.*

Section 4.1
polar molecule (114)
solvated (115)
electrolyte (115)
nonelectrolyte (117)

Section 4.2
molecular equation (118)
total ionic equation (118)
spectator ion (119)
net ionic equation (119)

Section 4.3
precipitation reaction (119)
precipitate (119)
metathesis reaction (120)

Section 4.4
acid-base reaction (123)
neutralization reaction (123)
acid (123)
base (123)
salt (125)
titration (126)
equivalence point (126)
end point (127)

Section 4.5
oxidation-reduction (redox) reaction (129)
oxidation (130)
reduction (130)
oxidizing agent (131)
reducing agent (131)
oxidation number (O.N.) (or oxidation state) (131)

Section 4.6
activity series of the metals (136)

• BRIEF SOLUTIONS TO *FOLLOW-UP PROBLEMS* *Compare your own solutions to these calculation steps and answers.*

4.1 (a) $KClO_4(s) \xrightarrow{H_2O} K^+(aq) + ClO_4^-(aq)$;
2 mol of K^+ and 2 mol of ClO_4^-

(b) $Mg(C_2H_3O_2)_2(s) \xrightarrow{H_2O} Mg^{2+}(aq) + 2C_2H_3O_2^-(aq)$;
2.49 mol of Mg^{2+} and 4.97 mol of $C_2H_3O_2^-$

(c) $(NH_4)_2CrO_4(s) \xrightarrow{H_2O} 2NH_4^+(aq) + CrO_4^{2-}(aq)$;
6.24 mol of NH_4^+ and 3.12 mol of CrO_4^{2-}

(d) $NaHSO_4(s) \xrightarrow{H_2O} Na^+(aq) + HSO_4^-(aq)$;
0.73 mol of Na^+ and 0.73 mol of HSO_4^-

4.2 (a) $Fe^{3+}(aq) + 3Cl^-(aq) + 3Cs^+(aq) + PO_4^{3-}(aq) \longrightarrow FePO_4(s) + 3Cl^-(aq) + 3Cs^+(aq)$

$Fe^{3+}(aq) + PO_4^{3-}(aq) \longrightarrow FePO_4(s)$

(b) $2Na^+(aq) + 2OH^-(aq) + Cd^{2+}(aq) + 2NO_3^-(aq) \longrightarrow 2Na^+(aq) + 2NO_3^-(aq) + Cd(OH)_2(s)$

$2OH^-(aq) + Cd^{2+}(aq) \longrightarrow Cd(OH)_2(s)$

(c) No reaction occurs

(d) $2Ag^+(aq) + SO_4^{2-}(aq) + Ba^{2+}(aq) + 2Cl^-(aq) \longrightarrow 2AgCl(s) + BaSO_4(s)$

Total and net ionic equations are identical.

4.3 (a) Beaker A contains a solution of $Zn(NO_3)_2$.

(b) Beaker B contains a solution of $Ba(OH)_2$.

(c) The precipitate is zinc hydroxide, and the spectator ions are Ba^{2+} and NO_3^-.

Molecular: $Zn(NO_3)_2(aq) + Ba(OH)_2(aq) \longrightarrow Zn(OH)_2(s) + Ba(NO_3)_2(aq)$

Total ionic:
$Zn^{2+}(aq) + 2NO_3^-(aq) + Ba^{2+}(aq) + 2OH^-(aq) \longrightarrow Zn(OH)_2(s) + Ba^{2+}(aq) + 2NO_3^-(aq)$

Net ionic: $Zn^{2+}(aq) + 2OH^-(aq) \longrightarrow Zn(OH)_2(s)$

(d) The OH^- ion is limiting.

Mass (g) of $Zn(OH)_2$

$$= 6\ \text{OH}^-\ \text{particles} \times \frac{0.050\ \text{mol OH}^-\ \text{ions}}{1\ \text{OH}^-\ \text{particle}} \times \frac{1\ \text{mol Zn(OH)}_2}{2\ \text{mol OH}^-\ \text{ions}} \times \frac{99.43\ \text{g Zn(OH)}_2}{1\ \text{mol Zn(OH)}_2}$$
$$= 15\ \text{g Zn(OH)}_2$$

4.4 Moles of H^+
$$= 451\ \text{mL} \times \frac{1\ \text{L}}{10^3\ \text{mL}} \times \frac{3.20\ \text{mol HBr}}{1\ \text{L soln}} \times \frac{1\ \text{mol H}^+}{1\ \text{mol HBr}}$$
$$= 1.44\ \text{mol H}^+$$

4.5 $Ca(OH)_2(aq) + 2HNO_3(aq) \longrightarrow Ca(NO_3)_2(aq) + 2H_2O(l)$

$Ca^{2+}(aq) + 2OH^-(aq) + 2H^+(aq) + 2NO_3^-(aq) \longrightarrow Ca^{2+}(aq) + 2NO_3^-(aq) + 2H_2O(l)$

$H^+(aq) + OH^-(aq) \longrightarrow H_2O(l)$

4.6 $Ba(OH)_2(aq) + 2HCl(aq) \longrightarrow BaCl_2(aq) + 2H_2O(l)$

Volume (L) of soln
$$= 50.00\ \text{mL HCl soln} \times \frac{1\ \text{L}}{10^3\ \text{mL}} \times \frac{0.1016\ \text{mol HCl}}{1\ \text{L soln}} \times \frac{1\ \text{mol Ba(OH)}_2}{2\ \text{mol HCl}} \times \frac{1\ \text{L soln}}{0.1292\ \text{mol Ba(OH)}_2}$$
$$= 0.01966\ \text{L}$$

4.7 (a) O.N. of Sc = +3; O.N. of O = −2

(b) O.N. of Ga = +3; O.N. of Cl = −1

(c) O.N. of H = +1; O.N. of P = +5; O.N. of O = −2

(d) O.N. of I = +3; O.N. of F = −1

4.8 (a) Fe is the reducing agent; Cl_2 is the oxidizing agent.

(b) C_2H_6 is the reducing agent; O_2 is the oxidizing agent.

(c) CO is the reducing agent; I_2O_5 is the oxidizing agent.

4.9 (a) Combination:

$S_8(s) + 16F_2(g) \longrightarrow 8SF_4(g)$

S_8 is the reducing agent; F_2 is the oxidizing agent.

(b) Displacement:

$2CsI(aq) + Cl_2(aq) \longrightarrow 2CsCl(aq) + I_2(aq)$

Cl_2 is the oxidizing agent; CsI is the reducing agent.

$2Cs^+(aq) + 2I^-(aq) + Cl_2(aq) \longrightarrow 2Cs^+(aq) + 2Cl^-(aq) + I_2(aq)$

$2I^-(aq) + Cl_2(aq) \longrightarrow 2Cl^-(aq) + I_2(aq)$

(c) Displacement:

$3Ni(NO_3)_2(aq) + 2Cr(s) \longrightarrow 3Ni(s) + 2Cr(NO_3)_3(aq)$

$3Ni^{2+}(aq) + 6NO_3^-(aq) + 2Cr(s) \longrightarrow 3Ni(s) + 2Cr^{3+}(aq) + 6NO_3^-(aq)$

$3Ni^{2+}(aq) + 2Cr(s) \longrightarrow 3Ni(s) + 2Cr^{3+}(aq)$

Cr is the reducing agent; $Ni(NO_3)_2$ is the oxidizing agent.

PROBLEMS

*Problems with **colored** numbers are answered in Appendix E. Sections match the text and provide the numbers of relevant sample problems. Bracketed problems are grouped in pairs (indicated by a short rule) that cover the same concept. Comprehensive Problems are based on material from any section or previous chapter.*

The Role of Water as a Solvent

(Sample Problem 4.1)

4.1 What two factors cause water to be polar?

4.2 What must be present in an aqueous solution for it to conduct an electric current? What general classes of compounds form solutions that conduct?

4.3 What occurs on the molecular level when an ionic compound dissolves in water?

4.4 Which of the following best represents how the ions occur in aqueous solutions of (a) $CaCl_2$, (b) Li_2SO_4, and (c) NH_4Br?

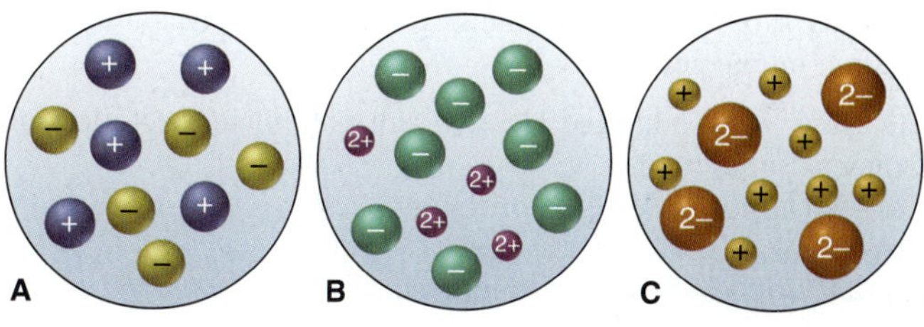

4.5 Which of the following best represents a volume from a solution of magnesium nitrate?

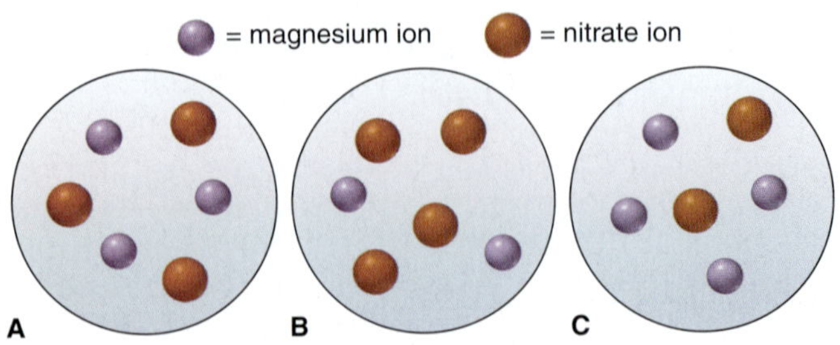

4.6 Why are some ionic compounds soluble in water and others are not?

4.7 Some covalent compounds dissociate into ions when they dissolve in water. What atom do these compounds have in their structures? What type of aqueous solution do they form? Name three examples of such an aqueous solution.

4.8 State whether each of the following substances is likely to be very soluble in water. Explain.
(a) Benzene, C_6H_6 (b) Sodium hydroxide
(c) Ethanol, CH_3CH_2OH (d) Potassium acetate

4.9 State whether each of the following substances is likely to be very soluble in water. Explain.
(a) Lithium nitrate (b) Glycine, H_2NCH_2COOH
(c) Pentane (d) Ethylene glycol, $HOCH_2CH_2OH$

4.10 State whether an aqueous solution of each of the following substances conducts an electric current. Explain your reasoning.
(a) Cesium bromide (b) Hydrogen iodide

4.11 State whether an aqueous solution of each of the following substances conducts an electric current. Explain your reasoning.
(a) Potassium sulfate (b) Sucrose, $C_{12}H_{22}O_{11}$

4.12 How many total moles of ions are released when each of the following samples dissolves completely in water?
(a) 0.75 mol of K_3PO_4 (b) 6.88×10^{-3} g of $NiBr_2 \cdot 3H_2O$
(c) 2.23×10^{22} formula units of $FeCl_3$

4.13 How many total moles of ions are released when each of the following samples dissolves completely in water?
(a) 0.734 mol of Na_2HPO_4 (b) 3.86 g of $CuSO_4 \cdot 5H_2O$
(c) 8.66×10^{20} formula units of $NiCl_2$

4.14 How many moles and numbers of ions of each type are present in the following aqueous solutions?
(a) 130 mL of 0.45 *M* aluminum chloride
(b) 9.80 L of a solution containing 2.59 g lithium sulfate/L
(c) 245 mL of a solution containing 3.68×10^{22} formula units of potassium bromide per liter

4.15 How many moles and numbers of ions of each type are present in the following aqueous solutions?
(a) 88 mL of 1.75 *M* magnesium chloride
(b) 321 mL of a solution containing 0.22 g aluminum sulfate/L
(c) 1.65 L of a solution containing 8.83×10^{21} formula units of cesium nitrate per liter

4.16 How many moles of H^+ ions are present in the following aqueous solutions?
(a) 1.40 L of 0.25 *M* perchloric acid
(b) 6.8 mL of 0.92 *M* nitric acid
(c) 2.6 L of 0.085 *M* hydrochloric acid

4.17 How many moles of H^+ ions are present in the following aqueous solutions?
(a) 1.4 mL of 0.75 *M* hydrobromic acid
(b) 2.47 mL of 1.98 *M* hydriodic acid
(c) 395 mL of 0.270 *M* nitric acid

4.18 To study a marine organism, a biologist prepares a 1.00-kg sample to simulate the ion concentrations in seawater. She mixes 26.5 g of NaCl, 2.40 g of $MgCl_2$, 3.35 g of $MgSO_4$, 1.20 g of $CaCl_2$, 1.05 g of KCl, 0.315 g of $NaHCO_3$, and 0.098 g of NaBr in distilled water. (a) If the density of this solution is 1.025 g/cm^3, what is the molarity of each ion? (b) What is the total molarity of alkali metal ions? (c) What is the total molarity of alkaline earth metal ions? (d) What is the total molarity of anions?

4.19 Water "softeners" remove metal ions such as Ca^{2+} and Fe^{3+} by replacing them with enough Na^+ ions to maintain the same number of positive charges in the solution. If 1.0×10^3 L of "hard" water is 0.015 *M* Ca^{2+} and 0.0010 *M* Fe^{3+}, how many moles of Na^+ are needed to replace these ions?

Writing Equations for Aqueous Ionic Reactions

4.20 Write two sets of equations (both molecular and total ionic) with different reactants that have the same net ionic equation as the following equation:

$$Ba(NO_3)_2(aq) + Na_2CO_3(aq) \longrightarrow BaCO_3(s) + 2NaNO_3(aq)$$

Precipitation Reactions

(Sample Problems 4.2 and 4.3)

4.21 Why do some pairs of ions precipitate and others do not?

4.22 Use Table 4.1 to determine which of the following combinations leads to a reaction. How can you identify the spectator ions in the reaction?
(a) Calcium nitrate(aq) + sodium chloride(aq) $\longrightarrow$
(b) Potassium chloride(aq) + lead(II) nitrate(aq) $\longrightarrow$

4.23 The beakers represent the aqueous reaction of $AgNO_3$ and NaCl. Silver ions are gray. What colors are used to represent NO_3^-, Na^+, and Cl^-? Write molecular, total ionic, and net ionic equations for the reaction.

4.24 Complete the following precipitation reactions with balanced molecular, total ionic, and net ionic equations:
(a) $Hg_2(NO_3)_2(aq) + KI(aq) \longrightarrow$
(b) $FeSO_4(aq) + Sr(OH)_2(aq) \longrightarrow$

4.25 Complete the following precipitation reactions with balanced molecular, total ionic, and net ionic equations:
(a) $CaCl_2(aq) + Cs_3PO_4(aq) \longrightarrow$
(b) $Na_2S(aq) + ZnSO_4(aq) \longrightarrow$

4.26 When each of the following pairs of aqueous solutions is mixed, does a precipitation reaction occur? If so, write balanced molecular, total ionic, and net ionic equations:

(a) Sodium nitrate + copper(II) sulfate
(b) Ammonium bromide + silver nitrate

4.27 When each of the following pairs of aqueous solutions is mixed, does a precipitation reaction occur? If so, write balanced molecular, total ionic, and net ionic equations:
(a) Potassium carbonate + barium hydroxide
(b) Aluminum nitrate + sodium phosphate

4.28 If 38.5 mL of lead(II) nitrate solution reacts completely with excess sodium iodide solution to yield 0.628 g of precipitate, what is the molarity of lead(II) ion in the original solution?

4.29 If 25.0 mL of silver nitrate solution reacts with excess potassium chloride solution to yield 0.842 g of precipitate, what is the molarity of silver ion in the original solution?

4.30 With ions shown as spheres and solvent molecules omitted for clarity, the circle *(right)* illustrates the solid formed when a solution containing K^+, Mg^{2+}, Ag^+, or Pb^{2+} (blue) is mixed with one containing ClO_4^-, NO_3^-, or SO_4^{2-} (yellow). (a) Identify the solid. (b) Write a balanced net ionic equation for the reaction. (c) If each sphere represents 5.0×10^{-4} mol of ion, what mass of product forms?

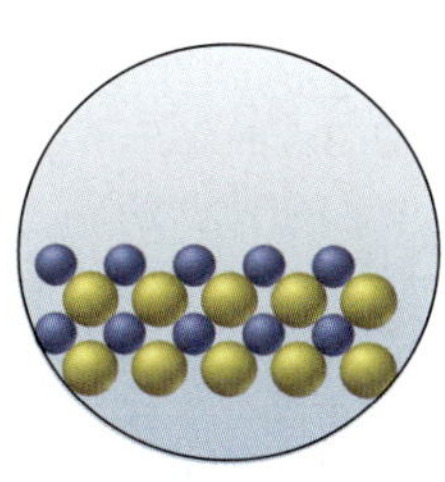

4.31 The precipitation reaction between 25.0 mL of a solution containing a cation (purple) and 35.0 mL of a solution containing an anion (green) is depicted below (with ions shown as spheres and solvent molecules omitted for clarity).

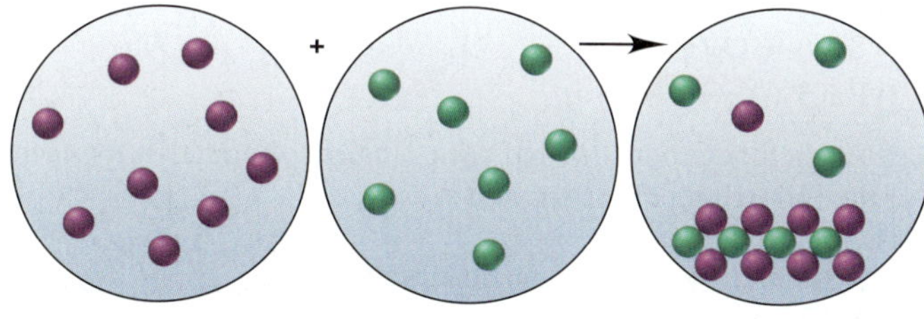

(a) Given the following choices of reactants, write balanced total ionic and net ionic equations that best represent the reaction:
(1) $KNO_3(aq) + CuCl_2(aq) \longrightarrow$
(2) $NaClO_4(aq) + CaCl_2(aq) \longrightarrow$
(3) $Li_2SO_4(aq) + AgNO_3(aq) \longrightarrow$
(4) $NH_4Br(aq) + Pb(CH_3COO)_2(aq) \longrightarrow$
(b) If each sphere represents 2.5×10^{-3} mol of ion, find the total number of ions present. (c) What is the mass of solid formed?

4.32 The mass percent of Cl^- in a seawater sample is determined by titrating 25.00 mL of seawater with $AgNO_3$ solution, causing a precipitation reaction. An indicator is used to detect the end point, which occurs when free Ag^+ ion is present in solution after all the Cl^- has reacted. If 53.63 mL of 0.2970 *M* $AgNO_3$ is required to reach the end point, what is the mass percent of Cl^- in the seawater (*d* of seawater = 1.024 g/mL)?

4.33 Aluminum sulfate, known as *cake alum,* has a wide range of uses, from dyeing leather and cloth to purifying sewage. In aqueous solution, it reacts with base to form a white precipitate. (a) Write balanced total and net ionic equations for its reaction with aqueous NaOH. (b) What mass of precipitate forms when 185.5 mL of 0.533 *M* NaOH is added to 627 mL of a solution that contains 15.8 g of aluminum sulfate per liter?

Acid-Base Reactions

(Sample Problems 4.4 to 4.6)

4.34 Is the total ionic equation the same as the net ionic equation when $Sr(OH)_2(aq)$ and $H_2SO_4(aq)$ react? Explain.

4.35 (a) Name three common strong acids. (b) Name three common strong bases. (c) What is a characteristic behavior of a strong acid or a strong base?

4.36 (a) Name three common weak acids. (b) Name one common weak base. (c) What is the major difference between a weak acid and a strong acid or between a weak base and a strong base, and what experiment would you perform to observe it?

4.37 The net ionic equation for the aqueous neutralization reaction between acetic acid and sodium hydroxide is different from that for the reaction between hydrochloric acid and sodium hydroxide. Explain by writing balanced net ionic equations.

4.38 Complete the following acid-base reactions with balanced molecular, total ionic, and net ionic equations:
(a) Potassium hydroxide(*aq*) + hydrobromic acid(*aq*) $\longrightarrow$
(b) Ammonia(*aq*) + hydrochloric acid(*aq*) $\longrightarrow$

4.39 Complete the following acid-base reactions with balanced molecular, total ionic, and net ionic equations:
(a) Cesium hydroxide(*aq*) + nitric acid(*aq*) $\longrightarrow$
(b) Calcium hydroxide(*aq*) + acetic acid(*aq*) $\longrightarrow$

4.40 Limestone (calcium carbonate) is insoluble in water but dissolves when a hydrochloric acid solution is added. Write balanced total ionic and net ionic equations, showing hydrochloric acid as it actually exists in water and the reaction as a proton-transfer process. [*Hint:* The H_2CO_3 that forms decomposes to H_2O and gaseous CO_2.]

4.41 Zinc hydroxide is insoluble in water but dissolves when a nitric acid solution is added. Write balanced total ionic and net ionic equations, showing nitric acid as it actually exists in water and the reaction as a proton-transfer process.

4.42 If 25.98 mL of a standard 0.1180 *M* KOH solution reacts with 52.50 mL of CH_3COOH solution, what is the molarity of the acid solution?

4.43 If 26.25 mL of a standard 0.1850 *M* NaOH solution is required to neutralize 25.00 mL of H_2SO_4, what is the molarity of the acid solution?

4.44 An auto mechanic spills 88 mL of 2.6 *M* H_2SO_4 solution from a rebuilt auto battery. How many milliliters of 1.6 *M* $NaHCO_3$ must be poured on the spill to react completely with the sulfuric acid? [*Hint:* H_2O and CO_2 are among the products.]

4.45 One of the first steps in the enrichment of uranium for use in nuclear power plants involves a displacement reaction between UO_2 and aqueous HF:

$$UO_2(s) + HF(aq) \longrightarrow UF_4(s) + H_2O(l) \text{ [unbalanced]}$$

How many liters of 2.40 *M* HF will react with 2.15 kg of UO_2?

4.46 An unknown amount of acid can often be determined by adding an excess of base and then "back-titrating" the excess. A 0.3471-g sample of a mixture of oxalic acid, which has two ionizable protons, and benzoic acid, which has one, is treated with 100.0 mL of 0.1000 *M* NaOH. The excess NaOH is titrated with 20.00 mL of 0.2000 *M* HCl. Find the mass % of benzoic acid.

4.47 A mixture of bases can sometimes be the active ingredient in antacid tablets. If 0.4826 g of a mixture of $Al(OH)_3$ and $Mg(OH)_2$ is neutralized with 17.30 mL of 1.000 *M* HNO_3, what is the mass % of $Al(OH)_3$ in the mixture?

Oxidation-Reduction (Redox) Reactions

(Sample Problems 4.7 and 4.8)

4.48 Why must every redox reaction involve an oxidizing agent and a reducing agent?

4.49 In which of the following equations does sulfuric acid act as an oxidizing agent? In which does it act as an acid? Explain.
(a) $4H^+(aq) + SO_4^{2-}(aq) + 2NaI(s) \longrightarrow 2Na^+(aq) + I_2(s) + SO_2(g) + 2H_2O(l)$
(b) $BaF_2(s) + 2H^+(aq) + SO_4^{2-}(aq) \longrightarrow 2HF(aq) + BaSO_4(s)$

4.50 Give the oxidation number of nitrogen in the following:
(a) NH_2OH (b) N_2F_4 (c) NH_4^+ (d) HNO_2

4.51 Give the oxidation number of sulfur in the following:
(a) $SOCl_2$ (b) H_2S_2 (c) H_2SO_3 (d) Na_2S

4.52 Give the oxidation number of arsenic in the following:
(a) AsH_3 (b) $H_3AsO_4^-$ (c) $AsCl_3$

4.53 Give the oxidation number of phosphorus in the following:
(a) $H_2P_2O_7^{2-}$ (b) PH_4^+ (c) PCl_5

4.54 Give the oxidation number of manganese in the following:
(a) MnO_4^{2-} (b) Mn_2O_3 (c) $KMnO_4$

4.55 Give the oxidation number of chromium in the following:
(a) CrO_3 (b) $Cr_2O_7^{2-}$ (c) $Cr_2(SO_4)_3$

4.56 Identify the oxidizing and reducing agents in the following:
(a) $5H_2C_2O_4(aq) + 2MnO_4^-(aq) + 6H^+(aq) \longrightarrow 2Mn^{2+}(aq) + 10CO_2(g) + 8H_2O(l)$
(b) $3Cu(s) + 8H^+(aq) + 2NO_3^-(aq) \longrightarrow 3Cu^{2+}(aq) + 2NO(g) + 4H_2O(l)$

4.57 Identify the oxidizing and reducing agents in the following:
(a) $Sn(s) + 2H^+(aq) \longrightarrow Sn^{2+}(aq) + H_2(g)$
(b) $2H^+(aq) + H_2O_2(aq) + 2Fe^{2+}(aq) \longrightarrow 2Fe^{3+}(aq) + 2H_2O(l)$

4.58 Identify the oxidizing and reducing agents in the following:
(a) $8H^+(aq) + 6Cl^-(aq) + Sn(s) + 4NO_3^-(aq) \longrightarrow SnCl_6^{2-}(aq) + 4NO_2(g) + 4H_2O(l)$
(b) $2MnO_4^-(aq) + 10Cl^-(aq) + 16H^+(aq) \longrightarrow 5Cl_2(g) + 2Mn^{2+}(aq) + 8H_2O(l)$

4.59 Identify the oxidizing and reducing agents in the following:
(a) $8H^+(aq) + Cr_2O_7^{2-}(aq) + 3SO_3^{2-}(aq) \longrightarrow 2Cr^{3+}(aq) + 3SO_4^{2-}(aq) + 4H_2O(l)$
(b) $NO_3^-(aq) + 4Zn(s) + 7OH^-(aq) + 6H_2O(l) \longrightarrow 4Zn(OH)_4^{2-}(aq) + NH_3(aq)$

4.60 Discuss each conclusion from a study of redox reactions:
(a) The sulfide ion functions only as a reducing agent.
(b) The sulfate ion functions only as an oxidizing agent.
(c) Sulfur dioxide functions as an oxidizing or a reducing agent.

4.61 Discuss each conclusion from a study of redox reactions:
(a) The nitride ion functions only as a reducing agent.
(b) The nitrate ion functions only as an oxidizing agent.
(c) The nitrite ion functions as an oxidizing or a reducing agent.

4.62 A person's blood alcohol (C_2H_5OH) level can be determined by titrating a sample of blood plasma with a potassium dichromate solution. The balanced equation is

$$16H^+(aq) + 2Cr_2O_7^{2-}(aq) + C_2H_5OH(aq) \longrightarrow 4Cr^{3+}(aq) + 2CO_2(g) + 11H_2O(l)$$

If 35.46 mL of 0.05961 M $Cr_2O_7^{2-}$ is required to titrate 28.00 g of plasma, what is the mass percent of alcohol in the blood?

Elements in Redox Reactions

(Sample Problem 4.9)

4.63 Which type of redox reaction leads to the following?
(a) An increase in the number of substances
(b) A decrease in the number of substances
(c) No change in the number of substances

4.64 Why do decomposition reactions typically have compounds as reactants, whereas combination and displacement reactions have one or more elements?

4.65 Which of the three types of reactions discussed in this section commonly produce one or more compounds?

4.66 Balance each of the following redox reactions and classify it as a combination, decomposition, or displacement reaction:
(a) $Sb(s) + Cl_2(g) \longrightarrow SbCl_3(s)$
(b) $AsH_3(g) \longrightarrow As(s) + H_2(g)$
(c) $Zn(s) + Fe(NO_3)_2(aq) \longrightarrow Zn(NO_3)_2(aq) + Fe(s)$

4.67 Balance each of the following redox reactions and classify it as a combination, decomposition, or displacement reaction:
(a) $Mg(s) + H_2O(g) \longrightarrow Mg(OH)_2(s) + H_2(g)$
(b) $Cr(NO_3)_3(aq) + Al(s) \longrightarrow Al(NO_3)_3(aq) + Cr(s)$
(c) $PF_3(g) + F_2(g) \longrightarrow PF_5(g)$

4.68 Predict the product(s) and write a balanced equation for each of the following redox reactions:
(a) $N_2(g) + H_2(g) \longrightarrow$
(b) $NaClO_3(s) \xrightarrow{\Delta}$
(c) $Ba(s) + H_2O(l) \longrightarrow$

4.69 Predict the product(s) and write a balanced equation for each of the following redox reactions:
(a) $Fe(s) + HClO_4(aq) \longrightarrow$
(b) $S_8(s) + O_2(g) \longrightarrow$
(c) $BaCl_2(l) \xrightarrow{\text{electricity}}$

4.70 Predict the product(s) and write a balanced equation for each of the following redox reactions:
(a) Cesium + iodine $\longrightarrow$
(b) Aluminum + aqueous manganese(II) sulfate $\longrightarrow$
(c) Sulfur dioxide + oxygen $\longrightarrow$
(d) Butane and oxygen $\longrightarrow$
(e) Write a balanced net ionic equation for (b).

4.71 Predict the product(s) and write a balanced equation for each of the following redox reactions:
(a) Pentane (C_5H_{12}) + oxygen $\longrightarrow$
(b) Phosphorus trichloride + chlorine $\longrightarrow$
(c) Zinc + hydrobromic acid $\longrightarrow$
(d) Aqueous potassium iodide + bromine $\longrightarrow$
(e) Write a balanced net ionic equation for (d).

4.72 How many grams of O_2 can be prepared from the thermal decomposition of 4.27 kg of HgO? Name and calculate the mass (in kg) of the other product.

4.73 How many grams of chlorine gas can be produced from the electrolytic decomposition of 874 g of calcium chloride? Name and calculate the mass (in g) of the other product.

4.74 In a combination reaction, 1.62 g of lithium is mixed with 6.50 g of oxygen. (a) Which reactant is present in excess? (b) How many moles of product are formed? (c) After reaction, how many grams of each reactant and product are present?

4.75 In a combination reaction, 2.22 g of magnesium is heated with 3.75 g of nitrogen. (a) Which reactant is present in excess? (b) How many moles of product are formed? (c) After reaction, how many grams of each reactant and product are present?

4.76 A mixture of $CaCO_3$ and CaO weighing 0.693 g was heated to produce gaseous CO_2. After heating, the remaining solid weighed 0.508 g. Assuming all the $CaCO_3$ broke down to CaO and CO_2, calculate the mass percent of $CaCO_3$ in the original mixture.

4.77 Before arc welding was developed, a displacement reaction involving aluminum and iron(III) oxide was commonly used to produce molten iron (the thermite process). This reaction was used, for example, to connect sections of iron railroad track. Calculate the mass of molten iron produced when 1.50 kg of aluminum reacts with 25.0 mol of iron(III) oxide.

Comprehensive Problems

Problems with an asterisk (*) are more challenging.

4.78 Nutritional biochemists have known for decades that acidic foods cooked in cast-iron cookware can supply significant amounts of dietary iron (ferrous ion).
(a) Write a balanced net ionic equation, with oxidation numbers, that supports this fact.
(b) Measurements show an increase from 3.3 mg of iron to 49 mg of iron per $\frac{1}{2}$-cup (125-g) serving during the slow preparation of tomato sauce in a cast-iron pot. How many ferrous ions are present in a 26-oz (737-g) jar of the tomato sauce?

4.79 The brewing industry uses yeast microorganisms to convert glucose to ethanol for wine and beer. The baking industry uses the carbon dioxide produced to make bread rise:

$$C_6H_{12}O_6(s) \xrightarrow{\text{yeast}} 2C_2H_5OH(l) + 2CO_2(g)$$

How many grams of ethanol can be produced from 10.0 g of glucose? What volume of CO_2 is produced? (Assume 1 mol of gas occupies 22.4 L at the conditions used.)

*** 4.80** A chemical engineer determines the mass percent of iron in an ore sample by converting the Fe to Fe^{2+} in acid and then titrating the Fe^{2+} with MnO_4^-. A 1.1081-g sample was dissolved in acid and then titrated with 39.32 mL of 0.03190 *M* $KMnO_4$. The balanced equation is

$$8H^+(aq) + 5Fe^{2+}(aq) + MnO_4^-(aq) \longrightarrow 5Fe^{3+}(aq) + Mn^{2+}(aq) + 4H_2O(l)$$

Calculate the mass percent of iron in the ore.

4.81 You are given solutions of HCl and NaOH and must determine their concentrations. You use 27.5 mL of NaOH to titrate 100. mL of HCl and 18.4 mL of NaOH to titrate 50.0 mL of 0.0782 *M* H_2SO_4. Find the unknown concentrations.

4.82 The flask *(right)* represents the products of the titration of 25 mL of sulfuric acid with 25 mL of sodium hydroxide.
(a) Write balanced molecular, total ionic, and net ionic equations for the reaction.
(b) If each orange sphere represents 0.010 mol of sulfate ion, how many moles of acid and of base reacted?
(c) What are the molarities of the acid and the base?

4.83 On a lab exam, you have to find the concentrations of the monoprotic (one proton per molecule) acids HA and HB. You

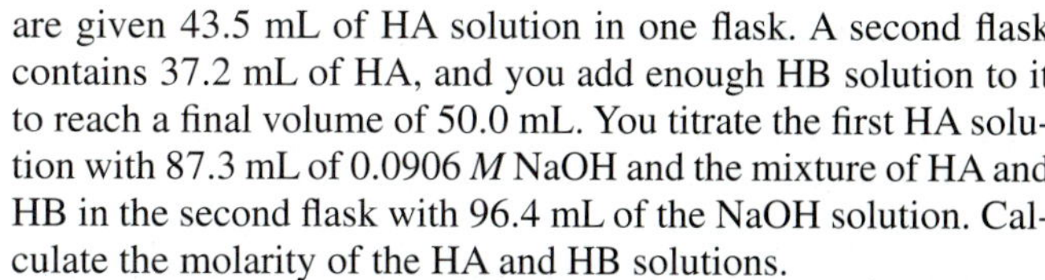

are given 43.5 mL of HA solution in one flask. A second flask contains 37.2 mL of HA, and you add enough HB solution to it to reach a final volume of 50.0 mL. You titrate the first HA solution with 87.3 mL of 0.0906 *M* NaOH and the mixture of HA and HB in the second flask with 96.4 mL of the NaOH solution. Calculate the molarity of the HA and HB solutions.

4.84 Nitric acid, a major industrial and laboratory acid, is produced commercially by the multistep Ostwald process, which begins with the oxidation of ammonia:

Step 1. $4NH_3(g) + 5O_2(g) \longrightarrow 4NO(g) + 6H_2O(l)$
Step 2. $2NO(g) + O_2(g) \longrightarrow 2NO_2(g)$
Step 3. $3NO_2(g) + H_2O(l) \longrightarrow 2HNO_3(l) + NO(g)$

(a) What are the oxidizing and reducing agents in each step?
(b) Assuming 100% yield in each step, what mass (in kg) of ammonia must be used to produce 3.0×10^4 kg of HNO_3?

4.85 For the following aqueous reactions, complete and balance the molecular equation and write a net ionic equation:
(a) Manganese(II) sulfide + hydrobromic acid
(b) Potassium carbonate + strontium nitrate
(c) Potassium nitrite + hydrochloric acid
(d) Calcium hydroxide + nitric acid
(e) Barium acetate + iron(II) sulfate
(f) Barium hydroxide + hydrocyanic acid
(g) Copper(II) nitrate + hydrosulfuric acid
(h) Magnesium hydroxide + chloric acid
(i) Potassium chloride + ammonium phosphate

4.86 There are various methods for finding the composition of an alloy (a metal-like mixture). Show that calculating the mass % of Mg in a magnesium-aluminum alloy ($d = 2.40$ g/cm^3) gives the same answer (within rounding) using each of these methods: (a) a 0.263-g sample of alloy (*d* of Mg = 1.74 g/cm^3; *d* of Al = 2.70 g/cm^3); (b) an identical sample reacting with excess aqueous HCl forms 1.38×10^{-2} mol of H_2; (c) an identical sample reacting with excess O_2 forms 0.483 g of oxide.

4.87 Sodium peroxide (Na_2O_2) is often used in self-contained breathing devices, such as those used in fire emergencies, because it reacts with exhaled CO_2 to form Na_2CO_3 and O_2. How many liters of respired air can react with 80.0 g of Na_2O_2 if each liter of respired air contains 0.0720 g of CO_2?

4.88 Magnesium is used in airplane bodies and other lightweight alloys. The metal is obtained from seawater in a process that includes precipitation, neutralization, evaporation, and electrolysis. How many kilograms of magnesium can be obtained from 1.00 km^3 of seawater if the initial Mg^{2+} concentration is 0.13% by mass (*d* of seawater = 1.04 g/mL)?

4.89 Physicians who specialize in sports medicine routinely treat athletes and dancers. Ethyl chloride, a local anesthetic commonly used for simple injuries, is the product of the combination of ethylene with hydrogen chloride:

$$C_2H_4(g) + HCl(g) \longrightarrow C_2H_5Cl(g)$$

If 0.100 kg of C_2H_4 and 0.100 kg of HCl react:
(a) How many molecules of gas (reactants plus products) are present when the reaction is complete?
(b) How many moles of gas are present when half the product forms?

4.90 Carbon dioxide is removed from the atmosphere of space capsules by reaction with a solid metal hydroxide. The products are water and the metal carbonate.

(a) Calculate the mass of CO_2 that can be removed by reaction with 3.50 kg of lithium hydroxide.
(b) How many grams of CO_2 can be removed by 1.00 g of each of the following: lithium hydroxide, magnesium hydroxide, and aluminum hydroxide?

* **4.91** Calcium dihydrogen phosphate, $Ca(H_2PO_4)_2$, and sodium hydrogen carbonate, $NaHCO_3$, are ingredients of baking powder that react with each other to produce CO_2, which causes dough or batter to rise:

$$Ca(H_2PO_4)_2(s) + NaHCO_3(s) \longrightarrow CO_2(g) + H_2O(g) + CaHPO_4(s) + Na_2HPO_4(s) \text{ [unbalanced]}$$

If the baking powder contains 31% $NaHCO_3$ and 35% $Ca(H_2PO_4)_2$ by mass:
(a) How many moles of CO_2 are produced from 1.00 g of baking powder?
(b) If 1 mol of CO_2 occupies 37.0 L at 350°F (a typical baking temperature), what volume of CO_2 is produced from 1.00 g of baking powder?

4.92 In a titration of HNO_3, you add a few drops of phenolphthalein indicator to 50.00 mL of acid in a flask. You quickly add 20.00 mL of 0.0502 *M* NaOH but overshoot the end point, and the solution turns deep pink. Instead of starting over, you add 30.00 mL of the acid, and the solution turns colorless. Then, it takes 3.22 mL of the NaOH to reach the end point. (a) What is the concentration of the HNO_3 solution? (b) How many moles of NaOH were in excess after the first addition?

* **4.93** The active compound in Pepto-Bismol contains C, H, O, and Bi.
(a) When 0.22105 g of it was burned in excess O_2, 0.1422 g of bismuth(III) oxide, 0.1880 g of carbon dioxide, and 0.02750 g of water were formed. What is the empirical formula of this compound?
(b) Given a molar mass of 1086 g/mol, determine the molecular formula.
(c) Complete and balance the acid-base reaction between bismuth(III) hydroxide and salicylic acid ($HC_7H_5O_3$), which is used to form this compound.
(d) A dose of Pepto-Bismol contains 0.600 mg of the active ingredient. If the yield of the reaction in part (c) is 88.0%, what mass (in mg) of bismuth(III) hydroxide is required to prepare one dose?

4.94 Two aqueous solutions contain the ions indicated below.

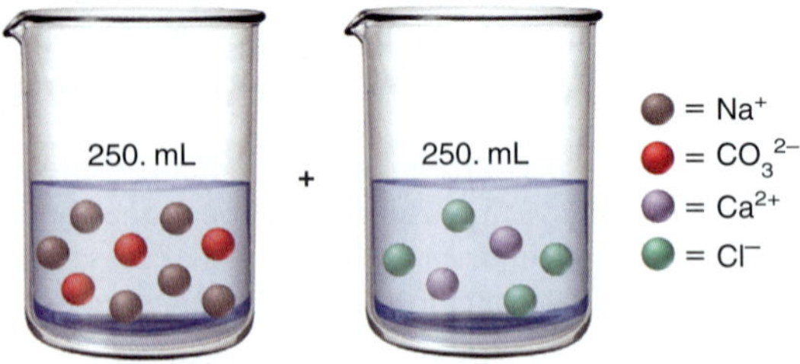

(a) Write balanced molecular, total ionic, and net ionic equations for the reaction that occurs when the solutions are mixed. (b) If each sphere represents 0.050 mol of ion, what mass (in g) of precipitate forms, assuming 100% reaction. (c) What is the concentration of each ion in solution after reaction?

* **4.95** In 1997, at the United Nations Conference on Climate Change, the major industrial nations agreed to expand their research efforts to develop renewable sources of carbon-based fuels. For more than a decade, Brazil has been engaged in a program to replace gasoline with ethanol derived from the root crop manioc (cassava).
(a) Write separate balanced equations for the complete combustion of ethanol (C_2H_5OH) and of gasoline (represented by the formula C_8H_{18}).
(b) What mass of oxygen is required to burn completely 1.00 L of a mixture that is 90.0% gasoline (d = 0.742 g/mL) and 10.0% ethanol (d = 0.789 g/mL) by volume?
(c) If 1.00 mol of O_2 occupies 22.4 L, what volume of O_2 is needed to burn 1.00 L of the mixture?
(d) Air is 20.9% O_2 by volume. What volume of air is needed to burn 1.00 L of the mixture?

* **4.96** In a car engine, gasoline (represented by C_8H_{18}) does not burn completely, and some CO, a toxic pollutant, forms along with CO_2 and H_2O. If 5.0% of the gasoline forms CO:
(a) What is the ratio of CO_2 to CO molecules in the exhaust?
(b) What is the mass ratio of CO_2 to CO?
(c) What percentage of the gasoline must form CO for the mass ratio of CO_2 to CO to be exactly 1/1?

* **4.97** The amount of ascorbic acid (vitamin C, $C_6H_8O_6$) in tablets is determined by reaction with bromine and then titration of the hydrobromic acid with standard base:

$$C_6H_8O_6 + Br_2 \longrightarrow C_6H_6O_6 + 2HBr$$
$$HBr + NaOH \longrightarrow NaBr + H_2O$$

A certain tablet is advertised as containing 500 mg of vitamin C. One tablet was dissolved in water and reacted with Br_2. The solution was then titrated with 43.20 mL of 0.1350 *M* NaOH. Did the tablet contain the advertised quantity of vitamin C?

4.98 In the process of *pickling,* rust is removed from newly produced steel by washing the steel in hydrochloric acid:
(1) $6HCl(aq) + Fe_2O_3(s) \longrightarrow 2FeCl_3(aq) + 3H_2O(l)$
During the process, some iron is lost as well:
(2) $2HCl(aq) + Fe(s) \longrightarrow FeCl_2(aq) + H_2(g)$
(a) Which reaction, if either, is a redox process? (b) If reaction 2 did not occur and all the HCl were used, how many grams of Fe_2O_3 could be removed and $FeCl_3$ produced in a 2.50×10^3-L bath of 3.00 *M* HCl? (c) If reaction 1 did not occur and all the HCl were used, how many grams of Fe could be lost and $FeCl_2$ produced in a 2.50×10^3-L bath of 3.00 *M* HCl? (d) If 0.280 g of Fe is lost per gram of Fe_2O_3 removed, what is the mass ratio of $FeCl_2$ to $FeCl_3$?

4.99 At liftoff, the space shuttle uses a solid mixture of ammonium perchlorate and aluminum powder to obtain great thrust from the volume change of solid to gas. In the presence of a catalyst, the mixture forms solid aluminum oxide and aluminum trichloride and gaseous water and nitrogen monoxide. (a) Write a balanced equation for the reaction, and identify the reducing and oxidizing agents. (b) How many total moles of gas (water vapor and nitrogen monoxide) are produced when 50.0 kg of ammonium perchlorate reacts with a stoichiometric amount of Al? (c) What is the volume change from this reaction? (d of NH_4ClO_4 = 1.95 g/cc, Al = 2.70 g/cc, Al_2O_3 = 3.97 g/cc, and $AlCl_3$ = 2.44 g/cc; assume 1 mol of gas occupies 22.4 L.)

Thermochemistry: Energy Flow and Chemical Change

Key Principles
to focus on while studying this chapter

- Chemical or physical change is *always* accompanied by a change in the energy content of the matter *(Introduction).*
- To study a *change in energy* (ΔE), scientists conceptually divide the universe into *system* (the part being studied) and *surroundings* (everything else). All energy changes occur as *heat* (q) and/or *work* (w), transferred either from the surroundings to the system or from the system to the surroundings ($\Delta E = q + w$). Thus, the total energy of the universe is constant (*law of conservation of energy* or *first law of thermodynamics*) *(Section 6.1).*
- The magnitude of ΔE is the same no matter how a given change in energy occurs. Because E is a *state function*—a property that depends *only* on the current state of the system—ΔE depends only on the difference between the initial and final values of E *(Section 6.1).*
- *Enthalpy* (H) is another state function and is related to E. The *change in enthalpy* (ΔH) equals the heat transferred at constant pressure, q_P. Most laboratory, environmental, and biological changes occur at constant P, so ΔH is more relevant than ΔE and easier to measure *(Section 6.2).*
- The change in the enthalpy of a reaction, called the *heat of reaction* (ΔH_{rxn}), is negative (<0) if the reaction releases heat *(exothermic)* and positive (>0) if it absorbs heat *(endothermic);* for example, the combustion of methane is exothermic ($\Delta H_{rxn} < 0$) and the melting of ice is endothermic ($\Delta H_{rxn} > 0$) *(Section 6.2).*
- The more heat a substance absorbs, the higher its temperature becomes, but each substance has its own *capacity* for absorbing heat. Knowing this capacity and measuring the change in temperature in a *calorimeter,* we can find ΔH_{rxn} *(Section 6.3).*
- The quantity of heat lost or gained in a reaction is related *stoichiometrically* to the amounts of reactants and products *(Section 6.4).*
- Because H is a state function, we can find ΔH of any reaction by imagining the reaction occurring as the sum of other reactions whose heats of reaction we know or can measure *(Hess's law of heat summation) (Section 6.5).*
- Chemists have specified a set of conditions, called *standard states,* to determine heats of different reactions. Each substance has a *standard heat of formation* (ΔH°_f), the heat of reaction when the substance is formed from its elements under these conditions, and the ΔH°_f values for each substance in a reaction are used to calculate the *standard heat of reaction* (ΔH°_{rxn}) *(Section 6.6).*

Heat Out and Heat In *Combustion releases heat and melting absorbs it.*

Outline

Concepts & Skills to Review before studying this chapter

- energy and its interconversion (Section 1.1)
- distinction between heat and temperature (Section 1.4)
- nature of chemical bonding (Section 2.7)
- calculations of reaction stoichiometry (Section 3.4)
- properties of the gaseous state (Section 5.1)
- relation between kinetic energy and temperature (Section 5.6)

Whenever matter changes, the energy content of the matter changes also. For example, during the chemical change that occurs when a candle burns, the wax and oxygen reactants contain more energy than the gaseous CO_2 and H_2O products, and this difference in energy is *released* as heat and light. In contrast, some of the energy in a flash of lightning is *absorbed* when lower energy N_2 and O_2 in the air react to form higher energy NO. Energy content changes during a physical change, too. For example, energy is *absorbed* when snow melts and is *released* when water vapor condenses.

As you probably already know, the production and utilization of energy have an enormous impact on society. Some of the largest industries manufacture products that release, absorb, or change the flow of energy. Common fuels—oil, wood, coal, and natural gas—release energy for heating and for powering combustion engines and steam turbines. Fertilizers enhance the ability of crops to absorb solar energy and convert it to the chemical energy of food, which our bodies then convert into other forms of energy. Metals are often used to increase the flow of energy, while plastic, fiberglass, and ceramic materials serve as insulators that limit the flow of energy.

Thermodynamics is the branch of physical science concerned with heat and its transformations to and from other forms of energy. It will be our focus here and again in Chapter 20. In this chapter, we highlight **thermochemistry,** which deals with the *heat* involved in chemical and physical changes.

6.1 FORMS OF ENERGY AND THEIR INTERCONVERSION

In Chapter 1, we discussed the facts that all energy is either potential or kinetic and that these forms are convertible from one to the other. An object has potential energy by virtue of its position and kinetic energy by virtue of its motion. The potential energy of a weight raised above the ground is converted to kinetic energy as it falls (see Figure 1.3, p. 7). When the weight hits the ground, it transfers some of that kinetic energy to the soil and pebbles, causing them to move, and thereby doing *work.* In addition, some of the transferred kinetic energy appears as *heat,* as it slightly warms the soil and pebbles. Thus, the potential energy of the weight is converted to kinetic energy, which is transferred to the ground as work and as heat.

Modern atomic theory allows us to consider other forms of energy—solar, electrical, nuclear, and chemical—as examples of potential and kinetic energy on the atomic and molecular scales. No matter what the details of the situation, *when energy is transferred from one object to another, it appears as work and/or as heat.* In this section, we examine this idea in terms of the loss or gain of energy that takes place during a chemical or physical change.

The System and Its Surroundings

In order to observe and measure a change in energy, we must first define the **system,** the part of the universe that we are going to focus on. The moment we define the system, everything else relevant to the change is defined as the **surroundings.** For example, in a flask containing a solution, if we define the system as the contents of the flask, then the flask itself, other nearby equipment, and perhaps the rest of the laboratory are the surroundings. In principle, the rest of the universe is the surroundings, but in practice, we need to consider only the portions of the universe relevant to the system. That is, it's not likely that a thunderstorm in central Asia or a methane blizzard on Neptune will affect the contents of the flask, but the temperature, pressure, and humidity of the lab might.

THINK OF IT THIS WAY
Wherever You Look, There Is a System

If we define a weight falling to the ground as the system, then the soil and pebbles that are moved and warmed when it lands are the surroundings. An astronomer may define a galaxy as the system and nearby galaxies as the surroundings. An ecologist in Africa can define a zebra herd as the system and the animals, plants, and water supplies that the herd has contact with as the surroundings. A microbiologist may define a bacterial cell as the system and the extracellular solution as the surroundings. Thus, in general, the nature of the experiment and the focus of the experimenter define system and surroundings.

Energy Flow to and from a System

Each particle in a system has potential energy and kinetic energy, and the sum of these energies for all the particles in the system is the **internal energy, *E*,** of the system (some texts use the symbol *U*). When a chemical system changes from reactants to products and the products return to the starting temperature, the internal energy has changed. To determine this change, ΔE, we measure the difference between the system's internal energy *after* the change (E_{final}) and *before* the change (E_{initial}):

$$\Delta E = E_{\text{final}} - E_{\text{initial}} = E_{\text{products}} - E_{\text{reactants}} \qquad (6.1)$$

where Δ (Greek *delta*) means "change (or difference) in." Note especially that Δ refers to the *final state of the system* ***minus*** *the initial state.*

Because the total energy must be conserved, *a change in the energy of the system is always accompanied by an* ***opposite*** *change in the energy of the surroundings.* We often represent this change with an *energy diagram* in which the final and initial states are horizontal lines on a vertical energy axis. The change in internal energy, ΔE, is the difference between the heights of the two lines. A system can change its internal energy in one of two ways:

1. By losing some energy *to* the surroundings, as shown in Figure 6.1A:

$$E_{\text{final}} < E_{\text{initial}} \qquad \Delta E < 0 \qquad (\Delta E \text{ is negative})$$

2. By gaining some energy *from* the surroundings, as shown in Figure 6.1B:

$$E_{\text{final}} > E_{\text{initial}} \qquad \Delta E > 0 \qquad (\Delta E \text{ is positive})$$

Note that the change in energy is always an energy *transfer* from system to surroundings, or vice versa.

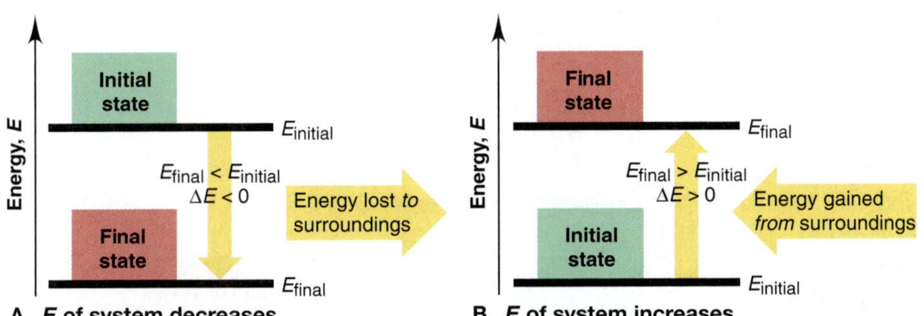

FIGURE 6.1 Energy diagrams for the transfer of internal energy (*E*) between a system and its surroundings. A, When the internal energy of a system *decreases,* the change in energy (ΔE) is lost *to* the surroundings; therefore, ΔE of the system ($E_{\text{final}} - E_{\text{initial}}$) is negative. **B,** When the system's internal energy *increases,* ΔE is gained *from* the surroundings and is positive. Note that the vertical yellow arrow, which signifies the direction of the change in energy, *always* has its tail at the initial state and its head at the final state.

Heat and Work: Two Forms of Energy Transfer

Just as we saw when a weight hits the ground, energy transfer outward from the system or inward from the surroundings can appear in two forms, heat and work. **Heat** (or *thermal energy,* symbol ***q***) is the energy transferred between a system and its surroundings as a result of a difference in their temperatures only. Energy in the form of heat is transferred from hot soup (system) to the bowl, air, and table (surroundings) because the surroundings have a lower temperature. All other forms of energy transfer (mechanical, electrical, and so on) involve some type of **work (*w*),** the energy transferred when an object is moved by a force. When you (system) kick a football (surroundings), energy is transferred as work to move the ball. When you inflate the ball, the inside air (system) exerts a force on the inner wall of the ball (surroundings) and does work to move it outward.

The total change in a system's internal energy is the sum of the energy transferred as heat and/or work:

$$\Delta E = q + w \tag{6.2}$$

The numerical values of q and w (and thus ΔE) can be either positive or negative, depending on the change the *system* undergoes. In other words, *we define the sign of the energy transfer from the* ***system's*** *perspective.* Energy coming *into* the system is *positive.* Energy going *out from* the system is *negative.* Of the innumerable changes possible in the system's internal energy, we'll examine the four simplest—two that involve only heat and two that involve only work.

Energy Transfer as Heat Only For a system that does no work but transfers energy only as heat (q), we know that $w = 0$. Therefore, from Equation 6.2, we have $\Delta E = q + 0 = q$. There are two possibilities:

1. *Heat flowing* ***out from*** *a system.* Suppose a sample of hot water is the system; then, the beaker containing it and the rest of the lab are the surroundings. The water transfers energy as heat to the surroundings until the temperature of the water equals that of the surroundings. The system's energy decreases as heat flows *out from* the system, so the final energy of the system is less than its initial energy. Heat was lost by the system, so *q is negative,* and therefore *ΔE is negative* (Figure 6.2A).

2. *Heat flowing* ***into*** *a system.* If the system consists of ice water, it gains energy as heat from the surroundings until the temperature of the water equals that of the surroundings. In this case, energy is transferred *into* the system, so the final energy of the system is higher than its initial energy. Heat was gained by the system, so *q is positive,* and therefore *ΔE is positive* (Figure 6.2B).

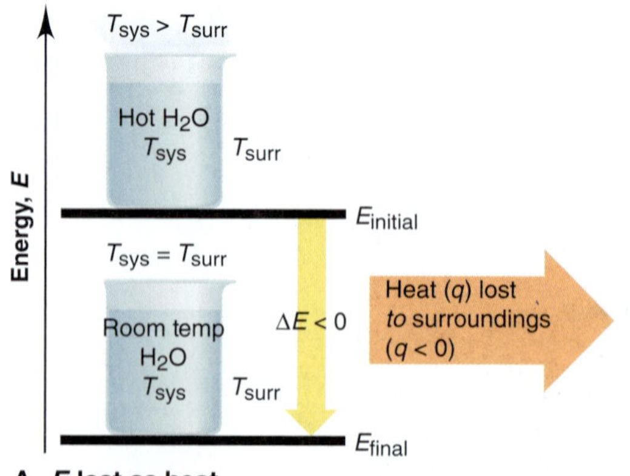

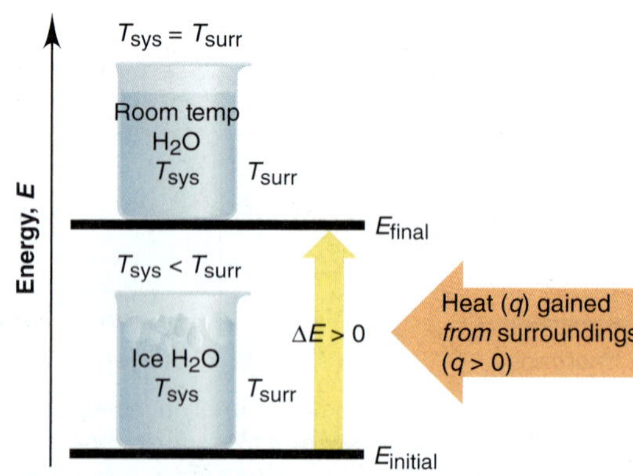

FIGURE 6.2 A system transferring energy as heat only. A, Hot water (the system, sys) transfers energy as heat (q) *to* the surroundings (surr) until $T_{sys} = T_{surr}$. Here $E_{initial} > E_{final}$ and $w = 0$, so $\Delta E < 0$ and the sign of q is negative. **B,** Ice water gains energy as heat (q) *from* the surroundings until $T_{sys} = T_{surr}$. Here $E_{initial} < E_{final}$ and $w = 0$, so $\Delta E > 0$ and the sign of q is positive.

THINK OF IT THIS WAY
Thermodynamics in the Kitchen

The functioning of two familiar appliances can clarify the loss or gain of heat and the sign of q. The air in a refrigerator (surroundings) has a lower temperature than a newly added piece of food (system), so the food loses energy as heat to the refrigerator air, or $q < 0$. The air in a hot oven (surroundings) has a higher temperature than a newly added piece of food (system), so the food gains energy as heat from the oven air, or $q > 0$.

Energy Transfer as Work Only For a system that transfers energy only as work (w), $q = 0$; therefore, $\Delta E = 0 + w = w$. Once again, there are two possibilities:

1. *Work done* **by** *a system.* Consider the reaction between zinc and hydrochloric acid as it takes place in an insulated container attached to a piston-cylinder assembly. We define the system as the atoms that make up the substances. In the initial state, the system's internal energy is that of the atoms in the form of the reactants, metallic Zn and aqueous H^+ and Cl^- ions. In the final state, the system's internal energy is that of the same atoms in the form of the products, H_2 gas and aqueous Zn^{2+} and Cl^- ions:

$$Zn(s) + 2H^+(aq) + 2Cl^-(aq) \longrightarrow H_2(g) + Zn^{2+}(aq) + 2Cl^-(aq)$$

As the H_2 gas forms, some of the internal energy is used *by* the system to do work *on* the surroundings and push the piston outward. Energy is lost by the system as work, so *w is negative* and *ΔE is negative,* as you see in Figure 6.3. The H_2 gas is doing **pressure-volume work (*PV* work),** the type of work in which a volume changes against an external pressure. The work done here is not very useful because it simply pushes back the piston and outside air. But, if the system is a ton of burning coal and O_2, and the surroundings are a locomotive engine, much of the internal energy lost from the system does the work of moving a train.

2. *Work done* **on** *a system.* If we increase the external pressure on the piston in Figure 6.3, the system gains energy because work is done *on* the system *by* the surroundings: *w is positive,* so *ΔE is positive.*

Table 6.1 summarizes the sign conventions for q and w and their effect on the sign of ΔE.

Animation: Energy Flow

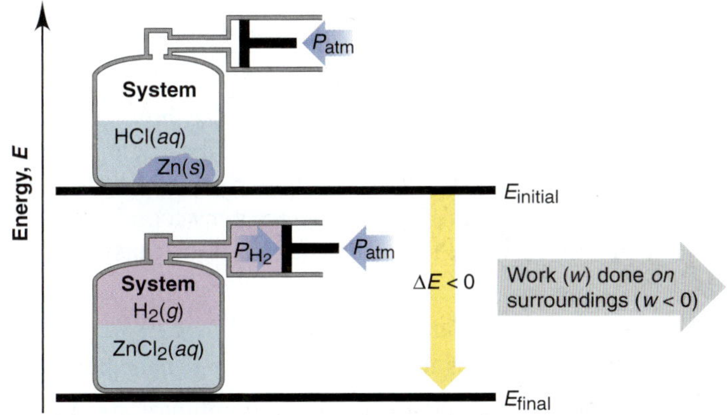

FIGURE 6.3 A system losing energy as work only. The internal energy of the system decreases as the reactants form products because the $H_2(g)$ does work (w) *on* the surroundings by pushing back the piston. The reaction vessel is insulated, so $q = 0$. Here $E_{initial} > E_{final}$, so $\Delta E < 0$ and the sign of w is negative.

Table 6.1 The Sign Conventions* for q, w, and ΔE

q	+	w	=	ΔE
+ (heat *gained*)		+ (work done *on*)		+ (energy *increased*)
+ (heat *gained*)		− (work done *by*)		Depends on ***sizes*** of q and w
− (heat *lost*)		+ (work done *on*)		Depends on ***sizes*** of q and w
− (heat *lost*)		− (work done *by*)		− (energy *decreased*)

*From the perspective of the system.

The Law of Energy Conservation

As you've seen, when a system gains energy, the surroundings lose it, and when a system loses energy, the surroundings gain it. Energy can be converted from one form to another as these transfers take place, but it cannot simply appear or disappear—it cannot be created or destroyed. The **law of conservation of energy** restates this basic observation as follows: *the total energy of the universe is constant.* This law is also known as the **first law of thermodynamics.**

Conservation of energy applies everywhere. As gasoline burns in a car engine, the released energy appears as an equivalent amount of heat and work. The heat warms the car parts, passenger compartment, and surrounding air. The work appears as mechanical energy to turn the car's wheels and belts. That energy is converted further into the electrical energy of the clock and radio, the radiant energy of the headlights, the chemical energy of the battery, the heat due to friction, and so forth. If you took the sum of all these energy forms, you would find that it equals the change in energy between the reactants and products as the gasoline is burned.

Complex biological processes also obey conservation of energy. Through photosynthesis, green plants convert radiant energy from the Sun into chemical energy, transforming low-energy CO_2 and H_2O into high-energy carbohydrates (such as wood) and O_2. When the wood is burned in air, those low-energy compounds form again, and the energy difference is released to the surroundings.

Thus, energy transfers between system and surroundings can be in the forms of heat and/or various types of work—mechanical, electrical, radiant, chemical—but *the energy of the system plus the energy of the surroundings remains constant: energy is conserved.* A mathematical expression of the law of conservation of energy (first law of thermodynamics) is

$$\Delta E_{\text{universe}} = \Delta E_{\text{system}} + \Delta E_{\text{surroundings}} = 0 \qquad \textbf{(6.3)}$$

This profound idea pertains to all systems, from a burning match to the movement of continents, from the inner workings of your heart to the formation of the Solar System.

Units of Energy

The SI unit of energy is the **joule (J),** a derived unit composed of three base units:

$$1\ \text{J} = 1\ \text{kg·m}^2/\text{s}^2$$

Both heat and work are expressed in joules. Let's see how these units arise in the case of work. The work (w) done on a mass is the force (F) times the distance (d) that the mass moves: $w = F \times d$. A *force* changes the velocity of (accelerates) a mass. Velocity has units of meters per second (m/s), so acceleration (a) has units of m/s^2. Force, therefore, has units of mass (m, in kilograms) times acceleration:

$$F = m \times a \quad \text{in units of} \quad \text{kg·m/s}^2$$

Therefore, $$w = F \times d \quad \text{has units of} \quad (\text{kg·m/s}^2) \times \text{m} = \text{kg·m}^2/\text{s}^2 = \text{J}$$

Potential energy, kinetic energy, and PV work are combinations of the same physical quantities and are also expressed in joules.

The **calorie (cal)** is an older unit that was defined originally as the quantity of energy needed to raise the temperature of 1 g of water by 1°C (from 14.5°C to 15.5°C). The calorie is now defined in terms of the joule:

$$1\ \text{cal} \equiv 4.184\ \text{J} \quad \text{or} \quad 1\ \text{J} = \frac{1}{4.184}\ \text{cal} = 0.2390\ \text{cal}$$

Because the quantities of energy involved in chemical reactions are usually quite large, chemists use the unit the kilojoule (kJ), or sometimes the kilocalorie (kcal):

$$1 \text{ kJ} = 1000 \text{ J} = 0.2390 \text{ kcal} = 239.0 \text{ cal}$$

The nutritional Calorie (note the capital C), the unit used to measure the energy available from food, is actually a kilocalorie. The *British thermal unit (Btu),* a unit in engineering that you may have seen used to indicate energy output of appliances, is the quantity of energy required to raise the temperature of 1 lb of water by 1°F and is equivalent to 1055 J. In general, the SI unit (J or kJ) is used throughout this text.

SAMPLE PROBLEM 6.1 Determining the Change in Internal Energy of a System

Problem When gasoline burns in an automobile engine, the heat released causes the products CO_2 and H_2O to expand, which pushes the pistons outward. Excess heat is removed by the car's cooling system. If the expanding gases do 451 J of work on the pistons and the system loses 325 J to the surroundings as heat, calculate the change in energy (ΔE) in J, kJ, and kcal.

Plan We must define system and surroundings, assign signs to q and w, and then calculate ΔE with Equation 6.2. The system is the reactants and products, and the surroundings are the pistons, the cooling system, and the rest of the car. Heat is released by the system, so q is negative. Work is done by the system to push the pistons outward, so w is also negative. We obtain the answer in J and then convert it to kJ and kcal.

Solution Calculating ΔE (from Equation 6.2) in J:

$$q = -325 \text{ J}$$
$$w = -451 \text{ J}$$
$$\Delta E = q + w = -325 \text{ J} + (-451 \text{ J})$$
$$= -776 \text{ J}$$

Converting from J to kJ:

$$\Delta E = -776 \text{ J} \times \frac{1 \text{ kJ}}{1000 \text{ J}}$$
$$= -0.776 \text{ kJ}$$

Converting from kJ to kcal:

$$\Delta E = -0.776 \text{ kJ} \times \frac{1 \text{ kcal}}{4.184 \text{ kJ}}$$
$$= -0.185 \text{ kcal}$$

Check The answer is reasonable: combustion releases energy from the system, so $E_{final} < E_{initial}$ and ΔE should be negative. Given that 4 kJ $\approx$ 1 kcal, with rounding, nearly 0.8 kJ should be nearly 0.2 kcal.

FOLLOW-UP PROBLEM 6.1 In a reaction, gaseous reactants form a liquid product. The heat absorbed by the surroundings is 26.0 kcal, and the work done on the system is 15.0 Btu. Calculate ΔE (in kJ).

State Functions and the Path Independence of the Energy Change

An important point to understand is that there is no particular sequence by which the internal energy (E) of a system must change. This is because E is a **state function,** a property dependent *only* on the current state of the system (its composition, volume, pressure, and temperature), *not* on the path the system took to reach that state; the current state depends only on the *difference* between the final and initial states.

THINK OF IT THIS WAY
Your Financial State Function

The balance in your checkbook is a state function of your personal financial system. For example, you can open a new account with a birthday gift of \$50, or you can open a new account with a deposit of a \$100 paycheck and then write two \$25 checks. The two paths to the balance are different, but the balance (current state) is the same.

Thus, the energy change of a system can occur by any one of countless combinations of heat (q) and work (w). No matter what the combination, however, the same overall energy change occurs, because ΔE *does* ***not*** *depend on how the change takes place.* As an example, let's define a system in its initial state as 1 mol of octane (a component of gasoline) together with enough O_2 to burn it. In its final state, the system is the CO_2 and H_2O that form (a fractional coefficient is needed for O_2 because we specified 1 mol of octane):

$$\underset{\text{initial state } (E_{\text{initial}})}{C_8H_{18}(l) + \tfrac{25}{2}O_2(g)} \longrightarrow \underset{\text{final state } (E_{\text{final}})}{8CO_2(g) + 9H_2O(g)}$$

Energy is released to warm the surroundings and/or do work on them, so ΔE is negative. Two of the ways the change can occur are shown in Figure 6.4. If we burn the octane in an open container, ΔE appears almost completely as heat (with a small amount of work done to push back the atmosphere). If we burn it in a car engine, a much larger portion (~30%) of ΔE appears as work that moves the car, with the rest used to heat the car, exhaust gases, and surrounding air. If we burn the octane in a lawn mower or a plane, ΔE appears as other combinations of work and heat.

Thus, for a given change, ΔE *(sum of q and w) is constant, even though q and w can vary.* Heat and work are *not* state functions because their values *do* depend on the path the system takes in undergoing the energy change. The pressure (P) of an ideal gas or the volume (V) of water in a beaker are other examples of state functions. This path independence means that *changes in state functions—ΔE, ΔP, and ΔV—depend only on their initial and final states.* (Note that symbols for state functions, such as E, P, and V, are capitalized.)

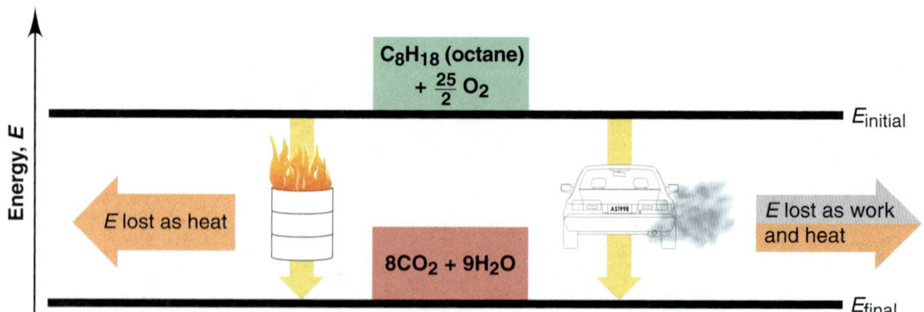

FIGURE 6.4 Two different paths for the energy change of a system. The change in internal energy when a given amount of octane burns in air is the same no matter how the energy is transferred. On the left, the fuel is burned in an open can, and the energy is lost almost entirely as heat. On the right, it is burned in a car engine; thus, a portion of the energy is lost as work to move the car, and less is lost as heat.

SECTION 6.1 SUMMARY

Energy is transferred as heat (q) when the system and surroundings are at different temperatures; energy is transferred as work (w) when an object is moved by a force. • Heat or work gained by a system ($q > 0$; $w > 0$) increases its internal energy (E); heat or work lost by the system ($q < 0$; $w < 0$) decreases E. The total change in the system's internal energy is the sum of the heat and work: $\Delta E = q + w$. • Heat and work are measured in joules (J). • Energy is always conserved: it changes from one form into another, moving into or out of the system, but the total quantity of energy in the universe (system *plus* surroundings) is constant. • Energy is a state function, which means that the same ΔE can occur through any combination of q and w.

6.2 ENTHALPY: HEATS OF REACTION AND CHEMICAL CHANGE

Most physical and chemical changes occur at virtually constant atmospheric pressure—a reaction in an open flask, the freezing of a lake, a drug response in an organism. In this section, we define a thermodynamic variable that makes it much easier to measure energy changes at constant pressure.

The Meaning of Enthalpy

To determine ΔE, we must measure both heat and work. The two most important types of chemical work are electrical work, the work done by moving charged particles (Chapter 21), and PV work, the work done by an expanding gas. We find the quantity of PV work done by multiplying the external pressure (P) by the change in volume of the gas (ΔV, or $V_{\text{final}} - V_{\text{initial}}$). In an open flask (or a cylinder with a weightless, frictionless piston), a gas does work by pushing back the atmosphere (Figure 6.5). Work done *on* the surroundings is a negative quantity because ΔV is positive; work done *on* the system is a positive quantity because ΔV is negative:

$$w = -P\Delta V \tag{6.4}$$

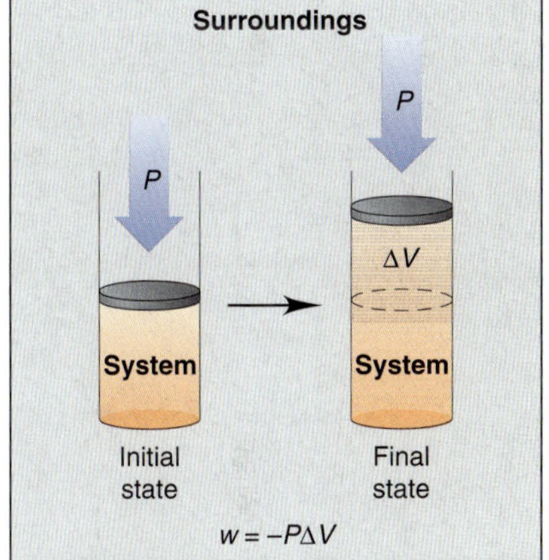

FIGURE 6.5 Pressure-volume work. When the volume (V) of a system increases by an amount ΔV against an external pressure (P), the system pushes back, and thus does PV work *on* the surroundings ($w = -P\Delta V$).

For reactions at constant pressure, a thermodynamic variable called **enthalpy (H)** eliminates the need to consider PV work separately. The enthalpy of a system is defined as the internal energy *plus* the product of the pressure and volume:

$$H = E + PV$$

The **change in enthalpy (ΔH)** is the change in internal energy *plus* the product of the constant pressure and the change in volume:

$$\Delta H = \Delta E + P\Delta V \tag{6.5}$$

Combining Equations 6.2 ($\Delta E = q + w$) and 6.4 leads to a key point about ΔH:

$$\Delta E = q + w = q + (-P\Delta V) = q - P\Delta V$$

At constant pressure, we denote q as q_P and solve for it:

$$q_P = \Delta E + P\Delta V$$

Notice the right side of this equation is identical to the right side of Equation 6.5:

$$q_P = \Delta E + P\Delta V = \Delta H \tag{6.6}$$

Thus, *the change in enthalpy equals the heat gained or lost at constant pressure*. Since most changes occur at constant pressure, ΔH is more relevant than ΔE *and* easier to find: *to find* ΔH, *measure* q_P. We discuss the laboratory method for measuring the heat involved in a chemical or physical change in Section 6.3.

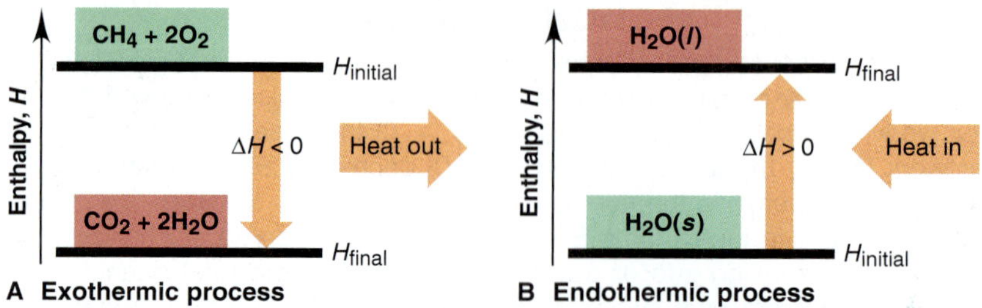

FIGURE 6.6 Enthalpy diagrams for exothermic and endothermic processes. A, Methane burns with a decrease in enthalpy because heat *leaves* the system. Therefore, $H_{final} < H_{initial}$, and the process is exothermic: $\Delta H < 0$. **B,** Ice melts with an increase in enthalpy because heat *enters* the system. Therefore, $H_{final} > H_{initial}$, and the process is endothermic: $\Delta H > 0$.

Exothermic and Endothermic Processes

Because E, P, and V are state functions, H is also a state function, which means that ΔH depends only on the *difference* between H_{final} and $H_{initial}$. The enthalpy change of a reaction, also called the **heat of reaction, ΔH_{rxn},** *always refers to H_{final} minus $H_{initial}$*:

$$\Delta H = H_{final} - H_{initial} = H_{products} - H_{reactants}$$

Therefore, because $H_{products}$ can be either more or less than $H_{reactants}$, the sign of ΔH indicates whether heat is absorbed or released in the change. We determine the sign of ΔH *by imagining the heat as a "reactant" or "product."* When methane burns in air, for example, we know that heat is produced, so we show it as a product (on the right):

$$CH_4(g) + 2O_2(g) \longrightarrow CO_2(g) + 2H_2O(g) + heat$$

Because heat is released to the surroundings, the products (1 mol of CO_2 and 2 mol of H_2O) must have less enthalpy than the reactants (1 mol of CH_4 and 2 mol of O_2). Therefore, ΔH ($H_{final} - H_{initial}$) is negative, as the **enthalpy diagram** in Figure 6.6A shows. An **exothermic** ("heat out") **process** *releases* heat and results in a *decrease* in the enthalpy of the system:

$$\text{Exothermic:} \quad H_{final} < H_{initial} \quad \Delta H < 0 \quad (\Delta H \text{ is negative})$$

An **endothermic** ("heat in") **process** *absorbs* heat and results in an *increase* in the enthalpy of the system. When ice melts, for instance, heat flows *into* the ice from the surroundings, so we show the heat as a reactant (on the left):

$$heat + H_2O(s) \longrightarrow H_2O(l)$$

Because heat is absorbed, the enthalpy of the liquid water is higher than that of the solid water, as Figure 6.6B shows. Therefore, ΔH ($H_{water} - H_{ice}$) is positive:

$$\text{Endothermic:} \quad H_{final} > H_{initial} \quad \Delta H > 0 \quad (\Delta H \text{ is positive})$$

In general, the value of an enthalpy change refers to reactants and products at the same temperature.

SAMPLE PROBLEM 6.2 Drawing Enthalpy Diagrams and Determining the Sign of ΔH

Problem In each of the following cases, determine the sign of ΔH, state whether the reaction is exothermic or endothermic, and draw an enthalpy diagram:

(a) $H_2(g) + \frac{1}{2}O_2(g) \longrightarrow H_2O(l) + 285.8\text{ kJ}$

(b) $40.7\text{ kJ} + H_2O(l) \longrightarrow H_2O(g)$

Plan From each equation, we see whether heat is a "product" (exothermic; $\Delta H < 0$) or a "reactant" (endothermic; $\Delta H > 0$). For exothermic reactions, reactants are above products on the enthalpy diagram; for endothermic reactions, reactants are below products. The ΔH arrow *always* points from reactants to products.

Solution (a) Heat is a product (on the right), so $\Delta H < 0$ and the reaction is exothermic. The enthalpy diagram appears in the margin *(top)*.

(b) Heat is a reactant (on the left), so $\Delta H > 0$ and the reaction is endothermic. The enthalpy diagram appears in the margin *(bottom)*.

Check Substances that are on the same side of the equation as the heat have less enthalpy than substances on the other side, so make sure they are placed on the lower line of the diagram.

Comment ΔH values depend on conditions. In (b), for instance, $\Delta H = 40.7$ kJ at 1 atm and 100°C; at 1 atm and 25°C, $\Delta H = 44.0$ kJ.

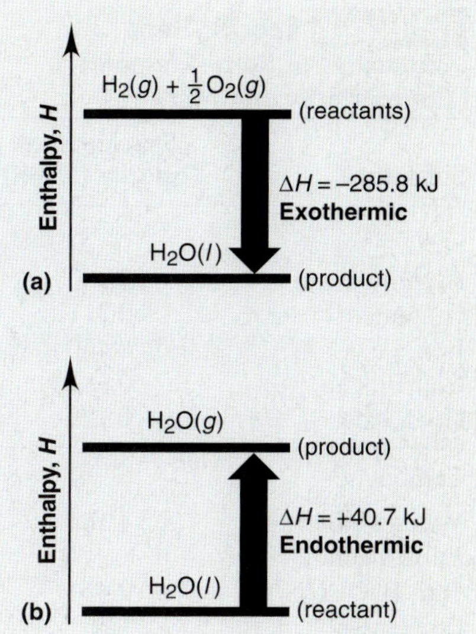

FOLLOW-UP PROBLEM 6.2 When 1 mol of nitroglycerine decomposes, it causes a violent explosion and releases 5.72×10^3 kJ of heat:

$$C_3H_5(NO_3)_3(l) \longrightarrow 3CO_2(g) + \tfrac{5}{2}H_2O(g) + \tfrac{1}{4}O_2(g) + \tfrac{3}{2}N_2(g)$$

Is the reaction exothermic or endothermic? Draw an enthalpy diagram for it.

SECTION 6.2 SUMMARY

The change in enthalpy, ΔH, is equal to the heat lost or gained during a chemical or physical change that occurs at constant pressure, q_P. • A change that releases heat is exothermic ($\Delta H < 0$); a change that absorbs heat is endothermic ($\Delta H > 0$).

6.3 CALORIMETRY: LABORATORY MEASUREMENT OF HEATS OF REACTION

Data about energy content and use are everywhere—the calories per serving of a slice of bread, the energy efficiency rating of a washing machine, the gas mileage of a new car, and so forth. How do we measure the heat released (or absorbed) by a change? To determine the energy content of a teaspoon of sugar, for example, you might think we can simply measure the enthalpies of the reactants (sucrose and O_2) and subtract them from the enthalpies of the products (CO_2 and H_2O). The problem is that the enthalpy (H) of a system in a given state cannot be measured because we have no starting point with which to compare it, no zero enthalpy. However, we *can* measure the *change* in enthalpy (ΔH) of a system. In this section, we'll see how ΔH values are determined.

To measure q_P, which is equal to ΔH, we construct "surroundings" that retain the heat, and we observe the temperature change. Then, we relate the quantity of heat released (or absorbed) to that temperature change through a physical property called the *specific heat capacity.*

Specific Heat Capacity

You know from everyday experience that the more you heat an object, the higher its temperature; that is, the quantity of heat (q) absorbed by an object is proportional to its temperature change:

$$q \propto \Delta T \qquad \text{or} \qquad q = \text{constant} \times \Delta T \qquad \text{or} \qquad \frac{q}{\Delta T} = \text{constant}$$

Every object has its own **heat capacity,** the quantity of heat required to change its temperature by 1 K. Heat capacity is the proportionality constant in the preceding equation:

$$\text{Heat capacity} = \frac{q}{\Delta T} \quad \text{[in units of J/K]}$$

Table 6.2 Specific Heat Capacities of Some Elements, Compounds, and Mixtures

Substance	Specific Heat Capacity (J/g·K)*
Elements	
Aluminum, Al	0.900
Graphite, C	0.711
Iron, Fe	0.450
Copper, Cu	0.387
Gold, Au	0.129
Compounds	
Water, $H_2O(l)$	4.184
Ethyl alcohol, $C_2H_5OH(l)$	2.46
Ethylene glycol, $(CH_2OH)_2(l)$	2.42
Carbon tetrachloride, $CCl_4(l)$	0.862
Solid mixtures	
Wood	1.76
Cement	0.88
Glass	0.84
Granite	0.79
Steel	0.45

*At 298 K (25°C).

A related property is **specific heat capacity (*c*),** the quantity of heat required to change the temperature of 1 *gram* of a substance by 1 K:*

$$\text{Specific heat capacity } (c) = \frac{q}{\text{mass} \times \Delta T} \quad \text{[in units of J/g·K]}$$

If we know *c* of the substance being heated (or cooled), we can measure its mass and temperature change and calculate the heat absorbed or released:

$$q = c \times \text{mass} \times \Delta T \qquad (6.7)$$

Notice that when an object gets hotter, ΔT (that is, $T_{\text{final}} - T_{\text{initial}}$) is positive. The object gains heat, so $q > 0$, as we expect. Similarly, when an object gets cooler, ΔT is negative; so $q < 0$ because heat is lost. Table 6.2 lists the specific heat capacities of some common substances and mixtures.

Closely related to the specific heat capacity is the **molar heat capacity (*C*;** note capital letter), the quantity of heat required to change the temperature of 1 *mole* of a substance by 1 K:

$$\text{Molar heat capacity } (C) = \frac{q}{\text{moles} \times \Delta T} \quad \text{[in units of J/mol·K]}$$

The specific heat capacity of liquid water is 4.184 J/g·K, so

$$C \text{ of } H_2O(l) = 4.184 \frac{\text{J}}{\text{g·K}} \times \frac{18.02 \text{ g}}{1 \text{ mol}} = 75.40 \frac{\text{J}}{\text{mol·K}}$$

SAMPLE PROBLEM 6.3 Finding Quantity of Heat from Specific Heat Capacity

Problem A layer of copper welded to the bottom of a skillet weighs 125 g. How much heat is needed to raise the temperature of the copper layer from 25°C to 300.°C? The specific heat capacity (*c*) of Cu is 0.387 J/g·K.

Plan We know the mass and *c* of Cu and can find ΔT in °C, which equals ΔT in K. We use this ΔT and Equation 6.7 to solve for the heat.

Solution Calculating ΔT and *q*:

$$\Delta T = T_{\text{final}} - T_{\text{initial}} = 300.°\text{C} - 25°\text{C} = 275°\text{C} = 275 \text{ K}$$

$$q = c \times \text{mass (g)} \times \Delta T = 0.387 \text{ J/g·K} \times 125 \text{ g} \times 275 \text{ K} = 1.33\times10^4 \text{ J}$$

Check Heat is absorbed by the copper bottom (system), so *q* is positive. Rounding shows that the arithmetic seems reasonable: $q \approx 0.4 \text{ J/g·K} \times 100 \text{ g} \times 300 \text{ K} = 1.2\times10^4 \text{ J}$.

FOLLOW-UP PROBLEM 6.3 Find the heat transferred (in kJ) when 5.50 L of ethylene glycol (d = 1.11 g/mL; see Table 6.2 for *c*) in a car radiator cools from 37.0°C to 25.0°C.

The Practice of Calorimetry

The **calorimeter** is used to measure the heat released (or absorbed) by a physical or chemical process. This apparatus is the "surroundings" that change temperature when heat is transferred to or from the system. Two common types are the constant-pressure and constant-volume calorimeters.

Constant-Pressure Calorimetry A constant-pressure calorimeter, simulated in the lab by a "coffee-cup" calorimeter (Figure 6.7), is often used to measure the heat transferred (q_P) in processes open to the atmosphere. One common use is to find the specific heat capacity of a solid that does not react with or dissolve in water. The solid (system) is weighed, heated to some known temperature, and added to a sample of water (surroundings) of known temperature and mass in the

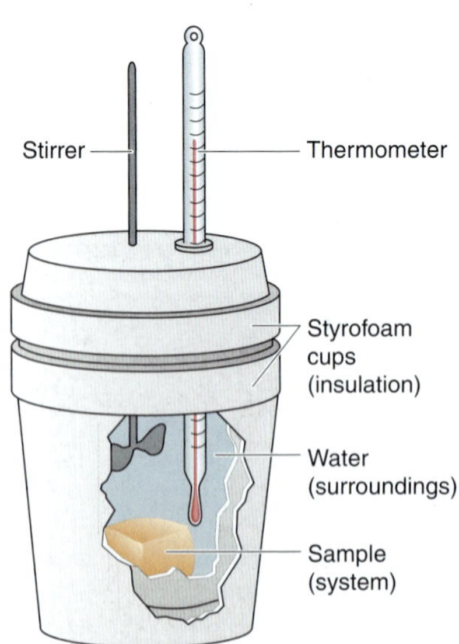

FIGURE 6.7 Coffee-cup calorimeter. This apparatus is used to measure the heat at constant pressure (q_P).

*Some texts use the term *specific heat* in place of *specific heat capacity.* This usage is very common but somewhat incorrect. *Specific heat* is the ratio of the heat capacity of 1 g of a substance to the heat capacity of 1 g of H_2O and therefore has no units.

calorimeter. With stirring, the final water temperature, which is also the final temperature of the solid, is measured.

The heat lost by the system ($-q_{sys}$, or $-q_{solid}$) is equal in magnitude but opposite in sign to the heat gained by the surroundings ($+q_{surr}$, or $+q_{H_2O}$):

$$-q_{solid} = q_{H_2O}$$

Substituting Equation 6.7 for each side of this equality gives

$$-(c_{solid} \times mass_{solid} \times \Delta T_{solid}) = c_{H_2O} \times mass_{H_2O} \times \Delta T_{H_2O}$$

All the quantities are known or measured except c_{solid}:

$$c_{solid} = -\frac{c_{H_2O} \times mass_{H_2O} \times \Delta T_{H_2O}}{mass_{solid} \times \Delta T_{solid}}$$

For example, suppose you heat a 25.64-g solid in a test tube to 100.00°C and carefully add it to 50.00 g of water in a coffee-cup calorimeter. The water temperature changes from 25.10°C to 28.49°C, and you want to find the specific heat capacity of the solid. Converting ΔT directly from °C to K, we know ΔT_{H_2O} = 3.39 K (28.49°C − 25.10°C) and ΔT_{solid} = −71.51 K (28.49°C − 100.00°C). Then, assuming all the heat lost by the solid is gained by the water, we have

$$c_{solid} = -\frac{c_{H_2O} \times mass_{H_2O} \times \Delta T_{H_2O}}{mass_{solid} \times \Delta T_{solid}} = -\frac{4.184\ \text{J/g·K} \times 50.00\ \text{g} \times 3.39\ \text{K}}{25.64\ \text{g} \times (-71.51\ \text{K})} = 0.387\ \text{J/g·K}$$

Follow-up Problem 6.4 applies this calculation, but Sample Problem 6.4 first shows how to find the heat of a reaction that takes place in a coffee-cup calorimeter.

SAMPLE PROBLEM 6.4 Determining the Heat of a Reaction

Problem You place 50.0 mL of 0.500 *M* NaOH in a coffee-cup calorimeter at 25.00°C and carefully add 25.0 mL of 0.500 *M* HCl, also at 25.00°C. After stirring, the final temperature is 27.21°C. Calculate q_{soln} (in J) and ΔH_{rxn} (in kJ/mol). (Assume the total volume is the sum of the individual volumes and that the final solution has the same density and specific heat capacity as water: d = 1.00 g/mL and c = 4.184 J/g·K.)

Plan We first find the heat given off to the solution (q_{soln}) for the amounts given and then use the equation to find the heat per mole of reaction. We know the solution volumes (25.0 mL and 50.0 mL), so we can find their masses from the given density (1.00 g/mL). Multiplying their total mass by the change in T and the given c, we can find q_{soln}. Then, writing the balanced net ionic equation for the acid-base reaction, we use the volumes and the concentrations (0.500 *M*) to find moles of reactants (H^+ and OH^-) and, thus, product (H_2O). Dividing q_{soln} by the moles of water formed gives ΔH_{rxn} in kJ/mol.

Solution Finding $mass_{soln}$ and ΔT_{soln}:

$$\text{Total mass (g) of solution} = (25.0\ \text{mL} + 50.0\ \text{mL}) \times 1.00\ \text{g/mL} = 75.0\ \text{g}$$

$$\Delta T = 27.21°\text{C} - 25.00°\text{C} = 2.21°\text{C} = 2.21\ \text{K}$$

Finding q_{soln}:

$$q_{soln} = c_{soln} \times mass_{soln} \times \Delta T_{soln} = (4.184\ \text{J/g·K})(75.0\ \text{g})(2.21\ \text{K}) = 693\ \text{J}$$

Writing the net ionic equation:

$$HCl(aq) + NaOH(aq) \longrightarrow H_2O(l) + NaCl(aq)$$

$$H^+(aq) + OH^-(aq) \longrightarrow H_2O(l)$$

Finding moles of reactants and products:

$$\text{Moles of } H^+ = 0.500\ \text{mol/L} \times 0.0250\ \text{L} = 0.0125\ \text{mol } H^+$$

$$\text{Moles of } OH^- = 0.500\ \text{mol/L} \times 0.0500\ \text{L} = 0.0250\ \text{mol } OH^-$$

Therefore, H^+ is limiting, so 0.0125 mol of H_2O is formed.

Finding ΔH_{rxn}: Heat gained by the water was lost by the reaction; that is,

$$q_{soln} = -q_{rxn} = 693\ \text{J} \quad \text{so} \quad q_{rxn} = -693\ \text{J}$$

$$\Delta H_{rxn}\ (\text{kJ/mol}) = \frac{q_{rxn}}{\text{mol } H_2O} \times \frac{1\ \text{kJ}}{1000\ \text{J}} = \frac{-693\ \text{J}}{0.0125\ \text{mol}} \times \frac{1\ \text{kJ}}{1000\ \text{J}} = -55.4\ \text{kJ/mol}$$

Check Rounding to check q_{soln} gives 4 J/g·K × 75 g × 2 K = 600 J. The volume of H^+ is half the volume of OH^-, so moles of H^+ determines moles of product. Taking the negative of q_{soln} to find ΔH_{rxn} gives −600 J/0.012 mol = -5×10^4 J/mol, or −50 kJ/mol.

FOLLOW-UP PROBLEM 6.4 In a purity check for industrial diamonds, a 10.25-carat (1 carat = 0.2000 g) diamond is heated to 74.21°C and immersed in 26.05 g of water in a constant-pressure calorimeter. The initial temperature of the water is 27.20°C. Calculate ΔT of the water and of the diamond ($c_{diamond}$ = 0.519 J/g·K).

Constant-Volume Calorimetry In the coffee-cup calorimeter, we assume all the heat is gained by the water, but some must be gained by the stirrer, thermometer, and so forth. For more precise work, as in constant-volume calorimetry, the *heat capacity of the entire calorimeter* must be known. One type of constant-volume apparatus is the *bomb calorimeter,* designed to measure very precisely the heat released in a combustion reaction. As Sample Problem 6.5 will show, this need for greater precision requires that we know (or determine) the heat capacity of the calorimeter.

Figure 6.8 depicts the preweighed combustible sample in a metal-walled chamber (the bomb), which is filled with oxygen gas and immersed in an insulated water bath fitted with motorized stirrer and thermometer. A heating coil connected to an electrical source ignites the sample, and the heat evolved raises the temperature of the bomb, water, and other calorimeter parts. Because we know the mass of the sample and the heat capacity of the entire calorimeter, we can use the measured ΔT to calculate the heat released.

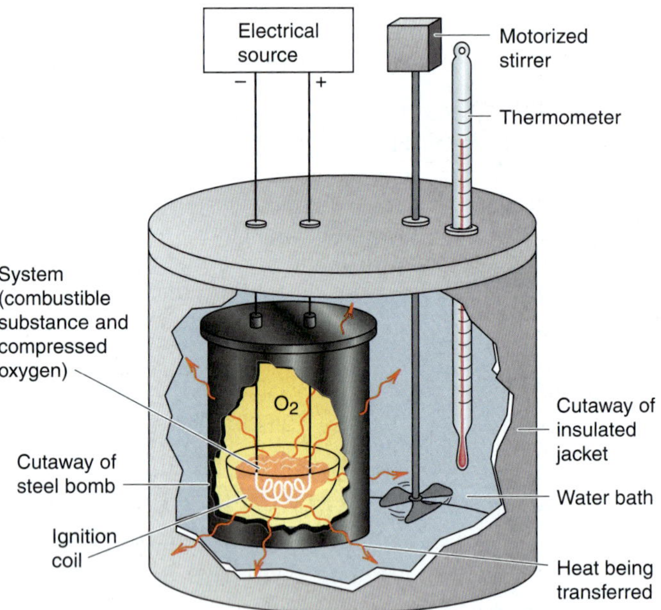

FIGURE 6.8 A bomb calorimeter. This device (not drawn to scale) is used to measure heat of combustion at constant volume (q_V).

SAMPLE PROBLEM 6.5 Calculating the Heat of a Combustion Reaction

Problem A manufacturer claims that its new dietetic dessert has "fewer than 10 Calories per serving." To test the claim, a chemist at the Department of Consumer Affairs places one serving in a bomb calorimeter and burns it in O_2 (heat capacity of the calorimeter = 8.151 kJ/K). The temperature increases 4.937°C. Is the manufacturer's claim correct?

Plan When the dessert burns, the heat released is gained by the calorimeter:

$$-q_{\text{sample}} = q_{\text{calorimeter}}$$

To find the heat, we multiply the given heat capacity of the calorimeter (8.151 kJ/K) by ΔT (4.937°C).

Solution Calculating the heat gained by the calorimeter:

$$q_{\text{calorimeter}} = \text{heat capacity} \times \Delta T = 8.151 \text{ kJ/K} \times 4.937 \text{ K} = 40.24 \text{ kJ}$$

Recall that 1 Calorie = 1 kcal = 4.184 kJ. Therefore, 10 Calories = 41.84 kJ, so the claim is correct.

Check A quick math check shows that the answer is reasonable: 8 kJ/K × 5 K = 40 kJ.

Comment With the volume of the steel bomb fixed, $\Delta V = 0$, and thus $P\Delta V = 0$. Thus, the energy change measured is the *heat at constant volume* (q_V), which equals ΔE, not ΔH:

$$\Delta E = q + w = q_V + 0 = q_V$$

However, in most cases, ΔH is usually very close to ΔE. For example, ΔH is only 0.5% larger than ΔE for the combustion of H_2 and only 0.2% smaller for the combustion of octane.

FOLLOW-UP PROBLEM 6.5 A chemist burns 0.8650 g of graphite (a form of carbon) in a new bomb calorimeter, and CO_2 forms. If 393.5 kJ of heat is released per mole of graphite and T increases 2.613 K, what is the heat capacity of the bomb calorimeter?

SECTION 6.3 SUMMARY

We calculate ΔH of a process by measuring the heat at constant pressure (q_P). To do this, we determine ΔT of the surroundings and relate it to q_P through the mass of the substance and its specific heat capacity (c), the quantity of energy needed to raise the temperature of 1 g of the substance by 1 K. • Calorimeters measure the heat released from a system either at constant pressure ($q_P = \Delta H$) or at constant volume ($q_V = \Delta E$).

6.4 STOICHIOMETRY OF THERMOCHEMICAL EQUATIONS

A **thermochemical equation** is a balanced equation that includes the heat of reaction (ΔH_{rxn}). Keep in mind that the ΔH_{rxn} value shown refers to the *amounts (moles) of substances* ***and*** *their states of matter in that specific equation.* The enthalpy change of any process has two aspects:

1. *Sign.* The sign of ΔH depends on whether the reaction is exothermic (−) or endothermic (+). A forward reaction has the *opposite* sign of the reverse reaction.

   Decomposition of 2 mol of water to its elements (endothermic):

   $$2H_2O(l) \longrightarrow 2H_2(g) + O_2(g) \qquad \Delta H_{\text{rxn}} = 572 \text{ kJ}$$

   Formation of 2 mol of water from its elements (exothermic):

   $$2H_2(g) + O_2(g) \longrightarrow 2H_2O(l) \qquad \Delta H_{\text{rxn}} = -572 \text{ kJ}$$

2. *Magnitude.* The magnitude of ΔH is *proportional to the amount of substance* reacting.

   Formation of 1 mol of water from its elements (half the amount in the preceding equation):

   $$H_2(g) + \tfrac{1}{2}O_2(g) \longrightarrow H_2O(l) \qquad \Delta H_{\text{rxn}} = -286 \text{ kJ}$$

Note that, in thermochemical equations, we often use fractional coefficients to specify the magnitude of ΔH_{rxn} for a *particular amount of substance.* Moreover, *in a particular reaction,* a certain amount of substance is thermochemically equivalent to a certain quantity of energy. In the reaction just shown,

286 kJ is thermochemically equivalent to 1 mol of $H_2(g)$
286 kJ is thermochemically equivalent to $\frac{1}{2}$ mol of $O_2(g)$
286 kJ is thermochemically equivalent to 1 mol of $H_2O(l)$

Just as we use stoichiometrically equivalent molar ratios to find amounts of substances, we use thermochemically equivalent quantities to find the heat of reaction for a given amount of substance. Also, just as we use molar mass (in g/mol of substance) to convert moles of a substance to grams, we use the heat of reaction (in kJ/mol of substance) to convert moles of a substance to an equivalent quantity of heat (in kJ). Figure 6.9 shows this new relationship, and Sample Problem 6.6 applies it.

FIGURE 6.9 **Summary of the relationship between amount (mol) of substance and the heat (kJ) transferred during a reaction.**

SAMPLE PROBLEM 6.6 Using the Heat of Reaction (ΔH_{rxn}) to Find Amounts

Problem The major source of aluminum in the world is bauxite (mostly aluminum oxide). Its thermal decomposition can be represented by

$$Al_2O_3(s) \xrightarrow{\Delta} 2Al(s) + \tfrac{3}{2}O_2(g) \quad \Delta H_{rxn} = 1676 \text{ kJ}$$

If aluminum is produced this way (see Comment), how many grams of aluminum can form when 1.000×10^3 kJ of heat is transferred?

Plan From the balanced equation and the enthalpy change, we see that 2 mol of Al forms when 1676 kJ of heat is absorbed. With this equivalent quantity, we convert the given kJ transferred to moles formed and then convert moles to grams.

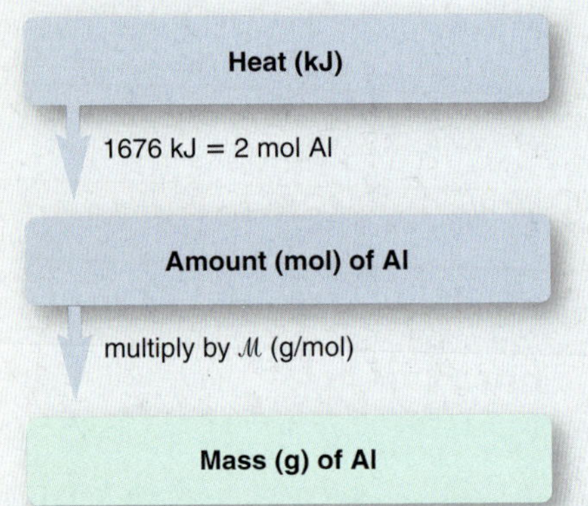

Solution Combining steps to convert from heat transferred to mass of Al:

$$\text{Mass (g) of Al} = (1.000\times10^3 \text{ kJ}) \times \frac{2 \text{ mol Al}}{1676 \text{ kJ}} \times \frac{26.98 \text{ g Al}}{1 \text{ mol Al}} = 32.20 \text{ g Al}$$

Check The mass of aluminum seems correct: ~1700 kJ forms about 2 mol of Al (54 g), so 1000 kJ should form a bit more than half that amount (27 g).

Comment In practice, aluminum is not obtained by heating but by supplying electrical energy (Chapter 21). Because ΔH is a state function, however, the total energy required for this chemical change is the same no matter how it occurs.

FOLLOW-UP PROBLEM 6.6 Organic hydrogenation reactions, in which H_2 and an "unsaturated" organic compound combine, are used in the food, fuel, and polymer industries. In the simplest case, ethene (C_2H_4) and H_2 form ethane (C_2H_6). If 137 kJ is given off per mole of C_2H_4 reacting, how much heat is released when 15.0 kg of C_2H_6 forms?

SECTION 6.4 SUMMARY

A thermochemical equation shows the balanced equation *and* its ΔH_{rxn}. The sign of ΔH for a forward reaction is opposite that for the reverse reaction. The magnitude of ΔH depends on the amount and physical state of the substance reacting and the ΔH per mole of substance. • We use the thermochemically equivalent amounts of substance and heat from the balanced equation as conversion factors to find the quantity of heat when a given amount of substance reacts.

6.5 HESS'S LAW OF HEAT SUMMATION

Many reactions are difficult, even impossible, to carry out separately. A reaction may be part of a complex biochemical process; or it may take place only under extreme environmental conditions; or it may require a change in conditions while it is occurring. Even if we can't run a reaction in the lab, it is still possible to find its enthalpy change. One of the most powerful applications of the state-function property of enthalpy (H) allows us to find the ΔH of *any* reaction for which we can write an equation.

This application is based on **Hess's law of heat summation:** *the enthalpy change of an overall process is the sum of the enthalpy changes of its individual steps.* To use Hess's law, we imagine an overall reaction as the sum of a series of reaction steps, whether or not it really occurs that way. Each step is chosen because its ΔH is known. Because the overall ΔH depends only on the initial and final states, Hess's law says that we add together the known ΔH values for the steps to get the unknown ΔH of the overall reaction. Similarly, if we know the ΔH values for the overall reaction and all but one of the steps, we can find the unknown ΔH of that step.

Let's see how we apply Hess's law in the case of the oxidation of sulfur to sulfur trioxide, the central process in the industrial production of sulfuric acid and in the formation of acid rain. (To introduce the approach, we'll simplify the equations by using S as the formula for sulfur, rather than the more correct S_8.) When we burn S in an excess of O_2, sulfur dioxide (SO_2) forms, *not* sulfur trioxide (SO_3). Equation 1 shows this step and its ΔH. If we change conditions and then add more O_2, we can oxidize SO_2 to SO_3 (Equation 2). In other words, we cannot put S and O_2 in a calorimeter and find ΔH for the overall reaction of S to SO_3 (Equation 3). But, we *can* find it with Hess's law. The three equations are

$$\text{Equation 1:} \quad S(s) + O_2(g) \longrightarrow SO_2(g) \quad \Delta H_1 = -296.8 \text{ kJ}$$

$$\text{Equation 2:} \quad 2SO_2(g) + O_2(g) \longrightarrow 2SO_3(g) \quad \Delta H_2 = -198.4 \text{ kJ}$$

$$\text{Equation 3:} \quad S(s) + \tfrac{3}{2}O_2(g) \longrightarrow SO_3(g) \quad \Delta H_3 = ?$$

Hess's law tells us that if we manipulate Equations 1 and/or 2 so that they add up to Equation 3, then ΔH_3 is the sum of the manipulated ΔH values of Equations 1 and 2.

First, we identify Equation 3 as our "target" equation, the one whose ΔH we want to find, and we carefully note the number of moles of each reactant and product in it. We also note that ΔH_1 and ΔH_2 are the values for Equations 1 and 2 *as written*. Now we manipulate Equations 1 and/or 2 as follows to make them add up to Equation 3:

- Equations 1 and 3 contain the same amount of S, so we leave Equation 1 unchanged.
- Equation 2 has twice as much SO_3 as Equation 3, so we multiply it by $\frac{1}{2}$, being sure to halve ΔH_2 as well.
- With the targeted amounts of reactants and products now present, we add Equation 1 to the halved Equation 2 and cancel terms that appear on both sides:

$$\text{Equation 1:} \quad S(s) + O_2(g) \longrightarrow SO_2(g) \quad \Delta H_1 = -296.8 \text{ kJ}$$

$$\tfrac{1}{2}(\text{Equation 2}): \quad SO_2(g) + \tfrac{1}{2}O_2(g) \longrightarrow SO_3(g) \quad \tfrac{1}{2}(\Delta H_2) = -99.2 \text{ kJ}$$

$$\text{Equation 3:} \quad S(s) + O_2(g) + \cancel{SO_2(g)} + \tfrac{1}{2}O_2(g) \longrightarrow \cancel{SO_2(g)} + SO_3(g) \quad \Delta H_3 = ?$$

$$\text{or} \quad S(s) + \tfrac{3}{2}O_2(g) \longrightarrow SO_3(g)$$

Adding the ΔH values gives

$$\Delta H_3 = \Delta H_1 + \tfrac{1}{2}(\Delta H_2) = -296.8 \text{ kJ} + (-99.2 \text{ kJ}) = -396.0 \text{ kJ}$$

Once again, the key point is that H is a state function, so the overall ΔH depends on the difference between the initial and final enthalpies only. Hess's law tells us that the difference between the enthalpies of the reactants (1 mol of S and $\frac{3}{2}$ mol of O_2) and that of the product (1 mol of SO_3) is the same, whether S is oxidized directly to SO_3 (impossible) or through the formation of SO_2 (actual).

To summarize, calculating an unknown ΔH involves three steps:

1. Identify the target equation, the step whose ΔH is unknown, and note the number of moles of each reactant and product.
2. Manipulate the equations with known ΔH values so that the target numbers of moles of reactants and products are on the correct sides. Remember to:
   - Change the sign of ΔH when you reverse an equation.
   - Multiply numbers of moles and ΔH by the same factor.
3. Add the manipulated equations to obtain the target equation. All substances except those in the target equation must cancel. Add their ΔH values to obtain the unknown ΔH.

SAMPLE PROBLEM 6.7 Using Hess's Law to Calculate an Unknown ΔH

Problem Two gaseous pollutants that form in auto exhaust are CO and NO. An environmental chemist is studying ways to convert them to less harmful gases through the following equation:

$$CO(g) + NO(g) \longrightarrow CO_2(g) + \tfrac{1}{2}N_2(g) \quad \Delta H = ?$$

Given the following information, calculate the unknown ΔH:

$$\text{Equation A:} \quad CO(g) + \tfrac{1}{2}O_2(g) \longrightarrow CO_2(g) \quad \Delta H_A = -283.0 \text{ kJ}$$

$$\text{Equation B:} \quad N_2(g) + O_2(g) \longrightarrow 2NO(g) \quad \Delta H_B = 180.6 \text{ kJ}$$

Plan We note the numbers of moles of each substance in the target equation, manipulate Equations A and/or B *and* their ΔH values, and then add them together to obtain the target equation and the unknown ΔH.

Solution Noting moles of substances in the target equation: There are 1 mol each of reactants CO and NO, 1 mol of product CO_2, and $\frac{1}{2}$ mol of product N_2.

Manipulating the given equations: Equation A has the same number of moles of CO and CO_2 as the target, so we leave it as written. Equation B has twice the needed amounts of N_2 and NO, and they are on the opposite sides from the target; therefore, we reverse Equation B, change the sign of ΔH_B, and multiply both by $\frac{1}{2}$:

$$\tfrac{1}{2}[2NO(g) \longrightarrow N_2(g) + O_2(g)] \quad \Delta H = -\tfrac{1}{2}(\Delta H_B) = -\tfrac{1}{2}(180.6 \text{ kJ})$$

or

$$NO(g) \longrightarrow \tfrac{1}{2}N_2(g) + \tfrac{1}{2}O_2(g) \quad \Delta H = -90.3 \text{ kJ}$$

Adding the manipulated equations to obtain the target equation:

$$\text{Equation A:} \quad CO(g) + \cancel{\tfrac{1}{2}O_2(g)} \longrightarrow CO_2(g) \quad \Delta H = -283.0 \text{ kJ}$$

$$\tfrac{1}{2}\text{(Equation B reversed):} \quad NO(g) \longrightarrow \tfrac{1}{2}N_2(g) + \cancel{\tfrac{1}{2}O_2(g)} \quad \Delta H = -90.3 \text{ kJ}$$

$$\text{Target:} \quad CO(g) + NO(g) \longrightarrow CO_2(g) + \tfrac{1}{2}N_2(g) \quad \Delta H = -373.3 \text{ kJ}$$

Check Obtaining the desired target equation is its own check. Be sure to remember to change the *sign* of ΔH for any equation you reverse.

FOLLOW-UP PROBLEM 6.7 Nitrogen oxides undergo many interesting reactions in the environment and in industry. Given the following information, calculate ΔH for the overall equation $2NO_2(g) + \frac{1}{2}O_2(g) \longrightarrow N_2O_5(s)$:

$$N_2O_5(s) \longrightarrow 2NO(g) + \tfrac{3}{2}O_2(g) \quad \Delta H = 223.7 \text{ kJ}$$

$$NO(g) + \tfrac{1}{2}O_2(g) \longrightarrow NO_2(g) \quad \Delta H = -57.1 \text{ kJ}$$

SECTION 6.5 SUMMARY

Because H is a state function, $\Delta H = H_{final} - H_{initial}$ and does not depend on how the reaction takes place. • Using Hess's law ($\Delta H_{total} = \Delta H_1 + \Delta H_2 + \cdots + \Delta H_n$), we can determine ΔH of any equation by manipulating the coefficients of other appropriate equations and their known ΔH values.

6.6 STANDARD HEATS OF REACTION (ΔH°_{rxn})

In this section, we see how Hess's law is used to determine the ΔH values of an enormous number of reactions. To begin we must take into account that thermodynamic variables, such as ΔH, vary somewhat with conditions. Therefore, to use heats of reaction, as well as other thermodynamic data that we will encounter in later chapters, chemists have established **standard states,** a set of specified conditions and concentrations:

- For a *gas,* the standard state is 1 atm* with the gas behaving ideally.
- For a substance in *aqueous solution,* the standard state is 1 M concentration.
- For a *pure substance* (element or compound), the standard state is usually the most stable form of the substance at 1 atm and the temperature of interest. In this text, that temperature is usually 25°C (298 K).†

We use the standard-state symbol (shown here as a degree sign) to indicate these standard states. In other words, when the heat of reaction, ΔH_{rxn}, has been measured with all the reactants and products in their standard states, it is referred to as the **standard heat of reaction, ΔH°_{rxn}.**

Formation Equations and Their Standard Enthalpy Changes

In a **formation equation,** 1 mole of a compound forms from its elements. The **standard heat of formation (ΔH°_f)** is the enthalpy change for the formation equation when all the substances are in their standard states. For instance, the formation equation for methane (CH_4) is

$$C(\text{graphite}) + 2H_2(g) \longrightarrow CH_4(g) \quad \Delta H^\circ_f = -74.9 \text{ kJ}$$

Thus, the standard heat of formation of methane is -74.9 kJ/mol. Some other examples are

$$Na(s) + \tfrac{1}{2}Cl_2(g) \longrightarrow NaCl(s) \quad \Delta H^\circ_f = -411.1 \text{ kJ}$$

$$2C(\text{graphite}) + 3H_2(g) + \tfrac{1}{2}O_2(g) \longrightarrow C_2H_5OH(l) \quad \Delta H^\circ_f = -277.6 \text{ kJ}$$

Standard heats of formation have been tabulated for many substances. Table 6.3 shows ΔH°_f values for several, and a much more extensive table appears in Appendix B.

The values in Table 6.3 were selected to make two points:

1. *An element in its standard state is assigned a ΔH°_f of zero.* For example, note that $\Delta H^\circ_f = 0$ for Na(s), but $\Delta H^\circ_f = 107.8$ kJ/mol for Na(g). These values mean that the gaseous state is *not* the most stable state of sodium at 1 atm and 298.15 K, and that heat is required to form Na(g). Note also that the standard state of chlorine is Cl_2 molecules, not Cl atoms. Several elements exist in different forms, only one of which is the standard state. Thus, the standard state of carbon is graphite, not diamond, so ΔH°_f of C(graphite) $= 0$. Similarly, the standard state of oxygen is dioxygen (O_2), not ozone (O_3), and the standard state of sulfur is S_8 in its rhombic crystal form, rather than its monoclinic form.
2. *Most compounds have a negative ΔH°_f.* That is, most compounds have exothermic formation reactions under standard conditions: *heat is given off when the compound forms.*

Table 6.3 Selected Standard Heats of Formation at 25°C (298 K)

Formula	ΔH°_f (kJ/mol)
Calcium	
Ca(s)	0
CaO(s)	−635.1
$CaCO_3$(s)	−1206.9
Carbon	
C(graphite)	0
C(diamond)	1.9
CO(g)	−110.5
CO_2(g)	−393.5
CH_4(g)	−74.9
CH_3OH(l)	−238.6
HCN(g)	135
CS_2(l)	87.9
Chlorine	
Cl(g)	121.0
Cl_2(g)	0
HCl(g)	−92.3
Hydrogen	
H(g)	218.0
H_2(g)	0
Nitrogen	
N_2(g)	0
NH_3(g)	−45.9
NO(g)	90.3
Oxygen	
O_2(g)	0
O_3(g)	143
H_2O(g)	−241.8
H_2O(l)	−285.8
Silver	
Ag(s)	0
AgCl(s)	−127.0
Sodium	
Na(s)	0
Na(g)	107.8
NaCl(s)	−411.1
Sulfur	
S_8(rhombic)	0
S_8(monoclinic)	0.3
SO_2(g)	−296.8
SO_3(g)	−396.0

*The definition of the standard state for gases has been changed to 1 bar, a slightly lower pressure than the 1 atm standard on which the data in this book are based (1 atm = 101.3 kPa = 1.013 bar). For most purposes, this makes *very* little difference in the standard enthalpy values.

†In the case of phosphorus, the most *common* form, white phosphorus (P_4), is chosen as the standard state, even though red phosphorus is more stable at 1 atm and 298 K.

SAMPLE PROBLEM 6.8 Writing Formation Equations

Problem Write balanced equations for the formation of 1 mole of each of the following compounds from their elements in their standard states, and include ΔH°_f.

(a) Silver chloride, AgCl, a solid at standard conditions
(b) Calcium carbonate, $CaCO_3$, a solid at standard conditions
(c) Hydrogen cyanide, HCN, a gas at standard conditions

Plan We write the elements as the reactants and 1 mol of the compound as the product, being sure all substances are in their standard states. Then, we balance the atoms and obtain the ΔH°_f values from Table 6.3 or Appendix B.

Solution (a) $Ag(s) + \frac{1}{2}Cl_2(g) \longrightarrow AgCl(s) \quad \Delta H^\circ_f = -127.0 \text{ kJ}$

(b) $Ca(s) + C(\text{graphite}) + \frac{3}{2}O_2(g) \longrightarrow CaCO_3(s) \quad \Delta H^\circ_f = -1206.9 \text{ kJ}$

(c) $\frac{1}{2}H_2(g) + C(\text{graphite}) + \frac{1}{2}N_2(g) \longrightarrow HCN(g) \quad \Delta H^\circ_f = 135 \text{ kJ}$

FOLLOW-UP PROBLEM 6.8 Write balanced equations for the formation of 1 mol of **(a)** $CH_3OH(l)$, **(b)** $CaO(s)$, and **(c)** $CS_2(l)$ from their elements in their standard states. Include ΔH°_f for each reaction.

Determining ΔH°_{rxn} from ΔH°_f Values of Reactants and Products

By applying Hess's law, we can use ΔH°_f values to determine ΔH°_{rxn} for any reaction. All we have to do is view the reaction as an imaginary two-step process.

Step 1. Each reactant decomposes to its elements. This is the *reverse* of the formation reaction for each *reactant,* so each standard enthalpy change is $-\Delta H^\circ_f$.

Step 2. Each product forms from its elements. This step *is* the formation reaction for each *product,* so each standard enthalpy change is ΔH°_f.

According to Hess's law, we add the enthalpy changes for these steps to obtain the overall enthalpy change for the reaction (ΔH°_{rxn}). Figure 6.10 depicts the conceptual process. Suppose we want ΔH°_{rxn} for

$$TiCl_4(l) + 2H_2O(g) \longrightarrow TiO_2(s) + 4HCl(g)$$

We write this equation as though it were the sum of four individual equations, one for each compound. The first two of these equations show the decomposition of the reactants to their elements (*reverse* of their formation), and the second two show the formation of the products from their elements:

$$\begin{array}{rcll} TiCl_4(l) & \longrightarrow & Ti(s) + 2Cl_2(g) & -\Delta H^\circ_f[TiCl_4(l)] \\ 2H_2O(g) & \longrightarrow & 2H_2(g) + O_2(g) & -2\Delta H^\circ_f[H_2O(g)] \\ Ti(s) + O_2(g) & \longrightarrow & TiO_2(s) & \Delta H^\circ_f[TiO_2(s)] \\ 2H_2(g) + 2Cl_2(g) & \longrightarrow & 4HCl(g) & 4\Delta H^\circ_f[HCl(g)] \end{array}$$

$$TiCl_4(l) + 2H_2O(g) + \cancel{Ti(s)} + \cancel{O_2(g)} + \cancel{2H_2(g)} + \cancel{2Cl_2(g)} \longrightarrow \cancel{Ti(s)} + \cancel{2Cl_2(g)} + \cancel{2H_2(g)} + \cancel{O_2(g)} + TiO_2(s) + 4HCl(g)$$

or $$TiCl_4(l) + 2H_2O(g) \longrightarrow TiO_2(s) + 4HCl(g)$$

It's important to realize that when titanium(IV) chloride and water react, the reactants don't *actually* decompose to their elements, which then recombine to form

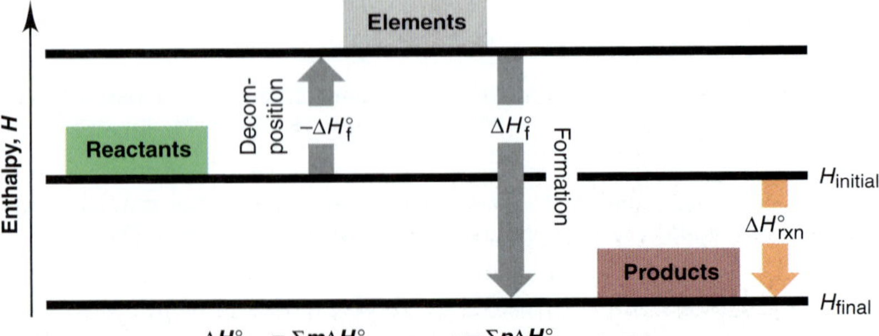

FIGURE 6.10 The general process for determining ΔH°_{rxn} from ΔH°_f values. For any reaction, ΔH°_{rxn} can be considered as the sum of the enthalpy changes for the decomposition of reactants to their elements $[-\Sigma n\Delta H^\circ_{f(reactants)}]$ and the formation of products from their elements $[+\Sigma m\Delta H^\circ_{f(products)}]$. [The factors m and n are the amounts (mol) of the products and reactants and equal the coefficients in the balanced equation, and Σ is the symbol for "sum of."]

the products. But that is the great usefulness of Hess's law and the state-function concept. Because ΔH°_{rxn} is the difference between two state functions, $\Delta H^\circ_{products}$ minus $\Delta H^\circ_{reactants}$, it doesn't matter *how* the change actually occurs. We simply add the individual enthalpy changes to find ΔH°_{rxn}:

$$\Delta H^\circ_{rxn} = \Delta H^\circ_f[TiO_2(s)] + 4\Delta H^\circ_f[HCl(g)] + \{-\Delta H^\circ_f[TiCl_4(l)]\} + \{-2\Delta H^\circ_f[H_2O(g)]\}$$
$$= \underbrace{\{\Delta H^\circ_f[TiO_2(s)] + 4\Delta H^\circ_f[HCl(g)]\}}_{\text{Products}} - \underbrace{\{\Delta H^\circ_f[TiCl_4(l)] + 2\Delta H^\circ_f[H_2O(g)]\}}_{\text{Reactants}}$$

By generalizing the result shown here, we see that *the standard heat of reaction is the sum of the standard heats of formation of the* ***products*** *minus the sum of the standard heats of formation of the* ***reactants*** (see Figure 6.10):

$$\Delta H^\circ_{rxn} = \Sigma m\Delta H^\circ_{f(products)} - \Sigma n\Delta H^\circ_{f(reactants)} \quad (6.8)$$

where the symbol Σ means "sum of," and m and n are the amounts (mol) of the products and reactants indicated by the coefficients from the balanced equation.

SAMPLE PROBLEM 6.9 Calculating the Heat of Reaction from Heats of Formation

Problem Nitric acid, whose worldwide annual production is nearly 10 billion kilograms, is used to make many products, including fertilizers, dyes, and explosives. The first step in the production process is the oxidation of ammonia:

$$4NH_3(g) + 5O_2(g) \longrightarrow 4NO(g) + 6H_2O(g)$$

Calculate ΔH°_{rxn} from ΔH°_f values.

Plan We use values from Table 6.3 (or Appendix B) and apply Equation 6.8 to find ΔH°_{rxn}.

Solution Calculating ΔH°_{rxn}:

$$\begin{aligned}\Delta H^\circ_{rxn} &= \Sigma m\Delta H^\circ_{f(products)} - \Sigma n\Delta H^\circ_{f(reactants)}\\ &= \{4\Delta H^\circ_f[NO(g)] + 6\Delta H^\circ_f[H_2O(g)]\} - \{4\Delta H^\circ_f[NH_3(g)] + 5\Delta H^\circ_f[O_2(g)]\}\\ &= (4\text{ mol})(90.3\text{ kJ/mol}) + (6\text{ mol})(-241.8\text{ kJ/mol})\\ &\quad -[(4\text{ mol})(-45.9\text{ kJ/mol}) + (5\text{ mol})(0\text{ kJ/mol})]\\ &= 361\text{ kJ} - 1451\text{ kJ} + 184\text{ kJ} - 0\text{ kJ} = -906\text{ kJ}\end{aligned}$$

Check One way to check is to write formation equations for the amounts of individual compounds in the correct direction and take their sum:

$$\begin{array}{llr} 4NH_3(g) \longrightarrow \cancel{2N_2(g)} + \cancel{6H_2(g)} & -4(-45.9\text{ kJ}) = & 184\text{ kJ}\\ \cancel{2N_2(g)} + 2O_2(g) \longrightarrow 4NO(g) & 4(90.3\text{ kJ}) = & 361\text{ kJ}\\ \cancel{6H_2(g)} + 3O_2(g) \longrightarrow 6H_2O(g) & 6(-241.8\text{ kJ}) = & -1451\text{ kJ}\\ \hline 4NH_3(g) + 5O_2(g) \longrightarrow 4NO(g) + 6H_2O(g) & & -906\text{ kJ}\end{array}$$

Comment In this problem, we know the individual ΔH°_f values and find the sum, ΔH°_{rxn}. In the follow-up problem, we know the sum and want to find an individual value.

FOLLOW-UP PROBLEM 6.9 Use the following information to find ΔH°_f of methanol [$CH_3OH(l)$]:

$$CH_3OH(l) + \tfrac{3}{2}O_2(g) \longrightarrow CO_2(g) + 2H_2O(g) \qquad \Delta H^\circ_{rxn} = -638.5\text{ kJ}$$
$$\Delta H^\circ_f\text{ of }CO_2(g) = -393.5\text{ kJ/mol} \qquad \Delta H^\circ_f\text{ of }H_2O(g) = -241.8\text{ kJ/mol}$$

Fossil Fuels and Climate Change

Out of the necessity to avoid catastrophe, the nations of the world are finally beginning to radically rethink the issue of energy use. No scientific challenge today is greater than reversing the climatic effects of our increasing dependence on the combustion of **fossil fuels**—coal, petroleum, and natural gas. Because these fuels form so much more slowly than we consume them, they are *nonrenewable.* In contrast, wood and other fuels derived from plant and animal matter are *renewable.*

All carbon-based fuels release CO_2 when burned, and in the past few decades it has become increasingly clear that our use of these fuels is changing Earth's

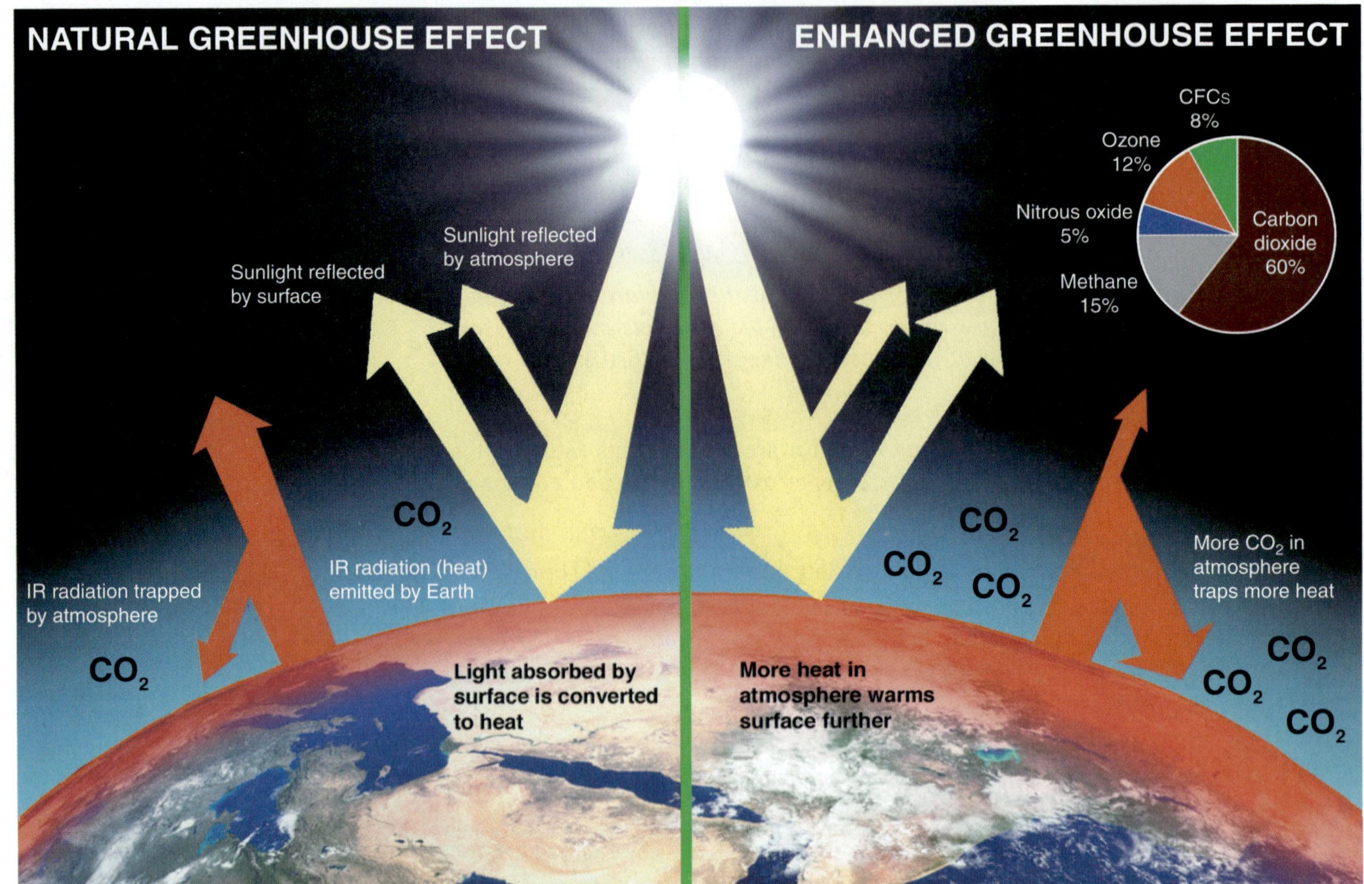

FIGURE 6.11 The trapping of heat by the atmosphere. Of the total sunlight reaching Earth, some is reflected by the atmosphere and some by the surface (especially snow, ice, and water). The remainder is absorbed by the surface and converted to IR radiation (heat). When this IR radiation is emitted by the surface, some is trapped by atmospheric components, especially CO_2. Without this *natural* greenhouse effect *(left)*, Earth's surface would have an average temperature of −18°C, far below water's freezing point, rather than its current average of 13°C. Since the early 19th century, and particularly in the past several decades, human activity has increased the amount of CO_2, along with several other greenhouse gases *(pie chart)*, and created an *enhanced* greenhouse effect *(right)*.

climate. The ability of CO_2 to absorb heat plays a key temperature-regulating role in the atmosphere. Much of the sunlight that shines on Earth is absorbed by the land and oceans and converted to heat. Like the glass of a greenhouse, atmospheric CO_2 does not absorb visible light from the Sun, but it traps some of the heat radiating back from Earth's surface and, thus, helps warm the atmosphere. This process is called the natural *greenhouse effect* (Figure 6.11, *left*).

Over several billion years, due largely to the spread of plant life, which uses CO_2 in photosynthesis, the amount of CO_2 originally present in Earth's atmosphere decreased to 0.028% by volume. However, today, as a result of the human use of fossil fuels for the past 200 years, this amount has increased to slighty over 0.036%. Thus, although the same amount of solar energy passes through the atmosphere, more is trapped as heat, which has created an *enhanced* greenhouse effect that is changing the climate through *global warming* (Figure 6.11, *right*). Based on current trends in fossil fuel use, the CO_2 concentration will increase to between 0.049% and 0.126% by 2100.

Computer-based models that simulate the climate's behavior are the best tools available for answering questions about how much the temperature will rise and how it will affect life on Earth. Even with ever-improving models, answers are difficult to obtain. Natural fluctuations in temperature and cyclic changes in solar activity must be taken into account. Moreover, as the amount of CO_2 increases from fossil-fuel burning, so does the amount of particulate matter, which may block sunlight and have a cooling effect. Water vapor also traps heat, and as temperatures rise, more water evaporates, which may thicken the cloud cover and also lead to cooling.

Despite these complicating factors, the best models predict a net warming of the atmosphere, and for the past several years, scientists have been documenting the predicted effects. The average temperature has increased by 0.6 ± 0.2°C since the late 19th century, of which 0.2–0.3°C has occurred over just the past 25 years. Globally, the 10 warmest years on record occurred in the last 15 years. Snow cover and glacier extent in the Northern Hemisphere and floating ice in the Arctic Ocean have decreased dramatically. Globally, sea level has risen 4–8 inches (10–20 cm) over the past century, and flooding and other extreme weather events have increased through much of the United States and Europe.

Today, the models predict a future temperature rise more than 50% higher than the 1.0–3.5°C rise predicted only 10 years ago. Such increases would significantly alter rainfall patterns and crop yields throughout the world and could increase sea level as much as 1 meter, thereby flooding low-lying regions, such as the Netherlands, half of Florida, much of southern Asia, and many Pacific island nations. To make matters worse, as we burn fossil fuels that *release* CO_2, we cut down the forests that *absorb* it.

In addition to developing alternative energy sources to reduce fossil-fuel consumption, researchers are studying *CO_2 sequestration* through large-scale tree planting and by liquefying CO_2 released from coal-fired power plants and burying it underground or injecting it deep into the oceans.

In 1997, the United Nations Conference on Climate Change in Kyoto, Japan, created an international treaty that set legally binding limits on release of greenhouse gases. It was ratified by 189 countries, but the largest emitter of CO_2, the United States, refused to do so. The 2005 conference in Montreal, Canada, presented overwhelming scientific evidence that confirmed the human impact on climate change, and the 2007 conference in Bali, Indonesia, issued a roadmap leading to a 2012 binding agreement on ways to address the effects of climate change and eventually reverse it.

SECTION 6.6 SUMMARY

Standard states are chosen conditions for substances. • When 1 mol of a compound forms from its elements with all substances in their standard states, the enthalpy change is ΔH_f°. • Hess's law allows us to picture a reaction as the decomposition of reactants to their elements, followed by the formation of products from their elements. • We use tabulated ΔH_f° values to find ΔH_{rxn}° or use known ΔH_{rxn}° and ΔH_f° values to find an unknown ΔH_f°. • As a result of increased fossil-fuel combustion, the amount of atmospheric CO_2 is climbing, which is seriously affecting Earth's climate.

CHAPTER REVIEW GUIDE

The following sections provide many aids to help you study this chapter. (Numbers in parentheses refer to pages, unless noted otherwise.)

• LEARNING OBJECTIVES *These are concepts and skills to review after studying this chapter.*

Related section (§), sample problem (SP), and end-of-chapter problem (EP) numbers are listed in parentheses.

1. Interconvert energy units; understand that ΔE of a system appears as the total heat and/or work transferred to or from its surroundings; understand the meaning of a state function (§ 6.1) (SP 6.1) (EPs 6.1–6.9)
2. Understand the meaning of H, why we measure ΔH, and the distinction between exothermic and endothermic reactions; draw enthalpy diagrams for chemical and physical changes (§ 6.2) (SP 6.2) (EPs 6.10–6.18)
3. Understand the relation between specific heat capacity and heat transferred in both constant-pressure (coffee-cup) and constant-volume (bomb) calorimeters (§ 6.3) (SPs 6.3–6.5) (EPs 6.19–6.32)
4. Understand the relation between heat of reaction and amount of substance (§ 6.4) (SP 6.6) (EPs 6.33–6.42)
5. Explain the importance of Hess's law and use it to find an unknown ΔH (§ 6.5) (SP 6.7) (EPs 6.43–6.48)
6. View a reaction as the decomposition of reactants followed by the formation of products; understand formation equations and how to use ΔH_f° values to find ΔH_{rxn}° (§ 6.6) (SPs 6.8, 6.9) (EPs 6.49–6.58)

• KEY TERMS *These important terms appear in boldface in the chapter and are defined again in the Glossary.*

thermodynamics (186)
thermochemistry (186)

Section 6.1
system (186)
surroundings (186)
internal energy (E) (187)
heat (q) (188)
work (w) (188)
pressure-volume work (PV work) (189)
law of conservation of energy (first law of thermodynamics) (190)
joule (J) (190)
calorie (cal) (191)
state function (191)

Section 6.2
enthalpy (H) (193)
change in enthalpy (ΔH) (193)
heat of reaction (ΔH_{rxn}) (194)
enthalpy diagram (194)
exothermic process (194)
endothermic process (194)

Section 6.3
heat capacity (195)
specific heat capacity (c) (196)
molar heat capacity (C) (196)
calorimeter (196)

Section 6.4
thermochemical equation (199)

Section 6.5
Hess's law of heat summation (201)

Section 6.6
standard states (203)
standard heat of reaction (ΔH°_{rxn}) (203)
formation equation (203)
standard heat of formation (ΔH°_f) (203)
fossil fuel (205)

• KEY EQUATIONS AND RELATIONSHIPS *Numbered and screened equations are listed for you to refer to or memorize.*

6.1 Defining the change in internal energy (187):

$$\Delta E = E_{final} - E_{initial} = E_{products} - E_{reactants}$$

6.2 Expressing the change in internal energy in terms of heat and work (188):

$$\Delta E = q + w$$

6.3 Stating the first law of thermodynamics (law of conservation of energy) (190):

$$\Delta E_{universe} = \Delta E_{system} + \Delta E_{surroundings} = 0$$

6.4 Determining the work due to a change in volume at constant pressure (PV work) (193):

$$w = -P\Delta V$$

6.5 Relating the enthalpy change to the internal energy change at constant pressure (193):

$$\Delta H = \Delta E + P\Delta V$$

6.6 Identifying the enthalpy change with the heat gained or lost at constant pressure (193):

$$q_P = \Delta E + P\Delta V = \Delta H$$

6.7 Calculating the heat absorbed or released when a substance undergoes a temperature change (196):

$$q = c \times \text{mass} \times \Delta T$$

6.8 Calculating the standard heat of reaction (205):

$$\Delta H^\circ_{rxn} = \Sigma m\Delta H^\circ_{f(products)} - \Sigma n\Delta H^\circ_{f(reactants)}$$

• BRIEF SOLUTIONS TO *FOLLOW-UP PROBLEMS* *Compare your own solutions to these calculation steps and answers.*

6.1 $\Delta E = q + w$

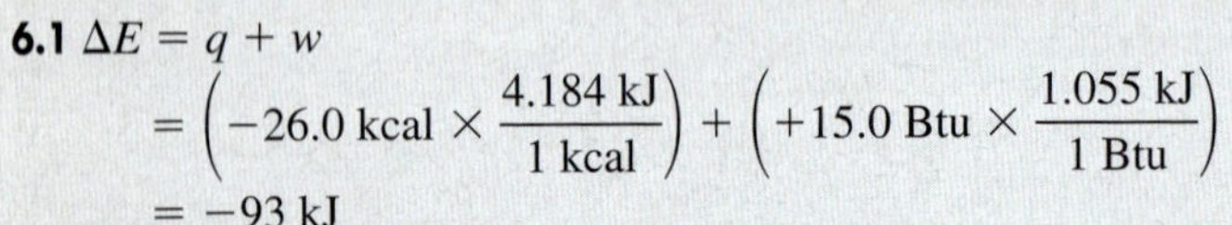

$$= \left(-26.0 \text{ kcal} \times \frac{4.184 \text{ kJ}}{1 \text{ kcal}}\right) + \left(+15.0 \text{ Btu} \times \frac{1.055 \text{ kJ}}{1 \text{ Btu}}\right)$$

$$= -93 \text{ kJ}$$

6.2 The reaction is exothermic.

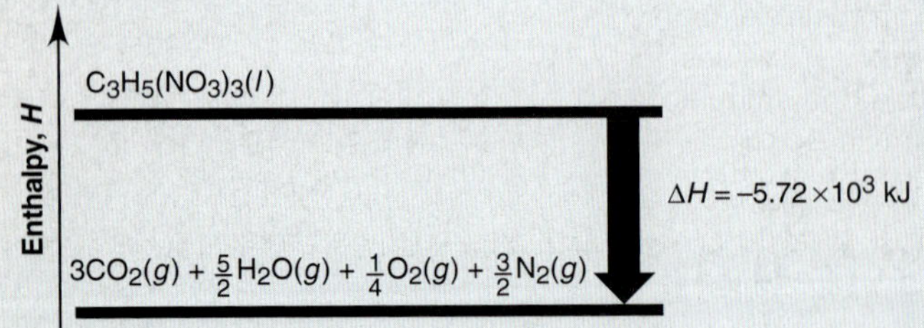

6.3 $\Delta T = 25.0°C - 37.0°C = -12.0°C = -12.0 \text{ K}$

$$\text{Mass (g)} = 1.11 \text{ g/mL} \times \frac{1000 \text{ mL}}{1 \text{ L}} \times 5.50 \text{ L} = 6.10\times10^3 \text{ g}$$

$$q = c \times \text{mass} \times \Delta T$$
$$= (2.42 \text{ J/g·K})\left(\frac{1 \text{ kJ}}{1000 \text{ J}}\right)(6.10\times10^3 \text{ g})(-12.0 \text{ K})$$
$$= -177 \text{ kJ}$$

6.4 $-q_{solid} = q_{water}$

$$-[(0.519 \text{ J/g·K})(2.050 \text{ g})(x - 74.21)] = [(4.184 \text{ J/g·K})(26.05 \text{ g})(x - 27.20)]$$
$$x = 27.65 \text{ K}$$
$$\Delta T_{diamond} = -46.56 \text{ K} \text{ and } \Delta T_{water} = 0.45 \text{ K}$$

6.5 $-q_{sample} = q_{calorimeter}$

$$-(0.8650 \text{ g C})\left(\frac{1 \text{ mol C}}{12.01 \text{ g C}}\right)(-393.5 \text{ kJ/mol C}) = (2.613 \text{ K})x$$
$$x = 10.85 \text{ kJ/K}$$

6.6 $C_2H_4(g) + H_2(g) \longrightarrow C_2H_6(g) + 137 \text{ kJ}$

$$\text{Heat (kJ)} = 15.0 \text{ kg} \times \frac{1000 \text{ g}}{1 \text{ kg}} \times \frac{1 \text{ mol } C_2H_6}{30.07 \text{ g } C_2H_6} \times \frac{137 \text{ kJ}}{1 \text{ mol}}$$
$$= 6.83\times10^4 \text{ kJ}$$

6.7

$$2NO(g) + \tfrac{3}{2}O_2(g) \longrightarrow N_2O_5(s) \qquad \Delta H = -223.7 \text{ kJ}$$
$$2NO_2(g) \longrightarrow 2NO(g) + O_2(g) \qquad \Delta H = 114.2 \text{ kJ}$$

$$\cancel{2NO(g)} + \tfrac{1}{2}\cancel{\tfrac{3}{2}}O_2(g) + 2NO_2(g) \longrightarrow N_2O_5(s) + \cancel{2NO(g)} + \cancel{O_2(g)}$$

$$2NO_2(g) + \tfrac{1}{2}O_2(g) \longrightarrow N_2O_5(s) \qquad \Delta H = -109.5 \text{ kJ}$$

6.8 (a) $C(\text{graphite}) + 2H_2(g) + \tfrac{1}{2}O_2(g) \longrightarrow CH_3OH(l) \qquad \Delta H^\circ_f = -238.6 \text{ kJ}$

(b) $Ca(s) + \tfrac{1}{2}O_2(g) \longrightarrow CaO(s) \qquad \Delta H^\circ_f = -635.1 \text{ kJ}$

(c) $C(\text{graphite}) + \tfrac{1}{4}S_8(\text{rhombic}) \longrightarrow CS_2(l) \qquad \Delta H^\circ_f = 87.9 \text{ kJ}$

6.9 ΔH°_f of $CH_3OH(l)$

$$= -\Delta H^\circ_{rxn} + 2\Delta H^\circ_f[H_2O(g)] + \Delta H^\circ_f[CO_2(g)]$$
$$= 638.5 \text{ kJ} + (2 \text{ mol})(-241.8 \text{ kJ/mol}) + (1 \text{ mol})(-393.5 \text{ kJ/mol})$$
$$= -238.6 \text{ kJ}$$

PROBLEMS

*Problems with **colored** numbers are answered in Appendix E. Sections match the text and provide the numbers of relevant sample problems. Bracketed problems are grouped in pairs (indicated by a short rule) that cover the same concept. Comprehensive Problems are based on material from any section or previous chapter.*

Forms of Energy and Their Interconversion

(Sample Problem 6.1)

6.1 If you feel warm after exercising, have you increased the internal energy of your body? Explain.

6.2 An *adiabatic* process is one that involves no heat transfer. What is the relationship between work and the change in internal energy in an adiabatic process?

6.3 Name a common device that is used to accomplish each energy change:
(a) Electrical energy to thermal energy
(b) Electrical energy to sound energy
(c) Electrical energy to light energy
(d) Mechanical energy to electrical energy
(e) Chemical energy to electrical energy

6.4 Imagine lifting your textbook into the air and dropping it onto a desktop. Describe all the energy transformations (from one form to another) that occur, moving backward in time from a moment after impact.

6.5 A system receives 425 J of heat and delivers 425 J of work to its surroundings. What is the change in internal energy of the system (in J)?

6.6 A system conducts 255 cal of heat to the surroundings while delivering 428 cal of work. What is the change in internal energy of the system (in cal)?

6.7 Complete combustion of 2.0 metric ton of coal (assuming pure carbon) to gaseous carbon dioxide releases 6.6×10^{10} J of heat. Convert this energy to (a) kilojoules; (b) kilocalories; (c) British thermal units.

6.8 Thermal decomposition of 5.0 metric tons of limestone to lime and carbon dioxide requires 9.0×10^{6} kJ of heat. Convert this energy to (a) joules; (b) calories; (c) British thermal units.

6.9 The nutritional calorie (Calorie) is equivalent to 1 kcal. One pound of body fat is equivalent to about 4.1×10^{3} Calories. Express this quantity of energy in joules and kilojoules.

Enthalpy: Heats of Reaction and Chemical Change

(Sample Problem 6.2)

6.10 Classify the following processes as exothermic or endothermic: (a) freezing of water; (b) boiling of water; (c) digestion of food; (d) a person running; (e) a person growing; (f) wood being chopped; (g) heating with a furnace.

6.11 Draw an enthalpy diagram for a general exothermic reaction; label axis, reactants, products, and ΔH with its sign.

6.12 Draw an enthalpy diagram for a general endothermic reaction; label axis, reactants, products, and ΔH with its sign.

6.13 Write a balanced equation and draw an approximate enthalpy diagram for each of the following: (a) the combustion of 1 mol of ethane in oxygen; (b) the freezing of liquid water.

6.14 Write a balanced equation and draw an approximate enthalpy diagram for each of the following: (a) the formation of 1 mol of sodium chloride from its elements (heat is released); (b) the conversion of liquid benzene to gaseous benzene.

6.15 Write a balanced equation and draw an approximate enthalpy diagram for each of the following changes: (a) the combustion of 1 mol of liquid methanol (CH_3OH); (b) the formation of 1 mol of nitrogen dioxide from its elements (heat is absorbed).

6.16 Write a balanced equation and draw an approximate enthalpy diagram for each of the following changes: (a) the sublimation of dry ice [conversion of $CO_2(s)$ directly to $CO_2(g)$]; (b) the reaction of 1 mol of sulfur dioxide with oxygen.

6.17 The circles below represent a phase change occurring at constant temperature:

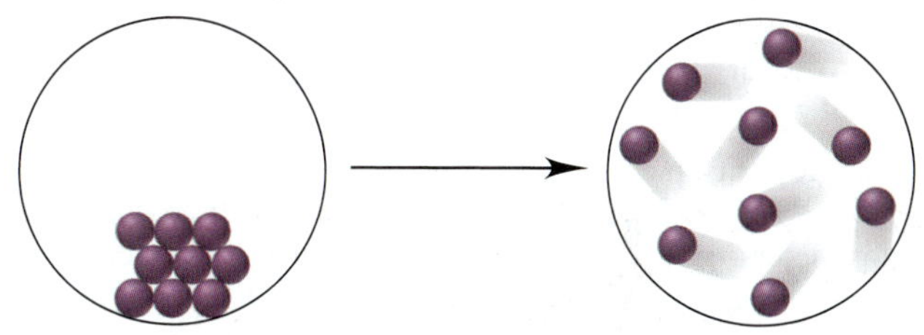

Is the value of each of the following positive (+), negative (−), or zero: (a) q_{sys}; (b) ΔE_{sys}; (c) ΔE_{univ}?

6.18 The piston-cylinder assemblies below represent a physical change occurring at constant pressure:

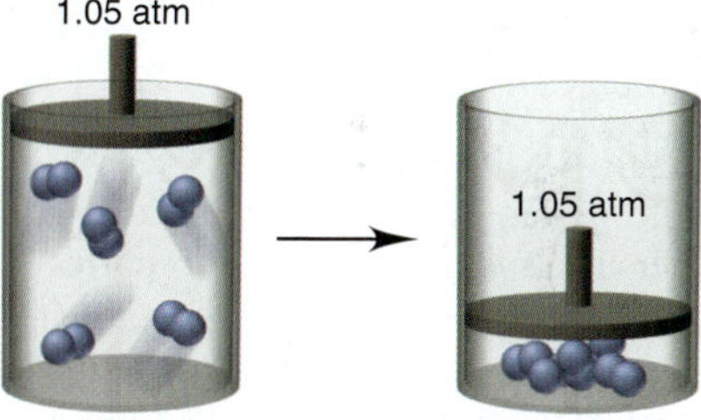

(a) Is w_{sys} +, −, or 0? (b) Is ΔH_{sys} +, −, or 0? (c) Can you determine whether ΔE_{surr} is +, −, or 0? Explain.

Calorimetry: Laboratory Measurement of Heats of Reaction

(Sample Problems 6.3 to 6.5)

6.19 Why can we measure only *changes* in enthalpy, not absolute enthalpy values?

6.20 What data do you need to determine the specific heat capacity of a substance?

6.21 Is the specific heat capacity of a substance an intensive or extensive property? Explain.

6.22 Calculate q when 22.0 g of water is heated from 25.°C to 100.°C.

6.23 Calculate q when 0.10 g of ice is cooled from 10.°C to −75°C (c_{ice} = 2.087 J/g·K).

6.24 A 295-g aluminum engine part at an initial temperature of 13.00°C absorbs 75.0 kJ of heat. What is the final temperature of the part (c of Al = 0.900 J/g·K)?

6.25 A 27.7-g sample of ethylene glycol, a car radiator coolant, loses 688 J of heat. What was the initial temperature of the ethylene glycol if the final temperature is 32.5°C (*c* of ethylene glycol = 2.42 J/g·K)?

6.26 Two iron bolts of equal mass—one at 100.°C, the other at 55°C—are placed in an insulated container. Assuming the heat capacity of the container is negligible, what is the final temperature inside the container (*c* of iron = 0.450 J/g·K)?

6.27 One piece of copper jewelry at 105°C has exactly twice the mass of another piece, which is at 45°C. Both pieces are placed inside a calorimeter whose heat capacity is negligible. What is the final temperature inside the calorimeter (*c* of copper = 0.387 J/g·K)?

6.28 When 155 mL of water at 26°C is mixed with 75 mL of water at 85°C, what is the final temperature? (Assume that no heat is lost to the surroundings; *d* of water = 1.00 g/mL.)

6.29 An unknown volume of water at 18.2°C is added to 24.4 mL of water at 35.0°C. If the final temperature is 23.5°C, what was the unknown volume? (Assume that no heat is lost to the surroundings; *d* of water = 1.00 g/mL.)

6.30 High-purity benzoic acid (C_6H_5COOH; ΔH_{rxn} for combustion = −3227 kJ/mol) is used as a standard for calibrating bomb calorimeters. A 1.221-g sample burns in a calorimeter (heat capacity = 1365 J/°C) that contains exactly 1.200 kg of water. What temperature change is observed?

6.31 Two aircraft rivets, one of iron and the other of copper, are placed in a calorimeter that has an initial temperature of 20.°C. The data for the metals are as follows:

	Iron	Copper
Mass (g)	30.0	20.0
Initial *T* (°C)	0.0	100.0
c (J/g·K)	0.450	0.387

(a) Will heat flow from Fe to Cu or from Cu to Fe?
(b) What other information is needed to correct any measurements that would be made in an actual experiment?
(c) What is the maximum final temperature of the system (assuming the heat capacity of the calorimeter is negligible)?

6.32 When 25.0 mL of 0.500 *M* H_2SO_4 is added to 25.0 mL of 1.00 *M* KOH in a coffee-cup calorimeter at 23.50°C, the temperature rises to 30.17°C. Calculate ΔH of this reaction. (Assume that the total volume is the sum of the individual volumes and that the density and specific heat capacity of the solution are the same as for pure water.)

Stoichiometry of Thermochemical Equations

(Sample Problem 6.6)

6.33 Would you expect $O_2(g) \longrightarrow 2O(g)$ to have a positive or a negative ΔH_{rxn}? Explain.

6.34 Is ΔH positive or negative when 1 mol of water vapor condenses to liquid water? Why? How does this value compare with that for the conversion of 2 mol of liquid water to water vapor?

6.35 Consider the following balanced thermochemical equation for a reaction sometimes used for H_2S production:

$$\tfrac{1}{8}S_8(s) + H_2(g) \longrightarrow H_2S(g) \quad \Delta H_{rxn} = -20.2\ \text{kJ}$$

(a) Is this an exothermic or endothermic reaction?
(b) What is ΔH_{rxn} for the reverse reaction?
(c) What is ΔH when 2.6 mol of S_8 reacts?
(d) What is ΔH when 25.0 g of S_8 reacts?

6.36 Consider the following balanced thermochemical equation for the decomposition of the mineral magnesite:

$$MgCO_3(s) \longrightarrow MgO(s) + CO_2(g) \quad \Delta H_{rxn} = 117.3\ \text{kJ}$$

(a) Is heat absorbed or released in the reaction?
(b) What is ΔH_{rxn} for the reverse reaction?
(c) What is ΔH when 5.35 mol of CO_2 reacts with excess MgO?
(d) What is ΔH when 35.5 g of CO_2 reacts with excess MgO?

6.37 When 1 mol of NO(*g*) forms from its elements, 90.29 kJ of heat is absorbed. (a) Write a balanced thermochemical equation for this reaction. (b) How much heat is involved when 3.50 g of NO decomposes to its elements?

6.38 When 1 mol of KBr(*s*) decomposes to its elements, 394 kJ of heat is absorbed. (a) Write a balanced thermochemical equation for this reaction. (b) How much heat is released when 10.0 kg of KBr forms from its elements?

6.39 Liquid hydrogen peroxide, an oxidizing agent in many rocket fuel mixtures, releases oxygen gas on decomposition:

$$2H_2O_2(l) \longrightarrow 2H_2O(l) + O_2(g) \quad \Delta H_{rxn} = -196.1\ \text{kJ}$$

How much heat is released when 652 kg of H_2O_2 decomposes?

6.40 Compounds of boron and hydrogen are remarkable for their unusual bonding (described in Section 14.5) and also for their reactivity. With the more reactive halogens, for example, diborane (B_2H_6) forms trihalides even at low temperatures:

$$B_2H_6(g) + 6Cl_2(g) \longrightarrow 2BCl_3(g) + 6HCl(g) \quad \Delta H_{rxn} = -755.4\ \text{kJ}$$

How much heat is released per kilogram of diborane that reacts?

6.41 Ethylene (C_2H_4) is the starting material for the preparation of polyethylene. Although typically made during the processing of petroleum, ethylene occurs naturally as a fruit-ripening hormone and as a component of natural gas.
(a) The heat of reaction for the combustion of C_2H_4 is −1411 kJ/mol. Write a balanced thermochemical equation for the combustion of C_2H_4.
(b) How many grams of C_2H_4 must burn to give 70.0 kJ of heat?

6.42 Sucrose ($C_{12}H_{22}O_{11}$, table sugar) is oxidized in the body by O_2 via a complex set of reactions that ultimately produces $CO_2(g)$ and $H_2O(g)$ and releases 5.64×10^3 kJ/mol sucrose.
(a) Write a balanced thermochemical equation for this reaction.
(b) How much heat is released per gram of sucrose oxidized?

Hess's Law of Heat Summation

(Sample Problem 6.7)

6.43 Express Hess's law in your own words.

6.44 Calculate ΔH_{rxn} for

$$Ca(s) + \tfrac{1}{2}O_2(g) + CO_2(g) \longrightarrow CaCO_3(s)$$

given the following set of reactions:

$$Ca(s) + \tfrac{1}{2}O_2(g) \longrightarrow CaO(s) \quad \Delta H = -635.1\ \text{kJ}$$
$$CaCO_3(s) \longrightarrow CaO(s) + CO_2(g) \quad \Delta H = 178.3\ \text{kJ}$$

6.45 Calculate ΔH_{rxn} for

$$2NOCl(g) \longrightarrow N_2(g) + O_2(g) + Cl_2(g)$$

given the following set of reactions:

$$\tfrac{1}{2}N_2(g) + \tfrac{1}{2}O_2(g) \longrightarrow NO(g) \quad \Delta H = 90.3\ \text{kJ}$$
$$NO(g) + \tfrac{1}{2}Cl_2(g) \longrightarrow NOCl(g) \quad \Delta H = -38.6\ \text{kJ}$$

6.46 Write the balanced overall equation for the following process (equation 3), calculate $\Delta H_{overall}$, and match the number of each equation with the letter of the appropriate arrow in Figure P6.46:

(1) $N_2(g) + O_2(g) \longrightarrow 2NO(g)$ $\quad \Delta H = 180.6$ kJ

(2) $2NO(g) + O_2(g) \longrightarrow 2NO_2(g)$ $\quad \Delta H = -114.2$ kJ

(3) $\quad \Delta H_{overall} = ?$

6.47 Write the balanced overall equation for the following process (equation 3), calculate $\Delta H_{overall}$, and match the number of each equation with the letter of the appropriate arrow in Figure P6.47:

(1) $P_4(s) + 6Cl_2(g) \longrightarrow 4PCl_3(g)$ $\quad \Delta H = -1148$ kJ

(2) $4PCl_3(g) + 4Cl_2(g) \longrightarrow 4PCl_5(g)$ $\quad \Delta H = -460$ kJ

(3) $\quad \Delta H_{overall} = ?$

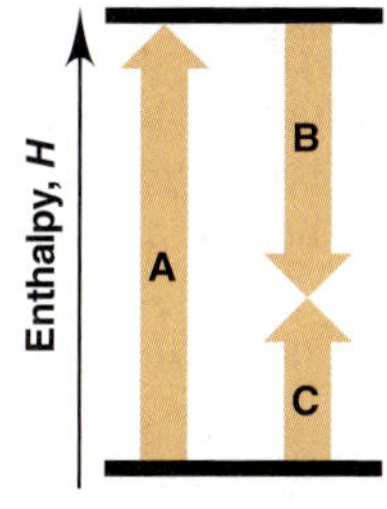

Figure P6.46

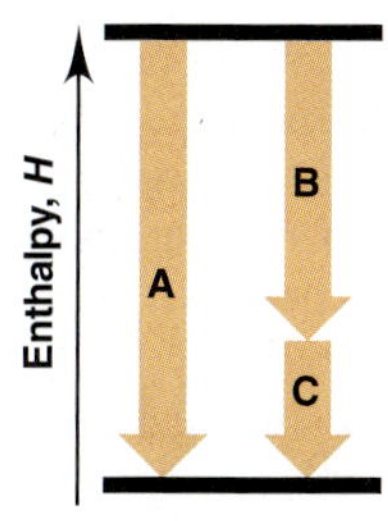

Figure P6.47

6.48 Diamond and graphite are two crystalline forms of carbon. At 1 atm and 25°C, diamond changes to graphite so slowly that the enthalpy change of the process must be obtained indirectly. Determine ΔH_{rxn} for

$$C(diamond) \longrightarrow C(graphite)$$

with equations from the following list:

(1) $C(diamond) + O_2(g) \longrightarrow CO_2(g)$ $\quad \Delta H = -395.4$ kJ

(2) $2CO_2(g) \longrightarrow 2CO(g) + O_2(g)$ $\quad \Delta H = 566.0$ kJ

(3) $C(graphite) + O_2(g) \longrightarrow CO_2(g)$ $\quad \Delta H = -393.5$ kJ

(4) $2CO(g) \longrightarrow C(graphite) + CO_2(g)$ $\quad \Delta H = -172.5$ kJ

Standard Heats of Reaction (ΔH°_{rxn})

(Sample Problems 6.8 and 6.9)

6.49 What is the difference between the standard heat of formation and the standard heat of reaction?

6.50 Make any changes needed in each of the following equations to make ΔH°_{rxn} equal to ΔH°_f for the compound present:

(a) $Cl(g) + Na(s) \longrightarrow NaCl(s)$

(b) $H_2O(g) \longrightarrow 2H(g) + \frac{1}{2}O_2(g)$

(c) $\frac{1}{2}N_2(g) + \frac{3}{2}H_2(g) \longrightarrow NH_3(g)$

6.51 Write balanced formation equations at standard conditions for each of the following compounds: (a) $CaCl_2$; (b) $NaHCO_3$; (c) CCl_4; (d) HNO_3.

6.52 Write balanced formation equations at standard conditions for each of the following compounds: (a) HI; (b) SiF_4; (c) O_3; (d) $Ca_3(PO_4)_2$.

6.53 Calculate ΔH°_{rxn} for each of the following:

(a) $2H_2S(g) + 3O_2(g) \longrightarrow 2SO_2(g) + 2H_2O(g)$

(b) $CH_4(g) + Cl_2(g) \longrightarrow CCl_4(l) + HCl(g)$ [unbalanced]

6.54 Calculate ΔH°_{rxn} for each of the following:

(a) $SiO_2(s) + 4HF(g) \longrightarrow SiF_4(g) + 2H_2O(l)$

(b) $C_2H_6(g) + O_2(g) \longrightarrow CO_2(g) + H_2O(g)$ [unbalanced]

6.55 Copper(I) oxide can be oxidized to copper(II) oxide:

$Cu_2O(s) + \frac{1}{2}O_2(g) \longrightarrow 2CuO(s)$ $\quad \Delta H^\circ_{rxn} = -146.0$ kJ

Given that ΔH°_f of $Cu_2O(s) = -168.6$ kJ/mol, what is ΔH°_f of $CuO(s)$?

6.56 Acetylene burns in air according to the following equation:

$C_2H_2(g) + \frac{5}{2}O_2(g) \longrightarrow 2CO_2(g) + H_2O(g)$

$\Delta H^\circ_{rxn} = -1255.8$ kJ

Given that ΔH°_f of $CO_2(g) = -393.5$ kJ/mol and that ΔH°_f of $H_2O(g) = -241.8$ kJ/mol, what is ΔH°_f of $C_2H_2(g)$?

6.57 Nitroglycerine, $C_3H_5(NO_3)_3(l)$, a powerful explosive used in mining, detonates to produce a hot gaseous mixture of nitrogen, water, carbon dioxide, and oxygen.

(a) Write a balanced equation for this reaction using the smallest whole-number coefficients.

(b) If $\Delta H^\circ_{rxn} = -2.29\times10^4$ kJ for the equation as written in part (a), calculate ΔH°_f of nitroglycerine.

6.58 The common lead-acid car battery produces a large burst of current, even at low temperatures, and is rechargeable. The reaction that occurs while recharging a "dead" battery is

$$2PbSO_4(s) + 2H_2O(l) \longrightarrow Pb(s) + PbO_2(s) + 2H_2SO_4(l)$$

(a) Use ΔH°_f values from Appendix B to calculate ΔH°_{rxn}.

(b) Use the following equations to check your answer in part (a):

(1) $Pb(s) + PbO_2(s) + 2SO_3(g) \longrightarrow 2PbSO_4(s)$

$\Delta H^\circ = -768$ kJ

(2) $SO_3(g) + H_2O(l) \longrightarrow H_2SO_4(l)$ $\quad \Delta H^\circ = -132$ kJ

Comprehensive Problems

Problems with an asterisk (*) are more challenging.

6.59 Stearic acid ($C_{18}H_{36}O_2$) is a typical fatty acid, a molecule with a long hydrocarbon chain and an organic acid group (COOH) at the end. It is used to make cosmetics, ointments, soaps, and candles and is found in animal tissue as part of many saturated fats. In fact, when you eat meat, chances are that you are ingesting some fats that contain stearic acid.

(a) Write a balanced equation for the complete combustion of stearic acid to gaseous products.

(b) Calculate ΔH°_{rxn} for this combustion ($\Delta H^\circ_f = -948$ kJ/mol).

(c) Calculate the heat (q) in kJ and kcal when 1.00 g of stearic acid is burned completely.

(d) The nutritional information for a candy bar states that one serving contains 11.0 g of fat and 100. Cal from fat (1 Cal = 1 kcal). Is this information consistent with your answer for part (c)?

6.60 A balloonist is preparing to make a trip in a helium-filled balloon. The trip begins in early morning at a temperature of 15°C. By midafternoon, the temperature has increased to 30.°C. Assuming the pressure remains constant at 1.00 atm, for each mole of helium, calculate:

(a) The initial and final volumes

(b) The change in internal energy, ΔE [*Hint:* Helium behaves like an ideal gas, so $E = \frac{3}{2}nRT$. Be sure the units of R are consistent with those of E.]

(c) The work (w) done by the helium (in J)

(d) The heat (q) transferred (in J)

(e) ΔH for the process (in J)

(f) Explain the relationship between the answers to (d) and (e).

6.61 In winemaking, the sugars in grapes undergo *fermentation* by yeast to yield CH_3CH_2OH and CO_2. During cellular *respiration,* sugar and ethanol are "burned" to water vapor and CO_2.

(a) Using $C_6H_{12}O_6$ for sugar, calculate ΔH°_{rxn} of fermentation and of respiration (combustion).
(b) Write a combustion reaction for ethanol. Which has a higher ΔH°_{rxn} for the combustion per mol of C, sugar or ethanol?

6.62 The following scenes represent a gaseous reaction between compounds of nitrogen (blue) and oxygen (red) at 298 K:

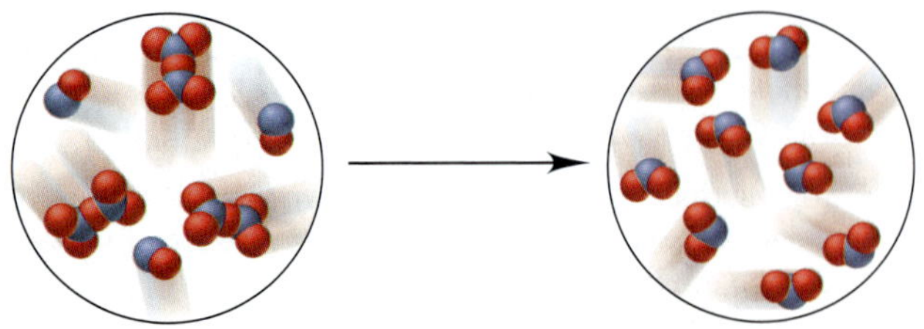

(a) Write a balanced equation and use Appendix B to calculate ΔH°_{rxn}. (b) If each molecule of product represents 1.50×10^{-2} mol, what quantity of heat (in J) is released or absorbed?

6.63 Iron metal is produced in a blast furnace through a complex series of reactions that involve reduction of iron(III) oxide with carbon monoxide.
(a) Write a balanced overall equation for the process, including the other product.
(b) Use the equations below to calculate ΔH°_{rxn} for the overall equation:
(1) $3Fe_2O_3(s) + CO(g) \longrightarrow 2Fe_3O_4(s) + CO_2(g)$ $\quad \Delta H^\circ = -48.5$ kJ
(2) $Fe(s) + CO_2(g) \longrightarrow FeO(s) + CO(g)$ $\quad \Delta H^\circ = -11.0$ kJ
(3) $Fe_3O_4(s) + CO(g) \longrightarrow 3FeO(s) + CO_2(g)$ $\quad \Delta H^\circ = 22$ kJ

6.64 Pure liquid octane (C_8H_{18}; $d = 0.702$ g/mL) is used as the fuel in a test of a new automobile drive train.
(a) How much energy (in kJ) is released by complete combustion of the octane in a 20.4-gal fuel tank to gases ($\Delta H^\circ_{rxn} = -5.45\times10^3$ kJ/mol)?
(b) The energy delivered to the wheels at 65 mph is 5.5×10^4 kJ/h. Assuming all the energy is transferred to the wheels, what is the cruising range (in km) of the car on a full tank?
(c) If the actual cruising range is 455 miles, explain your answer to part (b).

6.65 Four 50.-g samples of different liquids are placed in beakers at $T_{initial}$ of 25.00°C. Each liquid is heated until 450. J is absorbed; T_{final} is shown on each beaker below. Rank the liquids in order of increasing specific heat capacity.

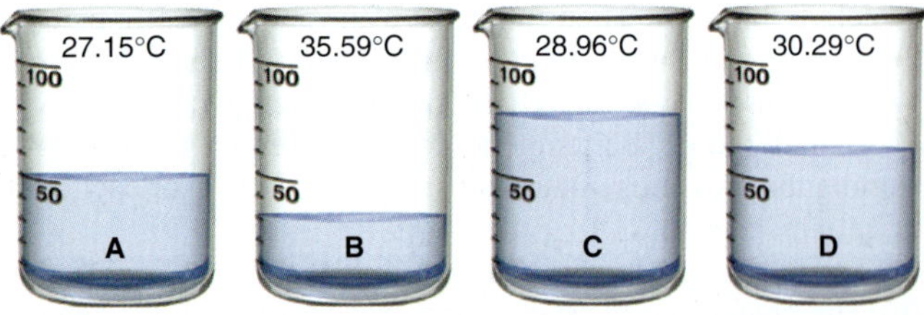

6.66 When simple sugars, called *monosaccharides,* link together, they form a variety of complex sugars and, ultimately, *polysaccharides,* such as starch, glycogen, and cellulose. Glucose and fructose have the same formula, $C_6H_{12}O_6$, but different arrangements of atoms. They link together to form a molecule of sucrose (table sugar) and a molecule of liquid water. The ΔH°_f values of glucose, fructose, and sucrose are −1273 kJ/mol, −1266 kJ/mol, and −2226 kJ/mol, respectively. Write a balanced equation for this reaction and calculate ΔH°_{rxn}.

6.67 Oxidation of gaseous ClF by F_2 yields liquid ClF_3, an important fluorinating agent. Use the following thermochemical equations to calculate ΔH°_{rxn} for the production of ClF_3:
(1) $2ClF(g) + O_2(g) \longrightarrow Cl_2O(g) + OF_2(g)$ $\quad \Delta H^\circ = 167.5$ kJ
(2) $2F_2(g) + O_2(g) \longrightarrow 2OF_2(g)$ $\quad \Delta H^\circ = -43.5$ kJ
(3) $2ClF_3(l) + 2O_2(g) \longrightarrow Cl_2O(g) + 3OF_2(g)$ $\quad \Delta H^\circ = 394.1$ kJ

6.68 Silver bromide is used to coat ordinary black-and-white photographic film, while high-speed film uses silver iodide.
(a) When 50.0 mL of 5.0 g/L $AgNO_3$ is added to a coffee-cup calorimeter containing 50.0 mL of 5.0 g/L NaI, with both solutions at 25°C, what mass of AgI forms?
(b) Use Appendix B to find ΔH°_{rxn}.
(c) What is ΔT_{soln} (assume the volumes are additive and the solution has the density and specific heat capacity of water)?

*** 6.69** Whenever organic matter is decomposed under oxygen-free (anaerobic) conditions, methane is one of the products. Thus, enormous deposits of natural gas, which is almost entirely methane, exist as a major source of fuel for home and industry.
(a) It is estimated that known sources of natural gas can produce 5600 EJ of energy (1 EJ = 10^{18} J). Current total global energy usage is 4.0×10^2 EJ per year. Find the mass (in kg) of known sources of natural gas (ΔH°_{rxn} for the combustion of CH_4 = −802 kJ/mol).
(b) For how many years could these sources supply the world's total energy needs?
(c) What volume (in ft^3) of natural gas, measured at STP, is required to heat 1.00 qt of water from 25.0°C to 100.0°C (d of H_2O = 1.00 g/mL; d of CH_4 at STP = 0.72 g/L)?
(d) The fission of 1 mol of uranium (about 4×10^{-4} ft^3) in a nuclear reactor produces 2×10^{13} J. What volume (in ft^3) of natural gas would produce the same amount of energy?

6.70 The heat of atomization (ΔH°_{atom}) is the heat needed to form separated gaseous atoms from a substance in its standard state. The equation for the atomization of graphite is

$$C(graphite) \longrightarrow C(g)$$

Use Hess's law to calculate ΔH°_{atom} of graphite from these data:
(1) ΔH°_f of CH_4 = −74.9 kJ/mol
(2) ΔH°_{atom} of CH_4 = 1660 kJ/mol
(3) ΔH°_{atom} of H_2 = 432 kJ/mol

6.71 A reaction is carried out in a steel vessel within a chamber filled with argon gas. Below are molecular views of the argon adjacent to the surface of the reaction vessel before and after the reaction. Was the reaction exothermic or endothermic? Explain.

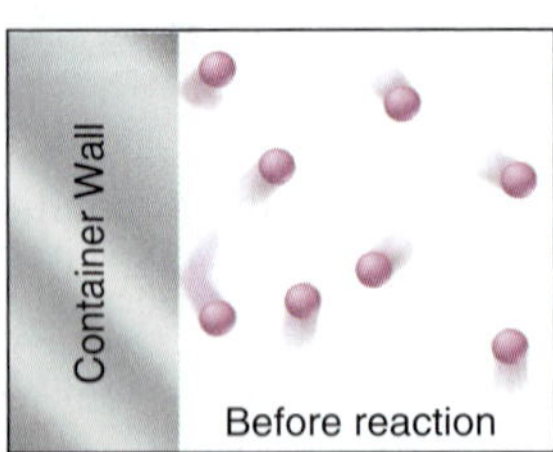

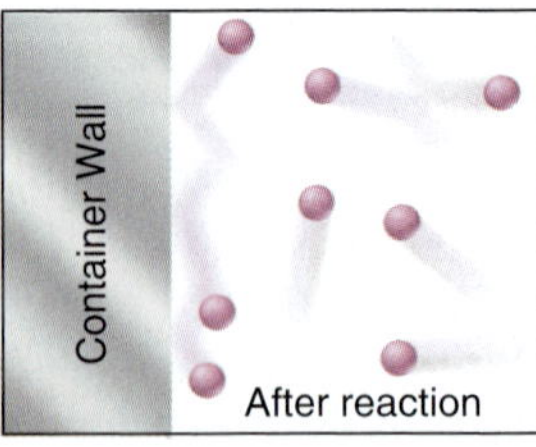

6.72 Benzene (C_6H_6) and acetylene (C_2H_2) have the same empirical formula, CH. Which releases more energy per mole of CH (ΔH°_f of gaseous C_6H_6 = 82.9 kJ/mol)?

6.73 An aqueous waste stream that has a maximum concentration of 0.50 M H_2SO_4 (d = 1.030 g/mL at 25°C) will be neutralized by controlled addition of 40% caustic soda (NaOH; d = 1.430 g/L) before it goes to the process sewer and then to the chemical plant waste treatment facility. However, a safety review finds that the waste stream could meet a small stream of an immiscible organic compound, which could form a flammable vapor in air at 40.°C. The maximum temperature of the caustic soda and the waste stream is 31°C. Could the temperature increase due to the heat of neutralization cause the vapor to explode? Assume the specific heat capacity of each solution is 4.184 J/g·K.

6.74 Kerosene, a common space-heater fuel, is a mixture of hydrocarbons whose "average" formula is $C_{12}H_{26}$.
(a) Write a balanced equation, using the simplest whole-number coefficients, for the complete combustion of kerosene to gases.
(b) If $\Delta H^\circ_{rxn} = -1.50\times10^4$ kJ for the combustion equation as written in part (a), determine ΔH°_f of kerosene.
(c) Calculate the heat produced by combustion of 0.50 gal of kerosene (d of kerosene = 0.749 g/mL).
(d) How many gallons of kerosene must be burned for a kerosene furnace to produce 1250. Btu (1 Btu = 1.055 kJ)?

* **6.75** Coal gasification is a multistep process to convert coal into cleaner-burning gaseous fuels. In one step, a certain coal sample reacts with superheated steam:
$C(coal) + H_2O(g) \longrightarrow CO(g) + H_2(g) \quad \Delta H^\circ_{rxn} = 129.7$ kJ
(a) Combine this reaction with the following two to write an overall reaction for the production of methane:
$CO(g) + H_2O(g) \longrightarrow CO_2(g) + H_2(g) \quad \Delta H^\circ_{rxn} = -41$ kJ
$CO(g) + 3H_2(g) \longrightarrow CH_4(g) + H_2O(g) \quad \Delta H^\circ_{rxn} = -206$ kJ
(b) Calculate ΔH°_{rxn} for this overall change.
(c) Using the value in (b) and calculating the ΔH°_{rxn} for combustion of methane, find the total heat for gasifying 1.00 kg of coal and burning the methane formed (assume water forms as a gas and $\mathcal{M}$ of coal = 12.00 g/mol).

6.76 Phosphorus pentachloride is used in the industrial preparation of organic phosphorus compounds. Equation 1 shows its preparation from PCl_3 and Cl_2:
(1) $PCl_3(l) + Cl_2(g) \longrightarrow PCl_5(s)$
Use equations 2 and 3 to calculate ΔH_{rxn} for equation 1:
(2) $P_4(s) + 6Cl_2(g) \longrightarrow 4PCl_3(l) \quad \Delta H = -1280$ kJ
(3) $P_4(s) + 10Cl_2(g) \longrightarrow 4PCl_5(s) \quad \Delta H = -1774$ kJ

6.77 A typical candy bar weighs about 2 oz (1.00 oz = 28.4 g).
(a) Assuming that a candy bar is 100% sugar and that 1.0 g of sugar is equivalent to about 4.0 Calories of energy, calculate the energy (in kJ) contained in a typical candy bar.
(b) Assuming that your mass is 58 kg and you convert chemical potential energy to work with 100% efficiency, how high would you have to climb to work off the energy in a candy bar? (Potential energy = mass × g × height, where g = 9.8 m/s^2.)
(c) Why is your actual conversion of potential energy to work less than 100% efficient?

6.78 Silicon tetrachloride is produced annually on the multikiloton scale for making transistor-grade silicon. It can be made directly from the elements (reaction 1) or, more cheaply, by heating sand and graphite with chlorine gas (reaction 2). If water is present in reaction 2, some tetrachloride may be lost in an unwanted side reaction (reaction 3):
(1) $Si(s) + 2Cl_2(g) \longrightarrow SiCl_4(g)$
(2) $SiO_2(s) + 2C(graphite) + 2Cl_2(g) \longrightarrow SiCl_4(g) + 2CO(g)$
(3) $SiCl_4(g) + 2H_2O(g) \longrightarrow SiO_2(s) + 4HCl(g)$
$\Delta H^\circ_{rxn} = -139.5$ kJ
(a) Use reaction 3 to calculate the heats of reaction of reactions 1 and 2. (b) What is the heat of reaction for the new reaction that is the sum of reactions 2 and 3?

6.79 Use the following information to find ΔH°_f of gaseous HCl:
$N_2(g) + 3H_2(g) \longrightarrow 2NH_3(g) \quad \Delta H^\circ_{rxn} = -91.8$ kJ
$N_2(g) + 4H_2(g) + Cl_2(g) \longrightarrow 2NH_4Cl(s) \quad \Delta H^\circ_{rxn} = -628.8$ kJ
$NH_3(g) + HCl(g) \longrightarrow NH_4Cl(s) \quad \Delta H^\circ_{rxn} = -176.2$ kJ

* **6.80** You want to determine ΔH° for the reaction
$$Zn(s) + 2HCl(aq) \longrightarrow ZnCl_2(aq) + H_2(g)$$
(a) To do so, you first determine the heat capacity of a calorimeter using the following reaction, whose ΔH is known:
$NaOH(aq) + HCl(aq) \longrightarrow NaCl(aq) + H_2O(l)$
$\Delta H^\circ = -57.32$ kJ
Calculate the heat capacity of the calorimeter from these data:
Amounts used: 50.0 mL of 2.00 M HCl and 50.0 mL of 2.00 M NaOH
Initial T of both solutions: 16.9°C
Maximum T recorded during reaction: 30.4°C
Density of resulting NaCl solution: 1.04 g/mL
c of 1.00 M NaCl(aq) = 3.93 J/g·K
(b) Use the result from part (a) and the following data to determine ΔH°_{rxn} for the reaction between zinc and HCl(aq):
Amounts used: 100.0 mL of 1.00 M HCl and 1.3078 g of Zn
Initial T of HCl solution and Zn: 16.8°C
Maximum T recorded during reaction: 24.1°C
Density of 1.0 M HCl solution = 1.015 g/mL
c of resulting $ZnCl_2(aq)$ = 3.95 J/g·K
(c) Given the values below, what is the error in your experiment?
$$\Delta H^\circ_f \text{ of } HCl(aq) = -1.652\times10^2 \text{ kJ/mol}$$
$$\Delta H^\circ_f \text{ of } ZnCl_2(aq) = -4.822\times10^2 \text{ kJ/mol}$$

* **6.81** One mole of nitrogen gas confined within a cylinder by a piston is heated from 0°C to 819°C at 1.00 atm.
(a) Calculate the work of expansion of the gas in joules (1 J = 9.87×10^{-3} atm·L). Assume all the energy is used to do work.
(b) What would be the temperature change if the gas were heated with the same amount of energy in a container of fixed volume? (Assume the specific heat capacity of N_2 is 1.00 J/g·K.)

6.82 The chemistry of nitrogen oxides is very versatile. Given the following reactions and their standard enthalpy changes,
(1) $NO(g) + NO_2(g) \longrightarrow N_2O_3(g) \quad \Delta H^\circ_{rxn} = -39.8$ kJ
(2) $NO(g) + NO_2(g) + O_2(g) \longrightarrow N_2O_5(g) \quad \Delta H^\circ_{rxn} = -112.5$ kJ
(3) $2NO_2(g) \longrightarrow N_2O_4(g) \quad \Delta H^\circ_{rxn} = -57.2$ kJ
(4) $2NO(g) + O_2(g) \longrightarrow 2NO_2(g) \quad \Delta H^\circ_{rxn} = -114.2$ kJ
(5) $N_2O_5(s) \longrightarrow N_2O_5(g) \quad \Delta H^\circ_{rxn} = 54.1$ kJ
calculate the heat of reaction for
$$N_2O_3(g) + N_2O_5(s) \longrightarrow 2N_2O_4(g)$$

* **6.83** Liquid methanol (CH_3OH) is used as an alternative fuel in truck engines. An industrial method for preparing it uses the catalytic hydrogenation of carbon monoxide:
$$CO(g) + 2H_2(g) \xrightarrow{\text{catalyst}} CH_3OH(l)$$
How much heat (in kJ) is released when 15.0 L of CO at 85°C and 112 kPa reacts with 18.5 L of H_2 at 75°C and 744 torr?

* **6.84** (a) How much heat is released when 25.0 g of methane burns in excess O_2 to form gaseous products?
(b) Calculate the temperature of the product mixture if the methane and air are both at an initial temperature of 0.0°C. Assume a stoichiometric ratio of methane to oxygen from the air, with air being 21% O_2 by volume (c of CO_2 = 57.2 J/mol·K; c of $H_2O(g)$ = 36.0 J/mol·K; c of N_2 = 30.5 J/mol·K).

7 Quantum Theory and Atomic Structure

Explaining the Spectacular *The light from neon signs and TV screens, as well as breathtaking firework and aurora displays, occur through changes in atomic energy levels, which you'll examine in this chapter.*

Outline

Key Principles
to focus on while studying this chapter

- All forms of *electromagnetic radiation* travel at the *speed of light* (*c*) in waves. The properties of a wave are its *wavelength* (λ, distance between corresponding points on adjacent waves), *frequency* (ν, number of cycles the wave undergoes per second), and *amplitude* (the height of the wave), which is related to the *intensity* (brightness) of the light. A region of the *electromagnetic spectrum* includes a range of wavelengths *(Section 7.1)*.
- In everyday experience, light is diffuse and matter is chunky, but certain phenomena—*blackbody radiation* (the light emitted by hot objects), the *photoelectric effect* (the flow of current when light strikes a metal), and *atomic spectra* (the specific colors seen when a substance is excited)—can only be explained if energy consists of "packets" *(quanta)* that occur in, and thus change by, *fixed amounts*. The energy of a quantum is related to its frequency *(Section 7.1)*.
- According to the *Bohr model,* an atomic spectrum consists of separate lines because an atom has certain allowable energy levels *(states)*. The energy of the atom changes when the electron moves from one *orbit* to another as the atom absorbs (or emits) light of a specific frequency *(Section 7.2)*.
- *Wave-particle duality* means that matter has wavelike properties (as shown by the *de Broglie wavelength* and *electron diffraction*) and energy has particle-like properties (as shown by *photons* of light behaving like particles with *momentum*). These properties are observable only on the atomic scale, and because of them, we can never know the position *and* speed of an electron in an atom *(uncertainty principle) (Section 7.3)*.
- According to the *quantum-mechanical model* of the H atom, each energy level of the atom is associated with an *atomic orbital (wave function),* a mathematical function describing the electron's position in three dimensions. We can know the *probability* of the electron being found within a particular tiny volume of space, but *not* its exact location. This probability decreases with distance from the nucleus *(Section 7.4)*.
- *Quantum numbers* denote the energy *(n, principal),* shape *(l, angular momentum),* and spatial orientation *(m_l, magnetic)* of each atomic orbital. An *energy level* consists of *sublevels,* which consist of *orbitals*. There is a *hierarchy* of quantum numbers, such that *n* limits *l*, which limits m_l *(Section 7.4)*.
- In the H atom, there is only one type of electrostatic interaction: attraction between nucleus and electron. For the H atom *only,* the energy levels depend exclusively on the principal quantum number (*n*) *(Section 7.4)*.

Over a few remarkable decades—from around 1890 to 1930—a revolution took place in how we view the makeup of the universe. But revolutions in science are not the violent upheavals of political overthrow. Rather, flaws appear in an established model as conflicting evidence mounts, a startling discovery or two widens the flaws into cracks, and the conceptual structure crumbles gradually from its inconsistencies. New insight, verified by experiment, then guides the building of a model more consistent with reality. So it was when Dalton's atomic theory established the idea of individual units of matter, and when Rutherford's nuclear model substituted atoms with rich internal structure for "plum puddings." In this chapter, you will see this process unfold again with the development of modern atomic theory.

Almost as soon as Rutherford proposed his nuclear model, a major problem arose. A nucleus and an electron attract each other, so if they are to remain apart, the energy of the electron's motion (kinetic energy) must balance the energy of attraction (potential energy). However, the laws of classical physics had established that a negative particle moving in a curved path around a positive one *must* emit radiation and thus lose energy. If this requirement applied to atoms, why didn't the orbiting electron lose energy continuously and spiral into the nucleus? Clearly, if electrons behaved the way classical physics predicted, all atoms would have collapsed eons ago! The behavior of subatomic matter seemed to violate real-world experience and accepted principles.

The breakthroughs that soon followed Rutherford's model forced a complete rethinking of the classical picture of matter and energy. In the macroscopic world, the two are distinct. Matter occurs in chunks you can hold and weigh, and you can change the amount of matter in a sample piece by piece. In contrast, energy is "massless," and its quantity changes in a continuous manner. Matter moves in specific paths, whereas light and other types of energy travel in diffuse waves. As you'll see in this chapter, however, as soon as 20th-century scientists probed the subatomic world, these clear distinctions between particulate matter and wavelike energy began to fade, revealing a much more amazing reality.

Concepts & Skills to Review before studying this chapter

- discovery of the electron and atomic nucleus (Section 2.4)
- major features of atomic structure (Section 2.5)
- changes in energy state of a system (Section 6.1)

7.1 THE NATURE OF LIGHT

Visible light is one type of **electromagnetic radiation** (also called *electromagnetic energy* or *radiant energy*). Other familiar types include x-rays, microwaves, and radio waves. All electromagnetic radiation consists of energy propagated by means of electric and magnetic fields that alternately increase and decrease in intensity as they move through space. This classical wave model distinguishes clearly between waves and particles; it is essential for understanding why rainbows form, how magnifying glasses work, why objects look distorted under water, and many other everyday observations. But, it cannot explain observations on the atomic scale because, in that unfamiliar realm, energy behaves as though it consists of particles!

The Wave Nature of Light

The wave properties of electromagnetic radiation are described by two interdependent variables, as Figure 7.1 shows:

- **Frequency** (ν, Greek *nu*) is the number of cycles the wave undergoes per second and is expressed in units of 1/second [s^{-1}; also called *hertz* (Hz)].
- **Wavelength** (λ, Greek *lambda*) is the distance between any point on a wave and the corresponding point on the next crest (or trough) of the wave, that is, the distance the wave travels during one cycle. Wavelength is expressed in meters and often, for very short wavelengths, in nanometers (nm, 10^{-9} m), picometers (pm, 10^{-12} m), or the non-SI unit angstroms (Å, 10^{-10} m).

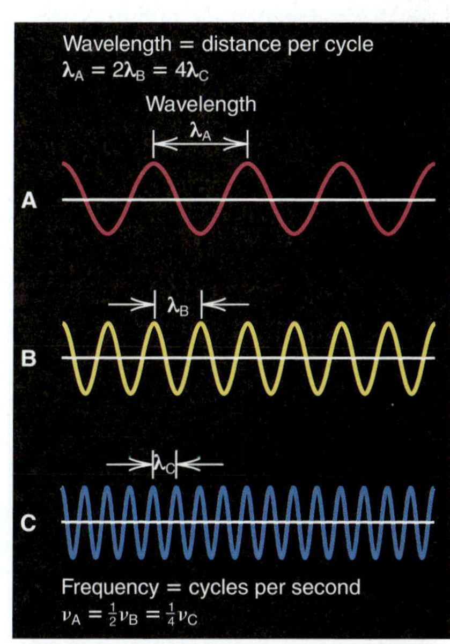

FIGURE 7.1 Frequency and wavelength. Three waves with different wavelengths (λ) and thus different frequencies (ν) are shown. Note that *as the wavelength decreases, the frequency increases, and vice versa.*

The speed of the wave, the distance traveled per unit time (in units of meters per second), is the product of its frequency (cycles per second) and its wavelength (meters per cycle):

$$\text{Units for speed of wave:} \quad \frac{\cancel{\text{cycles}}}{\text{s}} \times \frac{\text{m}}{\cancel{\text{cycle}}} = \frac{\text{m}}{\text{s}}$$

In a vacuum, all types of electromagnetic radiation travel at 2.99792458×10^8 m/s (3.00×10^8 m/s to three significant figures), which is a physical constant called the **speed of light (*c*):**

$$c = \nu \times \lambda \qquad (7.1)$$

As Equation 7.1 shows, the product of ν and λ is a constant. Thus, the individual terms have a reciprocal relationship to each other: *radiation with a high frequency has a short wavelength, and vice versa.*

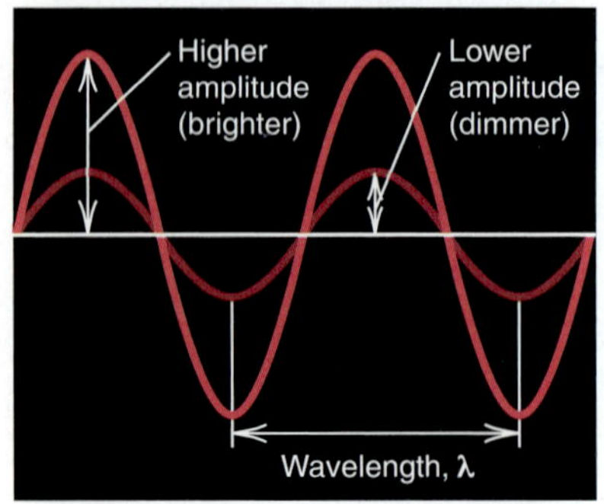

FIGURE 7.2 Amplitude (intensity) of a wave. Amplitude is represented by the height of the crest (or depth of the trough) of the wave. The two waves shown have the same wavelength (color) but different amplitudes and, therefore, different brightnesses (intensities).

Another characteristic of a wave is its **amplitude,** the height of the crest (or depth of the trough) of each wave (Figure 7.2). The amplitude of an electromagnetic wave is a measure of the strength of its electric and magnetic fields. Thus, amplitude is related to the *intensity* of the radiation, which we perceive as brightness in the case of visible light. Light of a particular color—fire-engine red, for instance—has a specific frequency and wavelength, but it can be dimmer (lower amplitude) or brighter (higher amplitude).

The Electromagnetic Spectrum Visible light represents a small portion of the continuum of radiant energy known as the **electromagnetic spectrum** (Figure 7.3). *All the waves in the spectrum travel at the same speed through a vacuum but differ in frequency and, therefore, wavelength.* Some regions of the spectrum are utilized by particular devices; for example, the long-wavelength, low-frequency radiation is used by microwave ovens and radios. Note that each region meets the next. For instance, the **infrared (IR)** region meets the microwave region on one end and the visible region on the other.

We perceive different wavelengths (or frequencies) of *visible* light as different colors, from red ($\lambda \approx 750$ nm) to violet ($\lambda \approx 400$ nm). Light of a single wavelength is called *monochromatic* (Greek, "one color"), whereas light of many wavelengths is *polychromatic.* White light is polychromatic. The region adjacent to visible light on the short-wavelength end consists of **ultraviolet (UV)** radiation (also called *ultraviolet light*). Still shorter wavelengths (higher frequencies) make up the x-ray and gamma (γ) ray regions. Thus, a TV signal, the green light from a traffic signal, and a gamma ray emitted by a radioactive element all travel at the same speed but differ in their frequency (and wavelength).

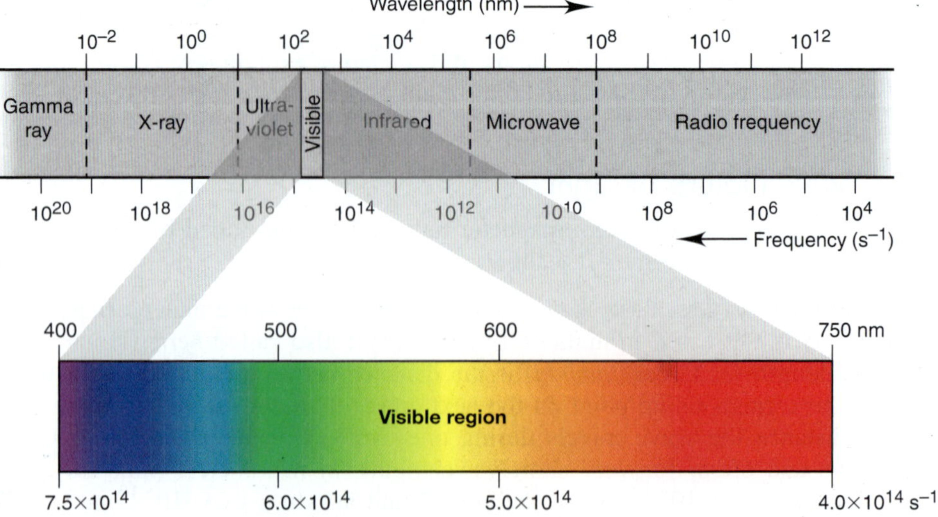

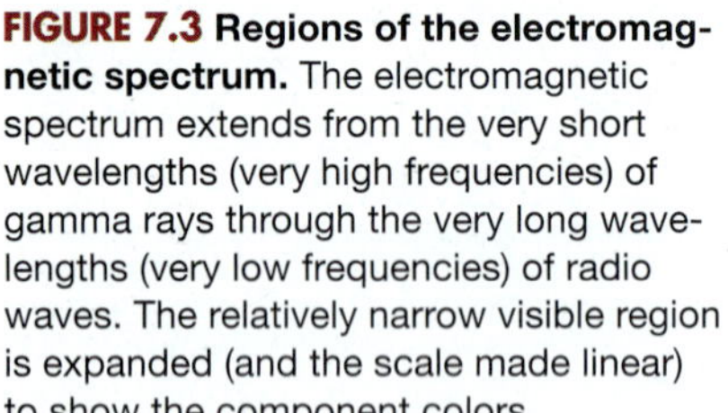

FIGURE 7.3 Regions of the electromagnetic spectrum. The electromagnetic spectrum extends from the very short wavelengths (very high frequencies) of gamma rays through the very long wavelengths (very low frequencies) of radio waves. The relatively narrow visible region is expanded (and the scale made linear) to show the component colors.

SAMPLE PROBLEM 7.1 Interconverting Wavelength and Frequency

Problem A dental hygienist uses x-rays (λ = 1.00 Å) to take a series of dental radiographs while the patient listens to a radio station (λ = 325 cm) and looks out the window at the blue sky (λ = 473 nm). What is the frequency (in s^{-1}) of the electromagnetic radiation from each source? (Assume that the radiation travels at the speed of light, 3.00×10^8 m/s.)

Plan We are given the wavelengths, so we use Equation 7.1 to find the frequencies. However, we must first convert the wavelengths to meters because c has units of m/s.

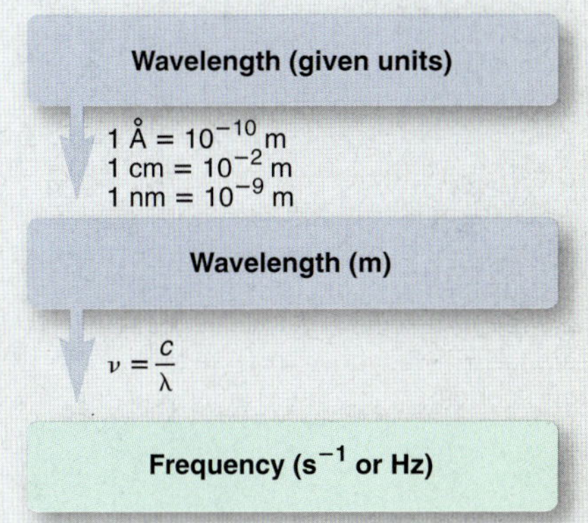

Solution For the x-rays: Converting from angstroms to meters,

$$\lambda = 1.00\ \text{Å} \times \frac{10^{-10}\ \text{m}}{1\ \text{Å}} = 1.00\times10^{-10}\ \text{m}$$

Calculating the frequency:

$$\nu = \frac{c}{\lambda} = \frac{3.00\times10^{8}\ \text{m/s}}{1.00\times10^{-10}\ \text{m}} = 3.00\times10^{18}\ \text{s}^{-1}$$

For the radio signal: Combining steps to calculate the frequency,

$$\nu = \frac{c}{\lambda} = \frac{3.00\times10^{8}\ \text{m/s}}{325\ \text{cm} \times \dfrac{10^{-2}\ \text{m}}{1\ \text{cm}}} = 9.23\times10^{7}\ \text{s}^{-1}$$

For the blue sky: Combining steps to calculate the frequency,

$$\nu = \frac{c}{\lambda} = \frac{3.00\times10^{8}\ \text{m/s}}{473\ \text{nm} \times \dfrac{10^{-9}\ \text{m}}{1\ \text{nm}}} = 6.34\times10^{14}\ \text{s}^{-1}$$

Check The orders of magnitude are correct for the regions of the electromagnetic spectrum (see Figure 7.3): x-rays (10^{19} to 10^{16} s^{-1}), radio waves (10^9 to 10^4 s^{-1}), and visible light (7.5×10^{14} to 4.0×10^{14} s^{-1}).

Comment The radio station here is broadcasting at 92.3×10^6 s^{-1}, or 92.3 million Hz (92.3 MHz), about midway in the FM range.

FOLLOW-UP PROBLEM 7.1 Some diamonds appear yellow because they contain nitrogen compounds that absorb purple light of frequency 7.23×10^{14} Hz. Calculate the wavelength (in nm and Å) of the absorbed light.

The Classical Distinction Between Energy and Matter In the everyday world, energy and matter behave very differently. Let's see how the behavior of light contrasts with the behavior of particles. Light of a given wavelength travels at different speeds through different transparent media—vacuum, air, water, quartz, and so forth. Therefore, when a light wave passes from one medium into another, say, from air to water, the speed of the wave changes. Figure 7.4A (next page) shows the phenomenon known as **refraction.** If the wave strikes the boundary between air and water, at an angle other than 90°, the change in speed causes a change in direction, and the wave continues at a different angle. The new angle (angle of refraction) depends on the materials on either side of the boundary and the wavelength of the light. In the process of *dispersion,* white light separates (disperses) into its component colors, as when it passes through a prism, because each incoming wave is refracted at a slightly different angle. Rainbows result when sunlight is dispersed through water droplets.

In contrast, a particle, like a pebble, does not undergo refraction when passing from one medium to another. Figure 7.4B shows that if you throw a pebble through the air into a pond, its speed changes abruptly and then it continues to slow down gradually in a curved path.

When a wave strikes the edge of an object, it bends around it in a phenomenon called **diffraction.** If the wave passes through a slit about as wide as its wavelength, it bends around both edges of the slit and forms a semicircular wave on the other side of the opening, as shown in Figure 7.4C.

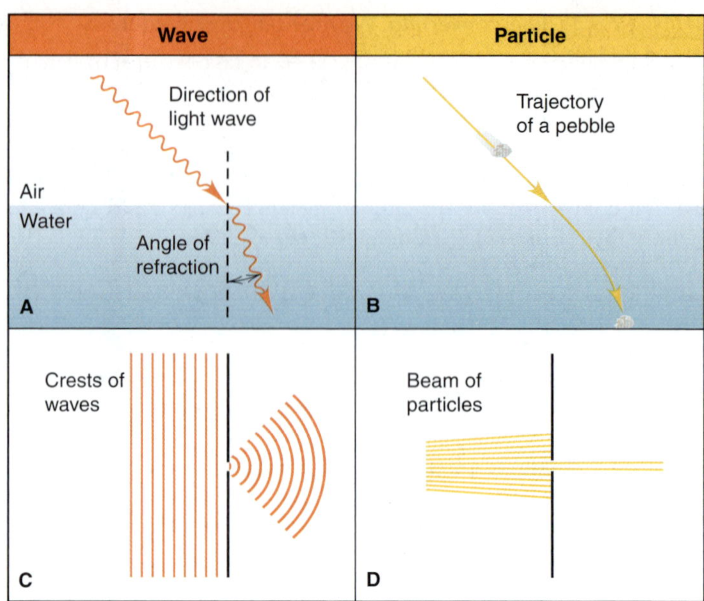

FIGURE 7.4 Different behaviors of waves and particles. A, A wave passing from air into water is *refracted* (bent at an angle). **B,** In contrast, a particle of matter (such as a pebble) entering a pond moves in a curved path, because gravity and the greater resistance (drag) of the water slow it down gradually. **C,** A wave is *diffracted* through a small opening, which gives rise to a circular wave on the other side. (The lines represent the crests of water waves as seen from above.) **D,** In contrast, when a collection of moving particles encounters a small opening, as when a handful of sand is thrown at a hole in a fence, some particles move through the opening and continue along their individual paths.

Once again, particles act very differently. Figure 7.4D shows that if you throw a collection of particles, like a handful of sand, at a small opening, some particles hit the edge, while others go through the opening and continue linearly in a narrower group.

If waves of light pass through two adjacent slits, the emerging circular waves interact with each other through the process of *interference.* If the crests of the waves coincide *(in phase),* they interfere *constructively* and the amplitudes add together. If the crests coincide with troughs *(out of phase),* they interfere *destructively* and the amplitudes cancel. The result is a diffraction pattern of brighter and darker regions (Figure 7.5). In contrast, particles passing through adjacent openings continue in straight paths, some colliding with each other and moving at different angles.

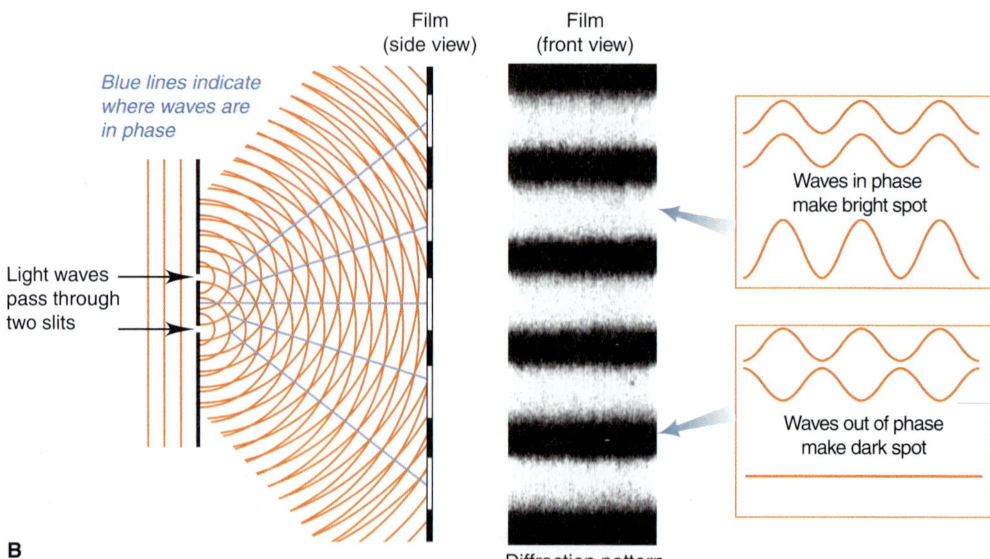

FIGURE 7.5 The diffraction pattern caused by waves passing through two adjacent slits. A, Constructive and destructive interference occurs as water waves viewed from above pass through two adjacent slits in a ripple tank. **B,** As light waves pass through two closely spaced slits, they also emerge as circular waves and interfere with each other. They create a diffraction (interference) pattern of bright and dark regions on a sheet of film. Bright regions appear where crests coincide and the amplitudes combine with each other (in phase); dark regions appear where crests meet troughs and the amplitudes cancel each other (out of phase).

At the end of the 19th century, all everyday and laboratory experience seemed to confirm these classical distinctions between the wave nature of energy and the particle nature of matter.

The Particle Nature of Light

Three phenomena involving matter and light confounded physicists at the turn of the 20th century: (1) blackbody radiation, (2) the photoelectric effect, and (3) atomic spectra. Explaining these phenomena required a radically new picture of energy. We discuss the first two here and the third in Section 7.2.

Blackbody Radiation and the Quantization of Energy When a solid object is heated to about 1000 K, it begins to emit visible light, as you can see in the soft red glow of smoldering coal. At about 1500 K, the light is brighter and more orange, like that from an electric heating coil. At temperatures greater than 2000 K, the light is still brighter and whiter, as from the filament of a lightbulb. These changes in intensity and wavelength of emitted light as an object is heated are characteristic of *blackbody radiation,* light given off by a hot *blackbody.** All attempts to account for these observed changes by applying classical electromagnetic theory failed. Then, in 1900, the German physicist Max Planck (1858–1947) made a radical assumption that eventually led to an entirely new view of energy. He proposed that the hot, glowing object could emit (or absorb) only *certain quantities* of energy:

$$E = nh\nu$$

where E is the energy of the radiation, ν is its frequency, n is a positive integer (1, 2, 3, and so on) called a **quantum number,** and h is a proportionality constant now known very precisely and called **Planck's constant.** With energy in joules (J) and frequency in s^{-1}, h has units of J·s:

$$h = 6.62606876\times10^{-34}\ \text{J·s} = 6.626\times10^{-34}\ \text{J·s} \qquad (4\ \text{sf})$$

Later interpretations of Planck's proposal stated that the hot object's radiation is emitted by the atoms contained within it. If an atom can *emit* only certain quantities of energy, it follows that *the atom itself can have only certain quantities of energy.* Thus, the energy of an atom is *quantized:* it exists only in certain fixed quantities, rather than being continuous. Each change in the atom's energy results from the gain or loss of one or more "packets," definite amounts, of energy. Each energy packet is called a **quantum** ("fixed quantity"; plural, *quanta*), and its energy is equal to $h\nu$. Thus, *an atom changes its energy state by emitting (or absorbing) one or more quanta,* and the energy of the emitted (or absorbed) radiation is equal to the *difference in the atom's energy states:*

$$\Delta E_{\text{atom}} = E_{\text{emitted (or absorbed) radiation}} = \Delta nh\nu$$

Because the atom can change its energy only by integer multiples of $h\nu$, the smallest change occurs when an atom in a given energy state changes to an adjacent state, that is, when $\Delta n = 1$:

$$\Delta E = h\nu \qquad \textbf{(7.2)}$$

The Photoelectric Effect and the Photon Theory of Light Despite the idea that energy is quantized, Planck and other physicists continued to picture the emitted energy as traveling in waves. However, the wave model could not explain the **photoelectric effect,** the flow of current when monochromatic light of sufficient frequency shines on a metal plate (Figure 7.6). The existence of the current was not puzzling: it could be understood as arising when the light transfers energy to the electrons at the metal surface, which break free and are collected by the

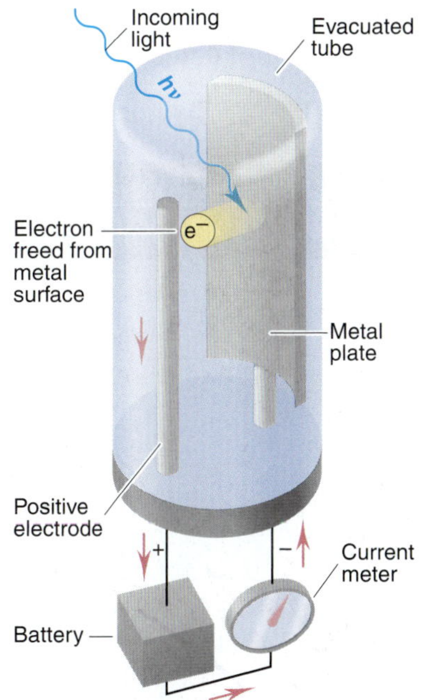

FIGURE 7.6 Demonstration of the photoelectric effect. When monochromatic light of high enough frequency strikes the metal plate, electrons are freed from the plate and travel to the positive electrode, creating a current.

*A blackbody is an idealized object that absorbs all the radiation incident on it. A hollow cube with a small hole in one wall approximates a blackbody.

positive electrode. However, the photoelectric effect had certain confusing features, in particular, the *presence of a threshold frequency* and the *absence of a time lag:*

1. *Presence of a threshold frequency.* Light shining on the metal must have a minimum *frequency,* or no current flows. (Different metals have different minimum frequencies.) The wave theory, however, associates the energy of the light with the *amplitude* (intensity) of the wave, not with its frequency (color). Thus, the wave theory predicts that an electron would break free when it absorbed enough energy from light of *any* color.
2. *Absence of a time lag.* Current flows the moment light of this minimum frequency shines on the metal, regardless of the light's intensity. The wave theory, however, predicts that in dim light there would be a time lag before the current flowed, because the electrons would have to absorb enough energy to break free.

Carrying Planck's idea of quantized energy further, the great physicist Albert Einstein proposed that light itself is particulate, that is, quantized into small "bundles" of electromagnetic energy, which were later called **photons.** In terms of Planck's work, we can say that each atom changes its energy whenever it absorbs or emits one photon, one "particle" of light, whose energy is fixed by its *frequency:*

$$E_{\text{photon}} = h\nu = \Delta E_{\text{atom}}$$

Let's see how Einstein's photon theory explains the photoelectric effect:

1. *Explanation of the threshold frequency.* According to the photon theory, a beam of light consists of an enormous number of photons. Light intensity (brightness) is related to the number of photons striking the surface per unit time, but *not* to their energy. Therefore, a photon of a certain minimum *energy* must be absorbed for an electron to be freed. Because energy depends on frequency ($h\nu$), the theory predicts a threshold frequency.
2. *Explanation of the time lag.* An electron cannot "save up" energy from several photons below the minimum energy until it has enough to break free. Rather, one electron breaks free the moment it absorbs one photon of *enough* energy. The current is weaker in dim light than in bright light because fewer photons of enough energy are present, so fewer electrons break free per unit time. But *some* current flows the moment photons reach the metal plate.

THINK OF IT THIS WAY
Ping-Pong Photons

Consider this analogy to help see why light of insufficient energy cannot free an electron from a metal surface. If one Ping-Pong ball does not have enough energy to knock a book off its shelf, neither does a series of Ping-Pong balls, because the book cannot save up the energy from the individual impacts. But one baseball traveling at the same speed does have enough energy to move the book. Whereas the energy of a ball is related to its mass and velocity, the energy of a photon is related to its frequency.

SAMPLE PROBLEM 7.2 Calculating the Energy of Radiation from Its Wavelength

Problem A cook uses a microwave oven to heat a meal. The wavelength of the radiation is 1.20 cm. What is the energy of one photon of this microwave radiation?
Plan We know λ in centimeters (1.20 cm) so we convert to meters, find the frequency with Equation 7.1, and then find the energy of one photon with Equation 7.2.
Solution Combining steps to find the energy:

$$E = h\nu = \frac{hc}{\lambda} = \frac{(6.626\times10^{-34}\ \text{J}\cdot\text{s})(3.00\times10^{8}\ \text{m/s})}{(1.20\ \text{cm})\left(\dfrac{10^{-2}\ \text{m}}{1\ \text{cm}}\right)} = 1.66\times10^{-23}\ \text{J}$$

Check Checking the order of magnitude gives $\dfrac{10^{-33}\ \text{J·s}\times10^{8}\ \text{m/s}}{10^{-2}\ \text{m}} = 10^{-23}\ \text{J}$.

FOLLOW-UP PROBLEM 7.2 Calculate the energies of one photon of ultraviolet ($\lambda = 1\times10^{-8}$ m), visible ($\lambda = 5\times10^{-7}$ m), and infrared ($\lambda = 1\times10^{-4}$ m) light. What do the answers indicate about the relationship between the wavelength and energy of light?

Planck's quantum theory and Einstein's photon theory assigned properties to energy that, until then, had always been reserved for matter: fixed quantity and discrete particles. These properties have since proved essential to explaining the interactions of matter and energy at the atomic level. But how can a particulate model of energy be made to fit the facts of diffraction and refraction, phenomena explained only in terms of waves? As you'll see shortly, the photon model does not *replace* the wave model. Rather, we have to accept *both* to understand reality. Before we discuss this astonishing notion, however, let's see how the new idea of quantized energy led to a key understanding about atomic behavior.

SECTION 7.1 SUMMARY

Electromagnetic radiation travels in waves of specific wavelength (λ) and frequency (ν). • All electromagnetic waves travel through a vacuum at the speed of light, c (3.00×10^{8} m/s), which is equal to $\nu \times \lambda$. The intensity (brightness) of a light wave is related to its amplitude. • The electromagnetic spectrum ranges from very long radio waves to very short gamma rays and includes the visible region [750 nm (red) to 400 nm (violet)]. • Refraction and diffraction indicate that electromagnetic radiation is wavelike, but blackbody radiation and the photoelectric effect indicate that it is particle-like. • Light exists as photons (quanta) that have an energy proportional to the frequency. • According to quantum theory, an atom has only certain quantities of energy ($E = nh\nu$), which it can change only by absorbing or emitting a photon.

7.2 ATOMIC SPECTRA

The third key observation about matter and energy that late 19th-century physicists could not explain involved the light emitted when an element is vaporized and then thermally or electrically excited, as you see in a neon sign. Figure 7.7A on the next page shows the result when light from excited hydrogen atoms passes through a narrow slit and is then refracted by a prism. Note that this light does not create a *continuous spectrum,* or rainbow, as sunlight does. Instead, it creates a **line spectrum,** a series of fine lines of individual colors separated by colorless (black) spaces.* The wavelengths of these spectral lines are characteristic of the element producing them (Figure 7.7B).

Spectroscopists studying the spectrum of atomic hydrogen had identified several series of such lines in different regions of the electromagnetic spectrum. Figure 7.8 on the next page shows three of these series of lines. Equations of the following general form, called the *Rydberg equation,* were found to predict the position and wavelength of any line in a given series:

$$\frac{1}{\lambda} = R\left(\frac{1}{n_1^2} - \frac{1}{n_2^2}\right) \qquad (7.3)$$

*The appearance of the spectrum as a series of lines results from the construction of the apparatus. If the light passed through a small hole, rather than a narrow slit, the spectrum would appear as a circular field of dots rather than a horizontal series of lines. The key point is that *the spectrum is discrete, rather than continuous.*

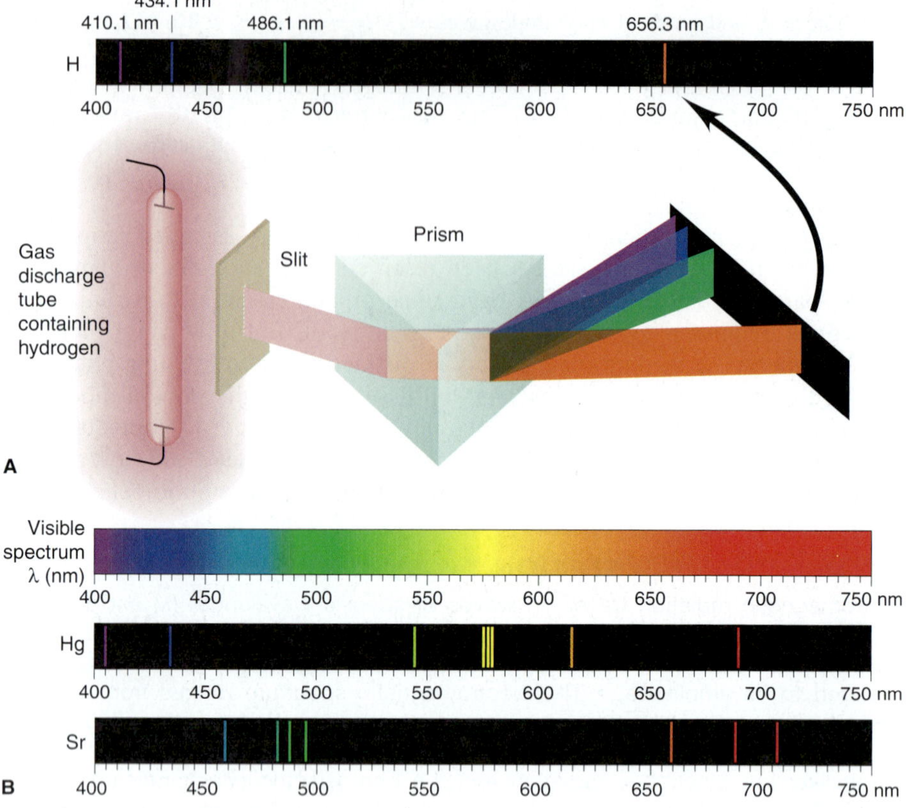

FIGURE 7.7 The line spectra of several elements. A, A sample of gaseous H_2 is dissociated into atoms and excited by an electric discharge. The emitted light passes through a slit and a prism, which disperses the light into individual wavelengths. The line spectrum of atomic H is shown *(top).* **B,** The continuous spectrum of white light is compared with the line spectra of mercury and strontium. Note that each line spectrum is different from the others.

Animation: Atomic Line Spectra

Animation: Emission Spectra

where λ is the wavelength of a spectral line, n_1 and n_2 are positive integers with $n_2 > n_1$, and R is the Rydberg constant ($1.096776 \times 10^7\ \text{m}^{-1}$). For the visible series of lines, $n_1 = 2$:

$$\frac{1}{\lambda} = R\left(\frac{1}{2^2} - \frac{1}{n_2^2}\right), \quad \text{with } n_2 = 3, 4, 5, \ldots$$

The Rydberg equation and the value of the constant are based on data rather than theory. No one knew *why* the spectral lines of hydrogen appear in this pattern. (Problems 7.19 and 7.20 are two of several at the end of the chapter that apply the Rydberg equation.)

The occurrence of line spectra did not correlate with classical theory for one major reason. As was mentioned in the chapter introduction, if an electron spiraled closer to the nucleus, it should emit radiation. Moreover, the frequency of the radiation should be related to the time of revolution. On the spiral path inward, that time should change smoothly, so the frequency of the radiation should change smoothly and create a continuous spectrum. Rutherford's nuclear model seemed totally at odds with atomic line spectra.

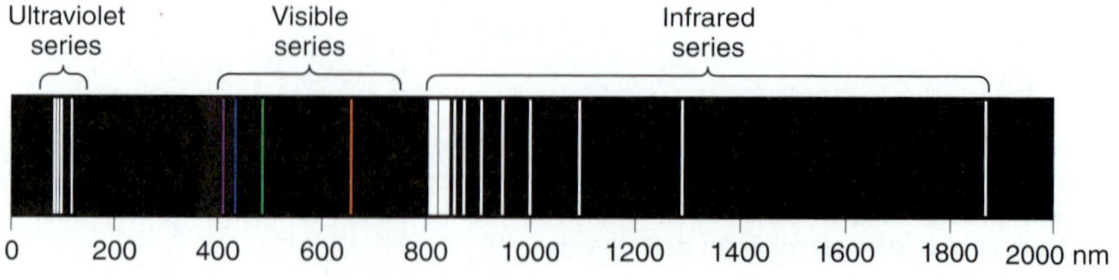

FIGURE 7.8 Three series of spectral lines of atomic hydrogen. These series appear in different regions of the electromagnetic spectrum. The hydrogen spectrum shown in Figure 7.7A is the visible series.

The Bohr Model of the Hydrogen Atom

Soon after the nuclear model was proposed, Niels Bohr (1885–1962), a young Danish physicist working in Rutherford's laboratory, suggested a model for the H atom that predicted the existence of line spectra. In his model, Bohr used Planck's and Einstein's ideas about quantized energy and proposed three postulates:

1. *The H atom has only certain allowable energy levels,* which Bohr called **stationary states.** Each of these states is associated with a fixed circular orbit of the electron around the nucleus.

2. *The atom does* **not** *radiate energy while in one of its stationary states.* That is, even though it violates the ideas of classical physics, the atom does not change energy while the electron moves *within* an orbit.

3. *The atom changes to another stationary state* (the electron moves to another orbit) *only by absorbing or emitting a photon whose energy equals the difference in energy between the two states:*

$$E_{\text{photon}} = E_{\text{state A}} - E_{\text{state B}} = h\nu$$

where the energy of state A is higher than that of state B. A spectral line results when a photon of specific energy (and thus specific frequency) is *emitted* as the electron moves from a higher energy state to a lower one. Therefore, Bohr's model explains that an atomic spectrum is not continuous because *the atom's energy has only certain discrete levels, or states.*

In Bohr's model, the quantum number n (1, 2, 3, . . .) is associated with the radius of an electron orbit, which is directly related to the electron's energy: *the lower the n value, the smaller the radius of the orbit, and the lower the energy level.* When the electron is in the first orbit ($n = 1$), the orbit closest to the nucleus, the H atom is in its lowest (first) energy level, called the **ground state.** If the H atom absorbs a photon whose energy equals the *difference* between the first and second energy levels, the electron moves to the second orbit ($n = 2$), the next orbit out from the nucleus. When the electron is in the second or any higher orbit, the atom is said to be in an **excited state.** If the H atom in the first excited state (the electron in the second orbit) emits a photon of that same energy, it returns to the ground state. Figure 7.9 shows a staircase analogy for this behavior.

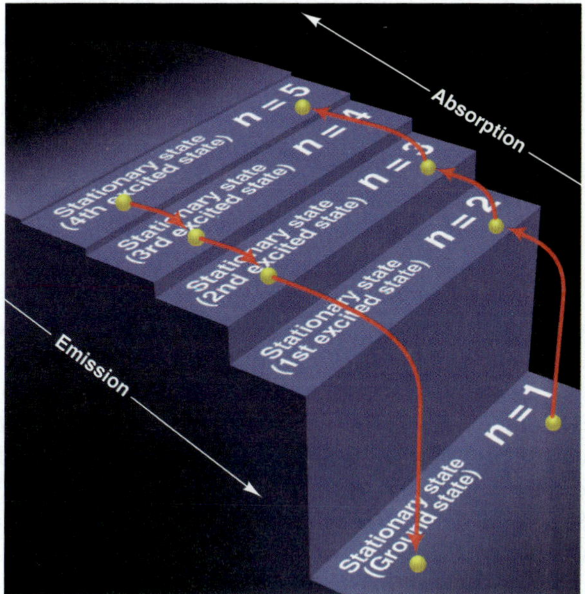

FIGURE 7.9 Quantum staircase. In this analogy for the energy levels of the hydrogen atom, an electron can absorb a photon and jump up to a higher "step" (stationary state) or emit a photon and jump down to a lower one. But the electron cannot lie between two steps.

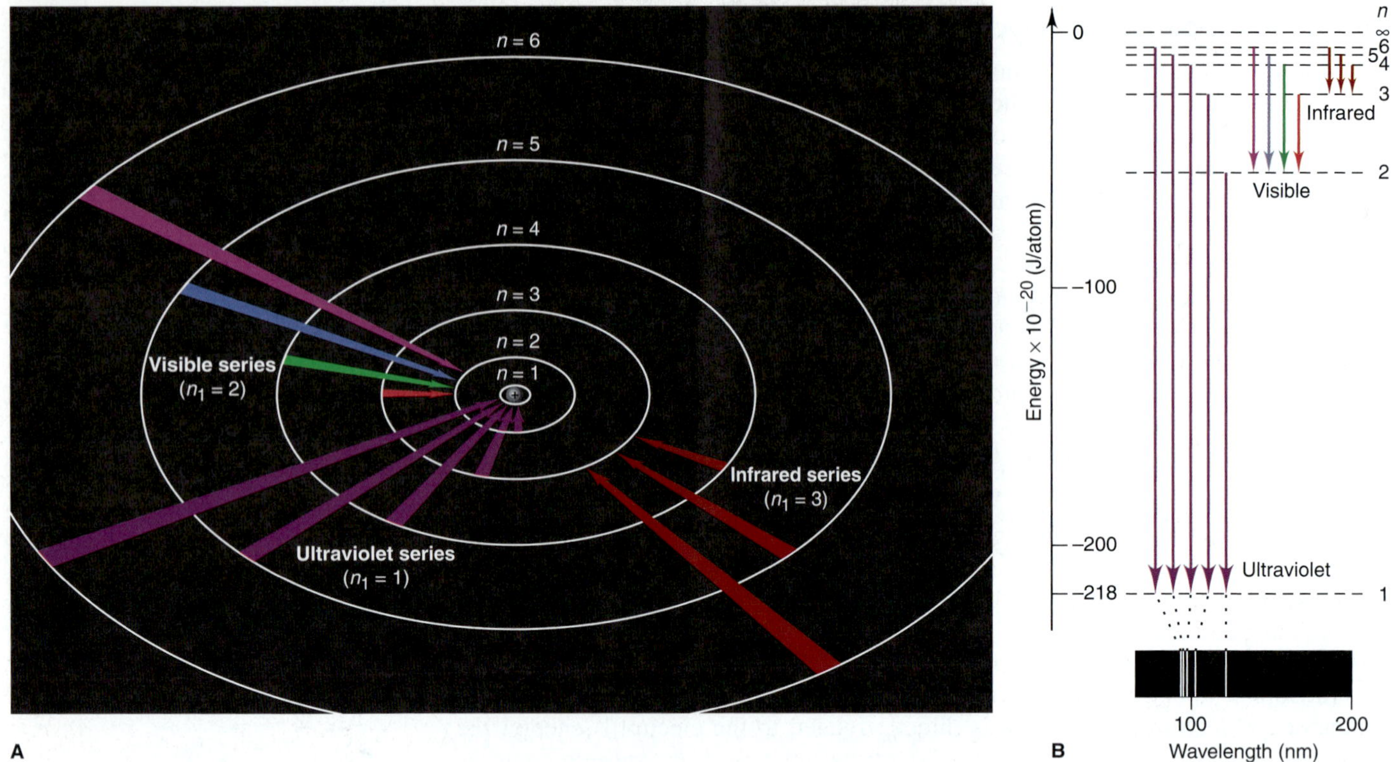

FIGURE 7.10 The Bohr explanation of three series of spectral lines. **A,** According to the Bohr model, when an electron drops from an outer orbit to an inner one, the atom emits a photon of specific energy that gives rise to a spectral line. In a given spectral series, each electron drop has the same inner orbit, that is, the same value of n_1 in the Rydberg equation (see Equation 7.3). (The orbit radius is proportional to n^2. Only the first six orbits are shown.) **B,** An energy diagram shows how the ultraviolet series arises. Within each series, the greater the *difference* in orbit radii, the greater the difference in energy levels (depicted as a downward arrow), and the higher the energy of the photon emitted. For example, in the ultraviolet series, in which $n_1 = 1$, a drop from $n = 5$ to $n = 1$ emits a photon with more energy (shorter λ, higher ν) than a drop from $n = 2$ to $n = 1$. [The axis shows negative values because $n = \infty$ (the electron completely separated from the nucleus) is *defined* as the atom with zero energy.]

Figure 7.10A shows how Bohr's model accounts for the three line spectra of hydrogen. When a sample of gaseous H atoms is excited, different atoms absorb different quantities of energy. Each atom has one electron, but so many atoms are present that all the energy levels (orbits) are populated by electrons. When the electrons drop from outer orbits to the $n = 3$ orbit (second excited state), the emitted photons create the infrared series of lines. The visible series arises when electrons drop to the $n = 2$ orbit (first excited state). Figure 7.10B shows that the ultraviolet series arises when electrons drop to the $n = 1$ orbit (ground state).

Despite its great success in accounting for the spectral lines of the H atom, the Bohr model failed to predict the spectrum of any other atom, even that of helium, the next simplest element. In essence, the Bohr model is a one-electron model. It works beautifully for the H atom and for other one-electron species, such as He^+ (Z = 2), Li^{2+} (Z = 3), and Be^{3+} (Z = 4), which have either been created in the lab or seen in the spectra of stars. But it fails for atoms with more than one electron because in these systems, electron-electron repulsions and additional nucleus-electron attractions are present as well. Moreover, as you'll soon see, electrons do not move in fixed orbits. As a picture of the atom, the Bohr model is incorrect, but we still use the terms "ground state" and "excited state" and retain one of Bohr's central ideas in our current model: *the energy of an atom occurs in discrete levels.*

The Energy States of the Hydrogen Atom

A very useful result from Bohr's work is an equation for calculating the energy levels of an atom, which he derived from the classical principles of electrostatic attraction and circular motion:

$$E = -2.18\times10^{-18}\,\text{J}\left(\frac{Z^2}{n^2}\right)$$

where Z is the charge of the nucleus. For the H atom, $Z = 1$, so we have

$$E = -2.18\times10^{-18}\,\text{J}\left(\frac{1^2}{n^2}\right) = -2.18\times10^{-18}\,\text{J}\left(\frac{1}{n^2}\right)$$

Therefore, the energy of the ground state ($n = 1$) is

$$E = -2.18\times10^{-18}\,\text{J}\left(\frac{1}{1^2}\right) = -2.18\times10^{-18}\,\text{J}$$

Don't be confused by the negative sign for the energy values (see the axis in Figure 7.10B). It appears because we *define* the zero point of the atom's energy when *the electron is completely removed from the nucleus.* Thus, $E = 0$ when $n = \infty$, so $E < 0$ for any smaller n.

THINK OF IT THIS WAY
A Book on a Desk and the H Atom's Energy

If you *define* the potential energy of a book-desk system as zero when the book rests on the desk, the system has negative energy when the book lies on the floor. Similarly, the H atom is defined as having zero energy when its electron is completely separated from the nucleus. Thus, its energy is negative when the electron is close enough to the nucleus to be attracted by it.

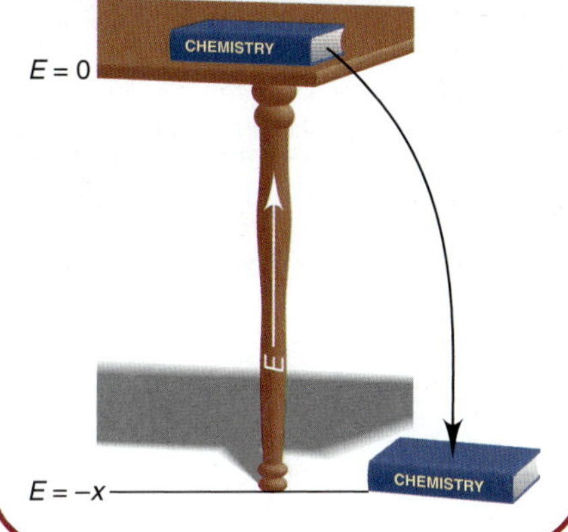

With n in the denominator of the energy equation, as the electron moves closer to the nucleus (n decreases), the atom becomes more stable (less energetic) and its energy becomes a *larger negative number.* As the electron moves away from the nucleus (n increases), the atom's energy increases (becomes a smaller negative number).

This equation is easily adapted to find the energy difference between any two levels:

$$\Delta E = E_{\text{final}} - E_{\text{initial}} = -2.18\times10^{-18}\,\text{J}\left(\frac{1}{n^2_{\text{final}}} - \frac{1}{n^2_{\text{initial}}}\right) \quad \textbf{(7.4)}$$

Using Equation 7.4, we can predict the wavelengths of the spectral lines of the H atom. Note that if we combine Equation 7.4 with Planck's expression for the change in an atom's energy (Equation 7.2), we obtain the Rydberg equation (Equation 7.3):

$$\Delta E = h\nu = \frac{hc}{\lambda} = -2.18\times10^{-18}\,\text{J}\left(\frac{1}{n^2_{\text{final}}} - \frac{1}{n^2_{\text{initial}}}\right)$$

Therefore,
$$\frac{1}{\lambda} = -\frac{2.18\times10^{-18}\,\text{J}}{hc}\left(\frac{1}{n^2_{\text{final}}} - \frac{1}{n^2_{\text{initial}}}\right)$$
$$= -\frac{2.18\times10^{-18}\,\text{J}}{(6.626\times10^{-34}\,\text{J·s})(3.00\times10^{8}\,\text{m/s})}\left(\frac{1}{n^2_{\text{final}}} - \frac{1}{n^2_{\text{initial}}}\right)$$
$$= -1.10\times10^{7}\,\text{m}^{-1}\left(\frac{1}{n^2_{\text{final}}} - \frac{1}{n^2_{\text{initial}}}\right)$$

where $n_{\text{final}} = n_2$, $n_{\text{initial}} = n_1$, and $1.10\times10^7\ \text{m}^{-1}$ is the Rydberg constant ($1.096776\times10^7\ \text{m}^{-1}$) to three significant figures. Thus, from classical relationships of charge and of motion combined with the idea that the H atom can have only certain values of energy, we obtain an equation from theory that leads directly to the empirical one! (Bohr's value for the Rydberg constant differed from the spectroscopists' value by only 0.05%!) In Sample Problem 7.3, we calculate the energy and wavelength of an electron jump in an H atom.

SAMPLE PROBLEM 7.3 Determining ΔE and λ of an Electron Transition

Problem A hydrogen atom absorbs a photon of visible light (see Figure 7.10), and its electron enters the $n = 4$ energy level. Calculate **(a)** the change in energy of the atom and **(b)** the wavelength (in nm) of the photon.

Plan (a) The H atom absorbs energy, so $E_{final} > E_{initial}$. We are given $n_{final} = 4$, and Figure 7.10 shows that $n_{initial} = 2$ because a visible photon is absorbed. We apply Equation 7.4 to find ΔE. **(b)** Once we know ΔE, we find the frequency with Equation 7.2 and the wavelength (in m) with Equation 7.1. Then we convert meters to nanometers.

Solution (a) Substituting the values into Equation 7.4:

$$\Delta E = -2.18\times10^{-18}\,\text{J}\left(\frac{1}{n_{final}^2} - \frac{1}{n_{initial}^2}\right)$$
$$= -2.18\times10^{-18}\,\text{J}\left(\frac{1}{4^2} - \frac{1}{2^2}\right) = -2.18\times10^{-18}\,\text{J}\left(\frac{1}{16} - \frac{1}{4}\right)$$
$$= 4.09\times10^{-19}\,\text{J}$$

(b) Combining Equations 7.2 and 7.1 and solving for λ:

$$\Delta E = h\nu = \frac{hc}{\lambda}$$

therefore, $$\lambda = \frac{hc}{\Delta E} = \frac{(6.626\times10^{-34}\,\text{J·s})(3.00\times10^{8}\,\text{m/s})}{4.09\times10^{-19}\,\text{J}} = 4.86\times10^{-7}\,\text{m}$$

Converting m to nm:

$$\lambda = 4.86\times10^{-7}\,\text{m} \times \frac{1\,\text{nm}}{10^{-9}\,\text{m}} = 486\,\text{nm}$$

Check In (a), the energy change is positive, which is consistent with an absorption. In (b), the wavelength is consistent with a visible photon (400–750 nm); it is blue-green.

Comment If ΔE is negative (the atom loses energy), we use its absolute value, $|\Delta E|$, because λ must have a positive value.

FOLLOW-UP PROBLEM 7.3 A hydrogen atom with its electron in the $n = 6$ energy level emits a photon of IR light. Calculate **(a)** the change in energy of the atom and **(b)** the wavelength (in Å) of the photon.

We can also use Equation 7.4 to find the quantity of energy needed to completely remove the electron from an H atom. In other words, what is ΔE for the following change?

$$H(g) \longrightarrow H^+(g) + e^-$$

We substitute $n_{final} = \infty$ and $n_{initial} = 1$ and obtain

$$\Delta E = E_{final} - E_{initial} = -2.18\times10^{-18}\,\text{J}\left(\frac{1}{\infty^2} - \frac{1}{1^2}\right) = -2.18\times10^{-18}\,\text{J}(0 - 1)$$
$$= 2.18\times10^{-18}\,\text{J}$$

ΔE is positive because energy is *absorbed* to remove the electron from the vicinity of the nucleus. For 1 mol of H atoms,

$$\Delta E = \left(2.18\times10^{-18}\,\frac{\text{J}}{\text{atom}}\right)\left(6.022\times10^{23}\,\frac{\text{atoms}}{\text{mol}}\right)\left(\frac{1\,\text{kJ}}{10^3\,\text{J}}\right) = 1.31\times10^3\,\text{kJ/mol}$$

This is the *ionization energy* of the H atom, the quantity of energy required to form 1 mol of gaseous H^+ ions from 1 mol of gaseous H atoms. We return to this idea in Chapter 8.

Spectral Analysis in the Laboratory

Analysis of the spectrum of the H atom led to the Bohr model, the first step toward our current model of the atom. From its use by 19th-century chemists as a means

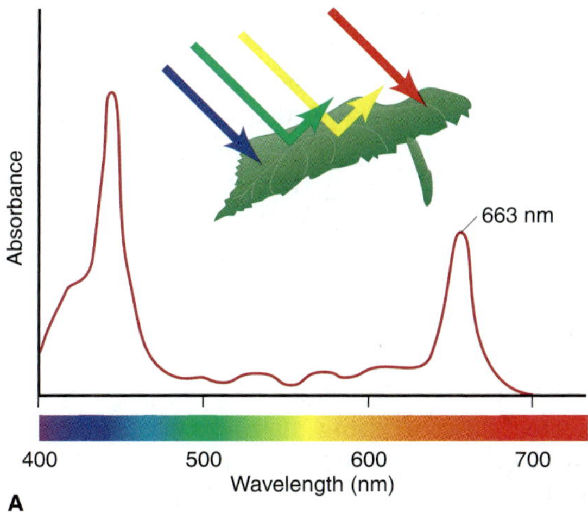

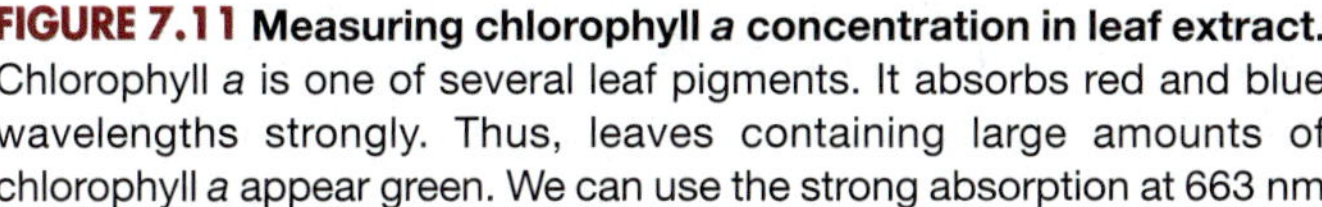

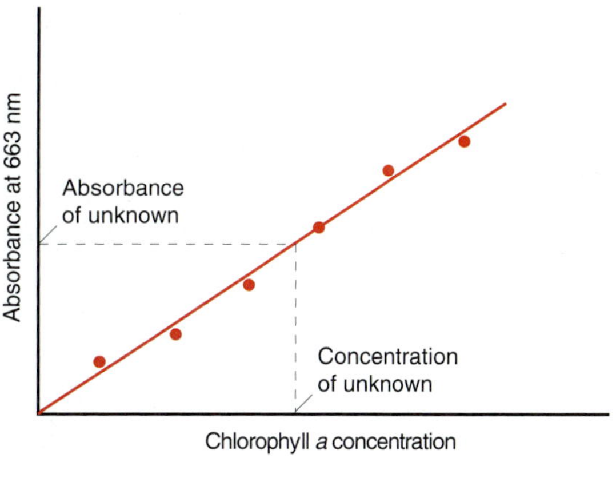

FIGURE 7.11 Measuring chlorophyll *a* concentration in leaf extract. Chlorophyll *a* is one of several leaf pigments. It absorbs red and blue wavelengths strongly. Thus, leaves containing large amounts of chlorophyll *a* appear green. We can use the strong absorption at 663 nm in the spectrum **(A)** to quantify the amount of chlorophyll *a* present in a plant extract by comparing that absorbance to a series of known standards **(B).**

of identifying elements and compounds, spectrometry has developed into a major tool of modern chemistry. The terms *spectroscopy, spectrometry,* and **spectrophotometry** refer to a large group of instrumental techniques that obtain spectra corresponding to a substance's atomic or molecular energy levels. The two types of spectra most often obtained are emission and absorption spectra:

- An **emission spectrum** is produced when atoms in an excited state emit photons characteristic of the element as they return to lower energy states. The characteristic colors of fireworks and sodium-vapor streetlights are due to one or a few prominent lines in the emission spectra of the atoms present.
- An **absorption spectrum** is produced when atoms absorb photons of certain wavelengths and become excited from lower to higher energy states. Therefore, the absorption spectrum of an element appears as dark lines against a bright background. Both the emission and absorption spectra of a substance are characteristic of that substance and used to identify it.

A spectrometer can also be used to measure the concentration of a substance in a solution because *the absorbance, the amount of light of a given wavelength absorbed by a substance, is proportional to the number of molecules.* Suppose, for example, you want to determine the concentration of chlorophyll in an ether solution of leaf extract. You select a strongly absorbed wavelength from the chlorophyll spectrum (such as 663 nm in Figure 7.11A), measure the absorbance of the leaf-extract solution, and compare it with the absorbances of a series of ether solutions with known chlorophyll concentrations (Figure 7.11B).

SECTION 7.2 SUMMARY

To explain the line spectrum of atomic hydrogen, Bohr proposed that the atom's energy is quantized because the electron's motion is restricted to fixed orbits. The electron can move from one orbit to another only if the atom absorbs or emits a photon whose energy equals the difference in energy levels (orbits). Line spectra are produced because these energy changes correspond to photons of specific wavelengths. • Bohr's model predicted the spectra of the H atom and other one-electron species, but not of any other atom. Despite this, Bohr's idea that atoms have quantized energy levels is a cornerstone of our current atomic model. • Spectrophotometry is an instrumental technique in which emission and absorption spectra are used to identify and measure concentrations of substances.

7.3 THE WAVE-PARTICLE DUALITY OF MATTER AND ENERGY

The early proponents of quantum theory demonstrated that *energy is particle-like.* Physicists who developed the theory turned this proposition upside down and showed that *matter is wavelike.* The sharp divisions we perceive between matter (chunky and massive) and energy (diffuse and massless) have been completely blurred. Strange as this idea may seem, it is the key to our modern atomic model.

The Wave Nature of Electrons and the Particle Nature of Photons

Bohr's efforts were a perfect case of fitting theory to data: he *assumed* that an atom has only certain allowable energy levels in order to *explain* the observed line spectrum. However, his assumption had no basis in physical theory. Then, in the early 1920s, a young French physics student named Louis de Broglie proposed a startling reason for fixed energy levels: *if energy is particle-like, perhaps matter is wavelike.* De Broglie had been thinking of other systems that display only certain allowed motions, such as the wave of a plucked guitar string. Figure 7.12 shows that, because the ends of the string are fixed, only certain vibrational frequencies (and wavelengths) are possible. De Broglie reasoned that *if electrons have wavelike motion* and are restricted to orbits of fixed radii, that would explain why they have only certain possible frequencies and energies.

Combining Einstein's famous equation for the quantity of energy equivalent to a given amount of mass ($E = mc^2$) with the equation for the energy of a photon ($E = h\nu = hc/\lambda$), de Broglie derived an equation for the wavelength of any particle of mass m—whether planet, baseball, or electron—moving at speed u:

$$\lambda = \frac{h}{mu} \tag{7.5}$$

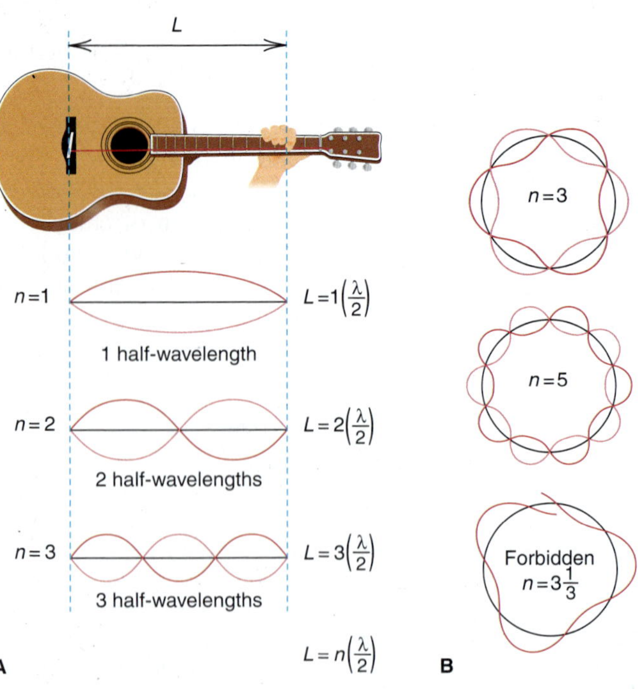

FIGURE 7.12 Wave motion in restricted systems. A, In a musical analogy to electron waves, one half-wavelength ($\lambda/2$) is the "quantum" of the guitar string's vibration. The string length L is fixed, so the only allowed vibrations occur when L is a whole-number multiple (n) of $\lambda/2$. **B,** If an electron occupies a circular orbit, only whole numbers of wavelengths are allowed ($n = 3$ and $n = 5$ are shown). A wave with a fractional number of wavelengths (such as $n = 3\frac{1}{3}$) is "forbidden" because it rapidly dies out through overlap of crests and troughs.

Table 7.1 The de Broglie Wavelengths (λ) of Several Objects

Substance	Mass (g)	Speed (m/s)	λ (m)
Slow electron	9×10^{-28}	1.0	7×10^{-4}
Fast electron	9×10^{-28}	5.9×10^{6}	1×10^{-10}
Alpha particle	6.6×10^{-24}	1.5×10^{7}	7×10^{-15}
One-gram mass	1.0	0.01	7×10^{-29}
Baseball	142	25.0	2×10^{-34}
Earth	6.0×10^{27}	3.0×10^{4}	4×10^{-63}

According to this equation for the **de Broglie wavelength,** *matter behaves as though it moves in a wave.* Note also that an object's wavelength is *inversely* proportional to its mass, so heavy objects such as planets and baseballs have wavelengths that are *many* orders of magnitude smaller than the object itself, as you can see in Table 7.1.

SAMPLE PROBLEM 7.4 Calculating the de Broglie Wavelength of an Electron

Problem Find the de Broglie wavelength of an electron with a speed of 1.00×10^{6} m/s (electron mass $= 9.11\times10^{-31}$ kg; $h = 6.626\times10^{-34}$ kg·m²/s).

Plan We know the speed (1.00×10^{6} m/s) and mass (9.11×10^{-31} kg) of the electron, so we substitute these into Equation 7.5 to find λ.

Solution

$$\lambda = \frac{h}{mu} = \frac{6.626\times10^{-34}\ \text{kg·m}^2\text{/s}}{(9.11\times10^{-31}\ \text{kg})(1.00\times10^{6}\ \text{m/s})} = 7.27\times10^{-10}\ \text{m}$$

Check The order of magnitude and units seem correct:

$$\lambda \approx \frac{10^{-33}\ \text{kg·m}^2\text{/s}}{(10^{-30}\ \text{kg})(10^{6}\ \text{m/s})} = 10^{-9}\ \text{m}$$

FOLLOW-UP PROBLEM 7.4 What is the speed of an electron that has a de Broglie wavelength of 100. nm?

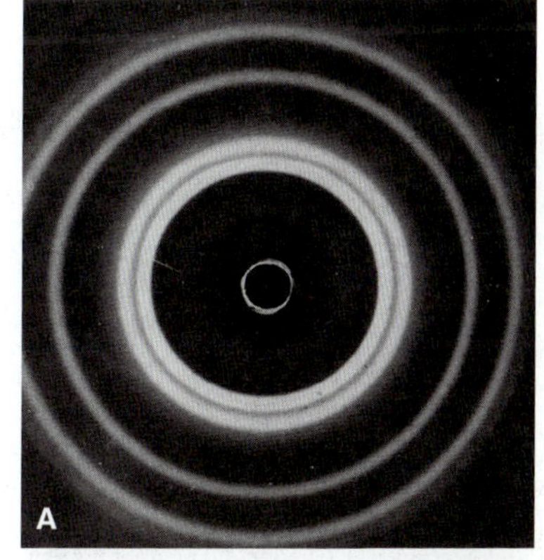

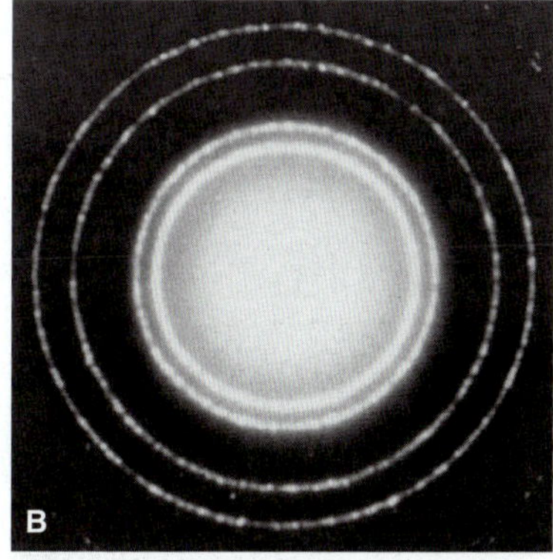

FIGURE 7.13 Comparing diffraction patterns of x-rays and electrons. A, X-ray diffraction pattern of aluminum. **B,** Electron diffraction pattern of aluminum. This behavior implies that both x-rays, which are electromagnetic radiation, and electrons, which are particles, travel in waves.

If particles travel in waves, electrons should exhibit diffraction and interference (see Section 7.1). A fast-moving electron has a wavelength of about 10^{-10} m, so perhaps a beam of electrons would be diffracted by the spaces about this size between atoms in a crystal. Indeed, in 1927, C. Davisson and L. Germer guided a beam of electrons at a nickel crystal and obtained a diffraction pattern. Figure 7.13 shows the diffraction patterns obtained when either x-rays or electrons impinge on aluminum foil. Apparently, electrons—particles with mass and charge—create diffraction patterns, just as electromagnetic waves do! (Indeed, the electron microscope—and its revolutionary impact on modern biology—depends on the wavelike behavior of electrons.) Even though electrons do not have orbits of fixed radius, as de Broglie thought, the energy levels of the atom *are* related to the wave nature of the electron.

If electrons have properties of energy, do photons have properties of matter? The de Broglie equation suggests that we can calculate the momentum (p), the product of mass and speed, for a photon of a given wavelength. Substituting the speed of light (c) for speed u in Equation 7.5 and solving for p gives

$$\lambda = \frac{h}{mc} = \frac{h}{p} \quad \text{and} \quad p = \frac{h}{\lambda}$$

Notice the inverse relationship between p and λ. This means that shorter wavelength (higher energy) photons have greater momentum. Thus, a decrease in a photon's momentum should appear as an increase in its wavelength. In 1923, Arthur Compton directed a beam of x-ray photons at a sample of graphite and observed that the wavelength of the reflected photons increased. This result means that the photons transferred some of their momentum to the electrons in the carbon atoms of the graphite, just as colliding billiard balls transfer momentum to one another. In this experiment, photons behave as particles with momentum!

To scientists of the time, these results were very unsettling. Classical experiments had shown matter to be particle-like and energy to be wavelike, but these new studies showed that, on the atomic scale, every characteristic trait used to define the one now also defined the other. Figure 7.14 summarizes the conceptual and experimental breakthroughs that led to this juncture.

The truth is that *both* matter and energy show *both* behaviors: each possesses both "faces." In some experiments, we observe one face; in other experiments, we observe the other face. The distinction between a particle and a wave is meaningful only in the macroscopic world, *not* in the atomic world. The distinction between matter and energy is in our minds and our limiting definitions, not inherent in nature. This dual character of matter and energy is known as the **wave-particle duality.**

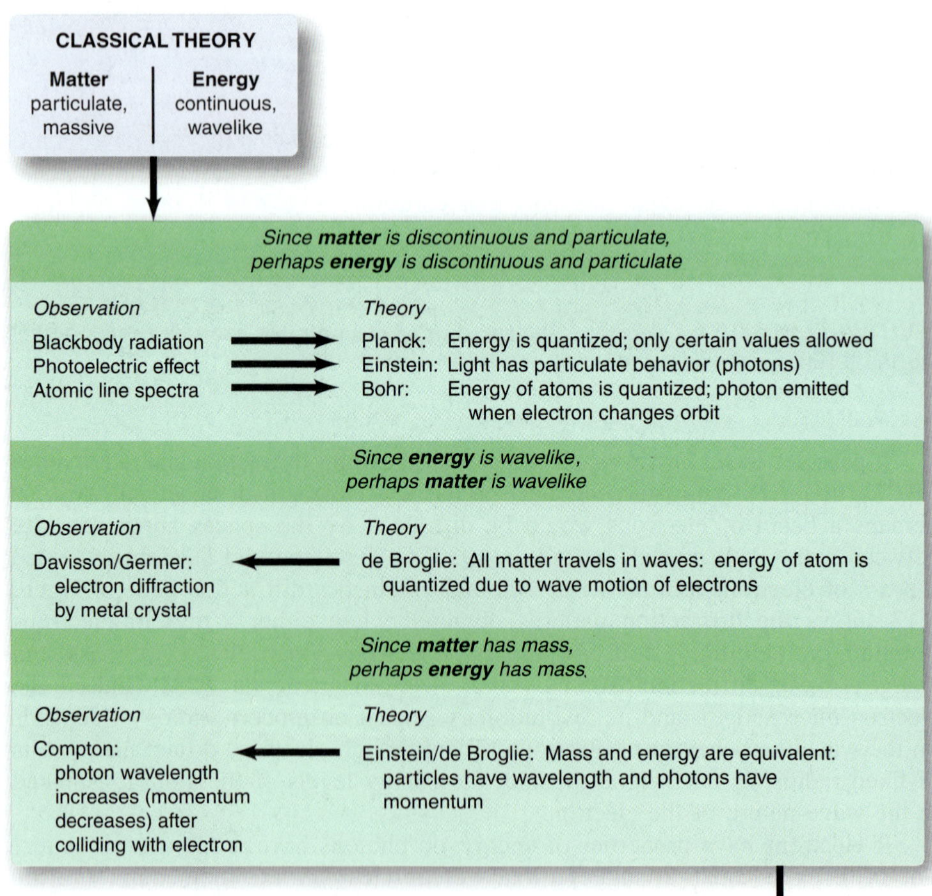

FIGURE 7.14 Summary of the major observations and theories leading from classical theory to quantum theory. As often happens in science, an observation (experiment) stimulates the need for an explanation (theory), and/or a theoretical insight provides the impetus for an experimental test.

The Heisenberg Uncertainty Principle

In the classical view of the world, a moving particle has a definite location at any instant, whereas a wave is spread out in space. If an electron has the properties of *both* a particle and a wave, what can we determine about its position in the atom? In 1927, the German physicist Werner Heisenberg postulated the **uncertainty principle,** which states that it is impossible to know the exact position *and* momentum (mass times speed) of a particle simultaneously. For a particle with constant mass m, the principle is expressed mathematically as

$$\Delta x \cdot m\Delta u \geq \frac{h}{4\pi} \qquad (7.6)$$

where Δx is the uncertainty in position and Δu is the uncertainty in speed. The more accurately we know the position of the particle (smaller Δx), the less accurately we know its speed (larger Δu), and vice versa.

By knowing the position and speed of a pitched baseball and using the classical laws of motion, we can predict its trajectory and whether it will be a strike or a ball. For a baseball, Δx and Δu are insignificant because its mass is enormous compared with $h/4\pi$. Knowing the position and speed of an electron, and from them its trajectory, is another situation entirely. For example, if we take an electron's speed as 6×10^6 m/s $\pm$ 1%, then Δu in Equation 7.6 is 6×10^4 m/s, and the uncertainty in the electron's position (Δx) is 10^{-9} m, which is about 10 times greater than the diameter of the entire atom (10^{-10} m)! Therefore, we have no precise idea where in the atom the electron is located.

The uncertainty principle has profound implications for an atomic model. It means that *we cannot assign fixed paths for electrons,* such as the circular orbits of Bohr's model. As you'll see next, the most we can ever hope to know is the *probability*—the odds—of finding an electron in a given region of space; but we are not *sure* it is there any more than a gambler is sure of the next roll of the dice.

SECTION 7.3 SUMMARY

As a result of Planck's quantum theory and Einstein's relativity theory, we no longer view matter and energy as distinct entities. • The de Broglie wavelength refers to the idea that electrons (and all matter) have wavelike motion. Allowed atomic energy levels are related to allowed wavelengths of the electron's motion. • Electrons exhibit diffraction patterns, as do waves of energy, and photons exhibit transfer of momentum, as do particles of mass. The wave-particle duality of matter and energy is observable only on the atomic scale. • According to the uncertainty principle, we cannot know simultaneously the position and speed of an electron.

7.4 THE QUANTUM-MECHANICAL MODEL OF THE ATOM

Acceptance of the dual nature of matter and energy and of the uncertainty principle culminated in the field of **quantum mechanics,** which examines the wave nature of objects on the atomic scale. In 1926, Erwin Schrödinger derived an equation that is the basis for the quantum-mechanical model of the hydrogen atom. The model describes an atom that has certain allowed quantities of energy due to the allowed frequencies of an electron whose behavior is wavelike and whose exact location is impossible to know.

The Atomic Orbital and the Probable Location of the Electron

The electron's matter-wave occupies the three-dimensional space near the nucleus and experiences a continuous, but varying, influence from the nuclear charge. The **Schrödinger equation** is quite complex but is represented as

$$\mathcal{H}\psi = E\psi$$

where E is the energy of the atom. The symbol ψ (Greek *psi,* pronounced "sigh") is called a **wave function,** a mathematical description of the electron's matter-wave in terms of position in three dimensions. The symbol $\mathcal{H}$, called the Hamiltonian operator, represents a set of mathematical operations that, when carried out on a particular ψ, yields an allowed energy value.*

Each solution to the equation (that is, each energy state of the atom) is associated with a given wave function, also called an **atomic orbital.** It's important to keep in mind that an "orbital" in the quantum-mechanical model *bears no resemblance* to an "orbit" in the Bohr model: an *orbit* was, supposedly, an electron's path around the nucleus, whereas an *orbital* is a mathematical function with no direct physical meaning.

We cannot know precisely where the electron is at any moment, but we can describe where it *probably* is, that is, where it is most likely to be found, or where it spends most of its time. Although the wave function (atomic orbital) has no direct physical meaning, the square of the wave function, ψ^2, is the *probability density,* a measure of the probability that the electron can be found within a particular tiny volume of the atom. (Whereas ψ can have positive or negative values, ψ^2 is always positive, which makes sense for a value that expresses a probability.) For a given energy level, we can depict this probability with an *electron probability density diagram,* or more simply, an **electron density diagram.** In Figure 7.15A, the value of ψ^2 for a given volume is represented pictorially by a certain density of dots: the greater the density of dots, the higher the probability of finding the electron within that volume.

Electron density diagrams are sometimes called **electron cloud** representations. If we *could* take a time-exposure photograph of the electron in wavelike motion around the nucleus, it would appear as a "cloud" of electron positions. The electron cloud is an *imaginary* picture of the electron changing its position rapidly over time; it does *not* mean that an electron is a diffuse cloud of charge. Note that *the electron probability density decreases with distance from the nucleus* along a line, r. The same concept is shown graphically in the plot of ψ^2 vs. r in Figure 7.15B. Note that due to the thickness of the printed line, the curve touches the axis; in reality, however, *the probability of the electron being far from the nucleus is very small, but not zero.*

The *total* probability of finding the electron at any distance r from the nucleus is also important. To find this, we mentally divide the volume around the nucleus into thin, concentric, spherical layers, like the layers of an onion (shown in cross section in Figure 7.15C), and ask in which *spherical layer* we are most likely to find the electron. This is the same as asking for the *sum of* ψ^2 *values* within each spherical layer. The steep falloff in probability density with distance (see Figure 7.15B) has an important effect. Near the nucleus, the volume of each layer increases faster than its probability density decreases. As a result, the *total* probability of finding the electron in the second layer is higher than in the first. Electron density drops off so quickly, however, that this effect soon diminishes with greater distance. Thus, even though the volume of each layer continues to increase, the total probability for a given layer gradually decreases. Because of these opposing effects of decreasing probability density and increasing layer volume, the total probability peaks in a layer some distance from the nucleus. Figure 7.15D shows this as a **radial probability distribution plot.**

---

*The complete form of the Schrödinger equation in terms of the three linear axes is

$$\left[-\frac{h^2}{8\pi^2 m_e}\left(\frac{d^2}{dx^2}+\frac{d^2}{dy^2}+\frac{d^2}{dz^2}\right)+V(x,y,z)\right]\psi(x,y,z)=E\psi(x,y,z)$$

where ψ is the wave function; m_e is the electron's mass; E is the total quantized energy of the atomic system; and V is the potential energy at point (x,y,z).

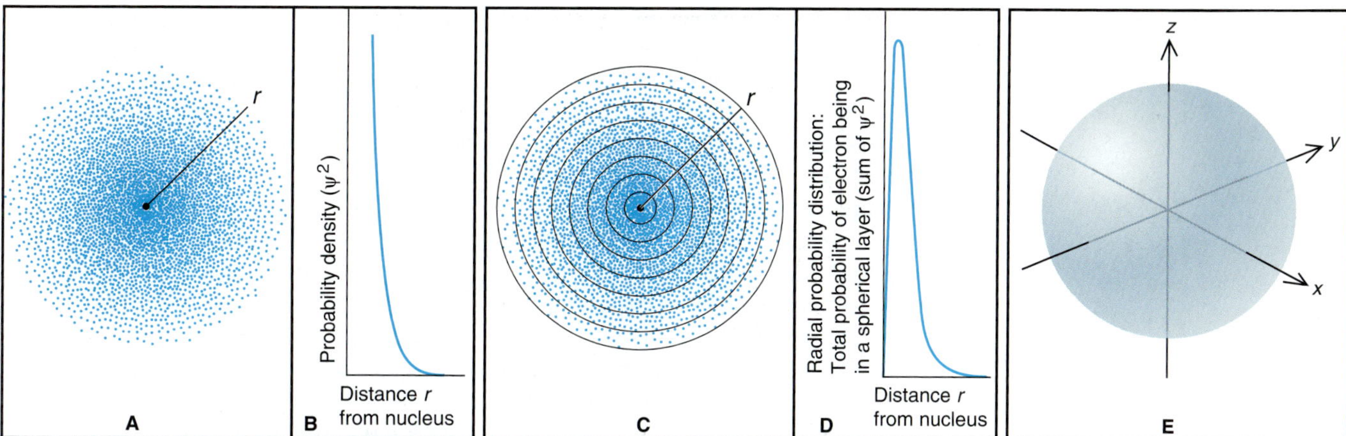

FIGURE 7.15 Electron probability density in the ground-state H atom. A, An electron density diagram shows a cross section of the H atom. The dots, each representing the probability of the electron being within a tiny volume, decrease along a line outward from the nucleus. **B,** A plot of the data in **A** shows that the probability density (ψ^2) decreases with distance from the nucleus but does not reach zero (the thickness of the line makes it appear to do so). **C,** Dividing the atom's volume into thin, concentric, spherical layers (shown in cross section) and counting the dots within each layer gives the total probability of finding the electron within that layer. **D,** A radial probability distribution plot shows total electron density in each spherical layer vs. *r*. Because electron density decreases more slowly than the volume of each concentric layer increases, the plot shows a peak. **E,** A 90% probability contour shows the ground state of the H atom (orbital of lowest energy) and represents the volume in which the electron spends 90% of its time.

THINK OF IT THIS WAY
A "Radial Probability Distribution" of Apples

Here's an analogy illustrating why the curve peaks in a radial probability distribution plot. Picture the fallen apples around the base of an apple tree: the density of apples is greatest near the trunk and decreases with distance. Divide the ground under the tree into foot-wide concentric rings and collect the apples within each ring. Apple density is greatest in the first ring; however, the area of the second ring is larger, so it contains a greater *total number* of apples. Farther out from the tree trunk, rings have more area but lower apple density, so the total number of apples decreases. A plot of "number of apples within a ring" vs. "distance of ring from trunk" shows a peak at some distance fairly close to the trunk.

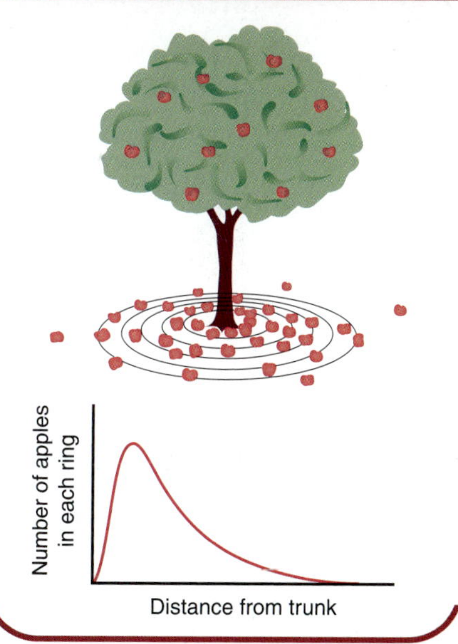

The peak of the radial probability distribution for the ground-state H atom appears at the same distance from the nucleus (0.529Å, or 5.29×10^{-11} m) as Bohr postulated for the closest orbit. Thus, at least for the ground state, the Schrödinger model predicts that the electron spends *most* of its time at the same distance that the Bohr model predicted it spent *all* of its time. The difference between "most" and "all" reflects the uncertainty of the electron's location in the Schrödinger model.

How far away from the nucleus can we find the electron? This is the same as asking "How large is the atom?" Recall from Figure 7.15B that the probability of finding the electron far from the nucleus is not zero. Therefore, we *cannot* assign a definite volume to an atom. However, we often visualize atoms with a 90% **probability contour,** such as in Figure 7.15E, which shows the volume within which the electron of the hydrogen atom spends 90% of its time.

Quantum Numbers of an Atomic Orbital

So far we have discussed the electron density for the *ground* state of the H atom. When the atom absorbs energy, it exists in an *excited* state and the region of space occupied by the electron is described by a different atomic orbital (wave function).

As you'll see, each atomic orbital has a distinctive radial probability distribution and 90% probability contour.

An atomic orbital is specified by three quantum numbers. One is related to the orbital's size, another to its shape, and the third to its orientation in space.* The quantum numbers have a hierarchical relationship: the size-related number limits the shape-related number, which limits the orientation-related number. Let's examine this hierarchy and then look at the shapes and orientations.

1. The **principal quantum number (n)** *is a positive integer* (1, 2, 3, and so forth). It indicates the relative *size* of the orbital and therefore the relative *distance from the nucleus* of the peak in the radial probability distribution plot. The principal quantum number specifies the *energy level* of the H atom: *the higher the n value, the higher the energy level.* When the electron occupies an orbital with $n = 1$, the H atom is in its ground state and has lower energy than when the electron occupies the $n = 2$ orbital (first excited state).

2. The **angular momentum quantum number (l)** *is an integer from* 0 *to* $n - 1$. It is related to the *shape* of the orbital and is sometimes called the *orbital-shape (or azimuthal) quantum number.* Note that the principal quantum number sets a limit on the values for the angular momentum quantum number; that is, n limits l. For an orbital with $n = 1$, l can have a value of only 0. For orbitals with $n = 2$, l can have a value of 0 or 1; for those with $n = 3$, l can be 0, 1, or 2; and so forth. Note that the number of possible l values equals the value of n.

3. The **magnetic quantum number (m_l)** *is an integer from* $-l$ *through* 0 *to* $+l$. It prescribes the *orientation* of the orbital in the space around the nucleus and is sometimes called the *orbital-orientation quantum number.* The possible values of an orbital's magnetic quantum number are set by its angular momentum quantum number; that is, l sets the possible values of m_l. An orbital with $l = 0$ can have only $m_l = 0$. However, an orbital with $l = 1$ can have any one of three m_l values, -1, 0, or $+1$; thus, there are three possible orbitals with $l = 1$, each with its own orientation. Note that the number of possible m_l values *equals* the number of orbitals, which is $2l + 1$ for a given l value.

Table 7.2 summarizes the hierarchy among the three quantum numbers. (In Chapter 8, we'll discuss a fourth quantum number that relates to a property of the electron itself.) The total number of orbitals for a given n value is n^2.

*For ease in discussion, we refer to the size, shape, and orientation of an "atomic orbital," although we really mean the size, shape, and orientation of an "atomic orbital's radial probability distribution." This usage is common in both introductory and advanced texts.

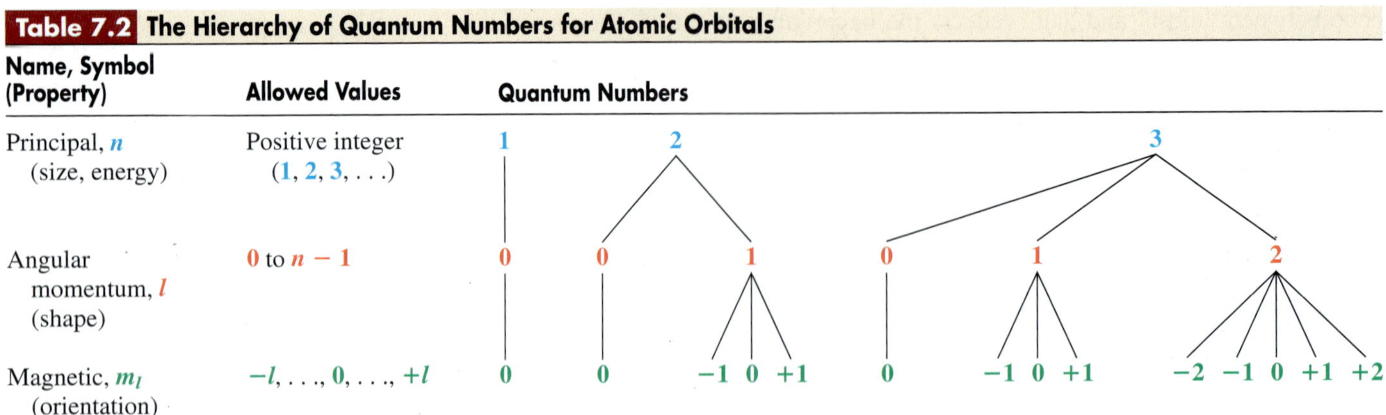

Table 7.2 The Hierarchy of Quantum Numbers for Atomic Orbitals

Name, Symbol (Property)	Allowed Values	Quantum Numbers								
Principal, n (size, energy)	Positive integer (1, 2, 3, . . .)	1	2		3		
Angular momentum, l (shape)	0 to $n - 1$	0	0	1	0	1	2
Magnetic, m_l (orientation)	$-l, \ldots, 0, \ldots, +l$	0	0	−1 0 +1	0	−1 0 +1	−2 −1 0 +1 +2

SAMPLE PROBLEM 7.5 Determining Quantum Numbers for an Energy Level

Problem What values of the angular momentum (l) and magnetic (m_l) quantum numbers are allowed for a principal quantum number (n) of 3? How many orbitals exist for $n = 3$?

Plan We determine allowable quantum numbers with the rules from the text: l values are integers from 0 to $n - 1$, and m_l values are integers from $-l$ to 0 to $+l$. One m_l value is assigned to each orbital, so the number of m_l values gives the number of orbitals.

Solution Determining l values: for $n = 3$, $l = 0, 1, 2$

Determining m_l for each l value:

For $l = 0$, $m_l = 0$

For $l = 1$, $m_l = -1, 0, +1$

For $l = 2$, $m_l = -2, -1, 0, +1, +2$

There are nine m_l values, so there are nine orbitals with $n = 3$.

Check Table 7.2 shows that we are correct. The total number of orbitals for a given n value is n^2, and for $n = 3$, $n^2 = 9$.

FOLLOW-UP PROBLEM 7.5 Specify the l and m_l values for $n = 4$.

The energy states and orbitals of the atom are described with specific terms and associated with one or more quantum numbers:

1. *Level.* The atom's energy **levels,** or *shells,* are given by the n value: the smaller the n value, the lower the energy level and the greater the probability of the electron being closer to the nucleus.

2. *Sublevel.* The atom's levels contain **sublevels,** or *subshells,* which designate the orbital shape. Each sublevel has a letter designation:

$l = 0$ is an s sublevel.
$l = 1$ is a p sublevel.
$l = 2$ is a d sublevel.
$l = 3$ is an f sublevel.

(The letters derive from the names of spectroscopic lines: *s*harp, *p*rincipal, *d*iffuse, and *f*undamental. Sublevels with l values greater than 3 are designated alphabetically: g sublevel, h sublevel, etc.) Sublevels are named by joining the n value and the letter designation. For example, the sublevel (subshell) with $n = 2$ and $l = 0$ is called the $2s$ sublevel.

3. *Orbital.* Each allowed combination of n, l, and m_l values specifies one of the atom's *orbitals.* Thus, the three quantum numbers that describe an orbital express its size (energy), shape, and spatial orientation. You can easily give the quantum numbers of the orbitals in any sublevel if you know the sublevel letter designation and the quantum number hierarchy. For example, the $2s$ sublevel has only one orbital, and its quantum numbers are $n = 2$, $l = 0$, and $m_l = 0$. The $3p$ sublevel has three orbitals: one with $n = 3$, $l = 1$, and $m_l = -1$; another with $n = 3$, $l = 1$, and $m_l = 0$; and a third with $n = 3$, $l = 1$, and $m_l = +1$.

SAMPLE PROBLEM 7.6 Determining Sublevel Names and Orbital Quantum Numbers

Problem Give the name, magnetic quantum numbers, and number of orbitals for each sublevel with the given quantum numbers:

(a) $n = 3$, $l = 2$ **(b)** $n = 2$, $l = 0$ **(c)** $n = 5$, $l = 1$ **(d)** $n = 4$, $l = 3$

Plan To name the sublevel (subshell), we combine the n value and l letter designation. We know l, so we can find the possible m_l values, whose total number equals the number of orbitals.

Solution

	n	l	Sublevel Name	Possible m_l Values	No. of Orbitals
(a)	3	2	$3d$	$-2, -1, 0, +1, +2$	5
(b)	2	0	$2s$	0	1
(c)	5	1	$5p$	$-1, 0, +1$	3
(d)	4	3	$4f$	$-3, -2, -1, 0, +1, +2, +3$	7

Check Check the number of orbitals in each sublevel using

$$\text{No. of orbitals} = \text{no. of } m_l \text{ values} = 2l + 1$$

FOLLOW-UP PROBLEM 7.6 What are the n, l, and possible m_l values for the $2p$ and $5f$ sublevels?

SAMPLE PROBLEM 7.7 Identifying Incorrect Quantum Numbers

Problem What is wrong with each of the following quantum number designations and/or sublevel names?

	n	l	m_l	Name
(a)	1	1	0	$1p$
(b)	4	3	+1	$4d$
(c)	3	1	−2	$3p$

Solution (a) A sublevel with $n = 1$ can have only $l = 0$, not $l = 1$. The only possible sublevel name is $1s$.
(b) A sublevel with $l = 3$ is an f sublevel, not a d sublevel. The name should be $4f$.
(c) A sublevel with $l = 1$ can have only m_l of $-1, 0, +1$, not -2.
Check Check that l is always less than n, and m_l is always $\geq -l$ and $\leq +l$.

FOLLOW-UP PROBLEM 7.7 Supply the missing quantum numbers and sublevel names.

	n	l	m_l	Name
(a)	?	?	0	$4p$
(b)	2	1	0	?
(c)	3	2	−2	?
(d)	?	?	?	$2s$

Shapes of Atomic Orbitals

Each sublevel of the H atom consists of a set of orbitals with characteristic shapes. As you'll see in Chapter 8, orbitals for the other atoms have similar shapes.

The *s* Orbital An orbital with $l = 0$ has a *spherical* shape with the nucleus at its center and is called an ***s* orbital.** The H atom's ground state, for example, has the electron in the $1s$ orbital, and *the electron probability density is highest at the nucleus.* Figure 7.16A shows this fact graphically *(top),* and an electron density *relief map (inset)* depicts this curve in three dimensions. The quarter-section of an electron cloud representation *(middle)* has the darkest shading at the nucleus. On the other hand, the radial probability distribution plot *(bottom),* which represents the probability of finding the electron (that is, where the electron spends most of its time), is highest slightly out from the nucleus. Both plots fall off smoothly with distance.

The $2s$ orbital (Figure 7.16B) has two regions of higher electron density. The radial probability distribution (Figure 7.16B, *bottom*) of the more distant region

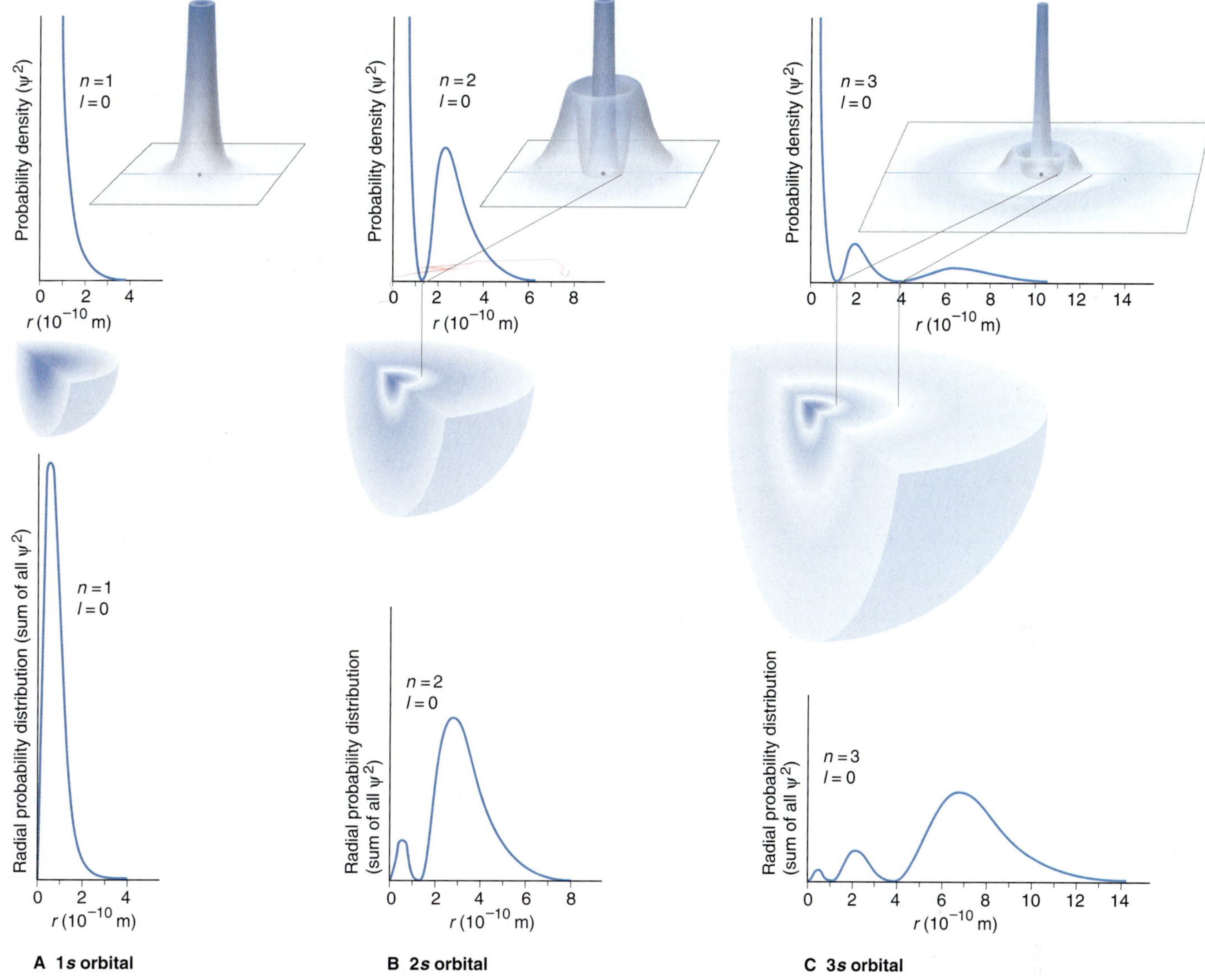

FIGURE 7.16 The 1s, 2s, and 3s orbitals. Information for each of the *s* orbitals is shown as a plot of probability density vs. distance (*top*, with the relief map (*inset*) showing the plot in three dimensions); as an electron cloud representation (*middle*), in which shading coincides with peaks in the plot above; and as a radial probability distribution (*bottom*) that shows where the electron spends its time. **A,** The 1*s* orbital. **B,** The 2*s* orbital. **C,** The 3*s* orbital. Nodes (regions of zero probability) appear in the 2*s* and 3*s* orbitals.

is *higher* than that of the closer one because the sum of its ψ^2 is taken over a much larger volume. Between the two regions is a spherical **node,** a shell-like region where the probability drops to zero ($\psi^2 = 0$ at the node, analogous to zero amplitude of a wave). Because the 2*s* orbital is larger than the 1*s*, an electron in the 2*s* spends more time *farther* from the nucleus than when it occupies the 1*s*.

The 3*s* orbital, shown in Figure 7.16C, has three regions of high electron density and two nodes. Here again, the highest radial probability is at the greatest distance from the nucleus because the sum of all ψ^2 is taken over a larger volume. This pattern of more nodes and higher probability with distance continues for *s* orbitals of higher *n* value. An *s* orbital has a spherical shape, so it can have only one orientation and, thus, only one value for the magnetic quantum number: for any *s* orbital, $m_l = 0$.

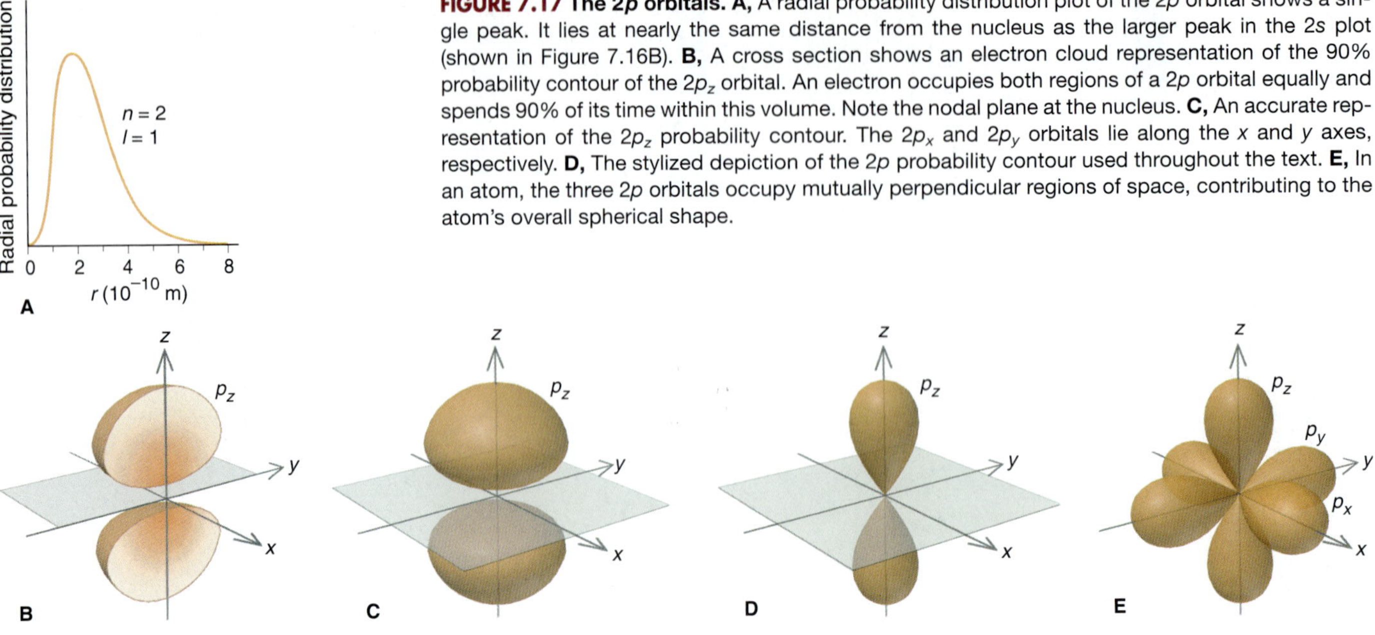

FIGURE 7.17 The 2*p* orbitals. A, A radial probability distribution plot of the 2*p* orbital shows a single peak. It lies at nearly the same distance from the nucleus as the larger peak in the 2*s* plot (shown in Figure 7.16B). **B,** A cross section shows an electron cloud representation of the 90% probability contour of the $2p_z$ orbital. An electron occupies both regions of a 2*p* orbital equally and spends 90% of its time within this volume. Note the nodal plane at the nucleus. **C,** An accurate representation of the $2p_z$ probability contour. The $2p_x$ and $2p_y$ orbitals lie along the *x* and *y* axes, respectively. **D,** The stylized depiction of the 2*p* probability contour used throughout the text. **E,** In an atom, the three 2*p* orbitals occupy mutually perpendicular regions of space, contributing to the atom's overall spherical shape.

The p Orbital An orbital with $l = 1$, called a ***p* orbital,** has two regions (lobes) of high probability, one on *either side* of the nucleus. Thus, as you can see in Figure 7.17, the *nucleus lies at the nodal plane* of this dumbbell-shaped orbital. The maximum value of l is $n - 1$, so only levels with $n = 2$ or higher can have a *p* orbital. Therefore, the lowest energy *p* orbital (the one closest to the nucleus) is the 2*p*. Keep in mind that *one p orbital consists of both lobes* and that the electron spends *equal* time in both. Similar to the pattern for *s* orbitals, a 3*p* orbital is larger than a 2*p* orbital, a 4*p* orbital is larger than a 3*p* orbital, and so forth.

Unlike an *s* orbital, each *p* orbital *does* have a specific orientation in space. The $l = 1$ value has three possible m_l values: -1, 0, and $+1$, which refer to three *mutually perpendicular p* orbitals. They are identical in size, shape, and energy, differing only in orientation. For convenience, we associate *p* orbitals with the *x*, *y*, and *z* axes (but there is no necessary relation between a spatial axis and a given m_l value): the p_x orbital lies along the *x* axis, the p_y along the *y* axis, and the p_z along the *z* axis.

The d Orbital An orbital with $l = 2$ is called a ***d* orbital.** There are five possible m_l values for the $l = 2$ value: -2, -1, 0, $+1$, and $+2$. Thus, a *d* orbital can have any one of five different orientations, as shown in Figure 7.18. Four of the five *d* orbitals have four lobes (a cloverleaf shape) prescribed by two mutually perpendicular nodal planes, with the nucleus lying at the junction of the lobes. Three of these orbitals lie in the mutually perpendicular *xy*, *xz*, and *yz* planes, with their lobes *between* the axes, and are called the d_{xy}, d_{xz}, and d_{yz} orbitals. A fourth, the $d_{x^2-y^2}$ orbital, also lies in the *xy* plane, but its lobes are directed *along* the axes. The fifth *d* orbital, the d_{z^2}, has a different shape: two major lobes lie along the *z* axis, and a donut-shaped region girdles the center. An electron associated with a given *d* orbital has equal probability of being in any of the orbital's lobes.

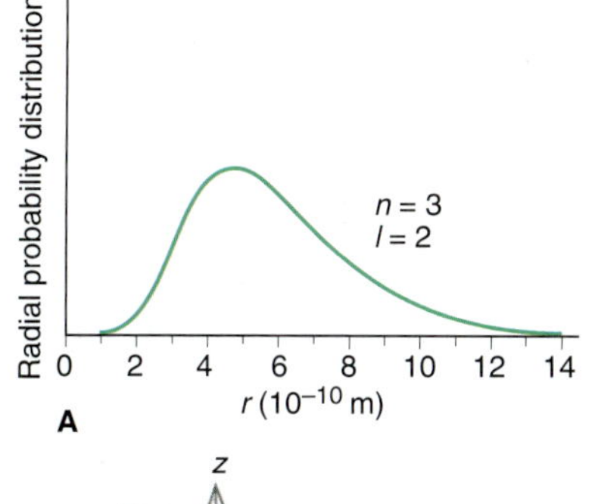

FIGURE 7.18 The 3*d* orbitals. A, A radial probability distribution plot. **B,** An electron cloud representation of the $3d_{yz}$ orbital in cross section. Note the mutually perpendicular nodal planes and the lobes lying *between* the axes. **C,** An accurate representation of the $3d_{yz}$ orbital probability contour. **D,** The stylized depiction of the $3d_{yz}$ orbital used throughout the text. **E,** The $3d_{xz}$ orbital. **F,** The $3d_{xy}$ orbital. **G,** The lobes of the $3d_{x^2-y^2}$ orbital lie *on* the *x* and *y* axes. **H,** The $3d_{z^2}$ orbital has two lobes and a central, donut-shaped region. **I,** A composite of the five 3*d* orbitals, which again contributes to an atom's overall spherical shape.

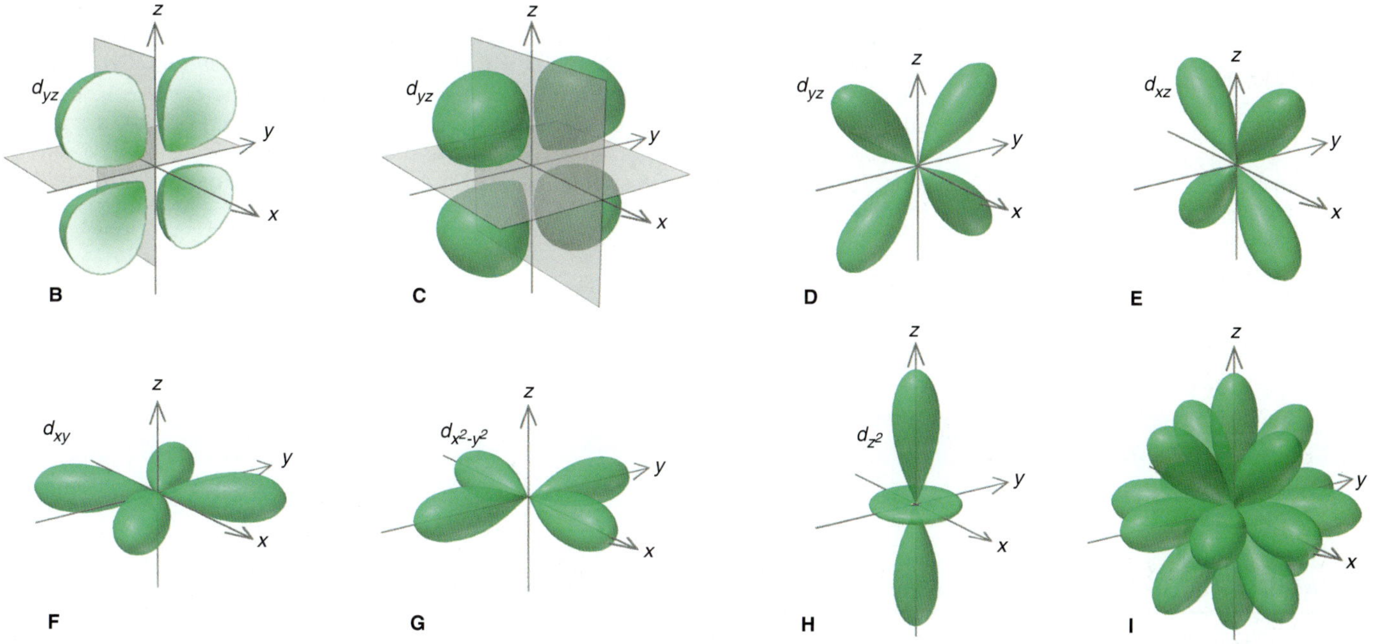

As we said for the *p* orbitals, an axis designation for a *d* orbital is *not* associated with a given m_l value. In keeping with the quantum number rules, a *d* orbital ($l = 2$) must have a principal quantum number of $n = 3$ or greater. The 4*d* orbitals extend farther from the nucleus than the 3*d* orbitals, and the 5*d* orbitals extend still farther.

Orbitals with Higher *l* Values Orbitals with $l = 3$ are *f* orbitals and must have a principal quantum number of at least $n = 4$. There are seven *f* orbitals ($2l + 1 = 7$), each with a complex, multilobed shape; Figure 7.19 shows one of them. Orbitals with $l = 4$ are *g* orbitals, but we will not discuss them further because they play no known role in chemical bonding.

The Special Case of the Hydrogen Atom

*The energy state of the H atom depends only on the principal quantum number **n**.* When an electron occupies an orbital with a higher *n* value, it occurs (on average) farther from the nucleus, so the atom is higher in energy. But the H atom is a special case because it has only one electron. In other words, *for the H atom only,* all four $n = 2$ orbitals (one 2*s* and three 2*p*) have the same energy, all nine $n = 3$ orbitals (one 3*s*, three 3*p*, and five 3*d*) have the same energy (Figure 7.20, next page), and so forth. Of course, atoms of all other elements have more than

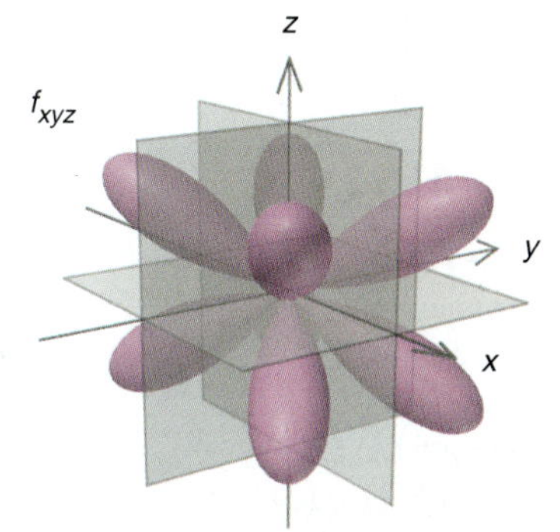

FIGURE 7.19 One of the seven possible 4*f* orbitals. The $4f_{xyz}$ orbital has eight lobes and three nodal planes. The other six 4*f* orbitals also have multilobed contours.

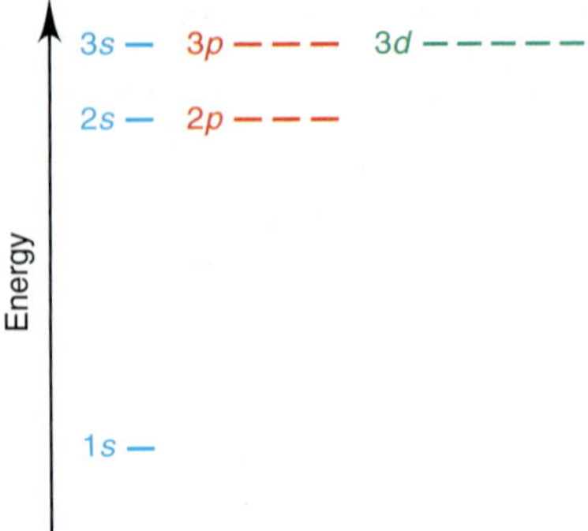

FIGURE 7.20 The energy levels in the H atom. The H atom is the only atom in which the energy level depends *only* on the *n* value of the sublevels. For example, the 2*s* and the three 2*p* sublevels (shown as short lines) all have the same energy.

one electron. As you'll see in Chapter 8, the additional attractions and repulsions that result make the energy states of all other atoms depend on *both* the *n* and *l* values of the occupied orbitals.

SECTION 7.4 SUMMARY

The electron's wave function (ψ, atomic orbital) is a mathematical description of the electron's wavelike behavior in an atom. Each wave function is associated with one of the atom's allowed energy states. • The probability density of finding the electron at a particular location is represented by ψ^2. An electron density diagram and a radial probability distribution plot show how the electron occupies the space near the nucleus for a particular energy level. • Three features of an atomic orbital are described by quantum numbers: size (n), shape (l), and orientation (m_l). Orbitals with the same n and l values constitute a sublevel; sublevels with the same n value constitute an energy level. • A sublevel with $l = 0$ has a spherical (*s*) orbital; a sublevel with $l = 1$ has three, two-lobed (*p*) orbitals; and a sublevel with $l = 2$ has five, multilobed (*d*) orbitals. • In the special case of the H atom, the energy levels depend on the *n* value only.

CHAPTER REVIEW GUIDE

The following sections provide many aids to help you study this chapter. (Numbers in parentheses refer to pages, unless noted otherwise.)

• LEARNING OBJECTIVES *These are concepts and skills you should know after studying this chapter.*

Related section (§), sample problem (SP), and end-of-chapter problem (EP) numbers are listed in parentheses.

1. Describe the relationships among frequency, wavelength, and energy of light, and know the meaning of amplitude; have a general understanding of the electromagnetic spectrum (§ 7.1) (SPs 7.1, 7.2) (EPs 7.1, 7.2, 7.6–7.14)
2. Understand how particles and waves differ and how the work of Planck (quantization of energy) and Einstein (photon theory) changed thinking about it (§ 7.1) (EPs 7.3–7.5)
3. Explain the Bohr theory and the importance of discrete atomic energy levels (§ 7.2) (SP 7.3) (EPs 7.15–7.28)
4. Describe the wave-particle duality of matter and energy and the theories and experiments that led to it (particle wavelength, electron diffraction, photon momentum, uncertainty principle) (§ 7.3) (SP 7.4) (EPs 7.29–7.36)
5. Distinguish between ψ (wave function) and ψ^2 (probability density); understand the meaning of electron density diagrams and radial probability distribution plots; describe the hierarchy of quantum numbers, the hierarchy of levels, sublevels, and orbitals, and the shapes and nodes of *s, p,* and *d* orbitals; and determine quantum numbers and sublevel designations (§ 7.4) (SPs 7.5–7.7) (EPs 7.37–7.49)

• KEY TERMS *These important terms appear in boldface in the chapter and are defined again in the Glossary.*

Section 7.1
electromagnetic radiation (215)
frequency (ν) (215)
wavelength (λ) (215)
speed of light (c) (216)
amplitude (216)
electromagnetic spectrum (216)
infrared (IR) (216)
ultraviolet (UV) (216)
refraction (217)
diffraction (217)
quantum number (219)
Planck's constant (h) (219)
quantum (219)
photoelectric effect (219)
photon (220)

Section 7.2
line spectrum (221)
stationary state (223)
ground state (223)
excited state (223)
spectrophotometry (227)
emission spectrum (227)
absorption spectrum (227)

Section 7.3
de Broglie wavelength (229)
wave-particle duality (230)
uncertainty principle (231)

Section 7.4
quantum mechanics (231)
Schrödinger equation (231)
wave function (atomic orbital) (232)
electron density diagram (232)
electron cloud (232)
radial probability distribution plot (232)
probability contour (233)
principal quantum number (n) (234)
angular momentum quantum number (l) (234)
magnetic quantum number (m_l) (234)
level (shell) (235)
sublevel (subshell) (235)
s orbital (236)
node (237)
p orbital (238)
d orbital (238)

KEY EQUATIONS AND RELATIONSHIPS *Numbered and screened concepts are listed for you to refer to or memorize.*

7.1 Relating the speed of light to its frequency and wavelength (216):

$$c = \nu \times \lambda$$

7.2 Determining the smallest change in an atom's energy (219):

$$\Delta E = h\nu$$

7.3 Calculating the wavelength of any line in the H atom spectrum (Rydberg equation) (221):

$$\frac{1}{\lambda} = R\left(\frac{1}{n_1^2} - \frac{1}{n_2^2}\right)$$

where n is a positive integer and $n_2 > n_1$

7.4 Finding the difference between two energy levels in the H atom (225):

$$\Delta E = E_{\text{final}} - E_{\text{initial}} = -2.18\times10^{-18}\text{ J}\left(\frac{1}{n_{\text{final}}^2} - \frac{1}{n_{\text{initial}}^2}\right)$$

7.5 Calculating the wavelength of any moving particle (de Broglie wavelength) (228):

$$\lambda = \frac{h}{mu}$$

7.6 Finding the uncertainty in position or speed of a particle (Heisenberg uncertainty principle) (231):

$$\Delta x \cdot m\Delta u \geq \frac{h}{4\pi}$$

BRIEF SOLUTIONS TO *FOLLOW-UP PROBLEMS* *Compare your own solutions to these calculation steps and answers.*

7.1 $\lambda\text{ (nm)} = \dfrac{3.00\times10^8\text{ m/s}}{7.23\times10^{14}\text{ s}^{-1}} \times \dfrac{10^9\text{ nm}}{1\text{ m}} = 415\text{ nm}$

$\lambda\text{ (Å)} = 415\text{ nm} \times \dfrac{10\text{ Å}}{1\text{ nm}} = 4150\text{ Å}$

7.2 UV: $E = hc/\lambda$

$$= \frac{(6.626\times10^{-34}\text{ J·s})(3.00\times10^8\text{ m/s})}{1\times10^{-8}\text{ m}} = 2\times10^{-17}\text{ J}$$

Visible: $E = 4\times10^{-19}$ J; IR: $E = 2\times10^{-21}$ J

As λ increases, E decreases.

7.3 With $n_{\text{final}} = 3$ for an IR photon:

$$\Delta E = -2.18\times10^{-18}\text{ J}\left(\frac{1}{n_{\text{final}}^2} - \frac{1}{n_{\text{initial}}^2}\right)$$

$$= -2.18\times10^{-18}\text{ J}\left(\frac{1}{3^2} - \frac{1}{6^2}\right)$$

$$= -2.18\times10^{-18}\text{ J}\left(\frac{1}{9} - \frac{1}{36}\right) = -1.82\times10^{-19}\text{ J}$$

$$\lambda = \frac{hc}{|\Delta E|} = \frac{(6.626\times10^{-34}\text{ J·s})(3.00\times10^8\text{ m/s})}{1.82\times10^{-19}\text{ J}} \times \frac{1\text{ Å}}{10^{-10}\text{ m}}$$

$$= 1.09\times10^4\text{ Å}$$

7.4 $u = \dfrac{h}{m\lambda} = \dfrac{6.626\times10^{-34}\text{ kg·m}^2\text{/s}}{(9.11\times10^{-31}\text{ kg})\left(100\text{ nm} \times \dfrac{1\text{ m}}{10^9\text{ nm}}\right)}$

$= 7.27\times10^3$ m/s

7.5 $n = 4$, so $l = 0, 1, 2, 3$. In addition to the m_l values in Sample Problem 7.5, we have those for $l = 3$:

$$m_l = -3, -2, -1, 0, +1, +2, +3$$

7.6 For $2p$: $n = 2, l = 1, m_l = -1, 0, +1$

For $5f$: $n = 5, l = 3, m_l = -3, -2, -1, 0, +1, +2, +3$

7.7 (a) $n = 4, l = 1$; (b) name is $2p$; (c) name is $3d$; (d) $n = 2, l = 0, m_l = 0$

PROBLEMS

*Problems with **colored** numbers are answered in Appendix E. Sections match the text and provide the numbers of relevant sample problems. Bracketed problems are grouped in pairs (indicated by a short rule) that cover the same concept. Comprehensive Problems are based on material from any section or previous chapter.*

The Nature of Light

(Sample Problems 7.1 and 7.2)

7.1 In what ways are microwave and ultraviolet radiation the same? In what ways are they different?

7.2 Consider the following types of electromagnetic radiation:

(1) microwave (2) ultraviolet (3) radio waves
(4) infrared (5) x-ray (6) visible

(a) Arrange them in order of increasing wavelength.
(b) Arrange them in order of increasing frequency.
(c) Arrange them in order of increasing energy.

7.3 In the mid-17th century, Isaac Newton proposed that light existed as a stream of particles, and the wave-particle debate continued for over 250 years until Planck and Einstein presented their revolutionary ideas. Give two pieces of evidence for the wave model and two for the particle model.

7.4 What new idea about energy did Planck use to explain blackbody radiation?

7.5 What new idea about light did Einstein use to explain the photoelectric effect? Why does the photoelectric effect exhibit a threshold frequency? Why does it *not* exhibit a time lag?

7.6 Portions of electromagnetic waves A, B, and C are represented below (not drawn to scale):

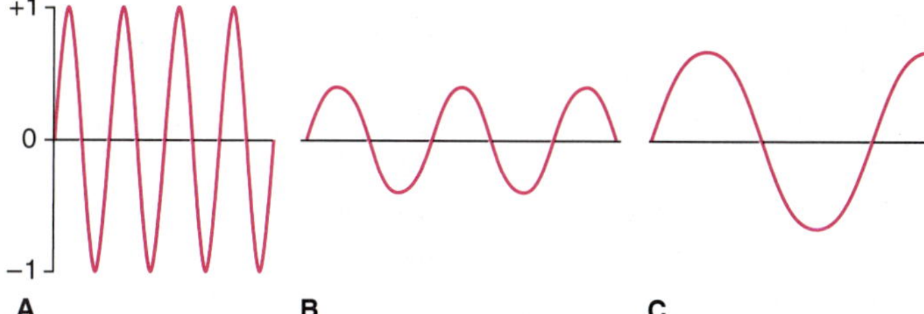

Rank them in order of (a) increasing frequency; (b) increasing energy; (c) increasing amplitude. (d) If wave B just barely fails to cause a current when shining on a metal, is wave A or C more likely to do so? (e) If wave B represents visible radiation, is wave A or C more likely to be IR radiation?

7.7 An AM station broadcasts rock music at "950 on your radio dial." Units for AM frequencies are given in kilohertz (kHz). Find the wavelength of the station's radio waves in meters (m), nanometers (nm), and angstroms (Å).

7.8 An FM station broadcasts classical music at 93.5 MHz (megahertz, or 10^6 Hz). Find the wavelength (in m, nm, and Å) of these radio waves.

7.9 A radio wave has a frequency of 3.6×10^{10} Hz. What is the energy (in J) of one photon of this radiation?

7.10 An x-ray has a wavelength of 1.3 Å. Calculate the energy (in J) of one photon of this radiation.

7.11 Rank the photons in terms of increasing energy: (a) blue ($\lambda = 453$ nm); (b) red ($\lambda = 660$ nm); (c) yellow ($\lambda = 595$ nm).

7.12 Rank the photons in terms of decreasing energy: (a) IR ($\nu = 6.5\times10^{13}\ s^{-1}$); (b) microwave ($\nu = 9.8\times10^{11}\ s^{-1}$); (c) UV ($\nu = 8.0\times10^{15}\ s^{-1}$).

7.13 Cobalt-60 is a radioactive isotope used to treat cancers of the brain and other tissues. A gamma ray emitted by an atom of this isotope has an energy of 1.33 MeV (million electron volts; 1 eV = 1.602×10^{-19} J). What is the frequency (in Hz) and the wavelength (in m) of this gamma ray?

7.14 (a) The first step in the formation of ozone in the upper atmosphere occurs when oxygen molecules absorb UV radiation of wavelengths $\leq$242 nm. Calculate the frequency and energy of the least energetic of these photons.
(b) Ozone absorbs light having wavelengths of 2200 to 2900 Å, thus protecting organisms on Earth's surface from this high-energy UV radiation. What are the frequency and energy of the most energetic of these photons?

Atomic Spectra

(Sample Problem 7.3)

7.15 How is n_1 in the Rydberg equation (Equation 7.3) related to the quantum number n in the Bohr model?

7.16 Distinguish between an absorption spectrum and an emission spectrum. With which did Bohr work?

7.17 Which of these electron transitions correspond to absorption of energy and which to emission?
(a) $n = 2$ to $n = 4$ (b) $n = 3$ to $n = 1$
(c) $n = 5$ to $n = 2$ (d) $n = 3$ to $n = 4$

7.18 Why could the Bohr model not predict line spectra for atoms other than hydrogen?

7.19 Use the Rydberg equation (Equation 7.3) to calculate the wavelength (in nm) of the photon emitted when a hydrogen atom undergoes a transition from $n = 5$ to $n = 2$.

7.20 Use the Rydberg equation to calculate the wavelength (in Å) of the photon absorbed when a hydrogen atom undergoes a transition from $n = 1$ to $n = 3$.

7.21 Calculate the energy difference (ΔE) for the transition in Problem 7.19 for 1 mol of H atoms.

7.22 Calculate the energy difference (ΔE) for the transition in Problem 7.20 for 1 mol of H atoms.

7.23 Arrange the following H atom electron transitions in order of *increasing* frequency of the photon absorbed or emitted:
(a) $n = 2$ to $n = 4$ (b) $n = 2$ to $n = 1$
(c) $n = 2$ to $n = 5$ (d) $n = 4$ to $n = 3$

7.24 Arrange the following H atom electron transitions in order of *decreasing* wavelength of the photon absorbed or emitted:
(a) $n = 2$ to $n = \infty$ (b) $n = 4$ to $n = 20$
(c) $n = 3$ to $n = 10$ (d) $n = 2$ to $n = 1$

7.25 The electron in a ground-state H atom absorbs a photon of wavelength 97.20 nm. To what energy level does the electron move?

7.26 An electron in the $n = 5$ level of an H atom emits a photon of wavelength 1281 nm. To what energy level does the electron move?

7.27 In addition to continuous radiation, fluorescent lamps emit sharp lines in the visible region from a mercury discharge within the tube. Much of this light has a wavelength of 436 nm. What is the energy (in J) of one photon of this light?

7.28 A Bohr-model representation of the H atom is shown below with several electron transitions depicted by arrows:

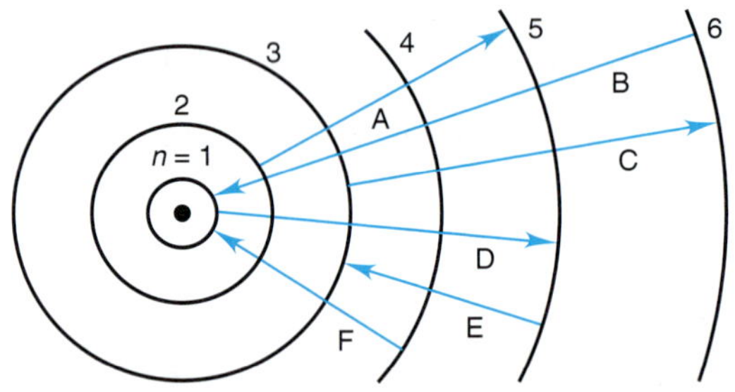

(a) Which transitions are absorptions and which are emissions? (b) Rank the emissions in terms of increasing energy. (c) Rank the absorptions in terms of increasing wavelength of light emitted.

The Wave-Particle Duality of Matter and Energy

(Sample Problem 7.4)

7.29 If particles have wavelike motion, why don't we observe that motion in the macroscopic world?

7.30 Why can't we overcome the uncertainty predicted by Heisenberg's principle by building more precise devices to reduce the error in measurements below the $h/4\pi$ limit?

7.31 A 232-lb fullback runs the 40-yd dash at a speed of 19.8 ± 0.1 mi/h. What is his de Broglie wavelength (in meters)?

7.32 An alpha particle (mass = 6.6×10^{-24} g) emitted by radium travels at $3.4\times10^7 \pm 0.1\times10^7$ mi/h. What is its de Broglie wavelength (in meters)?

7.33 How fast must a 56.5-g tennis ball travel in order to have a de Broglie wavelength that is equal to that of a photon of green light (5400 Å)?

7.34 How fast must a 142-g baseball travel in order to have a de Broglie wavelength that is equal to that of an x-ray photon with $\lambda = 100.$ pm?

7.35 A sodium flame has a characteristic yellow color due to emissions of wavelength 589 nm. What is the mass equivalence of one photon of this wavelength (1 J = 1 kg·m^2/s^2)?

7.36 A lithium flame has a characteristic red color due to emissions of wavelength 671 nm. What is the mass equivalence of 1 mol of photons of this wavelength (1 J = 1 kg·m^2/s^2)?

The Quantum-Mechanical Model of the Atom

(Sample Problems 7.5 to 7.7)

7.37 What physical meaning is attributed to ψ^2, the square of the wave function?

7.38 Explain in your own words what the "electron density" in a particular tiny volume of space means.

7.39 What feature of an orbital is related to each of these quantum numbers: (a) principal quantum number (n); (b) angular momentum quantum number (l); (c) magnetic quantum number (m_l)?

7.40 How many orbitals in an atom can have each of the following designations: (a) 1s; (b) 4d; (c) 3p; (d) $n = 3$?

7.41 How many orbitals in an atom can have each of the following designations: (a) 5f; (b) 4p; (c) 5d; (d) $n = 2$?

7.42 Give all possible m_l values for orbitals that have each of the following: (a) $l = 2$; (b) $n = 1$; (c) $n = 4, l = 3$.

7.43 Give all possible m_l values for orbitals that have each of the following: (a) $l = 3$; (b) $n = 2$; (c) $n = 6, l = 1$.

7.44 For each of the following, give the sublevel designation, the allowable m_l values, and the number of orbitals:
(a) $n = 4, l = 2$ (b) $n = 5, l = 1$ (c) $n = 6, l = 3$

7.45 For each of the following, give the sublevel designation, the allowable m_l values, and the number of orbitals:
(a) $n = 2, l = 0$ (b) $n = 3, l = 2$ (c) $n = 5, l = 1$

7.46 For each of the following sublevels, give the n and l values and the number of orbitals: (a) 5s; (b) 3p; (c) 4f.

7.47 For each of the following sublevels, give the n and l values and the number of orbitals: (a) 6g; (b) 4s; (c) 3d.

7.48 Are the following quantum number combinations allowed? If not, show two ways to correct them:
(a) $n = 2$; $l = 0$; $m_l = -1$ (b) $n = 4$; $l = 3$; $m_l = -1$
(c) $n = 3$; $l = 1$; $m_l = 0$ (d) $n = 5$; $l = 2$; $m_l = +3$

7.49 Are the following quantum number combinations allowed? If not, show two ways to correct them:
(a) $n = 1$; $l = 0$; $m_l = 0$ (b) $n = 2$; $l = 2$; $m_l = +1$
(c) $n = 7$; $l = 1$; $m_l = +2$ (d) $n = 3$; $l = 1$; $m_l = -2$

Comprehensive Problems

Problems with an asterisk (*) are more challenging.

7.50 The photoelectric effect is illustrated in a plot of the kinetic energies of electrons ejected from the surface of potassium metal or silver metal at different frequencies of incident light.

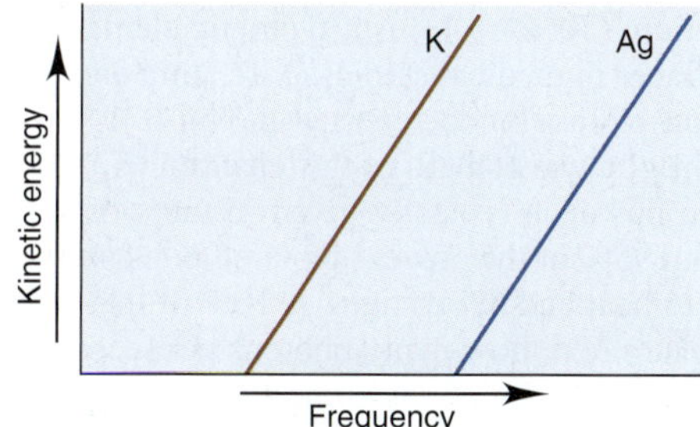

(a) Why don't the lines begin at the origin?
(b) Why don't the lines begin at the same point?
(c) From which metal will light of a shorter wavelength eject an electron?
(d) Why are the slopes of the lines equal?

7.51 The human eye is a complex sensing device for visible light. The optic nerve needs a minimum of 2.0×10^{-17} J of energy to trigger a series of impulses that eventually reach the brain.
(a) How many photons of red light (700. nm) are needed?
(b) How many photons of blue light (475 nm)?

7.52 One reason carbon monoxide (CO) is toxic is that it binds to the blood protein hemoglobin more strongly than oxygen does. The bond between hemoglobin and CO absorbs radiation of 1953 cm^{-1}. (The units are the reciprocal of the wavelength in centimeters.) Calculate the wavelength (in nm and Å) and the frequency (in Hz) of the absorbed radiation.

7.53 A metal ion M^{n+} has a single electron. The highest energy line in its emission spectrum occurs at a frequency of 2.961×10^{16} Hz. Identify the ion.

7.54 TV and radio stations transmit in specific frequency bands of the radio region of the electromagnetic spectrum.
(a) TV channels 2 to 13 (VHF) broadcast signals between the frequencies of 59.5 and 215.8 MHz, whereas FM radio stations broadcast signals with wavelengths between 2.78 and 3.41 m. Do these bands of signals overlap?
(b) AM radio signals have frequencies between 550 and 1600 kHz. Which has a broader transmission band, AM or FM?

7.55 To explain the threshold frequency in the photoelectric effect, Einstein reasoned that the absorbed photon must have the minimum energy required to dislodge an electron from the metal surface. This energy is called the *work function* (ϕ) of that metal. What is the longest wavelength of radiation (in nm) that could cause the photoelectric effect in each of these metals: (a) calcium, $\phi = 4.60\times10^{-19}$ J; (b) titanium, $\phi = 6.94\times10^{-19}$ J; (c) sodium, $\phi = 4.41\times10^{-19}$ J?

7.56 You have three metal samples—A, B, and C—that are tantalum (Ta), barium (Ba), and tungsten (W), but you don't know which is which. Metal A emits electrons in response to visible light; metals B and C require UV light. (a) Identify metal A, and find the longest wavelength that removes an electron. (b) What range of wavelengths would distinguish B and C? [The work functions are Ta (6.81×10^{-19} J), Ba (4.30×10^{-19} J), and W (7.16×10^{-19} J); work function is explained in Problem 7.55.]

7.57 A laser (*l*ight *a*mplification by *s*timulated *e*mission of *r*adiation) provides a coherent (in-phase) nearly monochromatic source of high-intensity light. Lasers are used in eye surgery, CD/DVD players, basic research, etc. Some modern dye lasers can be "tuned" to emit a desired wavelength. Fill in the blanks in the following table of the properties of some common lasers:

Type	λ (nm)	ν (s^{-1})	E (J)	Color
He-Ne	632.8	?	?	?
Ar	?	6.148×10^{14}	?	?
Ar-Kr	?	?	3.499×10^{-19}	?
Dye	663.7	?	?	?

7.58 As space exploration increases, means of communication with humans and probes on other planets are being developed. (a) How much time (in s) does it take for a radio wave of frequency 8.93×10^{7} s^{-1} to reach Mars, which is 8.1×10^{7} km from Earth? (b) If it takes this radiation 1.2 s to reach the Moon, how far (in m) is the Moon from Earth?

* **7.59** A ground-state H atom absorbs a photon of wavelength 94.91 nm, and its electron attains a higher energy level. The atom then emits two photons: one of wavelength 1281 nm to reach an intermediate level, and a second to reach the ground state.
(a) What higher level did the electron reach?
(b) What intermediate level did the electron reach?
(c) What was the wavelength of the second photon emitted?

7.60 Use the relative size of the 3s orbital represented on the next page to answer the following questions about orbitals A–D.

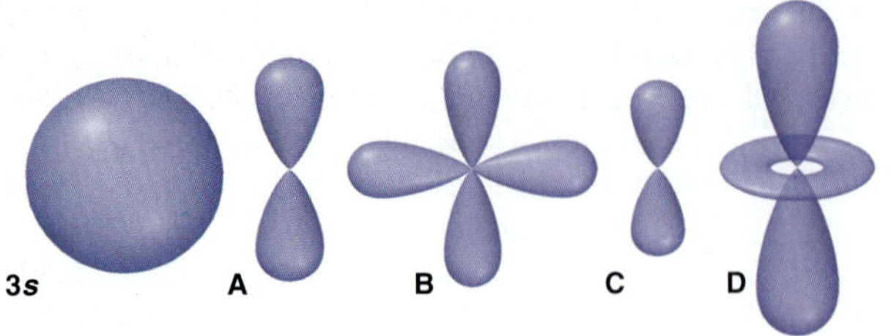

(a) Which orbital has the highest value of n? (b) Which orbital(s) have a value of $l = 1$? $l = 2$? (c) How many other orbitals with the same value of n have the same shape as orbital B? Orbital C? (d) Which orbital has the highest energy? Lowest energy?

*** 7.61** Why do the spaces between spectral lines within a series decrease as the wavelength becomes shorter?

7.62 Enormous numbers of microwave photons are needed to warm macroscopic samples of matter. A portion of soup containing 252 g of water is heated in a microwave oven from 20.°C to 98°C, with radiation of wavelength 1.55×10^{-2} m. How many photons are absorbed by the water in the soup?

*** 7.63** The quantum-mechanical treatment of the hydrogen atom gives an expression for the wave function, ψ, of the 1s orbital:

$$\psi = \frac{1}{\sqrt{\pi}}\left(\frac{1}{a_0}\right)^{3/2} e^{-r/a_0}$$

where r is the distance from the nucleus and a_0 is 52.92 pm. The electron probability density is the probability of finding the electron in a tiny volume at distance r from the nucleus and is proportional to ψ^2. The radial probability distribution is the total probability of finding the electron at all points at distance r from the nucleus and is proportional to $4\pi r^2\psi^2$. Calculate the values (to three significant figures) of ψ, ψ^2, and $4\pi r^2\psi^2$ to fill in the following table, and sketch plots of these quantities versus r. Compare the latter two plots with those in Figure 7.16A:

r (pm)	ψ (pm$^{-3/2}$)	ψ^2 (pm^{-3})	$4\pi r^2\psi^2$ (pm^{-1})
0			
50			
100			
200			

7.64 Photoelectron spectroscopy applies the principle of the photoelectric effect to study orbital energies of atoms and molecules. High-energy radiation (usually UV or x-ray) is absorbed by a sample and an electron is ejected. By knowing the energy of the radiation and measuring the energy of the electron lost, the orbital energy can be calculated. The following energy differences were determined for several electron transitions:

$$\Delta E_{2\longrightarrow 1} = 4.098\times10^{-17}\text{ J}$$
$$\Delta E_{3\longrightarrow 1} = 4.854\times10^{-17}\text{ J}$$
$$\Delta E_{5\longrightarrow 1} = 5.242\times10^{-17}\text{ J}$$
$$\Delta E_{4\longrightarrow 2} = 1.024\times10^{-17}\text{ J}$$

Calculate the energy change and the wavelength of a photon emitted in the following transitions.

(a) Level 3 $\longrightarrow$ 2 (b) Level 4 $\longrightarrow$ 1 (c) Level 5 $\longrightarrow$ 4

7.65 An electron microscope focuses electrons through magnetic lenses to observe objects at higher magnification than is possible with a light microscope. For any microscope, the smallest object that can be observed is one-half the wavelength of the radiation used. Thus, for example, the smallest object that can be observed with light of 400 nm is 2×10^{-7} m. (a) What is the smallest object observable with an electron microscope using electrons moving at 5.5×10^4 m/s? (b) At 3.0×10^7 m/s?

7.66 In a typical fireworks device, the heat of the reaction between a strong oxidizing agent, such as $KClO_4$, and an organic compound excites certain salts, which emit specific colors. Strontium salts have an intense emission at 641 nm, and barium salts have one at 493 nm. (a) What colors do these emissions produce? (b) What is the energy (in kJ) of these emissions for 5.00 g each of the chloride salts of Sr and Ba? (Assume that all the heat released is converted to light emitted.)

*** 7.67** Atomic hydrogen produces well-known series of spectral lines in several regions of the electromagnetic spectrum. Each series fits the Rydberg equation with its own particular n_1 value. Calculate the value of n_1 (by trial and error if necessary) that would produce a series of lines in which:

(a) The *highest* energy line has a wavelength of 3282 nm.

(b) The *lowest* energy line has a wavelength of 7460 nm.

7.68 Fish-liver oil is a good source of vitamin A, which is measured spectrophotometrically at a wavelength of 329 nm.

(a) Suggest a reason for using this wavelength.

(b) In what region of the spectrum does this wavelength lie?

(c) When 0.1232 g of fish-liver oil is dissolved in 500. mL of solvent, the absorbance is 0.724 units. When 1.67×10^{-3} g of vitamin A is dissolved in 250. mL of solvent, the absorbance is 1.018 units. Calculate the vitamin A concentration in the fish-liver oil.

7.69 Many calculators use photocells to provide their energy. Find the maximum wavelength needed to remove an electron from silver ($\phi = 7.59\times10^{-19}$ J). Is silver a good choice for a photocell that uses visible light?

7.70 In a game of "Clue," Ms. White is killed in the conservatory. You have a device in each room to help you find the murderer—a spectrometer that emits the entire visible spectrum to indicate who is in that room. For example, if someone wearing yellow is in a room, light at 580 nm is reflected. The suspects are Col. Mustard, Prof. Plum, Mr. Green, Ms. Peacock (blue), and Ms. Scarlet. At the time of the murder, the spectrometer in the dining room recorded a reflection at 520 nm, those in the lounge and study recorded reflections of lower frequencies, and the one in the library recorded a reflection of the shortest possible wavelength. Who killed Ms. White? Explain.

7.71 Technetium (Tc; $Z = 43$) is a synthetic element used as a radioactive tracer in medical studies. A Tc atom emits a beta particle (electron) with a kinetic energy (E_k) of 4.71×10^{-15} J. What is the de Broglie wavelength of this electron ($E_k = \frac{1}{2}mv^2$)?

7.72 Electric power is typically given in units of watts (1 W = 1 J/s). About 95% of the power output of an incandescent bulb is converted to heat and 5% to light. If 10% of that light shines on your chemistry text, how many photons per second shine on the book from a 75-W bulb? (Assume the photons have a wavelength of 550 nm.)

*** 7.73** The net change in the multistep biochemical process of photosynthesis is that CO_2 and H_2O form glucose ($C_6H_{12}O_6$) and O_2. Chlorophyll absorbs light in the 600 to 700 nm region. (a) Write a balanced thermochemical equation for formation of 1.00 mol of glucose. (b) What is the minimum number of photons with $\lambda = 680.$ nm needed to prepare 1.00 mol of glucose?

7.74 In contrast to the situation for an electron, calculate the uncertainty in the position of a 142-g baseball, pitched during the 2007 Red Sox–Rockies World Series, that moved at 100.0 mi/h ± 1.00%.

Electron Configuration and Chemical Periodicity 8

Key Principles
to focus on while studying this chapter

- Arranging the elements by atomic number exhibits a periodic recurrence of similar properties *(periodic law) (Section 8.1)*.
- The *electron configuration* of an atom, the distribution of electrons into its energy levels and sublevels, ultimately determines the behavior of the element *(Section 8.2)*.
- Three new features are important for atoms with more than one electron (all elements except hydrogen): a fourth quantum number (m_s) specifies *electron spin;* an orbital holds no more than two electrons *(exclusion principle);* and interactions between electrons and between nucleus and electrons (*shielding* and *penetration*) cause energy levels to split into sublevels of different energy *(Section 8.2)*.
- The periodic table is "*built up*" by adding one electron (and one proton and one or more neutrons) to each preceding atom. This method results in vertical groups of elements having *identical outer electron configurations* and, thus, similar behavior. The few seemingly *anomalous configurations* are explained by factors that affect the order of sublevel energies *(Section 8.3)*.
- Three atomic properties—*atomic size, ionization energy* (energy involved in removing an electron from an atom), and *electron affinity* (energy involved in adding an electron to an atom)—exhibit recurring trends throughout the periodic table *(Section 8.4)*.
- These atomic properties have a profound effect on many macroscopic properties, including *metallic behavior, acid-base behavior of oxides, ionic behavior,* and *magnetic behavior* of the elements and their compounds *(Section 8.5)*.

Patterns in Nature *Recurring patterns, such as tidal action, occur at the macroscopic level and, as you'll see in this chapter, at the atomic level as well.*

Outline

Concepts & Skills to Review before studying this chapter

- format of the periodic table (Section 2.6)
- characteristics of metals and nonmetals (Section 2.6)
- application of Coulomb's law to electrostatic attraction (Section 2.7)
- characteristics of acids and bases (Section 4.4)
- rules for assigning quantum numbers (Section 7.4)

In Chapter 7, you saw the outpouring of scientific creativity by early 20th-century physicists that led to a new understanding of matter and energy, which in turn led to the quantum-mechanical model of the atom. But you can be sure that late 19th-century chemists were not sitting idly by, waiting for their colleagues in physics to develop that model. They were exploring the nature of electrolytes, establishing the kinetic-molecular theory, and developing chemical thermodynamics. The fields of organic chemistry and biochemistry were born, as were the fertilizer, explosives, glassmaking, soapmaking, bleaching, and dyestuff industries. And, for the first time, chemistry became a university subject in Europe and America. Superimposed on this activity was the accumulation of an enormous body of facts about the elements, which became organized into the periodic table.

The goal of this chapter is to show how the organization of the table, condensed from countless hours of laboratory work, was explained perfectly by the new quantum-mechanical atomic model. This model answers one of the central questions in chemistry: why do the elements behave as they do? Or, rephrasing the question to fit the main topic of this chapter: how does the **electron configuration** of an element—*the distribution of electrons within the orbitals of its atoms*—relate to its chemical and physical properties?

8.1 DEVELOPMENT OF THE PERIODIC TABLE

An essential requirement for the amazing growth in theoretical and practical chemistry in the second half of the 19th century was the ability to organize the facts known about element behavior. In Chapter 2, you saw that the most successful organizing scheme was proposed by the Russian chemist Dmitri Mendeleev. In 1870, he arranged the 65 elements then known into a *periodic table* of rows and columns and summarized their behavior in the **periodic law:** when arranged by atomic mass, the elements exhibit a periodic recurrence of similar properties. It is a curious quirk of history that Mendeleev and the German chemist Julius Lothar Meyer arrived at virtually the same organization simultaneously, yet independently. Mendeleev focused on chemical properties and Meyer on physical properties. The greater credit has gone to Mendeleev because he was able to *predict* the properties of several as-yet-undiscovered elements, for which he had left blank spaces in his table.

Today's periodic table, which appears on the inside front cover of the text, resembles Mendeleev's in most details, although it includes 51 elements that were unknown in 1870. The only substantive change is that the elements are now arranged in order of *atomic number* (number of protons) rather than atomic mass.

SECTION 8.1 SUMMARY

The periodic law gave rise to the periodic table, in which the elements display similar properties when they are arranged by atomic number into rows and columns.

8.2 CHARACTERISTICS OF MANY-ELECTRON ATOMS

Like the Bohr model, the Schrödinger equation does not give *exact* solutions for many-electron atoms. However, unlike the Bohr model, the Schrödinger equation gives very good *approximate* solutions. These solutions show that the atomic orbitals of many-electron atoms resemble those of the H atom, which means we can use the same quantum numbers that we used for the H atom to describe the orbitals of other atoms.

Nevertheless, the existence of more than one electron in an atom requires that we consider three features that were not relevant in the case of hydrogen:

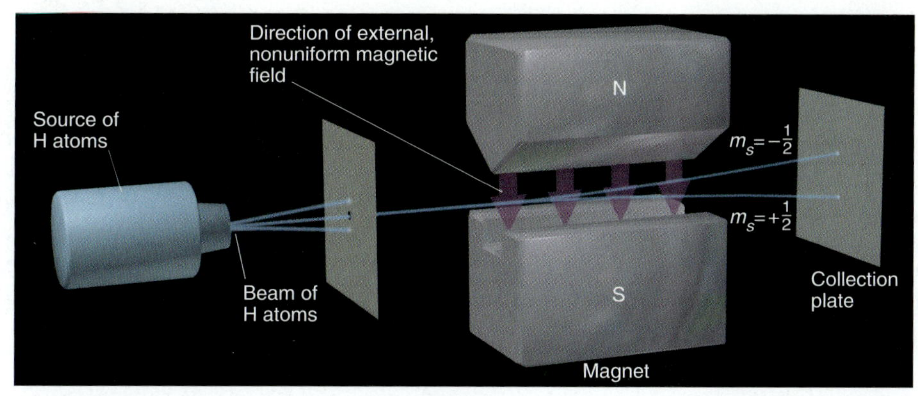

FIGURE 8.1 Observing the effect of electron spin. A nonuniform magnetic field, created by magnet faces with different shapes, splits a beam of hydrogen atoms in two. The split beam results from the two possible values of electron spin within each atom.

(1) the need for a fourth quantum number, (2) a limit on the number of electrons allowed in a given orbital, and (3) a more complex set of orbital energy levels. Let's examine these new features and then go on to determine the electron configuration for each element.

The Electron-Spin Quantum Number

Recall from Chapter 7 that the three quantum numbers n, l, and m_l describe the size (energy), shape, and orientation, respectively, of an atomic orbital. However, an additional quantum number is needed to describe a property of the electron itself, called *spin,* which is not a property of the orbital. Electron spin becomes important when more than one electron is present.

When a beam of H atoms passes through a nonuniform magnetic field, as shown in Figure 8.1, it splits into two beams that bend away from each other. The explanation of the split beam is that an electron generates a tiny magnetic field, as though it were a spinning charge. The single electron in each H atom can have one of two possible values of *spin,* each of which generates a tiny magnetic field. These two fields have opposing directions, so half of the electrons are attracted into the large external magnetic field and the other half are repelled by it. As a result, the beam of H atoms splits.

Like charge, spin is an intrinsic property of the electron, and the **spin quantum number (m_s)** has values of either $+\frac{1}{2}$ or $-\frac{1}{2}$. Thus, *each electron in an atom is described completely by a set of* ***four*** *quantum numbers: the first three describe its orbital, and the fourth describes its spin.* The quantum numbers are summarized in Table 8.1.

Table 8.1 Summary of Quantum Numbers of Electrons in Atoms

Name	Symbol	Permitted Values	Property
Principal	n	Positive integers (1, 2, 3, …)	Orbital energy (size)
Angular momentum	l	Integers from 0 to $n - 1$	Orbital shape (The l values 0, 1, 2, and 3 correspond to *s, p, d,* and *f* orbitals, respectively.)
Magnetic	m_l	Integers from $-l$ to 0 to $+l$	Orbital orientation
Spin	m_s	$+\frac{1}{2}$ or $-\frac{1}{2}$	Direction of e^- spin

Now we can write a set of four quantum numbers for any electron in the ground state of any atom. For example, the set of quantum numbers for the lone electron in hydrogen (H; $Z = 1$) is $n = 1$, $l = 0$, $m_l = 0$, and $m_s = +\frac{1}{2}$. (The spin quantum number for this electron could just as well have been $-\frac{1}{2}$, but by convention, we assign $+\frac{1}{2}$ for the first electron in an orbital.)

The Exclusion Principle

The element after hydrogen is helium (He; $Z = 2$), the first with atoms having more than one electron. The first electron in the He ground state has the same set of quantum numbers as the electron in the H atom, but the second He electron does not. Based on observations of the excited states of atoms, the Austrian physicist Wolfgang Pauli formulated the **exclusion principle:** *no two electrons in the same atom can have the same four quantum numbers.* That is, each electron must have a unique "identity" as expressed by its set of quantum numbers. Therefore, the second He electron occupies the same orbital as the first but has an opposite spin: $n = 1$, $l = 0$, $m_l = 0$, and $m_s = -\frac{1}{2}$.

THINK OF IT THIS WAY
Baseball Quantum Numbers

The unique set of quantum numbers that describes an electron is analogous to the unique location of a box seat at a baseball game. The stadium (atom) is divided into section (n, level), box (l, sublevel), row (m_l, orbital), and seat (m_s, spin). Only one person can have this particular set of stadium "quantum numbers," just as each electron in an atom has its own particular set.

Because the spin quantum number (m_s) can have only two values, the major consequence of the exclusion principle is that *an atomic orbital can hold a maximum of two electrons and they must have opposing spins.* We say that the 1*s* orbital in He is *filled* and that the electrons have *paired spins.* Thus, a beam of He atoms is not split in an experiment like that in Figure 8.1.

Electrostatic Effects and Energy-Level Splitting

Electrostatic effects play a major role in determining the energy states of many-electron atoms. Recall that the energy state of the H atom is determined *only* by the n value of the occupied orbital. In other words, in the H atom, all sublevels of a given level, such as the 2*s* and 2*p*, have the same energy. The reason is that the only electrostatic interaction is the attraction between nucleus and electron. On the other hand, the energy states of many-electron atoms arise not only from nucleus-electron attractions, but also electron-electron repulsions. One major consequence of these additional interactions is *the splitting of energy levels into sublevels of differing energies: the energy of an orbital in a many-electron atom depends mostly on its n value (size) and to a lesser extent on its l value (shape).* Thus, a more complex set of energy states exists for a many-electron atom than for the H atom. In fact, even helium (He), which has only one more electron, displays many more spectral lines than hydrogen, thus indicating more sublevel energies in its excited states.

Our first encounter with energy-level splitting in a ground-state configuration occurs with lithium (Li; $Z = 3$). By definition, the electrons of an atom in its ground state occupy the orbitals of lowest energy, so the first two electrons in the ground state of Li fill the 1*s* orbital. Then, the third Li electron must go into the $n = 2$ level. But this level has 2*s* and 2*p* sublevels: which sublevel is of lower energy, that is, which does the third electron enter? As you'll see, the 2*s* is lower in energy than the 2*p*. The reasons for this energy difference are based on three factors—*nuclear charge, electron repulsions,* and *orbital shape* (more specifically, radial probability distribution). Their interplay leads to the phenomena of *shielding* and *penetration*, which occur in all the atoms in the periodic table—except hydrogen.

The Effect of Nuclear Charge (Z) on Orbital Energy Nuclear protons create an ever-present pull on the electrons. You know that higher charges attract each other more strongly than lower charges (Coulomb's law, Section 2.7). Therefore, *higher nuclear charge lowers orbital energy (stabilizes the system) by increasing nucleus-electron attractions.* We can see this effect clearly by comparing the 1*s* orbital energies of three species with one electron—the H atom ($Z = 1$), He^+ ion ($Z = 2$), and Li^{2+} ion ($Z = 3$). The H 1*s* orbital is the least stable (highest energy), and the Li^{2+} 1*s* orbital is the most stable.

Shielding: The Effect of Electron Repulsions on Orbital Energy In many-electron atoms, each electron "feels" not only the attraction to the nucleus but also the repulsion from other electrons. This repulsion counteracts the nuclear attraction somewhat, making each electron easier to remove by, in effect, helping to push it away. We speak of each electron "shielding" the other electrons somewhat from the nucleus. **Shielding** (also called *screening*) reduces the full nuclear charge to an **effective nuclear charge (Z_{eff})**, the nuclear charge an electron *actually experiences. This lower nuclear charge makes the electron easier to remove.* We see the effect of shielding by electrons in the *same* orbital when we compare the He atom and He^+ ion: both have a 2+ nuclear charge, but He has two electrons in the 1*s* orbital and He^+ has only one. It takes less than half as much energy to remove an electron from He (2372 kJ/mol) than from He^+ (5250 kJ/mol) because the second electron in He repels the first, in effect shielding the first electron from the full nuclear charge (lowering Z_{eff}).

Much greater shielding is provided by inner electrons. Because they spend nearly all their time *between* the outer electrons and the nucleus, inner electrons shield outer electrons very effectively, in fact, *much more effectively* than do electrons in the same sublevel. *Shielding by inner electrons greatly lowers the Z_{eff} felt by outer electrons.*

Penetration: The Effect of Orbital Shape on Orbital Energy To resolve the question of why the third electron occupies the 2*s* orbital in the Li ground state, rather than the 2*p,* we have to consider orbital shapes, that is, radial probability distributions (Figure 8.2). At first, we might expect that the electron would enter the 2*p* orbital *(orange curve)* because it is slightly closer to the nucleus, on average, than the major portion of the 2*s* orbital *(blue curve).* But note that a minor portion of the 2*s* radial probability distribution appears within the 1*s* region. As a result, an electron in the 2*s* orbital spends part of its time "penetrating" very close to the nucleus. Charges attract more strongly if they are near each other than far apart (Coulomb's law, Section 2.7). Therefore, **penetration** by the 2*s* electron increases its overall attraction to the nucleus relative to that for a 2*p* electron. At the same time, penetration into the 1*s* region decreases the shielding of the 2*s* electron by the 1*s* electrons. Evidence shows that, indeed, the 2*s* orbital of Li *is* lower in energy than the 2*p* orbital, because it takes more energy to remove a 2*s* electron (520 kJ/mol) than a 2*p* (341 kJ/mol).

In general, *penetration and the resulting effects on shielding cause an energy level to split into sublevels of differing energy.* The lower the *l* value of

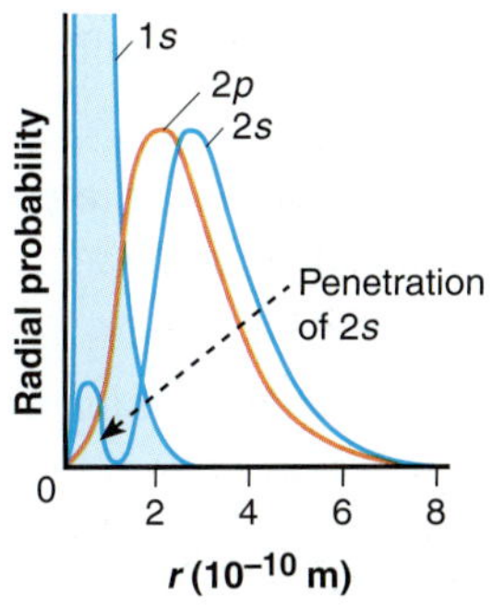

FIGURE 8.2 Penetration and orbital energy. Radial probability distributions show that a 2*s* electron spends most of its time slightly farther from the nucleus than does a 2*p* electron but penetrates near the nucleus for a small part of the time. Penetration by the 2*s* electron increases its overall attraction to the nucleus; thus, the 2*s* orbital is more stable (lower in energy) than the 2*p*.

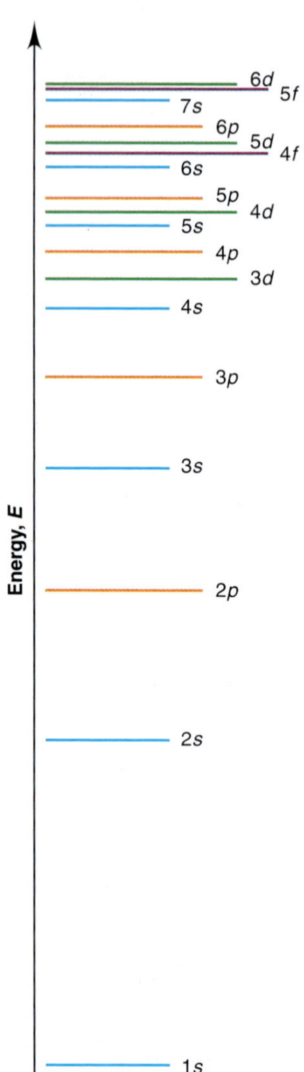

FIGURE 8.3 Order for filling energy sublevels with electrons. In many-electron atoms, energy levels split into sublevels. The relative energies of sublevels increase with principal quantum number n ($1 < 2 < 3$, etc.) and angular momentum quantum number l ($s < p < d < f$). As n increases, the energies become closer together. The penetration effect, together with this narrowing of energy differences, results in the overlap of some sublevels; for example, the 4s sublevel is slightly lower in energy than the 3d, so it is filled first. (Line color is by sublevel type; line lengths differ for ease in labeling.)

an orbital, the more its electrons penetrate, and so the greater their attraction to the nucleus. Therefore, *for a given n value, the lower the l value, the lower the sublevel energy:*

$$\text{Order of sublevel energies: } s < p < d < f \quad (8.1)$$

Thus, the 2s ($l = 0$) is lower in energy than the 2p ($l = 1$), the 3p ($l = 1$) is lower than the 3d ($l = 2$), and so forth.

Figure 8.3 shows the general energy order of levels (n value) and how they are split into sublevels (l values) of differing energies. (Compare this with the H atom energy levels in Figure 7.20, page 240.) Next, we use this energy order to construct a periodic table of ground-state atoms.

SECTION 8.2 SUMMARY

Identifying electrons in many-electron atoms requires four quantum numbers: three (n, l, m_l) describe the orbital, and a fourth (m_s) describes electron spin. • The exclusion principle requires each electron to have a unique set of four quantum numbers; therefore, an orbital can hold no more than two electrons, and their spins must be paired (opposite). • Electrostatic interactions determine orbital energies as follows:

1. Greater nuclear charge lowers orbital energy and makes electrons harder to remove.
2. Electron-electron repulsions raise orbital energy and make electrons easier to remove. Repulsions have the effect of *shielding* electrons from the full nuclear charge, reducing it to an effective nuclear charge, Z_{eff}. Inner electrons shield outer electrons most effectively.
3. Greater radial probability distribution near the nucleus (greater *penetration*) makes an electron harder to remove because it is attracted more strongly and shielded less effectively. As a result, an energy level (shell) is split into sublevels (subshells) with the energy order $s < p < d < f$.

8.3 THE QUANTUM-MECHANICAL MODEL AND THE PERIODIC TABLE

Quantum mechanics provides the theoretical foundation for the experimentally based periodic table. In this section, we fill the table with elements and determine their electron configurations—the distributions of electrons within their atoms' orbitals. Note especially the *recurring pattern in electron configurations, which is the basis for the recurring pattern in chemical behavior.*

Building Up Periods 1 and 2

A useful way to determine the electron configurations of the elements is to start at the beginning of the periodic table and add one electron per element to the *lowest-energy orbital available.* (Of course, one proton and one or more neutrons are also added to the nucleus.) This approach is based on the **aufbau principle** (German *aufbauen,* "to build up"), and it results in *ground-state* electron configurations. Let's assign sets of quantum numbers to the electrons in the ground state of the first 10 elements, those in the first two periods (horizontal rows).

For the electron in H, as you've seen, the set of quantum numbers is

$$\text{H } (Z = 1)\text{: } n = 1, l = 0, m_l = 0, m_s = +\tfrac{1}{2}$$

You also saw that the first electron in He has the same set as the electron in H, but the second He electron has opposing spin (exclusion principle):

$$\text{He } (Z = 2)\text{: } n = 1, l = 0, m_l = 0, m_s = -\tfrac{1}{2}$$

(As we go through each element in this discussion, the quantum numbers that follow refer to the element's *last added* electron.)

Here are two common ways to designate the orbital and its electrons:

1. *The electron configuration.* This shorthand notation consists of the principal energy level (n value), the letter designation of the sublevel (l value), and the number of electrons (#) in the sublevel, written as a superscript: $\boldsymbol{nl^{\#}}$. The electron configuration of H is $1s^1$ (spoken "one-ess-one"); that of He is $1s^2$ (spoken "one-ess-two," *not* "one-ess-squared"). This notation does *not* indicate electron spin but assumes you know that the two $1s$ electrons have paired (opposite) spins.

2. *The orbital diagram.* An **orbital diagram** consists of a box (or circle, or just a line) for each orbital in a given energy level, grouped by sublevel, with an arrow indicating an electron *and* its spin. (Traditionally, ↑ is $+\frac{1}{2}$ and ↓ is $-\frac{1}{2}$, but these are arbitrary; it is necessary only to be consistent. Throughout the text, orbital occupancy is also indicated by color intensity: an orbital with no color is empty, pale color means half-filled, and full color means filled.) The electron configurations and orbital diagrams for the first two elements are

H (Z = 1) $1s^1$ [↑] $1s$ He (Z = 2) $1s^2$ [↑↓] $1s$

The exclusion principle tells us that an orbital can hold only two electrons, so the $1s$ orbital in He is filled, and the $n = 1$ level is also filled. The $n = 2$ level is filled next, beginning with the $2s$ orbital, the next lowest in energy. As we said earlier, the first two electrons in Li fill the $1s$ orbital, and the last added Li electron has quantum numbers $n = 2$, $l = 0$, $m_l = 0$, $m_s = +\frac{1}{2}$. The electron configuration for Li is $1s^22s^1$. Note that the orbital diagram shows all the orbitals for $n = 2$, whether or not they are occupied:

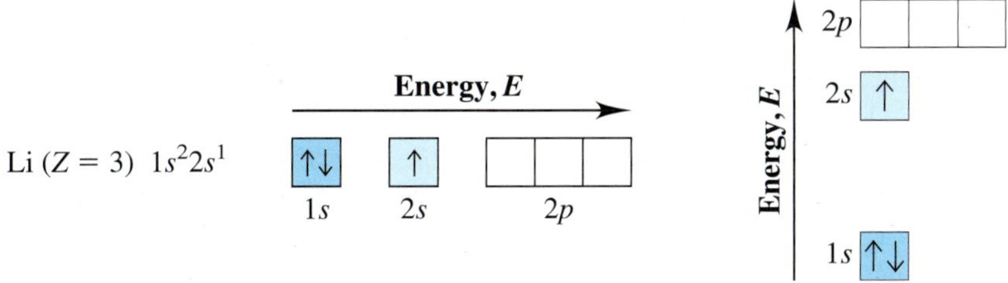

To save space on a page, orbital diagrams are often written horizontally, as shown at left. Note that *the energy of the sublevels increases from left to right.* The orbital diagram at right shows the sublevels vertically.

With the $2s$ orbital only half-filled in Li, the fourth electron of beryllium fills it with the electron's spin paired: $n = 2$, $l = 0$, $m_l = 0$, $m_s = -\frac{1}{2}$.

Be (Z = 4) $1s^22s^2$ [↑↓] $1s$ [↑↓] $2s$ [| |] $2p$

The next lowest energy sublevel is the $2p$. A p sublevel has $l = 1$, so the m_l (orientation) values can be -1, 0, or $+1$. The three orbitals in the $2p$ sublevel have *equal energy* (same n and l values), which means that the fifth electron of boron can go into *any one of the 2p orbitals.* For convenience, let's label the boxes from left to right, -1, 0, $+1$. By convention, we place the electron in the $m_l = -1$ orbital: $n = 2$, $l = 1$, $m_l = -1$, $m_s = +\frac{1}{2}$.

B (Z = 5) $1s^22s^22p^1$ [↑↓] $1s$ [↑↓] $2s$ [↑| |] $2p$ (boxes labeled -1 0 $+1$)

To minimize electron-electron repulsions, the last added (sixth) electron of carbon enters one of the *unoccupied* $2p$ orbitals; by convention, we place it in the $m_l = 0$ orbital. Experiment shows that the spin of this electron is *parallel* to (the same as) the spin of the other $2p$ electron: $n = 2$, $l = 1$, $m_l = 0$, $m_s = +\frac{1}{2}$.

C (Z = 6) $1s^22s^22p^2$

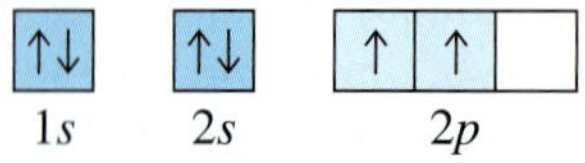

1s	2s	2p		
↑↓	↑↓	↑	↑	

This placement of electrons for carbon exemplifies **Hund's rule:** *when orbitals of equal energy are available, the electron configuration of lowest energy has the maximum number of unpaired electrons with parallel spins.* Based on Hund's rule, nitrogen's seventh electron enters the last empty $2p$ orbital, with its spin parallel to the two other $2p$ electrons: $n = 2$, $l = 1$, $m_l = +1$, $m_s = +\frac{1}{2}$.

N (Z = 7) $1s^22s^22p^3$

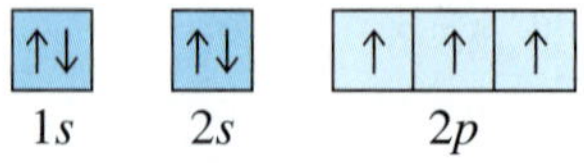

1s	2s	2p		
↑↓	↑↓	↑	↑	↑

The eighth electron in oxygen must enter one of these three half-filled $2p$ orbitals and "pair up" with (have opposing spin to) the electron already present. With the $2p$ orbitals all having the same energy, we place the electron in the orbital previously designated $m_l = -1$. The quantum numbers are $n = 2$, $l = 1$, $m_l = -1$, $m_s = -\frac{1}{2}$.

O (Z = 8) $1s^22s^22p^4$

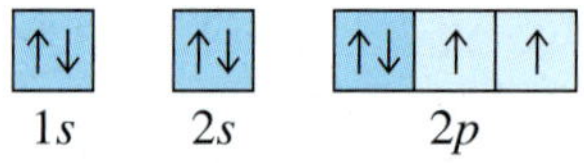

1s	2s	2p		
↑↓	↑↓	↑↓	↑	↑

Fluorine's ninth electron enters either of the two remaining half-filled $2p$ orbitals: $n = 2$, $l = 1$, $m_l = 0$, $m_s = -\frac{1}{2}$.

F (Z = 9) $1s^22s^22p^5$

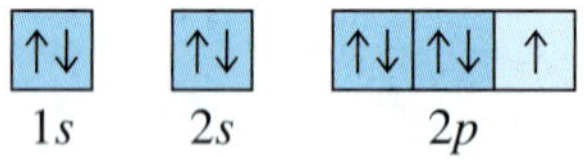

1s	2s	2p		
↑↓	↑↓	↑↓	↑↓	↑

Only one unfilled orbital remains in the $2p$ sublevel, so the tenth electron of neon occupies it: $n = 2$, $l = 1$, $m_l = +1$, $m_s = -\frac{1}{2}$. With neon, the $n = 2$ level is filled.

Ne (Z = 10) $1s^22s^22p^6$

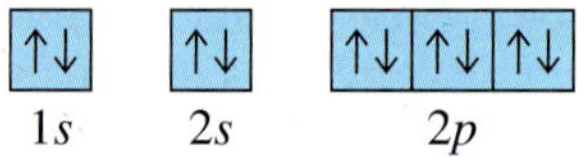

1s	2s	2p		
↑↓	↑↓	↑↓	↑↓	↑↓

SAMPLE PROBLEM 8.1 Determining Quantum Numbers from Orbital Diagrams

Problem Write a set of quantum numbers for the third electron and a set for the eighth electron of the F atom.

Plan Referring to the orbital diagram, we count to the electron of interest and note its level (n), sublevel (l), orbital (m_l), and spin (m_s).

Solution The third electron of the F atom is in the $2s$ orbital. The upward arrow indicates a spin of $+\frac{1}{2}$:

$$n = 2,\ l = 0,\ m_l = 0,\ m_s = +\tfrac{1}{2}$$

The eighth electron is in the first $2p$ orbital, which is designated $m_l = -1$, and has a downward arrow:

$$n = 2,\ l = 1,\ m_l = -1,\ m_s = -\tfrac{1}{2}$$

FOLLOW-UP PROBLEM 8.1 Use the periodic table to identify the element with the electron configuration $1s^22s^22p^4$. Write its orbital diagram, and give the quantum numbers of its sixth electron.

Table 8.2 Partial Orbital Diagrams and Electron Configurations* for the Elements in Period 3

Atomic Number	Element	Partial Orbital Diagram (3s and 3p Sublevels Only) 3s	3p	Full Electron Configuration	Condensed Electron Configuration
11	Na	↑	[ ][ ][ ]	$[1s^22s^22p^6]\ 3s^1$	$[Ne]\ 3s^1$
12	Mg	↑↓	[ ][ ][ ]	$[1s^22s^22p^6]\ 3s^2$	$[Ne]\ 3s^2$
13	Al	↑↓	[↑][ ][ ]	$[1s^22s^22p^6]\ 3s^23p^1$	$[Ne]\ 3s^23p^1$
14	Si	↑↓	[↑][↑][ ]	$[1s^22s^22p^6]\ 3s^23p^2$	$[Ne]\ 3s^23p^2$
15	P	↑↓	[↑][↑][↑]	$[1s^22s^22p^6]\ 3s^23p^3$	$[Ne]\ 3s^23p^3$
16	S	↑↓	[↑↓][↑][↑]	$[1s^22s^22p^6]\ 3s^23p^4$	$[Ne]\ 3s^23p^4$
17	Cl	↑↓	[↑↓][↑↓][↑]	$[1s^22s^22p^6]\ 3s^23p^5$	$[Ne]\ 3s^23p^5$
18	Ar	↑↓	[↑↓][↑↓][↑↓]	$[1s^22s^22p^6]\ 3s^23p^6$	$[Ne]\ 3s^23p^6$

*Colored type indicates the sublevel to which the last electron is added.

Building Up Period 3

The Period 3 elements, sodium through argon, lie directly under the Period 2 elements, lithium through neon. The sublevels of the $n = 3$ level are filled in the order 3*s*, 3*p*, 3*d*. Table 8.2 presents *partial* orbital diagrams (3*s* and 3*p* sublevels only) and electron configurations for the eight elements in Period 3 (with *filled inner levels* in brackets and the sublevel to which the last electron is added in colored type). Note the *group similarities in outer electron configuration* with the elements in Period 2 (refer to Figure 8.4).

In sodium (the second alkali metal) and magnesium (the second alkaline earth metal), electrons are added to the 3*s* sublevel, which contains the 3*s* orbital only, just as they filled the 2*s* sublevel in lithium and beryllium in Period 2. Then, just as for boron, carbon, and nitrogen in Period 2, the last electrons added to aluminum, silicon, and phosphorus in Period 3 half-fill the three 3*p* orbitals with spins parallel (Hund's rule). The last electrons added to sulfur, chlorine, and argon then successively enter the three half-filled 3*p* orbitals, filling the 3*p* sublevel. Argon, the next noble gas after helium and neon, ends Period 3. (As you'll see shortly, the 3*d* orbitals are filled in Period 4.)

The rightmost column of Table 8.2 shows the *condensed electron configuration.* In this simplified notation, the electron configuration of the previous noble gas is shown by its element symbol in brackets, and it is followed by the electron configuration of the energy level being filled. The condensed electron configuration of sulfur, for example, is $[Ne]\ 3s^23p^4$, where [Ne] stands for $1s^22s^22p^6$.

Electron Configurations Within Groups

One of the central points in all chemistry is that *similar outer electron configurations correlate with similar chemical behavior.* Figure 8.4 on the next page shows the condensed electron configurations of the first 18 elements. Note the similarities within each group.

Period	1A (1)	2A (2)	3A (13)	4A (14)	5A (15)	6A (16)	7A (17)	8A (18)
1	1 **H** $1s^1$							2 **He** $1s^2$
2	3 **Li** [He] $2s^1$	4 **Be** [He] $2s^2$	5 **B** [He] $2s^22p^1$	6 **C** [He] $2s^22p^2$	7 **N** [He] $2s^22p^3$	8 **O** [He] $2s^22p^4$	9 **F** [He] $2s^22p^5$	10 **Ne** [He] $2s^22p^6$
3	11 **Na** [Ne] $3s^1$	12 **Mg** [Ne] $3s^2$	13 **Al** [Ne] $3s^23p^1$	14 **Si** [Ne] $3s^23p^2$	15 **P** [Ne] $3s^23p^3$	16 **S** [Ne] $3s^23p^4$	17 **Cl** [Ne] $3s^23p^5$	18 **Ar** [Ne] $3s^23p^6$

FIGURE 8.4 Condensed ground-state electron configurations in the first three periods. The first 18 elements, H through Ar, are arranged in three periods containing two, eight, and eight elements. Each box shows the atomic number, atomic symbol, and condensed ground-state electron configuration. Note that elements in a group have similar outer electron configurations *(color).*

Here are some examples from just three groups:

- In Group 1A(1), lithium and sodium have the outer electron configuration ns^1 (where n is the quantum number of the outermost energy level), as do all other alkali metals (K, Rb, Cs, Fr). All are highly reactive metals that form ionic compounds with nonmetals with formulas such as MCl, M_2O, and M_2S (where M represents the alkali metal), and react vigorously with water to displace H_2.
- In Group 7A(17), fluorine and chlorine have the outer electron configuration ns^2np^5, as do the other halogens (Br, I, At). Little is known about rare, radioactive astatine (At), but all the others are reactive nonmetals that occur as diatomic molecules, X_2 (where X represents the halogen). All form ionic compounds with metals (KX, MgX_2), covalent compounds with hydrogen (HX) that yield acidic solutions in water, and covalent compounds with carbon (CX_4).
- In Group 8A(18), helium has the electron configuration ns^2, and all the others have the outer configuration ns^2np^6. Consistent with their *filled* energy levels, all of these elements are extremely unreactive monatomic gases.

To summarize the major connection between quantum mechanics and chemical periodicity: *orbitals are filled in order of increasing energy, which leads to outer electron configurations that recur periodically, which leads to chemical properties that recur periodically.*

The First *d*-Orbital Transition Series: Building Up Period 4

The $3d$ orbitals are filled in Period 4. Note, however, that *the 4s orbital is filled before the 3d.* This switch in filling order is due to the shielding and penetration effects that we discussed in Section 8.2. The radial probability distribution of the $3d$ orbital is greater outside the filled, inner $n = 1$ and $n = 2$ levels, so a $3d$ electron is shielded very effectively from the nuclear charge. In contrast, penetration by the $4s$ electron means that it spends a significant part of its time near the nucleus and feels a greater nuclear attraction. Thus, the $4s$ orbital is slightly *lower* in energy than the $3d$, and so fills first. Similarly, the $5s$ orbital fills before the $4d$, and the $6s$ fills before the $5d$. In general, *the ns sublevel fills before the* $(n - 1)d$ *sublevel.* As we proceed through the transition series, however, you'll see several exceptions to this pattern because the energies of the ns and $(n - 1)d$ sublevels become extremely close at higher values of n.

Table 8.3 shows the partial orbital diagrams and ground-state electron configurations for the 18 elements in Period 4 (again with filled inner levels in brackets and the sublevel to which the last electron has been added in colored type). The first two elements of the period, potassium and calcium, are the next alkali and alkaline earth metals, respectively, and their electrons fill the $4s$ sublevel. The third element, scandium ($Z = 21$), is the first of the **transition elements,** those in which d orbitals are being filled. The last electron in scandium occupies any one of the five $3d$ orbitals because they are equal in energy. Scandium has the electron configuration [Ar] $4s^23d^1$.

Table 8.3 Partial Orbital Diagrams and Electron Configurations* for the Elements in Period 4

Atomic Number	Element	Partial Orbital Diagram (4s, 3d, and 4p Sublevels Only): 4s	3d	4p	Full Electron Configuration	Condensed Electron Configuration
19	K	[↑]	[ ][ ][ ][ ][ ]	[ ][ ][ ]	$[1s^22s^22p^63s^23p^6]\ 4s^1$	$[Ar]\ 4s^1$
20	Ca	[↑↓]	[ ][ ][ ][ ][ ]	[ ][ ][ ]	$[1s^22s^22p^63s^23p^6]\ 4s^2$	$[Ar]\ 4s^2$
21	Sc	[↑↓]	[↑][ ][ ][ ][ ]	[ ][ ][ ]	$[1s^22s^22p^63s^23p^6]\ 4s^23d^1$	$[Ar]\ 4s^23d^1$
22	Ti	[↑↓]	[↑][↑][ ][ ][ ]	[ ][ ][ ]	$[1s^22s^22p^63s^23p^6]\ 4s^23d^2$	$[Ar]\ 4s^23d^2$
23	V	[↑↓]	[↑][↑][↑][ ][ ]	[ ][ ][ ]	$[1s^22s^22p^63s^23p^6]\ 4s^23d^3$	$[Ar]\ 4s^23d^3$
24	Cr	[↑]	[↑][↑][↑][↑][↑]	[ ][ ][ ]	$[1s^22s^22p^63s^23p^6]\ 4s^13d^5$	$[Ar]\ 4s^13d^5$
25	Mn	[↑↓]	[↑][↑][↑][↑][↑]	[ ][ ][ ]	$[1s^22s^22p^63s^23p^6]\ 4s^23d^5$	$[Ar]\ 4s^23d^5$
26	Fe	[↑↓]	[↑↓][↑][↑][↑][↑]	[ ][ ][ ]	$[1s^22s^22p^63s^23p^6]\ 4s^23d^6$	$[Ar]\ 4s^23d^6$
27	Co	[↑↓]	[↑↓][↑↓][↑][↑][↑]	[ ][ ][ ]	$[1s^22s^22p^63s^23p^6]\ 4s^23d^7$	$[Ar]\ 4s^23d^7$
28	Ni	[↑↓]	[↑↓][↑↓][↑↓][↑][↑]	[ ][ ][ ]	$[1s^22s^22p^63s^23p^6]\ 4s^23d^8$	$[Ar]\ 4s^23d^8$
29	Cu	[↑]	[↑↓][↑↓][↑↓][↑↓][↑↓]	[ ][ ][ ]	$[1s^22s^22p^63s^23p^6]\ 4s^13d^{10}$	$[Ar]\ 4s^13d^{10}$
30	Zn	[↑↓]	[↑↓][↑↓][↑↓][↑↓][↑↓]	[ ][ ][ ]	$[1s^22s^22p^63s^23p^6]\ 4s^23d^{10}$	$[Ar]\ 4s^23d^{10}$
31	Ga	[↑↓]	[↑↓][↑↓][↑↓][↑↓][↑↓]	[↑][ ][ ]	$[1s^22s^22p^63s^23p^6]\ 4s^23d^{10}4p^1$	$[Ar]\ 4s^23d^{10}4p^1$
32	Ge	[↑↓]	[↑↓][↑↓][↑↓][↑↓][↑↓]	[↑][↑][ ]	$[1s^22s^22p^63s^23p^6]\ 4s^23d^{10}4p^2$	$[Ar]\ 4s^23d^{10}4p^2$
33	As	[↑↓]	[↑↓][↑↓][↑↓][↑↓][↑↓]	[↑][↑][↑]	$[1s^22s^22p^63s^23p^6]\ 4s^23d^{10}4p^3$	$[Ar]\ 4s^23d^{10}4p^3$
34	Se	[↑↓]	[↑↓][↑↓][↑↓][↑↓][↑↓]	[↑↓][↑][↑]	$[1s^22s^22p^63s^23p^6]\ 4s^23d^{10}4p^4$	$[Ar]\ 4s^23d^{10}4p^4$
35	Br	[↑↓]	[↑↓][↑↓][↑↓][↑↓][↑↓]	[↑↓][↑↓][↑]	$[1s^22s^22p^63s^23p^6]\ 4s^23d^{10}4p^5$	$[Ar]\ 4s^23d^{10}4p^5$
36	Kr	[↑↓]	[↑↓][↑↓][↑↓][↑↓][↑↓]	[↑↓][↑↓][↑↓]	$[1s^22s^22p^63s^23p^6]\ 4s^23d^{10}4p^6$	$[Ar]\ 4s^23d^{10}4p^6$

*Colored type indicates sublevel(s) whose occupancy changes when the last electron is added.

The filling of 3*d* orbitals proceeds one at a time, as with *p* orbitals, except in two cases: chromium ($Z = 24$) and copper ($Z = 29$). Vanadium ($Z = 23$), the element before chromium, has three half-filled *d* orbitals ([Ar] $4s^23d^3$). Rather than having its last electron enter a fourth empty *d* orbital to give [Ar] $4s^23d^4$, chromium has one electron in the 4*s* sublevel and five in the 3*d* sublevel. Thus, both the 4*s* and the 3*d* sublevels are *half-filled* (see margin). In the next element, manganese ($Z = 25$), the 4*s* sublevel is filled again ([Ar] $4s^23d^5$).

Cr ($Z = 24$) [Ar] $4s^13d^5$

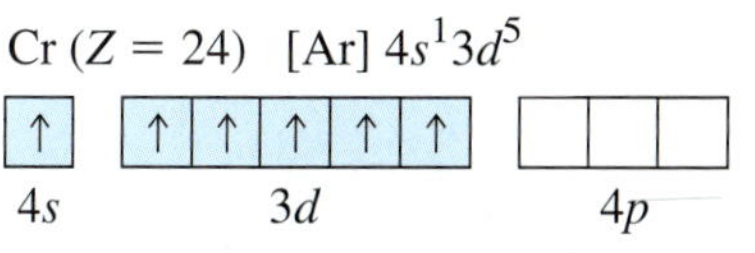

The other anomalous filling pattern occurs with copper. Following nickel ([Ar] $4s^23d^8$), copper would be expected to have the [Ar] $4s^23d^9$ configuration. Instead, the 4*s* orbital of copper is half-filled (1 electron), and the 3*d* orbitals are *filled* with 10 electrons (see margin). Then, in zinc ($Z = 30$), the 4*s* sublevel is filled ([Ar] $4s^23d^{10}$). The anomalous filling patterns in Cr and Cu lead us to conclude that *half-filled and filled sublevels are unexpectedly stable.* These are the first two cases of a pattern seen with many other elements.

Cu ($Z = 29$) [Ar] $4s^13d^{10}$

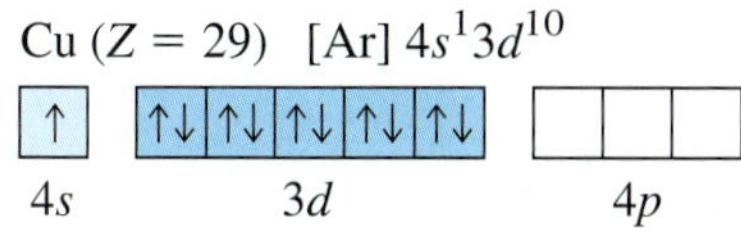

In zinc, both the 4*s* and 3*d* sublevels are completely filled, and the first transition series ends. As Table 8.3 shows, the 4*p* sublevel is then filled by the next six elements. Period 4 ends with krypton, the next noble gas.

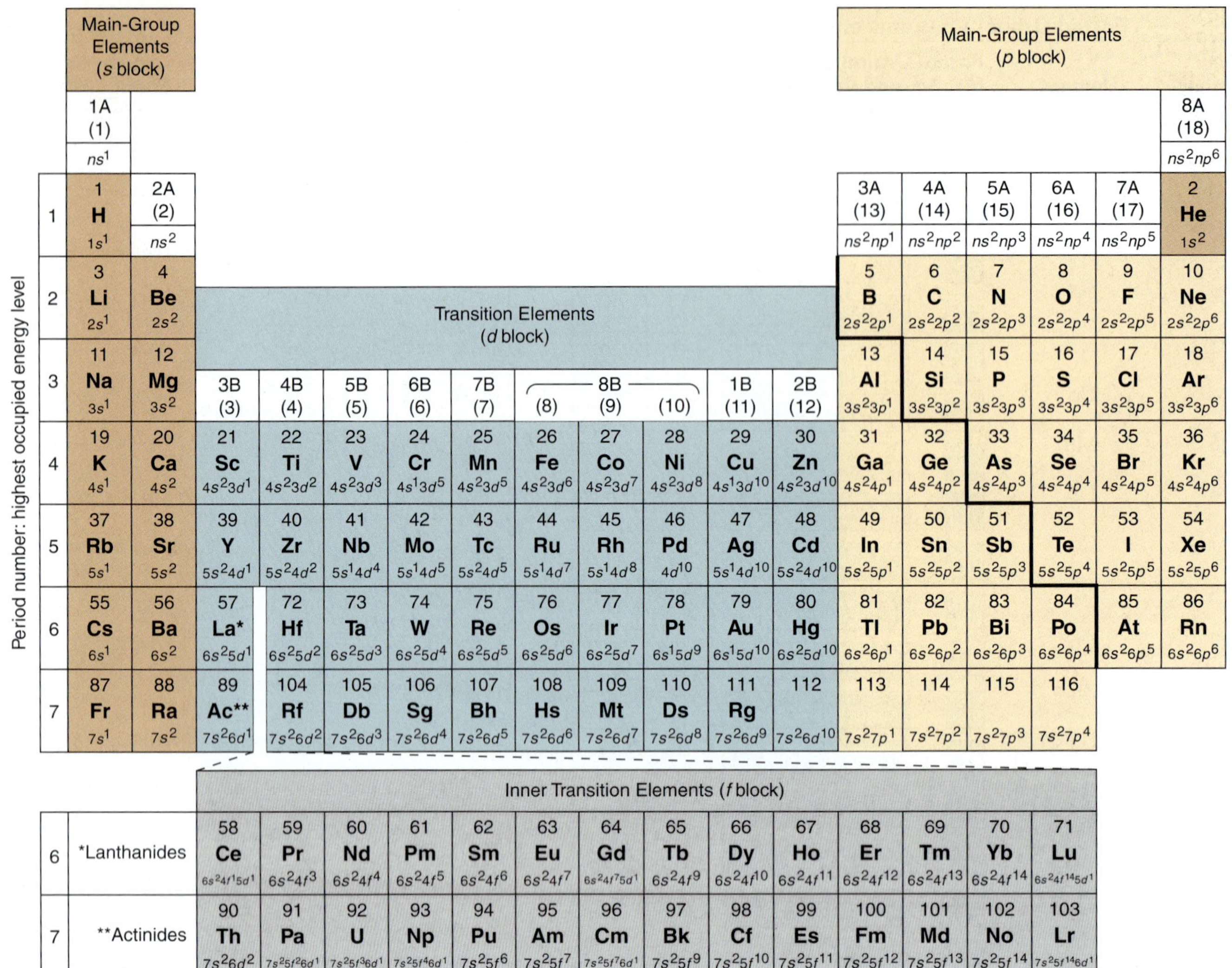

FIGURE 8.5 A periodic table of partial ground-state electron configurations. These ground-state electron configurations show the electrons beyond the previous noble gas in the sublevel block being filled (excluding filled inner sublevels). For main-group elements, the group heading identifies the general outer configuration. Anomalous electron configurations occur often among the *d*-block and *f*-block elements, with the first two appearing for Cr ($Z = 24$) and Cu ($Z = 29$). Helium is colored as an *s*-block element but placed with the other members of Group 8A(18). Configurations for elements 112 through 116 have not yet been confirmed.

General Principles of Electron Configurations

There are 80 known elements beyond the 36 we have considered. Let's survey the ground-state electron configurations to highlight some key ideas.

Similar Outer Electron Configurations Within a Group To repeat one of chemistry's central themes and the key to the usefulness of the periodic table, *elements in a group have similar chemical properties because they have similar outer electron configurations* (Figure 8.5). Among the main-group elements (A groups)—the *s*-block and *p*-block elements—outer electron configurations within a group are essentially identical, as shown by the group headings in Figure 8.5. Some variations in the transition elements (B groups, *d* block) and inner transition elements (*f* block) occur, as you'll see.

Orbital Filling Order When the elements are "built up" by filling their levels and sublevels in order of increasing energy, we obtain the actual sequence of elements in the periodic table (Figure 8.6).

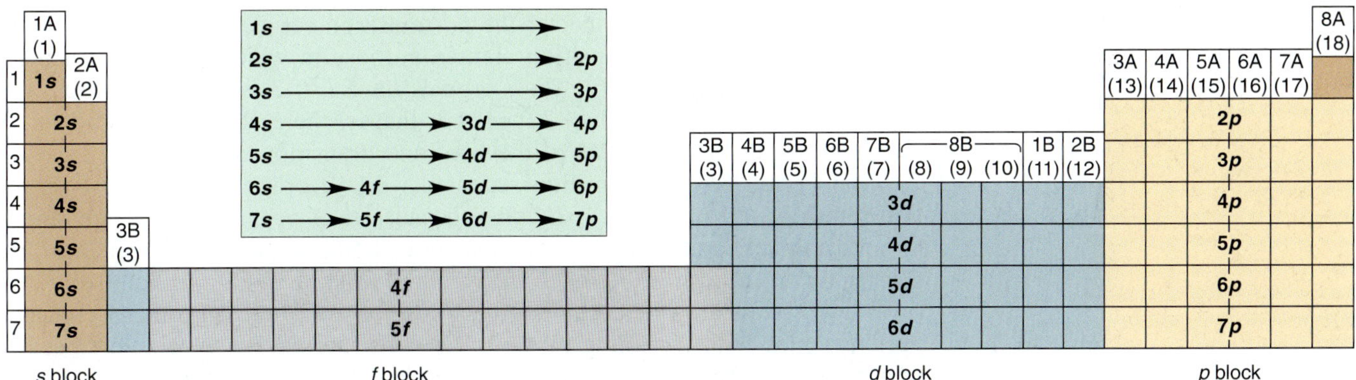

FIGURE 8.6 The relation between orbital filling and the periodic table. If we "read" the periods like the words on a page, the elements are arranged into sublevel blocks that occur in the order of increasing energy. This form of the periodic table shows the sublevel blocks. (The *f* blocks fit between the first and second elements of the *d* blocks in Periods 6 and 7.) *Inset:* A simple version of sublevel order.

THINK OF IT THIS WAY
The Filling Order of the Periodic Table

If you list the sublevels as shown at right and follow the direction of the arrows, starting from 1*s*, you obtain the sublevels in order of increasing energy. Note that
- the *n* value is constant horizontally,
- the *l* value is constant vertically, and
- the sum $n + l$ is constant diagonally.

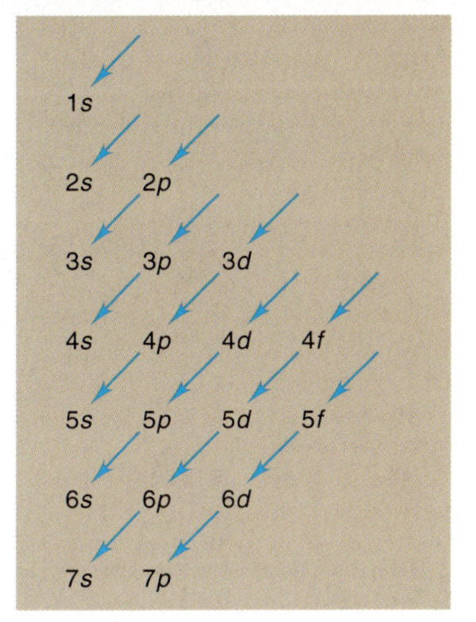

Categories of Electrons The elements have three categories of electrons:

1. **Inner (core) electrons** are those seen in the previous noble gas and any completed transition series. They fill all the *lower energy levels* of an atom.
2. **Outer electrons** are those in the *highest energy level* (highest *n* value). They spend most of their time farthest from the nucleus.
3. **Valence electrons** are those involved in forming compounds. *Among the main-group elements, the valence electrons* ***are*** *the outer electrons.* For the transition elements, all the $(n - 1)d$ electrons are counted among the valence electrons also, even though the elements Fe ($Z = 26$) through Zn ($Z = 30$) use only a few of them in bonding.

Group and Period Numbers Key information is embedded in the periodic table:

1. Among the main-group elements (A groups), *the group number equals the number of outer electrons* (those with the highest *n*): chlorine (Cl; Group **7**A) has 7 outer electrons, tellurium (Te; Group **6**A) has 6, and so forth.
2. *The period number is the n value of the highest energy level.* Thus, in Period 2, the $n = 2$ level has the highest energy; in Period 5, it is the $n = 5$ level.
3. The *n* value squared (n^2) gives the total number of *orbitals* in that energy level. Because an orbital can hold no more than two electrons (exclusion principle), $2n^2$ gives the maximum number of *electrons* (or elements) in the energy level. For example, for the $n = 3$ level, the number of orbitals is $n^2 = 9$: one 3*s*, three 3*p*, and five 3*d*. The number of electrons is $2n^2$, or 18: two 3*s* and six 3*p* electrons occur in the eight elements of Period 3, and ten 3*d* electrons are added in the ten transition elements of Period 4.

Unusual Configurations: Transition and Inner Transition Elements

Periods 4, 5, 6, and 7 incorporate the *d*-block transition elements. The general pattern, as you've seen, is that the $(n - 1)d$ orbitals are filled between the *ns* and *np* orbitals. Thus, Period 5 follows the same general pattern as Period 4. In Period 6, the 6*s* sublevel is filled in cesium (Cs) and barium (Ba), and then lanthanum

(La; $Z = 57$), the first member of the $5d$ transition series, occurs. At this point, the first series of **inner transition elements,** those in which f orbitals are being filled, intervenes (Figure 8.6). The f orbitals have $l = 3$, so the possible m_l values are -3, -2, -1, 0, $+1$, $+2$, and $+3$; that is, there are seven f orbitals, for a total of 14 elements in *each* of the two inner transition series.

The Period 6 inner transition series fills the $4f$ orbitals and consists of the **lanthanides** (or *rare earths*), so called because they occur after and are similar to lanthanum. The other inner transition series holds the **actinides,** which fill the $5f$ orbitals that appear in Period 7 after actinium (Ac; $Z = 89$). In both series, the $(n - 2)f$ orbitals are filled, after which filling of the $(n - 1)d$ orbitals proceeds. Period 6 ends with the filling of the $6p$ orbitals. Period 7 is incomplete because only four elements with $7p$ electrons have been synthesized at this time.

Several irregularities in filling pattern occur in both the d and f blocks. Two already mentioned occur in chromium (Cr) and copper (Cu) in Period 4. Silver (Ag) and gold (Au), the two elements under Cu in Group 1B(11), follow copper's pattern. Molybdenum (Mo) follows the pattern of Cr in Group 6B(6), but tungsten (W) does not. Other anomalous configurations appear among the transition elements in Periods 5 and 6. Note, however, that even though minor variations from the expected configurations occur, the sum of ns electrons and $(n - 1)d$ electrons always equals the *new* group number. For instance, despite variations in the electron configurations in Group 6B(**6**)—Cr, Mo, W, and Sg—the sum of ns and $(n - 1)d$ electrons is 6; in Group 8B(**10**)—Ni, Pd, Pt, and Ds—the sum is 10.

Whenever our observations differ from our expectations, remember that the fact always takes precedence over the model; in other words, the electrons don't "care" what orbitals *we* think they should occupy. As the atomic orbitals in larger atoms fill with electrons, sublevel energies differ very little, which results in these variations from the expected pattern.

SAMPLE PROBLEM 8.2 Determining Electron Configurations

Problem Using the periodic table on the inside front cover of the text (not Figure 8.5 or Table 8.3), give the full and condensed electron configurations, partial orbital diagrams showing valence electrons, and number of inner electrons for the following elements:
(a) Potassium (K; $Z = 19$) **(b)** Molybdenum (Mo; $Z = 42$) **(c)** Lead (Pb; $Z = 82$)

Plan The atomic number tells us the number of electrons, and the periodic table shows the order for filling sublevels. In the partial orbital diagrams, we include all electrons after those of the previous noble gas *except* those in *filled* inner sublevels. The number of inner electrons is the sum of those in the previous noble gas and in filled d and f sublevels.

Solution **(a)** For K ($Z = 19$), the full electron configuration is $1s^22s^22p^63s^23p^64s^1$.
The condensed configuration is [Ar] $4s^1$.
The partial orbital diagram for valence electrons is

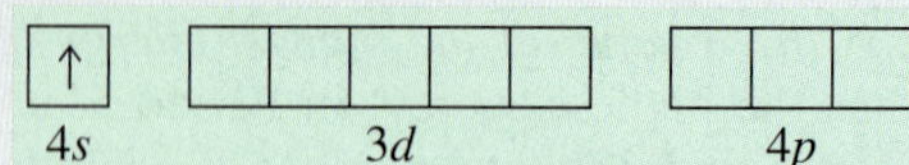

K is a main-group element in Group 1A(1) of Period 4, so there are 18 inner electrons.
(b) For Mo ($Z = 42$), we would expect the full electron configuration to be $1s^22s^22p^63s^23p^64s^23d^{10}4p^65s^24d^4$. However, Mo lies under Cr in Group 6B(6) and exhibits the same variation in filling pattern in the ns and $(n - 1)d$ sublevels:
$1s^22s^22p^63s^23p^64s^23d^{10}4p^65s^14d^5$.
The condensed electron configuration is [Kr] $5s^14d^5$.
The partial orbital diagram for valence electrons is

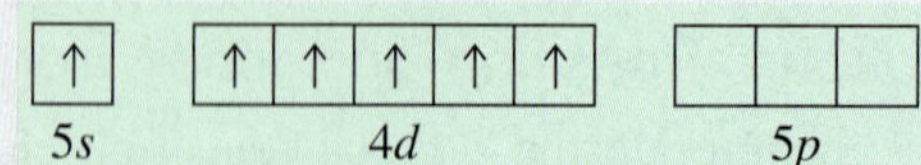

Mo is a transition element in Group 6B(6) of Period 5, so there are 36 inner electrons.

(c) For Pb ($Z = 82$), the full electron configuration is $1s^2 2s^2 2p^6 3s^2 3p^6 4s^2 3d^{10} 4p^6 5s^2 4d^{10} 5p^6 6s^2 4f^{14} 5d^{10} 6p^2$.

The condensed electron configuration is $[Xe]\ 6s^2 4f^{14} 5d^{10} 6p^2$.

The partial orbital diagram for valence electrons (no filled inner sublevels) is

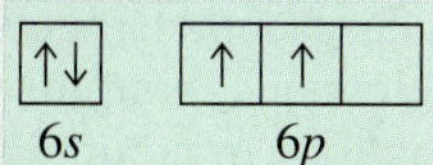

Pb is a main-group element in Group 4A(14) of Period 6, so there are 54 (in Xe) + 14 (in 4*f* series) + 10 (in 5*d* series) = 78 inner electrons.

Check Be sure the sum of the superscripts (electrons) in the full electron configuration equals the atomic number, and that the number of *valence* electrons in the condensed configuration equals the number of electrons in the partial orbital diagram.

FOLLOW-UP PROBLEM 8.2 Without referring to Table 8.3 or Figure 8.5, give full and condensed electron configurations, partial orbital diagrams showing valence electrons, and the number of inner electrons for the following elements:
(a) Ni ($Z = 28$) **(b)** Sr ($Z = 38$) **(c)** Po ($Z = 84$)

SECTION 8.3 SUMMARY

In the aufbau method, one electron is added to an atom of each successive element in accord with Pauli's exclusion principle (no two electrons can have the same set of quantum numbers) and Hund's rule (orbitals of equal energy become half-filled, with electron spins parallel, before any pairing occurs). • The elements of a group have similar outer electron configurations and thus similar chemical behavior. • For the main-group elements, valence electrons (those involved in reactions) are in the outer (highest energy) level only. • For transition elements, $(n-1)d$ electrons are also involved in reactions. In general, $(n-1)d$ orbitals fill after *ns* and before *np* orbitals. • In Periods 6 and 7, $(n-2)f$ orbitals fill between the first and second $(n-1)d$ orbitals.

8.4 TRENDS IN THREE KEY ATOMIC PROPERTIES

All physical and chemical behavior of the elements is based ultimately on the electron configurations of their atoms. In this section, we focus on three properties of atoms that are directly influenced by electron configuration and, thus, effective nuclear charge: atomic size, ionization energy (the energy required to remove an electron from a gaseous atom), and electron affinity (the energy change involved in adding an electron to a gaseous atom). These properties are *periodic:* they generally increase and decrease in a recurring manner throughout the periodic table. As a result, their relative magnitudes can often be predicted, and they often exhibit consistent changes, or *trends,* within a group or period that correlate with element behavior.

Trends in Atomic Size

In Chapter 7, we noted that an electron in an atom can lie relatively far from the nucleus, so we commonly represent atoms as spheres in which the electrons spend 90% of their time. However, we often *define* **atomic size** in terms of how closely one atom lies next to another. In practice, we measure the distance between identical, adjacent atomic nuclei in a sample of an element and divide that distance in half. (The technique is discussed in Chapter 12.) Because atoms do not have hard surfaces, the size of an atom in a compound depends somewhat on the atoms near it. In other words, *atomic size varies slightly from substance to substance.*

Figure 8.7 shows two common definitions of atomic size. The **metallic radius** is one-half the distance between nuclei of adjacent atoms in a crystal of the element; we typically use this definition for metals. For elements commonly

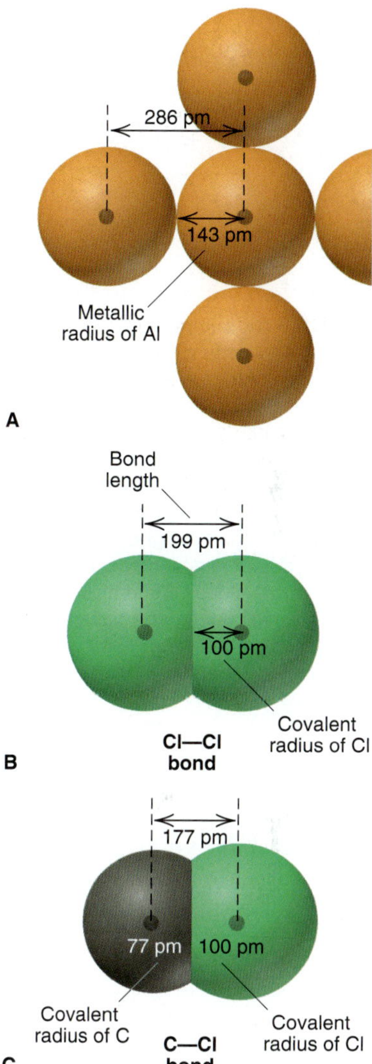

FIGURE 8.7 Defining metallic and covalent radii. A, The metallic radius is one-half the distance between nuclei of adjacent atoms in a crystal of the element, as shown here for aluminum. **B,** The covalent radius is one-half the distance between bonded nuclei in a molecule of the element, as shown here for chlorine. In effect, it is one-half the bond length. **C,** In a covalent compound, the bond length and known covalent radii are used to determine other radii. Here the C—Cl bond length (177 pm) and the covalent radius of Cl (100 pm) are used to find a value for the covalent radius of C (177 pm − 100 pm = 77 pm).

occurring as molecules, mostly nonmetals, we define atomic size by the **covalent radius,** one-half the distance between nuclei of identical covalently bonded atoms.

Trends Among the Main-Group Elements Atomic size greatly influences other atomic properties and is critical to understanding element behavior. Figure 8.8 shows the atomic radii of the main-group elements and most of the transition elements. Among the main-group elements, note that atomic size varies within both a group and a period. These variations in atomic size are the result of two opposing influences:

1. *Changes in n.* As the principal quantum number (n) increases, the probability that the outer electrons will spend more time farther from the nucleus increases as well; thus, the atoms are larger.
2. *Changes in Z_{eff}.* As the effective nuclear charge (Z_{eff})—the positive charge "felt" by an electron—increases, outer electrons are pulled closer to the nucleus; thus, the atoms are smaller.

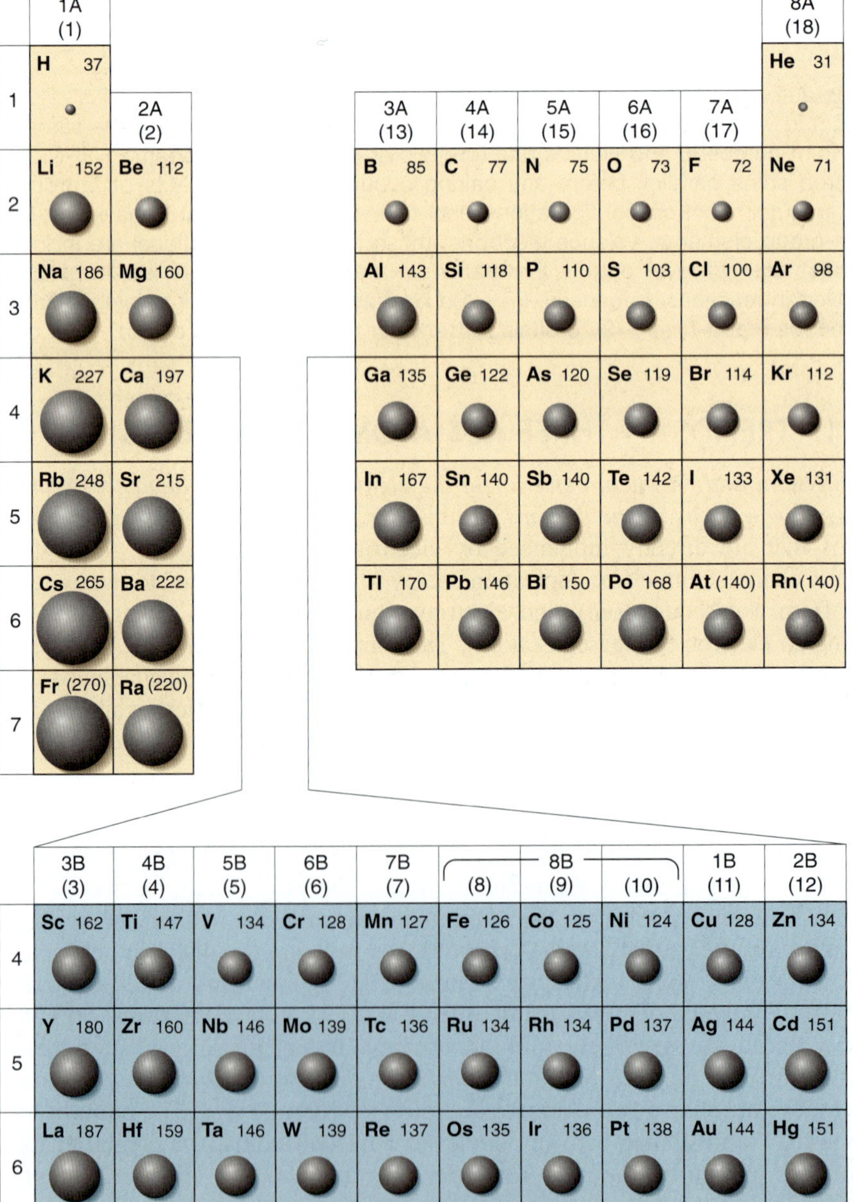

FIGURE 8.8 Atomic radii of the main-group and transition elements. Atomic radii (in picometers) are shown as half-spheres of proportional size for the main-group elements *(tan)* and the transition elements *(blue)*. Among the main-group elements, atomic radius generally *increases* from top to bottom and *decreases* from left to right. The transition elements do not exhibit these trends as consistently. (Values in parentheses have only two significant figures; values for the noble gases are based on quantum-mechanical calculations.)

The net effect of these influences depends on shielding of the increasing nuclear charge by inner electrons:

1. *Down a group, n dominates.* As we move down a main group, each member has one more level of *inner* electrons that shield the *outer* electrons very effectively. Even though calculations show Z_{eff} on the outer electrons rising moderately for each element in the group, the atoms get larger as a result of the increasing n value. *Atomic radius generally* ***increases*** *in a group from top to bottom.*
2. *Across a period, Z_{eff} dominates.* As we move across a period of main-group elements, electrons are added to the same outer level, so the shielding by inner electrons does not change. Because outer electrons shield each other poorly, Z_{eff} on the outer electrons rises significantly, and so they are pulled closer to the nucleus. *Atomic radius generally* ***decreases*** *in a period from left to right.*

Trends Among the Transition Elements As Figure 8.8 shows, these trends hold well for the main-group elements but *not* as consistently for the transition elements. As we move from left to right, size shrinks through the first two or three transition elements because of the increasing nuclear charge. But, from then on, *the size remains relatively constant* because shielding by the inner d electrons counteracts the usual increase in Z_{eff}. For instance, vanadium (V; $Z = 23$), the third Period 4 transition metal, has the same atomic radius as zinc (Zn; $Z = 30$), the last Period 4 transition metal. This pattern of atomic size shrinking also appears in Periods 5 and 6 in the d-block transition series and in both series of inner transition elements. The *lack* of a vertical size increase from the Period 5 to 6 transition metal is especially obvious.

This shielding by d electrons causes a major size decrease from Group 2A(2) to Group 3A(13), the two main groups that flank the transition series. The size decrease in Periods 4, 5, and 6 (*with* a transition series) is much greater than in Period 3 (*without* a transition series). Because electrons in the np orbitals penetrate more than those in the $(n - 1)d$ orbitals, the first np electron [Group 3A(13)] "feels" a Z_{eff} that has been increased by the protons added to all the intervening transition elements. The greatest change in size occurs in Period 4, in which calcium (Ca; $Z = 20$) is nearly 50% larger than gallium (Ga; $Z = 31$). In fact, shielding by the d orbitals in the transition series causes such a major size contraction that gallium is slightly *smaller* than aluminum (Al; $Z = 13$), even though Ga is below Al in the same group!

SAMPLE PROBLEM 8.3 Ranking Elements by Atomic Size

Problem Using only the periodic table (not Figure 8.8), rank each set of main-group elements in order of *decreasing* atomic size:

(a) Ca, Mg, Sr **(b)** K, Ga, Ca
(c) Br, Rb, Kr **(d)** Sr, Ca, Rb

Plan To rank the elements by atomic size, we find them in the periodic table. They are main-group elements, so size increases down a group and decreases across a period.

Solution (a) Sr > Ca > Mg. These three elements are in Group 2A(2), and size decreases up the group.

(b) K > Ca > Ga. These three elements are in Period 4, and size decreases across a period.

(c) Rb > Br > Kr. Rb is largest because it has one more energy level and is farthest to the left. Kr is smaller than Br because Kr is farther to the right in Period 4.

(d) Rb > Sr > Ca. Ca is smallest because it has one fewer energy level. Sr is smaller than Rb because it is farther to the right.

Check From Figure 8.8, we see that the rankings are correct.

FOLLOW-UP PROBLEM 8.3 Using only the periodic table, rank the elements in each set in order of *increasing* size: **(a)** Se, Br, Cl; **(b)** I, Xe, Ba.

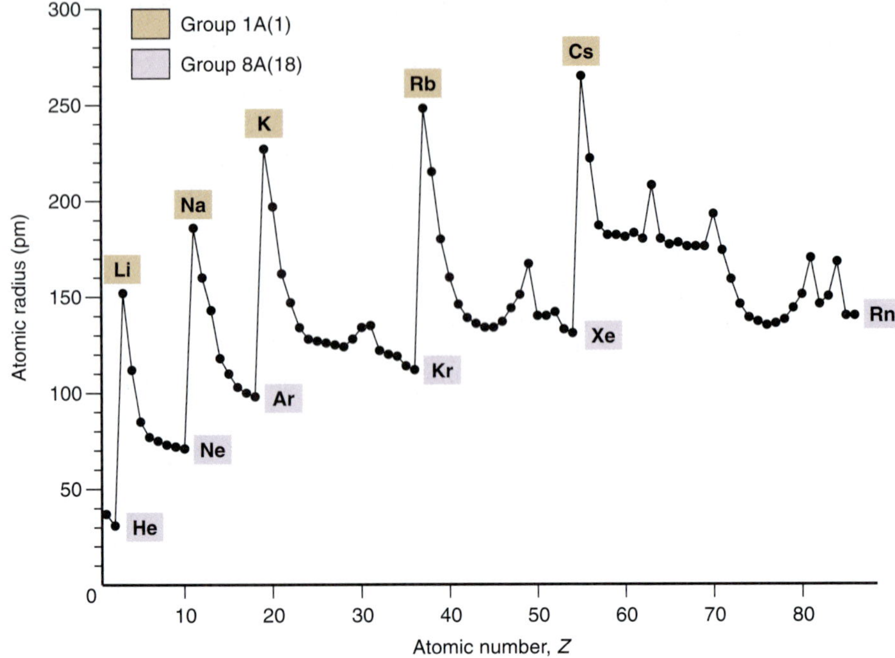

FIGURE 8.9 Periodicity of atomic radius. A plot of atomic radius vs. atomic number for the elements in Periods 1 through 6 shows a periodic change: the radius generally decreases through a period to the noble gas [Group 8A(18); *purple*] and then increases suddenly to the next alkali metal [Group 1A(1); *brown*]. Deviation from the general decrease occurs among the transition elements.

Figure 8.9 shows the overall variation in atomic size with increasing atomic number. Note the recurring up-and-down pattern as size drops across a period to the noble gas and then leaps up to the alkali metal that begins the next period. Also note how each transition series, beginning with that in Period 4 (K to Kr), throws off the smooth size decrease.

Trends in Ionization Energy

The **ionization energy (IE)** is the energy (in kJ) required for the *complete removal* of 1 mol of electrons from 1 mol of gaseous atoms or ions. Pulling an electron away from a nucleus *requires* energy to overcome the attraction. Because energy flows *into* the system, the ionization energy is always positive (like ΔH of an endothermic reaction).

In Chapter 7, you saw that the ionization energy of the H atom is the energy difference between $n = 1$ and $n = \infty$, the point at which the electron is completely removed. Many-electron atoms can lose more than one electron. The first ionization energy (IE_1) removes an outermost electron (highest energy sublevel) from the gaseous atom:

$$\text{Atom}(g) \longrightarrow \text{ion}^+(g) + e^- \qquad \Delta E = IE_1 > 0 \tag{8.2}$$

The second ionization energy (IE_2) removes a second electron. This electron is pulled away from a positively charged ion, so IE_2 is always larger than IE_1:

$$\text{Ion}^+(g) \longrightarrow \text{ion}^{2+}(g) + e^- \qquad \Delta E = IE_2 \text{ (always} > IE_1\text{)}$$

The first ionization energy is a key factor in an element's chemical reactivity because, as you'll see, *atoms with a low IE_1 tend to form cations during reactions, whereas those with a high IE_1 (except the noble gases) often form anions.*

Variations in First Ionization Energy The elements exhibit a periodic pattern in first ionization energy, as shown in Figure 8.10. By comparing this figure with Figure 8.9, you can see a roughly *inverse* relationship between IE_1 and atomic size: *as size decreases, it takes more energy to remove an electron.* This inverse relationship appears throughout the groups and periods of the table.

Let's examine the two trends and their exceptions.

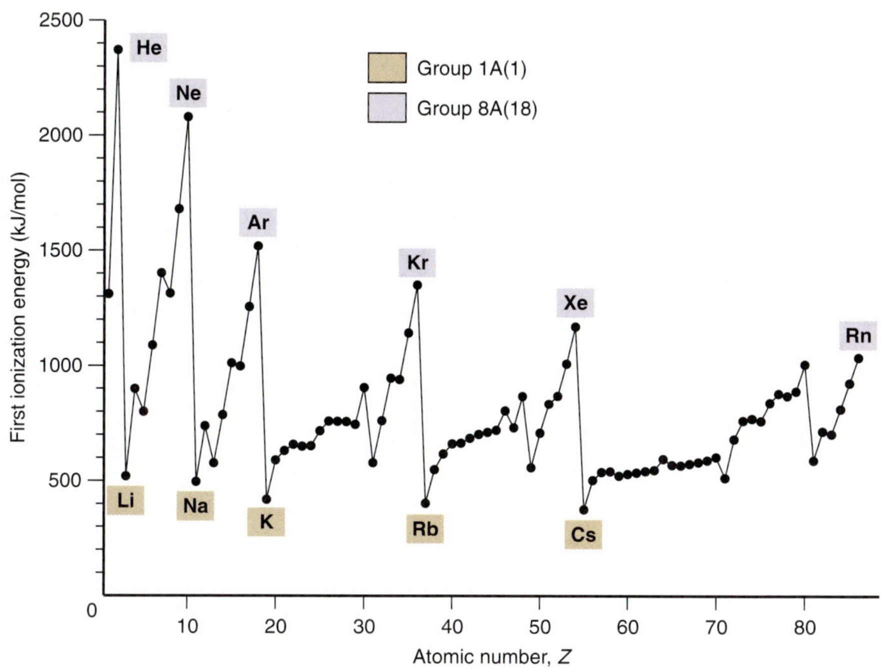

FIGURE 8.10 Periodicity of first ionization energy (IE_1). A plot of IE_1 vs. atomic number for the elements in Periods 1 through 6 shows a periodic pattern: the lowest values occur for the alkali metals *(brown)* and the highest for the noble gases *(purple)*. This is the *inverse* of the trend in atomic size (see Figure 8.9).

1. *Down a group.* As we move *down* a main group, the orbital's n value increases and so does atomic size. As the distance from nucleus to outermost electron increases, the attraction between them lessens, making the electron easier to remove. *Ionization energy generally* ***decreases down*** *a group* (Figure 8.11): it is easier to remove an outer electron from an element in Period 6 than one in Period 2.

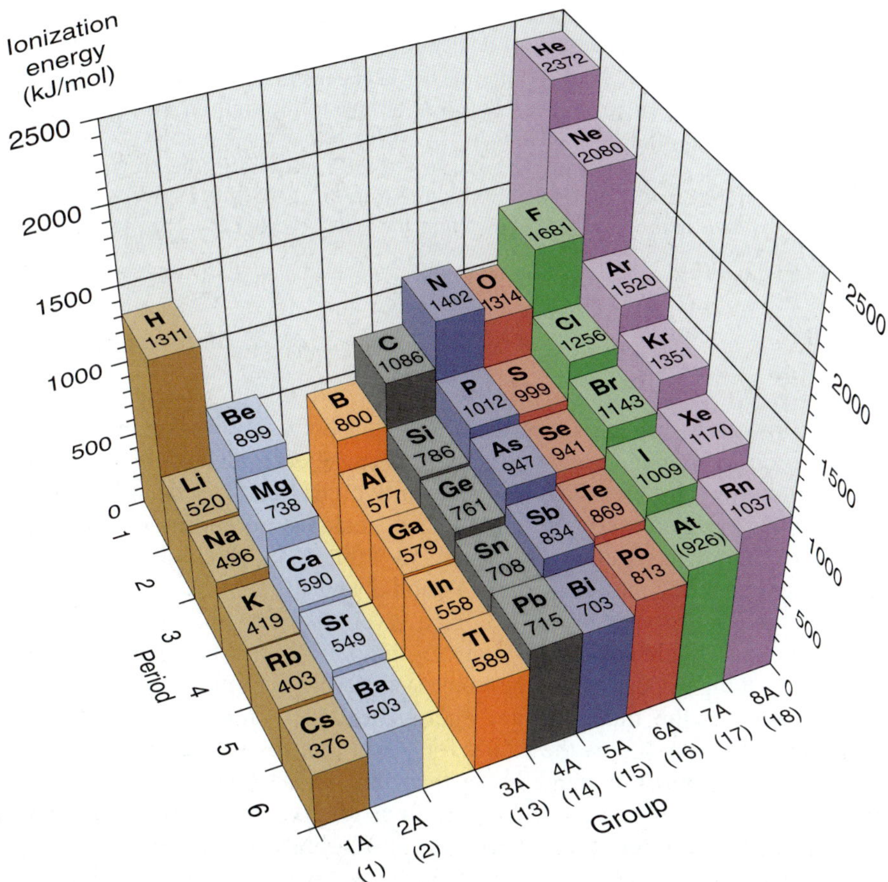

FIGURE 8.11 First ionization energies of the main-group elements. Values for IE_1 (in kJ/mol) of the main-group elements are shown as posts of varying height. Note the general increase within a period and decrease within a group. Thus, the lowest value is at the bottom of Group 1A(1), and the highest is at the top of Group 8A(18).

The only significant exception to this pattern occurs in Group 3A(13), right after the transition series, and is due to the effect of the series on atomic size: IE_1 decreases from boron (B) to aluminum (Al), but not for the rest of the group. Filling the *d* sublevels in Periods 4, 5, and 6 causes a greater than expected Z_{eff}, which holds the outer electrons more tightly in the larger Group 3A members.

2. *Across a period.* As we move left to right across a period, the orbital's *n* value stays the same, so Z_{eff} increases and atomic size decreases. As a result, the attraction between nucleus and outer electrons increases, which makes an electron harder to remove. *Ionization energy generally* ***increases across*** *a period:* it is easier to remove an outer electron from an alkali metal than from a noble gas.

There are several small "dips" in the otherwise smooth increase in ionization energy. These occur in Group 3A(13) for B and Al and in Group 6A(16) for O and S. The dips in Group 3A occur because these electrons are the first in the *np* sublevel. This sublevel is higher in energy than the *ns*, so the electron in it is pulled off more easily, leaving a stable, filled *ns* sublevel. The dips in Group 6A occur because the np^4 electron is the first to pair up with another *np* electron, and electron-electron repulsions raise the orbital energy. Removing the np^4 electron relieves the repulsions and leaves a stable, half-filled *np* sublevel; so the fourth *p* electron comes off more easily than the third one does.

SAMPLE PROBLEM 8.4 Ranking Elements by First Ionization Energy

Problem Using the periodic table only, rank the elements in each of the following sets in order of *decreasing* IE_1:
(a) Kr, He, Ar **(b)** Sb, Te, Sn **(c)** K, Ca, Rb **(d)** I, Xe, Cs

Plan As in Sample Problem 8.3, we first find the elements in the periodic table and then apply the general trends of decreasing IE_1 down a group and increasing IE_1 across a period.

Solution (a) He > Ar > Kr. These three are all in Group 8A(18), and IE_1 decreases down a group.

(b) Te > Sb > Sn. These three are all in Period 5, and IE_1 increases across a period.

(c) Ca > K > Rb. IE_1 of K is larger than IE_1 of Rb because K is higher in Group 1A(1). IE_1 of Ca is larger than IE_1 of K because Ca is farther to the right in Period 4.

(d) Xe > I > Cs. IE_1 of I is smaller than IE_1 of Xe because I is farther to the left. IE_1 of I is larger than IE_1 of Cs because I is farther to the right and in the previous period.

Check Because trends in IE_1 are generally the opposite of the trends in size, you can rank the elements by size and check that you obtain the reverse order.

FOLLOW-UP PROBLEM 8.4 Rank the elements in each of the following sets in order of increasing IE_1:
(a) Sb, Sn, I **(b)** Sr, Ca, Ba

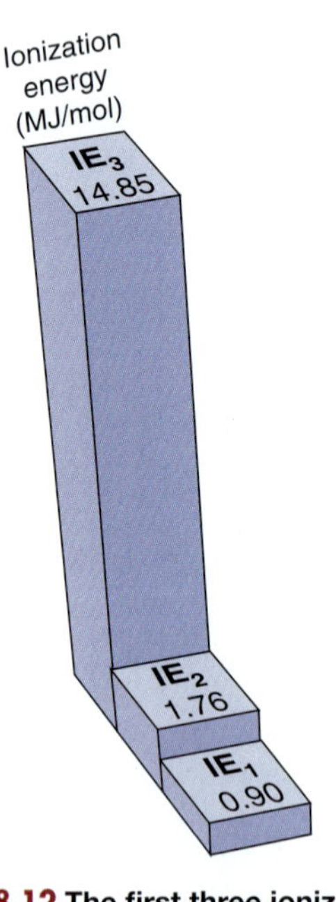

FIGURE 8.12 The first three ionization energies of beryllium (in MJ/mol). Successive ionization energies always increase, but an exceptionally large increase occurs when the first core electron is removed. For Be, this occurs with the third electron (IE_3). (Also see Table 8.4.)

Variations in Successive Ionization Energies Successive ionization energies (IE_1, IE_2, and so on) of a given element increase because each electron is pulled away from an ion with a progressively higher positive charge. Note from Figure 8.12, however, that this increase is not smooth, but includes an enormous jump.

A more complete picture is presented in Table 8.4, which shows successive ionization energies for the elements in Period 2 and the first element in Period 3. Move horizontally through the values for a given element, and you reach a point that separates relatively low from relatively high IE values (shaded area to right of line). This jump appears *after* the outer (valence) electrons have been removed and, thus, reflects the much greater energy needed to remove an inner (core) electron. For example, follow the values for boron (B): IE_1 is lower than IE_2, which is lower than IE_3, which is *much* lower than IE_4. Thus, boron has three electrons in the highest energy level ($1s^22s^22p^1$). Because of the significantly greater energy needed to remove core electrons, they are *not* involved in chemical reactions.

Table 8.4 Successive Ionization Energies of the Elements Lithium Through Sodium

Z	Element	Number of Valence Electrons	Ionization Energy (MJ/mol)*									
			IE_1	IE_2	IE_3	IE_4	IE_5	IE_6	IE_7	IE_8	IE_9	IE_{10}
3	Li	1	0.52	7.30	11.81							
4	Be	2	0.90	1.76	14.85	21.01			CORE ELECTRONS			
5	B	3	0.80	2.43	3.66	25.02	32.82					
6	C	4	1.09	2.35	4.62	6.22	37.83	47.28				
7	N	5	1.40	2.86	4.58	7.48	9.44	53.27	64.36			
8	O	6	1.31	3.39	5.30	7.47	10.98	13.33	71.33	84.08		
9	F	7	1.68	3.37	6.05	8.41	11.02	15.16	17.87	92.04	106.43	
10	Ne	8	2.08	3.95	6.12	9.37	12.18	15.24	20.00	23.07	115.38	131.43
11	Na	1	0.50	4.56	6.91	9.54	13.35	16.61	20.11	25.49	28.93	141.37

*MJ/mol, or megajoules per mole = 10^3 kJ/mol.

SAMPLE PROBLEM 8.5 Identifying an Element from Successive Ionization Energies

Problem Name the Period 3 element with the following ionization energies (in kJ/mol), and write its electron configuration:

IE_1	IE_2	IE_3	IE_4	IE_5	IE_6
1012	1903	2910	4956	6278	22,230

Plan We look for a large jump in the IE values, which occurs after all valence electrons have been removed. Then we refer to the periodic table to find the Period 3 element with this number of valence electrons and write its electron configuration.

Solution The exceptionally large jump occurs after IE_5, indicating that the element has five valence electrons and, thus, is in Group 5A(15). This Period 3 element is phosphorus (P; $Z = 15$). Its electron configuration is $1s^22s^22p^63s^23p^3$.

FOLLOW-UP PROBLEM 8.5 Element Q is in Period 3 and has the following ionization energies (in kJ/mol):

IE_1	IE_2	IE_3	IE_4	IE_5	IE_6
577	1816	2744	11,576	14,829	18,375

Name element Q and write its electron configuration.

Trends in Electron Affinity

The **electron affinity (EA)** is the energy change (in kJ) accompanying the *addition* of 1 mol of electrons to 1 mol of gaseous atoms or ions. As with ionization energy, there is a first electron affinity, a second, and so forth. The *first electron affinity* (EA_1) refers to the formation of 1 mol of monovalent (1−) gaseous anions:

$$\text{Atom}(g) + e^- \longrightarrow \text{ion}^-(g) \quad \Delta E = EA_1$$

In most cases, *energy is released when the first electron is added* because it is attracted to the atom's nuclear charge. Thus, EA_1 is usually negative (just as ΔH for an exothermic reaction is negative).* The second electron affinity (EA_2), however, is always positive because energy must be *absorbed* to overcome electrostatic repulsions and add another electron to a negative ion.

*Tables of first electron affinity often list them as positive if energy is *absorbed* to remove an electron from the anion. Keep this convention in mind when researching these values in reference texts. Electron affinities are difficult to measure, so values are frequently updated with more accurate data. Values for Group 2A(2) reflect recent changes.

FIGURE 8.13 Electron affinities of the main-group elements. The electron affinities (in kJ/mol) of the main-group elements are shown. Negative values indicate that energy is released when the anion forms. Positive values, which occur in Group 8A(18), indicate that energy is absorbed to form the anion; in fact, these anions are unstable and the values are estimated.

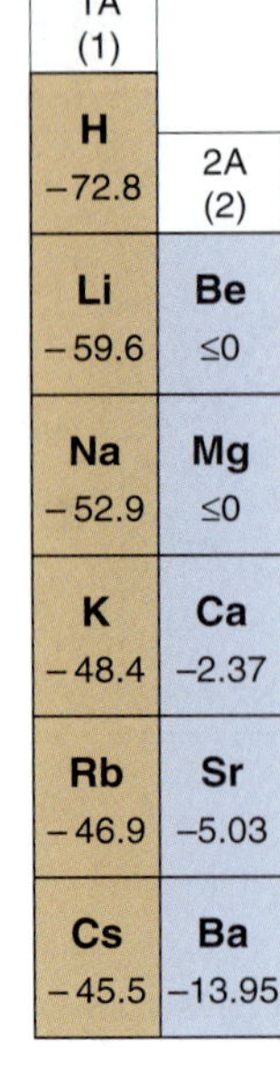

1A (1)	2A (2)	3A (13)	4A (14)	5A (15)	6A (16)	7A (17)	8A (18)
H −72.8							He (0.0)
Li −59.6	Be ≤0	B −26.7	C −122	N +7	O −141	F −328	Ne (+29)
Na −52.9	Mg ≤0	Al −42.5	Si −134	P −72.0	S −200	Cl −349	Ar (+35)
K −48.4	Ca −2.37	Ga −28.9	Ge −119	As −78.2	Se −195	Br −325	Kr (+39)
Rb −46.9	Sr −5.03	In −28.9	Sn −107	Sb −103	Te −190	I −295	Xe (+41)
Cs −45.5	Ba −13.95	Tl −19.3	Pb −35.1	Bi −91.3	Po −183	At −270	Rn (+41)

Factors other than Z_{eff} and atomic size affect electron affinities, so trends are not as regular as those for the previous two properties. For instance, we might expect electron affinities to decrease smoothly down a group (smaller negative number) because the nucleus is farther away from an electron being added. But, as Figure 8.13 shows, only Group 1A(1) exhibits this behavior. We might also expect a regular increase in electron affinities across a period (larger negative number) because size decreases and the increasing Z_{eff} should attract the electron being added more strongly. An overall left-to-right increase in magnitude is there, but we certainly cannot say that it is a regular increase. These exceptions arise from changes in sublevel energy and in electron-electron repulsion.

Despite the irregularities, three key points emerge when we examine the relative values of ionization energy and electron affinity:

1. *Reactive nonmetals.* The elements in Group 6A(16) and especially those in Group 7A(17) (halogens) have high ionization energies and highly negative (exothermic) electron affinities. These elements lose electrons with difficulty but attract them strongly. Therefore, *in their ionic compounds, they form negative ions.*
2. *Reactive metals.* The elements in Groups 1A(1) and 2A(2) have low ionization energies and slightly negative (exothermic) electron affinities. In both groups, the elements lose electrons readily but attract them only weakly, if at all. Therefore, *in their ionic compounds, they form positive ions.*
3. *Noble gases.* The elements in Group 8A(18) have very high ionization energies and slightly positive (endothermic) electron affinities. Therefore, *these elements tend* ***not*** *to lose or gain electrons.* In fact, only the larger members of the group (Kr, Xe, Rn) form any compounds at all.

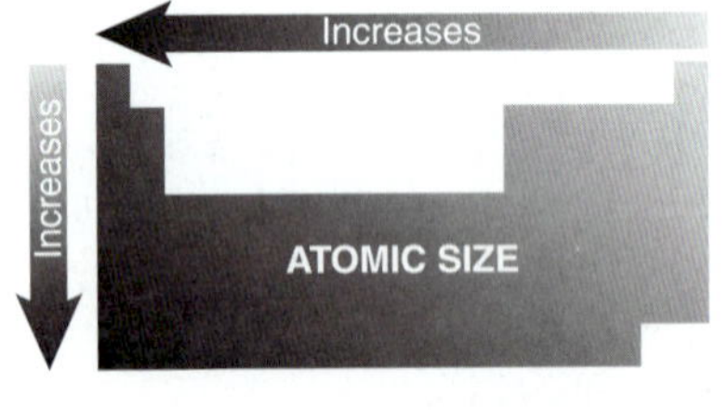

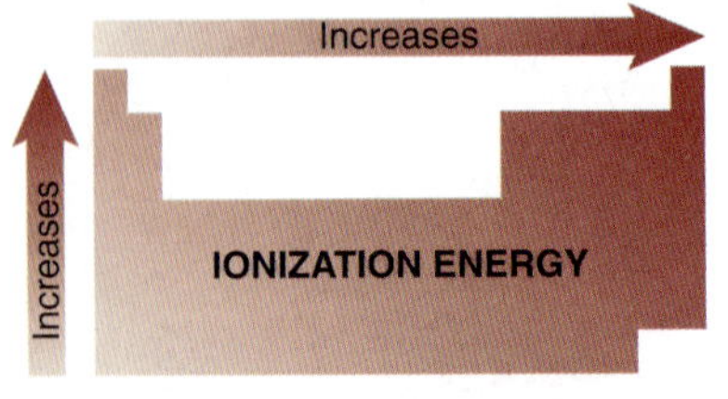

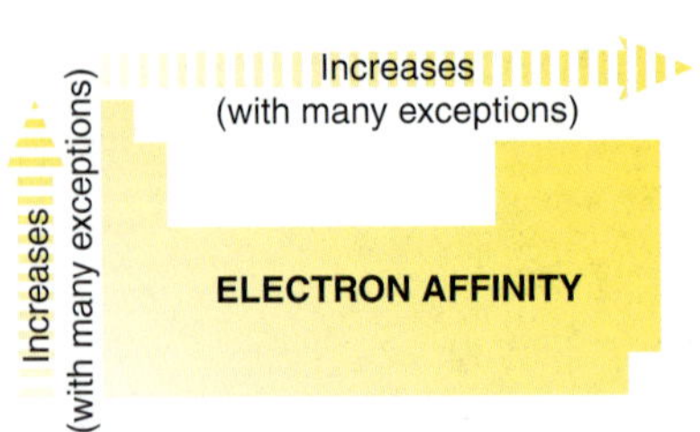

FIGURE 8.14 Trends in three atomic properties. Periodic trends are depicted as gradations in shading on miniature periodic tables, with arrows indicating the direction of general *increase* in a group or period. For electron affinity, Group 8A(18) is not shown, and the dashed arrows indicate the numerous exceptions to expected trends.

SECTION 8.4 SUMMARY

Trends in three atomic properties are summarized in Figure 8.14. Atomic size increases down a main group and decreases across a period. Across a transition series, size remains relatively constant. • First ionization energy (the energy required to remove the outermost electron from a mole of gaseous atoms) is inversely related to atomic size: IE_1 decreases down a main group and increases across a period. • An element's successive ionization energies show a very large increase when the first inner (core) electron is removed. • Electron affinity (the energy involved in adding an electron to a mole of gaseous atoms) shows many variations from expected trends. • Based on the relative sizes of IEs and EAs, in their ionic compounds, the Group 1A(1) and 2A(2) elements tend to form cations, and the Group 6A(16) and 7A(17) elements tend to form anions.

8.5 ATOMIC STRUCTURE AND CHEMICAL REACTIVITY

Our main purpose for discussing atomic properties is, of course, to see how they affect element behavior. In this section, you'll see how the properties we just examined influence metallic behavior and determine the type of ion an element can form, as well as how electron configuration relates to magnetic properties.

Trends in Metallic Behavior

Metals are located in the left and lower three-quarters of the periodic table. They are typically shiny solids with moderate to high melting points, are good thermal and electrical conductors, can be drawn into wires and rolled into sheets, and tend to lose electrons to nonmetals. *Nonmetals* are located in the upper right quarter of the table. They are typically not shiny, have relatively low melting points, are poor thermal and electrical conductors, are mostly crumbly solids or gases, and tend to gain electrons from metals. *Metalloids* are located in the region between the other two classes and have properties between them as well. Thus, *metallic behavior decreases left to right and increases top to bottom* in the periodic table (Figure 8.15).

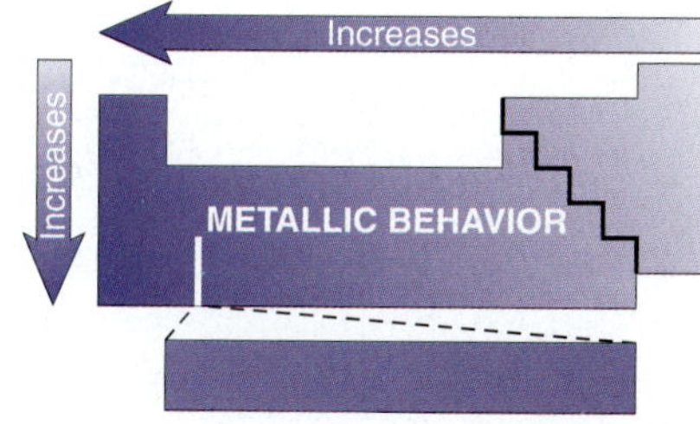

FIGURE 8.15 Trends in metallic behavior. The gradation in metallic behavior among the elements is depicted as a gradation in shading from bottom left to top right, with arrows showing the direction of increase. (Hydrogen appears next to helium in this periodic table.)

It's important to realize, however, that an element's properties may not fall neatly into our categories. For instance, the nonmetal carbon in the form of graphite is a good electrical conductor. Iodine, another nonmetal, is a shiny solid. Gallium and cesium are metals that melt at temperatures below body temperature, and mercury is a liquid at room temperature. And iron is quite brittle. Despite such exceptions, we can make several generalizations about metallic behavior.

Relative Tendency to Lose Electrons Metals tend to lose electrons during chemical reactions because they have low ionization energies compared to nonmetals. The increase in metallic behavior down a group is most obvious in the physical and chemical behavior of the elements in Groups 3A(13) through 6A(16), which contain more than one class of element. For example, consider the elements in Group 5A(15). Here, the change is so great that, with regard to monatomic ions, *elements at the top tend to form anions and those at the bottom tend to form cations.* Nitrogen (N) is a gaseous nonmetal, and phosphorus (P) is a solid nonmetal. Both occur occasionally as 3− anions in their compounds. Arsenic (As) and antimony (Sb) are metalloids, with Sb the more metallic of the two; neither forms ions readily. Bismuth (Bi), the largest member, is a typical metal, forming mostly ionic compounds in which it appears as a 3+ cation. Even in Group 2A(2), which consists entirely of metals, the tendency to form cations increases down the group. Thus, beryllium (Be) forms covalent compounds with nonmetals, whereas the compounds of barium (Ba) are ionic.

As we move across a period, it becomes more difficult to lose an electron (IE increases) and easier to gain one (EA becomes more negative). Therefore, with regard to monatomic ions, *elements at the left tend to form cations and those at the right tend to form anions.* The typical decrease in metallic behavior across a period is clear among the elements in Period 3. Sodium and magnesium are metals. Sodium is shiny when freshly cut under mineral oil, but it loses an electron so readily to O_2 that, if cut in air, its surface is coated immediately with a dull oxide. These metals exist naturally as Na^+ and Mg^{2+} ions in oceans, minerals, and organisms. Aluminum is metallic in its physical properties and forms the Al^{3+} ion in some compounds, but it bonds covalently in most others. Silicon (Si) is a shiny metalloid that does not occur as a monatomic ion. The most common form of phosphorus is a white, waxy nonmetal that, as noted earlier, forms the P^{3-} ion in a few compounds. Sulfur is a crumbly yellow nonmetal that forms the sulfide ion (S^{2-}) in many compounds. Diatomic chlorine (Cl_2) is a yellow-green, gaseous nonmetal that attracts electrons avidly and exists in nature as the Cl^- ion.

Acid-Base Behavior of the Element Oxides Metals are also distinguished from nonmetals by the acid-base behavior of their oxides in water:

- Most main-group metals *transfer* electrons to oxygen, so their *oxides are ionic. In water, the oxides act as bases,* producing OH^- ions from O^{2-} and reacting with acids.
- Nonmetals *share* electrons with oxygen, so *nonmetal oxides are covalent. In water, they act as acids,* producing H^+ ions and reacting with bases.

Some metals and many metalloids form oxides that are **amphoteric:** they can act as acids *or* as bases in water.

Figure 8.16 classifies the acid-base behavior of some common oxides, focusing on the elements in Group 5A(15) and Period 3. Note that *as the elements become more metallic down a group, their oxides become more basic.* Among oxides of Group 5A(15) elements, dinitrogen pentaoxide (N_2O_5) forms nitric acid, HNO_3, a strong acid, while tetraphosphorus decaoxide (P_4O_{10}) forms a weaker acid, H_3PO_4. The oxide of the metalloid arsenic (As_2O_5) is weakly acidic, whereas that of the metalloid antimony (Sb_2O_5) is weakly basic. Bismuth, the most metallic element of the group, forms an insoluble basic oxide (Bi_2O_3) that reacts with acid to form a salt and water.

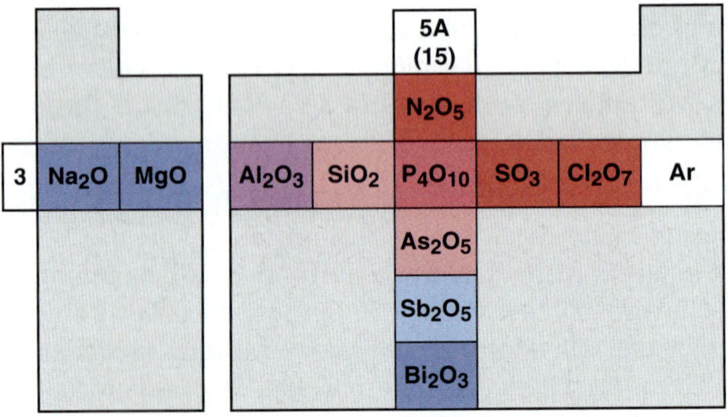

FIGURE 8.16 The trend in acid-base behavior of element oxides. The trend in acid-base behavior for some common oxides of Group 5A(15) and Period 3 elements is shown as a gradation in color (*red* = acidic; *blue* = basic). Note that the metals form basic oxides and the nonmetals form acidic oxides. Aluminum forms an oxide *(purple)* that can act as an acid or as a base. Thus, as atomic size increases, ionization energy decreases, and oxide basicity increases.

Note that *as the elements become less metallic across a period, their oxides become more acidic.* In Period 3, sodium and magnesium form the strongly basic oxides Na_2O and MgO. Metallic aluminum forms amphoteric aluminum oxide (Al_2O_3), which reacts with acid or with base, whereas silicon dioxide is weakly acidic. The common oxides of phosphorus, sulfur, and chlorine form acids of increasing strength: H_3PO_4, H_2SO_4, and $HClO_4$.

Properties of Monatomic Ions

So far we have focused on the reactants—the atoms—in the process of electron loss and gain. Now we focus on the products—the ions. We examine electron configurations, magnetic properties, and ionic radius relative to atomic radius.

Electron Configurations of Main-Group Ions In Chapter 2, you learned the symbols and charges of many monatomic ions. But *why* does an ion have a particular charge in its compounds? Why is a sodium ion Na^+ and not Na^{2+}, and why is a fluoride ion F^- and not F^{2-}? For elements at the left and right ends of the periodic table, the explanation concerns the very low reactivity of the noble gases. As we said earlier, because they have high IEs and positive (endothermic) EAs, the noble gases typically do not form ions but remain chemically stable with a *filled* outer energy level (ns^2np^6). *Elements in Groups 1A(1), 2A(2), 6A(16), and 7A(17) that readily form ions either lose or gain electrons to attain a filled outer level and thus a noble gas configuration.* Their ions are said to be **isoelectronic** (Greek *iso*, "same") with the nearest noble gas. Figure 8.17 shows this relationship.

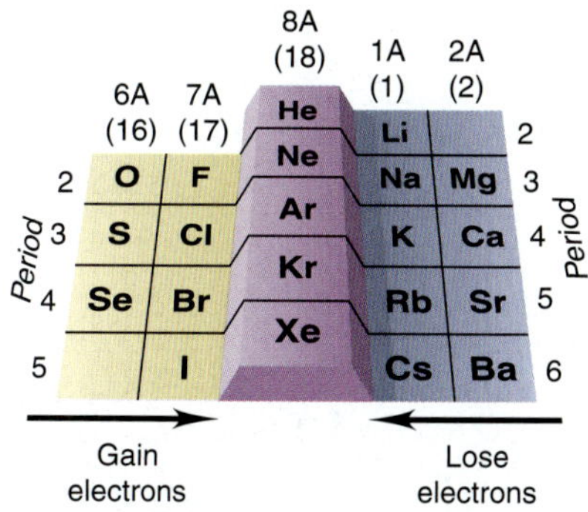

FIGURE 8.17 Main-group ions and the noble gas electron configurations. Most of the elements that form monatomic ions that are isoelectronic with a noble gas lie in the four groups that flank Group 8A(18), two on either side.

When an alkali metal atom [Group 1A(1)] loses its single valence electron, it becomes isoelectronic with the *previous* noble gas. The Na^+ ion, for example, is isoelectronic with neon (Ne):

$$\text{Na}\ (1s^22s^22p^63s^1) \longrightarrow \text{Na}^+\ (1s^22s^22p^6)\ [\text{isoelectronic with Ne}\ (1s^22s^22p^6)] + e^-$$

When a halogen atom [Group 7A(17)] adds a single electron to the five in its *np* sublevel, it becomes isoelectronic with the *next* noble gas. Bromide ion, for example, is isoelectronic with krypton (Kr):

$$\text{Br}\ ([\text{Ar}]\ 4s^23d^{10}4p^5) + e^- \longrightarrow \text{Br}^-\ ([\text{Ar}]\ 4s^23d^{10}4p^6)\ [\text{isoelectronic with Kr}\ ([\text{Ar}]\ 4s^23d^{10}4p^6)]$$

The energy needed to remove the electrons from metals to attain the previous noble gas configuration is supplied during their exothermic reactions with nonmetals. Removing more than one electron from Na to form Na^{2+} or more than two from Mg to form Mg^{3+} means removing core electrons, which requires more energy than is available in a reaction. This is the reason that $NaCl_2$ and MgF_3 do *not* exist. Similarly, adding two electrons to F to form F^{2-} or three to O to form O^{3-} means placing the extra electron into the next energy level. With 18 electrons acting as inner electrons and shielding the nuclear charge very effectively, adding an electron to the negative ion, F^- or O^{2-}, requires too much energy. Thus, we never see Na_2F or Mg_3O_2.

The larger metals of Groups 3A(13), 4A(14), and 5A(15) form cations through a different process, because it would be energetically impossible for them to lose enough electrons to attain a noble gas configuration. For example, tin (Sn; $Z = 50$) would have to lose 14 electrons—two 5*p*, ten 4*d*, and two 5*s*—to be isoelectronic with krypton (Kr; $Z = 36$), the previous noble gas. Instead, tin loses far fewer electrons and attains two different stable configurations. In the tin(IV) ion (Sn^{4+}), the metal atom empties its outer energy level and attains the stability of empty 5*s* and 5*p* sublevels and a filled inner 4*d* sublevel. This $(n-1)d^{10}$ configuration is called a **pseudo–noble gas configuration:**

$$\text{Sn}\ ([\text{Kr}]\ 5s^24d^{10}5p^2) \longrightarrow \text{Sn}^{4+}\ ([\text{Kr}]\ 4d^{10}) + 4e^-$$

Alternatively, in the more common tin(II) ion (Sn^{2+}), the atom loses the two 5*p* electrons only and attains the stability of filled 5*s* and 4*d* sublevels:

$$\text{Sn}\ ([\text{Kr}]\ 5s^24d^{10}5p^2) \longrightarrow \text{Sn}^{2+}\ ([\text{Kr}]\ 5s^24d^{10}) + 2e^-$$

The retained ns^2 electrons are sometimes called an *inert pair* because they seem difficult to remove. Thallium, lead, and bismuth, the largest and most metallic members of Groups 3A(13) to 5A(15), commonly form ions that retain the ns^2 pair of electrons: Tl^+, Pb^{2+}, and Bi^{3+}.

Excessively high energy cost is also the reason that some elements do not form monatomic ions in any of their reactions. For instance, carbon would have to lose four electrons to form C^{4+} and attain the He configuration, or gain four to form C^{4-} and attain the Ne configuration, but neither ion forms. (Such multivalent ions *are* observed in the spectra of stars, however, where temperatures exceed 10^6 K.) As you'll see in Chapter 9, carbon and other atoms that do not form ions attain a filled shell by *sharing* electrons through covalent bonding.

SAMPLE PROBLEM 8.6 Writing Electron Configurations of Main-Group Ions

Problem Using condensed electron configurations, write reactions for the formation of the common ions of the following elements:

(a) Iodine ($Z = 53$) **(b)** Potassium ($Z = 19$) **(c)** Indium ($Z = 49$)

Plan We identify the element's position in the periodic table and recall two general points:

- Ions of elements in Groups 1A(1), 2A(2), 6A(16), and 7A(17) are isoelectronic with the nearest noble gas.
- Metals in Groups 3A(13) to 5A(15) lose the *ns* and *np* electrons or just the *np* electrons.

Solution **(a)** Iodine is in Group 7A(17), so it gains one electron and is isoelectronic with xenon:

$$\text{I ([Kr] } 5s^24d^{10}5p^5) + e^- \longrightarrow \text{I}^- \text{ ([Kr] } 5s^24d^{10}5p^6) \quad \text{(same as Xe)}$$

(b) Potassium is in Group 1A(1), so it loses one electron and is isoelectronic with argon:

$$\text{K ([Ar] } 4s^1) \longrightarrow \text{K}^+ \text{ ([Ar])} + e^-$$

(c) Indium is in Group 3A(13), so it loses either three electrons to form In^{3+} (pseudo–noble gas configuration) or one to form In^+ (inert pair):

$$\text{In ([Kr] } 5s^24d^{10}5p^1) \longrightarrow \text{In}^{3+} \text{ ([Kr] } 4d^{10}) + 3e^-$$

$$\text{In ([Kr] } 5s^24d^{10}5p^1) \longrightarrow \text{In}^+ \text{ ([Kr] } 5s^24d^{10}) + e^-$$

Check Be sure that the number of electrons in the ion's electron configuration, plus those gained or lost to form the ion, equals Z.

FOLLOW-UP PROBLEM 8.6 Using condensed electron configurations, write reactions showing the formation of the common ions of the following elements:
(a) Ba (Z = 56) **(b)** O (Z = 8) **(c)** Pb (Z = 82)

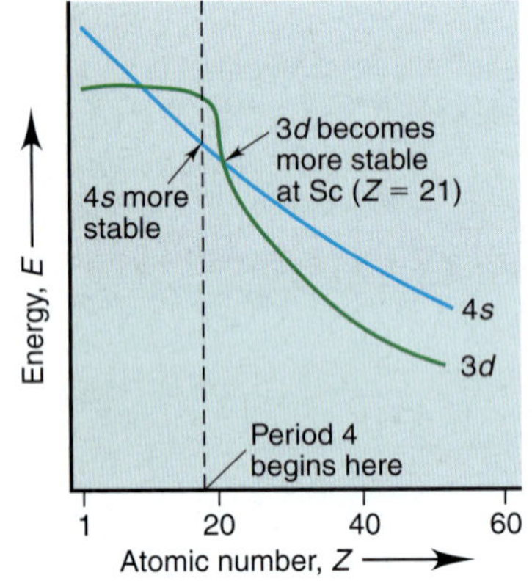

FIGURE 8.18 The Period 4 crossover in sublevel energies. The 3*d* orbitals are empty in elements at the beginning of Period 4. Because the 4*s* electron penetrates closer to the nucleus, the energy of the 4*s* orbital is lower in K and Ca; thus, the 4*s* fills before the 3*d*. But as the 3*d* orbitals fill, beginning with $Z = 21$, these inner electrons are attracted by the increasing nuclear charge, and they also shield the 4*s* electrons. As a result, there is an energy crossover, with the 3*d* sublevel becoming lower in energy than the 4*s*. For this reason, the 4*s* electrons are removed first when the transition metal ion forms. In other words,

- For a main-group metal ion, the highest *n* level of electrons is "last-in, first-out."
- For a transition metal ion, the highest *n* level of electrons is "first-in, first out."

Electron Configurations of Transition Metal Ions In contrast to most main-group ions, *transition metal ions rarely attain a noble gas configuration,* and the reason, once again, is that energy costs are too high. The exceptions in Period 4 are scandium, which forms Sc^{3+}, and titanium, which occasionally forms Ti^{4+} in some compounds. The typical behavior of a transition element is to *form more than one cation by losing all of its ns and some of its (n − 1)d electrons.* (We focus here on the Period 4 series, but these points hold for Periods 5 and 6 also.)

In the aufbau process of building up the ground-state atoms, Period 3 ends with the noble gas argon. At the beginning of Period 4, the radial probability distribution of the 4*s* orbital near the nucleus makes it more stable than the empty 3*d*. Therefore, the first and second electrons added in the period enter the 4*s* in K and Ca. But, as soon as we reach the transition elements and the 3*d* orbitals begin to fill, the increasing nuclear charge attracts their electrons more and more strongly. Moreover, the added 3*d* electrons fill inner orbitals, so they are not very well shielded from the increasing nuclear charge by the 4*s* electrons. As a result, the 3*d* orbital becomes *more stable* than the 4*s*. In effect, a *crossover in orbital energy* occurs as we enter the transition series (Figure 8.18). The effect on ion formation is critical: because the 3*d* orbitals are more stable, *the 4s electrons are lost before the 3d electrons to form the Period 4 transition metal ions.* Thus, the 4*s* electrons are added before the 3*d* to form the *atom* but lost before the 3*d* to form the *ion:* "first-in, first-out."

To summarize, *electrons with the highest n value are removed first.* Here are a few simple rules for forming the ion of any main-group or transition element:

- For main-group, *s*-block metals, remove all electrons with the highest *n* value.
- For main-group, *p*-block metals, remove *np* electrons before *ns* electrons.
- For transition (*d*-block) metals, remove *ns* electrons before $(n - 1)d$ electrons.
- For nonmetals, add electrons to the *p* orbitals of highest *n* value.

Magnetic Properties of Transition Metal Ions If we can't see electrons in orbitals, how do we know that a particular electron configuration is correct? Although analysis of atomic spectra is the most important method for determining configuration, the magnetic properties of an element and its compounds can support or refute conclusions from spectra. Recall that electron spin generates a tiny magnetic field, which causes a beam of H atoms to split in an external magnetic field (see Figure 8.1). Only chemical species (atoms, ions, or molecules)

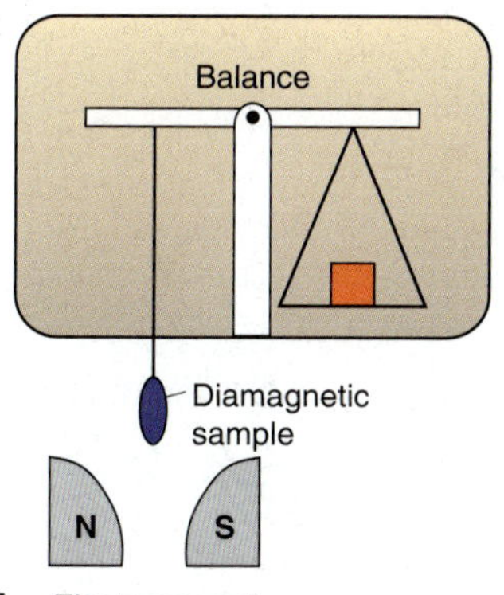

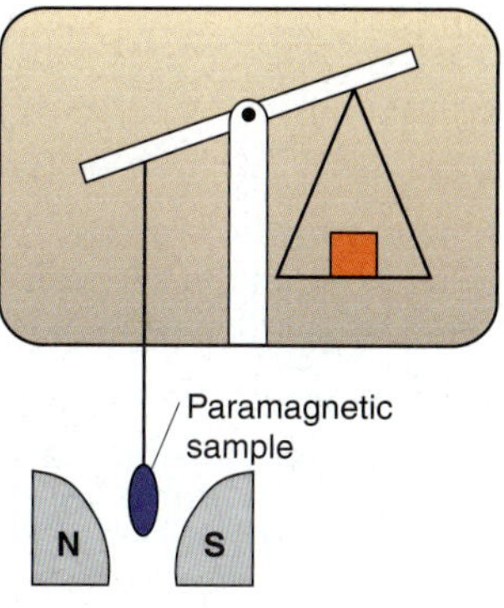

FIGURE 8.19 Apparatus for measuring the magnetic behavior of a sample. The substance is weighed on a very sensitive balance in the absence of an external magnetic field. **A,** If the substance is diamagnetic (has all *paired* electrons), its apparent mass is unaffected (or slightly reduced) when the magnetic field is "on." **B,** If the substance is paramagnetic (has *unpaired* electrons), its apparent mass increases when the field is "on" because the balance arm feels an additional force. This method is used to estimate the number of unpaired electrons in transition metal compounds.

with one or more *unpaired* electrons are affected by the external field. The species used in the original 1921 split-beam experiment was the silver atom:

Ag (Z = 47) [Kr] $5s^1 4d^{10}$ 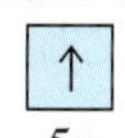 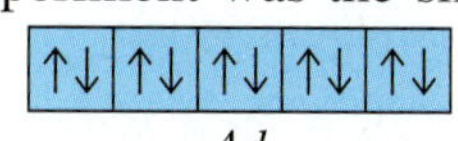 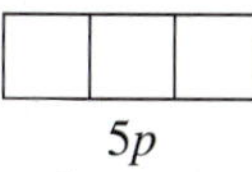

Note the unpaired $5s$ electron. A beam of cadmium atoms, the element after silver, is not split because their $5s$ electrons are *paired* (Cd: [Kr] $5s^2 4d^{10}$).

A species with unpaired electrons exhibits **paramagnetism:** it is attracted by an external magnetic field. A species with all electrons paired exhibits **diamagnetism:** it is not attracted (and, in fact, is slightly repelled) by a magnetic field. Figure 8.19 shows how this magnetic behavior is studied. Many transition metals and their compounds are paramagnetic because their atoms and ions have unpaired electrons.

Let's see how studies of paramagnetism might be used to provide additional evidence for a proposed electron configuration. Spectral analysis of the titanium atom yields the configuration [Ar] $4s^2 3d^2$. Experiment shows that Ti metal is paramagnetic, which is consistent with the presence of unpaired electrons in its atoms. Spectral analysis of the Ti^{2+} ion yields the configuration [Ar] $3d^2$, indicating loss of the two $4s$ electrons. Once again, experiment supports these findings by showing that Ti^{2+} compounds are paramagnetic. If Ti had lost its two $3d$ electrons during ion formation, its compounds would be diamagnetic because the $4s$ electrons are paired. Thus, the [Ar] $3d^2$ configuration supports the conclusion that electrons of highest n value are lost first:

$$\text{Ti ([Ar] } 4s^2 3d^2) \longrightarrow \text{Ti}^{2+} \text{ ([Ar] } 3d^2) + 2e^-$$

The partial orbital diagrams are

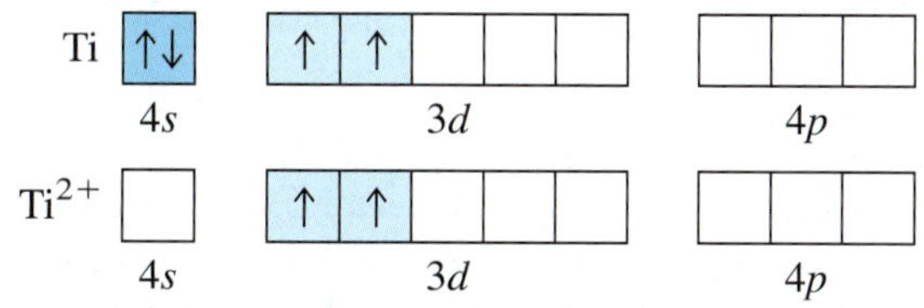

SAMPLE PROBLEM 8.7 Writing Electron Configurations and Predicting Magnetic Behavior of Transition Metal Ions

Problem Use condensed electron configurations to write the reaction for the formation of each transition metal ion, and predict whether the ion is paramagnetic:
(a) Mn^{2+} (Z = 25) **(b)** Cr^{3+} (Z = 24) **(c)** Hg^{2+} (Z = 80)

Plan We first write the condensed electron configuration of the atom, noting the irregularity for Cr in (b). Then we remove electrons, beginning with ns electrons, to attain the ion charge. If unpaired electrons are present, the ion is paramagnetic.

Solution **(a)** Mn ([Ar] $4s^2 3d^5$) $\longrightarrow$ Mn^{2+} ([Ar] $3d^5$) + $2e^-$

There are five unpaired e^-, so Mn^{2+} is paramagnetic.

(b) $Cr\ ([Ar]\ 4s^1 3d^5) \longrightarrow Cr^{3+}\ ([Ar]\ 3d^3) + 3e^-$

There are three unpaired e^-, so Cr^{3+} is paramagnetic.

(c) $Hg\ ([Xe]\ 6s^2 4f^{14} 5d^{10}) \longrightarrow Hg^{2+}\ ([Xe]\ 4f^{14} 5d^{10}) + 2e^-$

There are no unpaired e^-, so Hg^{2+} is *not* paramagnetic (is diamagnetic).

Check We removed the *ns* electrons first, and the sum of the lost electrons and those in the electron configuration of the ion equals *Z*.

FOLLOW-UP PROBLEM 8.7 Write the condensed electron configuration of each transition metal ion, and predict whether it is paramagnetic:
(a) V^{3+} ($Z = 23$) **(b)** Ni^{2+} ($Z = 28$) **(c)** La^{3+} ($Z = 57$)

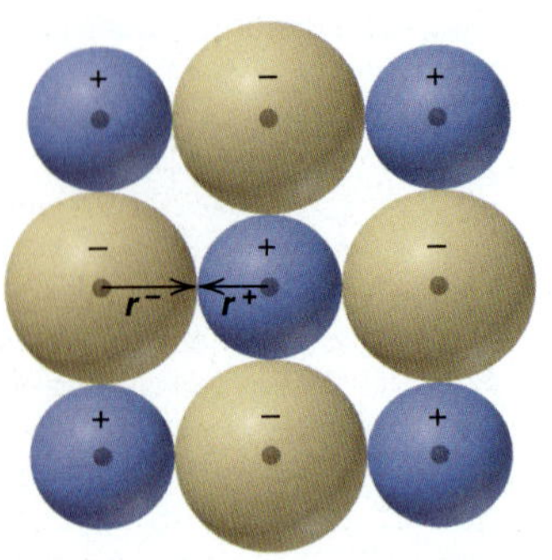

FIGURE 8.20 Depicting ionic radius. The cation radius (r^+) and the anion radius (r^-) each make up a portion of the total distance between the nuclei of adjacent ions in a crystalline ionic compound.

Ionic Size vs. Atomic Size The **ionic radius** is an estimate of the size of an ion in a crystalline ionic compound. You can picture it as one ion's portion of the distance between the nuclei of neighboring ions in the solid (Figure 8.20). From the relation between effective nuclear charge and atomic size, we can predict the size of an ion relative to its parent atom:

- *Cations are smaller than their parent atoms.* When a cation forms, electrons are *removed from* the outer level. The resulting decrease in electron repulsions allows the nuclear charge to pull the remaining electrons closer.
- *Anions are larger than their parent atoms.* When an anion forms, electrons are *added to* the outer level. The increase in repulsions causes the electrons to occupy more space.

Figure 8.21 shows the radii of some common main-group monatomic ions relative to their parent atoms. As you can see, *ionic size increases down a group* because the number of energy levels increases. Across a period, however, the pattern is more complex. Size decreases among the cations, then increases tremendously with the first of the anions, and finally decreases again among the anions.

This pattern results from changes in effective nuclear charge and electron-electron repulsions. In Period 3 (Na through Cl), for example, increasing Z_{eff} from left to right makes Na^+ larger than Mg^{2+}, which in turn is larger than Al^{3+}. The great jump in size from cations to anions occurs because we are *adding* electrons rather than removing them, so repulsions increase sharply. For instance, P^{3-} has eight more electrons than Al^{3+}. Then, the ongoing rise in Z_{eff} makes P^{3-} larger than S^{2-}, which is larger than Cl^-. These factors lead to some striking effects even among ions with the same number of electrons. Look at the ions within the dashed outline in Figure 8.21, which are all isoelectronic with neon. Even though the cations form from elements in the next period, the anions are still much larger. The pattern is

$$3- > 2- > 1- > 1+ > 2+ > 3+$$

When an element forms more than one cation, *the greater the ionic charge, the smaller the ionic radius.* Consider Fe^{2+} and Fe^{3+}. The number of protons is the same, but Fe^{3+} has one fewer electron, so electron repulsions are reduced somewhat. As a result, Z_{eff} increases, which pulls all the electrons closer, so Fe^{3+} is smaller than Fe^{2+}.

To summarize the main points,

- Ionic size increases down a group.
- Ionic size decreases across a period but increases from cations to anions.
- Ionic size decreases with increasing positive (or decreasing negative) charge in an isoelectronic series.
- Ionic size decreases as charge increases for different cations of a given element.

SAMPLE PROBLEM 8.8 Ranking Ions by Size

Problem Rank each set of ions in order of *decreasing* size, and explain your ranking:
(a) Ca^{2+}, Sr^{2+}, Mg^{2+} **(b)** K^+, S^{2-}, Cl^- **(c)** Au^+, Au^{3+}

Plan We find the position of each element in the periodic table and apply the ideas presented in the text.

Solution **(a)** Because Mg^{2+}, Ca^{2+}, and Sr^{2+} are all from Group 2A(2), they decrease in size up the group: $Sr^{2+} > Ca^{2+} > Mg^{2+}$.

(b) The ions K^+, S^{2-}, and Cl^- are isoelectronic. S^{2-} has a lower Z_{eff} than Cl^-, so it is larger. K^+ is a cation, and has the highest Z_{eff}, so it is smallest: $S^{2-} > Cl^- > K^+$.

(c) Au^+ has a lower charge than Au^{3+}, so it is larger: $Au^+ > Au^{3+}$.

FOLLOW-UP PROBLEM 8.8 Rank the ions in each set in order of *increasing* size:
(a) Cl^-, Br^-, F^- **(b)** Na^+, Mg^{2+}, F^- **(c)** Cr^{2+}, Cr^{3+}

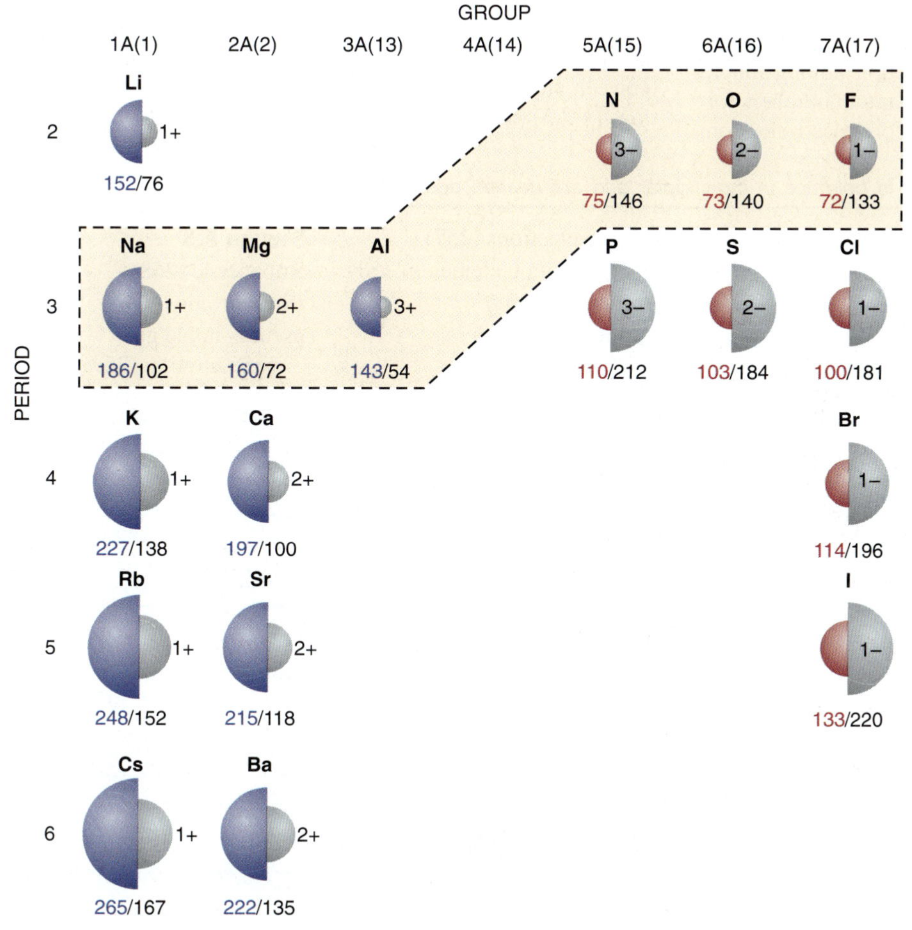

FIGURE 8.21 Ionic vs. atomic radii. The atomic radii *(colored half-spheres)* and ionic radii *(gray half-spheres)* of some main-group elements are arranged in periodic table format (with all radii values in picometers). Note that metal atoms *(blue)* form *smaller* positive ions, whereas nonmetal atoms *(red)* form *larger* negative ions. The dashed outline sets off ions of Period 2 nonmetals and Period 3 metals that are *isoelectronic* with neon. Note the size decrease from anions to cations.

Animation: Isoelectronic Series

SECTION 8.5 SUMMARY

Metallic behavior correlates with large atomic size and low ionization energy. Thus, metallic behavior increases down a group and decreases across a period. • Within the main groups, metal oxides are basic and nonmetal oxides acidic. Thus, oxides become more acidic across a period and more basic down a group. • Many main-group elements form ions that are isoelectronic with the nearest noble gas. Removing (or adding) more electrons than needed to attain the previous noble gas configuration requires a prohibitive amount of energy. • Metals in Groups 3A(13) to 5A(15) lose either their *np* electrons or both their *ns* and *np* electrons. • Transition metals lose *ns* electrons before $(n-1)d$ electrons and commonly form more than one ion. • Many transition metals and their compounds are paramagnetic because their atoms (or ions) have unpaired electrons. • Cations are smaller and anions larger than their parent atoms. Ionic radius increases down a group. Across a period, cationic and anionic radii decrease, but a large increase occurs from cations to anions.

CHAPTER REVIEW GUIDE

The following sections provide many aids to help you study this chapter. (Numbers in parentheses refer to pages, unless noted otherwise.)

LEARNING OBJECTIVES *These are concepts and skills to know after studying this chapter.*

Related section (§), sample problem (SP), and end-of-chapter problem (EP) numbers are listed in parentheses.

1. Understand the periodic law and the arrangement of elements by atomic number (§ 8.1) (EPs 8.1–8.3)
2. Describe the importance of the spin quantum number (m_s) and the exclusion principle for populating an orbital; understand how shielding and penetration lead to the splitting of energy levels into sublevels (§ 8.2) (EPs 8.4–8.13)
3. Understand orbital filling order, how outer configuration correlates with chemical behavior, and the distinction among inner, outer, and valence electrons; write the set of quantum numbers for any electron in an atom as well as full and condensed electron configurations and orbital diagrams for the atoms of any element (§ 8.3) (SPs 8.1, 8.2) (EPs 8.14–8.32)
4. Describe atomic radius, ionization energy, and electron affinity and their periodic trends; explain patterns in successive ionization energies and identify which electrons are involved in ion formation (to yield a noble gas or pseudo–noble gas electron configuration) (§ 8.4) (SPs 8.3–8.5) (EPs 8.33–8.47)
5. Describe the general properties of metals and nonmetals and understand how trends in metallic behavior relate to ion formation, oxide acidity, and magnetic behavior; understand the relation between atomic and ionic size and write ion electron configurations (§ 8.5) (SPs 8.6–8.8) (EPs 8.48–8.64)

KEY TERMS *These important terms appear in boldface in the chapter and are defined again in the Glossary.*

electron configuration (246)

Section 8.1

periodic law (246)

Section 8.2

spin quantum number (m_s) (247)
exclusion principle (248)
shielding (249)
effective nuclear charge (Z_{eff}) (249)
penetration (249)

Section 8.3

aufbau principle (250)
orbital diagram (251)
Hund's rule (252)
transition elements (254)
inner (core) electrons (257)
outer electrons (257)
valence electrons (257)
inner transition elements (258)
lanthanides (258)
actinides (258)

Section 8.4

atomic size (259)
metallic radius (259)
covalent radius (260)
ionization energy (IE) (262)
electron affinity (EA) (265)

Section 8.5

amphoteric (268)
isoelectronic (269)
pseudo–noble gas configuration (269)
paramagnetism (271)
diamagnetism (271)
ionic radius (272)

KEY EQUATIONS AND RELATIONSHIPS *Numbered and screened concepts are listed for you to refer to or memorize.*

8.1 Defining the energy order of sublevels in terms of the angular momentum quantum number (l value) (250):

Order of sublevel energies: $s < p < d < f$

8.2 Meaning of the first ionization energy (262):

$$\text{Atom}(g) \longrightarrow \text{ion}^+(g) + e^- \qquad \Delta E = IE_1 > 0$$

BRIEF SOLUTIONS TO *FOLLOW-UP PROBLEMS* *Compare your own solutions to these calculation steps and answers.*

8.1 The element has eight electrons, so $Z = 8$: oxygen.
Sixth electron: $n = 2$, $l = 1$, $m_l = 0$, $m_s = +\frac{1}{2}$

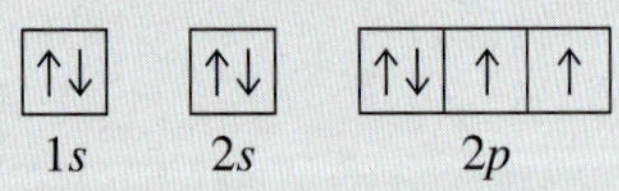

8.2 (a) For Ni, $1s^22s^22p^63s^23p^64s^23d^8$; [Ar] $4s^23d^8$

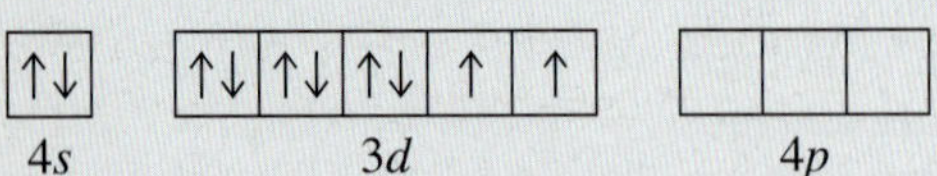

Ni has 18 inner electrons.
(b) For Sr, $1s^22s^22p^63s^23p^64s^23d^{10}4p^65s^2$; [Kr] $5s^2$

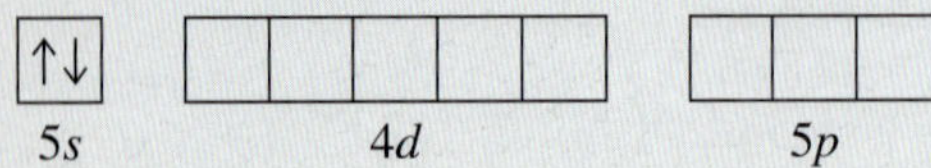

Sr has 36 inner electrons.

(c) For Po, $1s^22s^22p^63s^23p^64s^23d^{10}4p^65s^24d^{10}5p^66s^24f^{14}5d^{10}6p^4$; [Xe] $6s^24f^{14}5d^{10}6p^4$

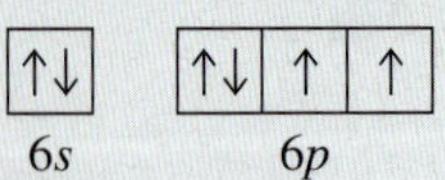

Po has 78 inner electrons.

8.3 (a) Cl < Br < Se; (b) Xe < I < Ba

8.4 (a) Sn < Sb < I; (b) Ba < Sr < Ca

8.5 Q is aluminum: $1s^22s^22p^63s^23p^1$

8.6 (a) Ba ([Xe] $6s^2$) $\longrightarrow$ Ba^{2+} ([Xe]) + $2e^-$
(b) O ([He] $2s^22p^4$) + $2e^-$ $\longrightarrow$ O^{2-} ([He] $2s^22p^6$) (same as Ne)
(c) Pb ([Xe] $6s^24f^{14}5d^{10}6p^2$) $\longrightarrow$ Pb^{2+} ([Xe] $6s^24f^{14}5d^{10}$) + $2e^-$
Pb ([Xe] $6s^24f^{14}5d^{10}6p^2$) $\longrightarrow$ Pb^{4+} ([Xe] $4f^{14}5d^{10}$) + $4e^-$

8.7 (a) V^{3+}: [Ar] $3d^2$; paramagnetic
(b) Ni^{2+}: [Ar] $3d^8$; paramagnetic
(c) La^{3+}: [Xe]; not paramagnetic (diamagnetic)

8.8 (a) $F^- < Cl^- < Br^-$; (b) $Mg^{2+} < Na^+ < F^-$;
(c) $Cr^{3+} < Cr^{2+}$

PROBLEMS

Problems with ***colored*** *numbers are answered in Appendix E. Sections match the text and provide the numbers of relevant sample problems. Bracketed problems are grouped in pairs (indicated by a short rule) that cover the same concept. Comprehensive Problems are based on material from any section or previous chapter.*

Development of the Periodic Table

8.1 What would be your reaction to a claim that a new element had been discovered, and it fit between tin (Sn) and antimony (Sb) in the periodic table?

8.2 Mendeleev arranged the elements in his periodic table by atomic mass. By what property are the elements now ordered in the periodic table? Give an example of a sequence of element order that would change if mass were still used.

8.3 Before Mendeleev published his periodic table, Johann Döbereiner grouped elements with similar properties into "triads," in which the unknown properties of one member could be predicted by averaging known values of the properties of the others. Predict the values of the following quantities:
(a) The atomic mass of K from the atomic masses of Na and Rb
(b) The melting point of Br_2 from the melting points of Cl_2 (−101.0°C) and I_2 (113.6°C) (actual value = −7.2°C)
(c) The boiling point of HBr from the boiling points of HCl (−84.9°C) and HI (−35.4°C) (actual value = −67.0°C)

Characteristics of Many-Electron Atoms

8.4 Summarize the rules for the allowable values of the four quantum numbers of an electron in an atom.

8.5 Which of the quantum numbers relate(s) to the electron only? Which relate(s) to the orbital?

8.6 State the exclusion principle. What does it imply about the number and spin of electrons in an atomic orbital?

8.7 What is the key distinction between sublevel energies in one-electron species, such as the H atom, and those in many-electron species, such as the C atom? What factors lead to this distinction? Would you expect the pattern of sublevel energies in Be^{3+} to be more like that in H or that in C? Explain.

8.8 Define *shielding* and *effective nuclear charge*. What is the connection between the two?

8.9 What is penetration? How is it related to shielding? Use the penetration effect to explain the difference in relative orbital energies of a 3*p* and a 3*d* electron in the same atom.

8.10 How many electrons in an atom can have each of the following quantum number or sublevel designations?
(a) $n = 2, l = 1$ (b) 3*d* (c) 4*s*

8.11 How many electrons in an atom can have each of the following quantum number or sublevel designations?
(a) $n = 2, l = 1, m_l = 0$ (b) 5*p* (c) $n = 4, l = 3$

8.12 How many electrons in an atom can have each of the following quantum number or sublevel designations?
(a) 4*p* (b) $n = 3, l = 1, m_l = +1$ (c) $n = 5, l = 3$

8.13 How many electrons in an atom can have each of the following quantum number or sublevel designations?
(a) 2*s* (b) $n = 3, l = 2$ (c) 6*d*

The Quantum-Mechanical Model and the Periodic Table

(Sample Problems 8.1 and 8.2)

8.14 State the periodic law, and explain its relation to electron configuration. (Use Na and K in your explanation.)

8.15 State Hund's rule in your own words, and show its application in the orbital diagram of the nitrogen atom.

8.16 How does the aufbau principle, in connection with the periodic law, lead to the format of the periodic table?

8.17 For main-group elements, are outer electron configurations similar or different within a group? Within a period? Explain.

8.18 Write a full set of quantum numbers for the following:
(a) The outermost electron in an Rb atom
(b) The electron gained when an S^- ion becomes an S^{2-} ion
(c) The electron lost when an Ag atom ionizes
(d) The electron gained when an F^- ion forms from an F atom

8.19 Write a full set of quantum numbers for the following:
(a) The outermost electron in an Li atom
(b) The electron gained when a Br atom becomes a Br^- ion
(c) The electron lost when a Cs atom ionizes
(d) The highest energy electron in the ground-state B atom

8.20 Write the full ground-state electron configuration for each:
(a) Rb (b) Ge (c) Ar

8.21 Write the full ground-state electron configuration for each:
(a) Br (b) Mg (c) Se

8.22 Draw an orbital diagram showing valence electrons, and write the condensed ground-state electron configuration for each:
(a) Ti (b) Cl (c) V

8.23 Draw an orbital diagram showing valence electrons, and write the condensed ground-state electron configuration for each:
(a) Ba (b) Co (c) Ag

8.24 Draw the partial (valence-level) orbital diagram, and write the symbol, group number, and period number of the element:
(a) [He] $2s^2 2p^4$ (b) [Ne] $3s^2 3p^3$

8.25 Draw the partial (valence-level) orbital diagram, and write the symbol, group number, and period number of the element:
(a) [Kr] $5s^2 4d^{10}$ (b) [Ar] $4s^2 3d^8$

8.26 From each partial (valence-level) orbital diagram, write the ground-state electron configuration and group number:

(a) [↑↓] [↑↓|↑↓|↑↓|↑↓|↑↓] [↑| |]
4*s* 3*d* 4*p*

(b) [↑↓] [↑↓|↑↓|↑↓]
2*s* 2*p*

8.27 From each partial (valence-level) orbital diagram, write the ground-state electron configuration and group number:

(a) [↑] [↑|↑|↑|↑|] [| |]
5*s* 4*d* 5*p*

(b) [↑↓] [↑|↑|↑]
2*s* 2*p*

8.28 How many inner, outer, and valence electrons are present in an atom of each of the following elements?
(a) O (b) Sn (c) Ca (d) Fe (e) Se

8.29 How many inner, outer, and valence electrons are present in an atom of each of the following elements?
(a) Br (b) Cs (c) Cr (d) Sr (e) F

8.30 Identify each element below, and give the symbols of the other elements in its group:
(a) [He] $2s^22p^1$ (b) [Ne] $3s^23p^4$ (c) [Xe] $6s^25d^1$

8.31 Identify each element below, and give the symbols of the other elements in its group:
(a) [Ar] $4s^23d^{10}4p^4$ (b) [Xe] $6s^24f^{14}5d^2$ (c) [Ar] $4s^23d^5$

8.32 One reason spectroscopists study excited states is to gain information about the energies of orbitals that are unoccupied in an atom's ground state. Each of the following electron configurations represents an atom in an excited state. Identify the element, and write its condensed ground-state configuration:
(a) $1s^22s^22p^63s^13p^1$ (b) $1s^22s^22p^63s^23p^44s^1$
(c) $1s^22s^22p^63s^23p^64s^23d^44p^1$ (d) $1s^22s^22p^53s^1$

Trends in Three Key Atomic Properties

(Sample Problems 8.3 to 8.5)

8.33 Explain the relationship between the trends in atomic size and in ionization energy within the main groups.

8.34 In what region of the periodic table will you find elements with relatively high IEs? With relatively low IEs?

8.35 Why do successive IEs of a given element always increase? When the difference between successive IEs of a given element is exceptionally large (for example, between IE_1 and IE_2 of K), what do we learn about its electron configuration?

8.36 Given the following partial (valence-level) electron configurations, (a) identify each element, (b) rank the four elements in order of increasing atomic size, and (c) rank them in order of increasing ionization energy:

A ↑↓ 3s; ↑ ↑ _ 3p
C ↑↓ 5s; _ _ _ 5p

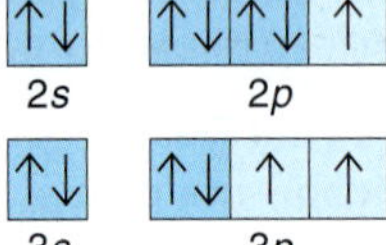

8.37 In a plot of IE_1 for the Period 3 elements (Figure 8.10, page 263), why do the values for elements in Groups 3A(13) and 6A(16) drop slightly below the generally increasing trend?

8.38 Which group in the periodic table has elements with high (endothermic) IE_1 and very negative (exothermic) first electron affinities (EA_1)? Give the charge on the ions these atoms form.

8.39 How does *d*-electron shielding influence atomic size among the Period 4 transition elements?

8.40 Arrange each set in order of *increasing* atomic size:
(a) Rb, K, Cs (b) C, O, Be (c) Cl, K, S (d) Mg, K, Ca

8.41 Arrange each set in order of *decreasing* atomic size:
(a) Ge, Pb, Sn (b) Sn, Te, Sr (c) F, Ne, Na (d) Be, Mg, Na

8.42 Arrange each set of atoms in order of *increasing* IE_1:
(a) Sr, Ca, Ba (b) N, B, Ne (c) Br, Rb, Se (d) As, Sb, Sn

8.43 Arrange each set of atoms in order of *decreasing* IE_1:
(a) Na, Li, K (b) Be, F, C (c) Cl, Ar, Na (d) Cl, Br, Se

8.44 Write the full electron configuration of the Period 2 element with the following successive IEs (in kJ/mol):
$IE_1 = 801$ $IE_2 = 2427$ $IE_3 = 3659$
$IE_4 = 25{,}022$ $IE_5 = 32{,}822$

8.45 Write the full electron configuration of the Period 3 element with the following successive IEs (in kJ/mol):
$IE_1 = 738$ $IE_2 = 1450$ $IE_3 = 7732$
$IE_4 = 10{,}539$ $IE_5 = 13{,}628$

8.46 Which element in each of the following sets would you expect to have the *highest* IE_2?
(a) Na, Mg, Al (b) Na, K, Fe (c) Sc, Be, Mg

8.47 Which element in each of the following sets would you expect to have the *lowest* IE_3?
(a) Na, Mg, Al (b) K, Ca, Sc (c) Li, Al, B

Atomic Structure and Chemical Reactivity

(Sample Problems 8.6 to 8.8)

8.48 List three ways in which metals and nonmetals differ.

8.49 Summarize the trend in metallic character as a function of position in the periodic table. Is it the same as the trend in atomic size? Ionization energy?

8.50 Summarize the acid-base behavior of the main-group metal and nonmetal oxides in water. How does oxide acidity in water change down a group and across a period?

8.51 What is a pseudo–noble gas configuration? Give an example of one ion from Group 3A(13) that has it.

8.52 The charges of a set of isoelectronic ions vary from 3+ to 3−. Place the ions in order of increasing size.

8.53 Which element would you expect to be *more* metallic?
(a) Ca or Rb (b) Mg or Ra (c) Br or I

8.54 Which element would you expect to be *less* metallic?
(a) S or Cl (b) In or Al (c) As or Br

8.55 Write the charge and full ground-state electron configuration of the monatomic ion most likely to be formed by each:
(a) Cl (b) Na (c) Ca

8.56 Write the charge and full ground-state electron configuration of the monatomic ion most likely to be formed by each:
(a) Rb (b) N (c) Br

8.57 How many unpaired electrons are present in the ground state of an atom from each of the following groups?
(a) 2A(2) (b) 5A(15) (c) 8A(18) (d) 3A(13)

8.58 How many unpaired electrons are present in the ground state of an atom from each of the following groups?
(a) 4A(14) (b) 7A(17) (c) 1A(1) (d) 6A(16)

8.59 Write the condensed ground-state electron configurations of these transition metal ions, and state which are paramagnetic:
(a) V^{3+} (b) Cd^{2+} (c) Co^{3+} (d) Ag^+

8.60 Write the condensed ground-state electron configurations of these transition metal ions, and state which are paramagnetic:
(a) Mo^{3+} (b) Au^+ (c) Mn^{2+} (d) Hf^{2+}

8.61 Palladium (Pd; $Z = 46$) is diamagnetic. Draw partial orbital diagrams to show which of the following electron configurations is consistent with this fact:
(a) [Kr] $5s^24d^8$ (b) [Kr] $4d^{10}$ (c) [Kr] $5s^14d^9$

8.62 Niobium (Nb; $Z = 41$) has an anomalous ground-state electron configuration for a Group 5B(5) element: [Kr] $5s^14d^4$. What is the expected electron configuration for elements in this group? Draw partial orbital diagrams to show how paramagnetic measurements could support niobium's actual configuration.

8.63 Rank the ions in each set in order of *increasing* size, and explain your ranking:
(a) Li^+, K^+, Na^+ (b) Se^{2-}, Rb^+, Br^- (c) O^{2-}, F^-, N^{3-}

8.64 Rank the ions in each set in order of *decreasing* size, and explain your ranking:
(a) Se^{2-}, S^{2-}, O^{2-} (b) Te^{2-}, Cs^+, I^- (c) Sr^{2+}, Ba^{2+}, Cs^+

Comprehensive Problems

Problems with an asterisk (*) are more challenging.

8.65 Name the element described in each of the following:
(a) Smallest atomic radius in Group 6A
(b) Largest atomic radius in Period 6
(c) Smallest metal in Period 3
(d) Highest IE_1 in Group 14
(e) Lowest IE_1 in Period 5
(f) Most metallic in Group 15
(g) Group 3A element that forms the most basic oxide
(h) Period 4 element with filled outer level
(i) Condensed ground-state electron configuration is [Ne] $3s^23p^2$
(j) Condensed ground-state electron configuration is [Kr] $5s^24d^6$
(k) Forms 2+ ion with electron configuration [Ar] $3d^3$
(l) Period 5 element that forms 3+ ion with pseudo–noble gas configuration
(m) Period 4 transition element that forms 3+ diamagnetic ion
(n) Period 4 transition element that forms 2+ ion with a half-filled *d* sublevel
(o) Heaviest lanthanide
(p) Period 3 element whose 2− ion is isoelectronic with Ar
(q) Alkaline earth metal whose cation is isoelectronic with Kr
(r) Group 5A(15) metalloid with the most acidic oxide

8.66 Rubidium and bromine atoms are depicted at right. (a) What monatomic ions do they form? (b) What electronic feature characterizes this pair of ions, and which noble gas are they related to? (c) Which pair best represents the relative ionic sizes?

* **8.67** When a nonmetal oxide reacts with water, it forms an oxoacid with the same nonmetal oxidation state. Give the name and formula of the oxide used to prepare each of these oxoacids: (a) hypochlorous acid; (b) chlorous acid; (c) chloric acid; (d) perchloric acid; (e) sulfuric acid; (f) sulfurous acid; (g) nitric acid; (h) nitrous acid; (i) carbonic acid; (j) phosphoric acid.

* **8.68** The energy difference between the 5*d* and 6*s* sublevels in gold accounts for its color. Assuming this energy difference is about 2.7 eV [1 electron volt (eV) $= 1.602\times10^{-19}$ J], explain why gold has a warm yellow color.

8.69 Write the formula and name of the compound formed from the following ionic interactions: (a) The 2+ ion and the 1− ion are both isoelectronic with the atoms of a chemically unreactive Period 4 element. (b) The 2+ ion and the 2− ion are both isoelectronic with the Period 3 noble gas. (c) The 2+ ion is the smallest with a filled *d* subshell; the anion forms from the smallest halogen. (d) The ions form from the largest and smallest ionizable atoms in Period 2.

8.70 The hot glowing gases around the Sun, the *corona*, can reach millions of degrees Celsius, high enough to remove many electrons from gaseous atoms. Iron ions with charges as high as 14+ have been observed in the corona. Which ions from Fe^+ to Fe^{14+} are paramagnetic? Which would be most attracted to a magnetic field?

8.71 Partial (valence-level) electron configurations for four different ions are shown below:

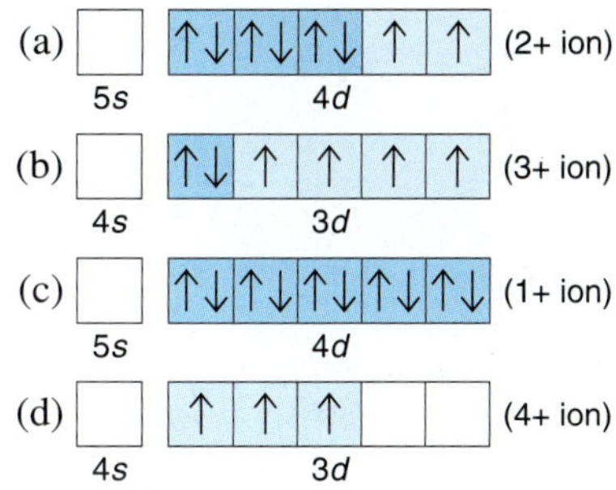

Identify the elements from which the ions are derived, and write the formula of the oxide each ion forms.

8.72 The bars in the graph at right represent the relative magnitudes of the first five ionization energies of an atom. Identify the element and write its complete electron configuration, assuming it comes from (a) Period 2; (b) Period 3; (c) Period 4.

* **8.73** On the planet Zog in the Andromeda galaxy, all of the stable elements have been studied. Data for some main-group elements are shown below (Zoggian units are unknown on Earth and, therefore, not shown). Limited communications with the Zoggians have indicated that balloonium is a monatomic gas with two positive charges in its nucleus. Use the data to deduce the names that Earthlings give to these elements.

Name	Atomic Radius	IE_1	EA_1
Balloonium	10	339	0
Inertium	24	297	+4.1
Allotropium	34	143	−28.6
Brinium	63	70.9	−7.6
Canium	47	101	−15.3
Fertilium	25	200	0
Liquidium	38	163	−46.4
Utilium	48	82.4	−6.1
Crimsonium	72	78.4	−2.9

Models of Chemical Bonding

Binding Atoms Together *The properties of substances, such as soft, shiny gold imbedded in hard, dull quartz, depend on how their atoms bind together.*

Outline

Key Principles
to focus on while studying this chapter

- Two classes of elements, metals and nonmetals, combine through three types of bonding: metal and nonmetal through *ionic bonding,* nonmetal and nonmetal through *covalent bonding,* and metal and metal through *metallic bonding (Section 9.1).*
- Ionic bonding is the attraction among the ions that are created when metal atoms *transfer* electrons to nonmetal atoms. Even though energy is required to form the ions, ionic compounds occur because much more energy is released when the ions attract each other to form a solid *(Section 9.2).*
- The strong attractions among their ions make ionic compounds hard, high-melting solids that conduct a current only when melted or dissolved *(Section 9.2).*
- A covalent bond is the attraction between the nuclei of two nonmetal atoms and the electron pair they share. Each covalent bond has specific *energy* and *length* that depend on the bonded atoms and an *order* that depends on the number of electron pairs shared *(Section 9.3).*
- Most covalent compounds consist of separate molecules, so they have low melting and boiling points. These physical changes disrupt the weak attractions *between* the molecules while leaving the strong covalent bonds *within* the molecules intact. Some substances have covalent bonds throughout, and they are very hard and high melting *(Section 9.3).*
- During a reaction, energy is *absorbed* to break certain bonds in the reactant molecules and is *released* to form other bonds that create the product molecules; the heat of reaction is the *difference* between the energy absorbed and the energy released *(Section 9.4).*
- Each atom in a covalent bond attracts the shared electron pair according to its *electronegativity (EN),* a property that, in general, is inversely related to atomic size. A covalent bond is *polar* if the two atoms have different EN values. The *ionic character* of a bond—from highly ionic to nonpolar covalent—varies with the difference in EN values of the atoms *(Section 9.5).*

Why do the substances around us behave as they do? That is, why is table salt (or any other ionic substance) a hard, brittle, high-melting solid that conducts a current only when molten or dissolved in water? Why is candle wax (along with most covalent substances) low melting, soft, and nonconducting, although diamond and a few other exceptions are high melting and extremely hard? And why is copper (and most other metallic substances) shiny, malleable, and able to conduct a current whether molten or solid? The answers lie in the *type of bonding within the substance.* In Chapter 8, we examined the properties of individual atoms and ions. Yet, in virtually all the substances in and around you, these particles are bonded to one another. As you'll see in this chapter, deeper insight comes as we discover how the properties of atoms influence the types of chemical bonds they form, because these are ultimately responsible for the behavior of substances.

Concepts & Skills to Review before studying this chapter

- characteristics of ionic and covalent bonding (Section 2.7)
- polar covalent bonds and the polarity of water (Section 4.1)
- Hess's law, ΔH°_{rxn}, and ΔH°_f (Sections 6.5 and 6.6)
- atomic and ionic electron configurations (Sections 8.3 and 8.5)
- trends in atomic properties and metallic behavior (Sections 8.4 and 8.5)

9.1 ATOMIC PROPERTIES AND CHEMICAL BONDS

Before we examine the types of chemical bonding, we should ask why atoms bond at all. In general terms, *bonding lowers the potential energy between positive and negative particles,* whether they are oppositely charged ions or nuclei and the electrons shared between them. Just as the electron configuration and the strength of the nucleus-electron attraction(s) determine the properties of an atom, the type and strength of chemical bonds determine the properties of a substance.

The Three Types of Chemical Bonding

On the atomic level, we distinguish a metal from a nonmetal on the basis of several properties that correlate with position in the periodic table (Figure 9.1 and inside the front cover). Recall from Chapter 8 that, in general, there is a gradation from more metal-like to more nonmetal-like behavior from left to right across a period and from bottom to top within most groups. Three types of bonding result from the three ways these two types of atoms can combine—metal with nonmetal, nonmetal with nonmetal, and metal with metal:

1. *Metal with nonmetal: electron transfer and ionic bonding* (Figure 9.2A, next page). We typically observe **ionic bonding** between atoms with large differences in their tendencies to lose or gain electrons. Such differences occur between reactive metals [Groups 1A(1) and 2A(2)] and nonmetals [Group 7A(17) and the top of

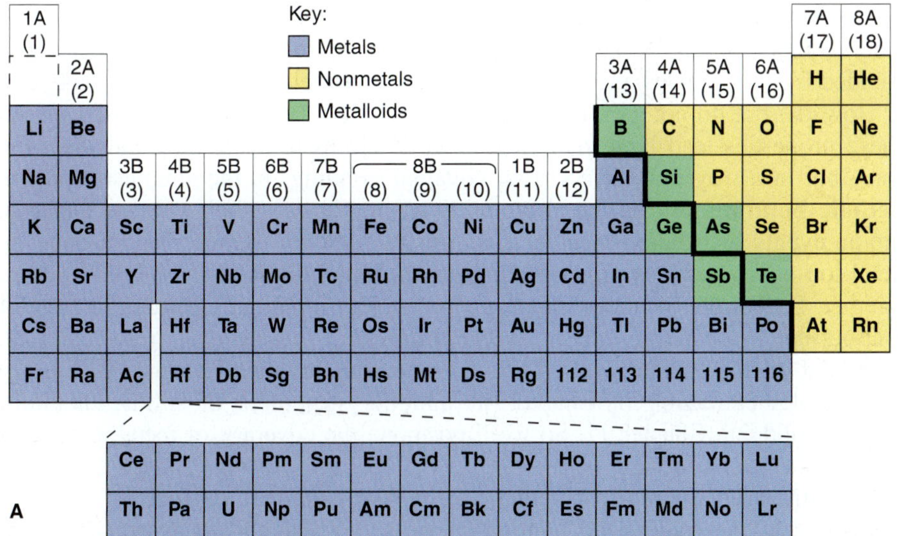

PROPERTY	METAL ATOM	NONMETAL ATOM
Atomic size	Larger	Smaller
Z_{eff}	Lower	Higher
IE	Lower	Higher
EA	Less negative	More negative

B

FIGURE 9.1 A general comparison of metals and nonmetals. A, The positions of metals, nonmetals, and metalloids within the periodic table. **B,** The relative magnitudes of some key atomic properties vary from left to right within a period and correlate with whether an element is metallic or nonmetallic.

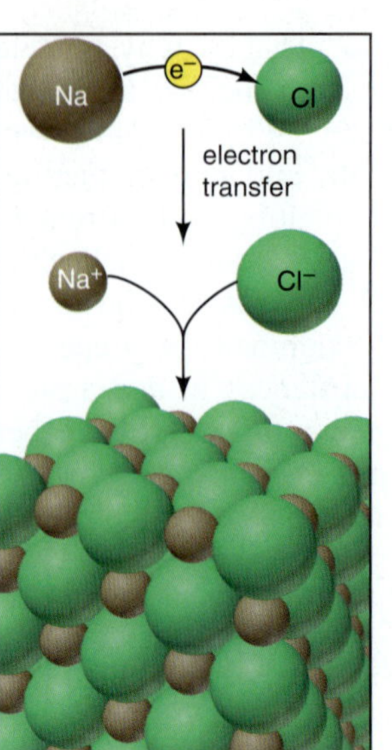

A Ionic bonding

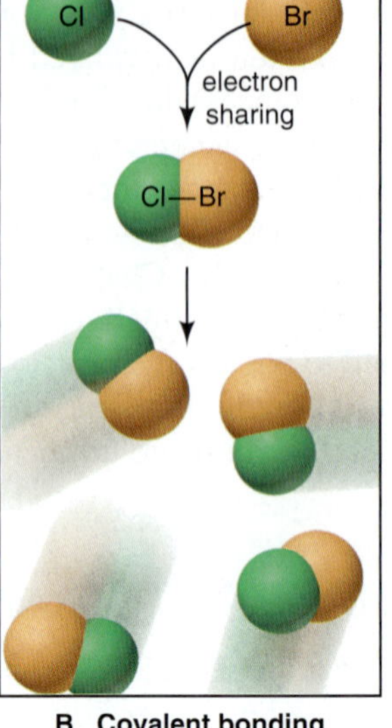

B Covalent bonding

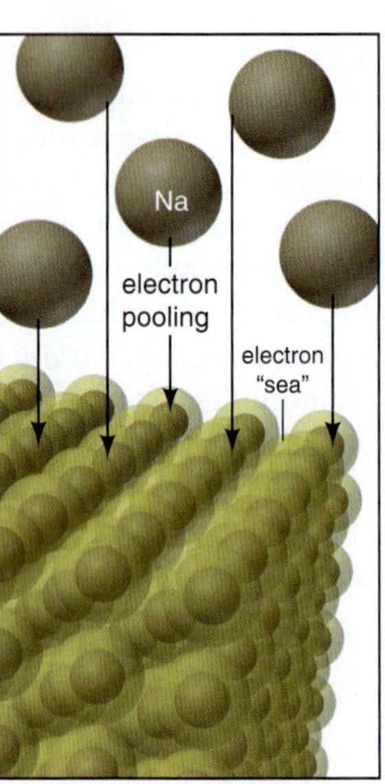

C Metallic bonding

FIGURE 9.2 The three models of chemical bonding. A, In ionic bonding, metal atoms transfer electron(s) to nonmetal atoms, forming oppositely charged ions that attract each other to form a solid. **B,** In covalent bonding, two atoms share an electron pair localized between their nuclei (shown here as a bond line). Most covalent substances consist of individual molecules, each made from two or more atoms. **C,** In metallic bonding, many metal atoms pool their valence electrons to form a delocalized electron "sea" that holds the metal-ion cores together.

Group 6A(16)]. The metal atom (low IE) loses its one or two valence electrons, and the nonmetal atom (highly negative EA) gains the electron(s). *Electron transfer* from metal to nonmetal occurs, and each atom forms an ion with a noble gas electron configuration. The electrostatic attraction between these positive and negative ions draws them into the three-dimensional array of an ionic solid, whose chemical formula represents the cation-to-anion ratio (empirical formula).

2. *Nonmetal with nonmetal: electron sharing and covalent bonding* (Figure 9.2B). When two atoms have a small difference in their tendencies to lose or gain electrons, we observe *electron sharing* and **covalent bonding.** This type of bonding most commonly occurs between nonmetal atoms (although a pair of metal atoms can sometimes form a covalent bond). Each nonmetal atom holds onto its own electrons tightly (high IE) and tends to attract other electrons as well (highly negative EA). The attraction of each nucleus for the valence electrons of the other draws the atoms together. A shared electron pair is considered to be *localized* between the two atoms because it spends most of its time there, linking them in a covalent bond of a particular length and strength. In most cases, separate molecules form when covalent bonding occurs, and the chemical formula reflects the actual numbers of atoms in the molecule (molecular formula).

3. *Metal with metal: electron pooling and metallic bonding* (Figure 9.2C). In general, metal atoms are relatively large, and their few outer electrons are well shielded by filled inner levels. Thus, they lose outer electrons comparatively easily (low IE) but do not gain them very readily (slightly negative or positive EA). These properties lead large numbers of metal atoms to share their valence electrons, but in a way that differs from covalent bonding. In the simplest model of **metallic bonding,** all the metal atoms in a sample *pool* their valence electrons into an evenly distributed "sea" of electrons that "flows" between and around the metal-ion cores (nucleus plus inner electrons), attracting them and holding them together. Unlike the localized electrons in covalent bonding, electrons in metallic bonding are *delocalized,* moving freely throughout the piece of metal. (For the remainder of this chapter, we'll focus on ionic and covalent bonding. We discuss electron delocalization in Chapter 11 and the structures of solids, including metallic solids, in Chapter 12. So we'll postpone the coverage of metallic bonding until then.)

It's important to remember that, in the world of real substances, there are exceptions to these idealized bonding models. You cannot always predict bond

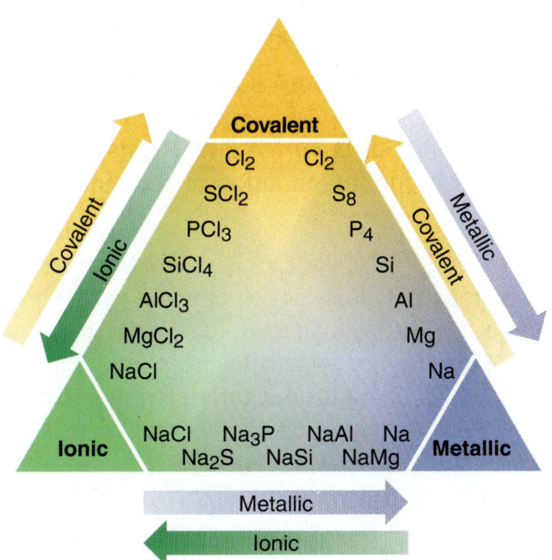

FIGURE 9.3 Gradation in bond type among the Period 3 elements. Along *the left side* of the triangle, compounds of each element with chlorine display a gradual change from ionic to covalent bonding. Along *the right side,* the elements themselves display a gradual change from covalent to metallic bonding. Along *the base,* compounds of each element with sodium display a gradual change from ionic to metallic bonding.

type solely from the elements' positions in the periodic table. For instance, all binary ionic compounds contain a metal and a nonmetal, but all metals do not form binary ionic compounds with all nonmetals. As just one example, when the metal beryllium [Group 2A(2)] combines with the nonmetal chlorine [Group 7A(17)], the bonding fits the covalent model better than the ionic model. In other words, just as we see a gradation in metallic behavior within groups and periods, we also see a gradation in bonding from one type to another (Figure 9.3).

Lewis Electron-Dot Symbols: Depicting Atoms in Chemical Bonding

Before turning to the two bonding models, let's discuss a method for depicting the valence electrons of interacting atoms. In the **Lewis electron-dot symbol** (named for the American chemist G. N. Lewis), the element symbol represents the nucleus *and* inner electrons, and the surrounding dots represent the valence electrons (Figure 9.4). The pattern of dots is the same for elements within a group.

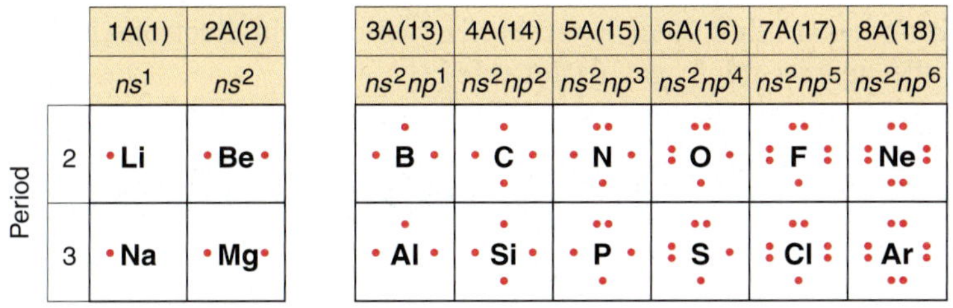

Period	1A(1)	2A(2)	3A(13)	4A(14)	5A(15)	6A(16)	7A(17)	8A(18)
	ns^1	ns^2	ns^2np^1	ns^2np^2	ns^2np^3	ns^2np^4	ns^2np^5	ns^2np^6
2	·Li	·Be·	·Ḃ·	·Ċ̣·	·N̤̈· (pair top, single bottom)	:Ö· (single bottom)	:F̈: (single bottom)	:Ne: (pairs top and bottom)
3	·Na	·Mg·	·Ȧl·	·Ṡi̤· (single top and bottom)	·P̈· (single bottom)	:S̈· (single bottom)	:C̈l: (single bottom)	:Ar: (pairs top and bottom)

FIGURE 9.4 Lewis electron-dot symbols for elements in Periods 2 and 3. The element symbol represents the nucleus and inner electrons, and the dots around it represent valence electrons, either paired or unpaired. The number of unpaired dots indicates the number of electrons a metal atom loses, or the number a nonmetal atom gains, or the number of covalent bonds a nonmetal atom usually forms.

It's easy to write the Lewis symbol for any main-group element:

1. Note its A-group number (1A to 8A), which equals the number of valence electrons.
2. Place one dot at a time on the four sides (top, right, bottom, left) of the element symbol.
3. Keep adding dots, pairing the dots until all are used up.

The specific placement of dots is not important; that is, in addition to the one shown in Figure 9.4, the Lewis symbol for nitrogen can *also* be written as

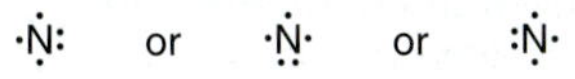

·Ṇ̇: or ·Ṅ·(pair bottom) or :Ṇ̇·

The Lewis symbol provides information about an element's bonding behavior:

- For a metal, the *total* number of dots is the maximum number of electrons an atom loses to form a cation.
- For a nonmetal, the number of *unpaired* dots equals either the number of electrons an atom gains in becoming an anion or the number it shares in forming covalent bonds.

To illustrate the last point, look at the Lewis symbol for carbon. Rather than one pair of dots and two unpaired dots, as its electron configuration ([He] $2s^22p^2$) would indicate, carbon has four unpaired dots because it forms four bonds. That is, in its compounds, carbon's four electrons are paired with four more electrons from its bonding partners for a total of eight electrons around carbon. (In Chapter 10, we'll see that larger nonmetals can sometimes form as many bonds as they have dots in the Lewis symbol.)

In his studies of bonding, Lewis generalized much of bonding behavior into the **octet rule:** *when atoms bond, they lose, gain, or share electrons to attain a filled outer level of eight (or two) electrons.* The octet rule holds for nearly all of the compounds of Period 2 elements and a large number of others as well.

SECTION 9.1 SUMMARY

Nearly all naturally occurring substances consist of atoms or ions bonded to others. Chemical bonding allows atoms to lower their energy. • Ionic bonding occurs when metal atoms transfer electrons to nonmetal atoms, and the resulting ions attract each other and form an ionic solid. • Covalent bonding most commonly occurs between nonmetal atoms and usually results in molecules. The bonded atoms share a pair of electrons, which remain localized between them. • Metallic bonding occurs when many metal atoms pool their valence electrons in a delocalized electron "sea" that holds all the atoms together. • The Lewis electron-dot symbol of an atom depicts the number of valence electrons for a main-group element. • In bonding, many atoms lose, gain, or share electrons to attain a filled outer level of eight (or two).

9.2 THE IONIC BONDING MODEL

The central idea of the ionic bonding model is the *transfer of electrons from metal atoms to nonmetal atoms to form ions that come together in a solid ionic compound.* For nearly every monatomic ion of a main-group element, the electron configuration has a filled outer level: either two or eight electrons, the same number as in the nearest noble gas (octet rule).

The transfer of an electron from a lithium atom to a fluorine atom is depicted in three ways in Figure 9.5. In each, Li loses its single outer electron and is left with a filled $n = 1$ level, while F gains a single electron to fill its $n = 2$ level. In

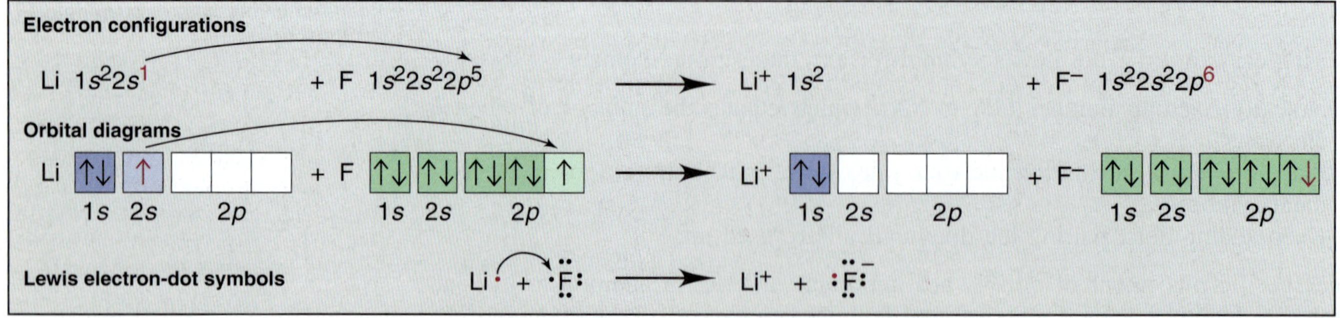

FIGURE 9.5 Three ways to represent the formation of Li^+ and F^- through electron transfer. The electron being transferred is indicated in red.

this case, each atom is one electron away from its nearest noble gas—He for Li and Ne for F—so the number of electrons lost by each Li equals the number gained by each F. Therefore, equal numbers of Li^+ and F^- ions form, as the formula LiF indicates. That is, in ionic bonding, *the total number of electrons lost by the metal atoms equals the total number of electrons gained by the nonmetal atoms.*

SAMPLE PROBLEM 9.1 Depicting Ion Formation

Problem Use partial orbital diagrams and Lewis symbols to depict the formation of Na^+ and O^{2-} ions from the atoms, and determine the formula of the compound the ions form.

Plan First, we draw the orbital diagrams and Lewis symbols for the Na and O atoms. To attain filled outer levels, Na loses one electron and O gains two. Thus, to make the number of electrons lost equal the number gained, two Na atoms are needed for each O atom.

Solution

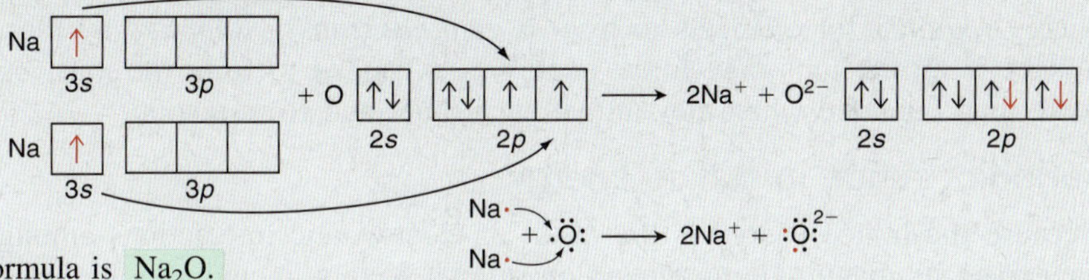

The formula is Na_2O.

FOLLOW-UP PROBLEM 9.1 Use condensed electron configurations and Lewis symbols to depict the formation of Mg^{2+} and Cl^- ions from the atoms, and write the formula of the ionic compound.

Energy Considerations in Ionic Bonding: The Importance of Lattice Energy

You may be surprised to learn that the electron-transfer process actually *absorbs* energy! So why does it occur? As you'll see, the reason ionic substances exist at all is because of the enormous *release* of energy that occurs when the ions come together and form a solid. Consider just the electron-transfer process for the formation of lithium fluoride, which involves two steps—a gaseous Li atom loses an electron, and a gaseous F atom gains it:

- The first ionization energy (IE_1) of Li is the energy change that occurs when 1 mol of gaseous Li atoms loses 1 mol of outer electrons:

$$Li(g) \longrightarrow Li^+(g) + e^- \qquad IE_1 = 520 \text{ kJ}$$

- The electron affinity (EA) of F is the energy change that occurs when 1 mol of gaseous F atoms gains 1 mol of electrons:

$$F(g) + e^- \longrightarrow F^-(g) \qquad EA = -328 \text{ kJ}$$

Note that the two-step electron-transfer process *by itself* requires energy:

$$Li(g) + F(g) \longrightarrow Li^+(g) + F^-(g) \qquad IE_1 + EA = 192 \text{ kJ}$$

The total energy needed for ion formation is even greater than this because metallic lithium and diatomic fluorine must first be converted to separate gaseous atoms, which also requires energy. Despite this, the standard heat of formation (ΔH_f°) of solid LiF is −617 kJ/mol; that is, 617 kJ is *released* when 1 mol of LiF(*s*) forms from its elements. The case of LiF is typical of many reactions between active metals and nonmetals: despite the endothermic electron transfer, ionic solids form readily, often vigorously. Figure 9.6 shows another example, the formation of NaBr.

Clearly, if the overall reaction of Li(*s*) and $F_2(g)$ to form LiF(*s*) releases energy, there must be some exothermic energy component large enough to overcome the endothermic steps. This component arises from the strong *attraction*

A

B

FIGURE 9.6 The reaction between sodium and bromine. A, Despite the endothermic electron-transfer process, all the Group 1A(1) metals react exothermically with any of the Group 7A(17) nonmetals to form solid alkali-metal halides. The reactants in the example shown are sodium (in beaker under mineral oil) and bromine. **B,** The reaction is usually rapid and vigorous.

among many oppositely charged ions. When 1 mol of $Li^+(g)$ and 1 mol of $F^-(g)$ form 1 mol of gaseous LiF molecules, a large quantity of heat is released:

$$Li^+(g) + F^-(g) \longrightarrow LiF(g) \qquad \Delta H^\circ = -755 \text{ kJ}$$

As you know, under ordinary conditions, LiF does not consist of gaseous molecules. *Much more energy is released when the gaseous ions coalesce into a crystalline solid.* That occurs because each ion attracts others of opposite charge:

$$Li^+(g) + F^-(g) \longrightarrow LiF(s) \qquad \Delta H^\circ = -1050 \text{ kJ}$$

The negative of this value, 1050 kJ, is the lattice energy of LiF. The **lattice energy** ($\boldsymbol{\Delta H^\circ_{lattice}}$) is the enthalpy change that occurs when 1 mol of ionic solid separates into gaseous ions. It indicates the strength of ionic interactions, which influence melting point, hardness, solubility, and other properties.

A key point to keep in mind is that *ionic solids exist only because the lattice energy exceeds the energy required for the electron transfer.* In other words, the energy *required* for elements to lose or gain electrons is *supplied* by the attraction between the ions they form: energy is expended to form the ions, but it is more than regained when they attract each other and form a solid.

Periodic Trends in Lattice Energy

Because the lattice energy is the result of electrostatic interactions among ions, we expect its magnitude to depend on several factors, including ionic size, ionic charge, and ionic arrangement in the solid. In Chapter 2, you were introduced to **Coulomb's law,** which states that the electrostatic energy between two charges (A and B) is directly proportional to the product of their magnitudes and inversely proportional to the distance between them:

$$\text{Electrostatic energy} \propto \frac{\text{charge A} \times \text{charge B}}{\text{distance}}$$

We can extend this relationship to the lattice energy ($\Delta H^\circ_{lattice}$) because it is directly proportional to the electrostatic energy. In an ionic solid, cations and anions lie as close to each other as possible, so the distance between them is the distance between their centers, or the sum of their radii (Figure 8.20, page 272):

$$\text{Electrostatic energy} \propto \frac{\text{cation charge} \times \text{anion charge}}{\text{cation radius} + \text{anion radius}} \propto \Delta H^\circ_{lattice} \qquad \textbf{(9.1)}$$

This relationship helps us predict trends in lattice energy and explain the effects of ionic size and charge:

1. *Effect of ionic size.* As we move down a group in the periodic table, the ionic radius increases. Therefore, the electrostatic energy between cations and anions decreases because the interionic distance is greater; thus, the lattice energies of their compounds should decrease as well. This prediction is borne out by the alkali-metal halides shown in Figure 9.7: note the regular decrease in lattice energy that occurs down a group whether we hold the cation constant (LiF to LiI) or the anion constant (LiF to RbF).

2. *Effect of ionic charge.* When we compare lithium fluoride with magnesium oxide, we find cations of about equal radii (Li^+ = 76 pm and Mg^{2+} = 72 pm) and anions of about equal radii (F^- = 133 pm and O^{2-} = 140 pm). Thus, the only significant difference is the ionic charge: LiF contains the singly charged Li^+ and F^- ions, whereas MgO contains the doubly charged Mg^{2+} and O^{2-} ions. The difference in their lattice energies is striking:

$$\Delta H^\circ_{lattice} \text{ of LiF} = 1050 \text{ kJ/mol} \qquad \text{and} \qquad \Delta H^\circ_{lattice} \text{ of MgO} = 3923 \text{ kJ/mol}$$

This nearly fourfold increase in $\Delta H^\circ_{lattice}$ reflects the fourfold increase in the product of the charges (1×1 vs. 2×2) in the numerator of Equation 9.1. The very large lattice energy of MgO more than compensates for the energy required to form the Mg^{2+} and O^{2-} ions. In fact, the lattice energy is the reason that compounds with 2+ cations and 2− anions even exist.

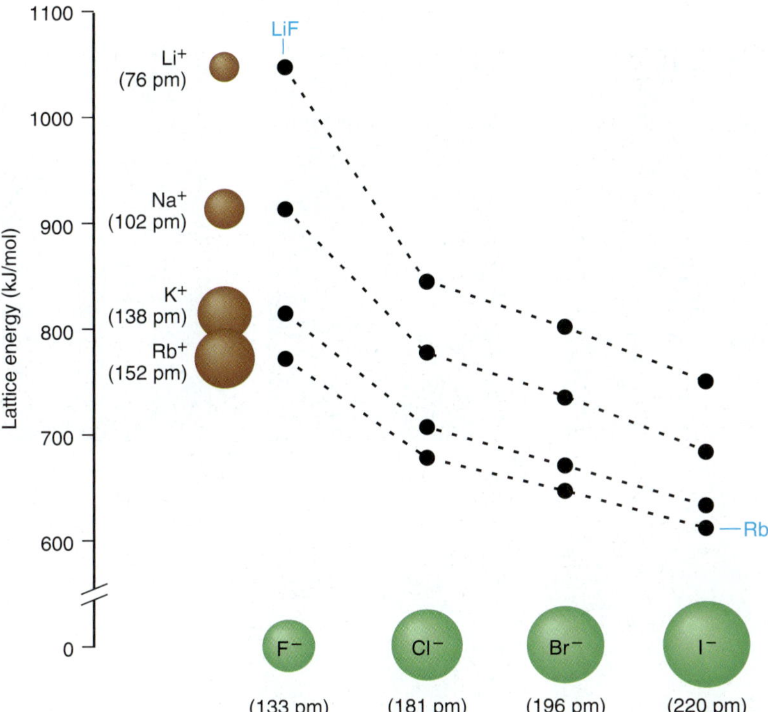

FIGURE 9.7 Trends in lattice energy. The lattice energies for many of the alkali-metal halides are shown. Each series of four points represents a given Group 1A(1) cation *(left side)* combining with each of the Group 7A(17) anions *(bottom)*. As ionic radii increase, the electrostatic attractions decrease, so the lattice energies of the compounds decrease as well. Thus, LiF (smallest ions shown) has the largest lattice energy, and RbI (largest ions) has the smallest.

How the Model Explains the Properties of Ionic Compounds

The first and most important job of any model is to explain the facts. By magnifying our view, we can see how the ionic bonding model accounts for the properties of ionic solids. You may have seen a piece of rock salt (NaCl). It is *hard* (does not dent), *rigid* (does not bend), and *brittle* (cracks without deforming). These properties are due to the powerful attractive forces that hold the ions *in specific positions* throughout the crystal. Moving the ions out of position requires overcoming these forces, so the sample resists denting and bending. If *enough* pressure is applied, ions of like charge are brought next to each other, and repulsive forces crack the sample suddenly (Figure 9.8).

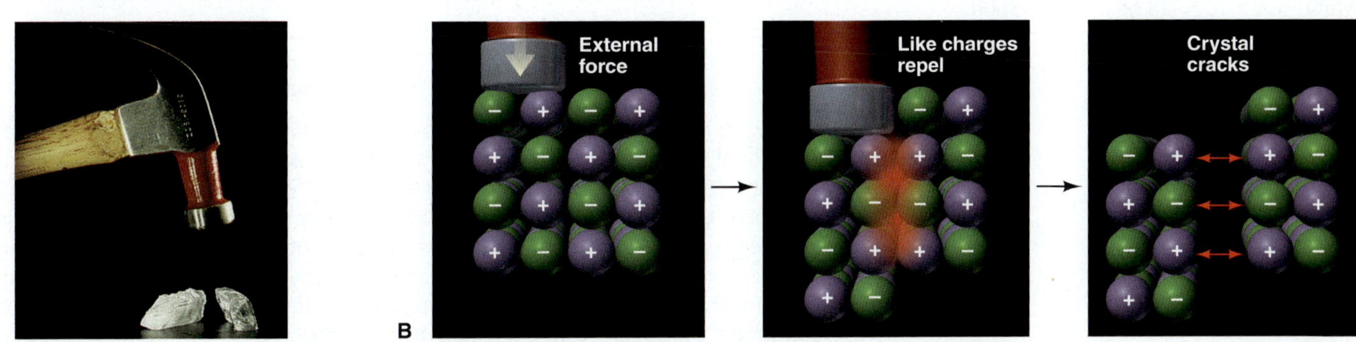

FIGURE 9.8 Electrostatic forces and the reason ionic compounds crack. A, Ionic compounds are hard and will crack, rather than bend, when struck with enough force. **B,** The positive and negative ions in the crystal are arranged to maximize their attractions. When an external force is applied, like charges move near each other, and the repulsions crack the piece apart.

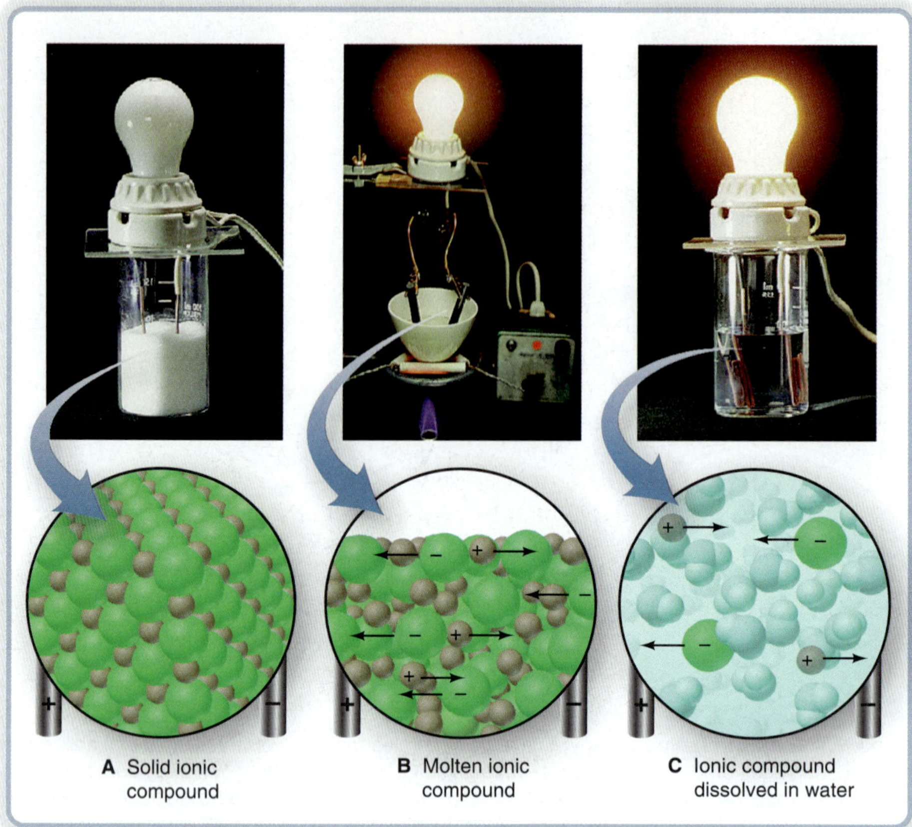

FIGURE 9.9 Electrical conductance and ion mobility. A, No current flows in the ionic solid because ions are immobile. **B,** In the molten compound, mobile ions flow toward the oppositely charged electrodes and carry a current. **C,** In an aqueous solution of the compound, mobile solvated ions carry a current.

Table 9.1 Melting and Boiling Points of Some Ionic Compounds

Compound	mp (°C)	bp (°C)
CsBr	636	1300
NaI	661	1304
KBr	734	1435
NaCl	801	1413
LiF	845	1676
KF	858	1505
MgO	2852	3600

Most ionic compounds *do not* conduct electricity in the solid state but *do* conduct it when melted or when dissolved in water. According to the ionic bonding model, the solid consists of immobilized ions. When it melts or dissolves, however, the ions are free to move and carry an electric current, as shown in Figure 9.9.

The model also explains that high temperatures are needed to melt and boil an ionic compound (Table 9.1) because freeing the ions from their positions (melting) requires large amounts of energy, and vaporizing them requires even more. In fact, the interionic attraction is so strong that the vapor consists of **ion pairs,** gaseous ionic molecules rather than individual ions. But keep in mind that in their ordinary (solid) state, ionic compounds consist of arrays of alternating ions that extend in all directions, and *no separate molecules exist.*

SECTION 9.2 SUMMARY

In ionic bonding, a metal transfers electrons to a nonmetal, and the resulting ions attract each other strongly to form a solid. • Main-group elements often attain a filled outer level of electrons (either eight or two) by forming ions with the electron configuration of the nearest noble gas. • Ion formation by itself *requires* energy. However, the lattice energy, the energy *absorbed* when the solid separates into gaseous ions, is large and is the major reason ionic solids exist. The lattice energy depends on ionic size and charge. • The ionic bonding model pictures oppositely charged ions held in position by strong electrostatic attractions and explains why ionic solids crack rather than bend and why they conduct electric current only when melted or dissolved. Gaseous ion pairs form when an ionic compound vaporizes, which requires very high temperatures.

9.3 THE COVALENT BONDING MODEL

Look through any large reference source of chemical compounds, such as the *Handbook of Chemistry and Physics,* and you'll find that the number of known covalent compounds dwarfs the number of known ionic compounds. Molecules held together by covalent bonds range from diatomic hydrogen to biological and synthetic macromolecules consisting of many hundreds or even thousands of atoms. We also find covalent bonds in many polyatomic ions. Without doubt, *sharing electrons is the principal way that atoms interact chemically.*

The Formation of a Covalent Bond

A sample of hydrogen gas consists of H_2 molecules. But why *do* the atoms bond to one another in pairs? Look at Figure 9.10 and imagine what happens to two isolated H atoms that approach each other from a distance (move *right* to *left* on the graph). When the atoms are far apart, each behaves as though the other were not present (point 1). As the distance between the nuclei decreases, each nucleus starts to attract the other atom's electron, which lowers the potential energy of the system. Attractions continue to draw the atoms closer, and the system becomes progressively lower in energy (point 2). As attractions increase, however, so do repulsions between the nuclei and between the electrons. At some internuclear distance, maximum attraction is achieved in the face of the increasing repulsion, and the system has its minimum energy (point 3, at the bottom of the energy "well"). Any shorter distance would increase repulsions and cause a rise in potential energy (point 4). Thus a **covalent bond,** such as the one that holds the atoms together in the H_2 molecule, arises from the balance between nucleus-electron attractions and electron-electron and nucleus-nucleus repulsions. Formation of a bond always results in *greater electron density between the nuclei.*

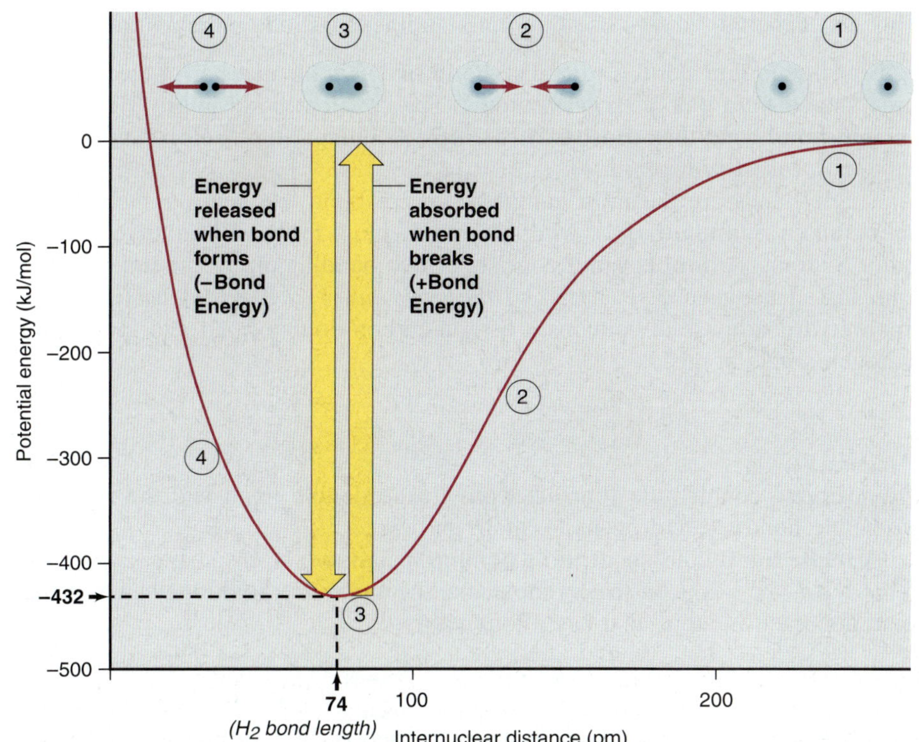

FIGURE 9.10 Covalent bond formation in H_2. The potential energy of a system of two H atoms is plotted against the distance between the nuclei, with a depiction of the atomic systems above. At point 1, the atoms are too far apart to attract each other. At 2, each nucleus attracts the other atom's electron. At 3, the combination of nucleus-electron attractions and electron-electron and nucleus-nucleus repulsions gives the minimum energy of the system. The energy difference between points 1 and 3 is the H_2 bond energy (432 kJ/mol). It is released when the bond forms and must be absorbed to break the bond. The internuclear distance at point 3 is the H_2 bond length (74 pm). If the atoms move closer, as at point 4, repulsions increase the system's energy and force the atoms apart to point 3 again.

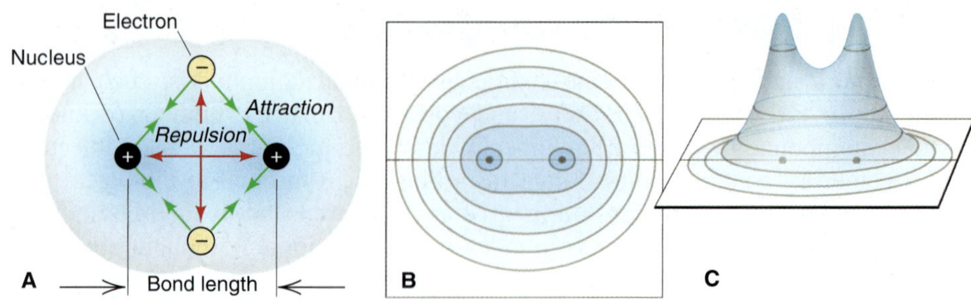

FIGURE 9.11 Distribution of electron density in H_2. A, At some optimum distance (bond length), attractions balance repulsions. Electron density *(blue shading)* is highest around and between the nuclei. **B,** This *contour map* shows a doubling of electron densities with each contour line; the dots represent the nuclei. **C,** This *relief map* depicts the varying electron densities of the contour map as peaks. The densest regions, by far, are around the nuclei (black dots on the "floor"), but the region between the nuclei—the bonding region—also has higher electron density.

Figure 9.11 depicts this fact in three ways: a cross-section of a space-filling model; an *electron density contour map,* with lines representing regular increments in electron density; and an *electron density relief map,* which portrays the contour map three-dimensionally as peaks of electron density.

Bonding Pairs and Lone Pairs In covalent bonding, as in ionic bonding, each atom achieves a full outer (valence) level of electrons, but this is accomplished by different means. *Each atom in a covalent bond "counts" the shared electrons as belonging entirely to itself.* Thus, the two electrons in the shared electron pair of H_2 simultaneously fill the outer level of *both* H atoms. The **shared pair,** or **bonding pair,** is represented by either a pair of dots or a line, H:H or H—H.

An outer-level electron pair that is *not* involved in bonding is called a **lone pair,** or **unshared pair.** The bonding pair in HF fills the outer level of the H atom *and,* together with three lone pairs, fills the outer level of the F atom as well:

bonding pair — lone pairs — H:F: or H—F:

In F_2 the bonding pair and three lone pairs fill the outer level of *each* F atom:

:F:F: or :F—F:

(This text generally shows bonding pairs as lines and lone pairs as dots.)

Types of Bonds and Bond Order The **bond order** is the number of electron pairs being shared by a pair of bonded atoms. The covalent bond in H_2, HF, or F_2 is a **single bond,** one that consists of a single bonding pair of electrons. A *single bond has a bond order of 1.*

Single bonds are the most common type of bond, but many molecules (and ions) contain multiple bonds. Multiple bonds most frequently involve C, O, N, and/or S atoms. A **double bond** consists of two bonding electron pairs, four electrons shared between two atoms, so *the bond order is 2.* Ethylene (C_2H_4) is a simple hydrocarbon that contains a carbon-carbon double bond and four carbon-hydrogen single bonds:

H:C::C:H (with H on each C) or $H_2C{=}CH_2$

Each carbon "counts" the four electrons in the double bond and the four in its two single bonds to hydrogens to attain an octet.

A **triple bond** consists of three bonding pairs; two atoms share six electrons, so *the bond order is 3.* In the N_2 molecule, the atoms are held together by a triple bond, and each N atom also has a lone pair:

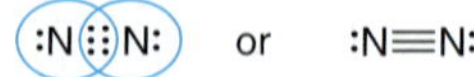

Six shared and two unshared electrons give *each* N atom an octet.

Properties of a Covalent Bond: Bond Energy and Bond Length

The strength of a covalent bond depends on the magnitude of the mutual attraction between bonded nuclei and shared electrons. The **bond energy (BE)** (also called *bond enthalpy* or *bond strength*) is the energy required to overcome this attraction and is defined as the standard enthalpy change for breaking the bond in 1 mol of gaseous molecules. *Bond breakage is an endothermic process, so the bond energy is always positive:*

$$\text{A—B}(g) \longrightarrow \text{A}(g) + \text{B}(g) \qquad \Delta H^\circ_{\text{bond breaking}} = \text{BE}_{\text{A—B}} \text{ (always} > 0)$$

Stated in another way, the bond energy is the difference in energy between the separated atoms and the bonded atoms (the potential energy difference between points 1 and 3 in Figure 9.10; the depth of the energy well). The energy absorbed to break the bond is released when the bond forms. *Bond formation is an exothermic process, so the sign of the enthalpy change is negative:*

$$\text{A}(g) + \text{B}(g) \longrightarrow \text{A—B}(g) \qquad \Delta H^\circ_{\text{bond forming}} = -\text{BE}_{\text{A—B}} \text{ (always} < 0)$$

Because bond energies depend on characteristics of the bonded atoms—their electron configurations, nuclear charges, and atomic radii—each type of bond has its own bond energy. Energies for some common bonds are listed in Table 9.2, along with each bond's length, which we discuss next. *Stronger bonds are lower in energy (have a deeper energy well); weaker bonds are higher in energy (have a shallower energy well).* The energy of a given bond varies slightly from molecule to molecule, and even within a molecule, so each tabulated value is an *average* bond energy.

A covalent bond has a **bond length,** the distance between the nuclei of two bonded atoms. In Figure 9.10, bond length is shown as the distance between the nuclei at the point of minimum energy, and Table 9.2 shows the lengths of some

Table 9.2 Average Bond Energies (kJ/mol) and Bond Lengths (pm)

	Bond	Energy	Length	Bond	Energy	Length	Bond	Energy	Length	Bond	Energy	Length
Single Bonds												
	H—H	432	74	N—H	391	101	Si—H	323	148	S—H	347	134
	H—F	565	92	N—N	160	146	Si—Si	226	234	S—S	266	204
	H—Cl	427	127	N—P	209	177	Si—O	368	161	S—F	327	158
	H—Br	363	141	N—O	201	144	Si—S	226	210	S—Cl	271	201
	H—I	295	161	N—F	272	139	Si—F	565	156	S—Br	218	225
				N—Cl	200	191	Si—Cl	381	204	S—I	~170	234
	C—H	413	109	N—Br	243	214	Si—Br	310	216			
	C—C	347	154	N—I	159	222	Si—I	234	240	F—F	159	143
	C—Si	301	186							F—Cl	193	166
	C—N	305	147	O—H	467	96	P—H	320	142	F—Br	212	178
	C—O	358	143	O—P	351	160	P—Si	213	227	F—I	263	187
	C—P	264	187	O—O	204	148	P—P	200	221	Cl—Cl	243	199
	C—S	259	181	O—S	265	151	P—F	490	156	Cl—Br	215	214
	C—F	453	133	O—F	190	142	P—Cl	331	204	Cl—I	208	243
	C—Cl	339	177	O—Cl	203	164	P—Br	272	222	Br—Br	193	228
	C—Br	276	194	O—Br	234	172	P—I	184	243	Br—I	175	248
	C—I	216	213	O—I	234	194				I—I	151	266
Multiple Bonds												
	C=C	614	134	N=N	418	122	C≡C	839	121	N≡N	945	110
	C=N	615	127	N=O	607	120	C≡N	891	115	N≡O	631	106
	C=O	745 (799 in CO_2)	123	O_2	498	121	C≡O	1070	113			

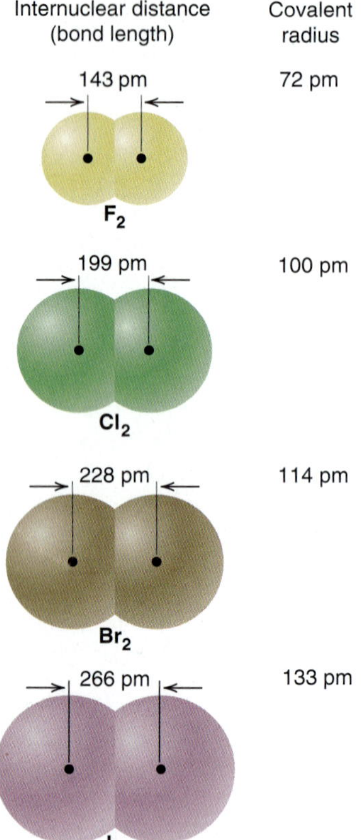

FIGURE 9.12 Bond length and covalent radius. Within a series of similar molecules, such as the diatomic halogen molecules, bond length increases as covalent radius increases.

covalent bonds. Like the bond energies, these values are *average* bond lengths for the given bond in different substances. Bond length is related to the sum of the radii of the bonded atoms. In fact, most atomic radii are calculated from measured bond lengths (see Figure 8.7C). Bond lengths for a series of similar bonds increase with atomic size, as shown in Figure 9.12 for the halogens.

A close relationship exists among bond order, bond length, and bond energy. Two nuclei are more strongly attracted to two shared electron pairs than to one: the atoms are drawn closer together *and* are more difficult to pull apart. Therefore, *for a given pair of atoms, a higher bond order results in a shorter bond length and a higher bond energy.* So, as Table 9.3 shows, for a given pair of atoms, *a shorter bond is a stronger bond.*

Table 9.3 The Relation of Bond Order, Bond Length, and Bond Energy

Bond	Bond Order	Average Bond Length (pm)	Average Bond Energy (kJ/mol)
C—O	1	143	358
C═O	2	123	745
C≡O	3	113	1070
C—C	1	154	347
C═C	2	134	614
C≡C	3	121	839
N—N	1	146	160
N═N	2	122	418
N≡N	3	110	945

In some cases, we can extend this relationship among atomic size, bond length, and bond strength by holding one atom in the bond constant and varying the other atom within a group or period. For example, the trend in carbon-halogen single bond lengths, C—I > C—Br > C—Cl, parallels the trend in atomic size, I > Br > Cl, and is opposite to the trend in bond energy, C—Cl > C—Br > C—I. Thus, for *single bonds,* longer bonds are usually weaker, and you can see many other examples of this relationship in Table 9.2.

SAMPLE PROBLEM 9.2 Comparing Bond Length and Bond Strength

Problem Without referring to Tables 9.2 and 9.3, rank the bonds in each set in order of *decreasing* bond length and bond strength:
(a) S—F, S—Br, S—Cl **(b)** C═O, C—O, C≡O

Plan In part (a), S is singly bonded to three different halogen atoms, so all members of the set have a bond order of 1. Bond length increases and bond strength decreases as the halogen's atomic radius increases, and that size trend is clear from the periodic table. In all the bonds in part (b), the same two atoms are involved, but the bond orders differ. In this case, bond strength increases and bond length decreases as bond order increases.

Solution (a) Atomic size increases down a group, so F < Cl < Br.

Bond length: S—Br > S—Cl > S—F

Bond strength: S—F > S—Cl > S—Br

(b) By ranking the bond orders, C≡O > C═O > C—O, we obtain

Bond length: C—O > C═O > C≡O

Bond strength: C≡O > C═O > C—O

Check From Tables 9.2 and 9.3, we see that the rankings are correct.

Comment Remember that for bonds involving pairs of different atoms, as in part (a), *the relationship between length and strength holds* only *for single bonds* and not in every case, so apply it carefully.

FOLLOW-UP PROBLEM 9.2 Rank the bonds in each set in order of *increasing* bond length and bond strength: **(a)** Si—F, Si—C, Si—O; **(b)** N═N, N—N, N≡N.

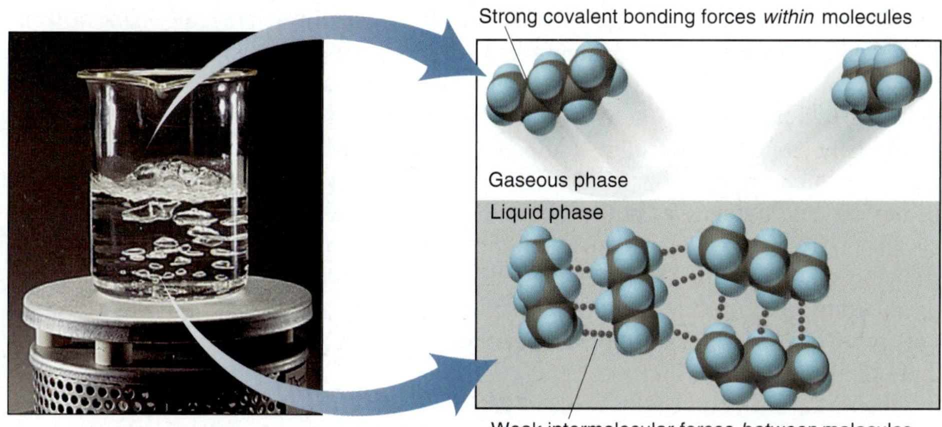

FIGURE 9.13 Strong forces within molecules and weak forces between them. When pentane boils, weak forces *between* molecules (intermolecular forces) are overcome, but the strong covalent bonds holding the atoms together within each molecule remain unaffected. Thus, the pentane molecules leave the liquid phase as intact units.

How the Model Explains the Properties of Covalent Substances

The covalent bonding model proposes that electron sharing between pairs of atoms leads to strong, localized bonds, usually within individual molecules. At first glance, however, it seems that the model is inconsistent with some of the familiar physical properties of covalent substances. After all, most are gases (such as methane and ammonia), liquids (such as benzene and water), or low-melting solids (such as sulfur and paraffin wax). If covalent bonds are so strong (~200 to 500 kJ/mol), why do covalent substances melt and boil at such low temperatures?

To answer this question, we must distinguish between two different sets of forces: (1) the *strong covalent bonding forces* holding the atoms together within the molecule (those we have been discussing), and (2) the *weak intermolecular forces* holding the separate molecules near each other in the macroscopic sample. It is these weak forces *between* the molecules, not the strong covalent bonds *within* each molecule, that are responsible for the physical properties of covalent substances. Consider, for example, what happens when pentane (C_5H_{12}) boils. As Figure 9.13 shows, the weak interactions *between* the pentane molecules are affected, not the strong C—C and C—H covalent bonds *within* each molecule.

Some covalent substances, called *network covalent solids,* do not consist of separate molecules. Rather, they are held together by covalent bonds that extend in three dimensions *throughout* the sample. The properties of these substances *do* reflect the strength of their covalent bonds. Two examples, quartz and diamond, are shown in Figure 9.14. Quartz (SiO_2) has silicon-oxygen covalent bonds that extend throughout the sample; no separate SiO_2 molecules exist. Quartz is very hard and melts at 1550°C. Diamond has covalent bonds connecting each of its carbon atoms to four others throughout the sample. It is the hardest natural substance known and melts at around 3550°C. Clearly, covalent bonds *are* strong, but because most covalent substances consist of separate molecules with weak forces between them, their physical properties do not reflect this bond strength. (We discuss intermolecular forces in detail in Chapter 12.)

Unlike ionic compounds, most covalent substances are poor electrical conductors, even when melted or when dissolved in water. An electric current is carried by either mobile electrons or mobile ions. In covalent substances the electrons are localized as either shared or unshared pairs, so they are not free to move, and no ions are present.

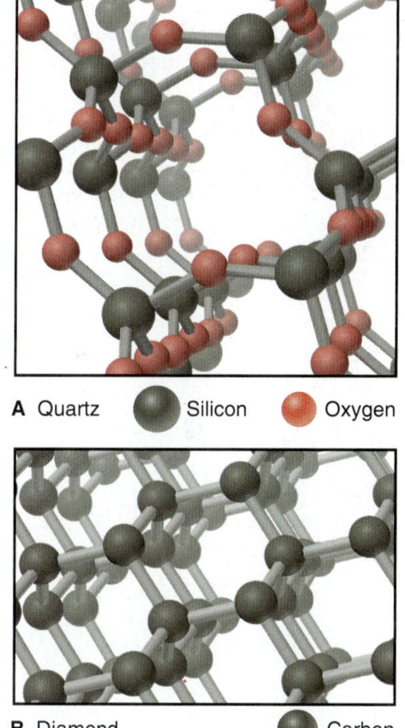

FIGURE 9.14 Covalent bonds of network covalent solids. A, In quartz (SiO_2), each Si atom is bonded covalently to four O atoms and each O atom is bonded to two Si atoms in a pattern that extends throughout the sample. Because no separate SiO_2 molecules are present, the melting point of quartz is very high, and it is very hard. **B,** In diamond, each C atom is covalently bonded to four other C atoms throughout the crystal. Diamond is the hardest natural substance known and has an extremely high melting point.

Chemists often study the types of covalent bonds in a molecule using a technique called **infrared (IR) spectroscopy.** All molecules, whether occurring as a gas, a liquid, or a solid, undergo continual vibrations. We can think of any covalent bond between two atoms, say, the C—C bond in ethane ($H_3C—CH_3$), as a spring that is continually stretching, twisting, and bending. Each motion occurs at a particular frequency, which depends on the "stiffness" of the spring (the bond energy), the type of motion, and the masses of the atoms. The frequencies of these vibrational motions correspond to the wavelengths of photons that lie within the IR region of the electromagnetic spectrum. Thus, the energies of these motions are quantized. And, just as an atom can absorb a photon of a particular energy and attain a different electron energy level (Chapter 7), a molecule can absorb an IR photon of a particular energy and attain a different vibrational energy level.

Each kind of bond (C—C, C═C, C—O, etc.) absorbs a characteristic range of IR wavelengths and quantity of radiation, which depends on the molecule's overall structure. The absorptions by all the bonds in a given molecule create a unique pattern of downward pointing peaks of varying depth and sharpness. Thus, *each compound has a characteristic IR spectrum* that can be used to identify it, much like a fingerprint is used to identify a person. As an example, consider the compounds 2-butanol and diethyl ether. These compounds have the same molecular formula ($C_4H_{10}O$) but different structural formulas and, therefore, are constitutional (structural) isomers. Figure 9.15 shows that they have very different IR spectra.

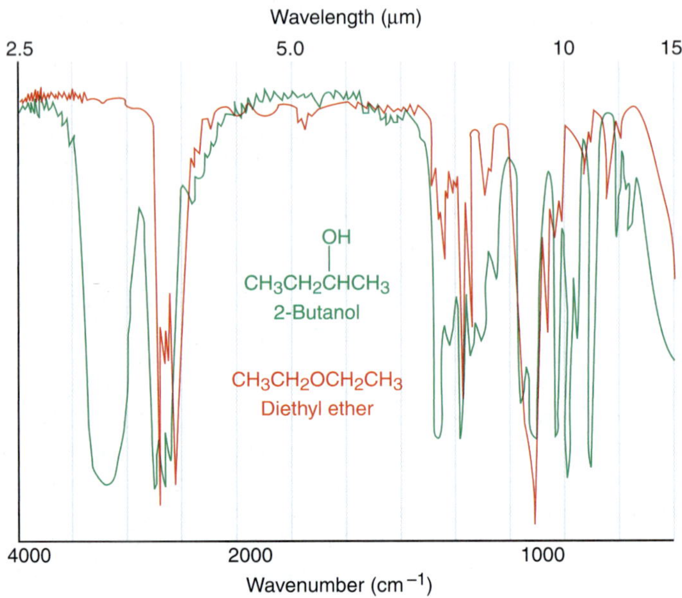

FIGURE 9.15 The infrared spectra of 2-butanol *(green)* and diethyl ether *(red)*.

SECTION 9.3 SUMMARY

A shared pair of valence electrons attracts the nuclei of two atoms and holds them together in a covalent bond, while filling each atom's outer level. The number of shared pairs between the two atoms is the bond order. • For a given type of bond, the bond energy is the average energy required to completely separate the bonded atoms; the bond length is the average distance between their nuclei. For a given pair of bonded atoms, bond order is directly related to bond energy and inversely related to bond length. • Substances that consist of separate molecules are generally soft and low melting because of the weak forces between molecules. Solids held together by covalent bonds extending throughout the sample are extremely hard and high melting. Most covalent substances have low electrical conductivity because electrons are localized and ions are absent. • The atoms in a covalent bond vibrate, and the energies of these vibrations can be studied with IR spectroscopy.

9.4 BOND ENERGY AND CHEMICAL CHANGE

The relative strengths of the bonds in reactants and products of a chemical change determine whether heat is released or absorbed. In fact, as you'll see in Chapter 20, bond strength is one of two essential factors determining whether the change occurs at all. In this section, we'll discuss the importance of bond energy in chemical change.

Changes in Bond Strengths: Where Does ΔH°_{rxn} Come From?

In Chapter 6, we discussed the heat involved in a chemical change (ΔH°_{rxn}), but we never stopped to ask a central question. When, for example, 1 mol of H_2 and 1 mol of F_2 react at 298 K, 2 mol of HF forms and 546 kJ of heat is released:

$$H_2(g) + F_2(g) \longrightarrow 2HF(g) + 546\text{ kJ}$$

Where does this heat come from? We find the answer through a very close-up view of the molecules and their energy components.

A system's internal energy has kinetic energy (E_k) and potential energy (E_p) components. Let's examine the contributions to these components to see which one changes during the reaction of H_2 and F_2 to form HF.

Of the various contributions to the kinetic energy, the most important come from the molecules moving through space, rotating, and vibrating and, of course, from the electrons moving within the atoms. Of the various contributions to the potential energy, the most important are electrostatic forces between the vibrating atoms, between nucleus and electrons (and between electrons) in each atom, between protons and neutrons in each nucleus, and, of course, between nuclei and shared electron pair in each bond.

The kinetic energy doesn't change during the reaction because the molecules' motion through space, rotation, and vibration are proportional to the temperature, which is constant at 298 K; and electron motion is not affected by a reaction. Of the potential energy contributions, those within the atoms and nuclei don't change, and vibrational forces vary only slightly as the bonded atoms change. The only significant change in potential energy is in the strength of attraction of the nuclei for the shared electron pair, that is, in the bond energy.

In other words, the answer to "Where does the heat come from?" is that it doesn't really "come from" anywhere: *the energy released or absorbed during a chemical change is due to differences between the reactant bond energies and the product bond energies.*

Using Bond Energies to Calculate ΔH°_{rxn}

We can think of any reaction as a two-step process in which *a quantity of heat is absorbed (ΔH° is positive) to break the reactant bonds and form separate atoms, and a different quantity is released (ΔH° is negative) when the atoms rearrange to form product bonds.* The sum (symbolized by Σ) of these enthalpy changes is the heat of reaction, ΔH°_{rxn}:

$$\Delta H^\circ_{rxn} = \Sigma\Delta H^\circ_{\text{reactant bonds broken}} + \Sigma\Delta H^\circ_{\text{product bonds formed}} \qquad (9.2)$$

- In an exothermic reaction, the total ΔH° for product bonds formed is *greater* than that for reactant bonds broken, so the sum, ΔH°_{rxn}, is *negative*.
- In an endothermic reaction, the total ΔH° for product bonds formed is *smaller* than that for reactant bonds broken, so the sum, ΔH°_{rxn}, is *positive*.

An equivalent form of Equation 9.2 uses bond energies:

$$\Delta H^\circ_{rxn} = \Sigma\text{BE}_{\text{reactant bonds broken}} - \Sigma\text{BE}_{\text{product bonds formed}}$$

The minus sign is needed because all bond energies are positive values (see Table 9.2).

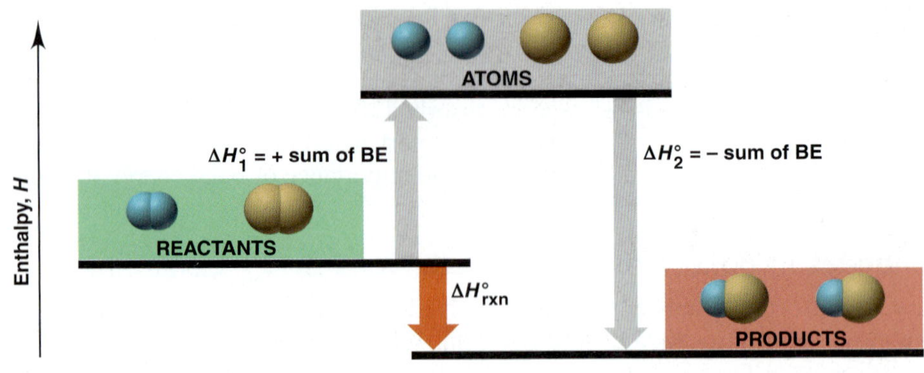

FIGURE 9.16 Using bond energies to calculate ΔH°_{rxn}. Any chemical reaction can be divided conceptually into two hypothetical steps: (1) reactant bonds break to yield separate atoms in a step that absorbs heat (+ sum of BE), and (2) the atoms combine to form product bonds in a step that releases heat (− sum of BE). When the total bond energy of the products is greater than that of the reactants, more energy is released than is absorbed, and the reaction is exothermic (as shown); ΔH°_{rxn} is negative. When the total bond energy of the products is less than that of the reactants, the reaction is endothermic; ΔH°_{rxn} is positive.

When 1 mol of H—H bonds and 1 mol of F—F bonds absorb energy and break, the 2 mol each of H and F atoms form 2 mol of H—F bonds, which releases energy (Figure 9.16). Recall that *weaker bonds (less stable, more reactive) are easier to break than stronger bonds (more stable, less reactive) because they are higher in energy.* Heat is released when HF forms because the bonds in H_2 and F_2 are weaker (less stable) than the bonds in HF (more stable). Put another way, the sum of the bond energies in 1 mol of H_2 and 1 mol of F_2 is *smaller* than the sum of the bond energies in 2 mol of HF.

We use bond energies to calculate ΔH°_{rxn} by assuming that ***all** the reactant bonds break to give individual atoms, from which **all** the product bonds form.* Even though, typically, only certain bonds break and form, Hess's law (see Section 6.5) allows us to imagine complete bond breakage and then sum the bond energies (with their appropriate signs) to arrive at the overall heat of reaction. (This method assumes all reactants and products are in the same physical state; when phase changes occur, additional heat must be taken into account. We address this topic in Chapter 12.)

Let's use bond energies to calculate ΔH°_{rxn} for the combustion of methane. Figure 9.17 shows that all the bonds in CH_4 and O_2 break, and the atoms form the bonds in CO_2 and H_2O. We find the bond energy values in Table 9.2, and use a positive sign for bonds broken and a negative sign for bonds formed:

Bonds broken

$$4 \times \text{C—H} = (4\ \text{mol})(413\ \text{kJ/mol}) = 1652\ \text{kJ}$$
$$2 \times O_2 = (2\ \text{mol})(498\ \text{kJ/mol}) = 996\ \text{kJ}$$
$$\Sigma\Delta H^\circ_{\text{reactant bonds broken}} = 2648\ \text{kJ}$$

Bonds formed

$$2 \times \text{C=O} = (2\ \text{mol})(-799\ \text{kJ/mol}) = -1598\ \text{kJ}$$
$$4 \times \text{O—H} = (4\ \text{mol})(-467\ \text{kJ/mol}) = -1868\ \text{kJ}$$
$$\Sigma\Delta H^\circ_{\text{product bonds formed}} = -3466\ \text{kJ}$$

Applying Equation 9.2 gives

$$\Delta H^\circ_{rxn} = \Sigma\Delta H^\circ_{\text{reactant bonds broken}} + \Sigma\Delta H^\circ_{\text{product bonds formed}}$$
$$= 2648\ \text{kJ} + (-3466\ \text{kJ}) = -818\ \text{kJ}$$

It is interesting to compare this value with the value obtained by calorimetry (Section 6.3), which is

$$CH_4(g) + 2O_2(g) \longrightarrow CO_2(g) + 2H_2O(g) \qquad \Delta H^\circ_{rxn} = -802\ \text{kJ}$$

Why is there a discrepancy between the bond energy value (−818 kJ) and the calorimetric value (−802 kJ)? Variations in experimental method always introduce small discrepancies, but there is a more basic reason in this case. Because bond energies are *average* values obtained from many different compounds, the energy of the bond *in a particular substance* is usually close, but not equal, to this average. For example, the tabulated C—H bond energy of 413 kJ/mol is the

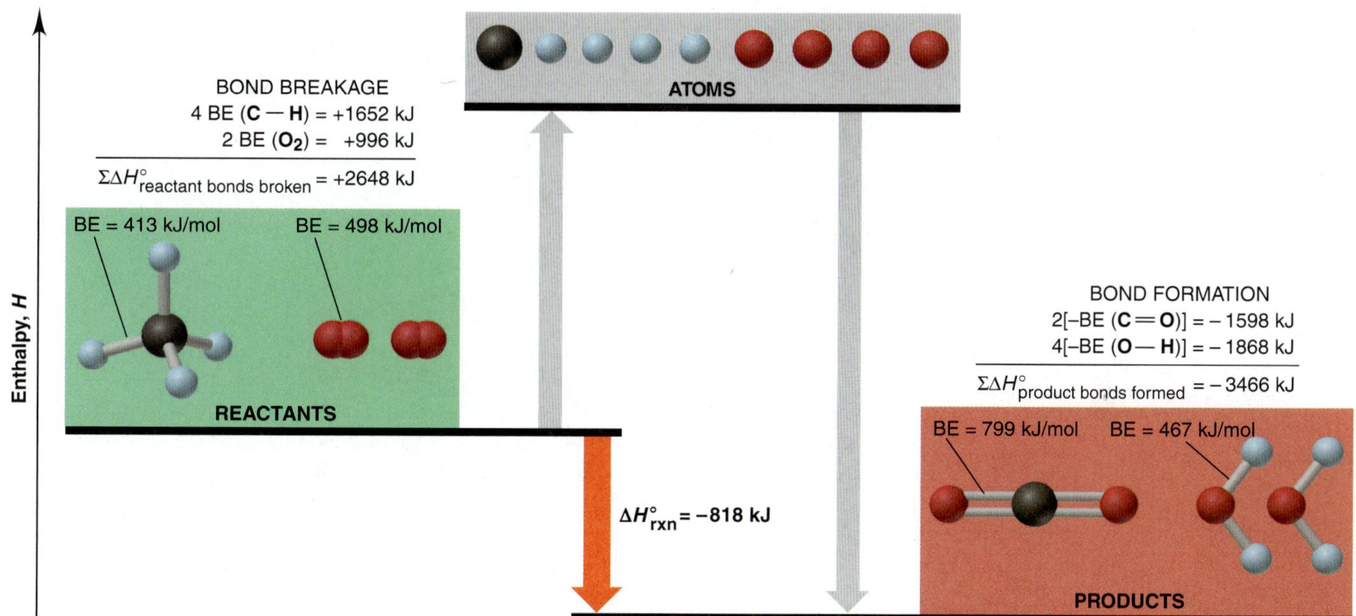

FIGURE 9.17 Using bond energies to calculate ΔH°_{rxn} of methane. Treating the combustion of methane as a hypothetical two-step process (see Figure 9.16) means breaking all the bonds in the reactants and forming all the bonds in the products.

average value of C—H bonds in many different molecules. In fact, 415 kJ is actually required to break 1 mol of C—H bonds in methane, or 1660 kJ for 4 mol of these bonds, which gives a ΔH°_{rxn} even closer to the calorimetric value. Thus, it isn't surprising to find a discrepancy between the two ΔH°_{rxn} values. What is surprising—and satisfying in its confirmation of bond theory—is that the values are so close.

SAMPLE PROBLEM 9.3 Using Bond Energies to Calculate ΔH°_{rxn}

Problem Calculate ΔH°_{rxn} for the chlorination of methane to form chloroform:

$$\mathrm{CH_4\ (H{-}C{-}H\ with\ H\ above\ and\ below)} + 3\,\mathrm{Cl{-}Cl} \longrightarrow \mathrm{Cl{-}C{-}Cl\ (H\ above,\ Cl\ below)} + 3\,\mathrm{H{-}Cl}$$

Plan We assume that, in the reaction, *all* the reactant bonds break and *all* the product bonds form. We find the bond energies in Table 9.2 and substitute the two sums, with correct signs, into Equation 9.2.

Solution Finding the standard enthalpy changes for bonds broken and for bonds formed: For bonds broken, the bond energy values are

$$\begin{aligned} 4 \times \mathrm{C{-}H} &= (4\ \mathrm{mol})(413\ \mathrm{kJ/mol}) = 1652\ \mathrm{kJ} \\ 3 \times \mathrm{Cl{-}Cl} &= (3\ \mathrm{mol})(243\ \mathrm{kJ/mol}) = 729\ \mathrm{kJ} \\ \hline \Sigma\Delta H^\circ_{\text{bonds broken}} &= 2381\ \mathrm{kJ} \end{aligned}$$

For bonds formed, the values are

$$\begin{aligned} 3 \times \mathrm{C{-}Cl} &= (3\ \mathrm{mol})(-339\ \mathrm{kJ/mol}) = -1017\ \mathrm{kJ} \\ 1 \times \mathrm{C{-}H} &= (1\ \mathrm{mol})(-413\ \mathrm{kJ/mol}) = -413\ \mathrm{kJ} \\ 3 \times \mathrm{H{-}Cl} &= (3\ \mathrm{mol})(-427\ \mathrm{kJ/mol}) = -1281\ \mathrm{kJ} \\ \hline \Sigma\Delta H^\circ_{\text{bonds formed}} &= -2711\ \mathrm{kJ} \end{aligned}$$

Calculating ΔH°_{rxn}:

$$\Delta H^\circ_{rxn} = \Sigma\Delta H^\circ_{\text{bonds broken}} + \Sigma\Delta H^\circ_{\text{bonds formed}} = 2381\ \mathrm{kJ} + (-2711\ \mathrm{kJ}) = -330\ \mathrm{kJ}$$

Check The signs of the enthalpy changes are correct: $\Sigma\Delta H^\circ_{\text{bonds broken}}$ should be >0, and $\Sigma\Delta H^\circ_{\text{bonds formed}} <0$. More energy is released than absorbed, so $\Delta H^\circ_{\text{rxn}}$ is negative:

$$\sim 2400 \text{ kJ} + [\sim(-2700 \text{ kJ})] = -300 \text{ kJ}$$

FOLLOW-UP PROBLEM 9.3 One of the most important industrial reactions is the formation of ammonia from its elements:

$$\text{N}{\equiv}\text{N} + 3\,\text{H—H} \longrightarrow 2\,\text{H—}\underset{\substack{|\\ \text{H}}}{\text{N}}\text{—H}$$

Use bond energies to calculate $\Delta H^\circ_{\text{rxn}}$.

SECTION 9.4 SUMMARY

The only component of internal energy that changes significantly during a reaction is the energy of the bonds in reactants and products; this change in bond energy appears as the heat of reaction, $\Delta H^\circ_{\text{rxn}}$. • A reaction involves breaking reactant bonds and forming product bonds. Applying Hess's law, we use tabulated bond energies to calculate $\Delta H^\circ_{\text{rxn}}$.

9.5 BETWEEN THE EXTREMES: ELECTRONEGATIVITY AND BOND POLARITY

Scientific models are idealized descriptions of reality. As we've discussed them so far, the ionic and covalent bonding models portray compounds as being formed by *either* complete electron transfer *or* complete electron sharing. In most real substances, however, the type of bonding lies somewhere between these extremes. Thus, the great majority of compounds have bonds that are more accurately thought of as "polar covalent," that is, partially ionic and partially covalent (Figure 9.18).

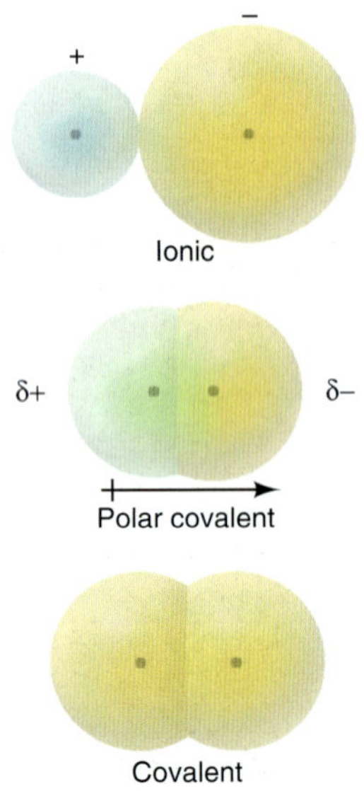

+ −

Ionic

δ+ δ−

Polar covalent

Covalent

FIGURE 9.18 The prevalence of electron sharing. The extremes of pure ionic *(top)* and pure covalent *(bottom)* bonding occur rarely, if ever. Much more common is the unequal electron sharing seen in polar covalent bonding.

Electronegativity

One of the most important concepts in chemical bonding is **electronegativity (EN),** the relative ability of a bonded atom to attract the shared electrons.* More than 50 years ago, the American chemist Linus Pauling developed the most common scale of relative EN values for the elements. Here is an example to show the basis of Pauling's approach. We might expect the bond energy of the HF bond to be the average of the energies of an H—H bond (432 kJ/mol) and an F—F bond (159 kJ/mol), or 296 kJ/mol. However, the actual bond energy of H—F is 565 kJ/mol, or 269 kJ/mol *higher* than the average. Pauling reasoned that this difference is due to an *electrostatic (charge) contribution* to the H—F bond energy. If F attracts the shared electron pair more strongly than H, that is, if F is more electronegative than H, the electrons will spend more time closer to F. This unequal sharing of electrons makes the F end of the bond partially negative and the H end partially positive, and the attraction between these partial charges *increases* the energy required to break the bond.

From similar studies with the remaining hydrogen halides and many other compounds, Pauling arrived at the scale of *relative EN values* shown in Figure 9.19. These values are not measured quantities but are based on Pauling's assignment of the highest EN value, 4.0, to fluorine.

Trends in Electronegativity Because the nucleus of a smaller atom is closer to the shared pair than that of a larger atom, it attracts the bonding electrons more

*Electronegativity is not the same as electron affinity (EA), although many elements with a high EN also have a highly negative EA. Electronegativity refers to a bonded atom attracting the shared electron pair; electron affinity refers to a separate atom in the gas phase gaining an electron to form a gaseous anion.

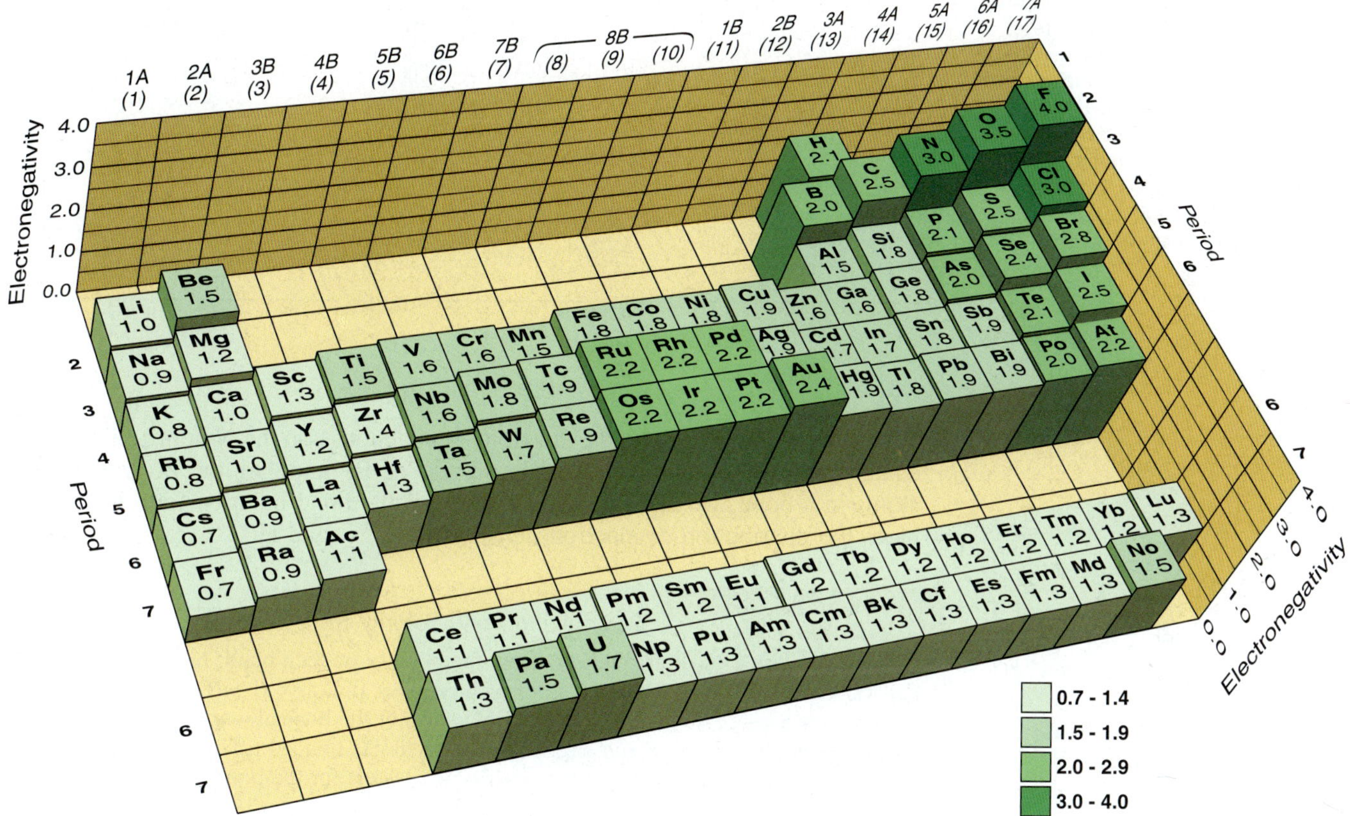

FIGURE 9.19 The Pauling electronegativity (EN) scale. The EN is shown by the height of the post with the value on top. The key indicates arbitrary EN cutoffs. In the main groups, EN generally *increases* from left to right and *decreases* from top to bottom. The noble gases are not shown. The transition and inner transition elements show relatively little change in EN. Hydrogen is shown near elements of similar EN.

strongly. So, in general, electronegativity is inversely related to atomic size. Thus, for the main-group elements, *electronegativity generally increases up a group and across a period.*

Electronegativity and Oxidation Number One important use of electronegativity is in determining an atom's oxidation number (O.N.; see Section 4.5):

1. The more electronegative atom in a bond is assigned *all* the *shared* electrons; the less electronegative atom is assigned *none.*
2. Each atom in a bond is assigned *all* of its *unshared* electrons.
3. The oxidation number is given by

$$\text{O.N.} = \text{no. of valence e}^- - (\text{no. of shared e}^- + \text{no. of unshared e}^-)$$

In HCl, for example, Cl is more electronegative than H. It has 7 valence electrons but is assigned 8 (2 shared + 6 unshared), so its oxidation number is $7 - 8 = -1$. The H atom has 1 valence electron and is assigned none, so its oxidation number is $1 - 0 = +1$.

Polar Covalent Bonds and Bond Polarity

Whenever atoms of different electronegativities form a bond, as in HF, the bonding pair is shared *un*equally. This unequal distribution of electron density gives the bond partially negative and positive poles. Such a **polar covalent bond** is depicted by a polar arrow (⟼) pointing toward the negative pole or by δ+ and δ− symbols, where the lowercase Greek letter delta (δ) represents a partial charge (see also the discussion of O—H bonds in water, Section 4.1):

⟼
H—F̤: or H—F̤:
δ+ δ−

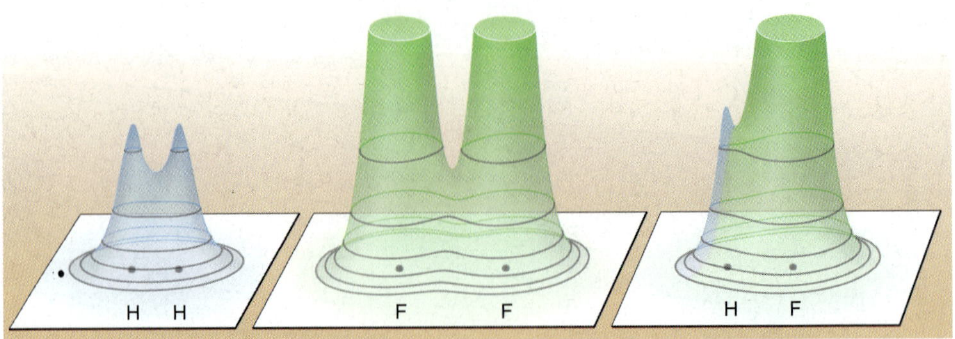

FIGURE 9.20 Electron density distributions in H_2, F_2, and HF. As these relief maps show, electron density is distributed equally around the two nuclei in the nonpolar covalent molecules H_2 and F_2. (The electron density around the F nuclei is so great that the peaks must be "cut off" to fit within the figure.) But, in polar covalent HF, the electron density is shifted away from H and toward F.

In the H—H and F—F bonds, the atoms are identical, so the bonding pair is shared equally, and a **nonpolar covalent bond** results. By knowing the EN values of the atoms in a bond, we can find the direction of the bond polarity. Figure 9.20 compares the distribution of electron density in H_2, F_2, and HF.

SAMPLE PROBLEM 9.4 Determining Bond Polarity from EN Values

Problem **(a)** Use a polar arrow to indicate the polarity of each bond: N—H, F—N, I—Cl. **(b)** Rank the following bonds in order of increasing polarity: H—N, H—O, H—C.

Plan **(a)** We use Figure 9.19 to find the EN values of the bonded atoms and point the polar arrow toward the negative end. **(b)** Each choice has H bonded to an atom from Period 2. EN increases across a period, so the polarity is greatest for the bond whose Period 2 atom is farthest to the right.

Solution **(a)** The EN of N = 3.0 and the EN of H = 2.1, so N is more electronegative than H: $\overset{\longleftarrow\!+}{\mathrm{N{-}H}}$

The EN of F = 4.0 and the EN of N = 3.0, so F is more electronegative: $\overset{\longleftarrow\!+}{\mathrm{F{-}N}}$

The EN of I = 2.5 and the EN of Cl = 3.0, so I is less electronegative: $\overset{+\!\longrightarrow}{\mathrm{I{-}Cl}}$

(b) The order of increasing EN is C < N < O, and each has a higher EN than H does. Therefore, O pulls most on the electron pair shared with H, and C pulls least; so the order of bond polarity is H—C < H—N < H—O.

Comment In Chapter 10, you'll see that the polarity of the bonds in a molecule contributes to the overall polarity of the molecule, which is a major factor determining the magnitudes of several physical properties.

FOLLOW-UP PROBLEM 9.4 Arrange each set of bonds in order of increasing polarity, and indicate bond polarity with $\delta+$ and $\delta-$ symbols:
(a) Cl—F, Br—Cl, Cl—Cl **(b)** Si—Cl, P—Cl, S—Cl, Si—Si

The Partial Ionic Character of Polar Covalent Bonds

As we've just seen, if you ask "Is an X—Y bond ionic or covalent?" the answer in almost every case is "Both, partially!" A better question is "*To what extent* is the bond ionic or covalent?" The **partial ionic character** of a bond is related directly to the **electronegativity difference (ΔEN),** the difference between the EN values of the bonded atoms: *a greater ΔEN results in larger partial charges and a higher partial ionic character.* Consider these three chlorine-containing molecules: ΔEN for LiCl(g) is 3.0 − 1.0 = 2.0; for HCl(g), it is 3.0 − 2.1 = 0.9; and for $Cl_2(g)$, it is 3.0 − 3.0 = 0. Thus, the bond in LiCl has more ionic character than the H—Cl bond, which has more than the Cl—Cl bond.

Various attempts have been made to classify the ionic character of bonds, but they all use arbitrary cutoff values, which is inconsistent with the gradation of

ΔEN	IONIC CHARACTER
>1.7	Mostly ionic
0.4-1.7	Polar covalent
<0.4	Mostly covalent
0	Nonpolar covalent

A

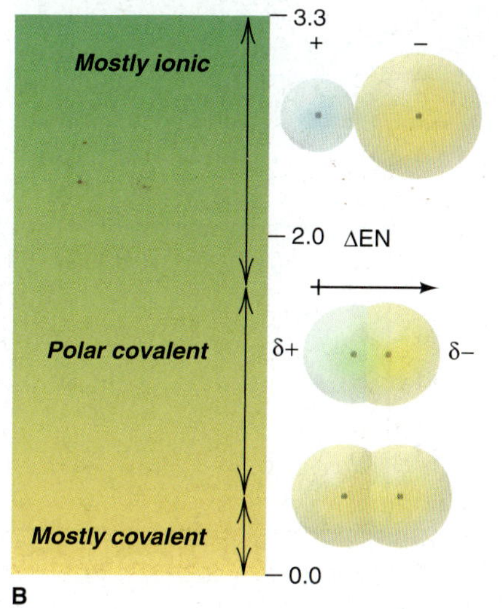

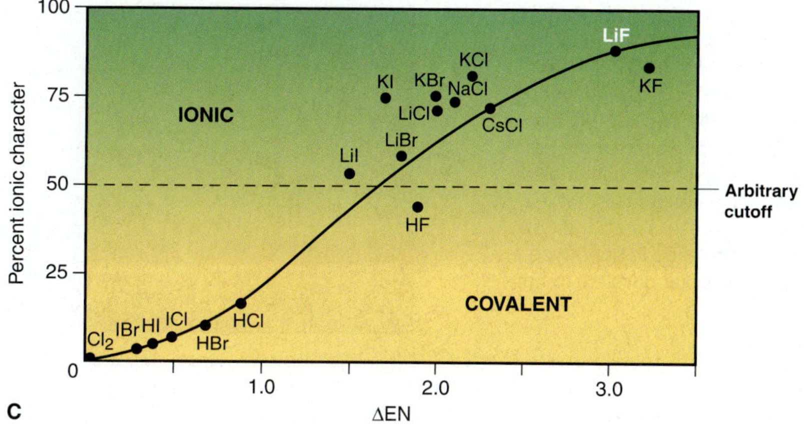

FIGURE 9.21 The ionic character of chemical bonds. A, The electronegativity difference (ΔEN) between bonded atoms shows cutoff values that act as a very general guide to a bond's ionic character. **B,** The gradation in ionic character across the entire bonding range is shown as shading from ionic *(green)* to covalent *(yellow)*. **C,** The percent ionic character is plotted against ΔEN for some gaseous diatomic molecules. Note that, in general, ΔEN correlates with ionic character. (The arbitrary cutoff for an ionic compound is >50% ionic character.)

ionic character observed experimentally. One approach uses ΔEN values to divide bonds into ionic, polar covalent, and nonpolar covalent. Based on a range of ΔEN values from 0 (completely nonpolar) to 3.3 (highly ionic), some *approximate* guidelines are given in Figure 9.21A and appear within a gradient in Figure 9.21B. Keep in mind that these ΔEN cutoff values are only useful for gaining a *general* idea of a compound's ionic character.

Another approach calculates the *percent ionic character* of a bond by comparing the behavior of a polar molecule in an electric field with the behavior it would show if the bonding pair shifted to the more electronegative atom completely (a pure ionic bond). A value of 50% ionic character divides substances we call "ionic" from those we call "covalent." Such methods show 43% ionic character for the H—F bond and expected decreases for the other hydrogen halides: H—Cl is 19% ionic, H—Br 11%, and H—I 4%. A plot of percent ionic character vs. ΔEN for a variety of gaseous diatomic molecules is shown in Figure 9.21C. The specific values are not important, but note that *percent ionic character generally increases with ΔEN.* Also note that some molecules, such as $Cl_2(g)$, have 0% ionic character, but none has 100% ionic character. Thus, *electron sharing occurs to some extent in every bond,* even one between an alkali metal and a halogen.

The Continuum of Bonding Across a Period

A metal and a nonmetal—elements from opposite sides of the periodic table—have a relatively large ΔEN and typically interact by electron transfer to form an ionic compound. Two nonmetals—elements from the same side of the table—have a small ΔEN and interact by electron sharing to form a covalent compound. When we combine the nonmetal chlorine with each of the other elements in Period 3, starting with sodium, we should observe a steady decrease in ΔEN and a gradation in bond type from ionic through polar covalent to nonpolar covalent.

FIGURE 9.22 Properties of the Period 3 chlorides. Samples of the compounds formed from each of the Period 3 elements with chlorine are shown in periodic table sequence in the photo. Note the trend in properties displayed in the bar graphs: as ΔEN decreases, both melting point and electrical conductivity (at the melting point) decrease. These trends are consistent with a change in bond type from ionic through polar covalent to nonpolar covalent.

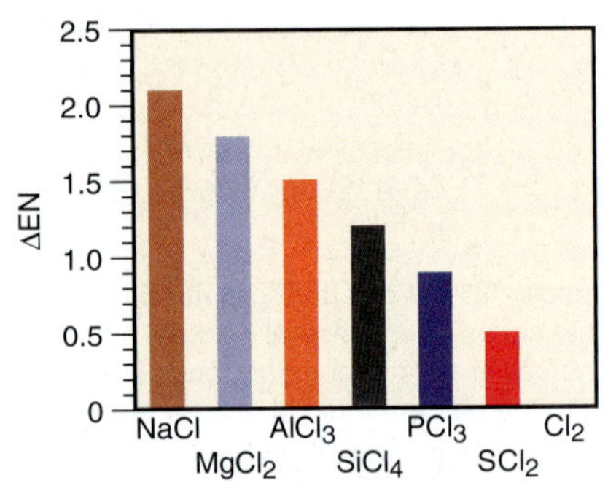

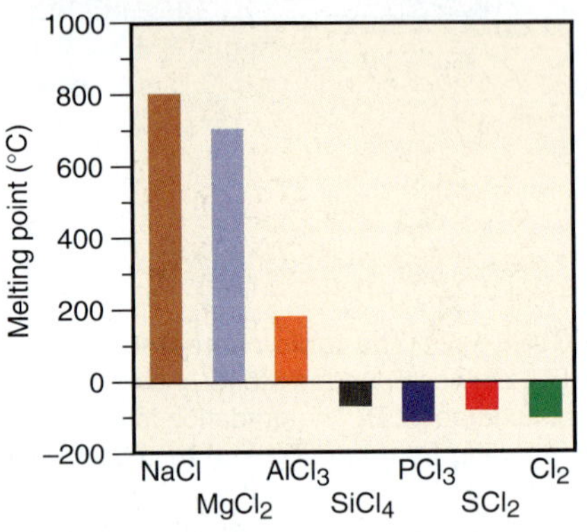

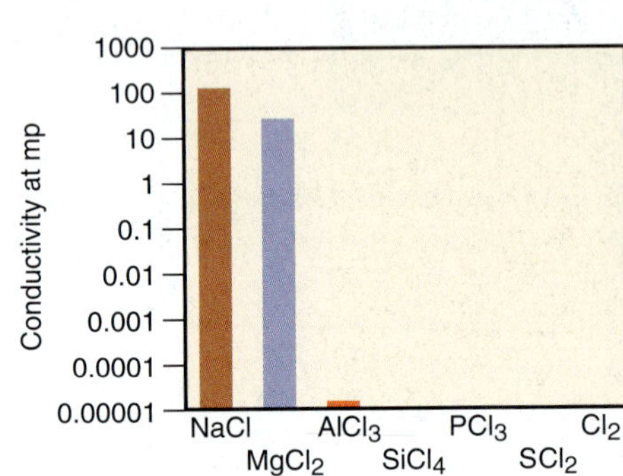

Figure 9.22 shows samples of the common Period 3 chlorides—NaCl, $MgCl_2$, $AlCl_3$, $SiCl_4$, PCl_3, and SCl_2, as well as Cl_2—with graphs showing their ΔEN values and two key macroscopic properties. With a ΔEN of 2.1, sodium chloride is the most ionic of this group, displaying the physical properties characteristic of an ionic solid: high melting point and high electrical conductivity when molten. $MgCl_2$, also a white crystalline solid, is also ionic, with ΔEN of 1.8, but it has a lower melting point and a slightly lower conductivity in the molten state.

The structure of $AlCl_3$ reflects its smaller ΔEN value of 1.5, which lies near the upper cutoff for polar covalent bonding. Instead of a three-dimensional lattice of Al^{3+} and Cl^- ions, $AlCl_3$ consists of extended layers of Al and Cl atoms connected through highly polar covalent bonds. Much weaker forces between the layers result in a significantly lower melting point for $AlCl_3$, and its very low conductivity when molten is consistent with a scarcity of ions.

The tendency toward less ionic and more covalent character in the bonds continues through $SiCl_4$, PCl_3, and SCl_2. Each of these compounds occurs as separate molecules, as indicated by the absence of any measurable conductivity. Moreover, the forces *between* the molecules are weak enough to lower the melting points for these compounds below room temperature. In Cl_2, the bond is nonpolar, and this is the only substance in this group that is a gas at room temperature.

Thus, *as ΔEN becomes smaller, the bond becomes more covalent,* and the macroscopic properties of the Period 3 chlorides and Cl_2 change from those of a solid consisting of ions to those of a gas consisting of individual molecules.

SECTION 9.5 SUMMARY

An atom's electronegativity refers to its ability to pull bonded electrons toward it, which generates partial charges at the ends of the bond. • Electronegativity increases across a period and decreases down a group, the reverse of the trends in atomic size. • The greater the ΔEN for the two atoms in a bond, the more polar the bond is and the greater its ionic character. • For chlorides of Period 3 elements, there is a gradation of bond type from ionic to polar covalent to nonpolar covalent.

CHAPTER REVIEW GUIDE

The following sections provide many aids to help you study this chapter. (Numbers in parentheses refer to pages, unless noted otherwise.)

LEARNING OBJECTIVES *These are concepts and skills you should know after studying this chapter.*

Related section (§), sample problem (SP), and end-of-chapter problem (EP) numbers are listed in parentheses.

1. Explain how differences in atomic properties lead to the three types of chemical bonding (§ 9.1) (EPs 9.1–9.7)
2. Depict main-group atoms with Lewis electron-dot symbols (§ 9.1) (EPs 9.8–9.11)
3. Understand the key features of ionic bonding, the significance of the lattice energy, and how the model explains the properties of ionic compounds (§ 9.2) (EPs 9.12–9.15, 9.20–9.22)
4. Depict the formation of binary ionic compounds with electron configurations, partial orbital diagrams, and Lewis electron-dot symbols (§ 9.2) (SP 9.1) (EPs 9.16–9.19)
5. Describe the formation of a covalent bond, the interrelationship among bond length, strength, and order, and how the model explains the properties of covalent compounds (§ 9.3) (SP 9.2) (EPs 9.23–9.31)
6. Understand how changes in bond energy account for ΔH°_{rxn} and be able to divide a reaction into bond-breaking and bond-forming steps (§ 9.4) (SP 9.3) (EPs 9.32–9.39)
7. Describe the trends in electronegativity, and understand how the polarity of a bond and the partial ionic character of a compound relate to ΔEN of the bonded atoms (§ 9.5) (SP 9.4) (EPs 9.40–9.55)

KEY TERMS *These important terms appear in boldface in the chapter and are defined again in the Glossary.*

Section 9.1
ionic bonding (279)
covalent bonding (280)
metallic bonding (280)
Lewis electron-dot symbol (281)
octet rule (282)

Section 9.2
lattice energy (284)
Coulomb's law (284)
ion pair (286)

Section 9.3
covalent bond (287)
bonding (shared) pair (288)
lone (unshared) pair (288)
bond order (288)
single bond (288)
double bond (288)
triple bond (288)
bond energy (BE) (289)
bond length (289)
infrared (IR) spectroscopy (292)

Section 9.5
electronegativity (EN) (296)
polar covalent bond (297)
nonpolar covalent bond (298)
partial ionic character (298)
electronegativity difference (ΔEN) (298)

KEY EQUATIONS AND RELATIONSHIPS *Numbered and screened concepts are listed for you to refer to or memorize.*

9.1 Relating the energy of attraction to the lattice energy (284):

$$\text{Electrostatic energy} \propto \frac{\text{cation charge} \times \text{anion charge}}{\text{cation radius} + \text{anion radius}} \propto \Delta H^\circ_{lattice}$$

9.2 Calculating heat of reaction from enthalpy changes or bond energies (293):

$$\Delta H^\circ_{rxn} = \Sigma\Delta H^\circ_{\text{reactant bonds broken}} + \Sigma\Delta H^\circ_{\text{product bonds formed}}$$

or $$\Delta H^\circ_{rxn} = \Sigma\text{BE}_{\text{reactant bonds broken}} - \Sigma\text{BE}_{\text{product bonds formed}}$$

BRIEF SOLUTIONS TO *FOLLOW-UP PROBLEMS* *Compare your own solutions to these calculation steps and answers.*

9.1 $Mg\ ([Ne]\ 3s^2) + 2Cl\ ([Ne]\ 3s^23p^5) \longrightarrow Mg^{2+}\ ([Ne]) + 2Cl^-\ ([Ne]\ 3s^23p^6)$

$\cdot Mg\cdot + 2\ :\ddot{C}l\cdot \longrightarrow Mg^{2+} + 2\ :\ddot{C}l:^-$ Formula: $MgCl_2$

9.2 (a) Bond length: Si—F < Si—O < Si—C
Bond strength: Si—C < Si—O < Si—F
(b) Bond length: N≡N < N═N < N—N
Bond strength: N—N < N═N < N≡N

9.3 $N{\equiv}N + 3\ H{-}H \longrightarrow 2\ H{-}\underset{\underset{H}{|}}{N}{-}H$

$\Sigma\Delta H^\circ_{\text{bonds broken}} = 1\ N{\equiv}N + 3\ H{-}H = 945\ kJ + 1296\ kJ = 2241\ kJ$

$\Sigma\Delta H^\circ_{\text{bonds formed}} = 6\ N{-}H = -2346\ kJ$

$\Delta H^\circ_{rxn} = -105\ kJ$

9.4 (a) $Cl{-}Cl < \overset{\delta+}{Br}{-}\overset{\delta-}{Cl} < \overset{\delta+}{Cl}{-}\overset{\delta-}{F}$

(b) $Si{-}Si < \overset{\delta+}{S}{-}\overset{\delta-}{Cl} < \overset{\delta+}{P}{-}\overset{\delta-}{Cl} < \overset{\delta+}{Si}{-}\overset{\delta-}{Cl}$

PROBLEMS

*Problems with **colored** numbers are answered in Appendix E. Sections match the text and provide the numbers of relevant sample problems. Bracketed problems are grouped in pairs (indicated by a short rule) that cover the same concept. Comprehensive Problems are based on material from any section or previous chapter.*

Atomic Properties and Chemical Bonds

9.1 In general terms, how does each of the following atomic properties influence the metallic character of the main-group elements in a period?
(a) Ionization energy
(b) Atomic radius
(c) Number of outer electrons
(d) Effective nuclear charge

9.2 What is the relationship between the tendency of a main-group element to form a monatomic ion and its position in the periodic table? In what part of the table are the main-group elements that typically form cations? Anions?

9.3 Three solids are represented below. What is the predominant type of intramolecular bonding in each?

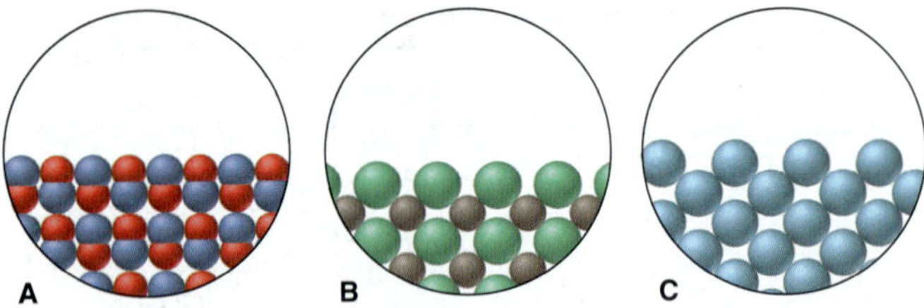

9.4 Which member of each pair is *more* metallic?
(a) Na or Cs (b) Mg or Rb (c) As or N

9.5 Which member of each pair is *less* metallic?
(a) I or O (b) Be or Ba (c) Se or Ge

9.6 State whether the type of bonding in the following compounds is best described as ionic or covalent: (a) CsF(*s*); (b) $N_2(g)$; (c) $H_2S(g)$.

9.7 State whether the type of bonding in the following compounds is best described as ionic or covalent: (a) $ICl_3(g)$; (b) $N_2O(g)$; (c) LiCl(*s*).

9.8 Draw a Lewis electron-dot symbol for each atom:
(a) Rb (b) Si (c) I

9.9 Draw a Lewis electron-dot symbol for each atom:
(a) Ba (b) Kr (c) Br

9.10 Give the group number and general electron configuration of an element with each electron-dot symbol:
(a) ·Ẍ̣: (b) Ẋ̣·

9.11 Give the group number and general electron configuration of an element with each electron-dot symbol:
(a) ·Ẋ̣: (b) ·Ẋ̣·

The Ionic Bonding Model
(Sample Problem 9.1)

9.12 If energy is required to form monatomic ions from metals and nonmetals, why do ionic compounds exist?

9.13 (a) In general, how does the lattice energy of an ionic compound depend on the charges and sizes of the ions? (b) Ion arrangements of three general salts are represented below. Rank them in order of increasing lattice energy.

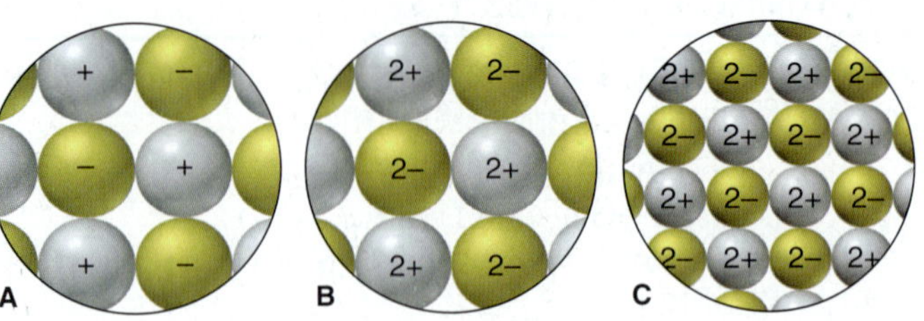

9.14 When gaseous Na^+ and Cl^- ions form gaseous NaCl ion pairs, 548 kJ/mol of energy is released. Why, then, does NaCl occur as a solid under ordinary conditions?

9.15 To form S^{2-} ions from gaseous sulfur atoms requires 214 kJ/mol, but these ions exist in solids such as K_2S. Explain.

9.16 Use condensed electron configurations and Lewis electron-dot symbols to depict the monatomic ions formed from each of the following atoms, and predict the formula of the compound the ions produce:
(a) Ba and Cl (b) Sr and O (c) Al and F (d) Rb and O

9.17 Use condensed electron configurations and Lewis electron-dot symbols to depict the monatomic ions formed from each of the following atoms, and predict the formula of the compound the ions produce:
(a) Cs and S (b) O and Ga (c) N and Mg (d) Br and Li

9.18 Identify the main group to which X belongs in each ionic compound formula: (a) X_2O_3; (b) XCO_3; (c) Na_2X.

9.19 Identify the main group to which X belongs in each ionic compound formula: (a) CaX_2; (b) Al_2X_3; (c) XPO_4.

9.20 For each pair, choose the compound with the lower lattice energy, and explain your choice:
(a) CaS or BaS (b) NaF or MgO

9.21 For each pair, choose the compound with the lower lattice energy, and explain your choice:
(a) NaF or NaCl (b) K_2O or K_2S

9.22 Aluminum oxide (Al_2O_3) is a widely used industrial abrasive (emery, corundum), for which the specific application depends on the hardness of the crystal. What does this hardness imply about the magnitude of the lattice energy? Would you have predicted from the chemical formula that Al_2O_3 is hard? Explain.

The Covalent Bonding Model
(Sample Problem 9.2)

9.23 Describe the interactions that occur between individual chlorine atoms as they approach each other and form Cl_2. What combination of forces gives rise to the energy holding the atoms together and to the final internuclear distance?

9.24 Define bond energy using the H—Cl bond as an example. When this bond breaks, is energy absorbed or released? Is the accompanying ΔH value positive or negative? How do the magnitude and sign of this ΔH value relate to the value that accompanies H—Cl bond formation?

9.25 For single bonds between similar types of atoms, how does the strength of the bond relate to the sizes of the atoms? Explain.

9.26 How does the energy of the bond between a given pair of atoms relate to the bond order? Why?

9.27 When liquid benzene (C_6H_6) boils, does the gas consist of molecules, ions, or separate atoms? Explain.

9.28 Using the periodic table only, arrange the members of each of the following sets in order of increasing bond *strength:*
(a) Br—Br, Cl—Cl, I—I (b) S—H, S—Br, S—Cl
(c) C=N, C—N, C≡N

9.29 Using the periodic table only, arrange the members of each of the following sets in order of increasing bond *length:*
(a) H—F, H—I, H—Cl (b) C—S, C=O, C—O
(c) N—H, N—S, N—O

9.30 Formic acid (HCOOH) has the structural formula shown at right. It is secreted by certain species of ants when they bite. Rank the relative strengths of (a) the C—O and C=O bonds, and (b) the H—C and H—O bonds. Explain these rankings.

```
     O
     ‖
H—C—O—H
```

9.31 In IR spectra, the stretching of a C=C bond appears at a shorter wavelength than that of a C—C bond. Would you expect the wavelength for the stretching of a C≡C bond to be shorter or longer than that for a C=C bond? Explain.

Bond Energy and Chemical Change

(Sample Problem 9.3)

9.32 Write a solution plan (without actual numbers, but including the bond energies you would use and how you would combine them algebraically) for calculating the total enthalpy change of the following reaction:

$$H_2(g) + O_2(g) \longrightarrow H_2O_2(g)\ (\text{H—O—O—H})$$

9.33 The text points out that, in general, a substance with weaker bonds is more reactive than one with stronger bonds. Why is this true?

9.34 Why is there often a discrepancy between a heat of reaction obtained from calorimetry and one obtained from bond energies?

9.35 Which gas has the greater heat of reaction per mole from combustion? Why?

```
 methane     or    formaldehyde

    H                  O
    |                  ‖
 H—C—H              H—C—H
    |
    H
```

9.36 Use bond energies to calculate the heat of reaction:

```
H       H                          H  H
  \   /                            |  |
   C=C     +  Cl—Cl  ——→       H—C—C—H
  /   \                            |  |
H       H                          Cl Cl
```

9.37 Use bond energies to calculate the heat of reaction:

```
                  H                       O
                  |                       ‖
O=C=O  +  2 N—H   ——→   H—N—C—N—H  +  H—O—H
                  |         |     |
                  H         H     H
```

9.38 An important industrial route to extremely pure acetic acid is the reaction of methanol with carbon monoxide:

```
    H                          H  O
    |                          |  ‖
H—C—O—H  +  C≡O  ——→      H—C—C—O—H
    |                          |
    H                          H
```

Use bond energies to calculate the heat of reaction.

9.39 Sports trainers treat sprains and soreness with ethyl bromide. It is manufactured by reacting ethylene with hydrogen bromide:

```
H       H                          H  H
  \   /                            |  |
   C=C     +  H—Br   ——→       H—C—C—Br
  /   \                            |  |
H       H                          H  H
```

Use bond energies to find the enthalpy change for this reaction.

Between the Extremes: Electronegativity and Bond Polarity

(Sample Problem 9.4)

9.40 Describe the vertical and horizontal trends in electronegativity (EN) among the main-group elements. According to Pauling's scale, what are the two most electronegative elements? The two least electronegative elements?

9.41 What is the general relationship between IE_1 and EN for the elements? Why?

9.42 Is the H—O bond in water nonpolar covalent, polar covalent, or ionic? Define each term, and explain your choice.

9.43 How does electronegativity differ from electron affinity?

9.44 How is the partial ionic character of a bond in a diatomic molecule related to ΔEN for the bonded atoms? Why?

9.45 Using the periodic table only, arrange the elements in each set in order of *increasing* EN: (a) S, O, Si; (b) Mg, P, As.

9.46 Using the periodic table only, arrange the elements in each set in order of *decreasing* EN: (a) I, Br, N; (b) Ca, H, F.

9.47 Use Figure 9.19 (p. 297) to indicate the polarity of each bond with a *polar arrow:* (a) N—B; (b) N—O; (c) C—S; (d) S—O; (e) N—H; (f) Cl—O.

9.48 Use Figure 9.19 (p. 297) to indicate the polarity of each bond with *partial charges:* (a) Br—Cl; (b) F—Cl; (c) H—O; (d) Se—H; (e) As—H; (f) S—N.

9.49 Which is the more polar bond in each of the following pairs from Problem 9.47: (a) or (b); (c) or (d); (e) or (f)?

9.50 Which is the more polar bond in each of the following pairs from Problem 9.48: (a) or (b); (c) or (d); (e) or (f)?

9.51 Are the bonds in each of the following substances ionic, nonpolar covalent, or polar covalent? Arrange the substances with polar covalent bonds in order of increasing bond polarity:
(a) S_8 (b) RbCl (c) PF_3 (d) SCl_2 (e) F_2 (f) SF_2

9.52 Are the bonds in each of the following substances ionic, nonpolar covalent, or polar covalent? Arrange the substances with polar covalent bonds in order of increasing bond polarity:
(a) KCl (b) P_4 (c) BF_3 (d) SO_2 (e) Br_2 (f) NO_2

9.53 Rank the members of each set of compounds in order of *increasing* ionic character of their bonds. Use a *polar arrow* to indicate the bond polarity of each:
(a) HBr, HCl, HI (b) H_2O, CH_4, HF (c) SCl_2, PCl_3, $SiCl_4$

9.54 Rank the members of each set of compounds in order of *decreasing* ionic character of their bonds. Use *partial charges* to indicate the bond polarity of each:
(a) PCl_3, PBr_3, PF_3 (b) BF_3, NF_3, CF_4 (c) SeF_4, TeF_4, BrF_3

9.55 The energy of the C—C bond is 347 kJ/mol, and that of the Cl—Cl bond is 243 kJ/mol. Which of the following values might you expect for the C—Cl bond energy? Explain.
(a) 590 kJ/mol (sum of the values given)
(b) 104 kJ/mol (difference of the values given)
(c) 295 kJ/mol (average of the values given)
(d) 339 kJ/mol (greater than the average of the values given)

Comprehensive Problems

Problems with an asterisk (*) are more challenging.

9.56 Geologists have a rule of thumb: when molten rock cools and solidifies, crystals of compounds with the smallest lattice energies appear at the bottom of the mass. Suggest a reason for this.

9.57 Acetylene gas (ethyne; HC≡CH) burns with oxygen in an oxyacetylene torch to produce carbon dioxide, water vapor, and the heat needed to weld metals. The heat of reaction for the combustion of acetylene is 1259 kJ/mol. (a) Calculate the C≡C bond energy, and compare your value with that in Table 9.2. (b) When 500.0 g of acetylene burns, how many kilojoules of heat are given off? (c) How many grams of CO_2 are produced? (d) How many liters of O_2 at 298 K and 18.0 atm are consumed?

* **9.58** Even though so much energy is required to form a metal cation with a 2+ charge, the alkaline earth metals form halides with general formula MX_2, rather than MX.
(a) Use the following data to calculate the ΔH°_f of MgCl:

$$Mg(s) \longrightarrow Mg(g) \qquad \Delta H^\circ = 148\ kJ$$
$$Cl_2(g) \longrightarrow 2Cl(g) \qquad \Delta H^\circ = 243\ kJ$$
$$Mg(g) \longrightarrow Mg^+(g) + e^- \qquad \Delta H^\circ = 738\ kJ$$
$$Cl(g) + e^- \longrightarrow Cl^-(g) \qquad \Delta H^\circ = -349\ kJ$$
$$\Delta H^\circ_{lattice} \text{ of MgCl} = 783.5\ kJ/mol$$

(b) Is MgCl favored energetically relative to its elements? Explain.
(c) Use Hess's law to calculate ΔH° for the conversion of MgCl to $MgCl_2$ and Mg (ΔH°_f of $MgCl_2 = -641.6$ kJ/mol).
(d) Is MgCl favored energetically relative to $MgCl_2$? Explain.

* **9.59** By using photons of specific wavelengths, chemists can dissociate gaseous HI to produce H atoms with certain speeds. When HI dissociates, the H atoms move away rapidly, whereas the relatively heavy I atoms move more slowly.
(a) What is the longest wavelength (in nm) that can dissociate a molecule of HI?
(b) If a photon of 254 nm is used, what is the excess energy (in J) over that needed for the dissociation?
(c) If all this excess energy is carried away by the H atom as kinetic energy, what is its speed (in m/s)?

9.60 Carbon dioxide is a linear molecule. Its vibrational motions include symmetrical stretching, bending, and asymmetrical stretching, and their frequencies are $4.02\times10^{13}\ s^{-1}$, $2.00\times10^{13}\ s^{-1}$, and $7.05\times10^{13}\ s^{-1}$, respectively. (a) In what region of the electromagnetic spectrum are these frequencies? (b) Calculate the energy (in J) of each vibration. Which occurs most readily (takes the least energy)?

9.61 In developing the concept of electronegativity, Pauling used the term *excess bond energy* for the difference between the actual bond energy of X—Y and the average bond energies of X—X and Y—Y (see text discussion for the case of HF). Based on the values in Figure 9.19, which of the following substances contains bonds with *no* excess bond energy?
(a) PH_3 (b) CS_2 (c) BrCl (d) BH_3 (e) Se_8

9.62 Without stratospheric ozone (O_3), harmful solar radiation would cause gene alterations. Ozone forms when O_2 breaks and each O atom reacts with another O_2 molecule. It is destroyed by reaction with Cl atoms that are formed when the C—Cl bond in synthetic chemicals breaks. Find the wavelengths of light that can break the C—Cl bond and the bond in O_2.

* **9.63** "Inert" xenon actually forms many compounds, especially with highly electronegative fluorine. The ΔH°_f values for xenon difluoride, tetrafluoride, and hexafluoride are −105, −284, and −402 kJ/mol, respectively. Find the average bond energy of the Xe—F bonds in each fluoride.

* **9.64** The HF bond length is 92 pm, 16% shorter than the sum of the covalent radii of H (37 pm) and F (72 pm). Suggest a reason for this difference. Similar calculations show that the difference becomes smaller down the group from HF to HI. Explain.

* **9.65** There are two main types of covalent bond breakage. In homolytic breakage (as in Table 9.2), each atom in the bond gets one of the shared electrons. In some cases, the electronegativity of adjacent atoms affects the bond energy. In heterolytic breakage, one atom gets both electrons and the other gets none; thus, a cation and an anion form.
(a) Why is the C—C bond in $H_3C—CF_3$ (423 kJ/mol) stronger than the bond in $H_3C—CH_3$ (376 kJ/mol)?
(b) Use bond energy and any other data to calculate the heat of reaction for the heterolytic cleavage of O_2.

9.66 Find the longest wavelengths of light that can cleave the bonds in elemental nitrogen, oxygen, and fluorine.

9.67 We can write equations for the formation of methane from ethane (C_2H_6) with its C—C bond, from ethene (C_2H_4) with its C═C bond, and from ethyne (C_2H_2) with its C≡C bond:

$$C_2H_6(g) + H_2(g) \longrightarrow 2CH_4(g) \qquad \Delta H^\circ_{rxn} = -65.07\ kJ/mol$$
$$C_2H_4(g) + 2H_2(g) \longrightarrow 2CH_4(g) \qquad \Delta H^\circ_{rxn} = -202.21\ kJ/mol$$
$$C_2H_2(g) + 3H_2(g) \longrightarrow 2CH_4(g) \qquad \Delta H^\circ_{rxn} = -376.74\ kJ/mol$$

Given that the average C—H bond energy in CH_4 is 415 kJ/mol, use values from Table 9.2 to calculate the average C—H bond energy in ethane, in ethene, and in ethyne.

9.68 Carbon-carbon bonds form the "backbone" of nearly every organic and biological molecule. The average bond energy of the C—C bond is 347 kJ/mol. Calculate the frequency and wavelength of the least energetic photon that can break an average C—C bond. In what region of the electromagnetic spectrum is this radiation?

9.69 In a future hydrogen-fuel economy, the cheapest source of H_2 will certainly be water. It takes 467 kJ to produce 1 mol of H atoms from water. What is the frequency, wavelength, and minimum energy of a photon that can free an H atom from water?

9.70 Dimethyl ether (CH_3OCH_3) and ethanol (CH_3CH_2OH) are constitutional isomers. (a) Use Table 9.2 to calculate ΔH°_{rxn} for the formation of each compound as a gas from methane and oxygen; water vapor also forms. (b) State which reaction is more exothermic. (c) Calculate ΔH°_{rxn} for the conversion of ethanol to dimethyl ether.

The Shapes of Molecules 10

Key Principles
to focus on while studying this chapter

- A *Lewis structure* shows the relative positions of the atoms in a molecule (or polyatomic ion), as well as the placement of all the shared and unshared electron pairs. It is generated from the molecular formula through a series of steps that often apply the octet rule *(Section 10.1)*.
- In many molecules or ions, one electron pair in a double bond spreads over an adjacent single bond, thereby *delocalizing* its charge and stabilizing the system. In such cases, more than one Lewis structure, each called a *resonance form,* can be drawn, and the species exists as a *resonance hybrid,* a mixture of the resonance forms *(Section 10.1)*.
- By assigning to each atom a *formal charge* based on the electrons belonging to the atom and shared by it, we can select the most important of the various resonance forms *(Section 10.1)*.
- According to VSEPR theory, each group of valence electrons, whether a bonding pair or a lone pair, around a central atom repels the others. These repulsions give rise to five geometric arrangements—*linear, trigonal planar, tetrahedral, trigonal bipyramidal,* and *octahedral.* Various molecular shapes, with characteristic bond angles, arise from these arrangements *(Section 10.2)*.
- A whole molecule may be polar or nonpolar, depending on its shape and the polarities of its bonds *(Section 10.3)*.

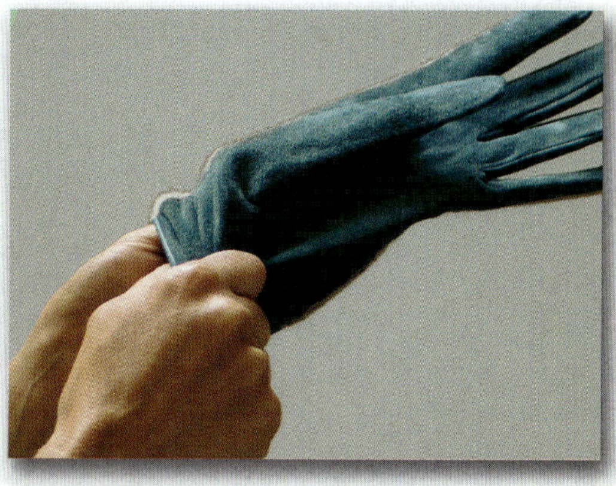

Fitting Together *In biological cells, the shape of one molecule changes the shape of another, just as a hand changes the shape of a glove.*

Outline

Concepts & Skills to Review before studying this chapter

- electron configurations of main-group elements (Section 8.3)
- electron-dot symbols (Section 9.1)
- the octet rule (Section 9.1)
- bond order, bond length, and bond energy (Sections 9.3 and 9.4)
- polar covalent bonds and bond polarity (Section 9.5)

The printed page, covered with atomic symbols, lines, and pairs of dots, makes it easy to forget the amazing, three-dimensional reality of molecular shape. In any molecule, each atom, bonding pair, and lone pair has its own position in space relative to the others, determined by the attractive and repulsive forces that govern all matter. With definite angles and distances between the nuclei, a molecule has a characteristic minute architecture, extending throughout its tiny volume of space. Whether we consider the details of simple reactions, the properties of synthetic materials, or the intricate life-sustaining processes of living cells, molecular shape is a crucial factor. In this chapter, we see how to depict molecules, first as two-dimensional drawings and then as three-dimensional objects.

10.1 DEPICTING MOLECULES AND IONS WITH LEWIS STRUCTURES

The first step toward visualizing what a molecule looks like is to convert its molecular formula to its **Lewis structure** (or **Lewis formula**).* This two-dimensional structural formula consists of electron-dot symbols that depict each atom and its neighbors, the bonding pairs that hold them together, and the lone pairs that fill each atom's outer level (valence shell). In many cases, the octet rule (Section 9.1) guides us in allotting electrons to the atoms in a Lewis structure; in many other cases, however, we set the rule aside.

Using the Octet Rule to Write Lewis Structures

To write a Lewis structure, we decide on the relative placement of the atoms in the molecule (or polyatomic ion)—that is, which atoms are adjacent and become bonded to each other—and distribute the total number of valence electrons as bonding and lone pairs. Let's begin by examining Lewis structures for species that "obey" the octet rule—those in which each atom fills its outer level with eight electrons (or two for hydrogen).

Lewis Structures for Molecules with Single Bonds First, we discuss the steps for writing Lewis structures for molecules that have only single bonds, using nitrogen trifluoride, NF_3, as an example. Figure 10.1 lays out the steps.

FIGURE 10.1 The steps in converting a molecular formula into a Lewis structure.

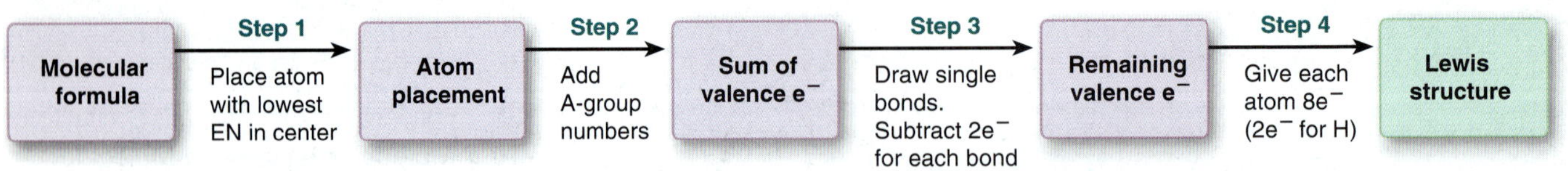

Step 1. Place the atoms relative to each other. For compounds of molecular formula AB_n, place the atom with *lower group number* in the center because it needs more electrons to attain an octet; usually, this is also the atom with the *lower electronegativity*. In NF_3, the N (Group 5A; EN = 3.0) has five electrons

*A Lewis *structure* may be more correctly called a Lewis *formula* because it provides information about the relative placement of atoms in a molecule or ion and shows which atoms are bonded to each other, but it does ***not*** indicate the three-dimensional shape. Nevertheless, use of the term Lewis *structure* is a convention that we follow.

and so needs three, whereas each F (Group 7A; EN = 4.0) has seven and so needs only one; thus, N goes in the center with the three F atoms around it:

```
    F

    N
 F     F
```

If the atoms have the same group number, as in SO_3 or ClF_3, place the atom with the *higher period number* in the center. H can form only one bond, so it is *never* a central atom.

Step 2. Determine the total number of valence electrons available. For molecules, add up the valence electrons of all the atoms. (Recall that the number of valence electrons equals the A-group number.) In NF_3, N has five valence electrons, and each F has seven:

$$[1 \times N(5e^-)] + [3 \times F(7e^-)] = 5e^- + 21e^- = 26 \text{ valence } e^-$$

For polyatomic ions, *add* one e^- for each negative charge of the ion, or *subtract* one e^- for each positive charge.

Step 3. Draw a single bond from each surrounding atom to the central atom, and subtract two valence electrons for each bond. There must be at least a single bond between bonded atoms:

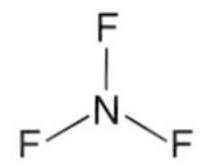

```
    F
    |
 F／N＼F
```

Subtract $2e^-$ for each single bond from the total number of valence electrons available (from step 2) to find the number remaining:

$$3 \text{ N—F bonds} \times 2e^- = 6e^- \quad \text{so} \quad 26e^- - 6e^- = 20e^- \text{ remaining}$$

Step 4. Distribute the remaining electrons in pairs so that each atom ends up with eight electrons (or two for H). First, place lone pairs on the *surrounding (more electronegative) atoms* to give each an octet. If any electrons remain, place them around the central atom. Then check that each atom has $8e^-$:

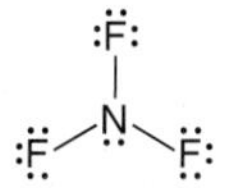

```
     :F̤̈:
      |
 :F̤̈／N̤＼F̤̈:
```

This is the Lewis structure for NF_3. Always check that the total number of electrons (bonds plus lone pairs) equals the sum of the valence electrons: $6e^-$ in three bonds plus $20e^-$ in ten lone pairs equals 26 valence electrons.

This particular arrangement of F atoms around an N atom resembles the molecular shape of NF_3 (Section 10.2). Because Lewis structures do not indicate shape, however, an equally correct Lewis structure for NF_3 is

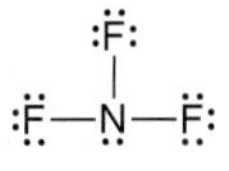

```
     :F̤̈:
      |
 :F̤̈—N̤—F̤̈:
```

or any other that retains the *same connections among the atoms*—a central N atom connected by single bonds to three surrounding F atoms.

Using these four steps, you can write a Lewis structure for any singly bonded molecule whose central atom is C, N, or O, as well as for some molecules with central atoms from higher periods. Remember that, in nearly all their compounds,

- Hydrogen atoms form one bond.
- Carbon atoms form four bonds.
- Nitrogen atoms form three bonds.
- Oxygen atoms form two bonds.
- Halogens form one bond when they are surrounding atoms; fluorine is always a surrounding atom.

SAMPLE PROBLEM 10.1 Writing Lewis Structures for Molecules with One Central Atom

Problem Write a Lewis structure for CCl_2F_2, one of the compounds responsible for the depletion of stratospheric ozone.

Step 1:

```
     Cl
  F  C  F
     Cl
```

Step 3:

```
     Cl
     |
  F—C—F
     |
     Cl
```

Step 4:

```
     :Cl:
      |
  :F—C—F:
      |
     :Cl:
```

Solution *Step 1.* Place the atoms relative to each other. In CCl_2F_2, carbon has the lowest group number and EN, so it is the central atom. The halogen atoms surround it, but their specific positions are not important (see margin).

Step 2. Determine the total number of valence electrons (from A-group numbers): C is in Group 4A, F is in Group 7A, and Cl is in Group 7A, too. Therefore, we have

$$[1 \times C(4e^-)] + [2 \times F(7e^-)] + [2 \times Cl(7e^-)] = 32 \text{ valence } e^-$$

Step 3. Draw single bonds to the central atom and subtract $2e^-$ for each bond (see margin). Four single bonds use $8e^-$, so $32e^- - 8e^-$ leaves $24e^-$ remaining.

Step 4. Distribute the remaining electrons in pairs, beginning with the surrounding atoms, so that each atom has an octet (see margin).

Check Counting the electrons shows that each atom has an octet. Remember that bonding electrons are counted as belonging to each atom in the bond. The total number of electrons in bonds (8) and lone pairs (24) equals 32 valence electrons. Note that, as expected, C has four bonds and the surrounding halogens have one each.

FOLLOW-UP PROBLEM 10.1 Write a Lewis structure for each of the following:
(a) H_2S **(b)** OF_2 **(c)** $SOCl_2$

A slightly more comple situation occurs when molecules have two or more central atoms bonded to each other, with the other atoms around them.

SAMPLE PROBLEM 10.2 Writing Lewis Structures for Molecules with More Than One Central Atom

Problem Write the Lewis structure for methanol, (molecular formula CH_4O), an important industrial alcohol that is being used as a gasoline alternative in car engines.

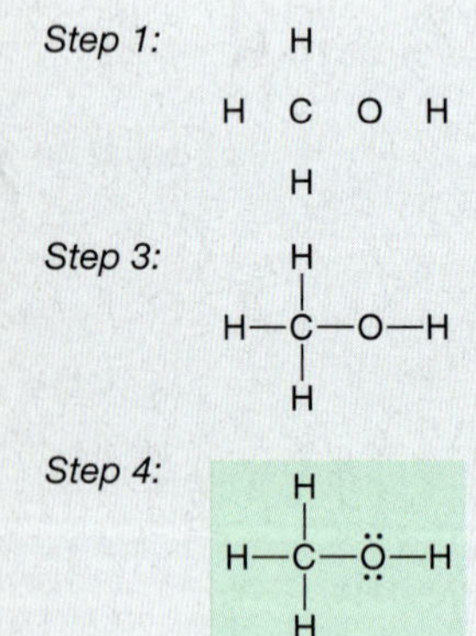

Solution *Step 1.* Place the atoms relative to each other. The H atoms can have only one bond, so C and O must be adjacent to each other. Recall that C has four bonds and O has two, so we arrange the H atoms to show this (see margin).

Step 2. Find the sum of valence electrons:

$$[1 \times C(4e^-)] + [1 \times O(6e^-)] + [4 \times H(1e^-)] = 14e^-$$

Step 3. Add single bonds and subtract $2e^-$ for each bond (see margin). Five bonds use $10e^-$, so $14e^- - 10e^-$ leaves $4e^-$ remaining.

Step 4. Add the remaining electrons in pairs. Carbon already has an octet, and each H shares two electrons with the C; so the four remaining valence electrons form two lone pairs on O. We now have the Lewis structure for methanol (see margin).

Check Each H atom has $2e^-$, and the C and O each have $8e^-$. The total number of valence electrons is $14e^-$, which equals $10e^-$ in bonds plus $4e^-$ in lone pairs. Also note that each H has one bond, C has four, and O has two.

FOLLOW-UP PROBLEM 10.2 Write a Lewis structure for each of the following:
(a) hydroxylamine (NH_3O) **(b)** dimethyl ether (C_2H_6O; no O—H bonds)

Lewis Structures for Molecules with Multiple Bonds Sometimes, you'll find that, after steps 1 to 4, there are not enough electrons for the central atom (or one of the central atoms) to attain an octet. This usually means that a multiple bond is present, and the following additional step is needed:

Step 5. Cases involving multiple bonds. If, after step 4, a central atom still does not have an octet, make a multiple bond by changing a lone pair from one of the surrounding atoms into a bonding pair to the central atom.

SAMPLE PROBLEM 10.3 Writing Lewis Structures for Molecules with Multiple Bonds

Problem Write Lewis structures for the following:
(a) Ethylene (C_2H_4), the most important reactant in the manufacture of polymers
(b) Nitrogen (N_2), the most abundant atmospheric gas

Plan We show the solution resulting from steps 1 to 4: placing the atoms, counting the total valence electrons, making single bonds, and distributing the remaining valence electrons in pairs to attain octets. Then we continue with step 5, if needed.

Solution (a) For C_2H_4. After steps 1 to 4, we have

H₂C—C̈H₂

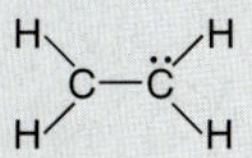

Step 5. Change a lone pair to a bonding pair. The C on the right has an octet, but the C on the left has only $6e^-$, so we convert the lone pair to another bonding pair between the two C atoms:

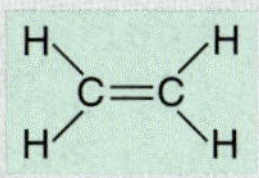

$H_2C{=}CH_2$

(b) For N_2. After steps 1 to 4, we have :N̈—N̈:
Step 5. Neither N has an octet, so we change a lone pair to a bonding pair: :N̈=N:
In this case, moving one lone pair to make a double bond still does not give the N on the right an octet, so we move a lone pair from the left N to make a triple bond: :N≡N:

Check In part (a), each C has four bonds and counts the $4e^-$ in the double bond as part of its own octet. The valence electron total is $12e^-$, all in six bonds. In part (b), each N has three bonds and counts the $6e^-$ in the triple bond as part of its own octet. The valence electron total is $10e^-$, which equals the electrons in three bonds and two lone pairs.

FOLLOW-UP PROBLEM 10.3 Write Lewis structures for each of the following:
(a) CO (the only common molecule in which C has only three bonds) **(b)** HCN **(c)** CO_2

Resonance: Delocalized Electron-Pair Bonding

We can often write more than one Lewis structure, each with the same relative placement of atoms, for a molecule or ion with *double bonds next to single bonds.* Consider ozone (O_3), a serious air pollutant at ground level but a life-sustaining absorber of harmful ultraviolet (UV) radiation in the stratosphere. Two valid Lewis structures (with lettered O atoms for clarity) are

I: O(A)=O(B)—O(C)   II: O(A)—O(B)=O(C)

In structure I, oxygen B has a double bond to oxygen A and a single bond to oxygen C. In structure II, the single and double bonds are reversed. These are *not* two different O_3 molecules, just different Lewis structures for the same molecule.

In fact, *neither* Lewis structure depicts O_3 accurately. Bond length and bond energy measurements indicate that the two oxygen-oxygen bonds in O_3 are identical, with properties that lie between those of an O—O bond and an O=O bond, something like a "one-and-a-half" bond. The molecule is shown more correctly with two Lewis structures, called **resonance structures** (or **resonance forms**), and a two-headed resonance arrow (⟷) between them. Resonance structures *have the same relative placement of atoms but different locations of bonding and lone electron pairs.* You can convert one resonance form to another by moving lone pairs to bonding positions, and vice versa:

I ⟷ II

Resonance structures are not real bonding depictions: O_3 does *not* change back and forth from structure I at one instant to structure II the next. The actual molecule is a **resonance hybrid,** an average of the resonance forms.

THINK OF IT THIS WAY
A Purple Mule, Not a Blue Horse and Then a Red Donkey

Consider these analogies for a resonance hybrid. A mule is a genetic mix, a hybrid, of a horse and a donkey; it is not a horse one instant and a donkey the next. Similarly, the color purple is a mix of two other colors, red and blue, not red one instant and blue the next. In the same sense, a resonance hybrid is one molecular species, not one resonance form this instant and another resonance form the next. The problem is that we cannot depict the hybrid accurately with a single Lewis structure.

Our need for more than one Lewis structure to depict the ozone molecule is the result of **electron-pair delocalization.** In a single, double, or triple bond, each electron pair is attracted by the nuclei of the two bonded atoms, and the electron density is greatest in the region between the nuclei: each electron pair is *localized.* In the resonance hybrid for O_3, however, two of the electron pairs (one bonding and one lone pair) are *delocalized:* their density is "spread" over the entire molecule. In O_3, this results in two identical bonds, each consisting of a single bond (the localized electron pair) and a *partial bond* (the contribution from one of the delocalized electron pairs). We draw the resonance hybrid with a curved dashed line to show the delocalized pairs:

[ozone resonance hybrid structure: O–O–O with curved dashed line]

Electron delocalization diffuses electron density over a greater volume, which reduces electron-electron repulsions and thus stabilizes the molecule. Resonance is very common, and many molecules (and ions) are best depicted as resonance hybrids. Benzene (C_6H_6), for example, has two important resonance forms in which alternating single and double bonds have different positions. The hybrid has six identical carbon-carbon bonds: there are six C—C bonds and three electron pairs delocalized over all six C atoms, often shown as a dashed (or solid) circle.

[benzene resonance structures]

resonance forms

or

[benzene resonance hybrid structure]

resonance hybrid

Partial bonding, such as that occurring in resonance hybrids, often leads to fractional bond orders. For O_3, we have

$$\text{Bond order} = \frac{3 \text{ electron pairs}}{2 \text{ bonded-atom pairs}} = 1\tfrac{1}{2}$$

The carbon-to-carbon bond order in benzene is 9 electron pairs/6 bonded-atom pairs, or $1\frac{1}{2}$ also. For the carbonate ion, CO_3^{2-}, three resonance structures can be drawn. Each has 4 electron pairs shared among 3 bonded-atom pairs, so the bond order is 4/3, or $1\frac{1}{3}$. One of the three resonance structures for CO_3^{2-} is

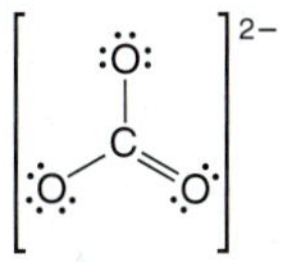

Note that *the Lewis structure of a polyatomic ion is shown in square brackets, with its charge as a right superscript outside the brackets.*

SAMPLE PROBLEM 10.4 Writing Resonance Structures

Problem Write resonance structures for the nitrate ion, NO_3^-.

Plan We write a Lewis structure, remembering to add $1e^-$ to the total number of valence electrons because of the 1− ionic charge. Then we move lone and bonding pairs to write other resonance forms and connect them with the resonance arrow.

Solution After steps 1 to 4, we have

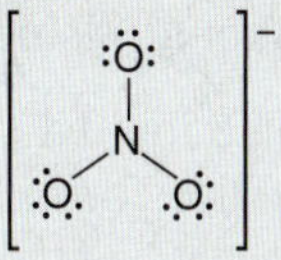

Step 5. Because N has only $6e^-$, we change one lone pair on an O atom to a bonding pair and form a double bond, which gives each atom an octet. All the O atoms are equivalent, however, so we can move a lone pair from any of the three O atoms and obtain three resonance structures:

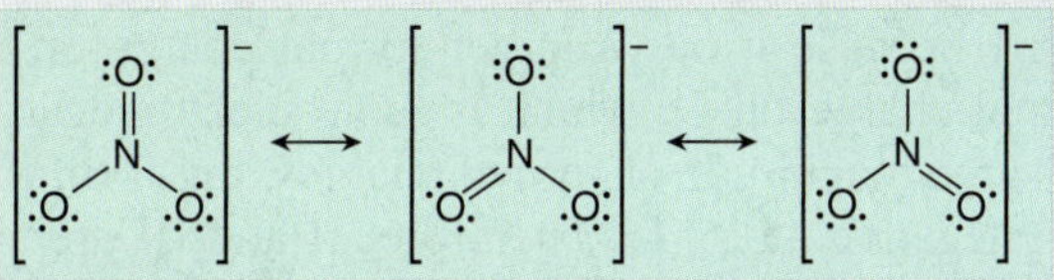

Check Each structure has the same relative placement of atoms, an octet around each atom, and $24e^-$ (the sum of the valence electron total and $1e^-$ from the ionic charge distributed in four bonds and eight lone pairs).

Comment Remember that no double bond actually exists in the NO_3^- ion. The ion is a resonance hybrid of these three structures with a bond order of $1\frac{1}{3}$. (You'll see in the upcoming discussion why N can have four bonds here.)

FOLLOW-UP PROBLEM 10.4 One of the three resonance structures for CO_3^{2-} was shown just before Sample Problem 10.4. Draw the other two.

Formal Charge: Selecting the More Important Resonance Structure

In the previous sample and follow-up problems, the resonance forms were mixed equally to form the resonance hybrid because the central atom (N or C) had surrounding atoms (O) that were all the same. When this is not the case, one resonance form may look more like the hybrid than the others. In other words, because the resonance hybrid is an average of the resonance forms, one form may contribute more and "weight" the average in its favor. We can often select the more important resonance form by determining each atom's **formal charge,** the charge it would have *if the bonding electrons were shared equally.*

An atom's formal charge is its total number of valence electrons minus *all* of its unshared valence electrons and *half* of its shared valence electrons. Thus,

$$\text{Formal charge of atom} = \text{no. of valence } e^- - (\text{no. of unshared valence } e^- + \tfrac{1}{2} \text{ no. of shared valence } e^-) \quad \textbf{(10.1)}$$

For example, in O_3, the formal charge of oxygen A in resonance form I is

$$6 \text{ valence } e^- - (4 \text{ unshared } e^- + \tfrac{1}{2} \text{ of } 4 \text{ shared } e^-) = 6 - 4 - 2 = 0$$

The formal charges of all the atoms in the two O_3 resonance forms are

$O_A[6 - 4 - \frac{1}{2}(4)] = 0$ (I), $O_A[6 - 6 - \frac{1}{2}(2)] = -1$ (II)

$O_B[6 - 2 - \frac{1}{2}(6)] = +1$ (I), $O_B[6 - 2 - \frac{1}{2}(6)] = +1$ (II)

$O_C[6 - 6 - \frac{1}{2}(2)] = -1$ (I), $O_C[6 - 4 - \frac{1}{2}(4)] = 0$ (II)

I ⟷ II

Forms I and II have the same formal charges but on different O atoms, so they contribute equally to the resonance hybrid. *Formal charges must sum to the actual charge on the species:* zero for a molecule and the ionic charge for an ion.

Note that, in form I, instead of the usual two bonds for oxygen, O_B has three bonds and O_C has one. Only when an atom has a zero formal charge does it have its usual number of bonds; the same holds for C in CO, N in NO_3^-, and so forth.

Three criteria help us choose the more important resonance structures:

- Smaller formal charges (positive or negative) are preferable to larger ones.
- Having the same nonzero formal charges on adjacent atoms is not preferred.
- A more negative formal charge should reside on a more electronegative atom.

Let's apply these criteria to the cyanate ion, NCO^-, which has two different atoms around the central one. Three resonance forms with formal charges are

	I		II		III
Formal charges:	(−2) (0) (+1)		(−1) (0) (0)		(0) (0) (−1)
Resonance forms:	[:N̤̈—C≡O:]⁻	⟷	[N̤̈=C=Ö̤]⁻	⟷	[:N≡C—Ö̤:]⁻

We eliminate form I because it has a larger formal charge on N than the others and a positive formal charge on O, which is more electronegative than N. Forms II and III have the same magnitude of formal charges, but form III has a −1 charge on the more electronegative atom, O. Therefore, II and III are significant contributors to the resonance hybrid of the cyanate ion, but III is the more important.

Note that formal charge (used to examine resonance structures) is *not* the same as oxidation number (used to monitor redox reactions):

- For a formal charge, bonding electrons are assigned *equally* to the atoms (as if the bonding were *nonpolar covalent*), so each atom has half of them:

  $$\text{Formal charge} = \text{valence } e^- - (\text{lone pair } e^- + \tfrac{1}{2} \text{ bonding } e^-)$$

- For an oxidation number, bonding electrons are assigned *completely* to the more electronegative atom (as if the bonding were *ionic*):

  $$\text{Oxidation number} = \text{valence } e^- - (\text{lone pair } e^- + \text{bonding } e^-)$$

For the three cyanate ion resonance structures,

Formal charges:	(−2) (0) (+1)		(−1) (0) (0)		(0) (0) (−1)
	[:N̤̈—C≡O:]⁻	⟷	[N̤̈=C=Ö̤]⁻	⟷	[:N≡C—Ö̤:]⁻
Oxidation numbers:	−3 +4 −2		−3 +4 −2		−3 +4 −2

Notice that the oxidation numbers *do not* change from one resonance form to another (because the electronegativities *do not* change), but the formal charges *do* change (because the numbers of bonding and lone pairs *do* change).

Lewis Structures for Exceptions to the Octet Rule

The octet rule is a useful guide for most molecules with Period 2 central atoms, but not for every one. Also, many molecules have central atoms from higher periods. As you'll see, some central atoms have fewer than eight electrons around them, and others have more. The most significant octet rule exceptions are for molecules containing electron-deficient atoms, odd-electron atoms, and especially atoms with expanded valence shells.

Electron-Deficient Molecules Gaseous molecules containing either beryllium or boron as the central atom are often **electron deficient;** that is, they have *fewer* than eight electrons around the Be or B atom. The Lewis structures, with formal charges, of gaseous beryllium chloride* and boron trifluoride are

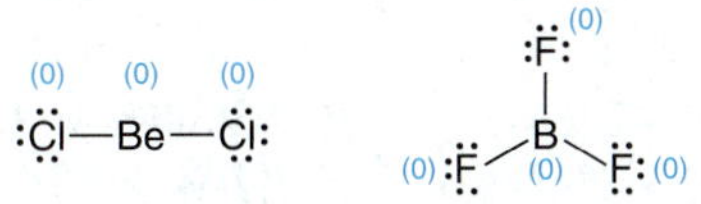

There are only four electrons around beryllium and six around boron. Why don't lone pairs from the surrounding halogen atoms form multiple bonds to the central atoms, thereby satisfying the octet rule? Because halogens are much more electronegative than beryllium or boron, formal charge rules make the following structures unlikely:

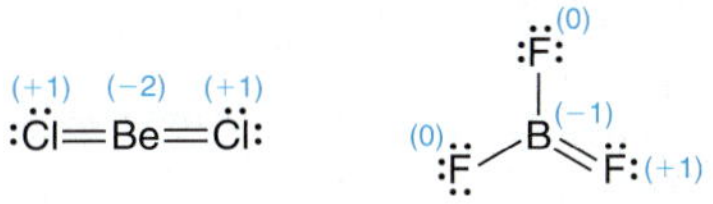

(Some data for BF_3 show a shorter than expected B—F bond. Shorter bonds indicate double-bond character, so the structure with the B=F bond may be a minor contributor to a resonance hybrid.) The main way electron-deficient atoms attain an octet is by forming additional bonds in reactions. When BF_3 reacts with ammonia, for instance, a compound forms in which boron attains its octet:

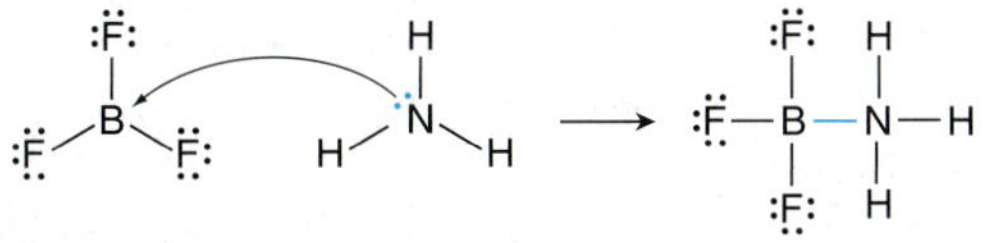

Odd-Electron Molecules A few molecules contain a central atom with an odd number of valence electrons, so they cannot possibly have all their electrons in pairs. Such species, called **free radicals,** contain a lone (unpaired) electron, which makes them paramagnetic (Section 8.5) and extremely reactive. Most odd-electron molecules have a central atom from an odd-numbered group, such as N [Group 5A(15)] or Cl [Group 7A(17)].

Consider nitrogen dioxide (NO_2) as an example. A major contributor to urban smog, it is formed when the NO in auto exhaust is oxidized. NO_2 has several resonance forms. Two involve an O atom that is doubly bonded, as in the case of ozone. Other resonance forms involve the location of the lone electron, and two of those are shown below. The lone electron is delocalized over the N and O atoms, but let's see if formal charges can help us decide where it occurs most of the time. The form with the lone electron on the singly bonded O (right) has zero formal charges:

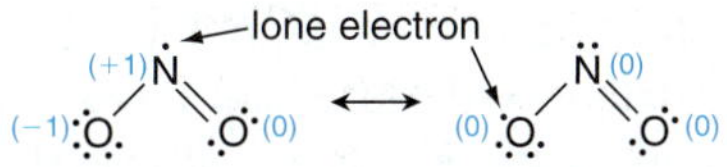

lone electron

But the form with the lone electron on N (left) may be more important because of the way NO_2 reacts. Free radicals react with each other to pair up their lone electrons. When two NO_2 molecules collide, the lone electrons pair up to form the N—N bond in dinitrogen tetraoxide (N_2O_4) and each N attains an octet:

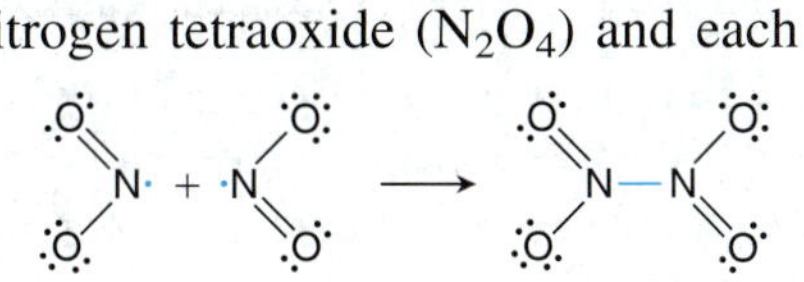

*Even though beryllium is an alkaline earth metal [Group 2A(2)], most of its compounds have properties consistent with covalent, rather than ionic, bonding (Chapter 14). For example, molten $BeCl_2$ does not conduct electricity, indicating the absence of ions.

Apparently, in this case, the lone electron spends most of its time on N, so formal charges are not very useful for picking the most important resonance form.

Expanded Valence Shells Many molecules and ions have more than eight valence electrons around the central atom. *An atom expands its valence shell to form more bonds,* a process that releases energy. Only larger central atoms can accommodate additional pairs, and empty *outer d* orbitals as well as occupied *s* and *p* orbitals are used. Therefore, **expanded valence shells** occur only with a *large central nonmetal atom in which d orbitals are available, that is, one from Period 3 or higher.*

One example is sulfur hexafluoride, SF_6, a remarkably dense and inert gas used as an insulator in electrical equipment. The central sulfur is surrounded by six single bonds, one to each fluorine, for a total of 12 electrons:

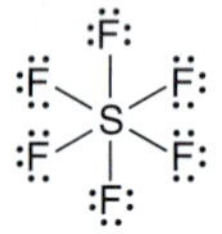

Another example is phosphorus pentachloride, PCl_5, a fuming yellow-white solid used in the manufacture of lacquers and films. PCl_5 forms when phosphorus trichloride, PCl_3, reacts with chlorine gas. The P in PCl_3 has an octet, but two more bonds to chlorine form and P expands its valence shell in PCl_5 to a total of 10 electrons. Note that when PCl_5 forms, *one* Cl—Cl bond breaks (left side of the equation), and *two* P—Cl bonds form (right side), for a net increase of one bond:

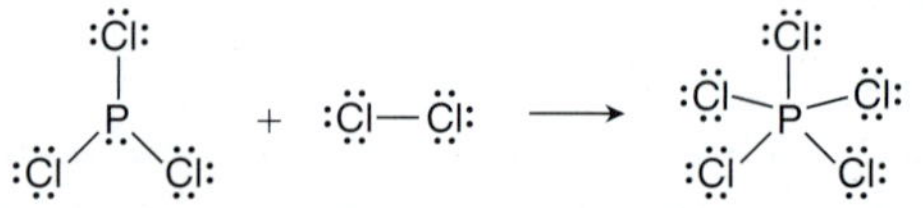

In SF_6 and PCl_5, the central atom forms bonds to *more than four* atoms. But, by applying formal charge rules, we can draw Lewis structures for many molecules with expanded valence shells in which the central atom bonds to *four or even fewer* atoms. Consider sulfuric acid, the industrial chemical produced in the greatest quantity. Two of the resonance forms for H_2SO_4, with formal charges, are

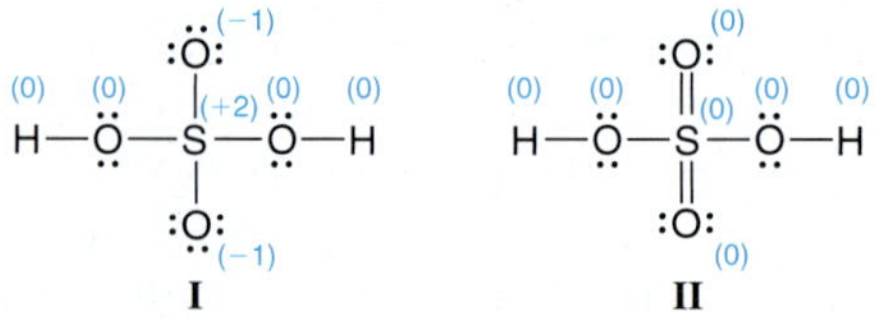

In form II, sulfur has an expanded valence shell of 12 electrons. Based on the formal charge rules, II contributes more than I to the resonance hybrid. More importantly, form II is consistent with observed bond lengths. In gaseous H_2SO_4, the two sulfur-oxygen bonds *with* H atoms attached to O are 157 pm long, whereas the two sulfur-oxygen bonds *without* H atoms attached to O are 142 pm long. This shorter bond length indicates double-bond character, which is shown in form II.

It's important to realize that determining formal charges is a useful, but not perfect, tool for assessing the importance of contributions to a resonance hybrid. You've already seen that it does not predict an important resonance form of NO_2. In fact, theoretical calculations indicate that, for many species with central atoms from Period 3 or higher, such as H_2SO_4, forms with expanded valence shells and zero formal charges (form II above) may be less important than forms with higher formal charges (form I). But we will continue to apply the formal charge rules because it is usually the simplest approach consistent with experimental data.

SAMPLE PROBLEM 10.5 Writing Lewis Structures for Octet-Rule Exceptions

Problem Write Lewis structures for (a) H_3PO_4 (pick the most likely structure); (b) $BFCl_2$.

Plan We write each Lewis structure and examine it for exceptions to the octet rule. In (a), the central atom is P, which is in Period 3, so it can use *d* orbitals to have more than an octet. Therefore, we can write more than one Lewis structure. We use formal charges to decide if one resonance form is more important. In (b), the central atom is B, which can have fewer than an octet of electrons.

Solution (a) For H_3PO_4, two possible Lewis structures, with formal charges, are

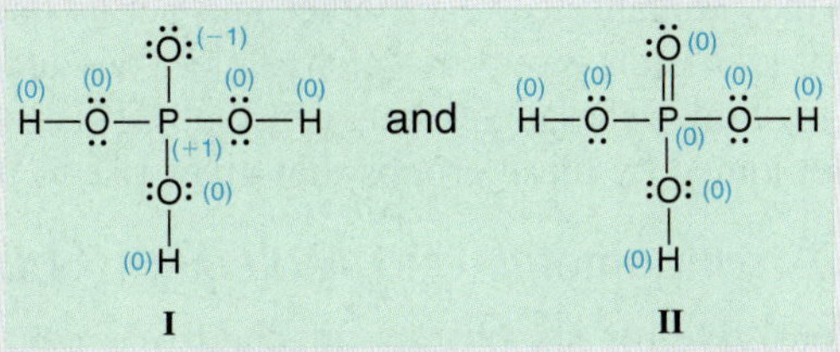

I and II

Structure II has lower formal charges, so it is the more important resonance form.

(b) For $BFCl_2$, the Lewis structure leaves B with only six electrons surrounding it:

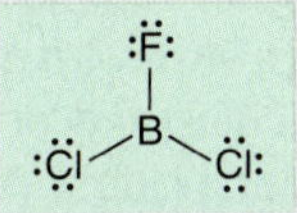

Comment In (a), structure II is also consistent with bond length measurements, which show one shorter (152 pm) phosphorus-oxygen bond and three longer (157 pm) ones. But, as we noted for H_2SO_4, calculations also indicate the importance of structure I.

FOLLOW-UP PROBLEM 10.5 Write the most likely Lewis structure for **(a)** $POCl_3$; **(b)** ClO_2; **(c)** XeF_4.

SECTION 10.1 SUMMARY

A stepwise process is used to convert a molecular formula into a Lewis structure, a *two-dimensional* representation of a molecule (or ion) that shows the relative placement of atoms and distribution of valence electrons among bonding and lone pairs. • When two or more Lewis structures can be drawn for the same relative placement of atoms, the actual structure is a hybrid of those resonance forms. • Formal charges are often useful for determining the most important contributor to the hybrid. • Electron-deficient molecules (central Be or B) and odd-electron species (free radicals) have less than an octet around the central atom but often attain an octet in reactions. • In a molecule (or ion) with a central atom from Period 3 or higher, the atom can hold more than eight electrons by using *d* orbitals to expand its valence shell.

10.2 VALENCE-SHELL ELECTRON-PAIR REPULSION (VSEPR) THEORY AND MOLECULAR SHAPE

Virtually every biochemical process hinges to a great extent on the shapes of interacting molecules. Every medicine you take, odor you smell, or flavor you taste depends on part or all of one molecule fitting physically together with another. This universal importance of molecular shape in the functioning of each organism carries over to the ecosystem. Biologists have found complex interactions regulating behaviors (such as mating, defense, navigation, and feeding) that depend on one molecule's shape matching up with that of another. In this section, we discuss a model for understanding and predicting molecular shape.

The Lewis structure of a molecule is something like the blueprint of a building: a flat drawing showing the relative placement of parts (atom cores), structural connections (groups of bonding valence electrons), and various attachments

(nonbonding, or lone, pairs of valence electrons). To construct the molecular shape from the Lewis structure, chemists employ **valence-shell electron-pair repulsion (VSEPR) theory.** Its basic principle is that *each group of valence electrons around a central atom is located as far away as possible from the others in order to minimize repulsions.* We define a "group" of electrons as any number of electrons that occupy a localized region around an atom. Thus, an electron group may consist of a single bond, a double bond, a triple bond, a lone pair, or even a lone electron. (The two electron pairs in a double bond or the three pairs in a triple bond occupy separate orbitals, so they remain near each other and act as one electron group, as you'll see in Chapter 11.) Each group of valence electrons around an atom repels the other groups to maximize the angles between them. It is the three-dimensional arrangement of nuclei joined by these groups that gives rise to the molecular shape.

Electron-Group Arrangements and Molecular Shapes

When two, three, four, five, or six objects attached to a central point maximize the space that each can occupy around that point, five geometric patterns result. Figure 10.2A depicts these patterns with balloons. If the objects are the valence-electron groups of a central atom, their repulsions maximize the space each occupies and give rise to the five *electron-group arrangements* of minimum energy seen in the great majority of molecules and polyatomic ions.

The electron-group arrangement is defined by the valence-electron groups, both bonding and nonbonding, around the central atom. On the other hand, the **molecular shape** is defined by the relative positions of the atomic nuclei. Figure 10.2B shows the molecular shapes that occur when *all* the surrounding electron groups are *bonding* groups. When some are *nonbonding* groups, different molecular shapes occur. Thus, *the same electron-group arrangement can give rise to different molecular shapes:* some with all bonding groups (as in Figure 10.2B) and others with bonding and nonbonding groups. To classify molecular shapes, we assign each a specific AX_mE_n designation, where m and n are integers, A is the central atom, X is a surrounding atom, and E is a nonbonding valence-electron group (usually a lone pair).

The **bond angle** is the angle formed by the nuclei of two surrounding atoms with the nucleus of the central atom at the vertex. The angles shown for the shapes in Figure 10.2B are *ideal* bond angles, those predicted by simple geometry alone. These are observed when all the bonding electron groups around a

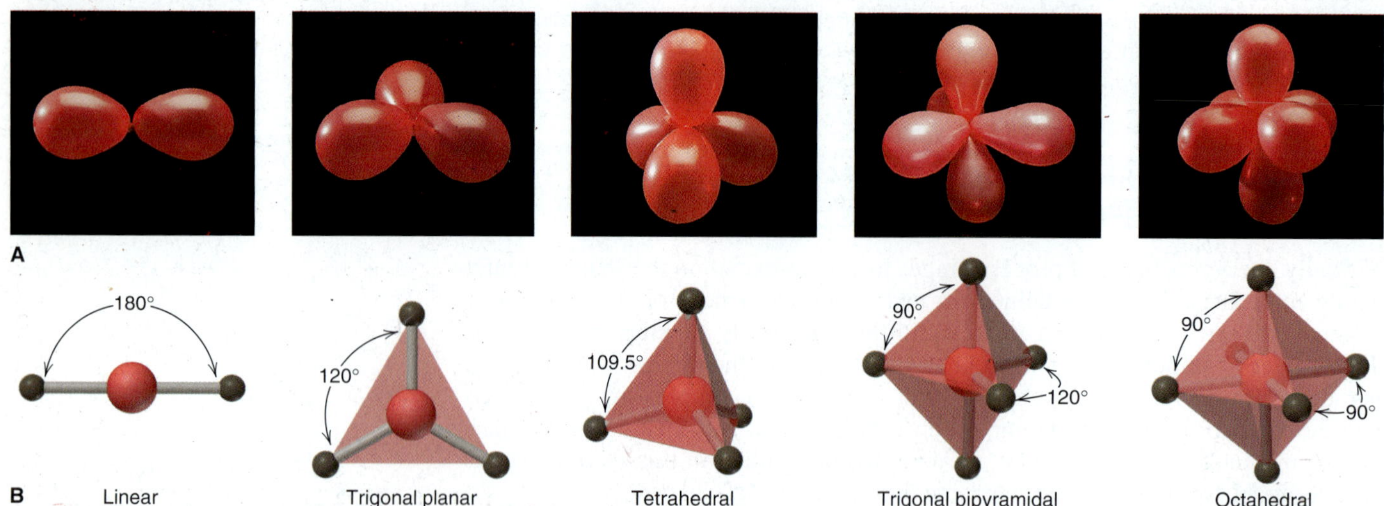

FIGURE 10.2 Electron-group repulsions and the five basic molecular shapes. A, As an analogy for electron-group arrangements, two to six attached balloons form five geometric orientations such that each balloon occupies as much space as possible. **B,** Mutually repelling electron groups attached to a central atom *(red)* occupy as much space as possible. If each is a bonding group to a surrounding atom *(dark gray),* these molecular shapes and bond angles are observed. The shape has the same name as the electron-group arrangement.

central atom are identical and are connected to atoms of the same element. When this is not the case, the bond angles deviate from the ideal angles, as you'll see shortly.

It's important to realize that we use the VSEPR model to account for the molecular shapes observed by means of various laboratory instruments. In almost every case, VSEPR predictions are in accord with actual observations. (We discuss some of these observational methods in Chapter 12.)

The Molecular Shape with Two Electron Groups (Linear Arrangement)

When two electron groups attached to a central atom are oriented as far apart as possible, they point in opposite directions. The **linear arrangement** of electron groups results in a molecule with a **linear shape** and a bond angle of 180°. Figure 10.3 shows the general form (top) and shape (middle) with VSEPR shape class (AX_2), and the formulas of some linear molecules.

Gaseous beryllium chloride ($BeCl_2$) is a linear molecule (AX_2). Gaseous beryllium compounds are electron deficient, with only two electron pairs around the central Be atom:

180°
:C̈l—Be—C̈l:

In carbon dioxide, the central C atom forms two double bonds with the O atoms:

180°
Ö═C═Ö

Each double bond acts as one electron group and is oriented 180° away from the other, so CO_2 is linear. Notice that the lone pairs on the O atoms of CO_2 or on the Cl atoms of $BeCl_2$ are not involved in the molecular shape: only electron groups around the *central* atom influence shape.

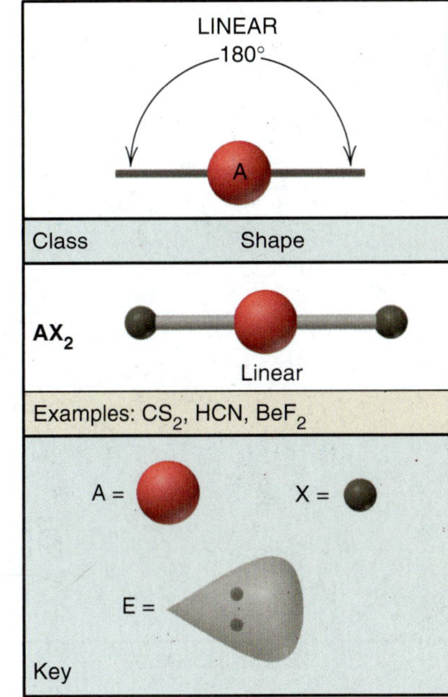

FIGURE 10.3 The single molecular shape of the linear electron-group arrangement. The key *(bottom)* for A, X, and E also refers to Figures 10.4, 10.5, 10.7, and 10.8.

Animation: VSEPR Theory and the Shape of Molecules

Molecular Shapes with Three Electron Groups (Trigonal Planar Arrangement)

Three electron groups around the central atom repel each other to the corners of an equilateral triangle, which gives the **trigonal planar arrangement,** shown in Figure 10.4, and an ideal bond angle of 120°. This arrangement has two possible molecular shapes, one with three surrounding atoms and the other with two atoms and one lone pair. It provides our first opportunity to see the effects of double bonds and lone pairs on bond angles.

When the three electron groups are bonding groups, the molecular shape is *trigonal planar* (AX_3). Boron trifluoride (BF_3), another electron-deficient molecule, is an example. It has six electrons around the central B atom in three single bonds to F atoms. The nuclei lie in a plane, and each F—B—F angle is 120°:

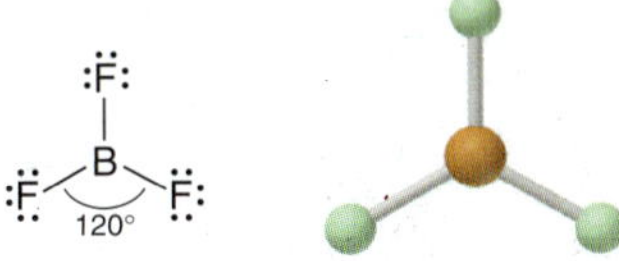

The nitrate ion (NO_3^-) is one of several polyatomic ions with the trigonal planar shape. One of three resonance forms of the nitrate ion (Sample Problem 10.4) is

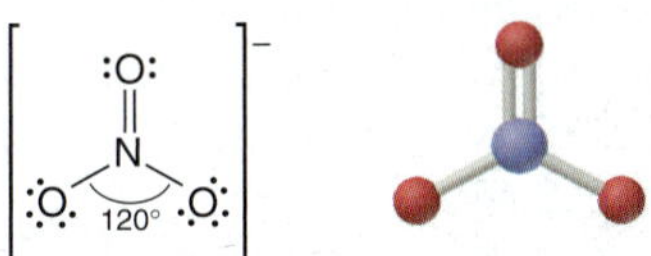

The resonance hybrid has three identical bonds of bond order $1\frac{1}{3}$, so the ideal bond angle is observed.

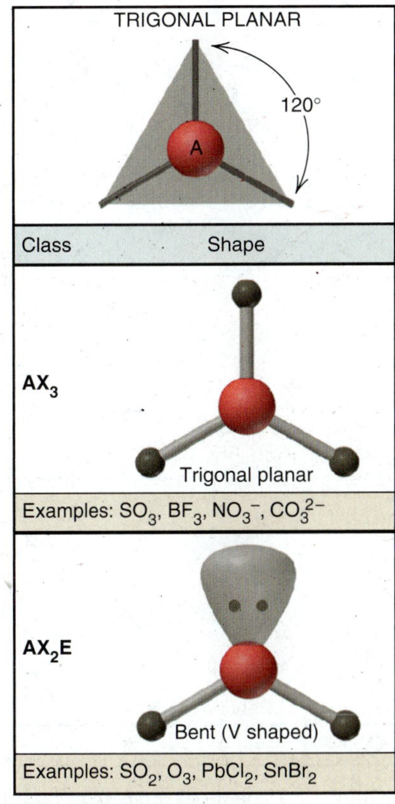

FIGURE 10.4 The two m... of the trigonal planar ... arrangement.

Effect of Double Bonds How do bond angles deviate from the ideal angles when the surrounding atoms and electron groups are not identical? Consider formaldehyde (CH_2O), a substance with many uses, including the manufacture of Formica countertops, the production of methanol, and the preservation of cadavers. Its trigonal planar shape is due to two types of surrounding atoms (O and H) and two types of electron groups (single and double bonds):

The actual bond angles deviate from the ideal because *the double bond, with its greater electron density, repels the two single bonds more strongly than they repel each other.* Note that the H—C—H bond angle is less than 120°.

Effect of Lone Pairs The molecular shape is defined *only* by the positions of the nuclei, so when one of the three electron groups is a lone pair (AX_2E), the shape is **bent,** or **V shaped,** not trigonal planar. Gaseous tin(II) chloride is an example, with the three electron groups in a trigonal plane and the lone pair at one of the triangle's corners. A lone pair can have a major effect on bond angle. Because a lone pair is held by only one nucleus, it is less confined and exerts stronger repulsions than a bonding pair. Thus, *a lone pair repels bonding pairs more strongly than bonding pairs repel each other.* This stronger repulsion *decreases* the angle between bonding pairs. Note the decrease from the ideal 120° angle in $SnCl_2$:

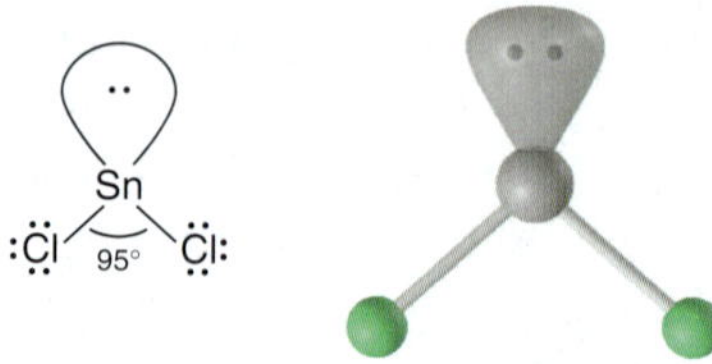

Molecular Shapes with Four Electron Groups (Tetrahedral Arrangement)

The shapes described so far have all been easy to depict in two dimensions, but four electron groups must use three dimensions to achieve maximal separation. Recall that *Lewis structures do **not** depict shape.* Consider methane. The Lewis structure shown below (left) indicates four bonds pointing to the corners of a square, which suggests a 90° bond angle. However, in three dimensions, the four electron groups move farther apart than 90° and point to the vertices of a tetrahedron, a polyhedron with four faces made of identical equilateral triangles. Methane has a bond angle of 109.5°. *Perspective drawings,* such as the one shown below (middle) for methane, indicate depth by using solid and dashed wedges for some of the bonds:

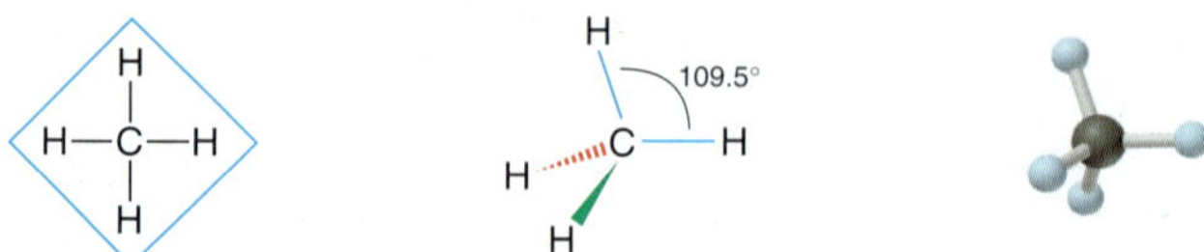

The normal bond lines (blue) represent bonds in the plane of the page; the solid wedge (green) is the bond between the atom in the plane of the page and a group lying toward you above the page; and the dashed wedge (red) is the bond to a group lying away from you below the page. The ball-and-stick model (right) shows the tetrahedral shape clearly.

*All molecules or ions with four electron groups around a central atom adopt the **tetrahedral arrangement*** (Figure 10.5). When all four electron groups are

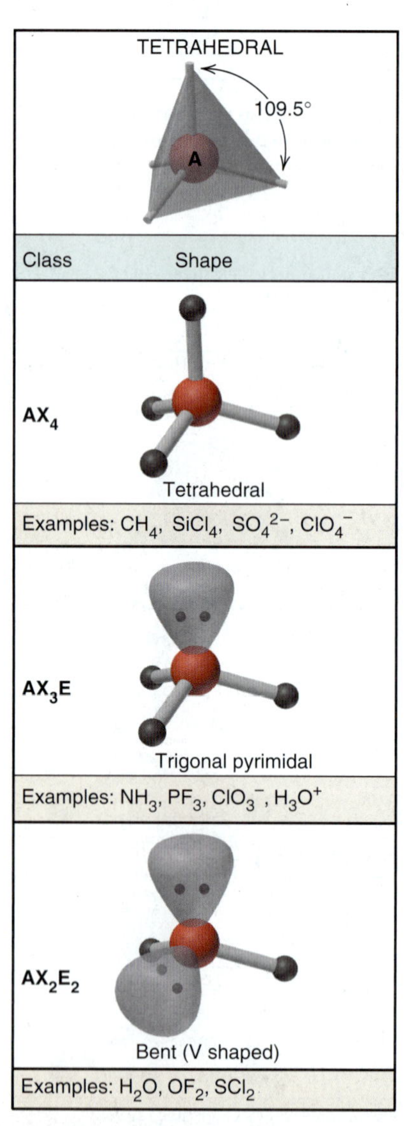

FIGURE 10.5 The three molecular shapes of the tetrahedral electron-group arrangement.

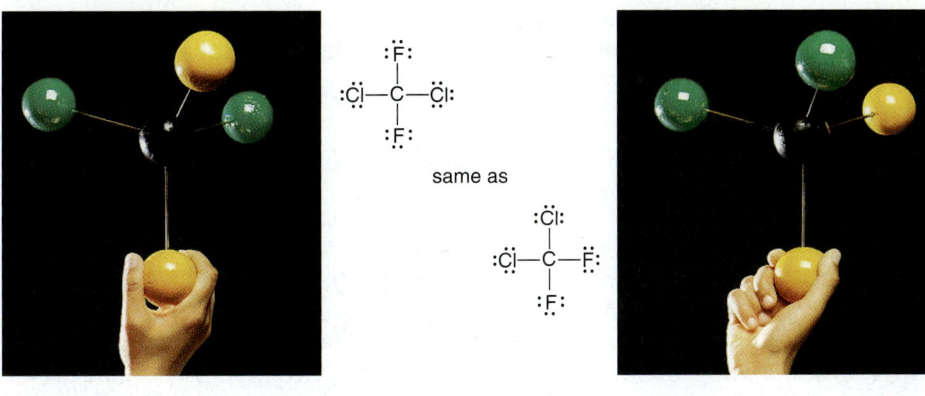

FIGURE 10.6 Lewis structures and molecular shapes. Lewis structures do not indicate geometry. For example, it may seem as if two different Lewis structures can be written for CCl_2F_2, but a twist of the model (Cl, *green;* F, *yellow*) shows that they represent the same molecule.

bonding groups, as in the case of methane, the molecular shape is also *tetrahedral* (AX_4), a very common geometry in organic molecules. In Sample Problem 10.1, we drew the Lewis structure for the tetrahedral molecule dichlorodifluoromethane (CCl_2F_2), without regard to how the halogen atoms surround the carbon atom. Because Lewis structures are flat, it may seem as if we can write two different structures for CCl_2F_2, but these actually represent the same molecule, as Figure 10.6 makes clear.

When one of the four electron groups in the tetrahedral arrangement is a lone pair, the molecular shape is that of a **trigonal pyramid** (AX_3E), a tetrahedron with one vertex "missing." Stronger repulsions due to the lone pair make the measured bond angle slightly less than the ideal 109.5°. In ammonia (NH_3), for example, the lone pair forces the N—H bonding pairs closer, and the H—N—H bond angle is 107.3°. NF_3 (see p. 307) also has a trigonal pyramidal shape.

Picturing molecular shapes is a great way to visualize what happens during a reaction. For instance, when ammonia accepts a proton from an acid, the lone pair on the N atom of trigonal pyramidal NH_3 forms a covalent bond to the H^+ and yields the ammonium ion (NH_4^+), one of many tetrahedral polyatomic ions. Note how the H—N—H bond angle expands from 107.3° in NH_3 to 109.5° in NH_4^+, as the lone pair becomes another bonding pair:

NH_3 (107.3°) $+ H^+ \longrightarrow [NH_4]^+$ (109.5°)

$+ H^+ \longrightarrow$

When the four electron groups around the central atom include two bonding and two nonbonding groups, the molecular shape is *bent, or V shaped* (AX_2E_2). [In the trigonal planar arrangement, the shape with two bonding groups and one lone pair (AX_2E) is also called bent, but its ideal bond angle is 120°, not 109.5°.] Water is the most important V-shaped molecule with the tetrahedral arrangement. We might expect the repulsions from its two lone pairs to have a *greater* effect on the bond angle than the repulsions from the single lone pair in NH_3. Indeed, the H—O—H bond angle is 104.5°, even less than the H—N—H angle in NH_3:

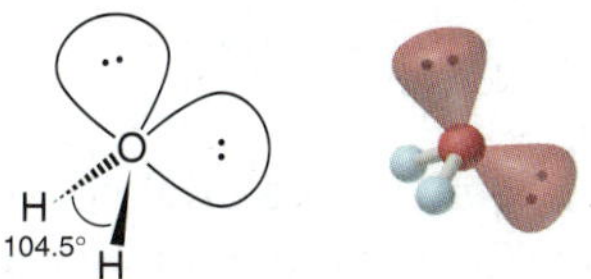

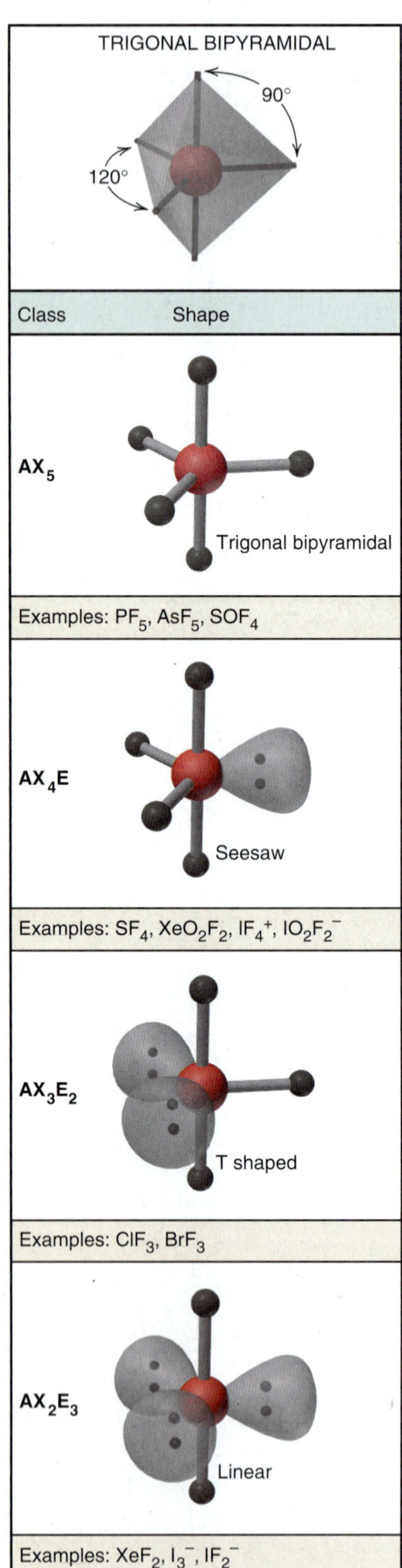

FIGURE 10.7 The four molecular shapes of the trigonal bipyramidal electron-group arrangement.

Thus, for similar molecules within a given electron-group arrangement, electron-pair repulsions cause deviations from ideal bond angles in the following order:

$$\text{Lone pair–lone pair} > \text{lone pair–bonding pair} > \text{bonding pair–bonding pair} \quad \textbf{(10.2)}$$

Molecular Shapes with Five Electron Groups (Trigonal Bipyramidal Arrangement)

All molecules with five or six electron groups have a central atom from Period 3 or higher because only these atoms have the *d* orbitals available to expand the valence shell beyond eight electrons.

When five electron groups maximize their separation, they form the **trigonal bipyramidal arrangement.** In a trigonal bipyramid, two trigonal pyramids share a common base, as shown in Figure 10.7. Note that, in a molecule with this arrangement, *there are two types of positions for surrounding electron groups and two ideal bond angles.* Three **equatorial groups** lie in a trigonal plane that includes the central atom, and two **axial groups** lie above and below this plane. Therefore, a 120° bond angle separates equatorial groups, and a 90° angle separates axial from equatorial groups. In general, the greater the bond angle, the weaker the repulsions, so *equatorial-equatorial (120°) repulsions are weaker than axial-equatorial (90°) repulsions.* The tendency of the electron groups to occupy *equatorial* positions, and thus minimize the stronger axial-equatorial repulsions, governs the four shapes of the trigonal bipyramidal arrangement.

With all five positions occupied by bonded atoms, the molecule has the *trigonal bipyramidal* shape (AX_5), as in phosphorus pentachloride (PCl_5):

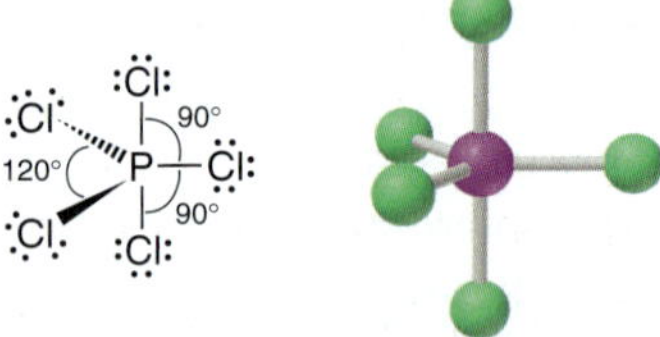

Three other shapes arise for molecules with lone pairs. Since lone pairs exert stronger repulsions than bonding pairs, *lone pairs occupy equatorial positions.* With one lone pair present at an equatorial position, the molecule has a **seesaw shape** (AX_4E). Sulfur tetrafluoride (SF_4), a powerful fluorinating agent, has this shape, shown here and in Figure 10.7 with the "seesaw" tipped up on an end. Note how the equatorial lone pair repels all four bonding pairs to reduce the bond angles:

F
S
101.5°
86.8°

The tendency of lone pairs to occupy equatorial positions causes molecules with three bonding groups and two lone pairs to have a **T shape** (AX_3E_2). Bromine trifluoride (BrF_3), one of many compounds with fluorine bonded to a larger halogen, has this shape. Note the predicted decrease from the ideal 90° F—Br—F bond angle:

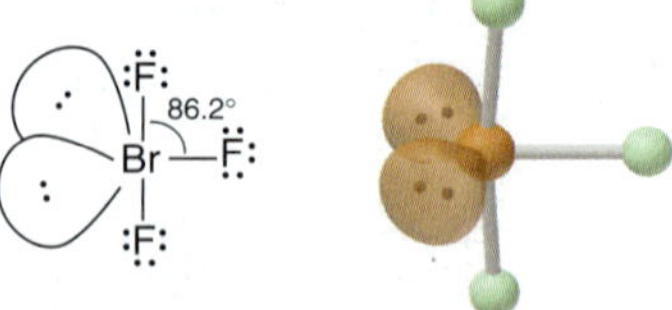

Molecules with three lone pairs in equatorial positions must have the two bonding groups in axial positions, which gives the molecule a *linear* shape (AX_2E_3) and a 180° axial-to-central-to-axial (X—A—X) bond angle. For example, the triiodide ion (I_3^-), which forms when I_2 dissolves in aqueous I^- solution, is linear:

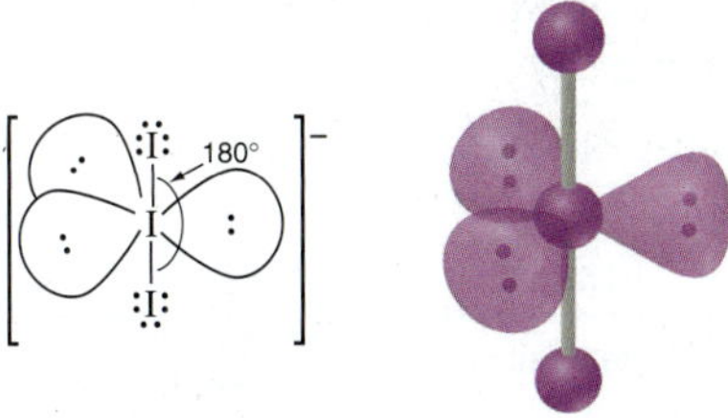

Molecular Shapes with Six Electron Groups (Octahedral Arrangement)

The last of the five major electron-group arrangements is the **octahedral arrangement.** An octahedron is a polyhedron with eight faces made of identical equilateral triangles and six identical vertices, as shown in Figure 10.8. In a molecule (or ion) with this arrangement, six electron groups surround the central atom and each points to one of the six vertices, which gives all the groups a 90° ideal bond angle. Three important molecular shapes occur with this arrangement.

With six bonding groups, the molecular shape is *octahedral* (AX_6), as in sulfur hexafluoride (SF_6):

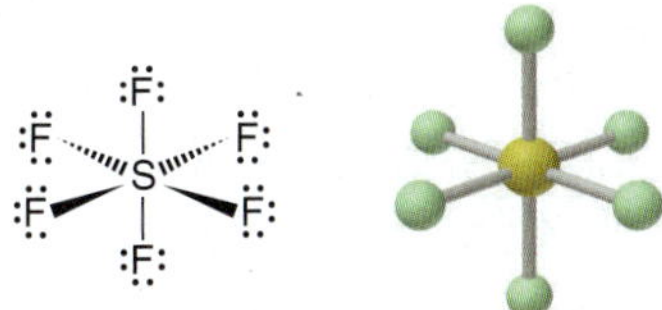

Because all six electron groups have the same ideal bond angle, it makes no difference which position one lone pair occupies. Five bonded atoms and one lone pair define the **square pyramidal shape** (AX_5E), as in iodine pentafluoride (IF_5):

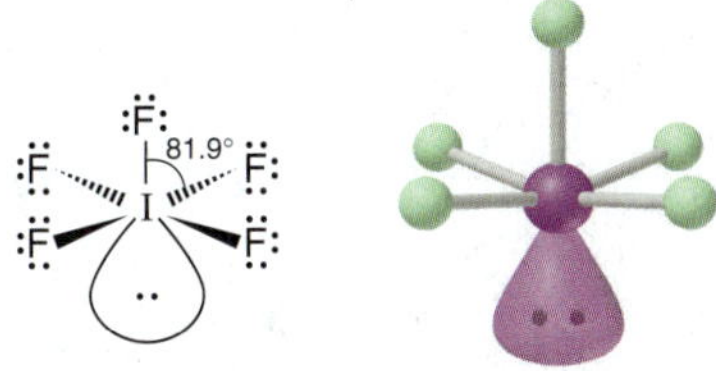

When a molecule has four bonded atoms and two lone pairs, however, the lone pairs always lie at *opposite vertices* to avoid the stronger 90° lone pair–lone pair repulsions. This positioning gives the **square planar shape** (AX_4E_2), as in xenon tetrafluoride (XeF_4):

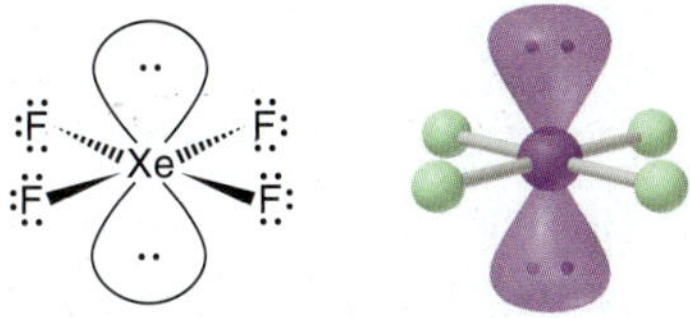

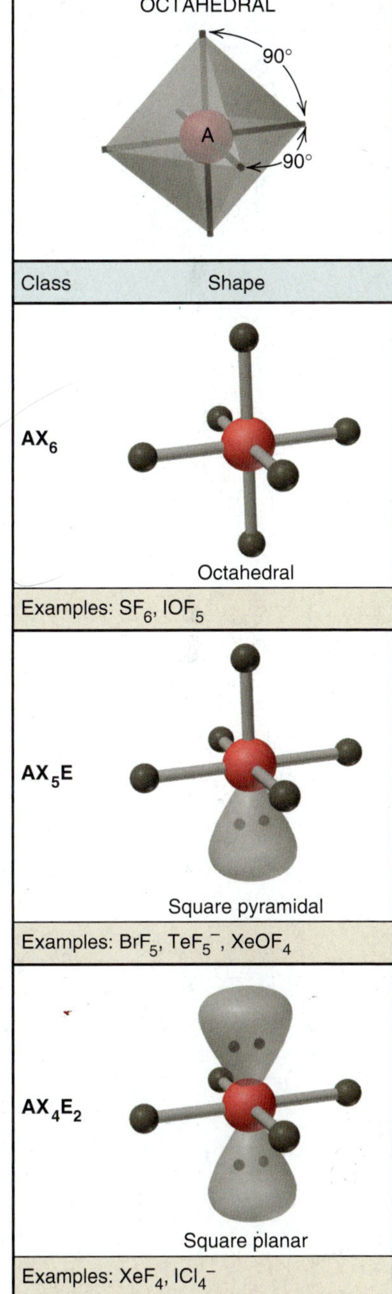

FIGURE 10.8 The three molecular shapes of the octahedral electron-group arrangement.

Using VSEPR Theory to Determine Molecular Shape

Let's apply a stepwise method for using the VSEPR theory to determine a molecular shape from a molecular formula:

Step 1. Write the Lewis structure from the molecular formula (Figure 10.1) to see the relative placement of atoms and the number of electron groups.

Step 2. Assign an electron-group arrangement by counting *all* electron groups around the central atom, bonding *plus* nonbonding.

Step 3. Predict the ideal bond angle from the electron-group arrangement and *the direction of any deviation* caused by lone pairs or double bonds.

Step 4. Draw and name the molecular shape by counting bonding groups and nonbonding groups separately.

Figure 10.9 summarizes these steps, and the next two sample problems apply them.

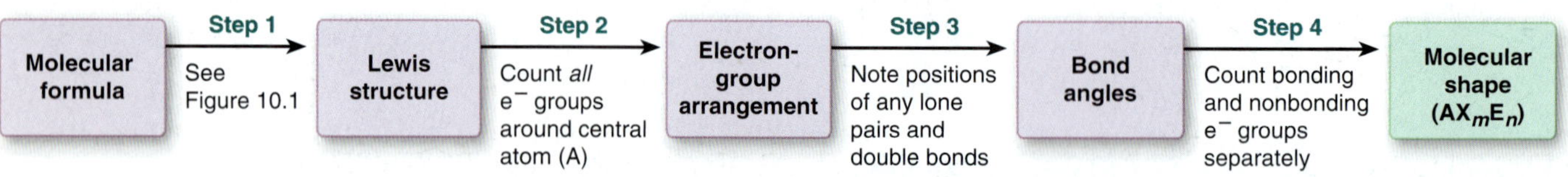

FIGURE 10.9 The steps in determining a molecular shape. Four steps are needed to convert a molecular formula to a molecular shape.

SAMPLE PROBLEM 10.6 Predicting Molecular Shapes with Two, Three, or Four Electron Groups

Problem Draw the molecular shapes and predict the bond angles (relative to the ideal angles) of **(a)** PF_3 and **(b)** $COCl_2$.

Solution (a) For PF_3.

Step 1. Write the Lewis structure from the formula (see below left).

Step 2. Assign the electron-group arrangement: Three bonding groups plus one lone pair give four electron groups around P and the *tetrahedral arrangement.*

Step 3. Predict the bond angle: For the tetrahedral electron-group arrangement, the ideal bond angle is 109.5°. There is one lone pair, so the actual bond angle should be less than 109.5°.

Step 4. Draw and name the molecular shape: With four electron groups, one of them a lone pair, PF_3 has a trigonal pyramidal shape (AX_3E):

4 e^- groups ⟹ Tetrahedral arrangement ⟹ 1 lone pair ⟹ <109.5° ⟹ 3 bonding groups ⟹ 96.3° AX_3E

(b) For $COCl_2$.

Step 1. Write the Lewis structure from the formula (see below left).

Step 2. Assign the electron-group arrangement: Two single bonds plus one double bond give three electron groups around C and the *trigonal planar arrangement.*

Step 3. Predict the bond angles: The ideal bond angle is 120°, but the double bond between C and O should compress the Cl—C—Cl angle to less than 120°.

Step 4. Draw and name the molecular shape: With three electron groups and no lone pairs, $COCl_2$ has a trigonal planar shape (AX_3):

3 e^- groups ⟹ Trigonal planar arrangement ⟹ 1 double bond ⟹ Cl—C—O >120°, Cl—C—Cl <120° ⟹ 3 bonding groups ⟹ 124.5°, 111° AX_3

Check We compare the answers with the information in Figures 10.4 and 10.5.

Comment Be sure the Lewis structure is correct because it determines the other steps.

FOLLOW-UP PROBLEM 10.6 Draw the molecular shapes and predict the bond angles (relative to the ideal angles) of **(a)** CS_2; **(b)** $PbCl_2$; **(c)** CBr_4; **(d)** SF_2.

SAMPLE PROBLEM 10.7 Predicting Molecular Shapes with Five or Six Electron Groups

Problem Determine the molecular shapes and predict the bond angles (relative to the ideal angles) of **(a)** SbF_5 and **(b)** BrF_5.

Plan We proceed as in Sample Problem 10.6, keeping in mind the need to minimize the number of 90° repulsions.

Solution **(a)** For SbF_5.

Step 1. Lewis structure (see below left).

Step 2. Electron-group arrangement: With five electron groups, this is the *trigonal bipyramidal* arrangement.

Step 3. Bond angles: All the groups and surrounding atoms are identical, so the bond angles are ideal: 120° between equatorial groups and 90° between axial and equatorial groups.

Step 4. Molecular shape: Five electron groups and no lone pairs give the trigonal bipyramidal shape (AX_5):

SbF_5 ⟹ (5 e^- groups) Trigonal bipyramidal arrangement ⟹ (no lone pairs or double bonds) Ideal bond angles ⟹ (5 bonding groups) 120°, 90°; AX_5

(b) For BrF_5.

Step 1. Lewis structure (see below left).

Step 2. Electron-group arrangement: Six electron groups give the *octahedral* arrangement.

Step 3. Bond angles: The lone pair should make all bond angles less than the ideal 90°.

Step 4. Molecular shape: With six electron groups and one of them a lone pair, BrF_5 has the square pyramidal shape (AX_5E):

BrF_5 ⟹ (6 e^- groups) Octahedral arrangement ⟹ (1 lone pair) <90° ⟹ (5 bonding groups) 84.8°; AX_5E

Check We compare our answers with Figures 10.7 and 10.8.

Comment We will encounter the linear, tetrahedral, square planar, and octahedral shapes in *coordination compounds,* which we discuss in Chapter 22.

FOLLOW-UP PROBLEM 10.7 Draw the molecular shapes and predict the bond angles (relative to the ideal angles) of **(a)** ICl_2^-; **(b)** ClF_3; **(c)** SOF_4.

Molecular Shapes with More Than One Central Atom

Many molecules, especially those in living systems, have more than one central atom. The shapes of these molecules are combinations of the molecular shapes for each central atom. For these molecules, we find the molecular shape around one central atom at a time. Consider ethane (CH_3CH_3; molecular formula C_2H_6), a component of natural gas (Figure 10.10A). With four bonding groups and no lone pairs around each of the two central carbons, ethane is shaped like two overlapping tetrahedra.

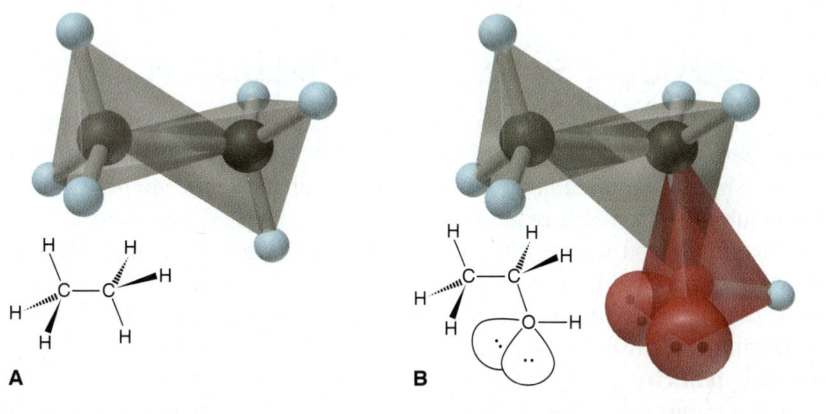

FIGURE 10.10 The tetrahedral centers of ethane and of ethanol. When a molecule has more than one central atom, the overall shape is a composite of the shape around each center. **A,** Ethane's shape can be viewed as two overlapping tetrahedra. **B,** Ethanol's shape can be viewed as three overlapping tetrahedral arrangements, with the shape around the O atom bent (V shaped) because of its two lone pairs.

Ethanol (CH_3CH_2OH; molecular formula C_2H_6O), the intoxicating substance in beer and wine, has three central atoms (Figure 10.10B). The CH_3— group is tetrahedrally shaped, and the —CH_2— group has four bonding groups around its central C atom, so it is tetrahedrally shaped also. The O atom has four electron groups and two lone pairs around it, which gives the V shape (AX_2E_2).

SAMPLE PROBLEM 10.8 Predicting Molecular Shapes with More Than One Central Atom

Problem Determine the shape around each of the central atoms in acetone, $(CH_3)_2C{=}O$.

Plan There are three central atoms, all C, two of which are in CH_3— groups. We determine the shape around one central atom at a time.

Solution

Step 1. Lewis structure (see below left).

Step 2. Electron-group arrangement: Each CH_3— group has four electron groups around its central C, so its electron-group arrangement is *tetrahedral.* The third C atom has three electron groups around it, so it has the *trigonal planar* arrangement.

Step 3. Bond angles: The H—C—H angle in the CH_3— groups should be near the ideal tetrahedral angle of 109.5°. The C=O double bond should compress the C—C—C angle to less than the ideal 120°.

Step 4. Shapes around central atoms: With four electron groups and no lone pairs, the shapes around the two C atoms in the CH_3— groups are tetrahedral (AX_4). With three electron groups and no lone pairs, the shape around the middle C atom is trigonal planar (AX_3):

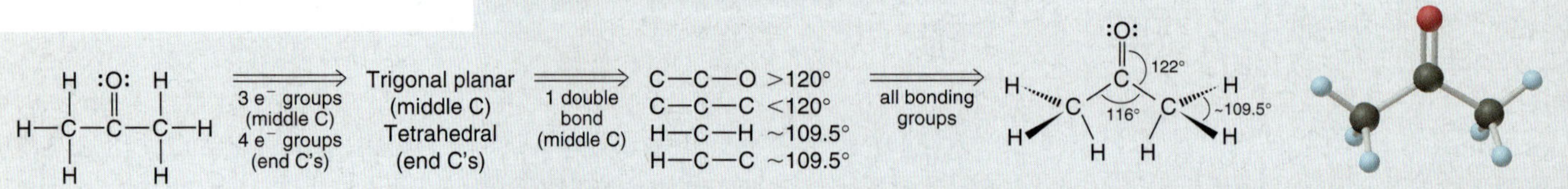

3 e⁻ groups (middle C), 4 e⁻ groups (end C's) ⟹ Trigonal planar (middle C), Tetrahedral (end C's) ⟹ 1 double bond (middle C) ⟹ C—C—O >120°, C—C—C <120°, H—C—H ~109.5°, H—C—C ~109.5° ⟹ all bonding groups ⟹ 122°, 116°, ~109.5°

FOLLOW-UP PROBLEM 10.8 Determine the shape around each central atom and predict any deviations from ideal bond angles in the following: **(a)** H_2SO_4; **(b)** propyne (C_3H_4; there is one C≡C bond); **(c)** S_2F_2.

SECTION 10.2 SUMMARY

The VSEPR theory proposes that each group of electrons (single bond, multiple bond, lone pair, or lone electron) around a central atom remains as far away from the others as possible. • One of five electron-group arrangements results when two, three, four, five, or six electron groups surround a central atom. Each arrangement is associated with one or more molecular shapes. • Ideal bond angles are prescribed by the regular geometric shapes; deviations from these angles occur when the surrounding atoms or electron groups are not identical. • Lone pairs and double bonds exert greater repulsions than single bonds. • Larger molecules have shapes that are composites of the shapes around each of their central atoms.

10.3 MOLECULAR SHAPE AND MOLECULAR POLARITY

Knowing the shape of a substance's molecules is a key to understanding its physical and chemical behavior. One of the most important and far-reaching effects of molecular shape is molecular polarity, which can influence melting and boiling points, solubility, chemical reactivity, and even biological function.

Animation: Influence of Shape on Polarity

In Chapter 9, you learned that a covalent bond is *polar* when it joins atoms of different electronegativities because the atoms share the electrons unequally. In diatomic molecules, such as HF, where there is only one bond, the bond polarity causes the molecule itself to be polar. Molecules with a net imbalance of charge have a **molecular polarity.** In molecules with more than two atoms, *both shape and bond polarity determine molecular polarity.* In an electric field, polar molecules become oriented, on average, with their partial charges pointing toward the oppositely charged electric plates, as shown for HF in Figure 10.11. The **dipole moment (μ)** is the product of these partial charges and the distance between them. It is typically measured in *debye* (D) units; using the SI units of charge (coulomb, C) and length (meter, m), 1 D = 3.34×10^{-30} C·m. [The unit is named for Peter Debye (1884–1966), the Dutch American chemist and physicist who won the Nobel Prize in 1936 for his major contributions to our understanding of molecular structure and solution behavior.]

To determine molecular polarity, we also consider shape because the presence of polar bonds does not *always* lead to a polar molecule. In carbon dioxide, for example, the large electronegativity difference between C (EN = 2.5) and O (EN = 3.5) makes each C═O bond quite polar. However, CO_2 is linear, so its bonds point 180° from each other. As a result, the two identical bond polarities are counterbalanced and give the molecule *no net dipole moment* (μ = 0 D). Note that the model shows regions of high negative charge *(red)* distributed equally on either side of the central region of high positive charge *(blue):*

O═C═O

Water also has identical atoms bonded to the central atom, but it *does* have a significant dipole moment (μ = 1.85 D). In each O—H bond, electron density is pulled toward the more electronegative O atom. Here, the bond polarities are *not* counterbalanced, because the water molecule is V shaped (also see Figure 4.1, page 114). Instead, the bond polarities are partially reinforced, and the O end of the molecule is more negative than the other end (the region between the H atoms), which the electron density model shows clearly:

(The molecular polarity of water has some amazing effects, from determining the composition of the oceans to supporting life itself, as you'll see in Chapter 12.)

In the two previous examples, molecular shape influences polarity. When different molecules have the same shape, the nature of the atoms surrounding the central atom can have a major effect on polarity. Consider carbon tetrachloride (CCl_4) and chloroform ($CHCl_3$), two tetrahedral molecules with very different polarities. In CCl_4, the surrounding atoms are all Cl atoms. Although each C—Cl bond is polar (ΔEN = 0.5), the molecule is nonpolar (μ = 0 D) because the individual bond polarities counterbalance each other. In $CHCl_3$, H substitutes for one Cl atom, disrupting the balance and giving chloroform a significant dipole moment (μ = 1.01 D):

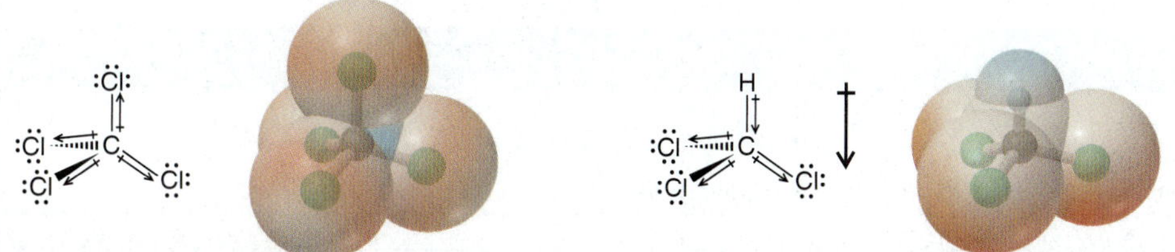

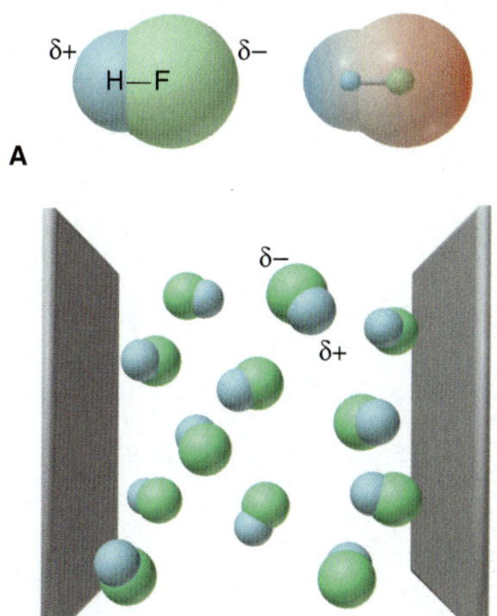

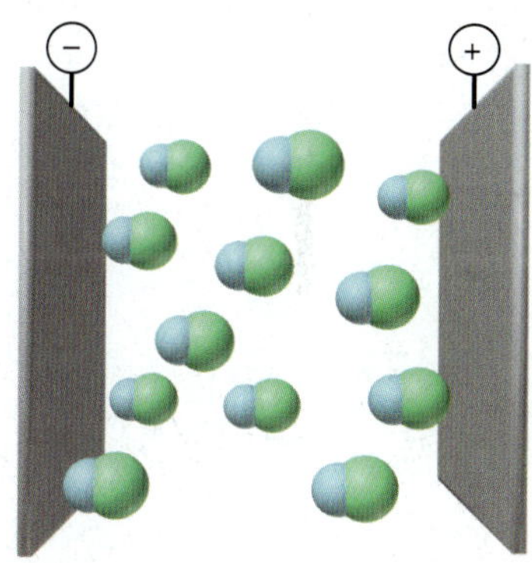

FIGURE 10.11 The orientation of polar molecules in an electric field.
A, A space-filling model of HF *(left)* shows the partial charges of this polar molecule. The electron density model *(right)* shows high electron density *(red)* associated with the F end and low electron density *(blue)* with the H end.
B, In the absence of an external electric field, HF molecules are oriented randomly.
C, In the presence of an electric field, the molecules, on average, become oriented with their partial charges pointing toward the oppositely charged plates.

SAMPLE PROBLEM 10.9 Predicting the Polarity of Molecules

Problem From electronegativity (EN) values and their periodic trends (Figure 9.19, page 297), predict whether each of the following molecules is polar and show the direction of bond dipoles and the overall molecular dipole when applicable:
(a) Ammonia, NH_3 **(b)** Boron trifluoride, BF_3
(c) Carbonyl sulfide, COS (atom sequence SCO)

Plan First, we draw and name the molecular shape. Then, using relative EN values, we decide on the direction of each bond dipole. Finally, we see if the bond dipoles balance or reinforce each other in the molecule as a whole.

Solution **(a)** For NH_3. The molecular shape is trigonal pyramidal. From Figure 9.19, we see that N (EN = 3.0) is more electronegative than H (EN = 2.1), so the bond dipoles point toward N. The bond dipoles partially reinforce each other, and thus the molecular dipole points toward N:

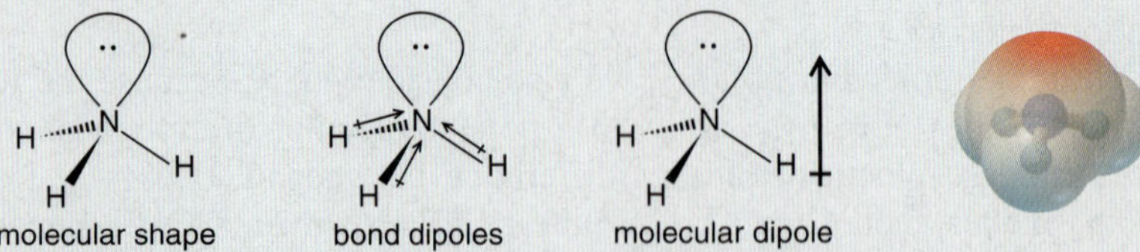

Therefore, ammonia is polar.

(b) For BF_3. The molecular shape is trigonal planar. Because F (EN = 4.0) is farther to the right in Period 2 than B (EN = 2.0), it is more electronegative; thus, each bond dipole points toward F. However, the bond angle is 120°, so the three bond dipoles counterbalance each other, and BF_3 has no molecular dipole:

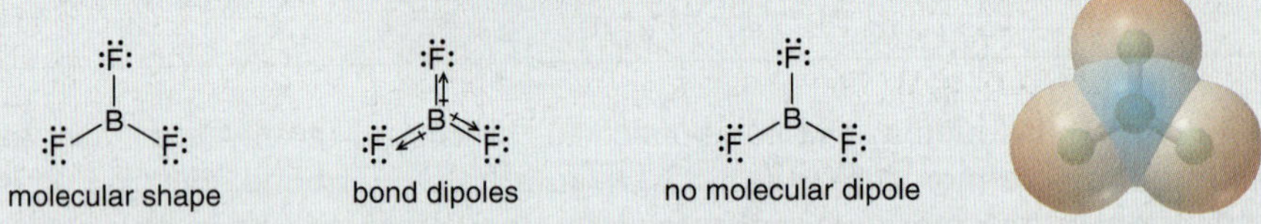

Therefore, boron trifluoride is nonpolar.

(c) For COS. The molecular shape is linear. With C and S having the same EN, the C=S bond is nonpolar, but the C=O bond is quite polar (ΔEN = 1.0), so there is a net molecular dipole toward the O:

S=C=O molecular shape | S=C=O bond dipole | S=C=O molecular dipole

Therefore, carbonyl sulfide is polar.

FOLLOW-UP PROBLEM 10.9 Show the bond dipoles and molecular dipole, if any, for **(a)** dichloromethane (CH_2Cl_2); **(b)** iodine oxide pentafluoride (IOF_5); **(c)** nitrogen tribromide (NBr_3).

SECTION 10.3 SUMMARY

Bond polarity and molecular shape determine molecular polarity, which is measured as a dipole moment. • When bond polarities counterbalance each other, the molecule is nonpolar; when they reinforce each other, even partially, the molecule is polar.

CHAPTER REVIEW GUIDE

The following sections provide many aids to help you study this chapter. (Numbers in parentheses refer to pages, unless noted otherwise.)

• LEARNING OBJECTIVES *These are concepts and skills to review after studying this chapter.*

Related section (§), sample problem (SP), and end-of-chapter problem (EP) numbers are listed in parentheses.

1. Use the octet rule to draw a Lewis structure from a molecular formula (§ 10.1) (SPs 10.1–10.3) (EPs 10.1, 10.5–10.8)
2. Understand how electron delocalization explains bond properties, and draw resonance structures (§ 10.1) (SP 10.4) (EPs 10.2, 10.9–10.12)
3. Describe the three types of exceptions to the octet rule, draw Lewis structures for such molecules, and use formal charges to select the most important resonance structure (§ 10.1) (SP 10.5) (EPs 10.3, 10.4, 10.13–10.24)
4. Describe the five electron-group arrangements and associated molecular shapes, predict molecular shapes from Lewis structures, and explain deviations from ideal bond angles (§ 10.2) (SPs 10.6–10.8) (EPs 10.25–10.50)
5. Understand how a molecule's polarity arises, and use molecular shape and EN values to predict the direction of a dipole (§ 10.3) (SP 10.9) (EPs 10.51–10.55)

• KEY TERMS *These important terms appear in boldface in the chapter and are defined again in the Glossary.*

Section 10.1
Lewis structure (Lewis formula) (306)
resonance structure (resonance form) (309)
resonance hybrid (309)
electron-pair delocalization (310)
formal charge (311)
electron deficient (313)
free radical (313)
expanded valence shell (314)

Section 10.2
valence-shell electron-pair repulsion (VSEPR) theory (316)
molecular shape (316)
bond angle (316)
linear arrangement (317)
linear shape (317)
trigonal planar arrangement (317)
bent shape (V shape) (318)
tetrahedral arrangement (318)
trigonal pyramidal shape (319)
trigonal bipyramidal arrangement (320)
equatorial group (320)
axial group (320)
seesaw shape (320)
T shape (320)
octahedral arrangement (321)
square pyramidal shape (321)
square planar shape (321)

Section 10.3
molecular polarity (324)
dipole moment (μ) (325)

• KEY EQUATIONS AND RELATIONSHIPS *Numbered and screened concepts are listed for you to refer to or memorize.*

10.1 Calculating the formal charge on an atom (311):
Formal charge of atom

$$= \text{no. of valence } e^- - (\text{no. of unshared valence } e^- + \tfrac{1}{2} \text{ no. of shared valence } e^-)$$

10.2 Ranking the effect of electron-pair repulsions on bond angle (320):
Lone pair–lone pair > lone pair–bonding pair > bonding pair–bonding pair

• BRIEF SOLUTIONS TO *FOLLOW-UP PROBLEMS* *Compare your own solutions to these calculation steps and answers.*

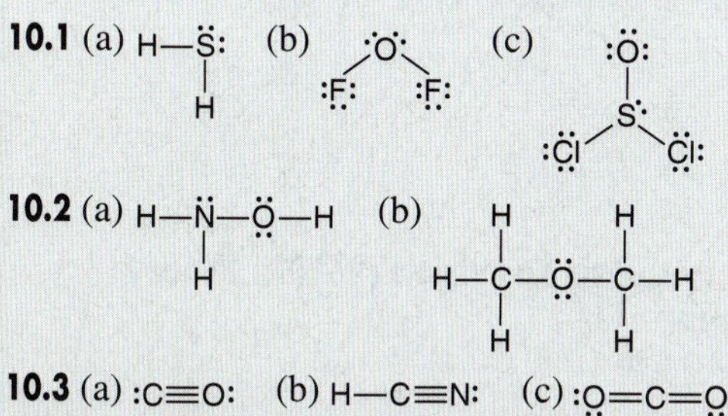

10.1 (a) (b) (c)

10.2 (a) (b)

10.3 (a) :C≡O: (b) H—C≡N: (c) :O=C=O:

10.4

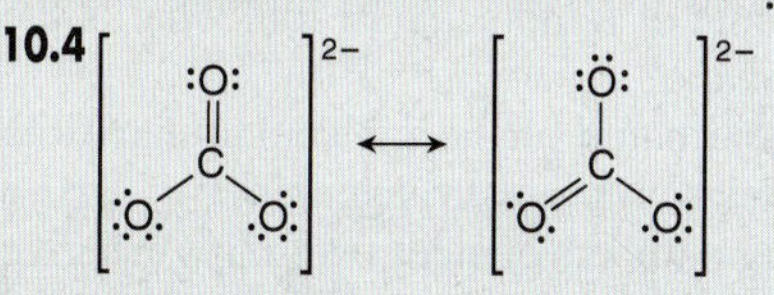

10.5 (a) (b) (c)

10.6 (a) S=C=S

Linear, 180°

(b) V shaped, <120°

(c) Tetrahedral, 109.5°

(d) V shaped, <109.5°

10.7

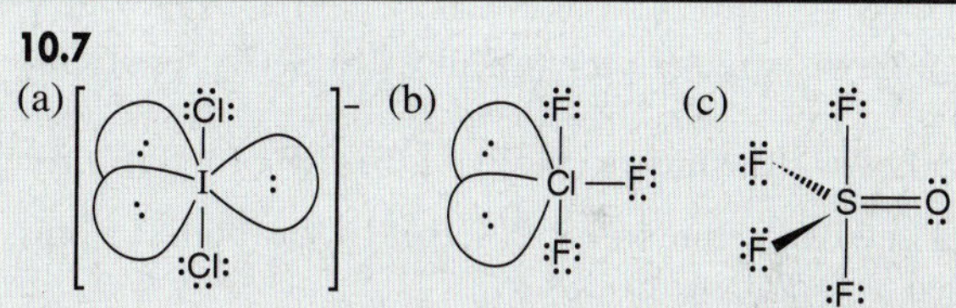

(a) Linear, 180°

(b) T shaped, <90°

(c) Trigonal bipyramidal
F_{eq}—S—F_{eq} angle <120°
F_{ax}—S—F_{eq} angle <90°

10.8 (a) S is tetrahedral; double bonds compress O—S—O angle to <109.5°. Each central O is V shaped; lone pairs compress H—O—S angle to <109.5°.

(b)

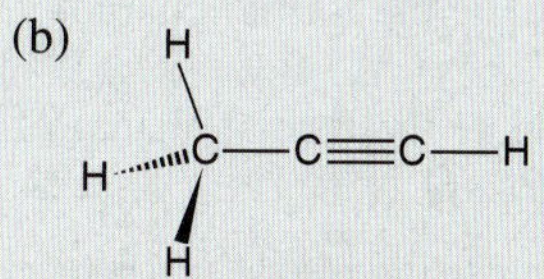

C in CH_3— is tetrahedral ~109.5°; other C atoms are linear, 180°.

(c) Each S is V shaped; F—S—S angle <109.5°.

10.9

(a) (b) (c)

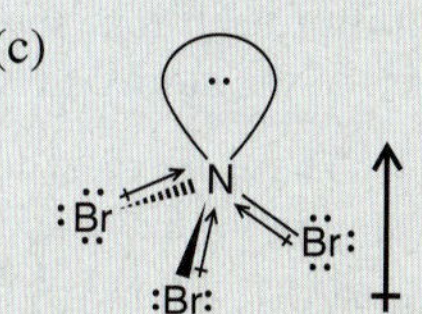

PROBLEMS

*Problems with **colored** numbers are answered in Appendix E. Sections match the text and provide the numbers of relevant sample problems. Bracketed problems are grouped in pairs (indicated by a short rule) that cover the same concept. Comprehensive Problems are based on material from any section or previous chapter.*

Depicting Molecules and Ions with Lewis Structures

(Sample Problems 10.1 to 10.5)

10.1 Which of these atoms *cannot* serve as a central atom in a Lewis structure: (a) O; (b) He; (c) F; (d) H; (e) P? Explain.

10.2 When is a resonance hybrid needed to adequately depict the bonding in a molecule? Using NO_2 as an example, explain how a resonance hybrid is consistent with the actual bond length, bond strength, and bond order.

10.3 In which of these bonding patterns does X obey the octet rule?

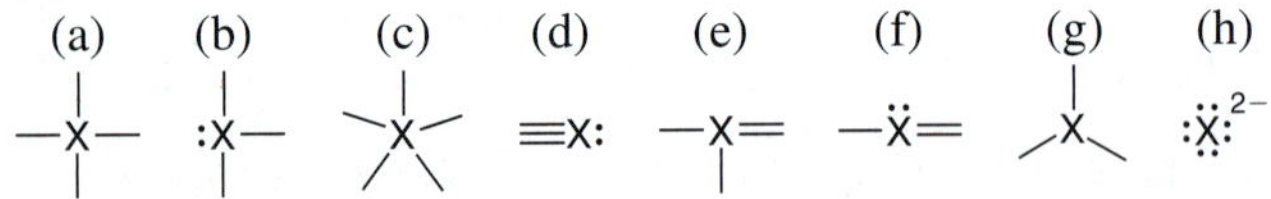

10.4 What is required for an atom to expand its valence shell? Which of the following atoms can expand its valence shell: F, S, H, Al, Se, Cl?

10.5 Draw a Lewis structure for (a) SiF_4; (b) $SeCl_2$; (c) COF_2 (C is central).

10.6 Draw a Lewis structure for (a) PH_4^+; (b) C_2F_4; (c) SbH_3.

10.7 Draw a Lewis structure for (a) PF_3; (b) H_2CO_3 (both H atoms are attached to O atoms); (c) CS_2.

10.8 Draw a Lewis structure for (a) CH_4S; (b) S_2Cl_2; (c) $CHCl_3$.

10.9 Draw Lewis structures of all the important resonance forms of (a) NO_2^+; (b) NO_2F (N is central).

10.10 Draw Lewis structures of all the important resonance forms of (a) HNO_3 ($HONO_2$); (b) $HAsO_4^{2-}$ ($HOAsO_3^{2-}$).

10.11 Draw Lewis structures of all the important resonance forms of (a) N_3^-; (b) NO_2^-.

10.12 Draw Lewis structures of all the important resonance forms of (a) HCO_2^- (H is attached to C); (b) $HBrO_4$ ($HOBrO_3$).

10.13 Draw a Lewis structure and calculate the formal charge of each atom in (a) IF_5; (b) AlH_4^-.

10.14 Draw a Lewis structure and calculate the formal charge of each atom in (a) COS (C is central); (b) NO.

10.15 Draw a Lewis structure for the most important resonance form of each ion, showing formal charges and oxidation numbers of the atoms: (a) BrO_3^-; (b) SO_3^{2-}.

10.16 Draw a Lewis structure for the most important resonance form of each ion, showing formal charges and oxidation numbers of the atoms: (a) AsO_4^{3-}; (b) ClO_2^-.

10.17 These species do not obey the octet rule. Draw a Lewis structure for each, and state the type of octet-rule exception:
(a) BH_3 (b) AsF_4^- (c) $SeCl_4$

10.18 These species do not obey the octet rule. Draw a Lewis structure for each, and state the type of octet-rule exception:
(a) PF_6^- (b) ClO_3 (c) H_3PO_3

10.19 These species do not obey the octet rule. Draw a Lewis structure for each, and state the type of octet-rule exception:
(a) BrF_3 (b) ICl_2^- (c) BeF_2

10.20 These species do not obey the octet rule. Draw a Lewis structure for each, and state the type of octet-rule exception:
(a) O_3^- (b) XeF_2 (c) SbF_4^-

10.21 Molten beryllium chloride reacts with chloride ion from molten NaCl to form the $BeCl_4^{2-}$ ion, in which the Be atom attains an octet. Show the net ionic reaction with Lewis structures.

10.22 Despite many attempts, the perbromate ion (BrO_4^-) was not prepared in the laboratory until about 1970. (Indeed, articles were published explaining theoretically why it could never be prepared!) Draw a Lewis structure for BrO_4^- in which all atoms have the lowest formal charges.

10.23 Cryolite (Na_3AlF_6) is an indispensable component in the electrochemical manufacture of aluminum. Draw a Lewis structure for the AlF_6^{3-} ion.

10.24 Phosgene is a colorless, highly toxic gas employed against troops in World War I and used today as a key reactant in certain organic syntheses. Use formal charges to select the most important of the following resonance structures:

A B C

Valence-Shell Electron-Pair Repulsion (VSEPR) Theory and Molecular Shape

(Sample Problems 10.6 to 10.8)

10.25 If you know the formula of a molecule or ion, what is the first step in predicting its shape?

10.26 In what situation is the name of the molecular shape the same as the name of the electron-group arrangement?

10.27 Which of the following numbers of electron groups can give rise to a bent (V-shaped) molecule: two, three, four, five, six? Draw an example for each case, showing the shape classification (AX_mE_n) and the ideal bond angle.

10.28 Name all the molecular shapes that have a tetrahedral electron-group arrangement.

10.29 Consider the following molecular shapes:

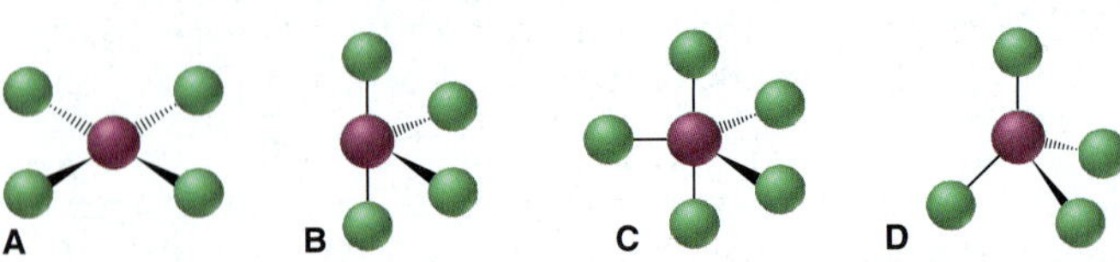

(a) Which has the most electron pairs (both shared and unshared) around the central atom? (b) Which has the most unshared pairs around the central atom? (c) Do any have only shared pairs around the central atom?

10.30 Why aren't lone pairs considered along with surrounding bonding groups when describing the molecular shape?

10.31 Use wedge-bond perspective drawings (if necessary) to sketch the atom positions in a general molecule of formula (not shape class) AX_n that has each of the following shapes:
(a) V shaped (b) trigonal planar (c) trigonal bipyramidal
(d) T shaped (e) trigonal pyramidal (f) square pyramidal

10.32 What would you expect the electron-group arrangement around atom A to be in each of the following cases? For each arrangement, give the ideal bond angle and the direction of any expected deviation:

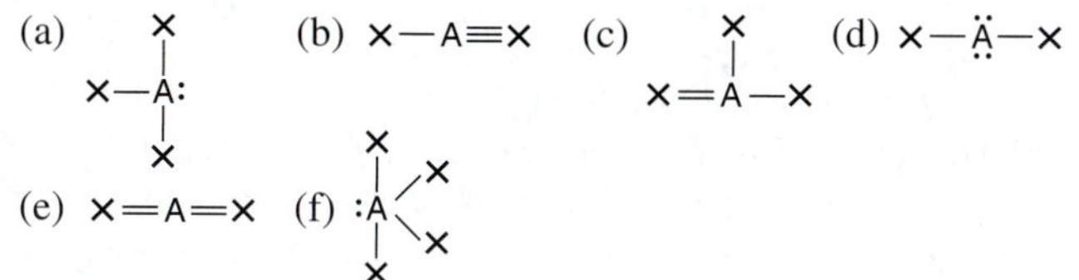

10.33 Determine the electron-group arrangement, molecular shape, and ideal bond angle(s) for each of the following:
(a) O_3 (b) H_3O^+ (c) NF_3

10.34 Determine the electron-group arrangement, molecular shape, and ideal bond angle(s) for each of the following:
(a) SO_4^{2-} (b) NO_2^- (c) PH_3

10.35 Determine the electron-group arrangement, molecular shape, and ideal bond angle(s) for each of the following:
(a) CO_3^{2-} (b) SO_2 (c) CF_4

10.36 Determine the electron-group arrangement, molecular shape, and ideal bond angle(s) for each of the following:
(a) SO_3 (b) N_2O (N is central) (c) CH_2Cl_2

10.37 Name the shape and give the AX_mE_n classification and ideal bond angle(s) for each of the following general molecules:

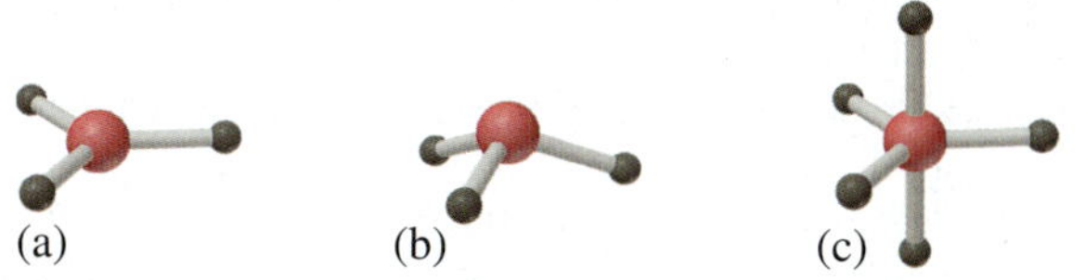

10.38 Name the shape and give the AX_mE_n classification and ideal bond angle(s) for each of the following general molecules:

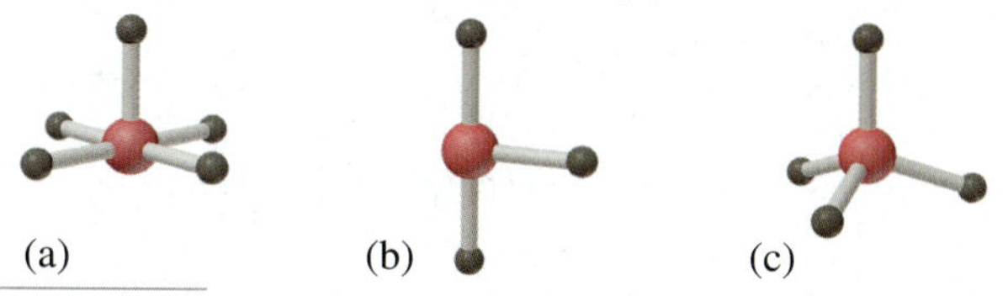

10.39 Determine the shape, ideal bond angle(s), and the direction of any deviation from these angles for each of the following:
(a) ClO_2^- (b) PF_5 (c) SeF_4 (d) KrF_2

10.40 Determine the shape, ideal bond angle(s), and the direction of any deviation from these angles for each of the following:
(a) ClO_3^- (b) IF_4^- (c) $SeOF_2$ (d) TeF_5^-

10.41 Determine the shape around each central atom in each molecule, and explain any deviation from ideal bond angles:
(a) CH_3OH (b) N_2O_4 (O_2NNO_2)

10.42 Determine the shape around each central atom in each molecule, and explain any deviation from ideal bond angles:
(a) H_3PO_4 (no H—P bond) (b) $CH_3—O—CH_2CH_3$

10.43 Determine the shape around each central atom in each molecule, and explain any deviation from ideal bond angles:
(a) CH_3COOH (b) H_2O_2

10.44 Determine the shape around each central atom in each molecule, and explain any deviation from ideal bond angles:
(a) H_2SO_3 (no H—S bond) (b) N_2O_3 ($ONNO_2$)

10.45 Arrange the following AF_n species in order of *increasing* F—A—F bond angles: BF_3, BeF_2, CF_4, NF_3, OF_2.

10.46 Arrange the following ACl_n species in order of *decreasing* Cl—A—Cl bond angles: SCl_2, OCl_2, PCl_3, $SiCl_4$, $SiCl_6^{2-}$.

10.47 State an ideal value for each of the bond angles in each molecule, and note where you expect deviations:

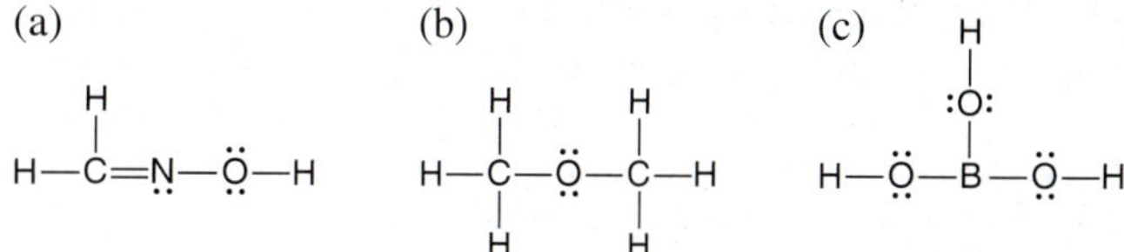

10.48 State an ideal value for each of the bond angles in each molecule, and note where you expect deviations:

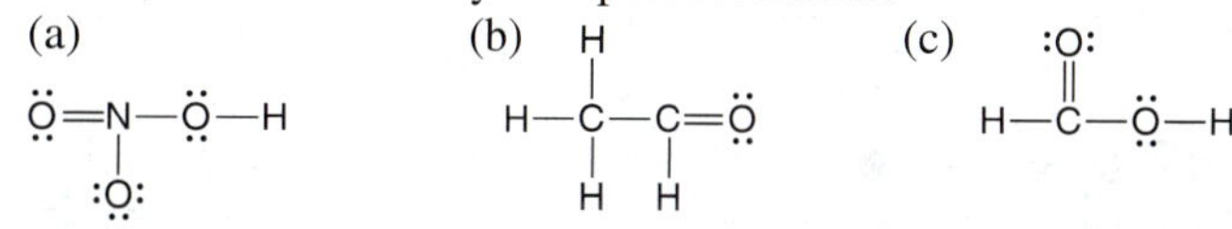

10.49 Because both tin and carbon are members of Group 4A(14), they form structurally similar compounds. However, tin exhibits a greater variety of structures because it forms several ionic species. Predict the shapes and ideal bond angles, including any deviations, for the following:
(a) $Sn(CH_3)_2$ (b) $SnCl_3^-$ (c) $Sn(CH_3)_4$
(d) SnF_5^- (e) SnF_6^{2-}

10.50 In the gas phase, phosphorus pentachloride exists as separate molecules. In the solid phase, however, the compound is composed of alternating PCl_4^+ and PCl_6^- ions. What change(s) in molecular shape occur(s) as PCl_5 solidifies? How does the Cl—P—Cl angle change?

Molecular Shape and Molecular Polarity

(Sample Problem 10.9)

10.51 How can a molecule with polar covalent bonds not be polar? Give an example.

10.52 Consider the molecules SCl_2, F_2, CS_2, CF_4, and BrCl.
(a) Which has bonds that are the most polar?
(b) Which have a molecular dipole moment?

10.53 Consider the molecules BF_3, PF_3, BrF_3, SF_4, and SF_6.
(a) Which has bonds that are the most polar?
(b) Which have a molecular dipole moment?

10.54 Which molecule in each pair has the greater dipole moment? Give the reason for your choice.
(a) SO_2 or SO_3 (b) ICl or IF
(c) SiF_4 or SF_4 (d) H_2O or H_2S

10.55 Which molecule in each pair has the greater dipole moment? Give the reason for your choice.
(a) ClO_2 or SO_2 (b) HBr or HCl
(c) $BeCl_2$ or SCl_2 (d) AsF_3 or AsF_5

Comprehensive Problems

Problems with an asterisk (*) are more challenging.

* **10.56** There are three different dichloroethylenes (molecular formula $C_2H_2Cl_2$), which we can designate X, Y, and Z. Compound X has no dipole moment, but compound Z does. Compounds X and Z each combine with hydrogen to give the same product:

$$C_2H_2Cl_2\text{ (X or Z)} + H_2 \longrightarrow ClCH_2{-}CH_2Cl$$

What are the structures of X, Y, and Z? Would you expect compound Y to have a dipole moment?

10.57 In addition to ammonia, nitrogen forms three other hydrides: hydrazine (N_2H_4), diazene (N_2H_2), and tetrazene (N_4H_4). (a) Use Lewis structures to compare the strength, length, and order of nitrogen-nitrogen bonds in hydrazine, diazene, and N_2. (b) Tetrazene (atom sequence H_2NNNNH_2) decomposes above 0°C to hydrazine and nitrogen gas. Draw a Lewis structure for tetrazene, and calculate ΔH°_{rxn} for this decomposition.

10.58 Both aluminum and iodine form chlorides, Al_2Cl_6 and I_2Cl_6, with "bridging" Cl atoms. The Lewis structures are

:Cl: Cl: Cl: / Al Al / :Cl: Cl: Cl: :Cl: Cl: Cl: / I I / :Cl: Cl: Cl:

(a) What is the formal charge on each atom? (b) Which of these molecules has a planar shape? Explain.

10.59 Nitrosyl fluoride (NOF) has an atom sequence in which all atoms have formal charges of zero. Write the Lewis structure consistent with this fact.

10.60 Consider the following reaction of silicon tetrafluoride:

$$SiF_4 + F^- \longrightarrow SiF_5^-$$

(a) Which depiction below best illustrates the change in molecular shape around Si? (b) Give the name and AX_mE_n designation of each shape in the depiction chosen in part (a).

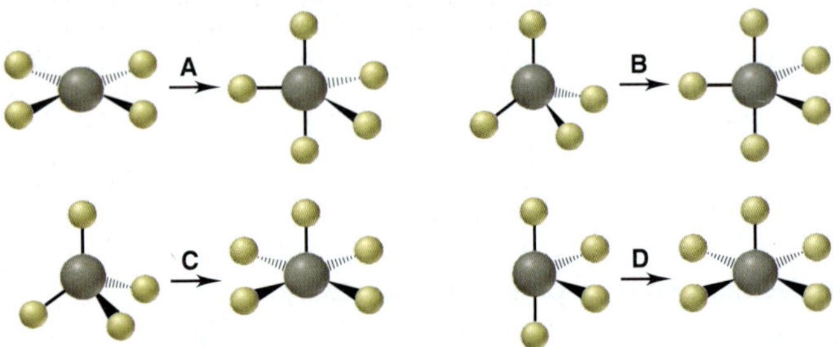

10.61 The VSEPR model was developed in the 1950s, before any xenon compounds had been prepared. Thus, in the early 1960s, these compounds provided an excellent test of the model's predictive power. What would you have predicted for the shapes of XeF_2, XeF_4, and XeF_6?

10.62 The actual bond angle in NO_2 is 134.3°, and in NO_2^- it is 115.4°, although the ideal bond angle is 120° in both. Explain.

10.63 "Inert" xenon actually forms several compounds, especially with the highly electronegative elements oxygen and fluorine. The simple fluorides XeF_2, XeF_4, and XeF_6 are all formed by direct reaction of the elements. As you might expect from the size of the xenon atom, the Xe—F bond is not a strong one. Calculate the Xe—F bond energy in XeF_6, given that the heat of formation is −402 kJ/mol.

10.64 Chloral, $Cl_3C{-}CH{=}O$, reacts with water to form the sedative and hypnotic agent chloral hydrate, $Cl_3C{-}CH(OH)_2$. Draw Lewis structures for these substances, and describe the change in molecular shape, if any, that occurs around each of the carbon atoms during the reaction.

* **10.65** Like several other bonds, carbon-oxygen bonds have lengths and strengths that depend on the bond order. Draw Lewis structures for the following species, and arrange them in order of increasing carbon-oxygen bond length and then by increasing carbon-oxygen bond strength: (a) CO; (b) CO_3^{2-}; (c) H_2CO; (d) CH_4O; (e) HCO_3^- (H attached to O).

10.66 A student isolates a product with the molecular shape shown at right (F is orange). (a) If the species is a neutral compound, can the black sphere represent selenium (Se)? (b) If the species is an anion, can the black sphere represent N? (c) If the black sphere represents Br, what is the charge of the species?

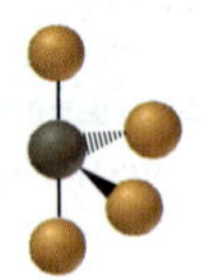

10.67 The four bonds of carbon tetrachloride (CCl_4) are polar, but the molecule is nonpolar because the bond polarity is canceled by the symmetric tetrahedral shape. When other atoms substitute for some of the Cl atoms, the symmetry is broken and the molecule becomes polar. Use Figure 9.19 (p. 297) to rank the following molecules from the least polar to the most polar: CH_2Br_2, CF_2Cl_2, CH_2F_2, CH_2Cl_2, CBr_4, CF_2Br_2.

* **10.68** Ethanol (CH_3CH_2OH) is being used as a gasoline additive or alternative in many parts of the world.
(a) Use bond energies to find the ΔH°_{rxn} for the combustion of gaseous ethanol. (Assume H_2O forms as a gas.)
(b) In its standard state at 25°C, ethanol is a liquid. Its vaporization requires 40.5 kJ/mol. Correct the value from part (a) to find the heat of reaction for the combustion of liquid ethanol.
(c) How does the value from part (b) compare with the value you calculate from standard heats of formation (Appendix B)?
(d) Methods for sustainable energy produce ethanol from corn and other plant material, but the main industrial method still involves hydrating ethylene from petroleum. Use Lewis structures and bond energies to calculate ΔH°_{rxn} for the formation of gaseous ethanol from ethylene gas with water vapor.

10.69 An oxide of nitrogen is 25.9% N by mass, has a molar mass of 108 g/mol, and contains no nitrogen-nitrogen or oxygen-oxygen bonds. Draw its Lewis structure, and name it.

* **10.70** An experiment requires 50.0 mL of 0.040 *M* NaOH for the titration of 1.00 mmol of acid. Mass analysis of the acid shows 2.24% hydrogen, 26.7% carbon, and 71.1% oxygen. Draw the Lewis structure of the acid.

10.71 When SO_3 gains two electrons, SO_3^{2-} forms. (a) Which depiction shown below best illustrates the change in molecular shape around S? (b) Does molecular polarity change during this reaction?

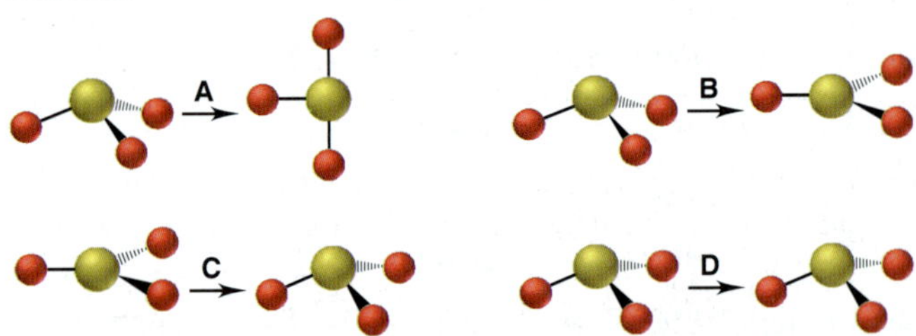

* **10.72** A major short-lived, neutral species in flames is OH.
(a) What is unusual about the electronic structure of OH?
(b) Use the standard heat of formation of OH(*g*) and bond energies to calculate the O—H bond energy in OH(*g*) [ΔH°_f of OH(*g*) = 39.0 kJ/mol].

(c) From the average value for the O—H bond energy in Table 9.2 and your value for the O—H bond energy in OH(g), find the energy needed to break the first O—H bond in water.

10.73 Pure HN_3 (atom sequence HNNN) is a very explosive compound. In aqueous solution, it is a weak acid (comparable to acetic acid) that yields the azide ion, N_3^-. Draw resonance structures to explain why the nitrogen-nitrogen bond lengths are equal in N_3^- but unequal in HN_3.

10.74 Except for nitrogen, the elements of Group 5A(15) all form pentafluorides, and most form pentachlorides. The chlorine atoms of PCl_5 can be replaced with fluorine atoms one at a time to give, successively, PCl_4F, PCl_3F_2, . . . , PF_5.
(a) Given the sizes of F and Cl, would you expect the first two F substitutions to be at axial or equatorial positions? Explain.
(b) Which of the five fluorine-containing molecules have no dipole moment?

10.75 Dinitrogen monoxide (N_2O), used as the anesthetic "laughing gas" in dental surgery, supports combustion in a manner similar to oxygen, with the nitrogen atoms forming N_2. Draw three resonance structures for N_2O (one N is central), and use formal charges to decide the relative importance of each. What correlation can you suggest between the most important structure and the observation that N_2O supports combustion?

10.76 Some scientists speculate that many organic molecules required for life on the young Earth arrived on meteorites. The Murchison meteorite that landed in Australia in 1969 contained 92 different amino acids, including 21 found in Earth organisms. A skeleton structure (single bonds only) of one of these extraterrestrial amino acids is

```
H3N—CH—C—O
    |   |
    CH2 O
    |
    CH3
```

Draw a Lewis structure, and identify any atoms with a nonzero formal charge.

10.77 When gaseous sulfur trioxide is dissolved in concentrated sulfuric acid, disulfuric acid forms:

$$SO_3(g) + H_2SO_4(l) \longrightarrow H_2S_2O_7(l)$$

Use bond energies (Table 9.2) to determine ΔH°_{rxn}. (The S atoms in $H_2S_2O_7$ are bonded through an O atom. Assume Lewis structures with zero formal charges; BE of S=O is 552 kJ/mol.)

10.78 In addition to propyne (see Follow-up Problem 10.8), there are two other constitutional isomers of formula C_3H_4. Draw a Lewis structure for each, determine the shape around each carbon, and predict any deviations from ideal bond angles.

10.79 A molecule of formula AY_3 is found experimentally to be polar. Which molecular shapes are possible and which impossible for AY_3?

* **10.80** In contrast to the cyanate ion (NCO^-), which is stable and found in many compounds, the fulminate ion (CNO^-), with its different atom sequence, is unstable and forms compounds with heavy metal ions, such as Ag^+ and Hg^{2+}, that are explosive. Like the cyanate ion, the fulminate ion has three resonance structures. Which is the most important contributor to the resonance hybrid? Suggest a reason for the instability of fulminate.

10.81 Consider the following molecular shapes:

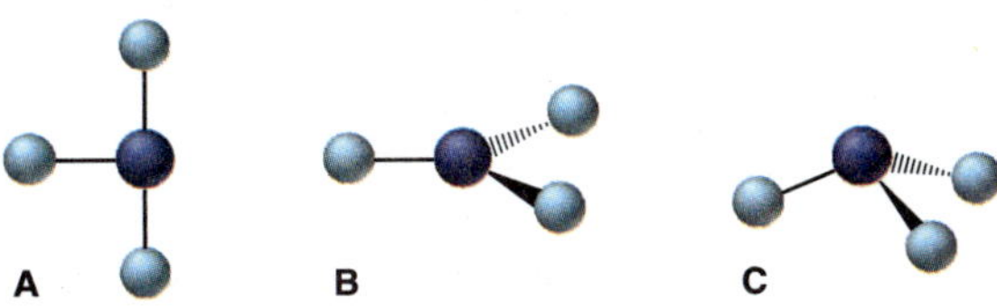

(a) Match each shape with one of the following species: XeF_3^+, $SbBr_3$, $GaCl_3$.
(b) Which, if any, is polar?
(c) Which has the most valence electrons around the cental atom?

10.82 Hydrogen cyanide (and organic nitriles, which contain the cyano group) can be catalytically reduced with hydrogen to form amines. Use Lewis structures and bond energies to determine ΔH°_{rxn} for

$$HCN(g) + 2H_2(g) \longrightarrow CH_3NH_2(g)$$

10.83 Ethylene, C_2H_4, and tetrafluoroethylene, C_2F_4, are used to make the polymers polyethylene and polytetrafluoroethylene (Teflon), respectively.
(a) Draw the Lewis structures for C_2H_4 and C_2F_4, and give the ideal H—C—H and F—C—F bond angles.
(b) The actual H—C—H and F—C—F bond angles are 117.4° and 112.4°, respectively. Explain these deviations.

10.84 Lewis structures of mescaline, a hallucinogenic compound in peyote cactus, and dopamine, a neurotransmitter in the mammalian brain, appear below. Suggest a reason for mescaline's ability to disrupt nerve impulses.

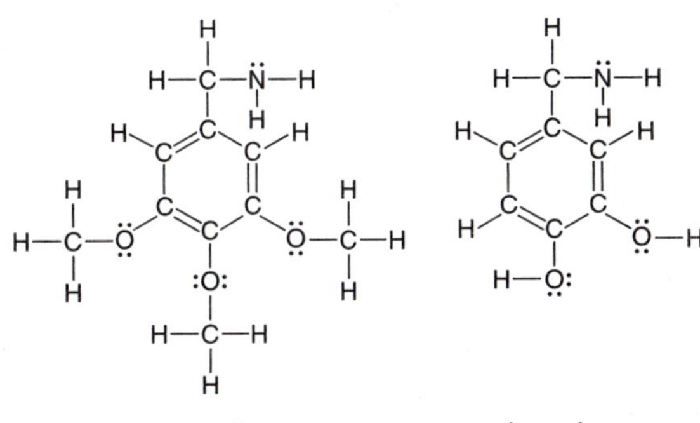

10.85 Phosphorus pentachloride, a key industrial compound with annual world production of about 2×10^7 kg, is used to make other compounds. It reacts with sulfur dioxide to produce phosphorus oxychloride ($POCl_3$) and thionyl chloride ($SOCl_2$). Draw a Lewis structure and name the molecular shape of each product.

Appendix B

STANDARD THERMODYNAMIC VALUES FOR SELECTED SUBSTANCES*

Substance or Ion	ΔH_f° (kJ/mol)	ΔG_f° (kJ/mol)	S° (J/mol·K)
$e^-(g)$	0	0	20.87
Aluminum			
$Al(s)$	0	0	28.3
$Al^{3+}(aq)$	−524.7	−481.2	−313
$AlCl_3(s)$	−704.2	−628.9	110.7
$Al_2O_3(s)$	−1676	−1582	50.94
Barium			
$Ba(s)$	0	0	62.5
$Ba(g)$	175.6	144.8	170.28
$Ba^{2+}(g)$	1649.9	—	—
$Ba^{2+}(aq)$	−538.36	−560.7	13
$BaCl_2(s)$	−806.06	−810.9	126
$BaCO_3(s)$	−1219	−1139	112
$BaO(s)$	−548.1	−520.4	72.07
$BaSO_4(s)$	−1465	−1353	132
Boron			
B(β-rhombohedral)	0	0	5.87
$BF_3(g)$	−1137.0	−1120.3	254.0
$BCl_3(g)$	−403.8	−388.7	290.0
$B_2H_6(g)$	35	86.6	232.0
$B_2O_3(s)$	−1272	−1193	53.8
$H_3BO_3(s)$	−1094.3	−969.01	88.83
Bromine			
$Br_2(l)$	0	0	152.23
$Br_2(g)$	30.91	3.13	245.38
$Br(g)$	111.9	82.40	174.90
$Br^-(g)$	−218.9	—	—
$Br^-(aq)$	−120.9	−102.82	80.71
$HBr(g)$	−36.3	−53.5	198.59
Cadmium			
$Cd(s)$	0	0	51.5
$Cd(g)$	112.8	78.20	167.64
$Cd^{2+}(aq)$	−72.38	−77.74	−61.1
$CdS(s)$	−144	−141	71
Calcium			
$Ca(s)$	0	0	41.6
$Ca(g)$	192.6	158.9	154.78
$Ca^{2+}(g)$	1934.1	—	—
$Ca^{2+}(aq)$	−542.96	−553.04	−55.2
$CaF_2(s)$	−1215	−1162	68.87
$CaCl_2(s)$	−795.0	−750.2	114
$CaCO_3(s)$	−1206.9	−1128.8	92.9
$CaO(s)$	−635.1	−603.5	38.2
$Ca(OH)_2(s)$	−986.09	−898.56	83.39
$Ca_3(PO_4)_2(s)$	−4138	−3899	263
$CaSO_4(s)$	−1432.7	−1320.3	107
Carbon			
C(graphite)	0	0	5.686
C(diamond)	1.896	2.866	2.439
$C(g)$	715.0	669.6	158.0
$CO(g)$	−110.5	−137.2	197.5
$CO_2(g)$	−393.5	−394.4	213.7
$CO_2(aq)$	−412.9	−386.2	121
$CO_3^{2-}(aq)$	−676.26	−528.10	−53.1
$HCO_3^-(aq)$	−691.11	587.06	95.0
$H_2CO_3(aq)$	−698.7	−623.42	191
$CH_4(g)$	−74.87	−50.81	186.1
$C_2H_2(g)$	227	209	200.85
$C_2H_4(g)$	52.47	68.36	219.22
$C_2H_6(g)$	−84.667	−32.89	229.5
$C_3H_8(g)$	−105	−24.5	269.9
$C_4H_{10}(g)$	−126	−16.7	310
$C_6H_6(l)$	49.0	124.5	172.8
$CH_3OH(g)$	−201.2	−161.9	238
$CH_3OH(l)$	−238.6	−166.2	127
$HCHO(g)$	−116	−110	219
$HCOO^-(aq)$	−410	−335	91.6
$HCOOH(l)$	−409	−346	129.0
$HCOOH(aq)$	−410	−356	164
$C_2H_5OH(g)$	−235.1	−168.6	282.6
$C_2H_5OH(l)$	−277.63	−174.8	161
$CH_3CHO(g)$	−166	−133.7	266
$CH_3COOH(l)$	−487.0	−392	160
$C_6H_{12}O_6(s)$	−1273.3	−910.56	212.1
$C_{12}H_{22}O_{11}(s)$	−2221.7	−1544.3	360.24
$CN^-(aq)$	151	166	118
$HCN(g)$	135	125	201.7
$HCN(l)$	105	121	112.8
$HCN(aq)$	105	112	129
$CS_2(g)$	117	66.9	237.79
$CS_2(l)$	87.9	63.6	151.0
$CH_3Cl(g)$	−83.7	−60.2	234
$CH_2Cl_2(l)$	−117	−63.2	179

*All values at 298 K.

(Continued)

Substance or Ion	ΔH_f° (kJ/mol)	ΔG_f° (kJ/mol)	S° (J/mol·K)
$CHCl_3(l)$	−132	−71.5	203
$CCl_4(g)$	−96.0	−53.7	309.7
$CCl_4(l)$	−139	−68.6	214.4
$COCl_2(g)$	−220	−206	283.74
Cesium			
$Cs(s)$	0	0	85.15
$Cs(g)$	76.7	49.7	175.5
$Cs^+(g)$	458.5	427.1	169.72
$Cs^+(aq)$	−248	−282.0	133
$CsF(s)$	−554.7	−525.4	88
$CsCl(s)$	−442.8	−414	101.18
$CsBr(s)$	−395	−383	121
$CsI(s)$	−337	−333	130
Chlorine			
$Cl_2(g)$	0	0	223.0
$Cl(g)$	121.0	105.0	165.1
$Cl^-(g)$	−234	−240	153.25
$Cl^-(aq)$	−167.46	−131.17	55.10
$HCl(g)$	−92.31	−95.30	186.79
$HCl(aq)$	−167.46	−131.17	55.06
$ClO_2(g)$	102	120	256.7
$Cl_2O(g)$	80.3	97.9	266.1
Chromium			
$Cr(s)$	0	0	23.8
$Cr^{3+}(aq)$	−1971	—	—
$CrO_4^{2-}(aq)$	−863.2	−706.3	38
$Cr_2O_7^{2-}(aq)$	−1461	−1257	214
Copper			
$Cu(s)$	0	0	33.1
$Cu(g)$	341.1	301.4	166.29
$Cu^+(aq)$	51.9	50.2	−26
$Cu^{2+}(aq)$	64.39	64.98	−98.7
$Cu_2O(s)$	−168.6	−146.0	93.1
$CuO(s)$	−157.3	−130	42.63
$Cu_2S(s)$	−79.5	−86.2	120.9
$CuS(s)$	−53.1	−53.6	66.5
Fluorine			
$F_2(g)$	0	0	202.7
$F(g)$	78.9	61.8	158.64
$F^-(g)$	−255.6	−262.5	145.47
$F^-(aq)$	−329.1	−276.5	−9.6
$HF(g)$	−273	−275	173.67
Hydrogen			
$H_2(g)$	0	0	130.6
$H(g)$	218.0	203.30	114.60
$H^+(aq)$	0	0	0
$H^+(g)$	1536.3	1517.1	108.83
Iodine			
$I_2(s)$	0	0	116.14
$I_2(g)$	62.442	19.38	260.58
$I(g)$	106.8	70.21	180.67
$I^-(g)$	−194.7	—	—
$I^-(aq)$	−55.94	−51.67	109.4
$HI(g)$	25.9	1.3	206.33
Iron			
$Fe(s)$	0	0	27.3
$Fe^{3+}(aq)$	−47.7	−10.5	−293
$Fe^{2+}(aq)$	−87.9	−84.94	−113
$FeCl_2(s)$	−341.8	−302.3	117.9
$FeCl_3(s)$	−399.5	−334.1	142
$FeO(s)$	−272.0	−251.4	60.75
$Fe_2O_3(s)$	−825.5	−743.6	87.400
$Fe_3O_4(s)$	−1121	−1018	145.3
Lead			
$Pb(s)$	0	0	64.785
$Pb^{2+}(aq)$	1.6	−24.3	21
$PbCl_2(s)$	−359	−314	136
$PbO(s)$	−218	−198	68.70
$PbO_2(s)$	−276.6	−219.0	76.6
$PbS(s)$	−98.3	−96.7	91.3
$PbSO_4(s)$	−918.39	−811.24	147
Lithium			
$Li(s)$	0	0	29.10
$Li(g)$	161	128	138.67
$Li^+(g)$	687.163	649.989	132.91
$Li^+(aq)$	−278.46	−293.8	14
$LiF(s)$	−616.9	−588.7	35.66
$LiCl(s)$	−408	−384	59.30
$LiBr(s)$	−351	−342	74.1
$LiI(s)$	−270	−270	85.8
Magnesium			
$Mg(s)$	0	0	32.69
$Mg(g)$	150	115	148.55
$Mg^{2+}(g)$	2351	—	—
$Mg^{2+}(aq)$	−461.96	−456.01	−118
$MgCl_2(s)$	−641.6	−592.1	89.630
$MgCO_3(s)$	−1112	−1028	65.86
$MgO(s)$	−601.2	−569.0	26.9
$Mg_3N_2(s)$	−461	−401	88
Manganese			
$Mn(s, \alpha)$	0	0	31.8
$Mn^{2+}(aq)$	−219	−223	−84
$MnO_2(s)$	−520.9	−466.1	53.1
$MnO_4^-(aq)$	−518.4	−425.1	190
Mercury			
$Hg(l)$	0	0	76.027
$Hg(g)$	61.30	31.8	174.87
$Hg^{2+}(aq)$	171	164.4	−32
$Hg_2^{2+}(aq)$	172	153.6	84.5
$HgCl_2(s)$	−230	−184	144
$Hg_2Cl_2(s)$	−264.9	−210.66	196
$HgO(s)$	−90.79	−58.50	70.27
Nitrogen			
$N_2(g)$	0	0	191.5
$N(g)$	473	456	153.2
$N_2O(g)$	82.05	104.2	219.7
$NO(g)$	90.29	86.60	210.65
$NO_2(g)$	33.2	51	239.9
$N_2O_4(g)$	9.16	97.7	304.3
$N_2O_5(g)$	11	118	346
$N_2O_5(s)$	−43.1	114	178
$NH_3(g)$	−45.9	−16	193
$NH_3(aq)$	−80.83	26.7	110
$N_2H_4(l)$	50.63	149.2	121.2

(Continued)

Substance or Ion	ΔH_f° (kJ/mol)	ΔG_f° (kJ/mol)	S° (J/mol·K)
$NO_3^-(aq)$	−206.57	−110.5	146
$HNO_3(l)$	−173.23	−79.914	155.6
$HNO_3(aq)$	−206.57	−110.5	146
$NF_3(g)$	−125	−83.3	260.6
$NOCl(g)$	51.71	66.07	261.6
$NH_4Cl(s)$	−314.4	−203.0	94.6
Oxygen			
$O_2(g)$	0	0	205.0
$O(g)$	249.2	231.7	160.95
$O_3(g)$	143	163	238.82
$OH^-(aq)$	−229.94	−157.30	−10.54
$H_2O(g)$	−241.826	−228.60	188.72
$H_2O(l)$	−285.840	−237.192	69.940
$H_2O_2(l)$	−187.8	−120.4	110
$H_2O_2(aq)$	−191.2	−134.1	144
Phosphorus			
$P_4(s,\text{ white})$	0	0	41.1
$P(g)$	314.6	278.3	163.1
$P(s,\text{ red})$	−17.6	−12.1	22.8
$P_2(g)$	144	104	218
$P_4(g)$	58.9	24.5	280
$PCl_3(g)$	−287	−268	312
$PCl_3(l)$	−320	−272	217
$PCl_5(g)$	−402	−323	353
$PCl_5(s)$	−443.5	—	—
$P_4O_{10}(s)$	−2984	−2698	229
$PO_4^{3-}(aq)$	−1266	−1013	−218
$HPO_4^{2-}(aq)$	−1281	−1082	−36
$H_2PO_4^-(aq)$	−1285	−1135	89.1
$H_3PO_4(aq)$	−1277	−1019	228
Potassium			
$K(s)$	0	0	64.672
$K(g)$	89.2	60.7	160.23
$K^+(g)$	514.197	481.202	154.47
$K^+(aq)$	−251.2	−282.28	103
$KF(s)$	−568.6	−538.9	66.55
$KCl(s)$	−436.7	−409.2	82.59
$KBr(s)$	−394	−380	95.94
$KI(s)$	−328	−323	106.39
$KOH(s)$	−424.8	−379.1	78.87
$KClO_3(s)$	−397.7	−296.3	143.1
$KClO_4(s)$	−432.75	−303.2	151.0
Rubidium			
$Rb(s)$	0	0	69.5
$Rb(g)$	85.81	55.86	169.99
$Rb^+(g)$	495.04	—	—
$Rb^+(aq)$	−246	−282.2	124
$RbF(s)$	−549.28	—	—
$RbCl(s)$	−435.35	−407.8	95.90
$RbBr(s)$	−389.2	−378.1	108.3
$RbI(s)$	−328	−326	118.0
Silicon			
$Si(s)$	0	0	18.0
$SiF_4(g)$	−1614.9	−1572.7	282.4
$SiO_2(s)$	−910.9	−856.5	41.5
Silver			
$Ag(s)$	0	0	42.702
$Ag(g)$	289.2	250.4	172.892
$Ag^+(aq)$	105.9	77.111	73.93
$AgF(s)$	−203	−185	84
$AgCl(s)$	−127.03	−109.72	96.11
$AgBr(s)$	−99.51	−95.939	107.1
$AgI(s)$	−62.38	−66.32	114
$AgNO_3(s)$	−45.06	19.1	128.2
$Ag_2S(s)$	−31.8	−40.3	146
Sodium			
$Na(s)$	0	0	51.446
$Na(g)$	107.76	77.299	153.61
$Na^+(g)$	609.839	574.877	147.85
$Na^+(aq)$	−239.66	−261.87	60.2
$NaF(s)$	−575.4	−545.1	51.21
$NaCl(s)$	−411.1	−384.0	72.12
$NaBr(s)$	−361	−349	86.82
$NaOH(s)$	−425.609	−379.53	64.454
$Na_2CO_3(s)$	−1130.8	−1048.1	139
$NaHCO_3(s)$	−947.7	−851.9	102
$NaI(s)$	−288	−285	98.5
Strontium			
$Sr(s)$	0	0	54.4
$Sr(g)$	164	110	164.54
$Sr^{2+}(g)$	1784	—	—
$Sr^{2+}(aq)$	−545.51	−557.3	−39
$SrCl_2(s)$	−828.4	−781.2	117
$SrCO_3(s)$	−1218	−1138	97.1
$SrO(s)$	−592.0	−562.4	55.5
$SrSO_4(s)$	−1445	−1334	122
Sulfur			
$S(\text{rhombic})$	0	0	31.9
$S(\text{monoclinic})$	0.3	0.096	32.6
$S(g)$	279	239	168
$S_2(g)$	129	80.1	228.1
$S_8(g)$	101	49.1	430.211
$S^{2-}(aq)$	41.8	83.7	22
$HS^-(aq)$	−17.7	12.6	61.1
$H_2S(g)$	−20.2	−33	205.6
$H_2S(aq)$	−39	−27.4	122
$SO_2(g)$	−296.8	−300.2	248.1
$SO_3(g)$	−396	−371	256.66
$SO_4^{2-}(aq)$	−907.51	−741.99	17
$HSO_4^-(aq)$	−885.75	−752.87	126.9
$H_2SO_4(l)$	−813.989	−690.059	156.90
$H_2SO_4(aq)$	−907.51	−741.99	17
Tin			
$Sn(\text{white})$	0	0	51.5
$Sn(\text{gray})$	3	4.6	44.8
$SnCl_4(l)$	−545.2	−474.0	259
$SnO_2(s)$	−580.7	−519.7	52.3
Zinc			
$Zn(s)$	0	0	41.6
$Zn(g)$	130.5	94.93	160.9
$Zn^{2+}(aq)$	−152.4	−147.21	−106.5
$ZnO(s)$	−348.0	−318.2	43.9
$ZnS(s,\text{ zinc blende})$	−203	−198	57.7

Appendix E

ANSWERS TO SELECTED PROBLEMS

Chapter 1

1.2 Gas molecules fill the entire container; the volume of a gas is the volume of the container. Solids and liquids have a definite volume. The volume of the container does not affect the volume of a solid or liquid. (a) gas (b) liquid (c) liquid **1.3** Physical property: a characteristic shown by a substance itself, without any interaction with or change into other substances. Chemical property: a characteristic of a substance that appears as it interacts with, or transforms into, other substances. (a) Color (yellow-green and silvery to white) and physical state (gas and metal to crystals) are physical properties. The interaction between chlorine gas and sodium metal is a chemical property. (b) Color and magnetism are physical properties. No chemical changes. **1.5**(a) Physical change; there is only a temperature change. (b) Chemical change; the change in appearance indicates an irreversible chemical change. (c) Physical change; there is only a change in size, not composition. (d) Chemical change; the wood (and air) become different substances with different compositions. **1.7**(a) fuel (b) wood **1.11** A well-designed experiment must have the following essential features: (1) There must be at least two variables that are expected to be related; (2) there must be a way to control all the variables, so that only one at a time may be changed; (3) the results must be reproducible. **1.14**(a) $(2.54\ \text{cm})^2/(1\ \text{in})^2$ and $(1\ \text{m})^2/(100\ \text{cm})^2$ (b) $(1000\ \text{m})^2/(1\ \text{km})^2$ and $(100\ \text{cm})^2/(1\ \text{m})^2$ (c) $(1.609\times10^3\ \text{m/mi})$ and $(1\ \text{h}/3600\ \text{s})$ (d) $(1000\ \text{g}/2.205\ \text{lb})$ and $(3.531\times10^{-5}\ \text{ft}^3/\text{cm}^3)$ **1.16** An extensive property depends on the amount of material present. An intensive property is the same regardless of how much material is present. (a) extensive property (b) intensive property (c) extensive property (d) intensive property **1.18**(a) increases (b) remains the same (c) decreases (d) increases (e) remains the same **1.21** 1.43 nm **1.23**(a) $2.07\times10^{-9}\ \text{km}^2$ (b) \$10.43 **1.25**(a) $5.52\times10^3\ \text{kg/m}^3$ (b) 345 lb/ft^3 **1.27**(a) $2.56\times10^{-9}\ \text{mm}^3$ (b) 10^{-10} L **1.29**(a) 9.626 cm^3 (b) 64.92 g **1.31** 2.70 g/cm^3 **1.33**(a) 22°C; 295 K (b) 109 K; −263°F (c) −273°C; −460.°F **1.37**(a) 2.47×10^{-7} m (b) 67.6 Å **1.43** Initial zeros are never significant; internal zeros are always significant; terminal zeros to the right of a decimal point are significant; terminal zeros to the left of a decimal point are significant only if they were measured. **1.44**(a) none (b) none (c) $0.041\underline{0}$ (d) $4.\underline{0}1\underline{00}\times10^4$ **1.46**(a) 1.34 m (b) 72.391 m^3 (c) 443 cm **1.48**(a) 1.310000×10^5 (b) 4.7×10^{-4} (c) 2.10006×10^5 (d) 2.1605×10^3 **1.50**(a) 5550 (b) 10,070 (c) 0.000000885 (d) 0.003004 **1.52**(a) 4.06×10^{-19} J (b) 1.61×10^{24} molecules (c) 1.82×10^5 J/mol **1.54**(a) Height measured, not exact. (b) Planets counted, exact. (c) Number of grams in a pound is not a unit definition, not exact. (d) Definition of "millimeter," exact. **1.56** 7.50 ± 0.05 cm **1.58**(a) $\text{I}_{avg} = 8.72$ g; $\text{II}_{avg} = 8.72$ g; $\text{III}_{avg} = 8.50$ g; $\text{IV}_{avg} = 8.56$ g; sets I and II are most accurate. (b) Set III is the most precise, but is the least accurate. (c) Set I has the best combination of high accuracy and high precision. (d) Set IV has both low accuracy and low precision. **1.61**(a)

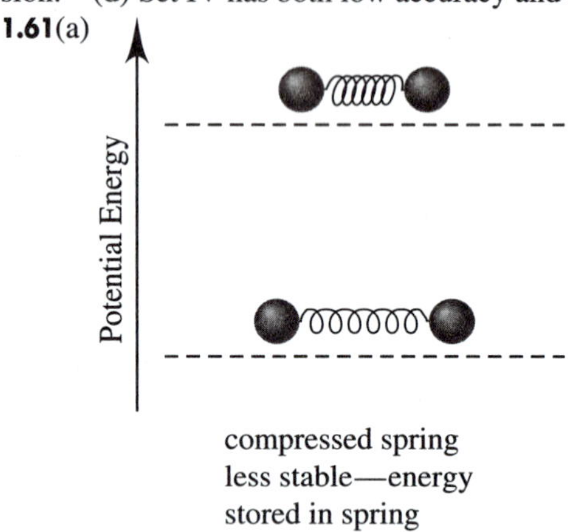

compressed spring
less stable—energy
stored in spring

(b)

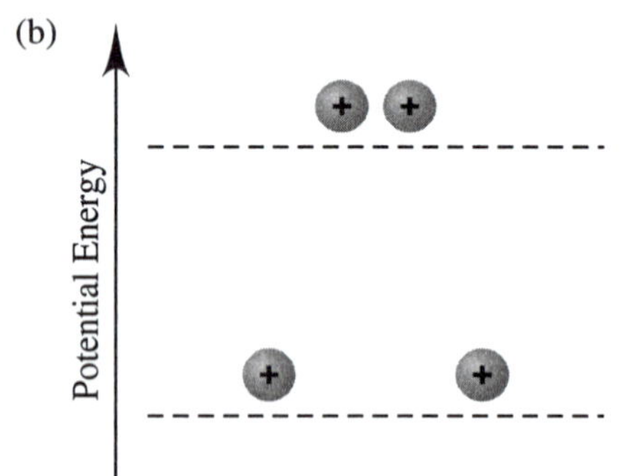

two charges near each other
less stable—repulsion of like
charges

1.65(a) \$19.40 before price increase; \$33.90 after price increase (b) 51.7 coins (c) 21.0 coins **1.68**(a) 0.21 g/L, will float (b) CO_2 is denser than air, will sink (c) 0.30 g/L, will float (d) O_2 is denser than air, will sink (e) 1.38 g/L, will sink (f) 0.50 g **1.71**(a) 8.0×10^{12} g (b) $4.1\times10^5\ \text{m}^3$ (c) 9.5×10^{13} dollars **1.73**(a) −195.79°C (b) −320.42°F (c) 5.05 L **1.76** 7.1×10^7 microparticles in the room; 1.2×10^3 microparticles in a breath **1.78** 2.3×10^{25} g oxygen; 1.4×10^{25} g silicon; 5×10^{15} g ruthenium (and rhodium) **1.80** freezing point = −3.7°X; boiling point = 63.3°X

Chapter 2

2.1 Compounds contain different types of atoms; there is only one type of atom in an element. **2.4**(a) The presence of more than one element makes pure calcium chloride a compound. (b) There is only one kind of atom, so sulfur is an element. (c) The presence of more than one compound makes baking powder a mixture. (d) The presence of more than one type of atom means cytosine cannot be an element. The specific, not variable, arrangement means it is a compound. **2.6**(a) elements, compounds, and mixtures (b) compounds (c) compounds **2.7**(a) Law of definite composition: the composition is the same regardless of its source. (b) Law of mass conservation: the total quantity of matter does not change. (c) Law of multiple proportions: two elements can combine to form two different compounds that have different proportions of those elements. **2.9**(a) No, the percent by mass of each element in a compound is fixed. (b) Yes, the *mass* of each element in a compound depends on the mass of the compound. (c) No, the percent by mass of each element in a compound is fixed. **2.10** The two experiments demonstrate the law of definite composition. The unknown compound decomposes the same way both times. The experiments also demonstrate the law of conservation of mass since the total mass before reaction equals the total mass after reaction. **2.12**(a) 1.34 g F (b) 0.514 Ca; 0.486 F (c) 51.4 mass % Ca; 48.6 mass % F **2.14** 3.498×10^6 g Cu; 1.766×10^6 g S **2.16** compound 1: 0.904 S/Cl; compound 2: 0.451 S/Cl; ratio: 2.00/1.00 **2.19** Coal A **2.20** Dalton postulated that atoms of an element are identical and that compounds result from the chemical combination of specific ratios of different elements. **2.21**(a) If you know the ratio of any two quantities and the value of one of them, the other can always be calculated; in this case, the charge and the charge-to-mass ratio were known. (b) The charge on each oil droplet has a common factor of -1.602×10^{-19} C, the charge of the electron. **2.24** All three isotopes have 18 protons and 18 electrons. Their respective mass numbers are 36, 38, and 40, with the respective numbers of neutrons being 18, 20, and 22. **2.26**(a) These have the same number of protons and electrons, but different numbers of neutrons; same *Z*. (b) These have the same number of neutrons, but different numbers of protons and electrons; same *N*. (c) These have different numbers of protons, neutrons, and electrons; same *A*. **2.28**(a) ${}^{38}_{18}\text{Ar}$ (b) ${}^{55}_{25}\text{Mn}$ (c) ${}^{109}_{47}\text{Ag}$ **2.30**(a) ${}^{48}_{22}\text{Ti}$ (b) ${}^{79}_{34}\text{Se}$ (c) ${}^{11}_{5}\text{B}$

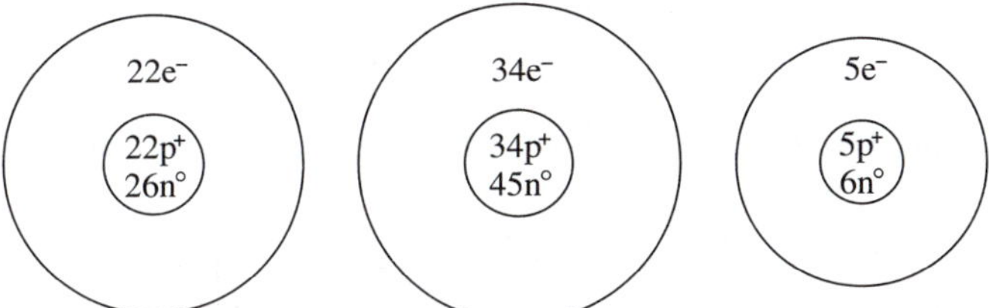

2.32 69.72 amu **2.34** ${}^{35}\text{Cl} = 75.774\%$, ${}^{37}\text{Cl} = 24.226\%$ **2.36**(a) In the modern periodic table, the elements are arranged in order of increasing atomic number. (b) Elements in a group (or family) have similar chemical properties. (c) Elements can be classified as metals, metalloids, or nonmetals. **2.39**(a) germanium; Ge; 4A(14); metalloid (b) phosphorus; P; 5A(15); nonmetal (c) helium; He; 8A(18); nonmetal (d) lithium; Li; 1A(1); metal (e) molybdenum; Mo; 6B(6); metal **2.41**(a) Ra; 88 (b) Si; 14 (c) Cu; 63.55 (d) Br; 79.90 **2.43** Atoms of these two kinds of substances will form ionic bonds, in which one or more electrons are transferred from the metal atom to the nonmetal atom to form a cation and an anion, respectively. **2.45** Coulomb's law states the energy of attraction in an ionic bond is directly proportional to the product of charges and inversely proportional to the distance between charges. The product of charges in MgO ($2+ \times 2-$) is greater than the product of charges in LiF ($1+ \times 1-$). Thus, MgO has stronger ionic bonding. **2.48** K^+; I^- **2.50**(a) oxygen; 17; 6A(16); 2 (b) fluorine; 19; 7A(17); 2 (c) calcium; 40; 2A(2); 4 **2.52** Lithium forms the Li^+ ion; oxygen forms the O^{2-} ion. Number of O^{2-} ions $= 4.2\times10^{21}$ O^{2-} ions. **2.54** NaCl **2.56** An empirical formula shows the simplest ratio of atoms of each element present in a compound, whereas a molecular formula describes the type and actual number of atoms of each element in a molecule of the compound. The empirical formula and the molecular formula can be the same. The molecular formula is always a whole-number multiple of the empirical formula. **2.57** The two samples are similar in that both contain 20 billion oxygen atoms and 20 billion hydrogen atoms. They differ in that they contain different types of molecules: H_2O_2 molecules in the hydrogen peroxide sample, and H_2 and O_2 molecules in the mixture. In addition, the mixture contains 20 billion molecules (10 billion H_2 and 10 billion O_2), while the hydrogen peroxide sample contains 10 billion molecules. **2.58**(a) NH_2 (b) CH_2O **2.60**(a) Na_3N, sodium nitride (b) SrO, strontium oxide (c) $AlCl_3$, aluminum chloride **2.62**(a) MgF_2, magnesium fluoride (b) ZnS, zinc sulfide (c) $SrCl_2$, strontium chloride **2.64**(a) $SnCl_4$ (b) iron(III) bromide (c) CuBr (d) manganese(III) oxide **2.66**(a) BaO (b) $Fe(NO_3)_2$ (c) MgS **2.68**(a) H_2SO_4; sulfuric acid (b) HIO_3; iodic acid (c) HCN; hydrocyanic acid (d) H_2S; hydrosulfuric acid **2.70** Disulfur tetrafluoride, S_2F_4 **2.72**(a) 12 oxygen atoms; 342.2 amu (b) 9 hydrogen atoms; 132.06 amu (c) 8 oxygen atoms; 344.6 amu **2.74**(a) $(NH_4)_2SO_4$; 132.15 amu (b) NaH_2PO_4; 119.98 amu (c) $KHCO_3$; 100.12 amu **2.76** disulfur dichloride; SCl; 135.04 amu **2.78**(a) SO_3, sulfur trioxide, 80.07 amu (b) N_2O, dinitrogen monoxide, 44.02 amu **2.80** Separating the components of a mixture requires physical methods only; that is, no chemical changes (no changes in composition) take place, and the components maintain their chemical identities and properties throughout. Separating the components of a compound requires a chemical change (change in composition). **2.83**(a) compound (b) homogeneous mixture (c) heterogeneous mixture (d) homogeneous mixture (e) homogeneous mixture **2.85**(a) fraction of volume $= 5.2\times10^{-13}$ (b) mass of nucleus $= 6.64466\times10^{-24}$ g; fraction of mass $= 0.999726$ **2.87**(a) I = NO; II = N_2O_3; III = N_2O_5 (b) I has 1.14 g O per 1.00 g N; II, 1.71 g O; III, 2.86 g O **2.89**(a) Formulas and masses in amu: ${}^{15}N_2{}^{18}O$, 48; ${}^{15}N_2{}^{16}O$, 46; ${}^{14}N_2{}^{18}O$, 46; ${}^{14}N_2{}^{16}O$, 44; ${}^{15}N^{14}N^{18}O$, 47; ${}^{15}N^{14}N^{16}O$, 45 (b) ${}^{15}N_2{}^{18}O$, least common; ${}^{14}N_2{}^{16}O$, most common **2.90**(a) Cl^-, 1.898 mass %; Na^+, 1.056 mass %; $SO_4{}^{2-}$, 0.265 mass %; Mg^{2+}, 0.127 mass %; Ca^{2+}, 0.04 mass %; K^+, 0.038 mass %; $HCO_3{}^-$, 0.014 mass % (b) 30.72% (c) Alkaline earth metal ions, total mass % = 0.17%; alkali metal ions, total mass % = 1.094% (d) Anions (2.177 mass %) make up a larger mass fraction than cations (1.26 mass %). **2.93** Molecular, $C_4H_6O_4$; empirical, $C_2H_3O_2$; molecular mass, 118.09 amu; 40.68% by mass C; 5.122% by mass H; 54.20% by mass O **2.94** B, C, D; 44.05 amu **2.96** 58.091 amu **2.98** 0.370 lb C; 0.0222 lb H; 0.423 lb O; 0.185 lb N **2.105**(1) chemical change (2) physical change (3) chemical change (4) chemical change (5) physical change

Chapter 3

3.2(a) 12 mol C atoms (b) 1.445×10^{25} C atoms **3.6**(a) left (b) left (c) left (d) neither **3.7**(a) 121.64 g/mol

(b) 76.02 g/mol (c) 106.44 g/mol (d) 152.00 g/mol **3.9**(a) 134.7 g/mol (b) 175.3 g/mol (c) 342.17 g/mol (d) 125.84 g/mol **3.11**(a) 1.1×10^2 g $KMnO_4$ (b) 0.188 mol O atoms (c) 1.5×10^{20} O atoms **3.13**(a) 9.73 g $MnSO_4$ (b) 44.6 mol $Fe(ClO_4)_3$ (c) 1.74×10^{21} N atoms **3.15**(a) 1.56×10^3 g Cu_2CO_3 (b) 0.0725 g N_2O_5 (c) 0.644 mol $NaClO_4$; 3.88×10^{23} formula units $NaClO_4$ (d) 3.88×10^{23} Na^+ ions; 3.88×10^{23} ClO_4^- ions; 3.88×10^{23} Cl atoms; 1.55×10^{24} O atoms **3.17**(a) 6.375 mass % H (b) 71.52 mass % O **3.19**(a) 0.9507 mol cisplatin (b) 3.5×10^{24} H atoms **3.21** $CO(NH_2)_2 > NH_4NO_3 > (NH_4)_2SO_4 > KNO_3$ **3.22**(a) 883 mol PbS (b) 1.88×10^{25} Pb atoms **3.24**(b) From the mass percent, determine the empirical formula. Add up the total number of atoms in the empirical formula, and divide that number into the total number of atoms in the molecule. The result is the multiplier that makes the empirical formula into the molecular formula. (c) (mass % expressed directly in grams)(1/molar mass) = moles of each element (e) Count the numbers of the various types of atoms in the structural formula and put these into a molecular formula. **3.25**(a) CH_2; 14.03 g/mol (b) CH_3O; 31.03 g/mol (c) N_2O_5; 108.02 g/mol (d) $Ba_3(PO_4)_2$; 601.8 g/mol (e) TeI_4; 635.2 g/mol **3.27**(a) C_3H_6 (b) N_2H_4 (c) N_2O_4 (d) $C_5H_5N_5$ **3.29**(a) Cl_2O_7 (b) $SiCl_4$ (c) CO_2 **3.31**(a) 1.20 mol F (b) 24.0 g M (c) calcium **3.33** $C_{21}H_{30}O_5$ **3.34** $C_{10}H_{20}O$ **3.36** b
3.37(a) $16Cu(s) + S_8(s) \longrightarrow 8Cu_2S(s)$
(b) $P_4O_{10}(s) + 6H_2O(l) \longrightarrow 4H_3PO_4(l)$
(c) $B_2O_3(s) + 6NaOH(aq) \longrightarrow 2Na_3BO_3(aq) + 3H_2O(l)$
(d) $4CH_3NH_2(g) + 9O_2(g) \longrightarrow 4CO_2(g) + 10H_2O(g) + 2N_2(g)$
3.39(a) $4Ga(s) + 3O_2(g) \longrightarrow 2Ga_2O_3(s)$
(b) $2C_6H_{14}(l) + 19O_2(g) \longrightarrow 12CO_2(g) + 14H_2O(g)$
(c) $3CaCl_2(aq) + 2Na_3PO_4(aq) \longrightarrow Ca_3(PO_4)_2(s) + 6NaCl(aq)$
3.42(a) 1.42×10^3 mol KNO_3 (b) 1.43×10^5 g KNO_3 **3.44** 195.8 g H_3BO_3; 19.16 g H_2 **3.46** 2.60×10^3 g Cl_2 **3.48**(a) 0.105 mol CaO (b) 0.175 mol CaO (c) calcium (d) 5.88 g CaO **3.50** 1.36 mol HIO_3, 239 g HIO_3; 44.9 g H_2O in excess **3.52** 4.40 g CO_2; 4.80 g O_2 in excess **3.54** 12.2 g $Al(NO_2)_3$, no NH_4Cl, 48.7 g $AlCl_3$, 30.7 g N_2, 39.5 g H_2O **3.56** 50.% **3.58** 90.5% **3.60** 24.0 g CH_3Cl **3.62** 39.7 g CF_4 **3.67**(a) C (b) B (c) C (d) B **3.68**(a) 7.85 g $Ca(C_2H_3O_2)_2$ (b) 0.254 *M* KI (c) 124 mol NaCN **3.70**(a) 0.0617 *M* KCl (b) 0.00363 *M* $(NH_4)_2SO_4$ (c) 0.138 *M* Na^+ **3.72**(a) 987 g HNO_3/L (b) 15.7 *M* HNO_3 **3.74** 845 mL **3.76** 0.88 g $BaSO_4$ **3.78**(a) Instructions: Be sure to wear goggles to protect your eyes! Pour approximately 2.0 gal of water into the container. Add to the water, slowly and with mixing, 0.90 gal of concentrated HCl. Dilute to 3.0 gal with more water. (b) 22.6 mL **3.79** $\chi = 3$ **3.80** ethane > propane > cetyl palmitate > ethanol > benzene **3.84** 89.8% **3.85**(a) $2AB_2 + B_2 \longrightarrow 2AB_3$ (b) AB_2 (c) 5.0 mol AB_3 (d) 0.5 mol B_2 **3.88**(a) C (b) B (c) D **3.92** 0.071 *M* KBr **3.95** (a) 586 g CO_2 (b) 10.5% CH_4 by mass **3.96** 10/0.66/1.0 **3.99** 32.7 mass % C **3.102**(a) 192.12 g/mol; $C_6H_8O_7$ (b) 0.549 mol **3.104**(a) 0.039 g heme (b) 6.3×10^{-5} mol heme (c) 3.5×10^{-3} g Fe (d) 4.1×10^{-2} g hemin **3.108**(a) 46.65 mass % N in urea; 31.98 mass % N in arginine; 21.04 mass % N in ornithine (b) 28.45 g N **3.109** A

Chapter 4

4.2 Ions must be present and they come from ionic compounds or from electrolytes such as acids and bases. **4.5** B **4.8**(a) Benzene is likely to be insoluble in water because it is nonpolar and water is polar. (b) Sodium hydroxide, an ionic compound, is likely to be very soluble in water. (c) Ethanol (CH_3CH_2OH) is likely to be soluble in water because the alcohol group (—OH) is polar. (d) Potassium acetate, an ionic compound, is likely to be very soluble in water. **4.10**(a) Yes, CsBr is a soluble salt. (b) Yes, HI is a strong acid. **4.12**(a) 3.0 mol (b) 7.57×10^{-5} mol (c) 0.148 mol **4.14**(a) 0.058 mol Al^{3+}; 3.5×10^{22} Al^{3+} ions; 0.18 mol Cl^-; 1.1×10^{23} Cl^- ions (b) 4.62×10^{-4} mol Li^+; 2.78×10^{20} Li^+ ions; 2.31×10^{-4} mol SO_4^{2-}; 1.39×10^{20} SO_4^{2-} ions (c) 1.50×10^{-2} mol K^+; 9.02×10^{21} K^+ ions; 1.50×10^{-2} mol Br^-; 9.02×10^{21} Br^- ions **4.16**(a) 0.35 mol H^+ (b) 6.3×10^{-3} mol H^+ (c) 0.22 mol H^+ **4.23** Assuming that the left beaker contains $AgNO_3$ (because it has gray Ag^+ ion), the right must contain NaCl. Then, NO_3^- is blue, Na^+ is brown, and Cl^- is green.
Molecular equation: $AgNO_3(aq) + NaCl(aq) \longrightarrow AgCl(s) + NaNO_3(aq)$
Total ionic equation: $Ag^+(aq) + NO_3^-(aq) + Na^+(aq) + Cl^-(aq) \longrightarrow AgCl(s) + Na^+(aq) + NO_3^-(aq)$
Net ionic equation: $Ag^+(aq) + Cl^-(aq) \longrightarrow AgCl(s)$
4.24(a) Molecular: $Hg_2(NO_3)_2(aq) + 2KI(aq) \longrightarrow Hg_2I_2(s) + 2KNO_3(aq)$
Total ionic: $Hg_2^{2+}(aq) + 2NO_3^-(aq) + 2K^+(aq) + 2I^-(aq) \longrightarrow Hg_2I_2(s) + 2K^+(aq) + 2NO_3^-(aq)$
Net ionic: $Hg_2^{2+}(aq) + 2I^-(aq) \longrightarrow Hg_2I_2(s)$
Spectator ions are K^+ and NO_3^-.
(b) Molecular: $FeSO_4(aq) + Sr(OH)_2(aq) \longrightarrow Fe(OH)_2(s) + SrSO_4(s)$
Total ionic: $Fe^{2+}(aq) + SO_4^{2-}(aq) + Sr^{2+}(aq) + 2OH^-(aq) \longrightarrow Fe(OH)_2(s) + SrSO_4(s)$
Net ionic: This is the same as the total ionic equation, because there are no spectator ions.
4.26(a) No precipitate will form. (b) A precipitate will form because silver ions, Ag^+, and bromide ions, Br^-, will combine to form a solid salt, silver bromide, AgBr. The ammonium and nitrate ions do not form a precipitate.
Molecular: $NH_4Br(aq) + AgNO_3(aq) \longrightarrow AgBr(s) + NH_4NO_3(aq)$
Total ionic: $NH_4^+(aq) + Br^-(aq) + Ag^+(aq) + NO_3^-(aq) \longrightarrow AgBr(s) + NH_4^+(aq) + NO_3^-(aq)$
Net ionic: $Ag^+(aq) + Br^-(aq) \longrightarrow AgBr(s)$
4.28 0.0354 *M* Pb^{2+} **4.30**(a) $PbSO_4$ (b) $Pb^{2+}(aq) + SO_4^{2-}(aq) \longrightarrow PbSO_4(s)$ (c) 1.5 g $PbSO_4$ **4.32** 2.206% Cl
4.38(a) Molecular equation: $KOH(aq) + HBr(aq) \longrightarrow KBr(aq) + H_2O(l)$
Total ionic equation: $K^+(aq) + OH^-(aq) + H^+(aq) + Br^-(aq) \longrightarrow K^+(aq) + Br^-(aq) + H_2O(l)$
Net ionic equation: $OH^-(aq) + H^+(aq) \longrightarrow H_2O(l)$
The spectator ions are $K^+(aq)$ and $Br^-(aq)$.
(b) Molecular equation: $NH_3(aq) + HCl(aq) \longrightarrow NH_4Cl(aq)$
Total ionic equation: $NH_3(aq) + H^+(aq) + Cl^-(aq) \longrightarrow NH_4^+(aq) + Cl^-(aq)$
NH_3 is a weak base so it is written as an intact molecule.

HCl, a strong acid, is written as dissociated ions. NH_4Cl is a soluble compound, because all ammonium compounds are soluble.
Net ionic equation: $NH_3(aq) + H^+(aq) \longrightarrow NH_4^+(aq)$
Cl^- is the only spectator ion.
4.40 Total ionic equation: $CaCO_3(s) + 2H^+(aq) + 2Cl^-(aq) \longrightarrow Ca^{2+}(aq) + 2Cl^-(aq) + H_2O(l) + CO_2(g)$
Net ionic equation: $CaCO_3(s) + 2H^+(aq) \longrightarrow Ca^{2+}(aq) + H_2O(l) + CO_2(g)$
4.42 0.05839 M CH_3COOH **4.49**(a) S has O.N. = +6 in SO_4^{2-} (i.e., H_2SO_4), and O.N. = +4 in SO_2, so S has been reduced (and I^- oxidized); H_2SO_4 acts as an oxidizing agent. (b) The oxidation numbers remain constant throughout; H_2SO_4 transfers a proton to F^- to produce HF, so it acts as an acid. **4.50**(a) −1 (b) +2 (c) −3 (d) +3 **4.52**(a) −3 (b) +5 (c) +3 **4.54**(a) +6 (b) +3 (c) +7 **4.56**(a) MnO_4^- is the oxidizing agent; $H_2C_2O_4$ is the reducing agent. (b) Cu is the reducing agent; NO_3^- is the oxidizing agent. **4.58**(a) Oxidizing agent is NO_3^-; reducing agent is Sn. (b) Oxidizing agent is MnO_4^-; reducing agent is Cl^-. **4.60** S is in Group 6A(16), so its highest possible O.N. is +6 and its lowest possible O.N. is 6 − 8 = −2. (a) S = −2. The S can only increase its O.N. (oxidize), so S^{2-} can function only as a reducing agent. (b) S = +6. The S can only decrease its O.N. (reduce), so SO_4^{2-} can function only as an oxidizing agent. (c) S = +4. The S can increase or decrease its O.N. Therefore, SO_2 can function as either an oxidizing or reducing agent.
4.66(a) $2Sb(s) + 3Cl_2(g) \longrightarrow 2SbCl_3(s)$; combination
(b) $2AsH_3(g) \longrightarrow 2As(s) + 3H_2(g)$; decomposition
(c) $Zn(s) + Fe(NO_3)_2(aq) \longrightarrow Zn(NO_3)_2(aq) + Fe(s)$; displacement
4.68(a) $N_2(g) + 3H_2(g) \longrightarrow 2NH_3(g)$
(b) $2NaClO_3(s) \xrightarrow{\Delta} 2NaCl(s) + 3O_2(g)$
(c) $Ba(s) + 2H_2O(l) \longrightarrow Ba(OH)_2(aq) + H_2(g)$
4.70(a) $2Cs(s) + I_2(s) \longrightarrow 2CsI(s)$
(b) $2Al(s) + 3MnSO_4(aq) \longrightarrow Al_2(SO_4)_3(aq) + 3Mn(s)$
(c) $2SO_2(g) + O_2(g) \longrightarrow 2SO_3(g)$
(d) $2C_4H_{10}(g) + 13O_2(g) \longrightarrow 8CO_2(g) + 10H_2O(g)$
(e) $2Al(s) + 3Mn^{2+}(aq) \longrightarrow 2Al^{3+}(aq) + 3Mn(s)$
4.72 315 g O_2; 3.95 kg Hg **4.74**(a) O_2 is in excess. (b) 0.117 mol Li_2O (c) 0 g Li, 3.49 g Li_2O, and 4.63 g O_2
4.77 2.79 kg Fe
4.78(a) $Fe(s) + 2H^+(aq) \longrightarrow Fe^{2+}(aq) + H_2(g)$
O.N.: 0 +1 +2 0
(b) 3.1×10^{21} Fe^{2+} ions
4.79 5.11 g C_2H_5OH; 2.49 L CO_2
4.84(a) Step 1: oxidizing agent is O_2; reducing agent is NH_3. Step 2: oxidizing agent is O_2; reducing agent is NO. Step 3: oxidizing agent is NO_2; reducing agent is NO_2. (b) 1.2×10^4 kg NH_3
4.87 627 L air **4.91** 3.0×10^{-3} mol CO_2 (b) 0.11 L CO_2
4.93(a) $C_7H_5O_4Bi$ (b) $C_{21}H_{15}O_{12}Bi_3$
(c) $Bi(OH)_3(s) + 3HC_7H_5O_3(aq) \longrightarrow Bi(C_7H_5O_3)_3(s) + 3H_2O(l)$
(d) 0.490 mg $Bi(OH)_3$
4.95(a) Ethanol: $C_2H_5OH(l) + 3O_2(g) \longrightarrow 2CO_2(g) + 3H_2O(l)$
Gasoline: $2C_8H_{18}(l) + 25O_2(g) \longrightarrow 16CO_2(g) + 18H_2O(g)$
(b) 2.50×10^3 g O_2 (c) 1.75×10^3 L O_2 (d) 8.38×10^3 L air
4.97 yes **4.98**(a) Reaction (2) is a redox process. (b) 2.00×10^5 g Fe_2O_3; 4.06×10^5 g $FeCl_3$ (c) 2.09×10^5 g Fe; 4.75×10^5 g $FeCl_2$ (d) 0.313

Chapter 6

6.5 0 J **6.7**(a) 6.6×10^7 kJ (b) 1.6×10^7 kcal (c) 6.3×10^7 Btu
6.10(a) exothermic (b) endothermic (c) exothermic (d) exothermic (e) endothermic (f) endothermic (g) exothermic
6.11

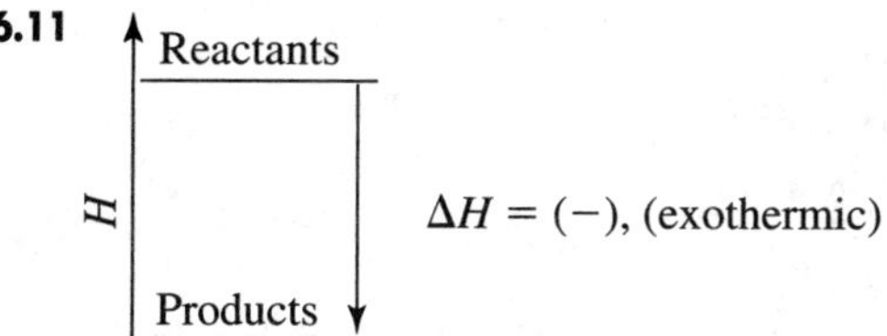

6.13(a) Combustion of ethane: $2C_2H_6(g) + 7O_2(g) \longrightarrow 4CO_2(g) + 6H_2O(g) + \text{heat}$

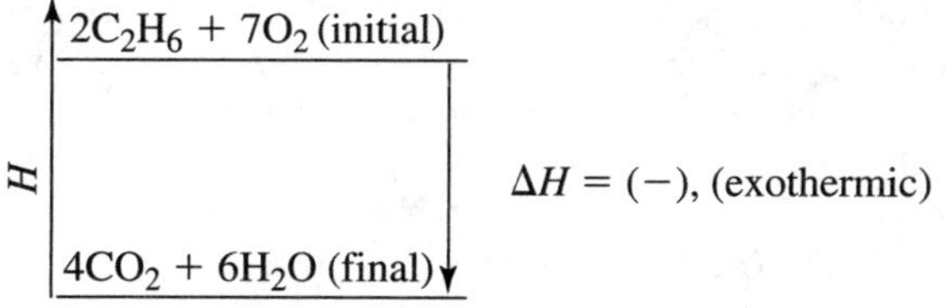

(b) Freezing of water: $H_2O(l) \longrightarrow H_2O(s) + \text{heat}$

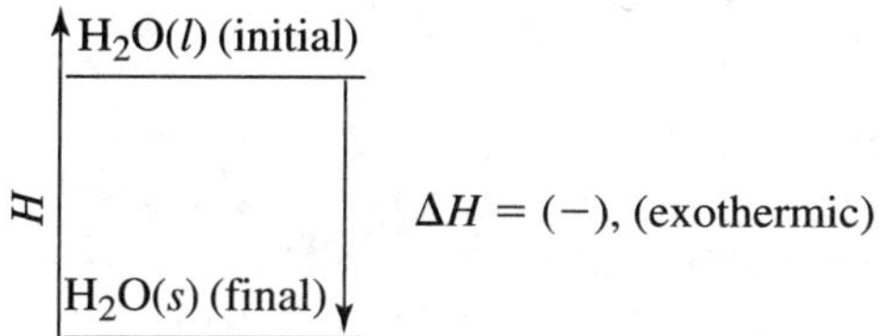

6.15(a) $2CH_3OH(l) + 3O_2(g) \longrightarrow 2CO_2(g) + 4H_2O(g) + \text{heat}$

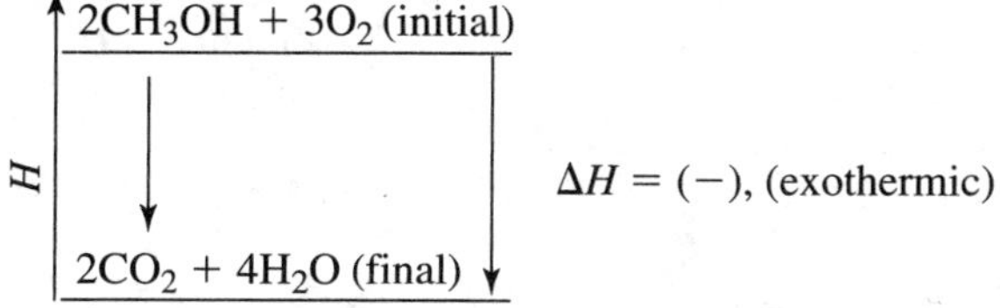

(b) $\frac{1}{2}N_2(g) + O_2(g) + \text{heat} \longrightarrow NO_2(g)$

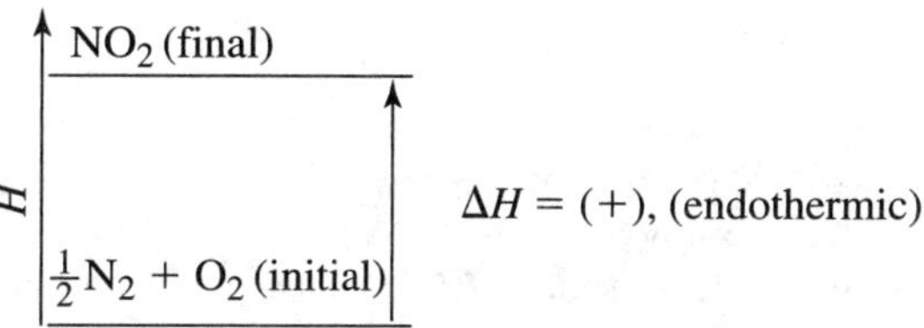

6.17(a) This is a phase change from the solid phase to the gas phase. Heat is absorbed by the system so q_{sys} is positive (+). (b) The volume of the system is expanding as more moles of gas are present after the phase change than were present before the phase change. So the system has done work of expansion, and w is negative. Since $\Delta E_{sys} = q + w$, q is positive, and w is negative, the sign of ΔE_{sys} cannot be predicted. It will be positive if $q > w$ and negative if $q < w$. (c) $\Delta E_{univ} = 0$. If the system loses energy, the surroundings gain an equal amount of energy. The sum of the energy of the system and the energy of the surroundings remains constant. **6.20** To determine the specific heat capacity of a substance, you need its mass, the heat added (or lost), and the change in temperature. **6.22** 6.9×10^3 J **6.24** 295°C **6.26** 77.5°C **6.28** 45°C **6.33** The reaction has a positive ΔH_{rxn},

because this reaction requires the input of energy to break the oxygen-oxygen bond. **6.34** ΔH is negative; it is opposite in sign and half of the value for the vaporization of 2 mol of H_2O.
6.35(a) exothermic (b) 20.2 kJ (c) -4.2×10^2 kJ (d) -15.7 kJ
6.37(a) $\frac{1}{2}N_2(g) + \frac{1}{2}O_2(g) \longrightarrow NO(g)$; $\Delta H = 90.29$ kJ
(b) -10.5 kJ
6.39 -1.88×10^6 kJ
6.41(a) $C_2H_4(g) + 3O_2(g) \longrightarrow 2CO_2(g) + 2H_2O(l)$;
$\Delta H_{rxn} = -1411$ kJ
(b) 1.39 g C_2H_4
6.44 -813.4 kJ
6.46 $N_2(g) + 2O_2(g) \longrightarrow 2NO_2(g)$; $\Delta H_{rxn} = +66.4$ kJ; A = 1, B = 2, C = 3 **6.49** The standard heat of reaction, ΔH°_{rxn}, is the enthalpy change for any reaction where all substances are in their standard states. The standard heat of formation, ΔH°_f, is the enthalpy change that accompanies the formation of one mole of a compound in its standard state from elements in their standard states.
6.50(a) $\frac{1}{2}Cl_2(g) + Na(s) \longrightarrow NaCl(s)$
(b) $H_2(g) + \frac{1}{2}O_2(g) \longrightarrow H_2O(l)$
(c) no changes
6.51(a) $Ca(s) + Cl_2(g) \longrightarrow CaCl_2(s)$
(b) $Na(s) + \frac{1}{2}H_2(g) + C(graphite) + \frac{3}{2}O_2(g) \longrightarrow NaHCO_3(s)$
(c) $C(graphite) + 2Cl_2(g) \longrightarrow CCl_4(l)$
(d) $\frac{1}{2}H_2(g) + \frac{1}{2}N_2(g) + \frac{3}{2}O_2(g) \longrightarrow HNO_3(l)$
6.53(a) -1036.8 kJ (b) -433 kJ **6.55** -157.3 kJ/mol
6.58(a) 503.9 kJ (b) $-\Delta H_1 + 2\Delta H_2 = 504$ kJ
6.59(a) $C_{18}H_{36}O_2(s) + 26O_2(g) \longrightarrow 18CO_2(g) + 18H_2O(g)$
(b) $-10,488$ kJ (c) -36.9 kJ; -8.81 kcal (d) 8.81 kcal/g $\times$ 11.0 g = 96.9 kcal **6.60**(a) 23.6 L/mol initial; 24.9 L/mol final
(b) 187 J (c) -1.2×10^2 J (d) 3.1×10^2 J (e) 310 J
(f) $\Delta H = \Delta E + P\Delta V = \Delta E - w = (q + w) - w = q_P$
6.70 721 kJ **6.78**(a) $\Delta H^\circ_{rxn1} = -657.0$ kJ; $\Delta H^\circ_{rxn2} = 32.9$ kJ
(b) -106.6 kJ **6.81**(a) -6.81×10^3 J (b) 243°C **6.82** -22.2 kJ
6.84(a) -1.25×10^3 kJ (b) 2.24×10^3°C

Chapter 7

7.2(a) x-ray < ultraviolet < visible < infrared < microwave < radio waves (b) radio < microwave < infrared < visible < ultraviolet < x-ray (c) radio < microwave < infrared < visible < ultraviolet < x-ray **7.4** The energy of an atom is not continuous, but quantized. It exists only in certain fixed amounts called *quanta*. **7.7** 316 m; 3.16×10^{11} nm; 3.16×10^{12} Å
7.9 2.4×10^{-23} J **7.11** b < c < a **7.14**(a) 1.24×10^{15} s^{-1}; 8.21×10^{-19} J (b) 1.4×10^{15} s^{-1}; 9.0×10^{-19} J **7.17**(a) absorption (b) emission (c) emission (d) absorption
7.19 434.17 nm **7.21** -2.76×10^5 J/mol **7.23** d < a < c < b
7.25 $n = 4$ **7.29** Macroscopic objects do exhibit a wavelike motion, but the wavelength is too small for humans to perceive.
7.31 7.10×10^{-37} m **7.33** 2.2×10^{-26} m/s **7.35** 3.75×10^{-36} kg
7.39(a) principal determinant of the electron's energy or distance from the nucleus (b) determines the shape of the orbital (c) determines the orientation of the orbital in three-dimensional space **7.40**(a) one (b) five (c) three (d) nine **7.42**(a) m_l: $-2, -1, 0, +1, +2$ (b) m_l: 0 (if $n = 1$, then $l = 0$) (c) m_l: $-3, -2, -1, 0, +1, +2, +3$

7.44

	Sublevel	Allowable m_l	No. o orbita
	(a) d ($l = 2$)	$-2, -1, 0, +1, +2$	5
	(b) p ($l = 1$)	$-1, 0, +1$	3
	(c) f ($l = 3$)	$-3, -2, -1, 0, +1, +2, +3$	7

7.46(a) $n = 5$ and $l = 0$; one orbital (b) $n = 3$ and $l = 1$; thre orbitals (c) $n = 4$ and $l = 3$; seven orbitals **7.48**(a) no; $n =$ $l = 1$, $m_l = -1$; $n = 2$, $l = 0$, $m_l = 0$ (b) allowed (c) allow (d) no; $n = 5$, $l = 3$, $m_l = +3$; $n = 5$, $l = 2$, $m_l = 0$
7.50(a) The attraction of the nucleus for the electrons must be overcome. (b) The electrons in silver are more tightly held by nucleus. (c) silver (d) Once the electron is freed from the ato its energy increases in proportion to the frequency of the light.
7.53 Li^{2+} **7.56**(a) Ba; 462 nm (b) 278 to 292 nm
7.58(a) 2.7×10^2 s (b) 3.6×10^8 m **7.62** 6.4×10^{27} photons
7.64(a) 7.56×10^{-18} J; 2.63×10^{-8} m (b) 5.122×10^{-17} J; 3.881×10^{-9} m (c) 1.2×10^{-18} J; 1.66×10^{-7} m
7.66(a) red; green (b) 5.89 kJ (Sr); 5.83 kJ (Ba)
7.68(a) This is the wavelength of maximum absorbance, so it gives the highest sensitivity. (b) ultraviolet region
(c) 1.93×10^{-2} g vitamin A/g oil **7.72** 1.0×10^{18} photons/s

Chapter 8

8.1 Elements are listed in the periodic table in an ordered, systematic way that correlates with a periodicity of their chemical and physical properties. The theoretical basis for the table in terms of atomic number and electron configuration does not allow for a "new element" between Sn and Sb.
8.3(a) predicted atomic mass = 54.23 amu (b) predicted mel ing point = 6.3°C (c) predicted boiling point = -60.2°C
8.5 The quantum number m_s relates to just the electron; all the others describe the orbital.
8.8 Shielding occurs when inner electrons protect, or shield, outer electrons from the full nuclear attraction. The effective nuclear charge is the nuclear charge an electron actually experiences. As the number of inner electrons increases, the effective nuclear charge decreases.
8.10(a) 6 (b) 10 (c) 2 **8.12**(a) 6 (b) 2 (c) 14
8.15 Electrons will occupy empty orbitals in the same sublevel before filling half-filled orbitals so there is the maximum numb of unpaired electrons with parallel spins. N: $1s^22s^22p^3$.

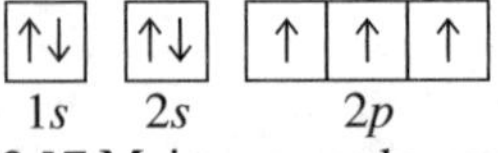

8.17 Main-group elements from the same group have similar outer electron configurations, and the (old) group number equa the number of outer electrons. Outer electron configurations va in a periodic manner within a period, with each succeeding ele ment having an additional electron.
8.18(a) $n = 5$, $l = 0$, $m_l = 0$, and $m_s = +\frac{1}{2}$ (b) $n = 3$, $l = 1$, $m_l = +1$, and $m_s = -\frac{1}{2}$ (c) $n = 5$, $l = 0$, $m_l = 0$, and $m_s = +$ (d) $n = 2$, $l = 1$, $m_l = +1$, and $m_s = -\frac{1}{2}$
8.20(a) Rb: $1s^22s^22p^63s^23p^64s^23d^{10}4p^65s^1$
(b) Ge: $1s^22s^22p^63s^23p^64s^23d^{10}4p^2$
(c) Ar: $1s^22s^22p^63s^23p^6$

8.22(a) Ti: [Ar] $4s^23d^2$

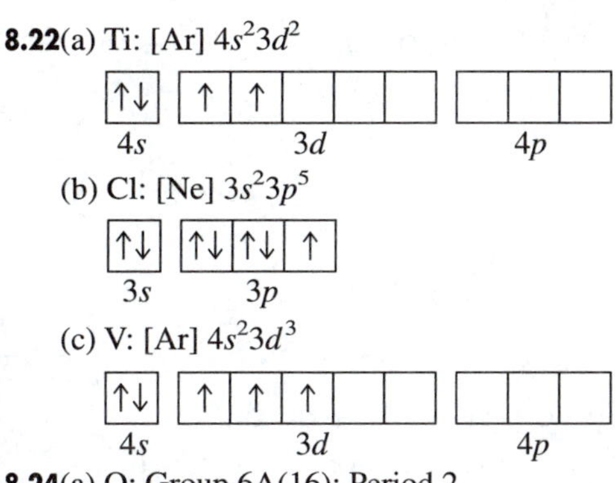

(b) Cl: [Ne] $3s^23p^5$

(c) V: [Ar] $4s^23d^3$

8.24(a) O; Group 6A(16); Period 2

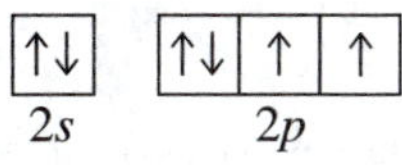

(b) P; Group 5A(15); Period 3

↑↓ | ↑ | ↑ | ↑
3s 3p

8.26(a) [Ar] $4s^23d^{10}4p^1$; Group 3A(13) (b) [He] $2s^22p^6$; Group 8A(18)

8.28

	Inner Electrons	Outer Electrons	Valence Electrons
(a) O	2	6	6
(b) Sn	46	4	4
(c) Ca	18	2	2
(d) Fe	18	2	8
(e) Se	28	6	6

8.30(a) B; Al, Ga, In, and Tl (b) S; O, Se, Te, and Po (c) La; Sc, Y, and Ac **8.33** Atomic size increases down a group. Ionization energy decreases down a group. These trends result because the outer electrons are more easily removed as the atom gets larger. **8.35** For a given element, successive ionization energies always increase. As each successive electron is removed, the positive charge on the ion increases, which results in a stronger attraction between the leaving electron and the ion. When a large jump between successive ionization energies is observed, the subsequent electron must come from a lower energy level. **8.38** A high IE_1 and a very negative EA_1 suggest that the elements are halogens, in Group 7A(17), which form 1− ions. **8.40**(a) K < Rb < Cs (b) O < C < Be (c) Cl < S < K (d) Mg < Ca < K **8.42**(a) Ba < Sr < Ca (b) B < N < Ne (c) Rb < Se < Br (d) Sn < Sb < As **8.44** $1s^22s^22p^1$ (boron, B) **8.46**(a) Na (b) Na (c) Be **8.48** (1) Metals conduct electricity, nonmetals do not. (2) When they form stable ions, metal ions tend to have a positive charge, nonmetal ions tend to have a negative charge. (3) Metal oxides are ionic and act as bases in water; nonmetal oxides are covalent and act as acids in water. **8.49** Metallic character increases down a group and decreases toward the right across a period. These trends are the same as those for atomic size and opposite those for ionization energy. **8.53**(a) Rb (b) Ra (c) I **8.55**(a) Cl^-: $1s^22s^22p^63s^23p^6$ (b) Na^+: $1s^22s^22p^6$ (c) Ca^{2+}: $1s^22s^22p^63s^23p^6$ **8.57**(a) 0 (b) 3 (c) 0 (d) 1 **8.59**(a) V^{3+}, [Ar] $3d^2$, paramagnetic (b) Cd^{2+}, [Kr] $4d^{10}$, diamagnetic (c) Co^{3+}, [Ar] $3d^6$, paramagnetic (d) Ag^+, [Kr] $4d^{10}$, diamagnetic **8.61** For palladium to be diamagnetic, all of its electrons must be paired. (a) You might first write the condensed electron configuration for Pd as [Kr] $5s^24d^8$. However, the partial orbital diagram is not consistent with diamagnetism.

↑↓ | ↑↓ | ↑↓ | ↑↓ | ↑ | ↑ | (5p empty)
5s 4d 5p

(b) This is the only configuration that supports diamagnetism, [Kr] $4d^{10}$.

(5s empty) | ↑↓ | ↑↓ | ↑↓ | ↑↓ | ↑↓ | (5p empty)
5s 4d 5p

(c) Promoting an s electron into the d sublevel still leaves two electrons unpaired.

↑ | ↑↓ | ↑↓ | ↑↓ | ↑↓ | ↑ | (5p empty)
5s 4d 5p

8.63(a) $Li^+ < Na^+ < K^+$ (b) $Rb^+ < Br^- < Se^{2-}$ (c) $F^- < O^{2-} < N^{3-}$ **8.67**(a) Cl_2O, dichlorine monoxide (b) Cl_2O_3, dichlorine trioxide (c) Cl_2O_5, dichlorine pentaoxide (d) Cl_2O_7, dichlorine heptaoxide (e) SO_3, sulfur trioxide (f) SO_2, sulfur dioxide (g) N_2O_5, dinitrogen pentaoxide (h) N_2O_3, dinitrogen trioxide (i) CO_2, carbon dioxide (j) P_2O_5, diphosphorus pentaoxide **8.69**(a) $SrBr_2$, strontium bromide (b) CaS, calcium sulfide (c) ZnF_2, zinc fluoride (d) LiF, lithium fluoride **8.70** All ions except Fe^{8+} and Fe^{14+} are paramagnetic; Fe^+ and Fe^{3+} would be most attracted.

Chapter 9

9.1(a) Greater ionization energy decreases metallic character. (b) Larger atomic radius increases metallic character. (c) Higher number of outer electrons decreases metallic character. (d) Larger effective nuclear charge decreases metallic character. **9.4**(a) Cs (b) Rb (c) As **9.6**(a) ionic (b) covalent (c) covalent **9.8**(a) Rb· (b) ·Ṣi· (c) :Ï· **9.10**(a) 6A(16); [noble gas] ns^2np^4 (b) 3(A)13; [noble gas] ns^2np^1 **9.13**(a) Because the lattice energy is the result of electrostatic attractions between oppositely charged ions, its magnitude depends on several factors, including ionic size and ionic charge. For a particular arrangement of ions, the lattice energy increases as the charge on the ions increases and as their radii decrease. (b) A < B < C **9.16**(a) Ba^{2+}, [Xe]; Cl^-, [Ne] $3s^23p^6$, :C̤̈l:$^-$; $BaCl_2$ (b) Sr^{2+}, [Kr]; O^{2-}, [He] $2s^22p^6$, :Ö:$^{2-}$; SrO (c) Al^{3+}, [Ne]; F^-, [He] $2s^22p^6$, :F̤̈:$^-$; AlF_3 (d) Rb^+, [Kr]; O^{2-}, [He] $2s^22p^6$, :Ö:$^{2-}$; Rb_2O. **9.18**(a) 3A(13) (b) 2A(2) (c) 6A(16) **9.20**(a) BaS; Ba^{2+} is larger than Ca^{2+}. (b) NaF; the charge on each ion is less than the charge on Mg and O. **9.23** When two chlorine atoms are far apart, there is no interaction between them. As the atoms move closer together, the nucleus of each atom attracts the electrons of the other atom. The closer the atoms, the greater this attraction; however, the repulsions of the two nuclei and two electrons also increase at the same time. The final internuclear distance is the distance at which maximum attraction is achieved in spite of the repulsion. **9.24** The bond energy is the energy required to break the bond between H atoms and Cl atoms in one mole of HCl molecules in the gaseous state. Energy is needed to break bonds, so bond energy is always endothermic and $\Delta H^\circ_{\text{bond breaking}}$ is positive. The amount of energy needed to break the bond is released upon its formation, so $\Delta H^\circ_{\text{bond forming}}$ has the same magnitude as $\Delta H^\circ_{\text{bond breaking}}$ but is opposite in sign (always exothermic and negative). **9.28**(a) I—I < Br—Br < Cl—Cl (b) S—Br < S—Cl < S—H (c) C—N < C=N < C≡N **9.30**(a) C—O < C=O; the C=O bond (bond order = 2) is stronger than the C—O bond (bond order = 1). (b) C—H < O—H; O is smaller than C so the O—H bond is shorter and stronger than the C—H bond. **9.33** Less energy is required to break weak bonds. **9.35** Both are one-carbon molecules. Since methane contains fewer C=O bonds, it will have the greater heat of combustion per mole. **9.36** −168 kJ **9.38** −22 kJ **9.39** −59 kJ **9.40** Electronegativity increases from left to right and increases from bottom to top within a group. Fluorine and oxygen are the two most electronegative elements. Cesium and francium are the two least electronegative elements. **9.42** The H—O bond in water is polar covalent. A nonpolar covalent bond occurs between two atoms with identical electronegativities. A polar covalent bond occurs when the atoms have differing electronegativities. Ionic bonds result from electron transfer between atoms. **9.45**(a) Si < S < O (b) Mg < As < P

9.47(a) N—B (b) N—O (c) C—S (none) (d) S—O (e) N—H (f) Cl—O

9.49 a, d, and e **9.51**(a) nonpolar covalent (b) ionic (c) polar covalent (d) polar covalent (e) nonpolar covalent (f) polar covalent; $SCl_2 < SF_2 < PF_3$

9.53(a) H—I < H—Br < H—Cl

(b) H—C < H—O < H—F

(c) S—Cl < P—Cl < Si—Cl

9.57(a) 800. kJ/mol, which is lower than the value in Table 9.2 (b) -2.417×10^4 kJ (c) 1690. g CO_2 (d) 65.2 L O_2 **9.58**(a) −125 kJ (b) yes, since ΔH_f° is negative (c) −392 kJ (d) No, ΔH_f° for $MgCl_2$ is much more negative than that for MgCl. **9.59**(a) 406 nm (b) 2.93×10^{-19} J (c) 1.87×10^4 m/s **9.62** C—Cl: 3.53×10^{-7} m; bond in O_2: 2.40×10^{-7} m **9.63** XeF_2: 132 kJ/mol; XeF_4: 150. kJ/mol; XeF_6: 146 kJ/mol **9.65**(a) The presence of the very electronegative fluorine atoms bonded to one of the carbons makes the C—C bond polar. This polar bond will tend to undergo heterolytic rather than homolytic cleavage. More energy is required to achieve heterolytic cleavage. (b) 1420 kJ **9.68** 8.70×10^{14} s^{-1}; 3.45×10^{-7} m, which is in the ultraviolet region of the electromagnetic spectrum. **9.70**(a) $CH_3OCH_3(g)$: −326 kJ; $CH_3CH_2OH(g)$: −369 kJ (b) The formation of gaseous ethanol is more exothermic. (c) 43 kJ

Chapter 10

10.1 He and H cannot serve as central atoms in a Lewis structure. Both can have no more than two valence electrons. Fluorine needs only one electron to complete its valence level, and it does not have *d* orbitals available to expand its valence level. Thus, it can bond to only one other atom. **10.3** All the structures obey the octet rule except c and g.

10.5(a) :F̤: / :F̤—Si—F̤: / :F̤: (b) :C̤l—S̤e—C̤l: (c) :F̤—C(=O:)—F̤:

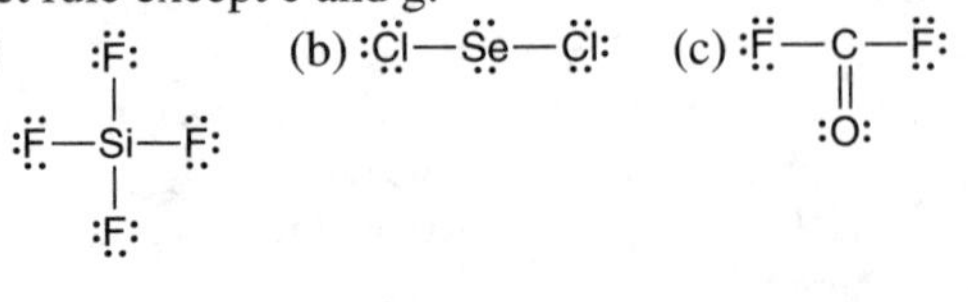

10.7(a) :F̤—P̤(—:F̤:)—F̤: (b) H—Ö—C(=:O:)—Ö—H (c) :S̈=C=S̈:

10.9(a) [Ö=N=Ö]⁺

(b) :F̤:—N(—Ö:)=Ö ⟷ :F̤:—N(=Ö)—Ö:

10.11(a) [:N̈=N=N̈:]⁻ ⟷ [:N̤—N≡N:]⁻ ⟷ [:N≡N—N̤:]⁻

(b) [:Ö=N̤—Ö:]⁻ ⟷ [:Ö—N̤=Ö:]⁻

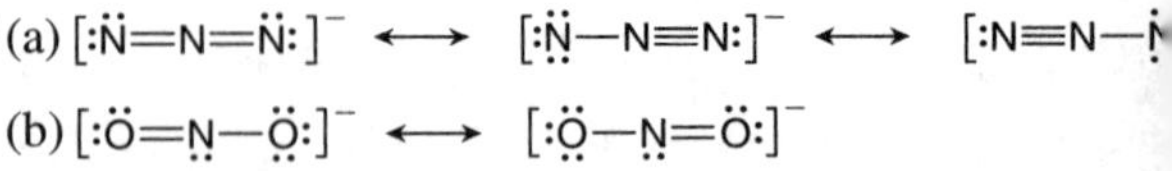

10.13(a) IF_5 (I with five F atoms and one lone pair) formal charges: I = 0, F = 0

(b) [Al with four H]⁻ formal charges: H = 0, Al = −1

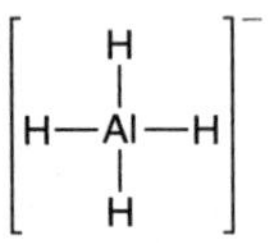

10.15(a) [Ö=B̈r(—:Ö:)=Ö]⁻ formal charges: Br = 0, doubly bonded O = 0, singly bonded O = −1 O.N.: Br = +5; O = −2

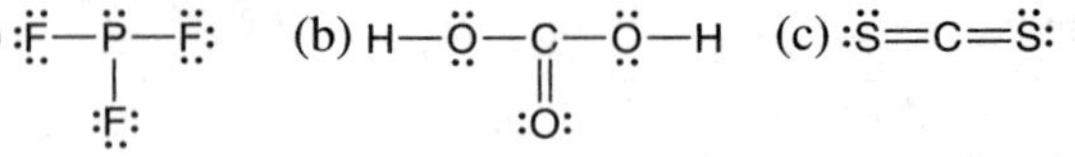

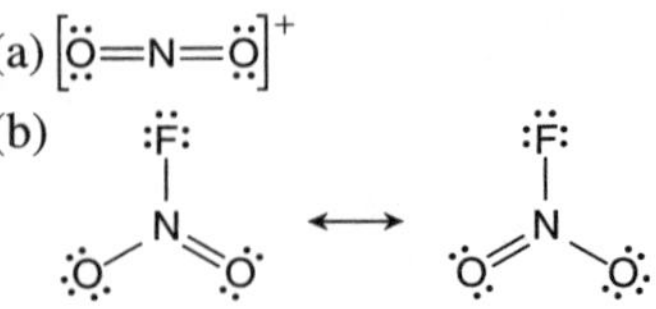

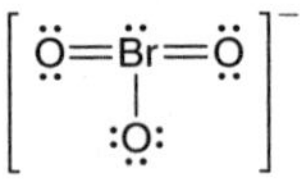

(b) [:Ö—S̈(=:O:)—Ö:]²⁻ formal charges: S = 0, singly bonded O = −1, doubly bonded O = 0 O.N.: S = +4; O = −2

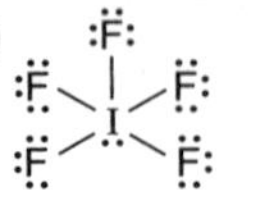

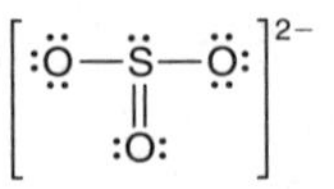

10.17(a) BH_3 has 6 valence electrons. (b) As has an expanded valence level with 10 electrons. (c) Se has an expanded valence level with 10 electrons.

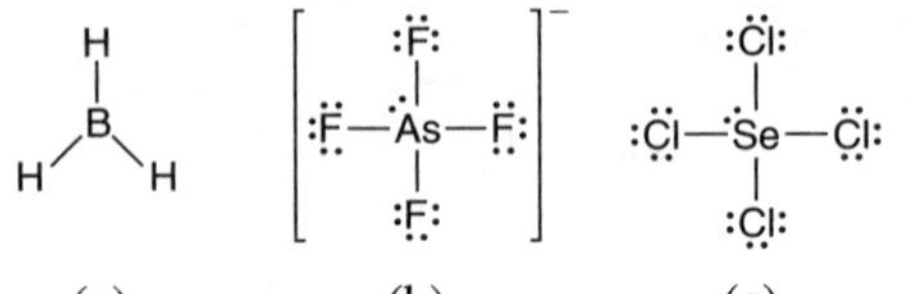

(a) H—B(—H)—H (b) [:F̤—A̤s(—:F̤:)(—:F̤:)—F̤:]⁻ (c) :C̤l—S̤e(—:C̤l:)(—:C̤l:)—C̤l:

10.19(a) Br expands its valence level to 10 electrons. (b) I has an expanded valence level of 10 electrons. (c) Be has only 4 valence shell electrons.

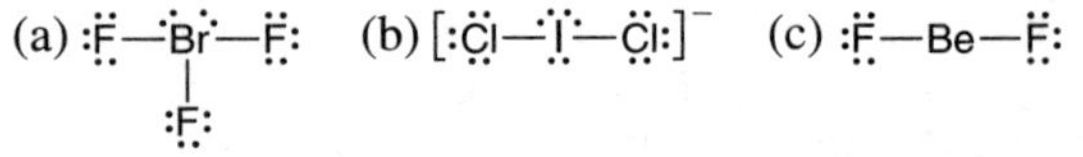

(a) :F̤—B̤r(—:F̤:)—F̤: (b) [:C̤l—I̤—C̤l:]⁻ (c) :F̤—Be—F̤:

10.21

:C̤l—Be—C̤l: + [:C̤l:]⁻ + [:C̤l:]⁻ ⟶ [:C̤l—Be(—:C̤l:)(—:C̤l:)—C̤l:]²⁻

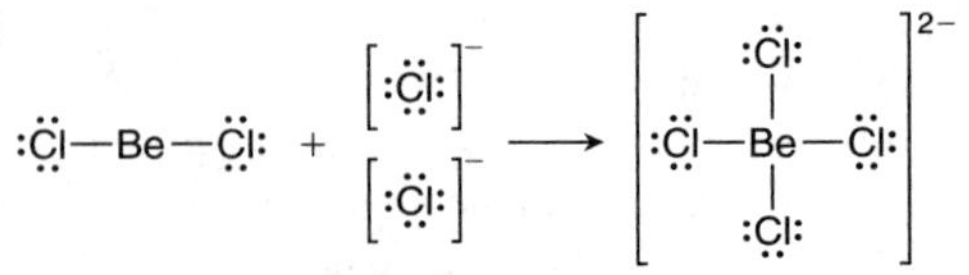

10.24 structure A **10.26** The molecular shape and the electron-group arrangement are the same when no lone pairs are present on the central atom. **10.28** tetrahedral, AX_4; trigonal pyramidal, AX_3E; bent or V shaped, AX_2E_2

10.31(a) X—A—X (bent) (b) A with three X (trigonal planar) (c) A with four X (tetrahedral) (d) X—A with X above and below (T shaped)

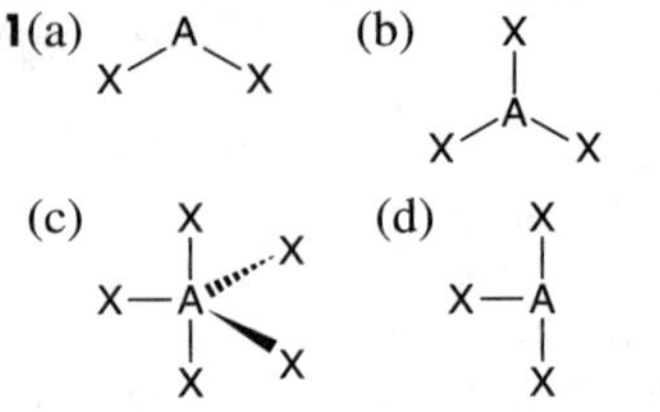

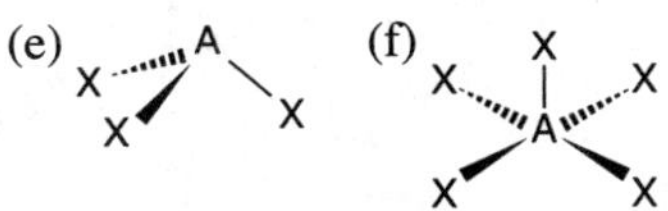

0.33(a) trigonal planar, bent, 120° (b) tetrahedral, trigonal yramidal, 109.5° (c) tetrahedral, trigonal pyramidal, 109.5° **0.35**(a) trigonal planar, trigonal planar, 120° (b) trigonal pla-ar, bent, 120° (c) tetrahedral, tetrahedral, 109.5° **0.37**(a) trigonal planar, AX_3, 120° (b) trigonal pyramidal, X_3E, 109.5° (c) trigonal bipyramidal, AX_5, 90° and 120° **0.39**(a) bent, 109.5°, less than 109.5° (b) trigonal bipyramidal, 0° and 120°, angles are ideal (c) see-saw, 90° and 120°, less han ideal (d) linear, 180°, angle is ideal **10.41**(a) C: tetrahe-ral, 109.5°; O: bent, $< 109.5°$ (b) N: trigonal planar, 120° **0.43**(a) C in CH_3: tetrahedral, 109.5°; C with C=O: trigonal lanar, 120°; O with H: bent, $< 109.5°$ (b) O: bent, $< 109.5°$ **0.45** $OF_2 < NF_3 < CF_4 < BF_3 < BeF_2$ **0.47**(a) The C and N each have three groups so the ideal angles re 120°, and the O has four groups so the ideal angle is 109.5°. he N and O have lone pairs so the angles are less than ideal. b) All central atoms have four pairs, so the ideal angles are 09.5°. The lone pairs on the O reduce this value. (c) The B has hree groups and an ideal bond angle of 120°. All the O's have our groups (ideal bond angles of 109.5°), two of which are lone airs that reduce the angle.

0.50 [Lewis structures of PCl_5, $[PCl_4]^+$, and $[PCl_6]^-$]

n the gas phase, PCl_5 is AX_5, so the shape is trigonal bipyrami-al, and the bond angles are 120° and 90°. The PCl_4^+ ion is AX o the shape is tetrahedral, and the bond angles are 109.5°. The Cl_6^- ion is AX_6, so the shape is octahedral, and the bond ngles are 90°. **10.52**(a) CF_4 (b) BrCl and SCl_2 **0.54**(a) SO_2, because it is polar and SO_3 is not. (b) IF has a reater electronegativity difference between its atoms. (c) SF ecause it is polar and SiF_4 is not. (d) H_2O has a greater elec-ronegativity difference between its atoms.

0.56 [Structures of ClHC=CHCl isomers X, Y, and Z]

X Y Z

Compound Y has a dipole moment

0.57 H—N̈—N̈—H (each N with one H below) H—N̈=N̈—H :N≡N:

Hydrazine Diazene Nitrogen

a) The single N—N bond (bond order = 1) is weaker and onger than the others. The triple bond (bond order = 3) is tronger and shorter than the others. The double bond (bond rder = 2) has an intermediate strength and length.
b) $\Delta H^\circ_{rxn} = -367$ kJ

H—N̈—N̈=N̈—N̈—H (H on first and last N) ⟶ H—N̈—N̈—H (H on each N) + :N≡N:

10.58(a) formal charges: Al = −1, end Cl = 0, bridging Cl = +1; I = −1, end Cl = 0, bridging Cl = +1 (b) The iodine atoms are each AX_4E_2 and the shape around each is square planar. Placing these square planar portions adjacent gives a planar molecule. **10.68**(a) −1267 kJ/mol (b) −1226 kJ/mol (c) −1234.8 kJ/mol. The two answers differ by less than 10 kJ/mol. This is very good agreement since average bond energies were used to calculate answers a and b. (d) −37 kJ

10.70 H—Ö—C—C—Ö—H (each C double-bonded to :O: below)

10.72 (a) The OH species only has 7 valence electrons, which is less than an octet, and 1 electron is unpaired. (b) 426 kJ (c) 508 kJ **10.74**(a) The F atoms will substitute at the axial positions first. (b) PF_5 and PCl_3F_2 **10.77** 22 kJ **10.79** Trigonal planar molecules are nonpolar, so AY_3 cannot be that shape. Trigonal pyramidal molecules and T-shaped molecules are polar, so either could represent AY_3.

Numbers in parentheses refer to the page(s) on which a term is introduced and/or discussed.

A

absolute scale (also *Kelvin scale*) The preferred temperature scale in scientific work, which has absolute zero (0 K, or −273.15°C) as the lowest temperature. (18) [See also *kelvin* (*K*).]
absorption spectrum The spectrum produced when atoms absorb specific wavelengths of incoming light and become excited from lower to higher energy levels. (227)
accuracy The closeness of a measurement to the actual value. (24)
acid In common laboratory terms, any species that produces H^+ ions when dissolved in water. (123) (See also *Arrhenius, Brønsted-Lowry,* and *Lewis acid-base definitions.*)
acid anhydride A compound, sometimes formed by a dehydration-condensation reaction of an oxoacid, that yields two molecules of the acid when it reacts with water. (489)
acid-base buffer (also *buffer*) A solution that resists changes in pH when a small amount of either strong acid or strong base is added. (632)
acid-base indicator A species whose color is different in acid and in base, which is used to monitor the equivalence point of a titration or the pH of a solution. (600)
acid-base reaction Any reaction between an acid and a base. (123) (See also *neutralization reaction.*)
acid-base titration curve A plot of the pH of a solution of acid (or base) versus the volume of base (or acid) added to the solution. (641)
acid-dissociation (acid-ionization) constant (K_a) An equilibrium constant for the dissociation of an acid (HA) in H_2O to yield the conjugate base (A^-) and H_3O^+:

$$K_a = \frac{[H_3O^+][A^-]}{[HA]} \qquad (594)$$

actinides The Period 7 elements that constitute the second inner transition series (5*f* block), which includes thorium (Th; $Z = 90$) through lawrencium (Lr; $Z = 103$). (258)
activated complex (See *transition state.*)
activation energy (E_a) The minimum energy with which molecules must collide to react. (527)
active site The region of an enzyme formed by specific amino acid side chains at which catalysis occurs. (542)
activity ($\mathcal{A}$) (also *decay rate*) The change in number of nuclei ($\mathcal{N}$) of a radioactive sample divided by the change in time (t). (793)
activity series of the metals A listing of metals arranged in order of their decreasing strength as reducing agents in aqueous reactions. (136)
actual yield The amount of product actually obtained in a chemical reaction. (97)
addition polymer (also *chain reaction,* or *chain-growth, polymer*) A polymer formed when monomers (usually containing C=C) combine through an addition reaction. (492)
addition reaction A type of organic reaction in which atoms linked by a multiple bond become bonded to more atoms. (481)
adduct The product of a Lewis acid-base reaction characterized by the formation of a new covalent bond. (621)
adenosine triphosphate (ATP) A high-energy molecule that serves most commonly as a store and source of energy in organisms. (693)
alcohol An organic compound (ending, *-ol*) that contains a $-\overset{|}{\underset{|}{C}}-\ddot{\underset{..}{O}}-H$ functional group. (484)
aldehyde An organic compound (ending, *-al*) that contains the carbonyl functional group ($C=\ddot{\underset{..}{O}}$) in which the carbonyl C is also bonded to H. (487)
alkane A hydrocarbon that contains only single bonds (general formula, C_nH_{2n+2}). (472)
alkene A hydrocarbon that contains at least one C=C bond (general formula, C_nH_{2n}). (477)
alkyl group A saturated hydrocarbon chain with one bond available. (481)
alkyl halide (See *haloalkane.*)
alkyne A hydrocarbon that contains at least one C≡C bond (general formula, C_nH_{2n-2}). (478)
allotrope One of two or more crystalline or molecular forms of an element. In general, one allotrope is more stable than another at a particular pressure and temperature. (444)
alloy A mixture with metallic properties that consists of solid phases of two or more pure elements, a solid-solid solution, or distinct intermediate phases. (404)
alpha (α) decay A radioactive process in which an alpha particle is emitted from a nucleus. (788)
alpha particle (α or $^4_2He^{2+}$) A positively charged particle, identical to a helium −4 nucleus, that is one of the common types of radioactive emissions. (786)
amide An organic compound that contains the $-\overset{:O:}{\overset{\|}{C}}-\overset{|}{\underset{..}{N}}-$ functional group. (488)
amine An organic compound (general formula, $-\overset{|}{\underset{|}{C}}-\overset{..}{\underset{|}{N}}-$) derived structurally by replacing one or more H atoms of ammonia with alkyl groups; a weak organic base. (485)
amino acid An organic compound [general formula, $H_2N-CH(R)-COOH$] with at least one carboxyl and one amine group on the same molecule; the monomer unit of a protein. (496)
amorphous solid A solid that occurs in different shapes because it lacks extensive molecular-level ordering of its particles. (379)
ampere (A) The SI unit of electric current; 1 ampere of current results when 1 coulomb flows through a conductor in 1 second. (747)
amphoteric Able to act as either an acid or a base. (268)
amplitude The height of the crest (or depth of the trough) of a wave; related to the intensity of the energy. (216)

angular momentum quantum number (l) (or *orbital-shape quantum number*) An integer from 0 to $n - 1$ that is related to the shape of an atomic orbital. (234)
anion A negatively charged ion. (49)
anode The electrode at which oxidation occurs in an electrochemical cell. Electrons are given up by the reducing agent and leave the cell at the anode. (709)
antibonding MO A molecular orbital formed when wave functions are subtracted from each other, which decreases electron density between the nuclei and leaves a node. Electrons occupying such an orbital destabilize the molecule. (344)
aqueous solution A solution in which water is the solvent. (61)
aromatic hydrocarbon A compound of C and H with one or more rings of C atoms (often drawn with alternating C—C and C═C bonds), in which there is extensive delocalization of π electrons. (480)
Arrhenius acid-base definition A model of acid-base behavior in which an acid is a substance that has H in its formula and produces H^+ in water, and a base is a substance that has OH in its formula and produces OH^- in water. (592)
Arrhenius equation An equation that expresses the exponential relationship between temperature and the rate constant: $k = Ae^{-E_a/RT}$. (527)
atmosphere (See *standard atmosphere.*)
atom The smallest particle of an element that retains the chemical nature of the element. A neutral, spherical entity composed of a positively charged central nucleus surrounded by one or more negatively charged electrons. (37)
atomic mass (also *atomic weight*) The average of the masses of the naturally occurring isotopes of an element weighted according to their abundances. (45)
atomic mass unit (amu) [also *dalton* (*Da*)] A mass exactly equal to $\frac{1}{12}$ the mass of a carbon-12 atom. (44)
atomic number (Z) The unique number of protons in the nucleus of each atom of an element (equal to the number of electrons in the neutral atom). An integer that expresses the positive charge of a nucleus in multiples of the electronic charge. (43)
atomic orbital (also *wave function*) A mathematical expression that describes the motion of the electron's matter-wave in terms of time and position in the region of the nucleus. The term is used qualitatively to mean the region of space in which there is a high probability of finding the electron. (232)
atomic size A term referring to the atomic radius, one-half the distance between nuclei of identical bonded elements. (259) (See also *covalent radius* and *metallic radius.*)
atomic solid A solid consisting of individual atoms held together by dispersion forces; the frozen noble gases are the only examples. (386)
atomic symbol (or *element symbol*) A one- or two-letter abbreviation for the English, Latin, or Greek name of an element. (43)
atomic weight (See *atomic mass.*)
ATP (See *adenosine triphosphate.*)
aufbau principle (or *building-up principle*) The conceptual basis of a process of building up atoms by adding one proton (and one or more neutrons) at a time to the nucleus and one electron around it to obtain the ground-state electron configurations of the elements. (250)
autoionization (also *self-ionization*) A reaction in which two molecules of a substance react to give ions. The most important example is for water:

$$2H_2O(l) \rightleftharpoons H_3O^+(aq) + OH^-(aq) \qquad (586)$$

average rate The change in concentration of reactants (or products) divided by a finite time period. (511)
Avogadro's law The gas law stating that, at fixed temperature and pressure, equal volumes of any ideal gas contain equal numbers of particles, and, therefore, the volume of a gas is directly proportional to its amount (mol): $V \propto n$. (154)
Avogadro's number A number (6.022×10^{23} to four significant figures) equal to the number of atoms in exactly 12 g of carbon-12; the number of atoms, molecules, or formula units in one mole of an element or compound. (72)
axial group A group (or atom) that lies above or below the trigonal plane of a trigonal bipyramidal molecule, or a similar structural feature in a molecule. (320)

B

background radiation Natural ionizing radiation, the most important form of which is cosmic radiation. (800)
balancing coefficient (also *stoichiometric coefficient*) A numerical multiplier of all the atoms in the formula immediately following it in a chemical equation. (85)
band of stability The narrow band of stable nuclides that appears on a plot of number of neutrons vs. number of protons for all nuclides. (790)
band theory An extension of molecular orbital (MO) theory that explains many properties of metals, in particular, the differences in electrical conductivity of conductors, semiconductors, and insulators. (389)
barometer A device used to measure atmospheric pressure. Most commonly, a tube open at one end, which is filled with mercury and inverted into a dish of mercury. (148)
base In common laboratory terms, any species that produces OH^- ions when dissolved in water. (123) (See also *Brønsted-Lowry, Arrhenius,* and *Lewis acid-base definitions.*)
base-dissociation (base-ionization) constant (K_b) An equilibrium constant for the reaction of a base (B) with H_2O to yield the conjugate acid (BH^+) and OH^-:

$$K_b = \frac{[BH^+][OH^-]}{[B]} \qquad (610)$$

base pair Two complementary bases in mononucleotides that are H bonded to each other; guanine (G) always pairs with cytosine (C), and adenine (A) always pairs with thymine (T) (or uracil, U). (500)
base unit (also *fundamental unit*) A unit that defines the standard for one of the seven physical quantities in the International System of Units (SI). (14)
battery A self-contained group of voltaic cells arranged in series. (732)
becquerel (Bq) The SI unit of radioactivity; 1 Bq = 1 d/s (disintegration per second). (793)
bent shape (also *V shape*) A molecular shape that arises when a central atom is bonded to two other atoms and has one or two lone pairs; occurs as the AX_2E shape class (bond angle $< 120°$) in the trigonal planar arrangement and as the AX_2E_2 shape class (bond angle $< 109.5°$) in the tetrahedral arrangement. (318)
β^- decay A radioactive process in which a beta particle is emitted from a nucleus. (788)
beta (β) decay A class of radioactive decay that includes β^- decay, β^+ emission, and e^- capture. (788)
beta particle (β, β^-, or ${}^{0}_{-1}\beta$) A negatively charged particle identified as a high-speed electron that is one of the common types of radioactive emissions. (786)

bimolecular reaction An elementary reaction involving the collision of two reactant species. (535)
binary covalent compound A compound that consists of atoms of two elements in which bonding occurs primarily through electron sharing. (57)
binary ionic compound A compound that consists of the oppositely charged ions of two elements. (49)
body-centered cubic unit cell A unit cell in which a particle lies at each corner and in the center of a cube. (380)
boiling point (bp or T_b) The temperature at which the vapor pressure of a gas equals the external (atmospheric) pressure. (365)
boiling point elevation (ΔT_b) The increase in the boiling point of a solvent caused by the presence of dissolved solute. (418)
bond angle The angle formed by the nuclei of two surrounding atoms with the nucleus of the central atom at the vertex. (316)
bond energy (BE) (or *bond strength*) The enthalpy change accompanying the breakage of a given bond in a mole of gaseous molecules. (289)
bond length The distance between the nuclei of two bonded atoms. (289)
bond order The number of electron pairs shared by two bonded atoms. (288)
bonding MO A molecular orbital formed when wave functions are added to each other, which increases electron density between the nuclei. Electrons occupying such an orbital stabilize the molecule. (344)
bonding pair (also *shared pair*) An electron pair shared by two nuclei; the mutual attraction between the nuclei and the electron pair forms a covalent bond. (288)
Boyle's law The gas law stating that, at constant temperature and amount of gas, the volume occupied by a gas is inversely proportional to the applied (external) pressure: $V \propto 1/P$. (150)
Brønsted-Lowry acid-base definition A model of acid-base behavior based on proton transfer, in which an acid and a base are defined, respectively, as species that donate and accept a proton. (600)
buffer (See *acid-base buffer.*)
buffer capacity A measure of the ability of a buffer to resist a change in pH; related to the total concentrations and relative proportions of buffer components. (637)
buffer range The pH range over which a buffer acts effectively; related to the relative component concentrations. (638)

C

calibration The process of correcting for systematic error of a measuring device by comparing it to a known standard. (24)
calorie (cal) A unit of energy defined as exactly 4.184 joules; originally defined as the heat needed to raise the temperature of 1 g of water 1°C (from 14.5°C to 15.5°C). (190)
calorimeter A device used to measure the heat released or absorbed by a physical or chemical process taking place within it. (196)
capillarity (or *capillary action*) A property that results in a liquid rising through a narrow space against the pull of gravity. (376)
carbonyl group The C=O grouping of atoms. (487)
carboxylic acid An organic compound (ending, *-oic acid*) that contains the $-\overset{\overset{\Large :\ddot{O}:}{\|}}{C}-\ddot{O}H$ group. (488)
catalyst A substance that increases the rate of a reaction without being used up in the process. (540)
cathode The electrode at which reduction occurs in an electrochemical cell. Electrons enter the cell and are acquired by the oxidizing agent at the cathode. (709)
cathode ray The ray of light emitted by the cathode (negative electrode) in a gas discharge tube; travels in straight lines, unless deflected by magnetic or electric fields. (39)
cation A positively charged ion. (49)
cell potential (E_{cell}) (also *electromotive force,* or *emf; cell voltage*) The potential difference between the electrodes of an electrochemical cell when no current flows. (715)
Celsius scale (formerly *centigrade scale*) A temperature scale in which the freezing and boiling points of water are defined as 0°C and 100°C, respectively. (18)
chain reaction In nuclear fission, a self-sustaining process in which neutrons released by splitting of one nucleus cause other nuclei to split, which releases more neutrons, and so on. (808)
change in enthalpy (ΔH) The change in internal energy plus the product of the constant pressure and the change in volume: $\Delta H = \Delta E + P\Delta V$; the heat lost or gained at constant pressure: $\Delta H = q_P$. (193)
charge density The ratio of the charge of an ion to its volume. (406)
Charles's law The gas law stating that at constant pressure, the volume occupied by a fixed amount of gas is directly proportional to its absolute temperature: $V \propto T$. (152)
chelate A complex ion in which the metal ion is bonded to a bidentate or polydentate ligand. (765)
chemical bond The force that holds two atoms together in a molecule (or formula unit). (48)
chemical change (also *chemical reaction*) A change in which a substance is converted into a substance with different composition and properties. (3)
chemical equation A statement that uses chemical formulas to express the identities and quantities of the substances involved in a chemical or physical change. (85)
chemical formula A notation of atomic symbols and numerical subscripts that shows the type and number of each atom in a molecule or formula unit of a substance. (52)
chemical kinetics The study of the rates and mechanisms of reactions. (508)
chemical property A characteristic of a substance that appears as it interacts with, or transforms into, other substances. (3)
chemical reaction (See *chemical change.*)
chemistry The scientific study of matter and the changes it undergoes. (2)
chiral molecule One that is not superimposable on its mirror image; an optically active molecule. In organic compounds, a chiral molecule typically contains a C atom bonded to four different groups (asymmetric C). (476)
chlor-alkali process An industrial method that electrolyzes concentrated aqueous NaCl and produces Cl_2, H_2, and NaOH. (742)
***cis-trans* isomers** (See *geometric isomers.*)
Clausius-Clapeyron equation An equation that expresses the relationship between vapor pressure P of a liquid and temperature T:

$$\ln P = \frac{-\Delta H_{vap}}{R}\left(\frac{1}{T}\right) + C, \text{ where } C \text{ is a constant} \quad (365)$$

colligative property A property of a solution that depends on the number, not the identity, of solute particles. (416) (See also

boiling point elevation, freezing point depression, osmotic pressure, and *vapor pressure lowering.*)
collision theory A model that explains reaction rate as the result of particles colliding with a certain minimum energy. (529)
combustion analysis A method for determining the formula of a compound from the amounts of its combustion products; used commonly for organic compounds. (82)
common-ion effect The shift in the position of an ionic equilibrium away from formation of an ion that is caused by the addition (or presence) of that ion. (633)
complex (See *coordination compound.*)
complex ion An ion consisting of a central metal ion bonded covalently to molecules and/or anions called ligands. (659, 763)
composition The types and amounts of simpler substances that make up a sample of matter. (2)
compound A substance composed of two or more elements that are chemically combined in fixed proportions. (32)
concentration A measure of the quantity of solute dissolved in a given quantity of solution. (99)
concentration cell A voltaic cell in which both compartments contain the same components but at different concentrations. (729)
condensation The process of a gas changing into a liquid. (358)
condensation polymer A polymer formed by monomers with two functional groups that are linked together in a dehydration-condensation reaction. (494)
conduction band In band theory, the empty, higher energy portion of the band of molecular orbitals into which electrons move when conducting heat and electricity. (390)
conductor A substance (usually a metal) that conducts an electric current well. (390)
conjugate acid-base pair Two species related to each other through the gain or loss of a proton; the acid has one more proton than its conjugate base. (601)
constitutional isomers (also *structural isomers*) Compounds with the same molecular formula but different arrangements of atoms. (474, 768)
controlled experiment An experiment that measures the effect of one variable at a time by keeping other variables constant. (8)
conversion factor A ratio of equivalent quantities that is equal to 1 and used to convert the units of a quantity. (10)
coordinate covalent bond A covalent bond formed when one atom donates both electrons to give the shared pair; once formed, it is identical to any covalent single bond. (770)
coordination compound (also *complex*) A substance containing at least one complex ion. (763)
coordination isomers Two or more coordination compounds with the same composition in which the complex ions have different ligand arrangements. (768)
coordination number In a crystal, the number of nearest neighbors surrounding a particle. (380) In a complex, the number of ligand atoms bonded to the central metal ion. (764)
core electrons (See *inner electrons.*)
corrosion The natural redox process that results in unwanted oxidation of a metal. (736)
coulomb (C) The SI unit of electric charge. One coulomb is the charge of 6.242×10^{18} electrons; one electron possesses a charge of 1.602×10^{-19} C. (715)
Coulomb's law A law stating that the electrostatic force associated with two charges A and B is directly proportional to the product of their magnitudes and inversely proportional to the square of the distance between them:

$$\text{electrostatic force} \propto \frac{\text{charge A} \times \text{charge B}}{(\text{distance})^2} \qquad (284)$$

counter ion A simple ion associated with a complex ion in a coordination compound. (763)
coupling of reactions The pairing of reactions of which one releases enough free energy for the other to occur. (692)
covalent bond A type of bond in which atoms are bonded through the sharing of two electrons; the mutual attraction of the nuclei and an electron pair that holds atoms together in a molecule. (50, 287)
covalent bonding The idealized bonding type that is based on localized electron-pair sharing between two atoms with little difference in their tendencies to lose or gain electrons (most commonly nonmetals). (280)
covalent compound A compound that consists of atoms bonded together by shared electron pairs. (48)
covalent radius One-half the distance between nuclei of identical covalently bonded atoms. (260)
critical mass The minimum mass needed to achieve a chain reaction. (808)
critical point The point on a phase diagram above which the vapor cannot be condensed to a liquid; the end of the liquid-gas curve. (367)
crystal field splitting energy (Δ) The difference in energy between two sets of metal-ion *d* orbitals that results from electrostatic interactions with the surrounding ligands. (774)
crystal field theory A model that explains the color and magnetism of coordination compounds based on the effects of ligands on metal-ion *d*-orbital energies. (772)
crystalline solid Solid with a well-defined shape because of the orderly arrangement of the atoms, molecules, or ions. (379)
cubic closest packing A crystal structure based on the face-centered cubic unit cell in which the layers have an *abcabc* . . . pattern. (382)
cubic meter (m^3) The SI derived unit of volume. (15)
curie (Ci) The most common unit of radioactivity, defined as the number of nuclei disintegrating each second in 1 g of radium-226; 1 Ci = 3.70×10^{10} d/s (disintegrations per second). (793)
cyclic hydrocarbon A hydrocarbon with one or more rings in its structure. (474)

D

***d* orbital** An atomic orbital with $l = 2$. (238)
dalton (Da) A unit of mass identical to *atomic mass unit.* (44)
Dalton's law of partial pressures A gas law stating that, in a mixture of unreacting gases, the total pressure is the sum of the partial pressures of the individual gases: $P_{total} = P_1 + P_2 + P_3 + \cdots$ (162)
data Pieces of quantitative information obtained by observation. (8)
de Broglie wavelength The wavelength of a moving particle obtained from the de Broglie equation: $\lambda = h/mu$. (229)
decay constant The rate constant *k* for radioactive decay. (794)
decay rate (See *activity.*)
decay series (also *disintegration series*) The succession of steps a parent nucleus undergoes as it decays into a stable daughter nucleus. (792)

dehydration-condensation reaction A reaction in which H and OH groups on two molecules react to form water as one of the products. (452)
delocalization (See *electron-pair delocalization.*)
density (*d*) An intensive physical property of a substance at a given temperature and pressure, defined as the ratio of the mass to the volume: $d = m/V$. (17)
deposition The process of changing directly from gas to solid. (359)
derived unit Any of various combinations of the seven SI base units. (14)
deuterons Nuclei of the stable hydrogen isotope deuterium, ^2H. (797)
diagonal relationship Physical and chemical similarities between a Period 2 element and one located diagonally down and to the right in Period 3. (438)
diamagnetism The tendency of a species not to be attracted (or to be slightly repelled) by a magnetic field as a result of its electrons being paired. (271)
diastereomers (See *geometric isomers.*)
diffraction The phenomenon in which a wave striking the edge of an object bends around it. A wave passing through a slit as wide as its wavelength forms a semicircular wave. (217)
diffusion The movement of one fluid through another. (173)
dimensional analysis (also *factor-label method*) A calculation method in which arithmetic steps are accompanied by the appropriate canceling of units. (11)
dipole-dipole force The intermolecular attraction between oppositely charged poles of nearby polar molecules. (370)
dipole–induced dipole force The intermolecular attraction between a polar molecule and the oppositely charged pole it induces in a nearby molecule. (401)
dipole moment (μ) A measure of molecular polarity; the magnitude of the partial charges on the ends of a molecule (in coulombs) times the distance between them (in meters). (325)
disaccharide An organic compound formed by a dehydration-condensation reaction between two simple sugars (monosaccharides). (495)
disintegration series (See *decay series.*)
dispersion force (also *London force*) The intermolecular attraction between all particles as a result of instantaneous polarizations of their electron clouds; the intermolecular force primarily responsible for the condensed states of nonpolar substances. (373)
disproportionation reaction A reaction in which a given substance is both oxidized and reduced. (450)
donor atom An atom that donates a lone pair of electrons to form a covalent bond, usually from ligand to metal ion in a complex. (764)
double bond A covalent bond that consists of two bonding pairs; two atoms sharing four electrons in the form of one σ and one π bond. (288)
double-displacement reaction (See *metathesis reaction.*)
double helix The two intertwined polynucleotide strands held together by H bonds that form the structure of DNA (deoxyribonucleic acid). (500)
Downs cell An industrial apparatus that electrolyzes molten NaCl to produce sodium and chlorine. (740)
dynamic equilibrium In a chemical or physical change, the condition at which the forward and reverse processes are taking place at the same rate, so there is no net change in the amounts of reactants or products. (364)

E

e_g orbitals The set of orbitals (composed of $d_{x^2-y^2}$ and d_{z^2}) that results when the energies of the metal-ion *d* orbitals are split by a ligand field. This set is higher in energy than the other (t_{2g}) set in an octahedral field of ligands and lower in energy in a tetrahedral field. (774)
effective collision A collision in which the particles meet with sufficient energy and an orientation that allows them to react. (530)
effective nuclear charge (Z_{eff}) The nuclear charge an electron actually experiences as a result of shielding effects due to the presence of other electrons. (249)
effusion The process by which a gas escapes from its container through a tiny hole into an evacuated space. (172)
electrochemical cell A system that incorporates a redox reaction to produce or use electrical energy. (705)
electrochemistry The study of the relationship between chemical change and electrical work. (705)
electrode The part of an electrochemical cell that conducts the electricity between the cell and the surroundings. (709)
electrolysis The nonspontaneous lysing (splitting) of a substance, often to its component elements, by supplying electrical energy. (740)
electrolyte A substance that conducts a current when it dissolves in water. (115, 417) A mixture of ions, in which the electrodes of an electrochemical cell are immersed, that conducts a current. (417)
electrolytic cell An electrochemical system that uses electrical energy to drive a nonspontaneous chemical reaction ($\Delta G > 0$). (709)
electromagnetic (EM) radiation (or *electromagnetic energy, radiant energy*) Oscillating, perpendicular electric and magnetic fields moving simultaneously through space as waves and manifested as visible light, x-rays, microwaves, radio waves, and so on. (215)
electromagnetic spectrum The continuum of wavelengths of radiant energy. (216)
electromotive force (emf) (See *cell potential.*) (715)
electron (e^-) A subatomic particle that possesses a unit negative charge (1.60218×10^{-19} C) and occupies the space around the atomic nucleus. (42)
electron affinity (EA) The energy change (in kJ) accompanying the addition of one mole of electrons to one mole of gaseous atoms or ions. (265)
electron (e^-) capture (EC) A type of radioactive decay in which a nucleus draws in an orbital electron, usually one from the lowest energy level, and releases energy. (788)
electron cloud An imaginary representation of an electron's rapidly changing position around the nucleus over time. (232)
electron configuration The distribution of electrons within the orbitals of the atoms of an element; also the notation for such a distribution. (246)
electron deficient Referring to a bonded atom, such as Be or B, that has fewer than eight valence electrons. (313)
electron density diagram (or *electron probability density diagram*) The pictorial representation for a given energy sublevel of the quantity ψ^2 (the probability density of the electron lying within a particular tiny volume) as a function of *r* (distance from the nucleus). (232)

electron volt (eV) The energy (in joules, J) that an electron acquires when it moves through a potential difference of 1 volt; 1 eV = 1.602×10^{-19} J. (805)
electronegativity (EN) The relative ability of a bonded atom to attract shared electrons. (296)
electronegativity difference (ΔEN) The difference in electronegativities between the atoms in a bond. (298)
electron-pair delocalization (also *delocalization*) The process by which electron density is spread over several atoms rather than remaining between two. (310)
electron-sea model A qualitative description of metallic bonding proposing that metal atoms pool their valence electrons into a delocalized "sea" of electrons in which the metal cores (metal ions) are submerged in an orderly array. (388)
element The simplest type of substance with unique physical and chemical properties. An element consists of only one kind of atom, so it cannot be broken down into simpler substances. (32)
elementary reaction (or *elementary step*) A simple reaction that describes a single molecular event in a proposed reaction mechanism. (535)
elimination reaction A type of organic reaction in which C atoms are bonded to fewer atoms in the product than in the reactant, which leads to multiple bonding. (481)
emission spectrum The line spectrum produced when excited atoms return to lower energy levels and emit photons characteristic of the element. (227)
empirical formula A chemical formula that shows the lowest relative numbers of atoms of elements in a compound. (52)
enantiomers (See *optical isomers.*)
end point The point in a titration at which the indicator changes color. (127, 642)
endothermic process One occurring with an absorption of heat from the surroundings and therefore an increase in the enthalpy of the system ($\Delta H > 0$). (194)
energy The capacity to do work, that is, to move matter. (6) [See also *kinetic energy* (E_k) and *potential energy* (E_p).]
enthalpy (*H*) A thermodynamic quantity that is the sum of the internal energy plus the product of the pressure and volume. (193)
enthalpy diagram A graphic depiction of the enthalpy change of a system. (194)
enthalpy of hydration (ΔH_{hydr}) (See *heat of hydration.*)
enthalpy of solution (ΔH_{soln}) (See *heat of solution.*)
entropy (*S*) A thermodynamic quantity related to the number of ways the energy of a system can be dispersed through the motions of its particles. (407, 673)
enzyme A biological macromolecule (usually a protein) that acts as a catalyst. (542)
equatorial group A group (or atom) that lies in the trigonal plane of a trigonal bipyramidal molecule, or a similar structural feature in a molecule. (320)
equilibrium (See *dynamic equilibrium.*)
equilibrium constant (*K*) The value obtained when equilibrium concentrations are substituted into the reaction quotient. (554)
equilibrium vapor pressure (See *vapor pressure.*)
equivalence point The point in a titration when the number of moles of the added species is stoichiometrically equivalent to the original number of moles of the other species. (126, 642)
ester An organic compound that contains the $-\overset{\overset{\displaystyle :\!O\!\cdot}{\|}}{C}-\ddot{\underset{\cdot\cdot}{O}}-\overset{|}{\underset{|}{C}}-$ group. (488)
exact number A quantity, usually obtained by counting or based on a unit definition, that has no uncertainty associated with it and, therefore, contains as many significant figures as a calculation requires. (23)
excited state Any electron configuration of an atom or molecule other than the lowest energy (ground) state. (223)
exclusion principle A principle developed by Wolfgang Pauli stating that no two electrons in an atom can have the same set of four quantum numbers. The principle arises from the fact that an orbital has a maximum occupancy of two electrons and their spins are paired. (248)
exothermic process One occurring with a release of heat to the surroundings and therefore a decrease in the enthalpy of the system ($\Delta H < 0$). (194)
expanded valence shell A valence level that can accommodate more than 8 electrons by using available *d* orbitals; occurs only for elements in Period 3 or higher. (314)
experiment A clear set of procedural steps that tests a hypothesis. (8)
extensive property A property, such as mass, that depends on the quantity of substance present. (17)

F

face-centered cubic unit cell A unit cell in which a particle occurs at each corner and in the center of each face of a cube. (380)
factor-label method (See *dimensional analysis.*)
Faraday constant (*F*) The physical constant representing the charge of 1 mol of electrons: $F = 96{,}485$ C/mol e^-. (724)
fatty acid A carboxylic acid that has a long hydrocarbon chain and is derived from a natural source. (489)
first law of thermodynamics (See *law of conservation of energy.*)
fission The process by which a heavier nucleus splits into lighter nuclei with the release of energy. (804)
formal charge The hypothetical charge on an atom in a molecule or ion, equal to the number of valence electrons minus the sum of all the unshared and half the shared valence electrons. (311)
formation constant (K_f) An equilibrium constant for the formation of a complex ion from the hydrated metal ion and ligands. (660)
formation equation An equation in which 1 mole of a compound forms from its elements. (203)
formula mass The sum (in amu) of the atomic masses of a formula unit of an ionic compound. (58)
formula unit The chemical unit of a compound that contains the number and type of atoms (or ions) expressed in the chemical formula. (53)
fossil fuel Any fuel, including coal, petroleum, and natural gas, derived from the products of the decay of dead organisms. (205)
fraction by mass (also *mass fraction*) The portion of a compound's mass contributed by an element; the mass of an element in a compound divided by the mass of the compound. (35)
free energy (*G*) A thermodynamic quantity that is the difference between the enthalpy and the product of the absolute temperature and the entropy: $G = H - TS$. (686)
free radical A molecular or atomic species with one or more unpaired electrons, which typically make it very reactive. (313)
freezing The process of cooling a liquid until it solidifies. (358)
freezing point depression (ΔT_f) A lowering of the freezing point of a solvent caused by the presence of dissolved solute particles. (420)

frequency (ν) The number of cycles a wave undergoes per second, expressed in units of 1/second, or s^{-1} [also called hertz (Hz)]. (215)
frequency factor (A) The product of the collision frequency Z and an orientation probability factor p that is specific for a reaction. (530)
fuel cell (or *flow battery*) A battery that is not self-contained and in which electricity is generated by the controlled oxidation of a fuel. (735)
functional group A specific combination of atoms, typically containing a carbon-carbon multiple bond and/or carbon-heteroatom bond, that reacts in a characteristic way no matter what molecule it occurs in. (469)
fundamental unit (See *base unit.*)
fusion (See *melting.*)
fusion (nuclear) The process by which light nuclei combine to form a heavier nucleus with the release of energy. (804)

galvanic cell (See *voltaic cell.*)
gamma emission The type of radioactive decay in which gamma rays are emitted from an excited nucleus. (789)
gamma (γ) ray A very high-energy photon. (786)
gas One of the three states of matter. A gas fills its container regardless of the shape. (4)
genetic code The set of three-base sequences that is translated into specific amino acids during the process of protein synthesis. (501)
geometric isomers (also *cis-trans isomers* or *diastereomers*) Stereoisomers in which the molecules have the same connections between atoms but differ in the spatial arrangements of the atoms. The *cis* isomer has similar groups on the same side of a structural feature; the *trans* isomer has them on opposite sides. (477–78, 768)
Graham's law of effusion A gas law stating that the rate of effusion of a gas is inversely proportional to the square root of its density (or molar mass):

$$\text{rate} \propto \frac{1}{\sqrt{\mathcal{M}}} \qquad (172)$$

gray (Gy) The SI unit of absorbed radiation dose; 1 Gy = 1 J/kg tissue. (799)
ground state The electron configuration of an atom or ion that is lowest in energy. (223)
group A vertical column in the periodic table. (46)

H

H bond (See *hydrogen bond.*)
Haber process An industrial process used to form ammonia from its elements. (582)
half-cell A portion of an electrochemical cell in which a half-reaction takes place. (711)
half-life ($t_{1/2}$) In chemical processes, the time required for half the initial reactant concentration to be consumed. (523) In nuclear processes, the time required for half the initial number of nuclei in a sample to decay. (794)
half-reaction method A method of balancing redox reactions by treating the oxidation and reduction half-reactions separately. (706)
haloalkane (also *alkyl halide*) A hydrocarbon with one or more halogen atoms (X) in place of H; contains a $-\overset{|}{\underset{|}{C}}-\ddot{\underset{\cdot\cdot}{X}}\!:$ group. (484)
heat (q) The energy transferred between objects because of differences in their temperatures only; thermal energy. (18, 188)
heat capacity The quantity of heat required to change the temperature of an object by 1 K. (195)
heat of fusion (ΔH°_{fus}) The enthalpy change occurring when 1 mol of a solid substance melts. (358)
heat of hydration (ΔH_{hydr}) (also *enthalpy of hydration*) The enthalpy change occurring when 1 mol of a gaseous species is hydrated. The sum of the enthalpies from separating water molecules and mixing the gaseous species with them. (406)
heat of reaction (ΔH°_{rxn}) The enthalpy change of a reaction. (194)
heat of solution (ΔH_{soln}) (also *enthalpy of solution*) The enthalpy change occurring when a solution forms from solute and solvent. The sum of the enthalpies from separating solute and solvent molecules and mixing them. (405)
heat of sublimation (ΔH°_{subl}) The enthalpy change occurring when 1 mol of a solid substance changes directly to a gas. The sum of the heats of fusion and vaporization. (359)
heat of vaporization (ΔH°_{vap}) The enthalpy change occurring when 1 mol of a liquid substance vaporizes. (358)
heating-cooling curve A plot of temperature vs. time for a substance when heat is absorbed or released by the system at a constant rate. (360)
Henderson-Hasselbalch equation An equation for calculating the pH of a buffer system:

$$\text{pH} = \text{p}K_a + \log\left(\frac{[\text{base}]}{[\text{acid}]}\right) \qquad (637)$$

Henry's law A law stating that the solubility of a gas in a liquid is directly proportional to the partial pressure of the gas above the liquid: $S_{gas} = k_H \times P_{gas}$. (411)
Hess's law of heat summation A law stating that the enthalpy change of an overall process is the sum of the enthalpy changes of the individual steps of the process. (201)
heteroatom Any atom in an organic compound other than C or H. (468)
heterogeneous catalyst A catalyst that occurs in a different phase from the reactants, usually a solid interacting with gaseous or liquid reactants. (541)
heterogeneous mixture A mixture that has one or more visible boundaries among its components. (61)
hexagonal closest packing A crystal structure based on the hexagonal unit cell in which the layers have an *abab* . . . pattern. (382)
high-spin complex Complex ion that has the same number of unpaired electrons as in the isolated metal ion; contains weak-field ligands. (776)
homogeneous catalyst A catalyst (gas, liquid, or soluble solid) that exists in the same phase as the reactants. (541)
homogeneous mixture (also *solution*) A mixture that has no visible boundaries among its components. (61)
homologous series A series of organic compounds in which each member differs from the next by a $-CH_2-$ (methylene) group. (472)
homonuclear diatomic molecule A molecule composed of two identical atoms. (346)
Hund's rule A principle stating that when orbitals of equal energy are available, the electron configuration of lowest energy has the maximum number of unpaired electrons with parallel spins. (252)

hybrid orbital An atomic orbital postulated to form during bonding by the mathematical mixing of specific combinations of nonequivalent orbitals in a given atom. (334)
hybridization A postulated process of orbital mixing to form hybrid orbitals. (334)
hydrate A compound in which a specific number of water molecules are associated with each formula unit. (55)
hydration Solvation in water. (406)
hydration shell The oriented cluster of water molecules that surrounds an ion in aqueous solution. (400)
hydrocarbon An organic compound that contains only H and C atoms. (469)
hydrogen bond (H bond) A type of dipole-dipole force that arises between molecules that have an H atom bonded to a small, highly electronegative atom with lone pairs, usually N, O, or F. (371)
hydrogenation The addition of hydrogen to a carbon-carbon multiple bond to form a carbon-carbon single bond. (542)
hydrolysis Cleaving a molecule by reaction with water, in which one part of the molecule bonds to the water —OH and the other to the water H. (489)
hydronium ion (H_3O^+) A proton covalently bonded to a water molecule. (591)
hypothesis A testable proposal made to explain an observation. If inconsistent with experimental results, a hypothesis is revised or discarded. (8)

I

ideal gas A hypothetical gas that exhibits linear relationships among volume, pressure, temperature, and amount (mol) at all conditions; approximated by simple gases at ordinary conditions. (150)
ideal gas law (or *ideal gas equation*) An equation that expresses the relationships among volume, pressure, temperature, and amount (mol) of an ideal gas: $PV = nRT$. (155)
ideal solution A solution whose vapor pressure equals the mole fraction of the solvent times the vapor pressure of the pure solvent; approximated only by very dilute solutions. (417) (See also *Raoult's law.*)
indicator (See *acid-base indicator.*)
infrared (IR) Radiation in the region of the electromagnetic spectrum between the microwave and visible regions. (216)
infrared (IR) spectroscopy An instrumental technique for determining the types of bonds in a covalent molecule by measuring the absorption of IR radiation. (292)
initial rate The instantaneous rate occurring as soon as the reactants are mixed, that is, at $t = 0$. (512)
inner electrons (also *core electrons*) Electrons that fill all the energy levels of an atom except the valence level; electrons also present in atoms of the previous noble gas and any completed transition series. (257)
inner transition elements The elements of the periodic table in which *f* orbitals are being filled; the lanthanides and actinides. (258)
instantaneous rate The reaction rate at a particular time, given by the slope of a tangent to a plot of reactant concentration vs. time. (511)
insulator A substance (usually a nonmetal) that does not conduct an electric current. (390)
integrated rate law A mathematical expression for reactant concentration as a function of time. (520)
intensive property A property, such as density, that does not depend on the quantity of substance present. (17)
intermolecular forces (or *interparticle forces*) The attractive and repulsive forces among the particles—molecules, atoms, or ions—in a sample of matter. (357)
internal energy (*E*) The sum of the kinetic and potential energies of all the particles in a system. (187)
ion A charged particle that forms from an atom (or covalently bonded group of atoms) when it gains or loses one or more electrons. (49)
ion-dipole force The intermolecular attractive force between an ion and a polar molecule (dipole). (370)
ion–induced dipole force The intermolecular attractive force between an ion and the dipole it induces in the electron cloud of a nearby particle. (401)
ion pair A pair of ions that form a gaseous ionic molecule; sometimes formed when a salt boils. (286)
ion-product constant for water (K_w) The equilibrium constant for the autoionization of water:

$$K_w = [H_3O^+][OH^-] \quad (596)$$

ionic atmosphere A cluster of ions of net opposite charge surrounding a given ion in solution. (424)
ionic bonding The idealized type of bonding based on the attraction of oppositely charged ions that arise through electron transfer between atoms with large differences in their tendencies to lose or gain electrons (typically metals and nonmetals). (279)
ionic compound A compound that consists of oppositely charged ions. (48)
ionic radius The size of an ion as measured by the distance between the centers of adjacent ions in a crystalline ionic compound. (272)
ionic solid A solid whose unit cell contains cations and anions. (386)
ionization The process by which a substance absorbs energy from high-energy radioactive particles and loses an electron to become ionized. (799)
ionization energy (IE) The energy (in kJ) required to remove completely one mole of electrons from one mole of gaseous atoms or ions. (262)
ionizing radiation The high-energy radiation that forms ions in a substance by causing electron loss. (799)
isoelectronic Having the same number and configuration of electrons as another species. (269)
isomer One of two or more compounds with the same molecular formula but different properties, often as a result of different arrangements of atoms. (83, 767)
isotopes Atoms of a given atomic number (that is, of a specific element) that have different numbers of neutrons and therefore different mass numbers. (43, 786)
isotopic mass The mass (in amu) of an isotope relative to the mass of the carbon-12 isotope. (45)

J

joule (J) The SI unit of energy; $1\ J = 1\ kg{\cdot}m^2/s^2$. (190)

kelvin (K) The SI base unit of temperature. The kelvin is the same size as the Celsius degree. (18)
Kelvin scale (See *absolute scale.*)

ketone An organic compound (ending, *-one*) that contains a carbonyl group bonded to two other C atoms, $-\overset{|}{\underset{|}{C}}-\overset{:O:}{\overset{\|}{C}}-\overset{|}{\underset{|}{C}}-$. (487)
kilogram (kg) The SI base unit of mass. (16)
kinetic energy (E_k) The energy an object has because of its motion. (6)
kinetic-molecular theory The model that explains gas behavior in terms of particles in random motion whose volumes and interactions are negligible. (167)

L

lanthanide contraction The additional decrease in atomic and ionic size, beyond the expected trend, caused by the poor shielding of the increasing nuclear charge by *f* electrons in the elements following the lanthanides. (760)
lanthanides (also *rare earths*) The Period 6 (4*f*) series of inner transition elements, which includes cerium (Ce; $Z = 58$) through lutetium (Lu; $Z = 71$). (258)
lattice The three-dimensional arrangement of points created by choosing each point to be at the same location within each particle of a crystal; thus, the lattice consists of all points with identical surroundings. (380)
lattice energy ($\Delta H^\circ_{\text{lattice}}$) The enthalpy change (always positive) that occurs when 1 mol of an ionic compound separates into gaseous ions, with all components in their standard states. (284)
law (See *natural law.*)
law of chemical equilibrium (also *law of mass action*) The law stating that when a system reaches equilibrium at a given temperature, the ratio of quantities that make up the reaction quotient has a constant numerical value. (556)
law of conservation of energy (also *first law of thermodynamics*) A basic observation that the total energy of the universe is constant: $\Delta E_{\text{universe}} = \Delta E_{\text{system}} + \Delta E_{\text{surroundings}} = 0$. (190)
law of definite (or constant) composition A mass law stating that, no matter what its source, a particular compound is composed of the same elements in the same parts (fractions) by mass. (35)
law of mass action (See *law of chemical equilibrium.*)
law of mass conservation A mass law stating that the total mass of substances does not change during a chemical reaction. (34)
law of multiple proportions A mass law stating that if elements A and B react to form two compounds, the different masses of B that combine with a fixed mass of A can be expressed as a ratio of small whole numbers. (36)
Le Châtelier's principle A principle stating that if a system in a state of equilibrium is disturbed, it will undergo a change that shifts its equilibrium position in a direction that reduces the effect of the disturbance. (573)
level (also *shell*) A specific energy state of an atom given by the principal quantum number *n*. (235)
Lewis acid-base definition A model of acid-base behavior in which acids and bases are defined, respectively, as species that accept and donate an electron pair. (621)
Lewis electron-dot symbol A notation in which the element symbol represents the nucleus and inner electrons, and surrounding dots represent the valence electrons. (281)
Lewis structure (or *Lewis formula*) A structural formula consisting of electron-dot symbols, with lines as bonding pairs and dots as lone pairs. (306)
ligand A molecule or anion bonded to a central metal ion in a complex ion. (659, 763)
like-dissolves-like rule An empirical observation stating that substances having similar kinds of intermolecular forces dissolve in each other. (400)
limiting reactant (or *limiting reagent*) The reactant that is consumed when a reaction occurs and therefore the one that determines the maximum amount of product that can form. (93)
line spectrum A series of separated lines of different colors representing photons whose wavelengths are characteristic of an element. (221) (See also *emission spectrum.*)
linear arrangement The geometric arrangement obtained when two electron groups maximize their separation around a central atom. (317)
linear shape A molecular shape formed by three atoms lying in a straight line, with a bond angle of 180° (shape class AX_2 or AX_2E_3). (317)
linkage isomers Coordination compounds with the same composition but with different ligand donor atoms linked to the central metal ion. (768)
lipid Any of a class of biomolecules, including fats and oils, that are soluble in nonpolar solvents. (489)
liquid One of the three states of matter. A liquid fills a container to the extent of its own volume and thus forms a surface. (4)
liter (L) A non-SI unit of volume equivalent to 1 cubic decimeter (0.001 m^3). (15)
London force (See *dispersion force.*)
lone pair (also *unshared pair*) An electron pair that is part of an atom's valence shell but not involved in covalent bonding. (288)
low-spin complex Complex ion that has fewer unpaired electrons than in the free metal ion because of the presence of strong-field ligands. (776)

M

macromolecule (See *polymer.*)
magnetic quantum number (m_l) (or *orbital-orientation quantum number*) An integer from $-l$ through 0 to $+l$ that specifies the orientation of an atomic orbital in the three-dimensional space about the nucleus. (234)
mass The quantity of matter an object contains. Balances are designed to measure mass. (16)
mass fraction (See *fraction by mass.*)
mass number (*A*) The total number of protons and neutrons in the nucleus of an atom. (43)
mass percent (also *mass %* or *percent by mass*) The fraction by mass expressed as a percentage. A concentration term [% (w/w)] expressed as the mass in grams of solute dissolved per 100. g of solution. (35, 413–14)
mass spectrometry An instrumental method for measuring the relative masses of particles in a sample by creating charged particles and separating them according to their mass-charge ratio. (44)
matter Anything that possesses mass and occupies volume. (2)
melting (also *fusion*) The change of a substance from a solid to a liquid. (358)
melting point (mp or T_f) The temperature at which the solid and liquid forms of a substance are at equilibrium. (366)

metal A substance or mixture that is relatively shiny and malleable and is a good conductor of heat and electricity. In reactions, metals tend to transfer electrons to nonmetals and form ionic compounds. (47)
metallic bonding An idealized type of bonding based on the attraction between metal ions and their delocalized valence electrons. (280) (See also *electron-sea model.*)
metallic radius One-half the distance between the nuclei of adjacent individual atoms in a crystal of an element. (259)
metallic solid A solid whose individual atoms are held together by metallic bonding. (387)
metalloid (also *semimetal*) An element with properties between those of metals and nonmetals. (47)
metathesis reaction (also *double-displacement reaction*) A reaction in which atoms or ions of two compounds exchange bonding partners. (120)
meter (m) The SI base unit of length. The distance light travels in a vacuum in 1/299,792,458 second. (15)
milliliter (mL) A volume (0.001 L) equivalent to 1 cm^3. (15)
millimeter of mercury (mmHg) A unit of pressure based on the difference in the heights of mercury in a barometer or manometer. Renamed the *torr* in honor of Evangelista Torricelli. (149)
miscible Soluble in any proportion. (399)
mixture A group of two or more elements and/or compounds that are physically intermingled. (33)
MO bond order One-half the difference between the numbers of electrons in bonding and antibonding MOs. (345)
model (also *theory*) A simplified conceptual picture based on experiment that explains how an aspect of nature occurs. (9)
molality (*m*) A concentration term expressed as number of moles of solute dissolved in 1000 g (1 kg) of solvent. (413)
molar heat capacity (*C*) The quantity of heat required to change the temperature of 1 mol of a substance by 1 K. (196)
molar mass ($\mathcal{M}$) (or *gram-molecular weight*) The mass of 1 mol of entities (atoms, molecules, or formula units) of a substance, in units of g/mol. (74)
molar solubility The solubility expressed in terms of amount (mol) of dissolved solute per liter of solution. (651)
molarity (*M*) A concentration term expressed as the moles of solute dissolved in 1 L of solution. (99)
mole (mol) The SI base unit for amount of a substance. The amount that contains a number of objects equal to the number of atoms in exactly 12 g of carbon-12. (72)
mole fraction (*X*) A concentration term expressed as the ratio of moles of one component of a mixture to the total moles present. (163, 414)
molecular equation A chemical equation showing a reaction in solution in which reactants and products appear as intact, undissociated compounds. (118)
molecular formula A formula that shows the actual number of atoms of each element in a molecule. (52)
molecular mass (or *molecular weight*) The sum (in amu) of the atomic masses of a formula unit of a compound. (58)
molecular orbital (MO) An orbital of given energy and shape that extends over a molecule and can be occupied by no more than two electrons. (343)
molecular orbital (MO) diagram A depiction of the relative energy and number of electrons in each MO, as well as the atomic orbitals from which the MOs form. (345)
molecular orbital (MO) theory A model that describes a molecule as a collection of nuclei and electrons in which the electrons occupy orbitals that extend over the entire molecule. (343)
molecular polarity The overall distribution of electronic charge in a molecule, determined by its shape and bond polarities. (324)
molecular shape The three-dimensional structure defined by the relative positions of the atomic nuclei in a molecule. (316)
molecular solid A solid held together by intermolecular forces between individual molecules. (386)
molecularity The number of reactant particles involved in an elementary step. (535)
molecule A structure consisting of two or more atoms that are chemically bound together and behave as an independent unit. (32)
monatomic ion An ion derived from a single atom. (49)
monomer A small molecule, linked covalently to others of the same or similar type to form a polymer, on which the repeat unit of the polymer is based. (492)
mononucleotide A monomer unit of a nucleic acid, consisting of an N-containing base, a sugar, and a phosphate group. (499)
monosaccharide A simple sugar; a polyhydroxy ketone or aldehyde with three to nine C atoms. (495)

N

natural law (also *law*) A summary, often in mathematical form, of a universal observation. (8)
Nernst equation An equation stating that the voltage of an electrochemical cell under any conditions depends on the standard cell voltage and the concentrations of the cell components:

$$E_{\text{cell}} = E^\circ_{\text{cell}} - \frac{RT}{nF}\ln Q \qquad (726)$$

net ionic equation A chemical equation of a reaction in solution in which spectator ions have been eliminated to show the actual chemical change. (119)
network covalent solid A solid in which all the atoms are bonded covalently. (387)
neutralization In the Arrhenius acid-base definition, the combination of the H^+ ion from the acid and the OH^- ion from the base to form H_2O. (592)
neutralization reaction An acid-base reaction that yields water and a solution of a salt; when a strong acid reacts with a stoichiometrically equivalent amount of a strong base, the solution is neutral. (123)
neutron (n^0) An uncharged subatomic particle found in the nucleus, with a mass slightly greater than that of a proton. (42)
nitrile An organic compound containing the $-C{\equiv}N$: group. (491)
node A region of an orbital where the probability of finding the electron is zero. (237)
nonelectrolyte A substance whose aqueous solution does not conduct an electric current. (117, 417)
nonmetal An element that lacks metallic properties. In reactions, nonmetals tend to bond with each other to form covalent compounds or accept electrons from metals to form ionic compounds. (47)
nonpolar covalent bond A covalent bond between identical atoms that share the bonding pair equally. (298)
nuclear binding energy The energy required to break 1 mol of nuclei of an element into individual nucleons. (805)

nuclear transmutation The induced conversion of one nucleus into another by bombardment with a particle. (797)
nucleic acid An unbranched polymer consisting of mononucleotides that occurs as two types, DNA and RNA (deoxyribonucleic and ribonucleic acids), which differ chemically in the nature of the sugar portion of the mononucleotides. (499)
nucleon A subatomic particle that makes up a nucleus; a proton or neutron. (786)
nucleus The tiny central region of the atom that contains all the positive charge and essentially all the mass. (41)
nuclide A nuclear species with specified numbers of protons and neutrons. (786)

observation A fact obtained with the senses, often with the aid of instruments. Quantitative observations provide data that can be compared objectively. (8)
octahedral arrangement The geometric arrangement obtained when six electron groups maximize their space around a central atom; when all six groups are bonding groups, the molecular shape is octahedral (AX_6; ideal bond angle = 90°). (321)
octet rule The observation that when atoms bond, they often lose, gain, or share electrons to attain a filled outer shell of eight electrons. (282)
optical isomers (also *enantiomers*) A pair of stereoisomers consisting of a molecule and its mirror image that cannot be superimposed on each other. (476, 769)
optically active Able to rotate the plane of polarized light. (477)
orbital diagram A depiction of electron number and spin in an atom's orbitals by means of arrows in a series of small boxes, lines, or circles. (251)
organic compound A compound in which carbon is nearly always bonded to at least one other carbon, to hydrogen, and often to other elements. (467)
osmosis The process by which solvent flows through a semipermeable membrane from a dilute to a concentrated solution. (421)
osmotic pressure (Π) The pressure that results from the inability of solute particles to cross a semipermeable membrane. The pressure required to prevent the net movement of solvent across the membrane. (421)
outer electrons Electrons that occupy the highest energy level (highest *n* value) and are, on average, farthest from the nucleus. (257)
overvoltage The additional voltage, usually associated with gaseous products, that is required above the standard cell voltage to accomplish electrolysis. (742)
oxidation The loss of electrons by a species, accompanied by an increase in oxidation number. (130)
oxidation number (O.N.) (also *oxidation state*) A number equal to the magnitude of the charge an atom would have if its shared electrons were held completely by the atom that attracts them more strongly. (131)
oxidation-reduction reaction (also *redox reaction*) A process in which there is a net movement of electrons from one reactant (reducing agent) to another (oxidizing agent). (129)
oxidation state (See *oxidation number.*)
oxidizing agent The substance that accepts electrons in a redox reaction and undergoes a decrease in oxidation number. (131)
oxoanion An anion in which an element is bonded to one or more oxygen atoms. (55)

P

***p* orbital** An atomic orbital with $l = 1$. (238)
packing efficiency The percentage of the available volume occupied by atoms, ions, or molecules in a unit cell. (382)
paramagnetism The tendency of a species with unpaired electrons to be attracted by an external magnetic field. (271)
partial ionic character An estimate of the actual charge separation in a bond (caused by the electronegativity difference of the bonded atoms) relative to complete separation. (298)
partial pressure The portion of the total pressure contributed by a gas in a mixture of gases. (162)
particle accelerator A device used to impart high kinetic energies to nuclear particles. (797)
pascal (Pa) The SI unit of pressure; 1 Pa = 1 N/m^2. (148)
penetration The process by which an outer electron moves through the region occupied by the core electrons to spend part of its time closer to the nucleus; penetration increases the average effective nuclear charge for that electron. (249)
percent by mass (mass %) (See *mass percent.*)
percent yield (% yield) The actual yield of a reaction expressed as a percentage of the theoretical yield. (97)
period A horizontal row of the periodic table. (46)
periodic law A law stating that when the elements are arranged by atomic number, they exhibit a periodic recurrence of properties. (246)
periodic table of the elements A table in which the elements are arranged by atomic number into columns (groups) and rows (periods). (46)
pH The negative common logarithm of $[H_3O^+]$. (597)
phase A physically distinct portion of a system. (357)
phase change A physical change from one phase to another, usually referring to a change in physical state. (357)
phase diagram A diagram used to describe the stable phases and phase changes of a substance as a function of temperature and pressure. (366)
photoelectric effect The observation that when monochromatic light of sufficient frequency shines on a metal, an electric current is produced. (219)
photon A quantum of electromagnetic radiation. (220)
physical change A change in which the physical form (or state) of a substance, but not its composition, is altered. (3)
physical property A characteristic shown by a substance itself, without interacting with or changing into other substances. (3)
pi (π) bond A covalent bond formed by sideways overlap of two atomic orbitals that has two regions of electron density, one above and one below the internuclear axis. (341)
pi (π) MO A molecular orbital formed by combination of two atomic (usually *p*) orbitals whose orientations are perpendicular to the internuclear axis. (347)
Planck's constant (*h*) A proportionality constant relating the energy and the frequency of a photon, equal to 6.626×10^{-34} J·s. (219)
polar covalent bond A covalent bond in which the electron pair is shared unequally, so the bond has partially negative and partially positive poles. (297)
polar molecule A molecule with an unequal distribution of charge as a result of its polar covalent bonds and shape. (114)
polarizability The ease with which a particle's electron cloud can be distorted. (372)

polyatomic ion An ion in which two or more atoms are bonded covalently. (51)
polymer (also *macromolecule*) An extremely large molecule that results from the covalent linking of many simpler molecular units (monomers). (492)
polyprotic acid An acid with more than one ionizable proton. (689)
polysaccharide A macromolecule composed of many simple sugars linked covalently. (495)
positron (β^+ or $^0_1\beta$) The antiparticle of an electron. (788)
positron (β^+) emission A type of radioactive decay in which a positron is emitted from a nucleus. (788)
potential energy (E_p) The energy an object has as a result of its position relative to other objects or because of its composition. (6)
precipitate The insoluble product of a precipitation reaction. (119)
precipitation reaction A reaction in which two soluble ionic compounds form an insoluble product, a precipitate. (119)
precision (also *reproducibility*) The closeness of a measurement to other measurements of the same phenomenon in a series of experiments. (24)
pressure (*P*) The force exerted per unit of surface area. (147)
pressure-volume work (*PV* work) A type of work in which a volume change occurs against an external pressure. (189)
principal quantum number (*n*) A positive integer that specifies the energy and relative size of an atomic orbital. (234)
probability contour A shape that defines the volume around an atomic nucleus within which an electron spends a given percentage of its time. (233)
product A substance formed in a chemical reaction. (85)
property A characteristic that gives a substance its unique identity. (2)
protein A natural, linear polymer composed of any of about 20 types of amino acid monomers linked together by peptide bonds. (496)
proton (p^+) A subatomic particle found in the nucleus that has a unit positive charge (1.60218×10^{-19} C). (42)
proton acceptor A substance that accepts an H^+ ion; a Brønsted-Lowry base. (600)
proton donor A substance that donates an H^+ ion; a Brønsted-Lowry acid. (600)
pseudo–noble gas configuration The $(n-1)d^{10}$ configuration of a *p*-block metal atom that has emptied its outer energy level. (269)

Q

quantum A packet of energy equal to $h\nu$. The smallest quantity of energy that can be emitted or absorbed. (219)
quantum mechanics The branch of physics that examines the wave motion of objects on the atomic scale. (231)
quantum number A number that specifies a property of an orbital or an electron. (219)

R

rad (radiation-absorbed dose) The quantity of radiation that results in 0.01 J of energy being absorbed per kilogram of tissue; 1 rad = 0.01 J/kg tissue = 0.01 Gy. (799)
radial probability distribution plot The graphic depiction of the total probability distribution (sum of ψ^2) of an electron in the region near the nucleus. (232)
radioactivity The emissions resulting from the spontaneous disintegration of an unstable nucleus. (785)
radioisotope An isotope with an unstable nucleus that decays through radioactive emissions. (796)
radioisotopic dating A method for determining the age of an object based on the rate of decay of a particular radioactive nuclide. (796)
random error Human error that occurs in all measurements and results in values *both* higher and lower than the actual value. (24)
Raoult's law A law stating that the vapor pressure of a solution is directly proportional to the mole fraction of solvent: $P_{solvent} = X_{solvent} \times P^\circ_{solvent}$. (417)
rare earths (See *lanthanides.*)
rate constant (*k*) The proportionality constant that relates the reaction rate to reactant (and product) concentrations. (514)
rate-determining step (or *rate-limiting step*) The slowest step in a reaction mechanism and therefore the step that limits the overall rate. (536)
rate law (or *rate equation*) An equation that expresses the rate of a reaction as a function of reactant (and product) concentrations. (514)
reactant A starting substance in a chemical reaction. (85)
reaction energy diagram A graph that shows the potential energy of a reacting system as it progresses from reactants to products. (532)
reaction intermediate A substance that is formed and used up during the overall reaction and therefore does not appear in the overall equation. (535)
reaction mechanism A series of elementary steps that sum to the overall reaction and is consistent with the rate law. (534)
reaction order The exponent of a reactant concentration in a rate law that shows how the rate is affected by changes in that concentration. (514)
reaction quotient (*Q*) A ratio of terms for a given reaction consisting of product concentrations multiplied together and divided by reactant concentrations multiplied together, each raised to the power of their balancing coefficient. The value of *Q* changes until the system reaches equilibrium, at which point it equals *K*. (556)
reaction rate The change in the concentrations of reactants (or products) with time. (510)
reactor core The part of a nuclear reactor that contains the fuel rods and generates heat from fission. (808)
redox reaction (See *oxidation-reduction reaction.*)
reducing agent The substance that donates electrons in a redox reaction and undergoes an increase in oxidation number. (131)
reduction The gain of electrons by a species, accompanied by a decrease in oxidation number. (130)
refraction A phenomenon in which a wave changes its speed and therefore its direction as it passes through a phase boundary. (217)
rem (*r*oentgen *e*quivalent for *m*an) The unit of radiation dosage for a human based on the product of the number of rads and a factor related to the biological tissue; 1 rem = 0.01 Sv. (799)
reproducibility (See *precision.*)
resonance hybrid The weighted average of the resonance structures of a molecule. (309)
resonance structure (or *resonance form*) One of two or more Lewis structures for a molecule that cannot be adequately depicted by a single structure. Resonance structures differ only in the position of bonding and lone electron pairs. (309)

rms (root-mean-square) speed (u_{rms}) The speed of a molecule having the average kinetic energy; very close to the most probable speed. (171)
round off The process of removing digits based on a series of rules to obtain an answer with the proper number of significant figures (or decimal places). (22)

S

***s* orbital** An atomic orbital with $l = 0$. (236)
salt An ionic compound that results from an Arrhenius acid-base reaction. (125)
salt bridge An inverted U tube containing a solution of nonreacting electrolyte that connects the compartments of a voltaic cell and maintains neutrality by allowing ions to flow between compartments. (713)
saturated hydrocarbon A hydrocarbon in which each C is bonded to four other atoms. (472)
saturated solution A solution that contains the maximum amount of dissolved solute at a given temperature in the presence of undissolved solute. (409)
Schrödinger equation An equation that describes how the electron matter-wave changes in space around the nucleus. Solutions of the equation provide allowable energy levels of the H atom. (231)
scientific method A process of creative thinking and testing aimed at objective, verifiable discoveries of the causes of natural events. (8)
second (s) The SI base unit of time. (20)
second law of thermodynamics A law stating that a process occurs spontaneously in the direction that increases the entropy of the universe. (676)
seesaw shape A molecular shape caused by the presence of one equatorial lone pair in a trigonal bipyramidal arrangement (AX_4E). (320)
self-ionization (See *autoionization.*)
semiconductor A substance whose electrical conductivity is poor at room temperature but increases significantly with rising temperature. (390)
semimetal (See *metalloid.*)
semipermeable membrane A membrane that allows solvent, but not solute, to pass through. (421)
shared pair (See *bonding pair.*)
shell (See *level.*)
shielding The ability of other electrons, especially inner ones, to lessen the nuclear attraction for an outer electron. (249)
SI unit A unit composed of one or more of the base units of the Système International d'Unités, a revised metric system. (13)
side reaction An undesired chemical reaction that consumes some of the reactant and reduces the overall yield of the desired product. (97)
sievert (Sv) The SI unit of human radiation dosage; 1 Sv = 100 rem. (799)
sigma (σ) bond A type of covalent bond that arises through end-to-end orbital overlap and has most of its electron density along the bond axis. (340)
sigma (σ) MO A molecular orbital that is cylindrically symmetrical about an imaginary line that runs through the nuclei of the component atoms. (345)
significant figures The digits obtained in a measurement. The greater the number of significant figures, the greater the certainty of the measurement. (21)
silicate A type of compound found throughout rocks and soil and consisting of repeating —Si—O groupings and, in most cases, metal cations. (446)
silicone A type of synthetic polymer containing —Si—O repeat units, with organic groups and crosslinks. (446)
simple cubic unit cell A unit cell in which a particle occurs at each corner of a cube. (380)
single bond A bond that consists of one shared electron pair. (288)
solid One of the three states of matter. A solid has a fixed shape that does not conform to the container shape. (4)
solubility (*S*) The maximum amount of solute that dissolves in a fixed quantity of a particular solvent at a specified temperature when excess solute is present. (399)
solubility-product constant (K_{sp}) An equilibrium constant for the dissolving of a slightly soluble ionic compound in water. (650)
solute The substance that dissolves in the solvent. (98, 399)
solution (See *homogeneous mixture.*)
solvated Surrounded closely by solvent molecules. (115)
solvation The process of surrounding a solute particle with solvent particles. (406)
solvent The substance in which the solute(s) dissolve. (98, 399)
***sp* hybrid orbital** An orbital formed by the mixing of one *s* and one *p* orbital of a central atom. (334)
sp^2 hybrid orbital An orbital formed by the mixing of one *s* and two *p* orbitals of a central atom. (336)
sp^3 hybrid orbital An orbital formed by the mixing of one *s* and three *p* orbitals of a central atom. (336)
sp^3d hybrid orbital An orbital formed by the mixing of one *s*, three *p*, and one *d* orbital of a central atom. (337)
sp^3d^2 hybrid orbital An orbital formed by the mixing of one *s*, three *p*, and two *d* orbitals of a central atom. (338)
specific heat capacity (*c*) The quantity of heat required to change the temperature of 1 gram of a substance by 1 K. (196)
spectator ion An ion that is present as part of a reactant but is not involved in the chemical change. (119)
spectrochemical series A ranking of ligands in terms of their ability to split *d*-orbital energies. (775)
spectrophotometry A group of instrumental techniques that create an electromagnetic spectrum to measure the atomic and molecular energy levels of a substance. (227)
speed of light (*c*) A fundamental constant giving the speed at which electromagnetic radiation travels in a vacuum: $c = 2.99792458 \times 10^8$ m/s. (216)
spin quantum number (m_s) A number, either $+\frac{1}{2}$ or $-\frac{1}{2}$, that indicates the direction of electron spin. (247)
spontaneous change A change that occurs by itself, that is, without an ongoing input of energy. (670)
square planar shape A molecular shape (AX_4E_2) caused by the presence of two axial lone pairs in an octahedral arrangement. (321)
square pyramidal shape A molecular shape (AX_5E) caused by the presence of one lone pair in an octahedral arrangement. (321)
standard atmosphere (atm) The average atmospheric pressure measured at sea level, defined as 1.01325×10^5 Pa. (148)
standard cell potential (E°_{cell}) The potential of a cell measured with all components in their standard states and no current flowing. (716)
standard electrode potential ($E^\circ_{half\text{-}cell}$) (also *standard half-cell potential*) The standard potential of a half-cell, with the half-reaction written as a reduction. (716)

standard entropy of reaction (ΔS°_{rxn}) The entropy change that occurs when all components are in their standard states. (682)
standard free energy change (ΔG°) The free energy change that occurs when all components are in their standard states. (687)
standard free energy of formation (ΔG°_f) The standard free energy change that occurs when 1 mol of a compound is made from its elements. (688)
standard half-cell potential (See *standard electrode potential.*)
standard heat of formation (ΔH°_f) The enthalpy change that occurs when 1 mol of a compound forms from its elements, with all substances in their standard states. (203)
standard heat of reaction (ΔH°_{rxn}) The enthalpy change that occurs during a reaction, with all substances in their standard states. (203)
standard hydrogen electrode (See *standard reference half-cell.*)
standard molar entropy (S°) The entropy of 1 mol of a substance in its standard state. (677)
standard molar volume The volume of 1 mol of an ideal gas at standard temperature and pressure: 22.4141 L. (154)
standard reference half-cell (also *standard hydrogen electrode*) A specially prepared platinum electrode immersed in 1 *M* $H^+(aq)$ through which H_2 gas at 1 atm is bubbled. $E^\circ_{half\text{-}cell}$ is defined as 0 V. (717)
standard states A set of specifications used to compare thermodynamic data: 1 atm for gases behaving ideally, 1 *M* for dissolved species, or the pure substance for liquids and solids. (203)
standard temperature and pressure (STP) The reference conditions for a gas:

$$0°\text{C } (273.15 \text{ K}) \text{ and } 1 \text{ atm } (760 \text{ torr}) \qquad (154)$$

state function A property of a system determined by its current state, regardless of how it arrived at that state. (191)
state of matter One of the three physical forms of matter: solid, liquid, or gas. (4)
stationary state In the Bohr model, one of the allowable energy levels of the atom in which it does not release or absorb energy. (223)
stereoisomers Molecules with the same connections of atoms but different orientations of groups in space. (476, 768) (See also *geometric isomers* and *optical isomers.*)
stoichiometric coefficient (See *balancing coefficient.*)
stoichiometry The study of the mass-mole-number relationships of chemical formulas and reactions. (72)
strong-field ligand A ligand that causes larger crystal field splitting energy and therefore is part of a low-spin complex. (774)
structural formula A formula that shows the actual numbers of atoms, their relative placement, and the bonds between them. (52)
structural isomers (See *constitutional isomers.*)
sublevel (or *subshell*) An energy substate of an atom within a level. Given by the *n* and *l* values, the sublevel designates the size and shape of the atomic orbitals. (235)
sublimation The process by which a solid changes directly into a gas. (359)
substance A type of matter, either an element or a compound, that has a fixed composition. (32)
substitution reaction An organic reaction that occurs when an atom (or group) from one reactant substitutes for one in another reactant. (481)
superconductivity The ability to conduct a current with no loss of energy to resistive heating. (391)
supersaturated solution An unstable solution in which more solute is dissolved than in a saturated solution. (409)
surface tension The energy required to increase the surface area of a liquid by a given amount. (375)
surroundings All parts of the universe other than the system being considered. (186)
system The defined part of the universe under study. (186)
systematic error A type of error producing values that are all either higher or lower than the actual value, often caused by faulty equipment or a consistent fault in technique. (24)

T

t_{2g} orbitals The set of orbitals (composed of d_{xy}, d_{yz}, and d_{xz}) that results when the energies of the metal-ion *d* orbitals are split by a ligand field. This set is lower in energy than the other (e_g) set in an octahedral field and higher in energy in a tetrahedral field. (774)
T shape A molecular shape caused by the presence of two equatorial lone pairs in a trigonal bipyramidal arrangement (AX_3E_2). (320)
temperature (*T*) A measure of how hot or cold a substance is relative to another substance. (18)
tetrahedral arrangement The geometric arrangement formed when four electron groups maximize their separation around a central atom; when all four groups are bonding groups, the molecular shape is tetrahedral (AX_4; ideal bond angle 109.5°). (318)
theoretical yield The amount of product predicted by the stoichiometrically equivalent molar ratio in the balanced equation. (97)
theory (See *model.*)
thermochemical equation A chemical equation that shows the heat of reaction for the amounts of substances specified. (199)
thermochemistry The branch of thermodynamics that focuses on the heat involved in chemical reactions. (186)
thermodynamics The study of heat (thermal energy) and its interconversions. (186)
thermometer A device for measuring temperature that contains a fluid that expands or contracts within a graduated tube. (18)
third law of thermodynamics A law stating that the entropy of a perfect crystal is zero at 0 K. (676)
titration A method of determining the concentration of a solution by monitoring its reaction with a solution of known concentration. (126)
torr A unit of pressure identical to 1 mmHg. (149)
total ionic equation A chemical equation for an aqueous reaction that shows all the soluble ionic substances dissociated into ions. (118)
tracer A radioisotope that signals the presence of the species of interest by emitting nonionizing radiation. (801)
transition element (or *transition metal*) An element that occupies the *d* block of the periodic table; one whose *d* orbitals are being filled. (254, 757)
transition state (also *activated complex*) An unstable species formed in an effective collision of reactants that exists momentarily when the system is highest in energy and that can either form products or re-form reactants. (531)
transition state theory A model that explains how the energy of reactant collision is used to form a high-energy transitional species that can change to reactant or product. (531)
transuranium element An element with atomic number higher than that of uranium ($Z = 92$). (798)

trigonal bipyramidal arrangement The geometric arrangement formed when five electron groups maximize their separation around a central atom. When all five groups are bonding groups, the molecular shape is trigonal bipyramidal (AX_5; ideal bond angles, axial-center-equatorial = 90° and equatorial-center-equatorial = 120°). (320)
trigonal planar arrangement The geometric arrangement formed when three electron groups maximize their separation around a central atom. (317)
trigonal planar shape A molecular shape (AX_3) formed when three atoms around a central atom lie at the corners of an equilateral triangle; ideal bond angle = 120°. (317)
trigonal pyramidal shape A molecular shape (AX_3E) caused by the presence of one lone pair in a tetrahedral arrangement. (319)
triple bond A covalent bond that consists of three bonding pairs, two atoms sharing six electrons; one σ and two π bonds. (288)
triple point The pressure and temperature at which three phases of a substance are in equilibrium. In a phase diagram, the point at which three phase-transition curves meet. (367)

U

ultraviolet (UV) Radiation in the region of the electromagnetic spectrum between the visible and the x-ray regions. (216)
uncertainty A characteristic of every measurement that results from the inexactness of the measuring device and the necessity of estimating when taking a reading. (20)
uncertainty principle The principle stated by Werner Heisenberg that it is impossible to know simultaneously the exact position and velocity of a particle; the principle becomes important only for particles of very small mass. (231)
unimolecular reaction An elementary reaction that involves the decomposition or rearrangement of a single particle. (535)
unit cell The smallest portion of a crystal that, if repeated in all three directions, gives the crystal. (380)
universal gas constant (*R*) A proportionality constant that relates the energy, amount of substance, and temperature of a system; $R = 0.0820578$ atm·L/mol·K = 8.31447 J/mol·K. (155)
unsaturated hydrocarbon A hydrocarbon with at least one carbon-carbon multiple bond; one in which at least two C atoms are bonded to fewer than four atoms. (477)
unsaturated solution A solution in which more solute can be dissolved at a given temperature. (409)
unshared pair (See *lone pair.*)

V

V shape (See *bent shape.*)
valence band In band theory, the lower energy portion of the band of molecular orbitals, which is filled with valence electrons. (390)
valence bond (VB) theory A model that attempts to reconcile the shapes of molecules with those of atomic orbitals through the concepts of orbital overlap and hybridization. (333)
valence electrons The electrons involved in compound formation; in main-group elements, the electrons in the valence (outer) level. (257)
valence-shell electron-pair repulsion (VSEPR) theory A model explaining that the shapes of molecules and ions result from minimizing electron-pair repulsions around a central atom. (316)
van der Waals equation An equation that accounts for the behavior of real gases. (176)
van der Waals radius One-half of the closest distance between the nuclei of identical nonbonded atoms. (369)
vapor pressure (also *equilibrium vapor pressure*) The pressure exerted by a vapor at equilibrium with its liquid in a closed system. (364)
vapor pressure lowering (ΔP) The lowering of the vapor pressure of a solvent caused by the presence of dissolved solute particles. (417)
vaporization The process of changing from a liquid to a gas. (358)
variable A quantity that can have more than a single value. (8) (See also *controlled experiment.*)
viscosity A measure of the resistance of a liquid to flow. (377)
volt (V) The SI unit of electric potential: 1 V = 1 J/C. (715)
voltage (See *cell potential.*)
voltaic cell (also *galvanic cell*) An electrochemical cell that uses a spontaneous reaction to generate electric energy. (709)
volume (*V*) The space occupied by a sample of matter. (15)
volume percent [% (v/v)] A concentration term defined as the volume of solute in 100. volumes of solution. (414)

W

wave function (See *atomic orbital.*)
wavelength (λ) The distance between any point on a wave and the corresponding point on the next wave, that is, the distance a wave travels during one cycle. (215)
wave-particle duality The principle stating that both matter and energy have wavelike and particle-like properties. (230)
weak-field ligand A ligand that causes smaller crystal field splitting energy and therefore is part of a high-spin complex. (774)
weight The force exerted by a gravitational field on an object. (16)
work (*w*) The energy transferred when an object is moved by a force. (188)

X

x-ray diffraction analysis An instrumental technique used to determine spatial dimensions of a crystal structure by measuring the diffraction patterns caused by x-rays impinging on the crystal. (384)

Photo Credits

Chapter 1 Figure 1.1a: © Paul Morrell/Stone/Getty Images; 1.1b, 1.5: © The McGraw-Hill Companies, Inc./Stephen Frisch, Photographer; 1.7b: © Hart Scientific, Inc.

Chapter 2 Opener: © Photolibrary/Alamy; Table 2.1a-c: © The McGraw-Hill Companies, Inc./Stephen Frisch, Photographer; Figure 2.2(top): © The Art Archive/Archaeological Museum Istanbul/Dagli Orti; 2.2(bottom): © Tim Ridley/DKimages.com; 2.11a(both), 2.11e: © The McGraw-Hill Companies, Inc./Stephen Frisch, Photographer; 2.15: © Dane S. Johnson/Visuals Unlimited; 2.19a-b: © The McGraw-Hill Companies, Inc./Stephen Frisch, Photographer.

Chapter 3 Opener: © Richard Megna/Fundamental Photographs; 3.1a-b, 3.2: © The McGraw-Hill Companies, Inc./Stephen Frisch, Photographer; 3.7(all): © Richard Megna/Fundamental Photographs; 3.11(both): © The McGraw-Hill Companies, Inc./Stephen Frisch, Photographer.

Chapter 4 Opener: © & Courtesy of Eastman Kodak Company; 4.3a-c, 4.4(both): © The McGraw-Hill Companies, Inc./Stephen Frisch, Photographer; 4.5: © The McGraw-Hill Companies, Inc./Richard Megna, Photographer; 4.6, Table 4.2(both), 4.7a-c, 4.12(all), 4.13(both): © The McGraw-Hill Companies, Inc./Stephen Frisch, Photographer.

Chapter 5 Opener: © k fotostudio/iStockphoto.com; 5.1a-c, 5.2a-b, 5.8: © The McGraw-Hill Companies, Inc./Stephen Frisch, Photographer.

Chapter 6 Opener: © Mike Goodwin/photographersdirect.com.

Chapter 7 Opener: © Ivan Cholakov/Shutterstock.com; 7.14a-b: PSSC Physics © 1965, Education Development Center, Inc.; D.C. Heath & Company/Education Development Center, Inc.

Chapter 8 Opener: © Jarvis Gray/Shutterstock.com.

Chapter 9 Opener: © Ken Lucas/Visuals Unlimited; 9.6a-b, 9.8a, 9.9a-c, 9.13, 9.22: © The McGraw-Hill Companies, Inc./Stephen Frisch, Photographer.

Chapter 10 Opener: © Robin Appleton Photography/photographersdirect.com; 10.2(all), 10.6(both): © The McGraw-Hill Companies, Inc./Stephen Frisch, Photographer.

Chapter 11 Opener: © Yoav Levy/Phototake.com; 11.21: © Richard Megna/Fundamental Photographs.

Chapter 12 Opener: © Kenneth G. Libbrecht/snowcrystals.net; p. 358: © Michael Lustbader/Photo Researchers, Inc.; 12.19a-b: © The McGraw-Hill Companies, Inc./Stephen Frisch, Photographer; 12.21b: © Scott Camazine/Photo Researchers, Inc.; 12.22a: © Jeffrey A. Scovil Photography (Wayne Thompson Minerals); 12.22b: © Mark Schneider/Visuals Unlimited; 12.22c: © Jeffrey A. Scovil Photography (Francis Benjamin Collection); 12.22d: © Jeffrey A. Scovil Photography; 12.25: © Michael Heron/Woodfin Camp & Associates; 12.33a: © The McGraw-Hill Companies, Inc./Stephen Frisch, Photographer; 12.36: © AT&T Bell Labs/SPL/Photo Researchers, Inc.

Chapter 13 Opener: © Melinda Podor Photography/photographersdirect.com; Table 13.1: © A. B. Dowsett/SPL/Photo Researchers, Inc.; 13.8a-c: © The McGraw-Hill Companies, Inc./Stephen Frisch, Photographer.

Chapter 14 Opener: © Tobias Machhaus/Shutterstock.com; Page 436(all), Page 439(all), Page 441(all), Page 443(all): © The McGraw-Hill Companies, Inc./Stephen Frisch, Photographer; 14.6a: © Gordon Vrdoljak, UC Berkeley Electron Microscope Lab; 14.6b: © Norbert Speicher/Alamy; Page 448(all), Page 453(all), Page 457(all), 14.16b(both), Page 460(all): © The McGraw-Hill Companies, Inc./Stephen Frisch, Photographer.

Chapter 15 Opener: © Alfred Pasieka/Photo Researchers, Inc.; 15.8: © Martin Bough/Fundamental Photographs.

Chapter 16 Opener: © JUPITER IMAGES/Creatas/Alamy RF.

Chapter 17 Opener: © Ken Lucas/Visuals Unlimited; 17.1a-d: © The McGraw-Hill Companies, Inc./Stephen Frisch, Photographer.

Chapter 18 Opener: © Richard Megna/Fundamental Photographs; 18.7a-b: © The McGraw-Hill Companies, Inc./Stephen Frisch, Photographer.

Chapter 19 Opener: © Tina Manley/North America/Alamy; 19.1a-b, 19.2a-b, 19.6, 19.10a-b, 19.11: © The McGraw-Hill Companies, Inc./Stephen Frisch, Photographer.

Chapter 20 Opener: NASA/Kennedy Space Center.

Chapter 21 Opener: © Gabriel Bouys/AFP/Getty Images; 21.1, 21.2, 21.4(both), 21.5b: © The McGraw-Hill Companies, Inc./Stephen Frisch, Photographer; 21.7: © Richard Megna/Fundamental Photographs; 21.9: © The McGraw-Hill Companies, Inc./Stephen Frisch, Photographer; 21.14: © Chris Sorensen Photography; 21.15: © The McGraw-Hill Companies, Inc./Pat Watson, Photographer; 21.17: © The McGraw-Hill Companies, Inc./Stephen Frisch, Photographer; 21.20a: © Alfred Dias/Photo Researchers, Inc.; 21.25, Page 744: © The McGraw-Hill Companies, Inc./Stephen Frisch, Photographer; 21.27b: © Tom Hollyman/Photo Researchers, Inc.

Chapter 22 Opener: © Bartlomiej Nowak/Shutterstock.com; 22.2(all), 22.5a-b, 22.6: © The McGraw-Hill Companies, Inc./Stephen Frisch, Photographer; 22.9a-b: © Richard Megna/Fundamental Photographs; 22.19: © The McGraw-Hill Companies, Inc./Pat Watson, Photographer; 22.20a-b: © The McGraw-Hill Companies, Inc./Stephen Frisch, Photographer.

Chapter 23 Opener: © U.S. Department of Energy/Photo Researchers, Inc.; 23.8: © Scott Camazine/Photo Researchers, Inc.; 23.9: © Dr. Robert Friedland/SPL/Photo Researchers, Inc.; 23.10: © Dr. Dennis Olson/Meat Lab, Iowa State University, Ames, IA; 23.14: © Albert Copley/Visuals Unlimited; 23.15: © Dietmar Krause/Princeton Plasma Physics Lab.

Index

Page numbers followed by *f* indicate figures; *n*, footnotes; and *t*, tables

D

E

F

G

H

N

U

V

W